12th International Symposium on the Science and Technology of Light Sources and the 3rd White LED Conference (LS-WLED 2010)

Eindhoven, The Netherlands
11 – 16 July 2010

Editors:

M. Haverlag
G.M.W. Kroesen
T. Taguchi

ISBN: 978-1-5108-4084-3

Printed from e-media with permission by:

Curran Associates, Inc.
57 Morehouse Lane
Red Hook, NY 12571

Some format issues inherent in the e-media version may also appear in this print version.

Printed by Curran Associates, Inc. (2017)

For permission requests, please contact Foundation for the Advancement of the Science & Technology of Light Sources (FAST-LS) at the address below.

Foundation for the Advancement of the Science & Technology of Light Sources
FAST-LS
Belmayne House
99 Clarkehouse Road
Sheffield, United Kingdom
S10 2LN

www.fast-ls.org

Additional copies of this publication are available from:

Curran Associates, Inc.
57 Morehouse Lane
Red Hook, NY 12571 USA
Phone: 845-758-0400
Fax: 845-758-2633
Email: curran@proceedings.com
Web: www.proceedings.com

LS - WLED 2010

LS-WLED 2010

Proceedings of the 12th International Symposium on the Science and Technology of light sources and the 3rd White LED conference

July 11-16, 2010, Eindhoven, The Netherlands

Edited by M. Haverlag, G.M.W. Kroesen and T. Taguchi

FAST-LS
Foundation for the Advancement of the Science & Technology of Light Sources

Prof. M. Haverlag
Secretary
Philips Lighting , LightLabs
P.O. Box 80020
5600 JM Eindhoven
The Netherlands

Prof. G.M.W. Kroesen
Chairman
Department of Physics
Eindhoven university of technology
P.O. Box 513
5600 MB Eindhoven
The Netherlands

Prof. T. Taguchi
Graduate School of Science and Engineering,
Yamaguchi University,
2-16-1 Tokiwadai, Ube,
Yamaguchi, 755-8611,
Japan

Published by FAST-LS Ltd.
Belmayne House, 99 Clarkehouse Rd.
Sheffield, S10 2LN, UK

www.FAST-LS.org

Produced by Canopus Publishing Limited
www.canopusbooks.com

LS12

White LED 3

TU/e Technische Universiteit **Eindhoven** University of Technology

The organisers of LS-12 / WhiteLED-3 gratefully acknowledge the important role of the following companies who have acted collectively as financial guarantors and/or supporters of the symposium:

PHILIPS

OSRAM

APL ENGINEERED MATERIALS, INC.

WelchAllyn®

SYLVANIA

GE Lighting

Heraeus

Preface

This volume contains both the full texts of the invited papers and the abstract of the accepted contributed posters that are to be presented on the 12th International Symposium on the Science and Technology of Light Sources (LS:12) and the 3rd WhiteLED conference. It is for the first time, but very likely not for the last time that these two conference series will be organized together. This fact reflects the growing importance of new light source technologies, e.g. LEDs and OLEDs in current and future lighting systems and applications.

LS-12 / WhiteLED-3 are held in the Auditorium building of the Eindhoven University of Technology from July 11th to the 16th, 2010. The joint conference is organized by the two international scientific committees of the LS- and WhiteLED series. All the organizers are dedicated to continue the special character of both these meetings, which is exemplified in the high quality of the meetings and open atmosphere in which new groups and companies are welcome to share results and ideas. Previous symposia in the LS-series were held in Loughborough (1975), Enschede (1979), Toulouse (1983), Karlsruhe (1986), York (1989), Budapest (1992), Kyoto (1995), Greifswald (1998), Ithaca (2001), Toulouse (2004) and Shanghai (2007). Previous symposia in the WhiteLED series were held in Tokyo (2007) and Taipei (2009).

The combined scope of the two conference series covers virtually all light source types and many different application areas, including special lighting applications, and covers both technical aspects and application aspect of light sources. The broad scope in terms of lighting technologies and applications makes this combined conference a must-be and must-see for everyone active in the field of light source technology and application. The joint symposium is set up in such a way that there is ample opportunity to meet and discuss in detail items of shared interest with other researchers active in the area of light sources and applications during the poster sessions, which are the core part of the symposium.

After the conference a selection of papers will be invited to submit full research papers to the Journal of Physics D : Applied Physics and the Journal of Light and Visual Environment. All these papers will be peer reviewed according to the normal procedures of these journals, and will appear in combined special issues.

G.M.W. Kroesen and M. Haverlag

Contents

Conference Chairs and Committees

Conference chairs of the joint LS12-WhiteLED3

Prof.dr.ir. G.M.W. Kroesen	Eindhoven University of Technology (NL)
Prof.dr. T. Taguchi	University of Yamaguchi (JP)

LS-12 co-chairs

Prof.dr. M. Haverlag	Philips Lighting (NL)
Prof.dr. K. Tachibana	Ehime University (JP)

WhiteLED-3 co-chairs

Prof.dr. I.T. Ferguson	Georgia Institute of Technology (US)
Prof.dr. C.C. Yang	National Taiwan University (TW)

Local organizing Committee

Prof.dr.ir. G.M.W. Kroesen (chair)	Eindhoven University of Technology (NL)
Prof.dr.ir. M. Haverlag (secretary)	Philips Lighting (NL)
Dr.ir. H.J. Gielen (treasurer)	Philips Lighting (NL)
Prof.dr. P. Awakowicz	Ruhr-University Bochum (DE)
Prof.dr. R. Coehoorn	Philips Research (DE)
Dr. R Geens	SLI-Sylvania (BE)
Dr. D. Eissler	OSRAM-OS (DE)
Dr. U. Hechtfischer	Philips Research Aachen (DE)
Dr.ir. H.T.J.M. Hintzen	Eindhoven University of Technology (NL)
Dr. Chr. Hoelen	Philips Lighting (NL)
Prof.dr. J.J.A.M. van der Mullen	Eindhoven University of Technology (NL)
Prof.dr. C. Ronda	Philips Research (DE)
TU/e Congress Office	Eindhoven University of Technology (NL)

International Symposium on the Science and Technology of Light Sources

International Committee

R. Devonshire (chair)	Sheffield University (UK)
G. Zissis (secretary)	Laplace Laboratory, Toulouse University (FR)
E. Arnold	Heraeus Noblelight (DE)
T. R. Brumleve	APL Engineered Materials, Inc (US)
J.T. Dakin	GE Lighting (US)
M. Haverlag	Philips Lighting (NL)
M. Jinno	Ehime University (JP)
J. Lawler	University of Wisconsin (US)
L. Brock	OSRAM Sylvania (US)
M.Q. Liu	Fudan University (CN)
J. Maya	Consultant (US)
H. Schöpp	INP Greifswald (DE)
C. Stewart	Welch Allyn Inc (US)
J. Stocks	Havells Sylvania UK Limited (UK)
T. Taguchi	University of Yamaguchi (JP)
Y. Yuge	Toshiba Lighting and Technology Corp (JP)
A. Corazza	SAES Getters (IT)
W. Lapatovich	Consultant (US)

International Conference on White LEDs and Solid State Lighting

International Committee

T. Taguchi (chair)	University of Yamaguchi (JP)
I. Ashdown	Philips (CA)
S.P. den Baars	UCSB (US)
R. Devonshire	Sheffield University (UK)
I.T. Ferguson	Georgia Institute of Technology (US)
G.M.W. Kroesen	Eindhoven University of Technology (NL)
V. Haerle	OSRAM-OS (DE)
M. Jinno	Ehime University (JP)
J. Kido	Yamagata University (CN)
W. Ling	SSL Alliance (US)
M. Liu	Fudan University (CN)
Y.S. Liu	National Tsing Hua University (TW)
Y. Luo	Tsinghua University (CN)
S.K. Min	Kyung Hee University (KR)
S. Nakamura	UCSB (US)
N. Narendran	Rensselaer Polytechnic Institute (US)
F. Orito	MCC (JP)
W.N. Wang	University of Bath (UK)
C.C. Yang	National Taiwan University (TW)
Y.M. Yu	KOPTI (KR)
G. Zissis	Laplace Laboratory, Toulouse University (FR)

This page intentionally left blank.

Prof. Zuquan Cai
(1924-2009)

Professor Zuquan Cai, the most respected and beloved pioneer of lighting science and technology in China, died on July 27th, 2009 at the age of 84. Professor Cai worked at Fudan University in Shanghai for 47 years.

His initial studies were in the fields of vacuum science and technology and vacuum devices, work that included, for example, a collaboration which led to the manufacture in 1953 of the first X-ray tube made in China. In 1960 he turned his attention to light sources and lighting and this field soon came to be his major and, in the event, his lifelong interest.

In 1978 Professor Cai founded The Institute for Electric Light Sources at Fudan University and, under his leadership, lamp research and test laboratories were established. Working in close collaboration with industry, the Institute was responsible for the development, and first introduction in China, of many major lamp types including high pressure mercury lamps in 1961, halogen lamps in 1963, xenon lamps in 1964, high pressure sodium lamps in 1968, and metal halide lamps in 1978. During this period the Institute produced also many spectroscopic lamps.

After his formal retirement Professor Cai remained as Honorary Director of the Institute and, until prevented from doing so by illness, he actively helped his successors. One of his important legacies is that many of the lighting researchers and engineers active currently in China were his students at Fudan University.

Professor Cai actively promoted scientific and technical exchange both nationally and internationally. He was one of the founders of the Shanghai Illuminating Engineering Society and the Chinese Illuminating Engineering Society. He sponsored and hosted the first Lux Pacific in 1980 in Shanghai and was a Member of the International Organizing Committee and, latterly, the International Advisory Committee, of the LS symposium series. During his final conversations he expressed his happiness to have fulfilled all of his wishes, foremost amongst them being that he had had the honour to host LS-11 in Shanghai in 2007.

Shaolong Zhu, Retired Director,
Institute for Light Sources,
Fudan University,

Rongqing Liang,
Director,
Institute for Light Sources,
Fudan University,

Muqing Liu
International Organizing Committee
LS Symposium Series

Prof. Dr. Jean-Jacques Damelincourt
(1939-2008)

Professor Jean-Jacques Damelincourt died in his hometown, Toulouse, on November 29th, 2008 at the age of 69. He was Professor in Toulouse 3 University and he acted for more than 40 years as a prominent promoter of light sources science and technology.

Born in Paris, on July 9th, 1939, he earned his bachelor degree in Physical Sciences in 1962 and a degree in Electrical Engineering from ENSEEIHT, the French National High School of Electrical and Electronics Engineering of Toulouse in 1964. He began his research activities in 1965 and obtained his first university position as an Assistant in ENSEEIHT in 1966. In 1967 he joined Toulouse 3 University in the Centre of Atomic Physics of Toulouse (CPAT) to work under the direction of Professor Daniel Blanc in the domain of light sources. After obtaining his PhD (Doctorat d'Etat) in 1973 for work on probe measurements of fluorescent lamps he progressed through the academic scales, eventually being appointed full professor in Toulouse 3 University in 1990.

In 1970 he established his own research group within CPAT known as the "High Intensity Photon Sources" team. In the period from 1970 to 2004 more than 100 master and more than 30 PhD students worked under his direction in the domain of light sources. Following his retirement in 2004 he was Professor Emeritus. During his academic career he published more than 200 papers, communications, book chapters, etc., in the domain of light sources.

He was a very active member of the French national community of illuminating engineers and acted as President for the Regional branch of the French Illuminating Engineering Society (AFE) for more than 6 years. For the services that he rendered to the community the AFE awarded him its highest honour: the Alfred Monnier medal.

He was also a dedicated member of the Symposia on the Science and Technology of Light Sources community. He participated in the symposia for the first time in 1979 when he attended LS:2 in Enschede; he then undertook the pivotal and demanding role of Secretary of the Local Organising Committee for LS:3 held in Toulouse in April, 1983. The meeting was a great success with the Carcassonne Excursion and Banquet being remembered to this day by participants. He applied his organisational skills again when he acted as Chairman of the Local Organising Committee for LS:10 held in Toulouse in 2004. He was a member of the LS International Scientific Committee until 1998 and served as a member of the LS Advisory Committee until the time of his death.

His passing means that the light sources community at large has lost a devoted member and his close collaborators an excellent advisor and colleague.

Professor Georges Zissis
International Organising Committee
LS Symposium series

Prof. Dr. Ryohei Itatani
(1930-2010)

The world–renowned plasma scientist Professor Dr. Ryohei Itatani died in Kyoto, Japan on the 7th of March, 2010, aged 79. Professor Itatani was born on April 5th, 1930, in Suma, Kobe.

After graduating from Kyoto University with a B.S. degree in 1953 he began work at NEC Corporation on fluorescent lamp R&D. In 1955 he returned to Kyoto University as a research associate and continued then firstly to gain a Ph.D. Degree and then, in 1968, to be appointed full professor. His entire research career at the university was dedicated to the development of plasma applications, in particular lighting plasmas, work for which he gained national and international recognition.

Amongst his many achievements, Professor Itatani invented and developed a variable-colour fluorescent lamp, work which he presented at LS:5 in Budapest in 1992. This innovative lamp was commercialized and is installed at many locations in Japan. He chaired the Local Organizing Committee for LS:7 held in Kyoto in 1995, a meeting which he and his colleagues promoted vigorously and successfully, overcoming initial concerns about whether or not there was an audience for these meetings outside Europe.

One of Professor Itatani's important legacies is that he acted as tutor and friend to numerous students and researchers, many of whom continue to work in the field of light sources either at universities or lighting companies in Japan. He was honoured to lead the Illuminating Engineering Institute of Japan as President from 1997 to 1999.

For his considerable, sustained, contributions to plasma science he received the Order of the Sacred Treasure (Class: Gold Rays with Neck Ribbon), a Japanese national order, in 2005 and the rank of "sho-shi-i (the major 4th rank)" in 2010.

His passing is a sad loss for both the light source and plasma research communities; he will be remembered as an outstanding researcher and educator in these fields.

Masafumi Jinno
International Organizing Committee
LS Symposium Series

Johan A.J.M. van Vliet
(1940-2007)

On July 4th 2007 Johan van Vliet passed away. He was an internationally recognized expert in the field of gas-discharge lamps and as such a prominent member of the international lighting community. During the greater part of his professional life he was active within the Research and Development laboratories of Philips Lighting in Eindhoven, The Netherlands.

Born in Waalwijk in the southern part of the Netherlands, he started work at Philips Research Laboratories in Eindhoven in 1962, where he became involved in electron-optical devices. Besides his daily tasks for Philips he undertook further education courses in physics and engineering at the Institute for Higher Professional Education in Eindhoven, from where he gained an engineering degree in 1967. Thereafter he studied physics at Eindhoven University of Technology from where he gained a Masters degree *cum laude* in 1972. This demanding combination of work and study indicated already his characteristic perseverance. After completing his studies, he continued his career as a physicist in the Central Development Laboratories of Philips Lighting, where he stayed until his early retirement due to health reasons in 1996. Following his retirement he maintained his involvement with Philips Lighting as a teacher in various educational projects.

Johan was an all-rounder in the field of gas-discharge lamp engineering and a major characteristic of his professional approach was the perfect balance he achieved between the important aspects of practical engineering and the scientific basis of gas-discharge physics. His interests included all of the major relevant fields including: discharge-ballast interaction, electrodes, ignition phenomena, modeling of high-pressure Na discharges, and the development of metal-halide lamps. This breadth of interest is evidenced by his having (co-) authored many papers, patents and conference contributions. His international reputation, however, is particularly connected with the standard work on high-pressure sodium lamps that he co-authored with J.J.(Sjef) de Groot. ("*The High-Pressure Sodium Lamp," Philips Technical Library, 1986*). The scope of this book gives ample evidence of his mastery of gas-discharge lamp engineering.

In person, Johan had an amiable and strong personality which he combined with great drive and an inspiring, infectious enthusiasm. Despite increasing health problems during his last years he always maintained a positive outlook. It was a privilege to have worked with him.

J.J. de Groot
W.J. van den Hoek
R.A.J.Keijser

List of Invited, Landmark and Contributed Papers

Sunday Lectures

Session One

Session Two

Session Four

Session Five

Session Seven

Session Eight

Sunday Lectures

Notes on the History of Incandescent Lamps

Wim van den Hoek [1]

Museum "*Philips Incandescent Lamp Factory of 1891*"
Eindhoven , The Netherlands
contact: wj.vandenhoek@planet.nl

ABSTRACT

The incandescent lamp[2] is one of the longest existing electro-technical products, its history comprises about 200 years. The essential start of this history is the experiment of Sir Humphrey Davy in 1802 whereby a Pt wire was heated to incandescence by the passage of an electric current from a Volta pile. The first historic milestone was reached with the *carbon filament lamp* of Edison in 1879. From then a series of inventions and developments gave rise around 1912 to the most important milestone the *gas-filled tungsten filament lamp*. This was due to the achievements in the tungsten field by Coolidge and the physical insights of Langmuir. Since then there have been no essential innovations in the physics of the incandescent lamp. Further innovations and developments belong primarily to the fields of technology and manufacturing. These have been numerous and impressive. Particularly important is the development of large-scale lamp production on high-speed machinery. The last essential milestone was the invention of the *halogen lamp* in 1959. At present the product-lifecycle of the incandescent lamp seems to come to an end, primarily because of the inherent low luminous efficacy. More efficient light sources are available and the expansion of environmental awareness has resulted in restrictive governmental legislation [1].

INTRODUCTION

It is important to realize that the study of the historical evolution of the incandescent lamp (and in general of any other technical product) concerns not only the *history of the technical aspects* of the product. Of decisive importance are the *scientific history* (involving the ideas of the mind of brilliant individuals and the realization and dissemination thereof, etc.) and the *economic history* (involving markets, entrepreneurs, companies, patents, competition, cartels, trusts etc.). These historic 'flows' are strongly interacting domains. These aspects, however, are not covered in the present paper. Here only some notes will be presented as what the author considers as important topics in the physics, chemistry and materials science of the historic development of the device.

The sources of the present paper are on the one hand the personal recollections of the author. On the other hand two books have been guiding. First, the historical studies on the Philips company by Heerding [2] ,which primarily contribute to the economic aspects and second, the authorative work by Howell and Schroeder [3] , which is a lucid and fascinating review of the early technical history[3]. A number of specific references will be given, from where the original literature can be traced. Furthermore, a huge amount of information can be obtained from the internet[4].

In Table I the various stages to be distinguished in the historal development are given. The indicated years are to a certain extent approximate, since there are differences in the points in time of invention, patent application, market introduction etc. Note especially the indication of the field of basic scientific insights, underlying the progression. In general it should be kept in mind that lamp science has a strong *empirical* basis.

---

[1] Physical chemist, from 1972-2004 employed in the Laboratories of Philips Lighting in Eindhoven.
[2] Here the term 'incandescent' implies incandescence due to ohmic heating. In historical context it is important to note that incandescence can also be brought about by gas heating or anodic arc heating etc.
[3] J.W.Howell was a collaborator of Thomas A. Edison
[4] Not included in the References.

TABLE I
The various stages in the history of the incandescent lamp

Period	characterization	lm/W	essential 'science'
1802-1879	Initial concepts		
1879-1905	Carbon filament lamp	~1.7-4.3	
1902-1913	Metal (tungsten) vacuum lamp	~7	*Materials science*
1913-	Tungsten gas-filled lamp	~10-15	*Physics*
1959-	Tungsten halogen lamp	~20-30	*Chemistry*

For further discussion a reminder of some basic facts of the black-body radiator is appropriate. Filament temperatures of incandescent lamps are in the range 2000-3000 K. In that range the luminous efficacy of a black-body radiator (with only radiative losses) runs from 1-20 lm/W. Because of its spectral selectivity the corresponding numbers for tungsten are 1-30 lm/W. A hypothetical black-body radiator which emits only in the visible region would have an approximate efficacy of 230 lm/W [5].

INITIAL CONCEPTS

The steps leading to the inventions culminating in the incandescent lamps by J. Swan (1878) and Th. Edison (1879) should be seen in in the light of the general expansion of science, technology and industrial activities in the beginning of the 19th century[6]. Several aspects are in particular to be considered. Basic are the experiments by Davy on incandescence (1802) and on the carbon arc (1809) which gave rise to the ideas on possible light sources. These ideas should be seen against the background of 19th century lighting practice, which consisted of candles, oil lamps and gas light[7]. Lighting constituted a market which, in principle, had a huge potential for adsorbing electric light sources. Following the steps by Davy, an impressive number of inventive individuals made attempts to construct electric incandescent lamps. Heerding [2] gives a list of 16 inventors from the whole of Europe and the USA in the period 1802-1879. So one cannot speak of individual inventors of the incandescent lamp, the principles became fairly well known during that period. It was however Edison, who made the first *lighting system* based on incandescent lamps. In addition, it should be noted that the invention and development of electric power generators ("dynamo's") took place in the same era.

Besides the developments in *electric* incandescent sources, attempts to make other types of lamps took place. Following the arc experiments of Davy (1809) a sophisticated arc lamp was developed by Jablochkoff (1877). Gas lighting was expanded by the invention of a gas flame surrounded by an incandescent ceramic mantle (C. Auer von Welsbach 1885) and later on *electric* ceramic incandescence was invented by W. Nernst in 1897. These lamps have disappeared due to the later

[5] Some of the older lm/W values have been recalculated from W/candle values, the unit used in the early years of lamp production. Note that one candle equals approximately 4π lumen.
[6] Some important historical data of the physics of light generation: Kirchhoff 1859, Stefan-Boltzmann 1884, Wien 1893, Planck 1900.
[7] A readable historical review is given by Stoer [4].

spectacular improvements in the incandescent lamp[8].

CARBON FILAMENT LAMPS

Although in the first experimental prototypes Pt wires were used (melting point 2043 K), it was soon recognized that materials allowing higher temperatures were needed. At that time only carbon, e.g. in the form of carbonized paper or cotton, seemed to be appropriate. Carbon has a sublimation temperature in excess of 3500 K. The lamps of J. Swan (1878) and Edison (1879) used filaments [9](i.e. thin incandescent bodies) of carbonized cotton wire. As already stated, the real milestone relates to the pioneering work of Edison, which started around 1877. At that time the principle of incandescent lighting had been established and amply demonstrated. As a man, Edison combined all the qualities of a brilliant inventor and entrepreneur. He had already made significant inventions and he ran a laboratory of about 100 people in Menlo Park, New Yersey. Moreover, he knew how to finance his activities. All his predecessors in the lamp field had made 'single' lamps, which at the most were operated in series. Edison however keenly recognized the application aspect, i.e. the chain: *electrical power supply, distribution and large amounts of lamps.* Edison's approach to this problem of '*subdividing*' the light was *operating the lamps parallel,* demanding high-resistance filaments. Today this seems trivial, but at that time it was revolutionary! It was this **system approach** that made the contribution of Edison to the history of lighting unique.

The very first lighting system was installed on a ship. This has a special historical significance, since it was around 1884 that the young dutch engineer Gerard Philips (1858-1942), working in the shipbuilding industry in Glasgow, got acquainted with electric lighting, which ultimately led to the foundation of the Philips company in Eindhoven in 1891. Edison's successful example generated a huge industrial activity in the USA as well as in Europe. Heerding [7] mentions for instance a number of 29 companies in western Europe in the year 1890 , one year before the start of Philips !

The precise technical history of the carbon filament lamp can be read in Howell & Schroeder [3]. There is at least one aspect that needs further comments. There has been a remarkable evolution in the preparation of the carbon filament. The filament is the heart of the lamp, so it is essentially decisive for the behavior of the lamp. Table II gives a summary. It is seen that over a period of 25 years the efficacy has increased significantly. This improvement can be explained in terms of modifications in the structure of the filament carbon[10]. The *carbonization* (up to ~1300 K) of (solid) organic materials leads to so-called '*defect structures*' of graphitic carbon. There is a large variety of these defect structures depending on the starting material and the particular carbonization conditions. Upon further heat treatment (>1300-3000 K) these defect structures will develop into the direction of 'perfect' graphite. This process is called '*graphitization*'. Depending on the particular defect structure, graphitization can be accompanied by changes in the macroscopic dimensions and also by further degassing, etc. (see the discussion on '*hard*' and '*soft*' carbons in Ref.[7]). So *optimum* filaments should have a structure before lamp mounting *'as near as possible to graphite'*, in order to avoid failure due to further graphitization. Graphite is strongly anisotropic and this has consequences also for its vaporization behavior and the emissivity, which both appear to be optimal in the c-direction. The so-called 'treating' , which is essentially the deposition of carbon [11] from gas-phase carbonization from e.g. CH_4 , results in a graphite layer with the c-axis perpendicular to the

---

[8] The principle of the Auer mantle and the Nernst lamp have attracted research attention to far into the 20th century [5,6]. However, these activities have never led to a lamp product.

[9] The term 'filament' originates from Edison.

[10] See Ref.[7] for detailed discussions and pertinent references. Around 1972 there has been a renewed interest in carbon filaments lamps within Philips, dictated by the predictions on the futural availability of lamp materials, in particular tungsten, in the reports of the "Club of Rome'. See Refs. [8,9].

[11] A historical case of *Chemical Vapor Deposition*! This also led to a 'flattened' temperature profile of the filament, since carbon deposition is enhanced at spots with higher temperatures (e.g. due to local differences in diameter).

filament axis (pyrolytic graphite [10]. In retrospect,it can be said that the filaments from 1904 (Whitney from GE) thus were optimally oriented and graphitized.

TABLE II
Historical evolution of carbon filament preparation

Year	Who & what	**lm/W**	structure	vapor.	emis.
< 1879	Carbonized 'paper'				
1877-1879	**Edison**: -carbonized cotton (sewing thread)				
1880-1891	-Japanese bamboo	1.7→ 2.3	-improved 'hardness' -improved degassing		
1879-1885	**Sawyer & Mann**: 'treating' gas-phase deposition		-pyrolytic graphite on surface -control of resistance -flattened temperature profile	+	+
1888-1894	**Swan, Lane Fox a.o.** squirted cellulose	3.4	-improved structure -improved control -freedom in size etc.		
1904	**Whitney** heating to > 3000 C	4.3	-optimally graphitized & oriented	+	+

Remarks:
-First three columns are based on Ref.[3], lm/W values are based on 600 h lamp life.
-Note that progress is not only due to filament improvement, but also vacuum technique improvements etc., etc.

It is important to note again that these developments had a purely *empirical* basis, the final elucidation of the structure of graphite had to wait until 1924. In is interesting to note that the idea behind Whitney's treatment to over 3000 C was the removal of impurities.

TUNGSTEN FILAMENT (VACUUM) LAMPS

From 1905 to 1918 the gradual transition from carbon filament lamps to tungsten filament lamps took place. From the insights gained from carbon incandescence, it was recognized that a significant step in the luminous efficacy was only to be obtained with a filament of a metal, which combined primarily a melting point far above 2000 K and low vapor pressure[12]. This pointed into the direction of the refractory metals, many of which had been discovered c.q. isolated during the 19th century (see Hammond [11]). The principles of potential technologies for metal-filament making like wire drawing (operational since mediaeval times), powder metallurgy (PM) and deposition from the gas phase (CVD) were essentially known, but these techniques had to be made operational for the specific

[12] For instance at 3200 K the ratio of the vapor pressures of carbon and of tungsten is about 3000 [7]. Some melting points: Os 3323 K, Ta 3250 K, W 3653 K.

metals.

The innovative activities took place in the *geographical* triangle "Vienna-Berlin-New York". Around 1898 the Austrian chemist Carl Auer von Welsbach (1858-1929) succeeded in making a powder-metallurgical Os filament. Such a lamp had about 5.9 lm/W which meant a significant step. The osmium filament lamp is only of historical interest. It never was a large activity. Os minerals appeared to be very rare in the earth crust. Moreover Os oxide is extremely toxic.

In 1902 the Russian chemist J.R. von Bolton (1868-1912) working in Berlin with the company Siemens & Halske, used powder metallurgy to make Ta filaments, starting from the oxide followed by reduction to Ta metal. Later on von Bolton succeeded in drawing Ta wires. Key factor in this process appeared to be the purity of the Ta. Tantalum lamps are reported to have an efficacy of 4.8 lm/W, so they were better or at least fully compatible with the carbon lamps. Ta filament lamps disappeared around 1913. Both the Os and the Ta filaments showed the *'sag'*-phenomenon, which later was also one the essential problems of tungsten.

Suggestions on the combination carbon-tungsten were made as early as 1887. Just and Hanaman in Vienna tried filaments of a carbon core with a tungsten deposition. This was effected by heating a C-wire in an atmosphere of (gaseous) tungsten oxidechloride. At incandescent temperatures W reacts with C to form tungsten carbides, so this was not a success. A second approach of these investigators was the preparation of filaments from squirting a paste of W-powder and an organic binder through a diamond die, followed by baking and sintering step in a H_2 / N_2 / H_2O atmosphere. Tungsten filament lamps based on this PM scheme came on the market in 1906. The luminous efficacy was 7.8 lm/W, which meant an enormous improvement [13].

The invention of the tungsten filament (vacuum) lamp changed the industrial landscape completely. Every lamp company had to make the transition to the new technology. Because of patent rights , the Edison companies (GE) could not use the powder metallurgical method for filament preparation. For this reason a project to come to drawn tungsten was started at GE. Before 1906 it was the general opinion among the experts that W could not be drawn, it was too brittle. William D. Coolidge (1873-1975) took up the task to tackle that problem. Coolidge found that purity (as compared with Ta drawing) was not the issue and he realized that brittleness is an inherent property of W. Subsequent studies revealed that ductility of W can be reached after strong mechanical deformation (hammering). Most metals exhibit work-hardening on deformation. With W the strongly deformed state appeared to give rise to ductility, thus allowing heavily worked tungsten to be drawn. This was a pioneering piece of research! A precise description of Coolidge's efforts can be found in Howell[12]. This work is the basis of the well-known process in the lighting industry, where powder metallurgical tungsten bars (ingots) are hammered, rolled and swaged, followed by series of drawing and annealing steps. The process is still in operation today. The work of Coolidge enabled the manufacture of drawn tungsten wire, in 1911 the first drawn tungsten-filament lamps came on the market. The USA had now definitely transferred its backlog in lamp technology into a significant advantage. This was again a revolution in the world of the lamp companies.

A further essential problem, however, in the filament application (2000-3000 K) of tungsten is the 'sagging' of the filament. Sag is the off-setting of crystallites along the grain boundaries due to gravity. It was found by Coolidge that 'sag' and 'non-sag' depended on the conditions of tungsten powder preparation. Further work showed that a small K-Al-silicate doping eliminated the problem (Pacz, 1917). The study of 'non-sag' tungsten has occupied the research in the lighting industry till the end of the 20th-century. Especially the availability of electron microscopy has revealed the existence of

[13] Note that the length of a C-filament was of the order of 0.1 m. Due to the different electrical properties of W (and also Os and Ta) the length of the filament became of the order of 1m. This necessitated an extensive system of filament support wires.

rows of *gaseous potassium bubbles* (Moon & Stickler 1971). Without potassium doping the initial fibrous microstructure of the drawn wire is re-crystallized during operation into an equi-axed grain structure. With potassium dope, an interlocking grain structure develops which impedes the off-setting slippage along crystal boundaries. The wealth of knowledge on tungsten for high-temperature applications has been summarized in the book by Pink and Bartha [13].

TUNGSTEN FILAMENT *GAS-FILLED* LAMPS

Besides the development of 'non-sag' tungsten, there was a second innovation from the laboratories of GE in the race for optimal life and increasing efficiency. Increasing the temperature of the filament leads to increased evaporation, thus wall-blackening. Preventing this by a gas filling will lead to unacceptable heat losses. This latter aspect was already known from the work of S.G. Lane-Fox (1856-1932). It was Irving Langmuir who proposed in 1912, based on insights into the heat transfer at the solid-gas interface, the decisive steps of coiling the filament in combination with an inert gas-filling. Coiling of the filament reduced the effective length of ~ 1 m to some 0,01 m, thus reducing the gas losses, while the inert gas impedes the evaporation. This made higher filament temperatures possible (up to 2750 K), which made efficacies of 10-15 lm/W the standard. Again a revolutionary step, which forced all the lamp companies either to adapt or to disappear. The versatile physicist Irving Langmuir (1881-1942, Nobel prize 1932) not only made important contributions to lamp physics, but later on also to plasma physics, adsorption etc. His list of publications is impressive, in the year 1913 alone he published 10 papers mainly on the chemistry and physics of incandescent lamps [14].

As already stated in the beginning: *since the work of Coolidge and Langmuir around 1912 no essential innovations in the physics of the incandescent lamp have taken place*. Most further developments, some of which were very innovative, belong to the fields of technology and manufacture. Typical aspects of technological improvements, evolutionary developments or optimizations are, for instance: type extensions, lamp components (e.g. getters), coiled coils, reflectors and coatings, automotive lamps, environmentally more acceptable materials, etc. A historical review on lamp manufacture has been given by Veldman in 1994 [15].

There has been a steady flow of papers on the science of incandescent lamps. Here two highlights will be mentioned relating to: heat-and mass transfer, and the end-of-life mechanism[14]. Langmuir himself modeled in 1912 the heat transfer by conduction over a stagnant gas layer, the *Langmuir layer*, followed by convective transport outside this layer. This theory was further developed by Fonda and later by Elenbaas [17]. A milestone in this area was made in 1974 by the work of Fischer and Fitzgerald, who calculated the temperature profile and the flow field from basic principles. The results were found to be in agreement with experimental results using holographic interferometry and visualization of the flow field using fine quartz particles. The availability of computer programs to solve the transport equations forms the basis of design programs, which every company has been using.

A second important aspect relates the end-of-life mechanisms. An informative review on the failure mechanisms is given by Covington[18]. These causes may be quite diverse, but most important in this connection is the so-called *'spot model'* from 1972 [19], which is based on the existence of local inhomogenities in the filament. These inhomogeneities may be either initially present or may develop during lamplife as a result of all kind of mass transport mechanisms. The simplest example is the intially present local constriction of the wire diameter. Ohmic dissipation is then locally enhanced, resulting in locally enhanced W transport, which gives a positive feedback to the process, ultimately resulting in local melting of the filament. Incorporation of such mechanisms in the design

[14] Both topics have been lucidly discussed in ref. [16], where also the main references can be found.

programs has led to quantitative lamp life prediction.

HALOGEN LAMPS

The final innovation (at least up to now) took place in 1959. It essentially implies the deliberate introduction of chemical reactions in the light bulb. It had been recognized since the end of the 19th century that blackening in carbon filament lamps was due to the equilibrium reactions

$$CO_2 + C \leftrightarrows 2\,CO\,.$$

The equilibrium is shifted to the right at filament temperatures and shifted to the left at wall temperatures. This gives rise to chemical transport[15] of carbon from filament to wall. The elimination of oxygen and water, thus avoiding or minimizing the reactions was already a key-factor in the processing of the first carbon filament lamps. Around 1912 it was again Langmuir, who elucidated the importance of W-transport due to the equilibrium

$$W + 2\,H_2O \leftrightarrows WO_2 + 4\,H\quad.$$

Based on these transport observations in lamps and inspired by purification techniques via metal iodides in inorganic chemistry (van Arkel-de Boer process) the idea arose to introduce iodine in the light bulb[16]. This appeared to give rise to transport reactions which transport tungsten from the wall to the filament, the reverse of the 'old' transport chemistry as discussed above.

Four scientists from GE are the inventors of tungsten-iodine lamp E.G. Zubler (1925-2004), F.A. Mosby, E.G. Fridrich and E.H. Wiley [22]. The tungsten-iodine reactions appeared to keep the bulb wall clean. This dissolving of deposited tungsten in the gas phase at wall temperatures allowed the application of higher filament temperatures, small bulbs and higher inert gas pressures, thus allowing to carry the incandescent lamp concept to its very limits. A notable further innovation was the use of CH_2Br_2 as the filling gas [23]. Technologically this constituted an important step forward in the manufacture of the lamp, since dosing of the gaseous CH_2Br_2 is easier than dosing iodine.

The high-temperature chemistry of the halogen lamp lends itself nicely for thermodynamic considerations. The first paper on this topic came from Kopelman and van Wormer [24]. This was followed by a plethora of papers full of partial pressure curves (see e.g. [25]). The chemistry appeared to be more complex than initially thought, for instance, for proper operation of the tungsten-iodine cycle the presence of minor amount of oxygen is necessary. The species WO_2I_2 was proven to play a major role.

A significant further step forward in efficiency improvement was the application of *infra-red reflecting optical filters* [21, 26]. These interference (or semiconductor) filters allow the visible radiation to be transmitted and the infra-red heat radiation to be reflected back to the filament, thus enhancing the efficacy. Especially around 1970 the possibilities of optical filter application became a hot R&D topic. In theory the possibilities are clear but the practical realization appeared to be very hard. The developments have resulted e.g. in halogen lamps with a filter on the outer surface. Typical data for such a coated lamp are 30W, 20 lm/W, 3000 h lamp life. Further important products in this connection are halogen lamps in coated reflectors [21].

QUEST FOR EFFICIENCY BREAKTHROUGH

[15] See Ref.[20] for a detailed account of chemical transport. Ref. [21] gives a brief introduction especially with respect to lamps. Lamps always involve large temperature gradients, which make them pre-eminently devices for chemical transport. Carbon transport processes in lamps have been studied in detail [7].

[16] Introduction of a halogen in the light bulb was already patented in 1882. See references in [7]

In the previous sections the essential historical steps have been discussed. During history there have been many attempts to come with 'new' concepts. This relates particularly to *new incandescent materials*. The quest for such materials continues to this very day. The use of selective thermal radiators available up to around 1970 has been ruled out by Kauer [6]. The use of tantalum carbide [27] (melting point over 4150 K) and other binary refractories remains to attract attention, but the incongruent vaporization behaviour of these materials will always pose a problem. However, at the end of the 20th century new ideas arose on surface-modified tungsten. A first suggestion on nano-scale structures has been made by Waymouth [28]. In recent years publications on so-called *photonic crystals* pop up (see e.g. [29, 30]), which all predict promising results. With whatever new innovative material, however, the essential problem will always remain, how to maintain these structures for a sufficiently long duration at high temperature[17].

FINAL REMARKS

The history of the incandescent lamp represents a unique chapter covering 200 years in the history of technology. The impact of the incandescent lamp on society in the broadest sense has been enormous. It is to be expected that with the end of the incandescent lamp era more extensive historical studies will be published. The present paper is only a small contribution related to the scientific aspects. The discussed topics, however, cover to the opinion of the author the basic issues.

REFERENCES

1. Official Journal of the European Union 24-3-2009 L76/3 Commission Regulation 244/2009 18 march 2009. See also : http://ec.europa.eu/energy/efficiency/ecodesign/lumen/doc/incandescent-bulbs-en.pdf.
2. A. Heerding , The History of N.V. Philips Gloeilampenfabrieken, Vol. 1 *The origin of the Dutch incandescent lamp industry,* Vol. 2 *A company of many parts,* (translation) Cambridge University Press 1986.
3. J.W. Howell and H. Schroeder, History of the Incandescent Lamp, The Maqua Company, Schenectady, 1927.
4. G.W. Stoer, History of Light and Lighting, published by Philips Lighting Division, Eindhoven, The Netherlands, 1986.
5. H.F. Ivey, Review article : Candoluminescence and radical-excited luminescence, J.Luminescence 8, 1974, 271-307.
6. E.Kauer, Generating lights with selective thermal radiators, Philips Techn. Rev. 26, 1965, 33.
7. W.J. van den Hoek, An evaluation of the carbon filament lamp, Philips Res.Repts. 31, 1976, 129-152.
8. L. Meadows, The limits to growth, Universe Books, New York, 1972.
9. L.M.Nijland, Lamp development and the scarcity of vital materials, CIE session, London, september 1975, p.54.
10. W.F. Knippenberg ,B.Lersmacher, H.Lydtin and A.W.Moore, Pyrolytic graphite, Philips Techn. Rev. 28, 1967, 231.
11. C.R. Hammond, The Elements, in Handbook of Chemistry and Physics 49th Ed. 1968., B97, The Chemical Rubber, Cleveland.
12. J.W. Howell, loc.cit . p.91-129.
13. E.Pink and L.Bartha, The metallurgy of Doped/Non-Sag Tungsten, Elsevier Appl.Sc. London, 1989.
14. C.G.Suits and M.J.Martin, Irving Langmuir (1881-1957) A biographical Memorial, 1974 Nat.Ac. Sci.
15. H.E.Veldman, Innovaties in de lampenfabricage bij Philips, 1900-1980, Thesis , Technical

[17] A typical time constant for surface-diffusion of W at 2500K is of the order of several hours.

University Eindhoven, 1994.
16. E.Fischer, J.Fitzgerald, W.Lechner and W.Lems, Transport phenomena in incandescent lamps and the burn-out mechanism, Philips Techn.Rev, 35,1975, 296.
17. W. Elenbaas, Light Sources, Macmillan, London, 1972.
18. E.J. Covington, The failure of incandescent lamps, in E.Pink and L.Bartha, loc.cit.p.267-283,
19. H.Hörster, E.Kauer, W.Lechner, The burn-out mechanism of incandescent lamps, Philips Techn.Rev. 32, 1971, 155.
20. H.Schäfer, Chemical Transport reactions, Academic Press, New York 1964.
21. W.J. van den Hoek, G.M.J.F. Luijks, J.J.de Groot Lamps, Ullmann's Encyclopedia of Industrial Chemistry, 7th Ed., 2006, Wiley, Weinheim,
22. E.G. Zubler and F.A. Mosby, Illum.Eng. 54, 1959, 734.
23. G.T'jampens and M.H.A. van de Weijer, Philips Techn. Rev.27, 1966, 173,
24. B.Kopelman, and G.H. van Wormer, Thermodynamic considerations on Tungsten Halogen Lamps, Illum.Eng. 1968, 176.
25. J.H.Dettingmeijer, G.Dittmer, A.Klopfer and J.Schröder, Regenerative cycleprocesses in tungsten-halogen lamps, Philips Tech. Rev. 35, 1975, 302.
26. H Köstlin, Application of thin semiconductor and metal films in energy technology, Festkörperprobleme 22, 1982, 229.
27. See Patents WO 2006/045273 and WO 2006/060995
28. J.F.Waymouth, Invited paper at Light Sources conference, 1989, York.
29. Sandia Nat.Labs. News release May 2002.
30. A.Y. Vorobyev, V.S.Makin , C.Guo, Phys.Rev.Lett. 102, 2009, 234301.

ACKNOWLEDGEMENTS
The author thanks his former colleagues Marco Haverlag, for suggesting and supporting the present work, and Peter Drop for stimulating discussions.

Session One

Light Source Science & Technology 2010: A New Paradigm Emerges

R. Devonshire

Department of Chemistry, University of Sheffield, Sheffield S3 7HF, UK

Contact: r.devonshire@sheffield.ac.uk

ABSTRACT

Lighting is in the midst of one of its most revolutionary and complex periods of change. The recent introduction of solid-state lighting (SSL) technologies is instigating major changes in the entire industry, from producer through to end-user. The impact of SSL is not a result simply of its adding to the existing technology mix but because its appearance has coincided with, and in part helped create, other strong influences on lighting. These influences include: changes to the structure of the supply chain; accommodation of much longer bulb operating lifetimes; measures of, and standards for, light quality and lamp performance; increasing requirements on, and opportunities for, lighting controls; increasing awareness and knowledge of human interactions with artificial light; new environmentally-related legislation; and recognition of the importance of lighting in global electricity energy consumption. The talk will address these separate themes and analyse their various interactions.

INTRODUCTION

These are extraordinary times to be involved in the lighting industry; the pace and complexity of the changes taking place are unprecedented. The extraordinary image shown in Figure 1 below is perhaps an allegory for what is happening.

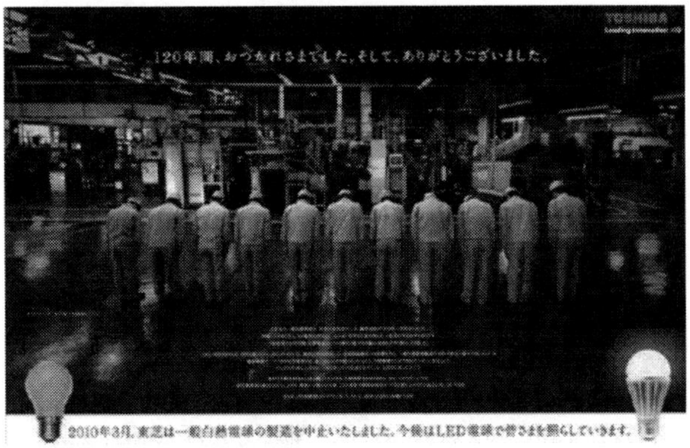

Fig. 1: The closure of an incandescent manufacturing line March 2010
(*Reproduced with permission of Toshiba Lighting Technology*)

The picture in Figure 1 was taken at the closure of a major Japanese manufacturer's last remaining incandescent lamp manufacturing line in March 2010, a closure that anticipates an instruction from the Japanese government to end Japanese domestic manufacture of such bulbs by 2012. Many other countries also have plans in place to remove low efficacy incandescent bulbs from their domestic markets [1]. The workers in the picture are showing their respect and appreciation for a record of continuous production that stretched back 120 years to 1890 when the company produced the first incandescent bulbs in Japan, at a rate of 10/day.

DISCUSSION

The general background situation in the lighting industry that has resulted in such closures is a complex one. The changes do not reflect simply the impact of the appearance of a more efficient technology, something that has happened many times previously in the history of the industry, but they are a response to more fundamental aspects of the manufacture and performance of the newly available, and imminent, light source technologies and to a variety of other influences on the industry, both internal and external.

Examples of the different influences that are being factored in to new industry models include:
(i) major changes to the structure and nature of the supply chain where the LED chips and modules are sourced from a small number of high volume LED manufacturers and supplied to a large, diverse and rapidly growing number of SSL system integrators;
(ii) the potentially much longer operating lifetimes of white LED sources in comparison with existing thermal and discharge sources is not only a challenge to important practical issues such as accelerated life testing but also impacts important economic models both for manufacturers (pricing) and end users (total cost of ownership);
(iii) major differences in the nature of the spectral power distributions from white LEDs, e.g., narrow-band LED plus phosphor, RGB LED, and RAGB LED, in comparison with existing thermal and discharge sources is necessitating the introduction of new measures of colour rendering performance and increasing attention being given to colour temperature;
(iv) increasing requirements for the control of lighting installations, in particular for lighting energy management and lighting maintenance in large commercial buildings and in large scale, integrated, street lighting systems, but also for residential lighting in relation to security, energy savings, etc.;
(v) increasing awareness and knowledge of human interactions with artificial light;
(vi) increasing legislation [2]. The industry in Europe is subject already to major, environmentally-related, EU Directives (RoHS; WEEE; EuP; and Exposure to Artificial Optical Radiation), however, increasingly sophisticated models of light source Life Cycle Analysis (LCA) are being developed (Figure 2; [3]) whose impacts on the industry are not yet clear.

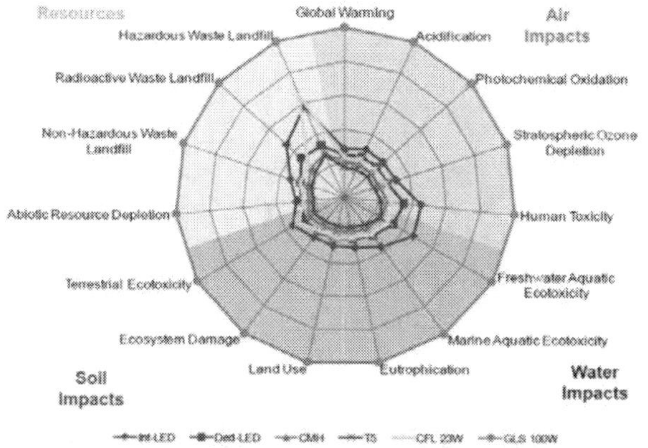

Fig. 2: An LCA analysis of different light source technologies using as a benchmark (i.e., perimeter) a 100W GLS lamp

and (vii) government and public awareness of the important contribution that artificial lighting makes towards total global electricity energy consumption. The open interface between consumers and residential lighting technologies, i.e., a market sector that is dominated by bulb replacement at end-of-life on the basis of standardised sockets, has played a major part in lighting becoming a focal point for public debate on global warming.

Taken together these various influences are not only changing the industry model but, of course, are changing the entire focus and scope of the supporting R&D. The LS:X symposia have provided a regular snap-shot of the lighting industry's R&D priorities and concerns since the series began in the 1970s. The LS12-WLED3 conference proceedings are likely to be a watershed in the series.

REFERENCES

[1] P. Waide, "*Phase Out on Incandescent Lamps*," IEA Information Paper, April 2010
[2] R. Devonshire, "*The Competitive Technology Environment for LED Lighting*," J. Light & Vis. Env., 32(3) 275-287 (2008)
[3] Defra Report SPMT08_069, "*Life Cycle Assessment of Ultra-Efficient Lamps*," Navigant Consulting Europe, May 2009.

On solder joint reliability in LEDs by Accelerated Life Testing

M. Erinc[1], J. Kloosterman [2], E.P. Veninga[1], W.D. van Driel[2], A.W.J. Gielen[1], G.Q. Zhang[2]

[1] *TNO Science and Industry, Materials Performance, NL-5600HE, Eindhoven, The Netherlands*
[2] *Philips Lighting, LightLabs, NL-5611BD, Eindhoven, The Netherlands*
Contact: Muge.Erinc@tno.nl

ABSTRACT

The increasing use of solid-state lighting brings an industrial need for efficient thermal characterization methods to assess long-term reliability [1]. Accelerated Life Test (ALT) is an empirical (or semi-phenomenological) method used to identify the limiting failure modes of a product and the stresses at which these failures occur. The motivation of this study is to predict long-term reliability of LEDs on a physics of failure basis, which requires a fundamental understanding and description of the damage mechanisms involved. The aspects to be covered in this study are material degradation (both continuum methods and fracture mechanics), failure analysis, damage modelling and reliability modelling of LEDs. Main goals of the study are:

1. Developing numerical tools to evaluate LED reliability with a physics of failure approach based on ALT.
2. Correlating ALT performance to the service conditions and developing numerical tools to incompatibility.

INTRODUCTION

The purpose of ALT is to induce field failure in the laboratory at a much faster rate by providing a harsher yet representative environment. Currently, the industrial trend is towards products with long-term reliability which have to be, contradictorily, developed in a shorter time to market. For instance, typical life times for automotive electronics is 15 years, LEDs 7,000 to 50,000-100,0000 hours and solar panels more than 20 years. These long term reliability requirements make the established 1.000-6,000 hours ALT schemes questionable.

ALT amplifies the rate of damage evolution (delamination, creep-fatigue) that should have happened at a slower pace. ALT schemes are generally validated by comparing the specimens failed in service with specimens failed in ALT. It is questionable whether the qualitative base for ALT is well substantiated in this manner; time dependent mechanisms, mixed interactions induced during accelerated testing such as time and temperature, temperature and humidity etc. are generally overlooked.

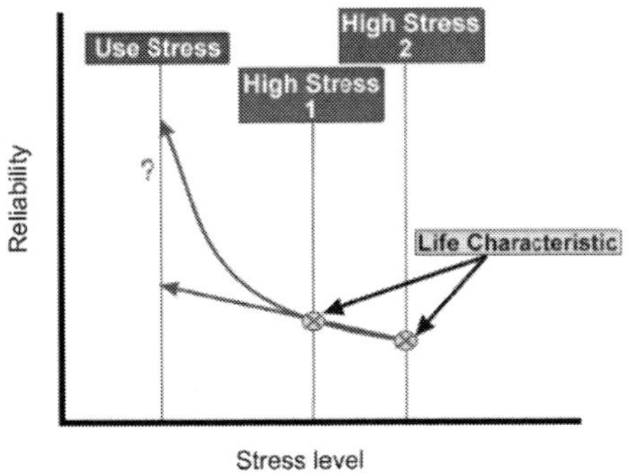

Figure 1. Schematic of stress level vs. life time for a hypothetical electronic product [2].

The reliability information obtained by ALT are fed into empirical or (semi) phenomenological models, and used to extrapolate the long-term reliability, of which the accuracy of the predictions and tolerances are questionable. Furthermore, from an engineering perspective, how to correlate the information obtained from ALT to the field use, quantitatively, is a remaining research question (Figure 1).

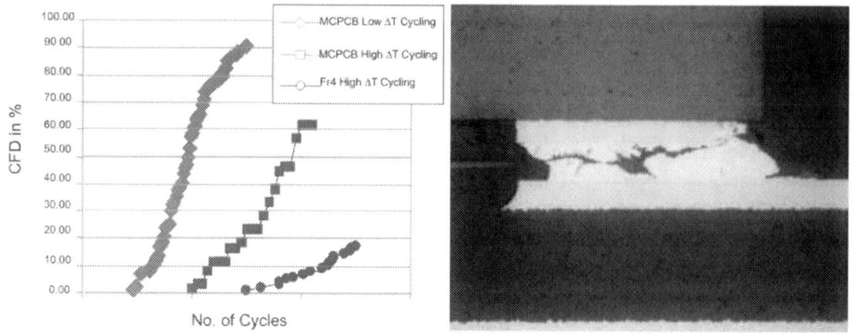

Figure 2. Cumulative failure distribution (left) and failure mode (right).

DEMONSTRATOR: SOLDER FATIGUE

At the LED-Package level the reliability challenge is dominantly solder joint fatigue and mixed creep-fatigue interactions. LED packages bring a variety of dissimilar materials together such as metals, polymers, ceramics etc. Especially for ceramic based LED packages, the amplified mismatch in material properties causes solder joint fatigue failures. In addition, long-term performance of SnAgCu solders is not well known and there is a knowledge gap in the literature with respect to long-term reliability. Factors influencing solder joint fatigue are:

- CTE, to a lesser extent Young's modulus of the substrate and component
- Component / Substrate thickness, dimensions / footprint
- Component lay-out (Leadless, Gullwing, J-lead, S-lead etc.)
- Type of solder, solder height
- Temperature cycling range, mean temperature and dwell times
- On / off frequency

In the ALT approach adopted here, the following factors are considered: Large LED versus small LED, different substrates (FR4, Metal Core PCB and polyimide based), solder type and the effect of temperature swing (5 different swings). Finite element models are created and validated to a selected number of test variations. This combined experimental - numerical approach is expected to provide a 95% confidence interval on solder joint fatigue predictions. High confidence intervals are crucial for accurate warranty period calculations and cost estimations in SSL systems. In Figure 2, different failure performances of an LED device mounted on MCPCB, subjected to a low temperature swing and a high temperature swing is shown.

FUTURE WORK

The goal of this study is to develop numerical tools for long-term reliability predictions of LEDs, based on ALT performance using a physics of failure approach. In order to achieve this goal, the following will be carried out:

1. Main damage mechanisms will be identified and described by constitutive relationships using specimens in-house subjected to ALT.
2. LEDs subjected to ALT will be simulated in a commercial finite element package. In this way the material and damage models will be validated based on dominant damage mechanisms and failure criteria.
3. Once a validated model is obtained, actual field conditions can be simulated promptly, and correlations between ALT and service use will be established. Thereby, acceleration factors will be calculated.

REFERENCES

[1] C.J.M. Lasance, 'Challenges in LED thermal characterization', In the Proc. of 10th Int. Conf. on Thermal, Mechanical and Multiphysics Simulation and Experiments in Micro-Electronics and Micro-Systems, EuroSimE (2009).

[2] Website: http://weibull.com/ . Access date: March 2010.

Highly efficient bulb-shaped ballast-integrated LED lamps replacing incandescent lamps for general lighting

Masao Segawa , Shigeru Osawa , Takeshi Hisayasu , Masahiko Kamata , Takuro Hiramatsu , Takeo Yasuda

Toshiba Lighting & Technology Corporation, 1-201-1, Funakoshi, Yokosuka-city, Japan, 237-8510
Contact: masao.segawa@tlt.co.jp

ABSTRACT

The progress of light emitting diode (LED) in luminous efficacy has been made extremely rapid, and the luminous output for one package of LED has also much more increased in these several years. Thus, the bulb-shaped ballast-integrated LED lamps, emitting total luminous flux of several-hundred-lumens, have been realized to replace incandescent lamps (ILs). In our developed LED lamps, the maximum value of the luminous efficacy reaches 93 lm/W, which already exceeds those of ballast-integrated compact fluorescent lamps (BCFLs).

INTRODUCTION

High efficacy is one of the most important characteristics of electric lamps for general lighting. It enables to reduce the emission of CO_2, considered as a global warming gas, at power stations, and it also simply reduces electric fees for individuals. ILs are now going to wane, and their uses have started to be banned in many countries because their efficacies are as low as approximately 10 to 20 lm/W. In spring of 2008, Japanese government announced the policy to abolish the production of ILs by 2012. As a result, the sales record of BCFLs in fiscal 2008 was 1.4 times increased for the previous year. BCFLs normally have the efficacies around 50 to 70 lm/W, and the shape and size are nearly the same as those of an IL in particular BCFL [1]. Consequently, these lamps were thought to be the most promising substitute for a while. However, the year of 2009 in Japan showed completely different situation in this field and would be memorized as the breaking year for LED lamps.

LAMP STRUCTURE

Figure 1 illustrates the typical structure of our developed LED lamp. The hemisphere-shaped diffusive cover is used to enlarge the vertical light distribution angle and simultaneously make the total lamp image resemble a popular IL bulb. The vertical light distribution angle is yet narrower than those of ILs (and BCFLs) as shown in Figure 2, which is an obvious problem required to be improved in the future. Accordingly, LED lamps are most convenient to be used in downlight fixtures so far.

Since the luminous output of LED decreases as the temperature increases, heat control is a key design parameter to achieve a high efficacy. Furthermore, it also affects the reliability and luminous maintenance characteristic during the long life of tens of thousand hours. The aluminum body is designed as a heat conductor and radiator. The number and thickness of fins, the width of ditches are optimized by experiments as well as the calculations of CFD (computational fluid dynamics), for natural air convection.

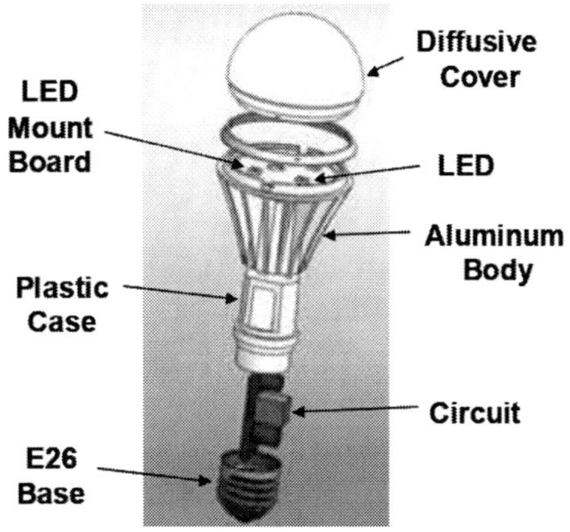

Figure 1. Typical structure of our developed LED lamp. The hemisphere-shaped diffusive cover is used to enlarge the light distribution angle and make the total lamp image resemble a popular IL bulb.

TECHNICAL DATA

The specifications of our developed LED lamps operated with 100 V input are shown in Table 1. Each type has two kinds of colors: 2800K and 5000 K. The luminous efficacy of the former is 24-37% lower than that of the latter in all lamps. This is due to the loss by the down-conversion from LED-chip blue to YAG-phosphor yellow, as often seen on other LED applications. The efficacies of 52 to 69 lm/W in 2800 K type are comparable to those of BCFLs; however, 76 to 93 lm/W in 5000 K are overwhelming.

The lifetime is determined according to the temperature of LED. Lifetime of the lamp named LDA3-E17 is half of those of the others. Though the consuming power is as low as only 3.4 W, the heat density is relatively high because of its smallness.

It is needless to say that the luminous run-up time of LED is extremely short, thus the run-up time of the LED lamp is determined by the boot-time of the circuit. It is still short within several tens of milli-seconds, while the time of BCFLs are several tens of seconds to several minutes. This is a significant difference, especially in case of use under cold circumstance.

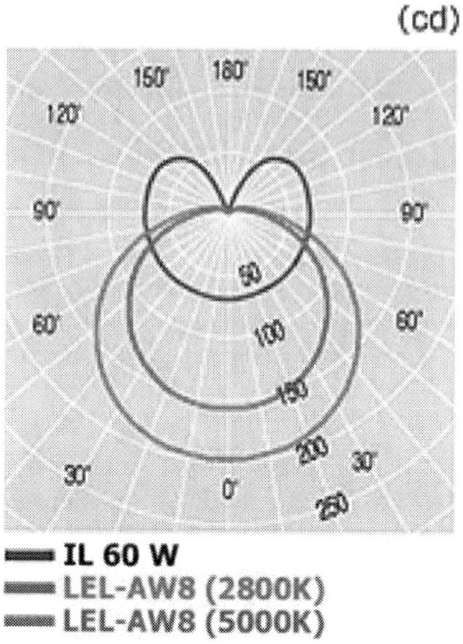

Figure 2. Vertical light distribution curves of IL 60 W and LEL-AW8 (2800K and 5000K). All lamps are operated without fixture on base-up position.

Table 1. Specifications of our developed 4 kinds of bulb-shaped ballast-integrated LED lamps. While these lamps cannot be used with dimmer, dimmable LED lamps have been also developed with slightly low efficacy.

Type Name	LDA3-E17		LDA5		LDA6		LEL-AW8	
Base	E17		E26		E26		E26	
Consuming Power (W)	3.4		4.6		6.4		8.7	
Color Temperature (K)	2800	5000	2800	5000	2800	5000	2800	5000
Luminous Flux (lm)	220	290	240	350	330	520	600	810
Efficacy (lm/W)	65	85	52	76	52	81	69	93
Life Time (hours)	20,000		40,000		40,000		40,000	
Appearance								
Outer Diameter (mm)	ϕ43		ϕ60		ϕ60		ϕ60	
Total Length (mm)	72		107		107		119	

SUMMARY

We have developed several kinds of LED lamps with the features of high efficacy and long lifetime. These lamps possibly replace ILs, especially used in downlight fixtures. The highest efficacy reaches 93 lm/W with the correlated color temperature of 5000 K, and this value has already exceeded that of BCFL. It is demanded by market to enlarge the vertical light distribution angle.

REFERENCES

[1] T. Yasuda et al., J. Light & Vis. Env. 33-3 (2009) 161.

Physics and Chemistry of the LiFi® Lamp

Light Emitting Plasma in a Resonant Dielectric Cavity

R. Gilliard, M. DeVincentis, A. Hafidi, D. O'Hare, G. Hollingsworth
Luxim Corporation, 1171 Borregas Ave., Sunnyvale, CA 94089 USA
Contact: rgilliard@luxim.com

ABSTRACT

The LiFi® lamp generates light from a thermal plasma in which electrical power is delivered through an electric field produced in a resonant cavity at high frequency rather than through electrodes. The result is a long-lived lamp without many of the failure mechanisms of conventional HID lamps. The flatter arc temperature profile results in a fuller spectral power distribution, generally resulting in high efficiency and improved color quality over standard metal halide lamps.

INTRODUCTION

Although a light emitting plasma in a resonant cavity produces light in much the same way as a standard HID discharge, there are subtle differences. For example, arc position is not controlled by electrode placement and chemical processes do not involve tungsten transport.

CONFIGURATION

The configuration of the LiFi® lamp exhibits similarities and differences when compared to a conventional HID lamp. In both cases, light is produced by an LTE discharge in which power is delivered by ohmic heating of the gas. The means for creating the electric current are different, however.

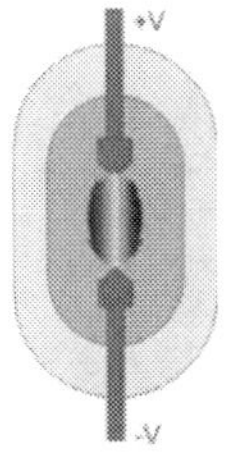

Figure 1: Configuration of Conventional HID Lamp

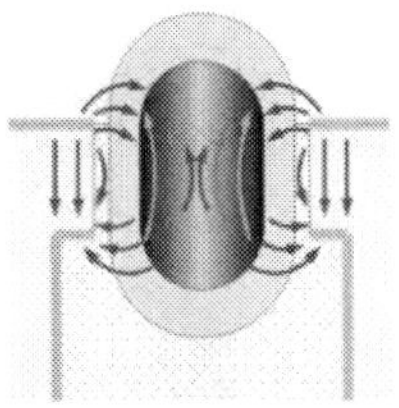

Figure 2: Configuration of LiFi® Lamp

In the case of the standard HID lamp the electric current is generated by an axial potential difference across the partially ionized gas. In the case of the LiFi® lamp, an electric field is produced in a dielectric material. A gap in the material enables a fringing field in which a filled capsule can be positioned. This results in an electric field configuration in generally the same axial direction as in a conventional arc tube.

The inherent advantages of operating the discharge without the need for electrodes are clear. Electrodes require a glass-to-metal seal which is generally the weakest part of the structure, an important consideration in high pressure operation. In addition, tungsten electrodes, operating at high temperature, are inherently reactive with typical fill materials and the arc tube material. Various transport processes deposit tungsten on the wall causing darkening and a diminution in light output. Finally, the typical voltage drops associated with electrodes themselves result in a power loss which ultimately lowers the efficiency of the lamp.

The fill materials for the capsule include the same set of materials as are currently used in HID lamps plus others that are normally excluded from electroded lamps because of their incompatibility with tungsten and other refractory metals.

Power delivery within the capsule does not result in the typical parabolic temperature profile common to an electroded HID lamp in a cylindrically symmetric capsule. In a LiFi® lamp, the temperature peaks at a radius typically greater than half the capsule radius rather than at the center. Unlike an electroded lamp in which the current flows through the core of the arc tube, the field is produced externally and its penetration is limited by the plasma itself resulting in temperature profiles shown below. Since the skin effect is frequency dependent, peak temperatures are at increasingly higher radii for higher operating frequencies.

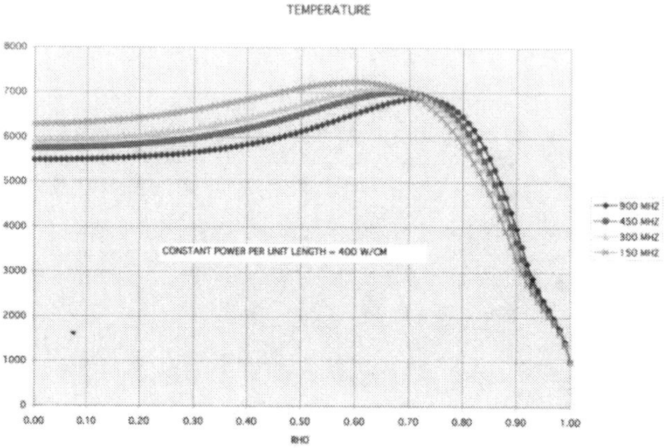

Figure 3: Temperature Profiles from Fringing Fields[1]

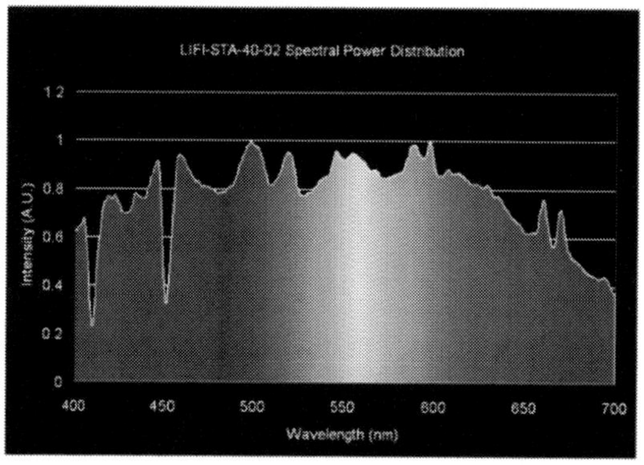

Figure 4: Example LiFi® Lamp Spectrum

CONCLUSIONS

The use of fringing fields from a resonator to produce a light emitting plasma has shown several advantages over standard HID technology. Besides the obvious benefits of electrodeless operation, the flatter plasma profile results in broad spectral power distributions in compact capsules for high brightness sources with excellent color quality. Typical lamp fills include rare earth and other metal halides. Efficiencies above 140 LPW have been achieved with excellent life characteristics.

REFERENCES

[1] John Waymouth, private communication
[2] J. F. Waymouth, *Applications of microwave discharges to high-power light sources, in Microwave Discharges: Fundamentals and Applications,* NATO ASI Series B, Vol. 302, eds. C. M. Ferreira and M. Moisan, Plenum, New York, 1993 pp. 427-443.

Photoluminescence of Oxygen Deficiency Centers in Fused Quartz - Effect of Thermal Treatment

M. Stamminger[1], B. Kühn[1], E. Arnold[2], H.-D. Witzke[2], F.-J. Schilling[2], A. Schreiber[1], C. Neumann[1]

[1] *Heraeus Quarzglas GmbH & Co. KG, Quarzstraße 8, D-63450 Hanau, Germany*
[2] *Heraeus Noblelight GmbH, Heraeusstrasse 12-14, D-63450 Hanau, Germany*
Contact: mark.stamminger@heraeus.com

ABSTRACT

The emission of low-pressure mercury lamps at 253.7 nm, mainly used for germicidal purposes, can be affected by an absorption band in the glass envelope centered at 245 nm. This absorption, due to oxygen deficiency centers (ODC) in fused quartz containing hydroxyl (SiOH), can be reduced by an annealing process which also removes the hydrogen from the glass. To investigate the kinetics of this process, a sample of flame-fused quartz glass was annealed in air at 1000°C, with hydrogen content, transmission, and photoluminescence (PL) being measured in regular intervals during the treatment to monitor the spatial and temporal development of the ODC content. The removal of hydrogen during thermal treatment is accompanied by a diffusion-influenced reduction in the oxygen deficiency center photoluminescence at 280 nm and 390 nm, though at a slower rate than for a reaction controlled solely by hydrogen diffusion.

INTRODUCTION

The transmission loss at 245 nm caused by oxygen deficiency centers (ODC) in natural fused silica can be improved by thermal treatment. In [1], a reaction (see Equation 1) involving SiOH with an oxygen-deficient

$$2SiOH + 2Si^{+3} \xrightarrow{1000°C} 2SiOSi + H_2 \uparrow \quad (1)$$

defect in the glass was assumed to be responsible for the transmission improvement seen in natural fused quartz containing SiOH. The ODC was represented in this older work as Si^{+3} to signify a reduced condition of the glass without specifying the exact nature of the center. Although there has been considerable research done on silicon and germanium ODCs in the intervening decades, there is still some controversy as to the exact nature of the defect responsible for the absorption at 245 nm and the PL at 280 and 390 nm (see, for example [2]). We will simply use the general notation ODC in this work. In order to shed more light on the kinetics of the annealing process for the natural fused quartz Heralux, a flame-fused sample of dimensions 10 mm by 10 mm by 55 mm with about 165 ppm OH and $2.5*10^{17}$ hydrogen molecules/cm^3 (before annealing) was treated for 103 hours at 1000°C in air. In regular intervals, measurements of the H_2, SiH, and SiOH content were made with a Raman spectrometer (Jobin Yvon). Transmission was measured with a Lambda 950 spectrometer (Perkin Elmer) and the photoluminescence (excitation at 250 nm) measurements with an Edinburgh Instruments 920 fluorescence spectrometer.

RESULTS

The photoluminescence intensities at 280 nm and 390 nm under 250 nm excitation correlate with the absorption at 245 nm in fused quartz and give an indication of the ODC content in the sample [3]. The PL intensities at 280 nm and 390 nm in the center of the sample, as well as the measured hydrogen content, were plotted versus annealing time at 1000°C (see Figure 1). The measured hydrogen content can be modeled with simple diffusion out of the sample at 1000°C, with a diffusion coefficient of $3.5*10^{-6}$ cm^2/s. This is within an order of magnitude of other literature values for the diffusion coefficient of hydrogen at 1000°C [4]. The development of the PL at 280 nm and 390 nm during the annealing cannot be modeled as well by simple diffusion, and the best fit simple diffusion curve is achieved with a diffusion coefficient of $0.9*10^{-6}$ cm^2/s for both PL wavelengths. The PL measurements have the advantage of greater sensitivity in comparison with the transmission measurements and the Raman measurements of hydrogen. After about 40 hours, further changes in transmission could not be detected, and after twenty hours the hydrogen content was below the detection limit of the Raman spectrometer. The PL at 280 nm and 390 nm, however, could still be measured after 103 hours. Figure 2 shows the PL at 280 nm measured over the thickness of the sample after 56 and 78 hours, respectively. The PL profiles show a typical diffusion-controlled form, with lower concentrations toward the edges of the sample. These results demonstrate that the ODC content continues to decrease even after transmission changes can no longer be measured. Given the correlation between the PL kinetics and the kinetics of hydrogen removal by diffusion, these PL profiles at 280 nm are likely indirect evidence of the very small hydrogen concentrations which still exist in the sample, and influence the ODC equilibrium, but cannot be measured by other means. If Reaction (1) is valid, then the decrease in ODC PL should be accompanied by a reduction in SiOH. In the first 4 hours of the thermal treatment, the SiH content of the

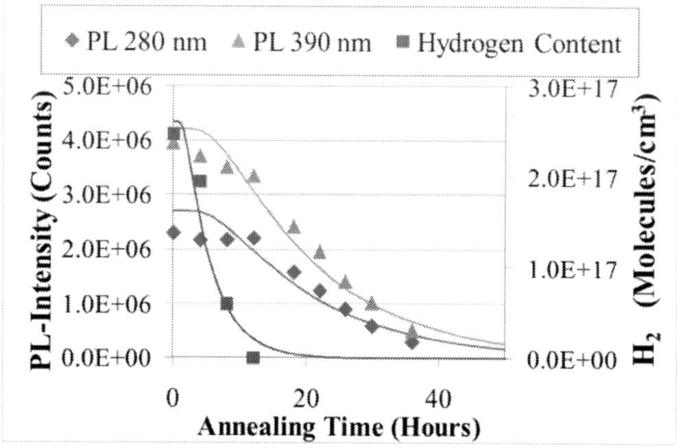

Figure 1: PL at 280 and 390 nm and H_2 Content in Heralux during Thermal Treatment at 1000°C

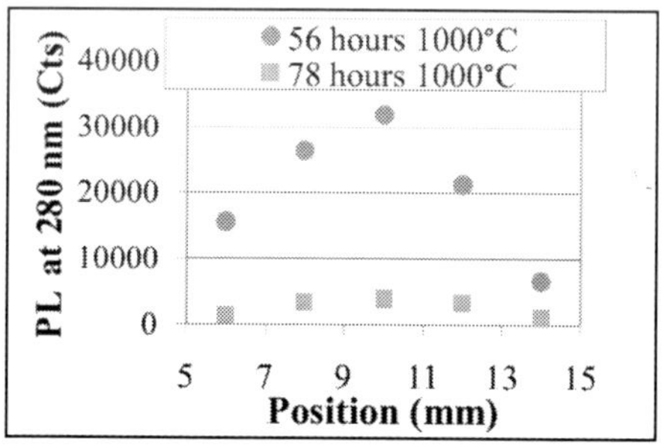

Figure 2: Photoluminescence at 280 nm across the thickness of the sample after 56 and 78 hours at 1000°C

sample decreased by $9.4*10^{17}/cm^3$ and the SiOH content decreased by $8.9*10^{17}/cm^3$. However, because these decreases are approximately the same, they can be fully accounted for by another reaction which occurs during the thermal treatment (Equation 2). A rough calculation of the initial ODC content based on Smakula's equation [2] gives at most about $5*10^{16}$ ODC /cm^3. By reaction (1), this would lead to a decrease in SiOH content of the same magnitude, which value is too small to be distinguished by our Raman method.

$$SiOH + SiH \xrightarrow{1000°C} SiOSi + H_2 \uparrow \qquad (2)$$

Thus, the proposed reaction (1) may still give a valid description of the annealing process.

CONCLUSION

For low-pressure mercury lamps with emission at 253.7 nm, thermal treatment can improve the transmission of the fused quartz envelope in the 245 nm region. In Heralux, the removal of hydrogen during this treatment is accompanied by a diffusion-influenced reduction in the ODC photoluminescence at 280 nm and 390 nm, though at a slower rate than for a reaction controlled solely by hydrogen diffusion. This leads to the conclusion that the kinetics of the ODC reduction reaction limits the overall rate of the process. A direct proof of the contribution of SiOH to the ODC reduction reaction was not found.

REFERENCES

[1] T. Bell, G. Hetherington, K.H. Jack, Physics and Chemistry of Glasses, Vol. 3 No. 5 October 1962.
[2] L. Skuja, Journal of Non-Crystalline Solids, 239 (1998) 16-48.
[3] Y. Sakurai, Journal of Non-Crystalline Solids, 271 (2000) 218-223.
[4] R.H. Doremus, Diffusion of Reactive Molecules in Solids and Melts, Wiley, 2002, p. 151.

High pulsed power UV and VUV excilamps

V. Tarasenko, E. Baksht, M. Erofeev, M. Lomaev, D. Rybka, D. Sorokin
High Current Electronics Institute, Akademicheskii ave., 2/3, Tomsk, 634055, Russia
Contact: VFT@loi.hcei.tsc.ru

ABSTRACT

This paper reports on the pulse sources of spontaneous UV and VUV radiation based on the non-equilibrium high-pressure gas discharge plasma. Emission characteristics of runaway electrons preionized diffuse discharge plasma of inert gases and mixtures of inert gases with halogens, and nitrogen under pressure up to 15 atm. and at high specific power of excitation (~ 100 MW/cm^3) have been investigated. The power of ~ 1 MW /cm^3 at duration of ~ 8 ns of Xe_2^* dimmers radiation was obtained at at pressure of 12 atm.

GENERAL

The characteristics of radiant emittance of the volume high-current discharge plasma in inert gases at pressure greater than 1 atm. was investigated by using a RADAN-220 generator with the voltage pulse amplitude of 220 kV and pulse duration at a matched load of ~ 2 ns. The width of the pulse leading edge was ~ 0.5 ns. A flat anode and a cathode with a small-curvature radius provided the electric field gain in the near-cathode region. Interelectrode distance ranged as 4 - 16 mm. The most homogeneous discharge was obtained in the high-pressure helium. The subnanosecond electron beam of fast electrons was registered in He at the pressure of 15 atm. The beam current pulse duration at a half-height was ~100 ps. A runaway preionized diffuse discharge in helium was realized at pressures up to 15 atm.

In xenon at pressure of 1.2 atm, the energy of spontaneous radiation in the full solid angle was ~ 45 mJ/cm^3, and the FWHM of a radiation pulse was ~ 110 ns. The spontaneous radiation power monotonic rise and decreasing of pulse duration in xenon was observed at pressures up to 12 atm. The power of ~ 1 MW /cm^3 at duration of ~ 8 ns of Xe_2^* dimmers radiation was obtained at this pressure, see Fig.1.

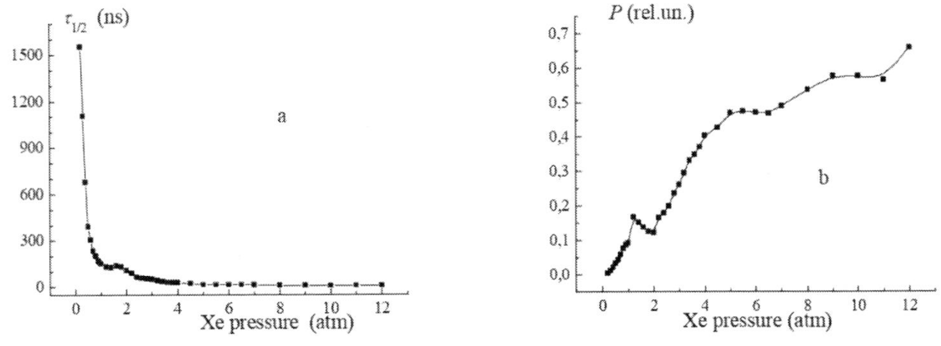

Fig.1. Duration (a) and power (b) of VUV radiation of Xe_2^* dimmers vs Xe pressure.

The maximum radiant power on Xe_2^* dimmers to the full spatial angle ~ 8 MW was obtained with set-up No. 2. This set-up forming a diffuse discharge without preionization, included a primary capacitive store, a pulse transformer, a coaxial line with a wave impedance of 10 Ω, a peaking spark gap and gas diode.

At a presence in argon and krypton small admix of xenon (~ 0.01%) was found high-power narrow-band emission at wavelength of ~ 147 nm to arise. The intensity of this band achieved its maximum in argon at pressure of 0.9 atm. (see Fig.2) and in krypton at pressure of 0.3 atm.

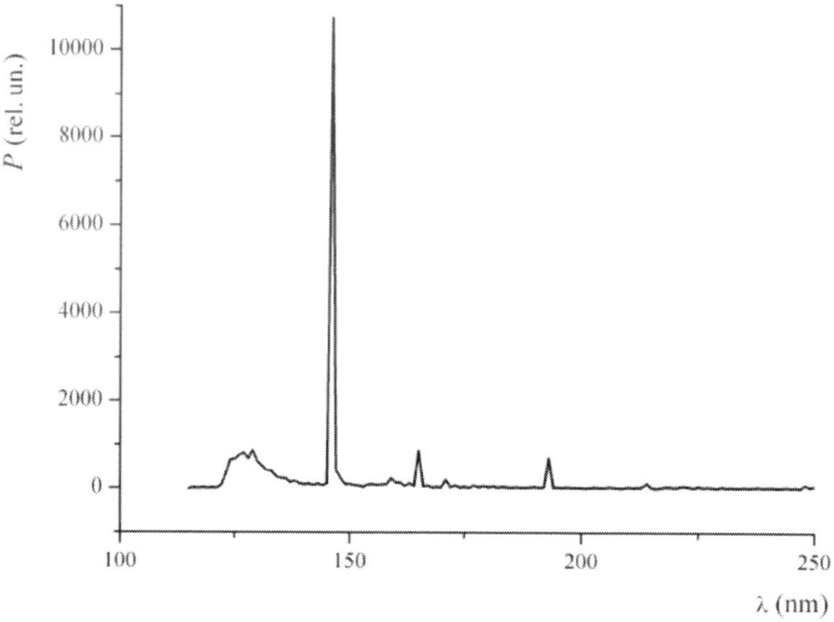

Fig.2. Emission spectrum in argon small admix of xenon (~ 0.01%).

The results of our experimental investigation of the radiation characteristics of nitrogen in a wide pressure range, p = 0.1–12 atm., upon excitation by a volume discharge initiated by a beam of runaway electrons. By optimizing the electrode system and reducing the inductance of the discharge chamber, the specific radiation power of transitions of the (2+) system of nitrogen was increased to ~ 50 kW/cm^3. In this case, the radiation power of the entire excited volume of the discharge plasma was ~ 120 kW. We determined the amplitude, time, and spectral characteristics of the (2+) system of nitrogen excited by volume and by contracted discharges. We showed that the time dynamics of the population of the $C^3\Pi u$ state at the trailing edge of the radiation pulse is determined by the radiative decay and collision quenching. By measuring the Stark broadening of spectral lines, we determined the electron concentration in the volume discharge plasma of helium in the pressure range 1–6 atm and in the discharge contraction channel in nitrogen. The estimation of the electron temperature in nitrogen at p ~ 1 atm. showed that the value of this parameter averaged per pulse is Te ~ 2.3 eV and that T_e rapidly decrease from ~3.5 to ~2 eV within a few nanoseconds after the breakdown of the gap.
Investigations of radiation of inert gas halogens excited by high voltage generator RADAN-150 (set-up No. 3) was carried out in the operating mixtures pressure range of 60–750 Torr at various inert gas/halogen. The opportunity of generation of a runaway preionized diffuse discharge in inert gas halogenides without preionization is also shown. Peak power of volume discharge radiation increases with increasing gap between electrodes. At low pressure (60–120 Torr) and discharge gap of 12 mm discharge had diffuse radiated cone-like form. With increasing of pressure discharge obtained form of the diffuse channel with diameter ~3 mm which contracted at the pressure of 500 Torr and than became in a spark. The optimal values of working pressures for the working gas mixtures Kr/Cl_2, Xe/Cl_2, Xe/Br_2 are 500 Torr at the ratio inert gas/halogen = 50/1. For Kr/Br_2 the working pressure was 750 Torr at Kr/Br_2 = 100:1. The highest pulsed power densities of KrCl*, XeCl*, XeBr*, and KrBr* molecules radiation were 3.7 kW/cm^2, 3.1 kW/cm^2, 4.5 kW/cm^2, and 2.1 kW/cm^2 at the efficiencies 5%, 4.8%, 5.5%, and 4%, respectively. In the optimal conditions, radiation pulse duration at a half-height was 30 – 40 ns. At this conditions input energy to discharge plasma was ~ 1 J. The radiation spectra of KrCl-, XeCl-, XeBr- excilamps possessing the high-pulsed power density consist of the narrow intensive bands of several nm at a half-height of B–X transitions of the respective molecules.

Colour rendition properties of solid-state lighting sources

A. Žukauskas[1] , R. Vaicekauskas[2] , M. S. Shur[3]
[1]*Institute of Applied Research, Vilnius University, Saulėtekio al. 9, bldg. III, Vilnius, LT-10222 Lithuania*
[2]*Department of Computer Science, Vilnius University Naugarduko g. 24, Vilnius, LT-03225 Lithuania*
[3]*Department of Electrical, Computer, and System Engineering, Rensselaer Polytechnic Institute, 110 8th Street, Troy, NY 12180 USA*
Contact: arturas.zukauskas@ff.vu.lt

ABSTRACT

Assessment of colour quality of solid-state lighting sources that contain narrow-band components requires approaches that are beyond the standard metric of colour rendering index (CRI). Neither CRI nor refined metrics that are based on measuring colour shifts for a small number of test colour samples are able to completely distinguish between two principle colour rendition properties of illumination: the ability to render object colours with high fidelity and the ability to increase chromatic contrast. To that end, the statistical approach, which is based on sorting a very large number of test colour samples in respect of just perceivable colour distortions of several kinds allows to reveal and assess three distinct types of solid state lighting sources: (i) high-fidelity lamps, which cover the entire spectrum with both the yellow and red spectral components present; (ii) colour-saturating lamps, which lack the yellow component; and (iii) colour-dulling lamps, which lack the red component.

INTRODUCTION

White light-emitting diodes (LEDs) have already outperformed common lamps in efficiency and are rapidly encroaching on the lighting market. However, assessing colour quality of illumination provided by solid-state lamps that emit due to narrow electroluminescence bands still presents an issue [1]. In particular, the conventional colour-quality metric, colour rendering index (CRI) [2], which has been introduced for ranking broad-band fluorescent lamps, appeared to be in a conflict with subjective visual preferences when a set of lighting sources under assessment contains white LEDs.

To resolve this conflict, here we address two colour rendition properties of illumination: the ability to make object colours appear "natural" (colour fidelity) and the ability to make colours appear "vivid" and easily distinguishable (colour saturating) [3]. We discuss the use of different approaches in assessing these two colour rendition properties and categorize solid-state lamps (white phosphor converted (pc) LEDs and multiwavelength LED clusters) into several groups depending on their colour-fidelity and colour-saturating characteristics.

COLOUR QUALITY METRICS

The most widely used metric for rating lighting sources in colour quality is CRI, which relies on numerically estimating colour differences for a set of test colour samples within the 1964 CIE colour space when a reference source is replaced by that under assessment. The integral figure of merit is the general colour rendering index R_a, which is the average of special CRIs derived form the colour differences for eight test colour samples. The main drawback of the CRI metric is disregarding the directions of the colour shifts, i.e. ignoring colour-saturating properties of a lamp. Other essential drawbacks are the lack of uniformity of the 1964 CIE colour space in respect of perceptually resolved colour differences and missing colour distortions for many hues due to the use of a small number of test colour samples.

An attempt to meet the need in improved assessing the colour rendition properties of LED-based lighting sources is colour quality scale (CQS) [4]. In particular, this metric employs a larger number of test-colour samples (15), a more equidistant colour space (CIELAB), improved averaging of colour shifts, and a more convenient rating scale. The main idea of CQS is in that an increase in chromatic saturation of a test colour sample is excluded from the colour shift. However, although within such an approach a lamp is not derated for colour-saturating effect, the CQS score still makes difficult distinguishing between lighting sources that render colours with high fidelity and those that render colours with increased chromatic saturation.

In order to reveal the colour-saturating effect, the CRI and CQS metrics can be complemented by gamut-area indexes, which are the relative change in the area of the polygon, defined by the coordinates of the test

colour samples in the colour space. However because of a small number of test colour samples used and both positive and negative contributions of different samples to the polygon area, the gamut area indices lack exactness. Also, they are difficult to collate with CRI and CQS scores because of the different format.
An alternative metric that allows revealing different colour rendition properties of lighting sources within a single format is the statistical approach [5]. The approach is based on computational sorting of a very large number of test colour samples (e.g. 1269 Munsell samples [6]) to several groups depending on the direction and magnitude of the colour shifts (colour rendition vectors, CRVs). In order to avoid analytical colour spaces, which are non-ideal approximations of perceptually uniform colour space, the behaviour of the CRVs is analysed in respect of the colour-discrimination shapes, which are built basing on the experimental data on just perceivable chromaticity differences (MacAdam ellipses) and luminance discrimination thresholds estimated individually for each colour test sample. In particular, the CRVs that do not escape from colour discrimination shapes are scored to the colour fidelity index (CFI) and those that have excess projections directed toward increased or decreased chroma are scored to the colour saturation index (CSI) or colour dulling index (CDI), respectively. Each index (the percentage of the scored samples) quantifies a particular colour rendition characteristic within the single statistical format.

SOLID-STATE LAMPS WITH DIFFERENT COLOUR RENDITION PROPERTIES

Different colour quality metrics (CRI and CQS supplemented by gamut-area indices, as well as statistical approach) were applied to the analysis of spectral power distributions (SPDs) of various solid-state lighting sources, such as white pc LEDs and LED clusters. The most comprehensive results were obtained for the statistical approach, which allowed to attribute the sources to three conditional groups of, namely, high-fidelity (CFI > 50%), colour-saturating (CSI > 50%), and colour-dulling lamps (CDI > 50%), respectively.
Examples of high-fidelity lamps are trichromatic warm white phosphor converted LED (CFI = 61%) and optimized multiwavelength LED clusters containing at least four coloured LEDs or a white pc LED supplemented by at least two coloured LEDs. Figure 1(a) shows an SPD of a tetrachromatic cluster composed of red, pc amber, green, and blue LEDs (RpcAGB cluster), which has CFI of 80%. Even higher colour-fidelity characteristics (CFI > 90%) can be obtained for pentachromatic coloured-LED clusters or a warm white LED complemented by three coloured LEDs.
Colour-saturating solid-state lamps are trichromatic LED clusters, which do not cover the spectral region between 530 nm and 610 nm [7]. Figure 1(b) shows an SPD of such a cluster composed of red, green, and blue LEDs (RGB cluster), which has CSI of 73%.
Examples of colour-dulling solid-state lamps are dichromatic daylight pc LED (CDI =53 %) and LED clusters, which have no red component. Figure 1(c) shows an SPD of such a cluster composed of amber, green, and blue LEDs (AGB cluster), which has CDI of 61%.

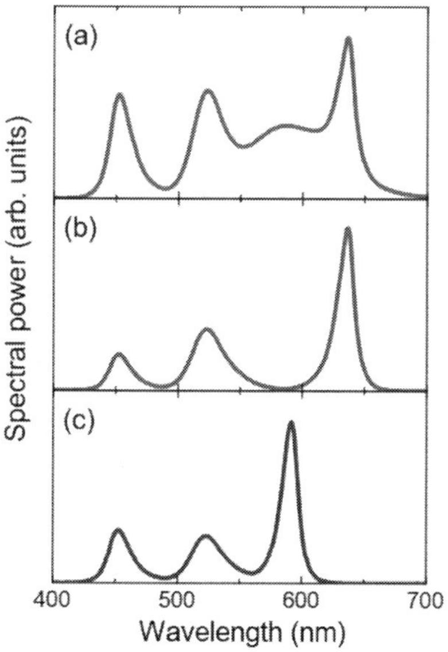

Fig. 1. Examples of SPDs of LED clusters with different colour rendition characteristics at a CCT of 4000 K: (a) high-fidelity RpcAGB cluster; (b) colour-saturating RGB cluster; (c) colour-dulling AGB cluster.

REFERENCES

[1] Commission Internationale de l'Éclairage, Pub. CIE 177 (2007).
[2] Commission Internationale de l'Éclairage, Pub. CIE 13.3 (1995).
[3] M.S. Rea, J.P. Freyssinier-Nova, Color Res. Appl., 33 (2008) 192.
[4] W. Davis, Y. Ohno, Proc. SPIE, 5941 (2005) 59411G.
[5] A. Žukauskas, R. Vaicekauskas, F. Ivanauskas, H. Vaitkevičius, P. Vitta, M.S. Shur, IEEE J. Sel. Top. Quantum Electron. 15 (2009) 1753.
[6] Spectral Database, University of Joensuu Color Group. Available: http://spectral.joensuu.fi/
[7] A. Žukauskas, R. Vaicekauskas, M. Shur, Opt. Express, 18 (2010) 2287.

Effect of Temperature and Current Variation on the Colour Quality of White Light Emitting Diodes

O. V. Khokhlev, O. V. Bord , M. V. Bogdanov, K. A. Bulashevich, M. S. Ramm, I. Yu. Evstratov and S. Yu. Karpov

STR Group – Soft-Impact, Ltd., 27 Engels ave., St.Petersburg, 194156 Russia

Contact: sergey.karpov@str-soft.com

ABSTRACT

Using simulations, we have examined effects of ambient temperature and operating current variation on the colour characteristics of white light obtained by phosphor-converted emission of a blue light emitting diode (LED). The temperature rise is found to primarily affect the correlated colour temperature (CCT), keeping the colour rendering index (CRI) practically unchanged, while the current variation is predicted to influence more the CRI at nearly the same CCT.

INTRODUCTION

The efficacy and light quality normally assessed in terms of CCT, colour coordinates, and CRI are the most important properties of white LEDs considered as the basic light sources for solid-state lighting. Generally, such devices should operate in a wide range of ambient temperatures and provide stable colour characteristics at the light dimming up and down. While the efficacy of white LEDs was the main focus of numerous studies for a long time, the stability of their colour under temperature and current variation has been investigated in insufficient detail.

This paper reports on the results of modelling research of the above issues. The general LED design considered in this study (see Fig.1a) is similar to that of Luxeon K2 produced by Philips Lumileds. For better predictability, experimental electrical and optical characteristics of blue LEDs are used in our simulations. To make interpretation of the modelling results more straightforward and clear, we consider here a single-phosphor light conversion with YAG:Ce^{3+}. The phosphorescent layer is assumed to be conformally deposited on the emitting surface of an LED die.

MODEL

Modelling of the LED operation has been done with the commercial SimuLAMP™ package [1]. The simulator predicts 3D temperature distribution in the device produced by both heat release in the LED chip and light absorption in the lamp units. 3D spectrally-resolved ray tracing, accounting for the light conversion in the phosphorescent medium, is performed to calculate the luminous flux outgoing from the LED, total emission spectrum, colour characteristics of the emitted light, and volume heat sources originated from the light absorption. Physics-based approximations for the current-voltage and optical power-current characteristics and the initial emission spectrum of blue LED were preliminarily derived from experimental data with account of governing parameters dependence on temperature and/or operating current. The temperature-dependent quantum efficiency of the YAG:Ce^{3+} phosphor [2,3] was included in the model to calculate accurately the resulting emission spectrum due to partial blue light conversion. The Henyey-Greenstein's scattering diagram with the asymmetry factor $g = 0.9$ [4] was attributed to the phosphor particles. The ratio of the scattering to absorption cross-sections of 4:1 was chosen in accordance with the Mie theory predictions for the particles of ~18 μm in diameter. In more detail, the simulation model will be described elsewhere.

RESULTS

In all the considered cases, a rather non-uniform angle-distribution of the emitted light colour is predicted by simulations (see Fig.1b), caused by different pathways of the rays passing the phosphorescent layer at nearly normal or glancing angles. Mean CCT is found to be most sensitive to the ambient temperature variation. It rises from ~5000 to more than 6000 K, while the temperature varies from –25°C to 150°C (see Fig.1f). Detailed examination of the emission spectra (Fig.1d) shows that the

CCT rise is caused by broadening of the blue LED spectrum at lower temperatures and thermal quenching of the phosphor quantum efficiency at higher temperatures, resulting in suppression of the yellow component of the integral emission spectrum. In contrast, CRI is predicted to vary just slightly, between 72 and 73, in the above temperature range (Fig.1c).

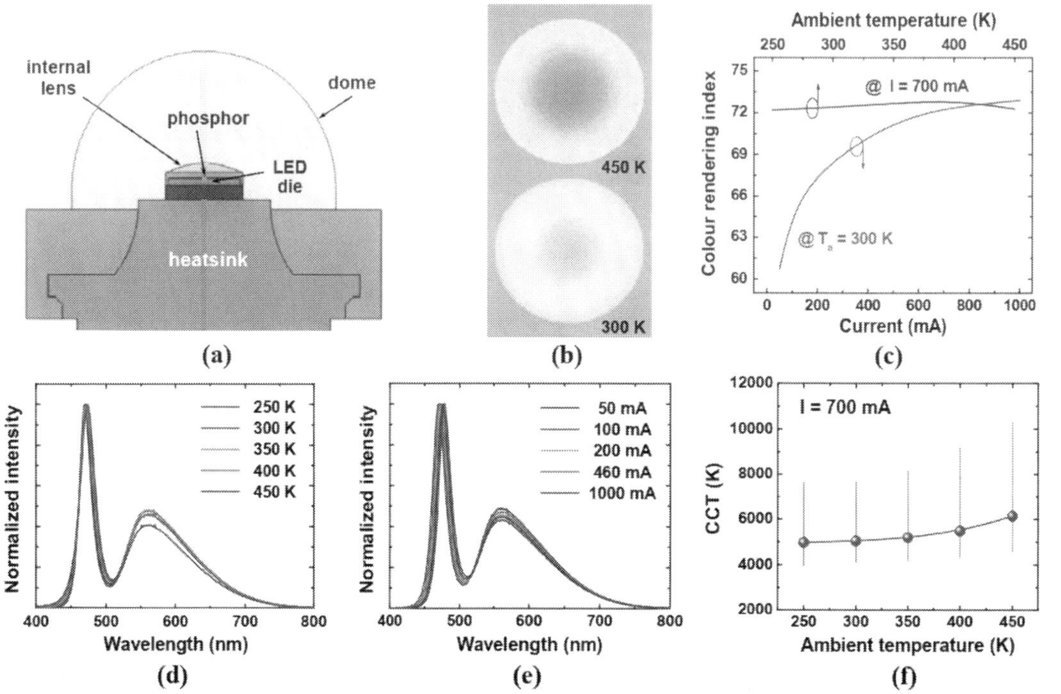

Fig.1 (a) schematic LED design; (b) far-field light colour distributions at different temperatures and current of 700 mA; (c) CRI as a function of temperature and operating current; integral emission spectra of LED (d) at different temperatures and (e) currents; (f) CCT as a function of ambient temperature, where bars indicate the CCT variation with observation angle in the far-field zone.

Increasing operating current affects slightly the yellow component in the total emission spectrum and produces a distinct, ~8 nm, blue shift of the LED emission spectrum (Fig.1e). As a result of good mutual balance between spectral components, the CCT remains nearly constant, about 4990-5070 K, in the whole range of the current variation. In contrast, CRI falls down well below 70, while the current becomes low enough (see Fig.1c) and the LED generates light with a longer peak wavelength. The reason for that is a dramatic increase of the *y*-color coordinate, moving the chromaticity point of emitted light far away from the black-body radiation locus.

CONCLUSION

This simulation study has revealed quite different behaviour of the white LED colour characteristics upon ambient temperature and operating current variations, which is important for development of solid-state luminaries. The variations of colour characteristic with temperature and current are, however, comparable with those observed in the far-field zone (see Fig.1b). The latter ones can be avoided, at least partly, by accurate optimization of the phosphorescent medium design, namely, its location inside the LED lamp, thickness, and phosphor particle parameters like mean size and concentration.

REFERENCES

[1] http://www.str-soft.com/products/SimuLED/SimuLAMP

[2] R.-J. Xie and N. Hirosaki, Sci. Technol. Adv. Mater. 8 (2007) 588.

[3] V. Bachmann, C. Ronda, and A. Meijerink, Chem. Mater. 21 (2009) 2077.

[4] N. Linder, D. Eisert, F. Jermann, and D. Berben, In: ***Nitride Semiconductor Devices: Principles and Simulation***, Ed. J. Piprek, Wiley-VCH Verlag, Weinheim (2007) Ch.15, pp.327-351.

Microwave excitation of high pressure metal halide lamps

M. Kettlitz, R. Kozakov, H. Schneidenbach, R. Methling
INP Greifswald, Felix-Hausdorff-Str. 2, D-17489 Greifswald, Germany
Contact: kettlitz@inp-greifswald.de

ABSTRACT

This work discusses the electrodeless high frequency power input and the plasma behavior in high pressure lamps. Therefore several lamp geometries and Hg-free lamp fillings were tested. The lamps were made of quartz containing Ar as an ignition gas and several metal halides as light emitting substances. The lamps were operated in a special cavity configuration at a frequency of 2.45 GHz. The main energy input was varied between 10 and 100 W. Input power, spectral distribution from the ultraviolet to the near infrared spectral range and vessel temperature were observed. Variations in the filling, the input power and the wall temperature lead to changes in the spectral output, color temperature and color rendering.

INTRODUCTION

High intensity discharge (HID) lamps as compact light sources are characterized by high efficiency, excellent color rendering and long lifetime. Usually the lifetime is mainly influenced by electrode erosion leading to wall blackening and a reduction of the luminous flux. Therefore it is of topical research interest to have an electrodeless power input into HID lamps. More than ten years ago there was a high power (above 1 kW) electrode-less sulfur lamp with a microwave excitation commercially available [1]. New trends with lower power HID units below 50 W including research on electrodeless HID lamps [2 - 4] could expand the field of application and break into new markets.

EXPERIMENT

Plasma was generated and investigated in a quartz envelope within the microwave cavity while excited by a magnetron. The plasma was investigated both during the ignition and warm-up phase as well as during the subsequent steady state. From ignition until steady state wide pressure and temperature ranges were encountered.

Several lamp geometries and Hg-free lamp fillings were tested. The lamps were made of quartz spheres containing some mbar Ar as an ignition gas and some mg of several metal halides as light emitting substances. The diameter of the lamps was varied between 5 and 20 mm. The lamps were ignited by an external high voltage pulse and operated in a cavity configuration [5] at a frequency of 2.45 GHz. The main energy input was varied between 10 and 100 W. The power was supplied by a coaxial cord from the magnetron (Sairem GMP20KED) to the cavity. Input and reflected power were determined by a power meter (Agilent E4419B).

Overview spectra from the ultraviolet to the near infrared spectral range were observed by a fiber spectrometer. Beside overview spectra, high-resolution spectra for diagnostic lines were measured in the ultraviolet and visible range. High accuracy is necessary in order to get well-resolved line profiles for temperature determination. The spectroscopic measurement of absolute side-on intensities was carried out with a 1 m-spectrograph (Acton AM510) with an intensified CCD camera (Proscan). The radiation was imaged by an optical system to the entrance slit of the spectrograph. The calibration of absolute intensities was done with a tungsten strip lamp. The vessel temperature was measured by a pyrometer (Kleiber KPE 660/5) at 5 μm.

RESULTS

Variations in filling, input power and wall temperature lead to changes in the spectral output, color temperature, color rendering and to changes in the coupling with the circuit. Lamps containing InI, InCl or InBr strongly emit In lines at 410 and 451 nm at low power. This leads to a blue impression of the lamps combined with high correlated color temperatures (CCT) above 15 000 K. There is an increase of continuous radiation for raising input powers shown in figure 1 for an InCl lamp with a

diameter of 15 mm. This is accompanied by a decrease of the CCT, a good color rendering and a white impression of the lamp. The continuous radiation is emitted from indium halide molecules. Measured spectra can be explained qualitatively by model calculations on the basis of a semi-empirical model of the molecular radiation [6]. A comparison of the spectra of InI, InCl and InBr shows that there is only a slight change of the intensity of the continuous part for different halides. But there seems to be no change of the spectral distribution if halides are exchanged.
The plasma temperature in these lamps was obtained by the Bartels method from the optically thick indium line at 451 nm [7, 8]. It is about 3 000 K in the arc centre of InI, InCl and InBr discharges at an input power of 40 W.

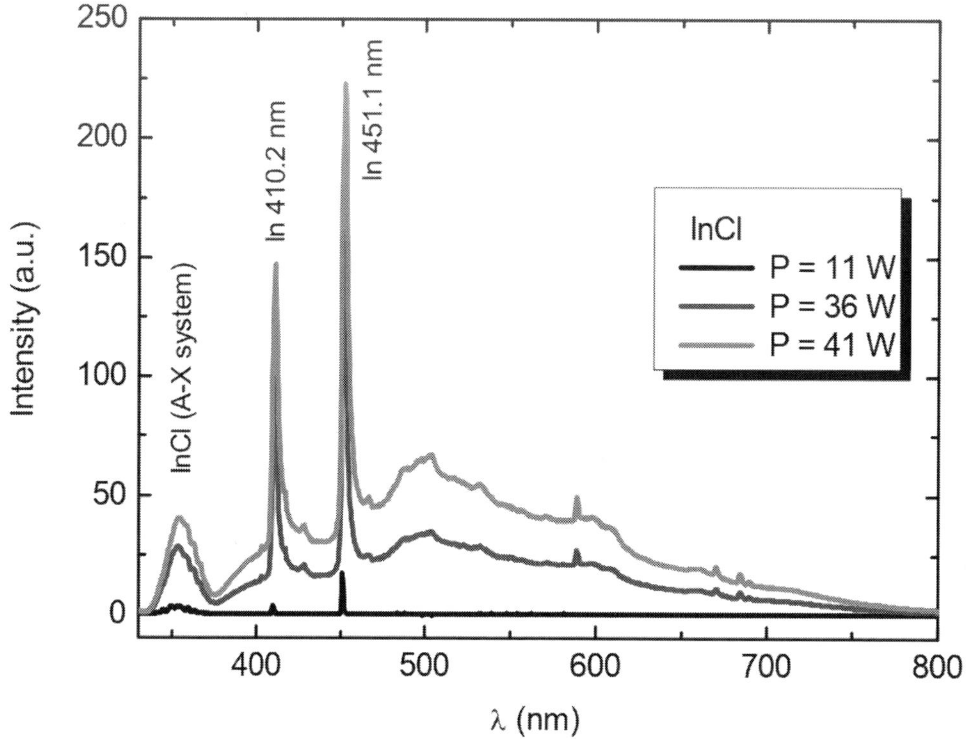

Figure 1: Overview spectra of a lamp with a diameter of 15 mm, filled with InCl for input powers from 11 to 41 W. The wall temperature increases with power from 450 to 970 °C.

Further experiments with an amplifier on semiconductor basis, other cavity configurations and lamp shapes are planned.

REFERENCES

[1] J. T. Dolan, M. G. Ury, and C. H. Wood, US patent 5,404,076, (1995).
[2] G. G. Lister, J. E. Lawler, W. P. Lapatovich, V. A. Godyak, Rev. Mod. Phys. **76** (2004) 541.
[3] V. Babykumar, A. Neate, E. Odell, B. Preston, A. Sadiq, R. Devonshire, Proc. LS 11, Shanghai, (2007) 109.
[4] M. Kettlitz, R. Kozakov, S. Kühn, R. Gesche, H. Porteanu, Plasma Science - Abstracts, ICOPS IEEE, 10.1109/PLASMA.2009.5227671 (2009).
[5] C. I. M. Beenakker, Spectrochim. Acta **31B** (1976) 483.
[6] S. Franke, R. Methling, H. Hess, H. Schneidenbach, H. Schöpp, L. Hitzschke, M. Käning, B. Schalk, J. Phys. D: Appl. Phys. **40** (2007) 3836.
[7] H. Bartels, Z. Phys. **127** (1950) 243 and Z. Phys. **128** (1950) 546.
[8] S. Franke, H. Schneidenbach, J. Phys. D: Appl. Phys. **41** (2008) 15.

The peculiarities of microwave gas discharge as UV-source

E.M.Barkhudarov[1], N.V.Denisova[2], I.A.Kossyi[1], M.A.Misakyan[1]
[1] *Prokhorov General Physics Institute of RAS, Moscow, Russia*
[2] *Institute of Theoretical and Applied Mechanics of RAS, Novosibirsk, Russia*
Contact: kossyi@fpl.gpi.ru

ABSTRACT

Results are presented from theoretical and experimental studies of microwave discharges excited in Ar-Hg mixtures and output optical radiation from such discharges. The emphasis is on microwave discharges when "plasma-resonance" conditions at low pressures are realized. Under these conditions with decreasing Ar pressure, the radiation intensity in the spectrum of Hg (including bactericidal UV radiation) increases sharply while the intensity of Ar lines decreases also abruptly. The radiation intensity is calculated using a developed collisional--radiative model. The results of these calculations are in qualitative agreement with experimental data. Based on the obtained results microwave UV source has been developed. Performed measurements manifested high degree of its perfection.

GENERAL

In recent years, a new type of microwave gas-discharge plasma sources based on the phenomenon known in the literature as the "plasma resonance" has attracted the attention of both theorists and experimentalists. This phenomenon consists in the totality of nonlinear processes occurring where the electron density of nonmagnetized plasma approaches the critical value $n_{ec} = \dfrac{m(\omega^2 + \nu_{eff}^2)}{4\pi e^2}$ on its essentially nonuniform profile [1,2]. Here, ν_{eff} is the effective frequency of electron—neutral collisions, ω is cyclic frequency of microwave radiation. In a fairly narrow layer adjacent to the resonance surface, the microwave energy can be absorbed and through the excited strong plasma Langmuir wave converted efficiently into the energy of plasma electrons, so that the presence of plasma resonance strongly affects the parameters of microwave plasma. This type of discharges has been much investigated both experimentally and theoretically.

In [2] conditions were attained wherein electrons accelerated at the plasma resonance mostly expended their energy for ionizing heavy particles, thereby producing a plasma "aureole" around the plasma-resonance regions. As a plasma source, a microwave coaxial plasmatron was chosen for these experiments. The results have been presented in [2] of experimental studies of the plasmatron parameters, namely, data of measurements of the plasma density and potential, and the dynamics of discharge glow. These experimental data allow us to affirm that, at relatively low pressures of the working gas (argon), conditions are attained wherein the presence of resonance regions in an inhomogeneous plasma and definitely nonlinear processes of the interaction between this region and microwaves play a crucial role in the overall process and formation of the parameters of the ionized gas medium in the reactor volume. In this case, the plasma density in the chamber far exceeds the critical density and the electron temperature is higher than 10—15 eV.

In this paper, we will study the characteristics of optical radiation from the plasma of the coaxial microwave plasmatron and will demonstrate the possibility of using the microwave discharge in Ar—Hg mixtures as an effective source of intense UV radiation.

It is shown experimentally that the microwave discharge excited by the coaxial plasmatron in argon at relatively low pressures, where the electron density and temperature sharply increase, demonstrates also the peculiar behavior of the optical radiation spectrum. Thus, with a small admixture of mercury to argon that does not alter the character of gas-discharge processes and does not affect the main plasma parameters, in the same region of low pressures ($p < 0.1$ Torr), the intensity of mercury lines (including in the UV spectral region) increases sharply, whereas the intensity of argon lines drops. Typical dependences of mercury and argon lines intensity as a function of gas pressure are presented on Figures 1 and 2.

To account for this effect in the collisional—radiative model of radiation from a microwave discharge in an Ar—Hg mixture [3], it is necessary to assume that (i) the electron temperature values are close to those measured in previous experiments and also predicted by the resonance discharge model and (ii) the values of electron density far exceed the critical density for microwaves used in these experiments.

The measurements and calculations reported in the paper are of obvious physical interest, as they pertain to the physics of microwave discharges. At the same time peculiarity of the radiative characteristics of the microwave discharge is important in technology of gas-discharge UV lamps and gives ways of reaching the high efficiency of microwave energy conversion into bactericidal UV radiation. Based on the obtained results microwave UV source destined for potable water disinfection has been developed. Performed measurements manifested high degree of its perfection.

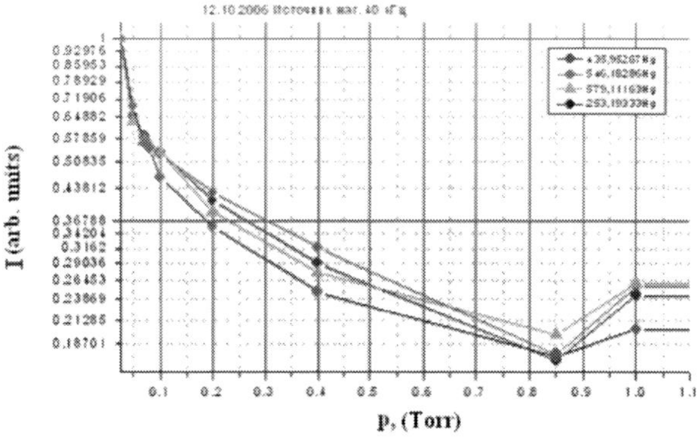

Figure 1. Dependence of the mercury line intensities (in arb. units) on the argon pressure at the operation of the coaxial microwave plasmatron in an Ar—Hg mixture.

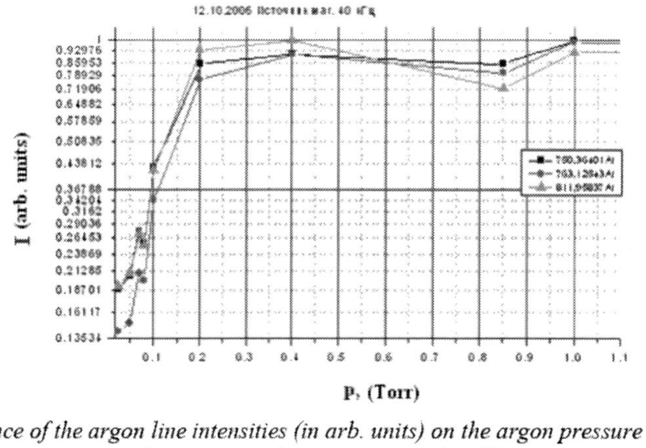

Figure 2. Dependence of the argon line intensities (in arb. units) on the argon pressure at the operation of the coaxial microwave plasmatron in an Ar—Hg mixture.

REFERENCES

[1] Generation of Nonlinear Waves and Quasi-Steady Currents in Plasma. Transact. GPI, vol. 16, Moscow: Nauka, 1988 (in Russian).

[2] Berezhetskaya N.K., Kop'ev V.A., Kossyi I.A., Misakyan M.A., Taktakishvili M.I., Temchin S. and Y.D.Lee.Eur. Phys. J. Appl. Phys., 2008, 42, 327.

[3] E.M.Barkhudarov, N.V.Denisova, I.A.Kossyi and M.A.Misakyan. "Resonance" Microwave Discharge as a Source of UV Radiation. Plasma Physics Reports, 2009, vol. 35, No 7, pp. 559-566.

Ultraviolet Radiation Emissions from Compact Fluorescent Lamps and Halogen Lamps

T. Cantwell[1] , L. Browne[1] , R. Dowdall[1]
[1] *School of Physics, Dublin Institute of Technology, Kevin Street, Dublin 8, Ireland*
Contact: tom.cantwell@dit.ie

ABSTRACT
Household incandescent lamps are being replaced with more efficient alternatives such as compact fluorescent lamps (CFLs), light emitting diodes (LEDs) and improved incandescent lamps with halogen technology (HLs). However, the potential adverse effects of ultraviolet radiation (UVR) from indoor lighting, especially CFLs which contain mercury, is becoming increasingly recognized. The UVR emitted from 107 CFLs [85 open (single envelope) and 22 encapsulated (double envelope)] and 15 HLs was measured with a double monochromator scanning spectroradiometer in the wavelength range, 250 nm to 400 nm. The biologically effective exposure from each lamp was assessed using the International Commission on Non-Ionizing Radiation Protection (ICNIRP) weighting function [1] and compared to the Exposure Limit values to evaluate potentially hazardous exposures. No HL or double envelope CFL exceeded the 8-hour limit value at 200 mm from the lamp. 9.4% of the single envelope CFLs exceeded the 8-hour limit value of 30 J m^{-2}.hhh

INTRODUCTION
On 13 April 2009, new legislation came into force, throughout the European Union, which is designed to replace household non-directional lamps with more energy-efficient types. Regulation EC/244/2009 [2] effects this change as part of the Implementing Measures of the Eco-Design Directive of Energy Using Products (Directive 2005/32/EC). This paper details results of measurements of UVR emitted from energy saving lamps available to Irish consumers from retail outlets. A number of recent studies [3-6] have measured UVR from household lamps and discussed their significance for exposure of the eye and skin with respect to the ICNIRP guidelines [1] and for photosensitive individuals.

PROCEDURE
The spectral irradiance in the wavelength range, 250 nm to 400 nm, was measured using a Bentham DMc 150 double monochromator scanning spectroradiometer with a bandwidth of 1 nm. The input optic consisted of a very high precision cosine response diffuser (f2 error < 1%) and the measurements were performed in 1 nm intervals. Calibration was carried out with reference to a quartz-halogen lamp traceable to the UK National Physical Laboratory. A total of 107 CFLs (85 single envelope and 22 double envelope) were measured; they were bought in 4 lots at 4-monthly intervals from retail outlets in Dublin over the period from September 2008 to January 2010. The lamps, ranging in power from 11W to 25W and from 8 different manufacturers, all had Class A energy rating. 15 HLs (Class C) were measured; these lamps, which were bought in January 2010, are currently available from 3 manufacturers and they vary from 42W to 105W. The measurements were performed at distances of 20 mm and 200 mm from the lamp to the diffuser.

RESULTS
Only very small quantities of UVA were emitted from the HLs. Because of their mercury content, the CFLs emitted significant quantities of UVA, especially at 365 nm. Many of the CFLs had sizeable outputs at 313 nm and, in some cases, at 254 nm – see Figure 1. The UVR emitted from double envelope CFLs was much reduced compared to single envelope CFLs. Further details are given in Table 1.

Table 1. Mean Total Irradiance and Effective Irradiance of the CFLs (mW m^{-2})

Distance = 20 mm	UVC (250-280 nm)	UVB (280-315 nm)	UVA (315–400 nm)	Mean Effective Irradiance
Single Envelope	0.52	570	7900	13.0
Double Envelope	0.37	20	2100	0.46
Distance = 200 mm				
Single Envelope	0.49	15	170	0.43
Double Envelope	0.36	0.87	66	0.09

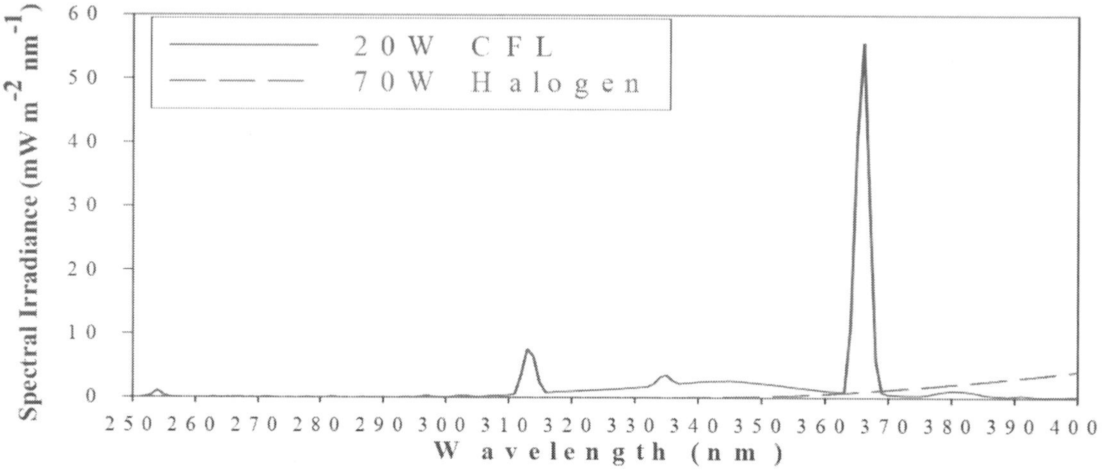

Figure 1. Spectral Irradiance at 200 mm of a 20W CFL and a 70W Halogen Lamp

Table 2. Time to exceed the ICNIRP Exposure Limit value of 30 J m^{-2}

Distance from long axis of CFL = 200 mm	**< 8 hours**	**8 – 10 hours**	**> 10 hours**
% Single Envelopes	9.4	5.6	85.0
% Double Envelopes	0.0	0.0	100.0

The biologically effective exposure from each lamp was assessed using the ICNIRP weighting function [1] and compared to the Exposure Limit values to evaluate potentially hazardous exposures. No HL or double envelope CFL exceeded the limit value of 30 J m^{-2} at 200 mm from the lamp within eight hours. However, 9.4% of the single envelope CFLs exceeded the ICNIRP limit value in a time less than eight hours (Table 2). The Artificial Optical Radiation Directive [7] will become law throughout EU member states on 27 April 2010. The Directive requires businesses, including those based in the home, to limit the exposure of workers to optical radiation, including exposure to UVR hazards from general lighting. Since the exposure limits in the Directive are based on the ICNIRP values, our results may be of significance in this regard.

REFERENCES

[1] The International Commission on Non-Ionizing Radiation Protection, *Guidelines on Limits of Exposure of Ultraviolet Radiation Wavelengths between 180 nm and 400 nm* (Incoherent Optical Radiation). Health Physics 87 (2): 171-186;2004

[2] COMMISSION REGULATION (EC) No 244/2009 of 18 March 2009 *Implementing Directive 2005/32/EC of the European Parliament and of the Council with regard to ecodesign requirements for non-directional household lamps,* OJ L76 of 24.3.2009 p. 3-16.

[3] Khazova M., J. B. O'Hagan, *Optical Radiation Emissions from Compact Fluorescent Lamps,*Radiation Protection Dosimetry , (2008) Vol 1 131, No. 4, pp. 521-525.

[4] Nuzum-Keim A.D., Sontheimer R.D., *Ultraviolet light output of compact fluorescent lamps: comparison to conventional incandescent and halogen residential lighting sources* Lupus (2009)18, 556-560

[5] Klein R.S., Werth V.P., Dowdy J.C., Sayre R.M., *Analysis of Compact Fluorescent Lights for Use by Patients with Photosensitive Conditions*, Photochemistry and Photobiology, 2009, 85: 1004-1010

[6] Eadie E., Fergson J. and Moseley H., *A preliminary investigation into the effect of exposure of photosensitive individuals to light from compact fluorescent lamps*, British Journal of Dermatology 2009 **160,** pp659-664

[7] Directive 2006/25/EC of the European Parliament and of the Council on the minimum Health and Safety Requirements Regarding the Exposure of Workers to Risks Arising from Physical Agents (Artificial Optical Radiation Directive). Official Journal of the European Union, L114/38-14/59, 27April 2006.

Investigation of low pressure discharges containing InBr

S. Briefi[1], U. Fantz[1,2]

[1]*Lehrstuhl für Experimentelle Plasmaphysik, Universität Augsburg, Universitätsstr. 1, 86135 Augsburg, Germany*
[2]*Max-Planck-Institut für Plasmaphysik, EURATOM Association, Boltzmannstr. 2, 85748 Garching, Germany*
Contact: stefan.briefi@physik.uni-augsburg.de

ABSTRACT

Metal halides such as InBr could be used as a substitute for mercury in low pressure discharge lamps. As InBr emits between 350 and 400 nm the Stokes shift can be reduced greatly. Therefore ICP discharges in Argon containing InBr were investigated spectroscopically. Emphasis is put on the understanding of the changing plasma parameters with increasing amount of InBr in the discharge which has a great influence on the intensity of the InBr band emission.

INTRODUCTION

One main goal of the current research in lighting technology is to find an efficient radiator as substitute for mercury in fluorescent lamps. Molecular radiation, especially from metal halides such as InBr, is under discussion to be a good alternative [1], [2]. InBr emits a broad and intense band spectrum in the near UV range (350 - 400 nm) which makes the conversion into visible light by a phosphor more efficient compared to mercury in fluorescent lamps. This emission originates from the transitions of the first two electronically excited states to the ground state (A $^3\Pi_{0^+} - X\ ^1\Sigma_0^+$ and B $^3\Pi_1 - X\ ^1\Sigma_0^+$). The present investigations focus on the understanding of the change of plasma parameters and InBr emission with increasing amount of InBr in the discharge. As diagnostic techniques emission and absorption spectroscopy are used which are supported by a simulation of the emission spectrum and the generation of molecular data such as Franck-Condon-Factors and vibrationally resolved transition probabilities [3]. The aim is to maximize the molecular radiation at low discharge power by varying the pressure and the type of the buffer gas as well as the amount of InBr in the plasma.

EXPERIMENTAL SETUP

As metal halides would quickly erode internal electrodes, one has to use electrodeless RF-coupling methods. Therefore the discharge is generated in sealed cylindrical quartz cells (length 18 cm, diameter 2.5 cm) via inductive coupling (13.56 MHz, $0-100$ W). The cell is filled with a few mg of InBr salt and a background gas at a few mbar. As InBr has a very low vapour pressure at ambient temperature, the whole cell is heated up to several hundred degrees centigrade using flowing hot air. The amount of InBr in the gas phase and in the plasma is determined by the coldest spot of the quartz wall. Emission and absorption spectra were recorded using a high resolution spectrometer ($\Delta\lambda_{FWHM} = 25$ pm). For absorption spectroscopy a high-pressure Xe-lamp is used. To achieve a low detection limit for the absorption spectroscopy, an axial line of sight is being used (radial centered).

RESULTS

With increasing heating temperature and therefore with increasing amount of InBr in the plasma the emission spectrum changes drastically. The discharge changes from a pure Argon discharge to an InBr/In discharge in which the Argon emission vanishes. Figure 1 shows measured densities of a representative excited Argon state and of the ground states of InBr and of In which is a result from the dissociation of the molecule. As soon as InBr is evaporated the Argon population decreases strongly. Since the detection limit of the InBr ground state is worse than the one of Indium, the In ground state density (which is a representative curve of the InBr evaporation due to the dissociation) can also be measured at lower heating temperatures. The decrease of the Argon population is explained as follows: as soon as InBr evaporates and dissociates in the plasma, Indium ions are created. The ionization energy of In is much lower compared to Argon (E_{Ion}(In) $=$ 5.79 eV, E_{Ion}(Ar) $= 15.76$ eV) which means the electron temperature decreases due to the ionization balance.

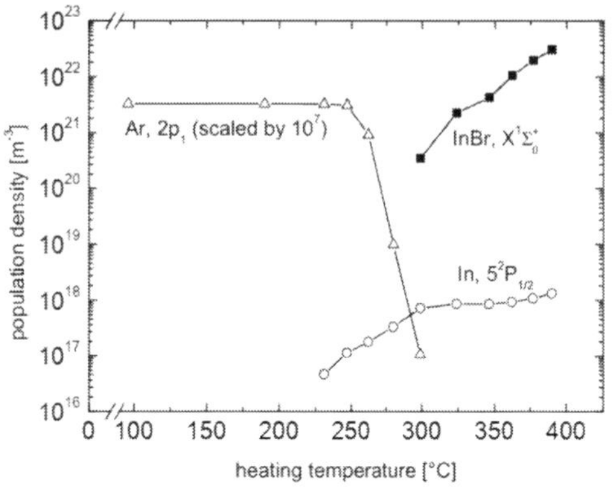

Figure 1. Densities of an excited state of Argon and the ground states of InBr and In with increasing heating temperature.

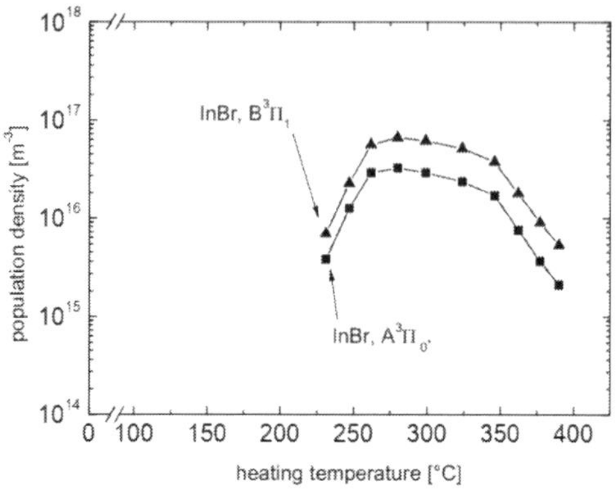

Figure 2. Population of the electronic A and B states of InBr with varying heating temperature.

This results in the decreasing population of the excited Argon states (excitation energy $E_{Exc}(\mathrm{Ar}) \approx 12$ eV, $E_{Exc}(\mathrm{InBr, In}) \approx 3$ eV).

The population of the excited states of InBr increases with increasing heating temperature of the quartz cell (see figure 2). If the heating temperature exceeds 300 °C, the InBr density in the plasma gets too high and re-absorption occurs in the intense sequences of the emission spectrum. This results in a decrease of the emission and therefore of the measured population at higher heating temperatures.

CONCLUSION AND OUTLOOK

Low pressure discharges containing InBr were investigated spectroscopically. The change of the plasma emission with increasing amount of InBr due to the changing plasma parameters and the peaking of the molecular radiation could be explained. In a next step a comparison of ICP and CCP will be carried out, as the capacitive coupling seems to allow to operate the discharge at reduced power. The emission of discharges containing other metal halides which have a higher vapour pressure is very sensitive on the cold spot temperature. Therefore a new experimental setup is under construction which allows a direct change of the coldest spot with high accuracy.

REFERENCES

[1] S. Kitsinelis, G. Zissis, E. Fokitis, J.Phys. D: Appl. Phys, 42 (2009) 045209.

[2] A. Körber, D. Hayashi, 28th ICPIG conference, Prague (2007).

[3] U. Fantz, S. Briefi, R. Friedl, contribution to this conference.

Driving of dielectric barrier discharge lamps with a pulsed transformer-less full-bridge topology

M. Meisser, M. Paravia, R. Kling, W. Heering
Light Technology Institute (LTI), Karlsruhe Institute of Technology (KIT),
Engesserstrasse 13, 76131 Karlsruhe, Germany
Contact: michael.meisser@kit.edu

ABSTRACT

This paper presents a new transformer-less topology and switching pattern to drive dielectric barrier discharge (DBD) lamps in pulsed operation mode. Supplied by a 450 V intermediate supply, pulses with peak voltage exceeding 1.5 kV are generated by a resonant principle. Driving a DBD lamp with C_{DBD} = 2.76 nF at a pulse frequency of 1.15 MHz, an electrical efficiency of more than 80 % is achieved.

DESCRIPTION OF THE FULL-BRIDGE TOPOLOGY

Dielectric barrier discharge (DBD) lamps have a low power factor and should be operated in pulsed mode to obtain high plasma efficiency. Furthermore, after the first ignition per pulse, an immediate subsequent second ignition with opposite current polarity is required for homogeneous discharge distribution.

Full-bridge configurations to drive DBDs are known from [1, 2]. In contrast, the presented topology does not store energy in the lamp capacitance between the pulses, but recovers energy after the pulses and thereby offers the feature of low voltage during pulse pauses. Because a transformer is avoided, differing from [3], exceptional high resonance frequencies can be achieved. The used full bridge configuration is advantageous because energy fed-in and energy recovery is possible regardless of actual lamp current direction. As depicted in Figure 1, the full-bridge excites a resonant tank, which is set up by an inductor L and the capacitance of the DBD C_{DBD}. This allows a multiple voltage magnification every time, energy is fed into the resonant circuit. The aim is to reach ignition voltage with a minimum amount of resonant swings prior to first ignition. This is because any AC voltage applied to the DBD before the first ignition leads to losses inside the DBD as well as it can impede homogeneity of the discharge. The number of resonant swings, which are necessary to generate the ignition voltage, depends on the ratio $V_{lamp,ign}/V_s$ and the circuit quality factor Q_s. After the voltage half-wave that leads to the first ignition, the following half-wave must be tuned to support an intensive second ignition resetting the internal charges by utilizing them for a second ignition. At times of ignitions a high current should flow, which, however, should commutate shortly after ignition to restrict glow phase losses. This can be achieved by a current cut-off that simultaneously recovers the energy, stored in the resonant tank, back to the intermediate supply. The body diodes of the high switches HS1 and HS2 are bypassed by zero-recovery silicon carbide diodes in order to prevent any return flow of the recovered energy into the circuit.

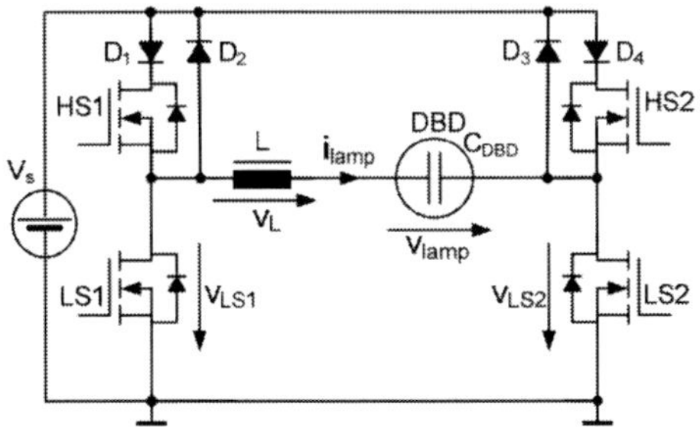

Figure 1: Schematic of the full-bridge topology.

Because of the missing transformer, the effective lamp capacitance can lie within the order of magnitude of the output capacitances of the switches. In this case zero-voltage switching (ZVS) transitions are no longer effective, as a significant percentage of circuit energy is required to charge the output capacitances. Therefore, for applications with low lamp capacitance and low pulse frequency, zero current switching (ZCS) is preferable. However, if lamp capacitance is large and pulse frequency high, ZVS should be applied.

EXPERIMENTAL RESULTS

Figure 2 shows experimental results obtained by driving a 2.5 nF flat DBD lamp with an additionally attached 1.87 nF capacitor and an electrical efficiency of more than 92 %. The corresponding infrared radiation, given in Figure 2A, indicates the first lamp ignition during the time of the second voltage half-wave. To achieve a strong third ignition, further energy is fed in by closing HS1 and LS2 for a second time. Energy recovery is initiated by keeping all switches off. Delayed by the reverse recovery time of the LS body

diodes, current leaves the circuit by passing alternately D_2 and D_3. This is indicated in Figure 2B by alternating peaks of V_{LS1} and V_{LS2}, respectively. As it can be derived from Figure 2A, 15 W of power is consumed in the first half-wave by a loss process inside the DBD. This is almost 6.7 % of total lamp power, which restricts lamp efficiency. The Power Factor (PF) of the second voltage half-wave, in which the first ignition occurs, has almost the double value compared to the following voltage half-waves. In contrast, the highest level of apparent power has to be distributed within the third half-wave to lead to the second ignition.
In Figure 3, results are given for a 1.15 MHz pulse generated to drive the lamp with attached 260 pF capacitor. The electrical efficiency exceeded 80 %. Due to lower V_S, three resonant swings were necessary to reach ignition voltage. To achieve a strong second ignition, energy recovery is deferred by closing LS1. The rising edge of V_{LS1} at $t = 2\mu s$ in Figure 3B indicates that energy recovery is then initiated by opening LS1.
Because measurements of lamp voltage had to be performed differentially, efficiency calculations from mathematically offset-corrected waveforms show a rather high uncertainty. However, thermal measurements prove in both cases an electrical efficiency well above 80 %. The efficiency is remarkably high, which is mainly due to low conduction losses and missing HV HF transformer losses.

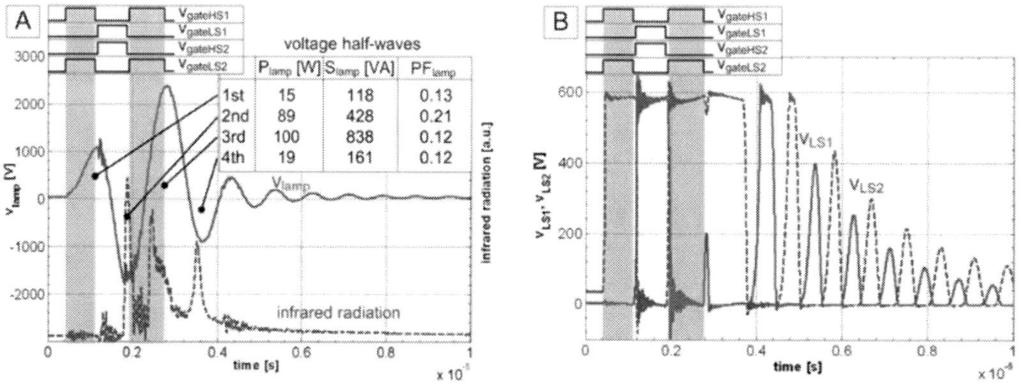

Figure 2: Experimental waveforms: $T_{rep} = 40\ \mu s, V_s = 600\ V, f_{pulse} = 500\ kHz, P_{lamp} = 228\ W,\ \eta_{ECG} > 92\ \%$.

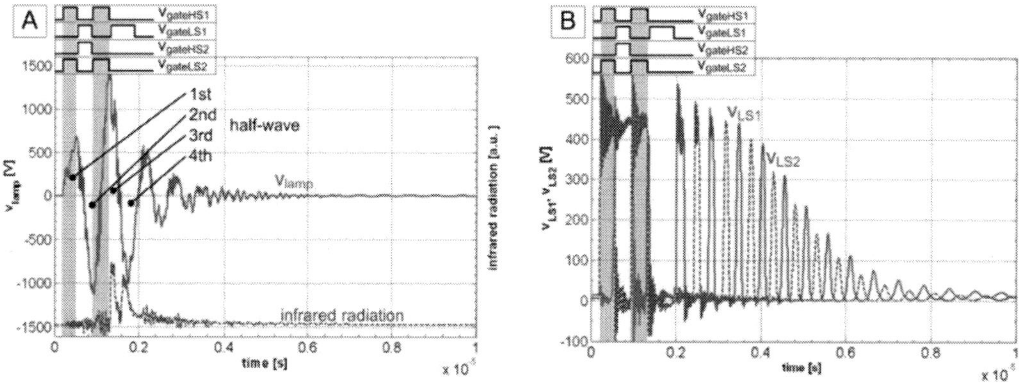

Figure 3: Experimental waveforms: $T_{rep} = 40\ \mu s, V_s = 450\ V, f_{pulse} = 1.15\ MHz, P_{lamp} = 195\ W,\ \eta_{ECG} > 80\ \%$.

CONCLUSION

The presented full-bridge topology and corresponding switching pattern offers a high frequency pulsed operation of DBD lamps. The implemented energy recovery feature allows to drive a 2.5 nF flat lamp with attached 260 nF capacitor at a pulse frequency of 1.15 MHz with an electrical efficiency exceeding 80 %. At 500 kHz and almost doubled lamp capacitance, an electrical efficiency of more than 92 % was obtained.

REFERENCES

1. K. Kyrberg, et al., *IEEE Trans. on Power Electronics*, vol. 22, no. 3: p. 926-933, 2007.
2. H. Piquet, et al., *8th Int. Conf. on Atomic and Molecular Pulsed Lasers*, 2007, p. 93810-93810.
3. O. Koudriavtsev, et al., *IEEE Trans. on Industry Appl.*, vol. 38, no. 2: p. 369-378, 2002.

Experimental Study of the effect of germicidal lamp power supply on the UV-C bacteria inactivation in water treatment

B. Mrabet[1] , M. Ben Said[2]
[1] *Electrical Department, College of Engeneering, Alkharj University, Alkharj, Saudi Arabia*
[2]*Laboratoire de Traitement et Recyclage des Eaux Usées, Research and Water Technologies Center, Tunisia.*
Contact: mrabet@ksu.edu.sa

ABSTRACT

The effectiveness of a UV system in water treatment to eliminate microbiological contamination is directly dependent on the UV dose applied to the water. UV-C rays (photons) produced by a discharge lamp penetrate microorganisms and become absorbed by the DNA of the pathogen in the water being treated. The DNA is altered in such a way that the pathogen cannot reproduce and is essentially killed and cannot cause infection. This process of DNA modification is called inactivation. This paper describes the influence of the current frequency of low pressure lamp on a bacterial strain of Pseudomonas aeruginosa ATCC 15422. To reach this objective, an experimental setup is carried out and a regulated-current electronic power supply is realized. A study of the kinetics of inactivation and mechanisms of bacterial reactivation are carried out.

INRODUCTION

The inactivation of microorganisms by UV radiation produced by a discharge lamp is allotted to photochemical reactions induced by the radiation UV-C. The germicidal properties of ultraviolet irradiation are due to absorption of UV light by DNA molecules, causing cross-linking between neighboring pyrimidine nucleosides (thymine and cytosine) within the same DNA strand [1]. The aim of this work is to study the effect of the discharge lamp power supply : electromagnetic at Low frequency (E1) and electronic ballast at High frequency (E2), on the kinetics of inactivation of bacterial strain ATCC 15422 and to study the adaptive response of this bacterial strain post-exposed to UV by highlighting the mechanism of reactivations and repairs adopted by this bacterial species.

EXPERIMENTAL

The electronic power supply of the lamp is presented in reference [2] and the electrical and geometrical characteristics of the germicidal lamp are (length = 400 mm, diameter = 15 mm, I_{arc}= 0.65 A, V_{arc}= 110 V).
The study relates to the bacterial strain named ATCC 15422 known by its environmental flexibility and its genetic complexity [3, 4]. The tests were carried out on a reactor built in CERTE. For the kinetics of inactivation, the strain will be exposed to UV during split time intervals in order to estimate the number of the viable bacteria before and after each UV exposure. For the mechanism of photoreactivation and the mechanism of obscure repair, the strain will be deposited during split rest periods in the presence of light then in the darkness. The UV emitted by the lamp was measured by a UV detector connected to a radiometer.

RESULTATS

The lamp current and the lamp voltage waves for 50 Hz and for 64kHz are shown in figure1 and figure2, respectively. In both cases the lamp power is maintained constant

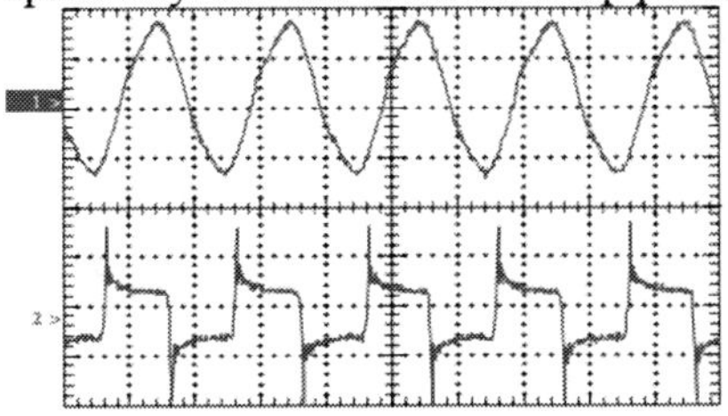

Figure 1 : *I_{arc} (wave 1 ; 500 mA/Div -10ms/Div), U_{arc} (wave 2; 200V/Div-1ms/Div)) at f = 50Hz*

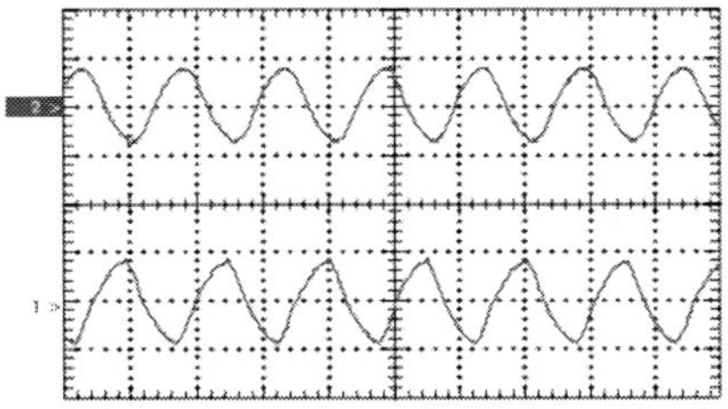

Figure 2 : *I_{arc} (wave 2 ; 1A/Div -10µs/Div), U_{arc} (wave 1 ; 200V/Div-10µs/Div)) at f = 64KHz*

The results of the kinetics inactivation, per standard enumeration carried out for the studied bacterial strain

ATCC 15422, are presented in figure3. The results of the photoreactivation mechanism, and the obscure repair mechanism are recapitulated in Figure 4.

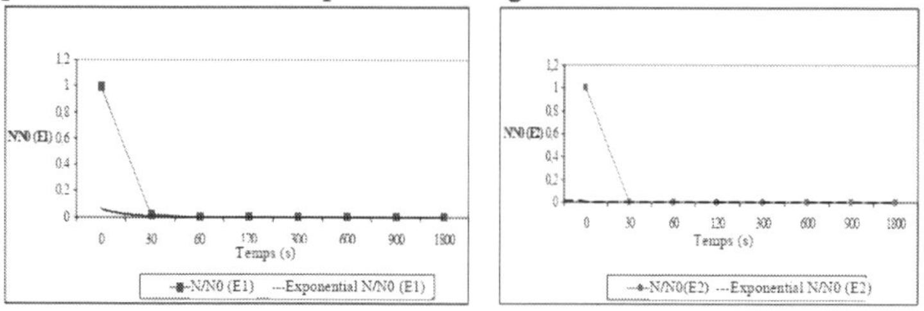

***Figure3** : kinetics of inactivation of ATCC 15422 at low frequency 50Hz (E1) and at high frequency (E2).*

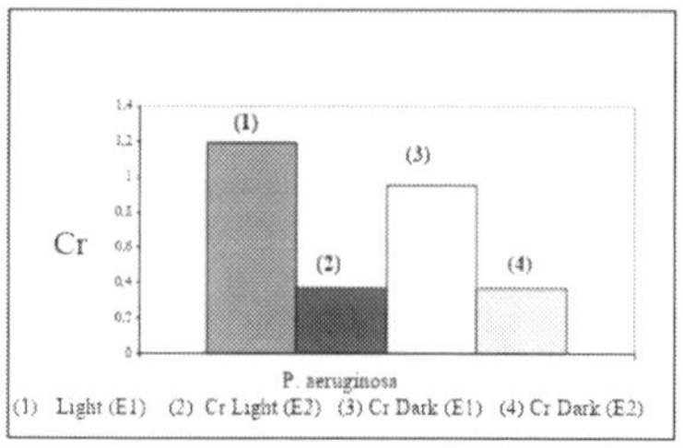

***Figure 4** : Reactivation of ATCC15422 in light and in darkness for E1 and E2*

The kinetics of inactivation of the bacterial strain of Pseudomonas aeruginosa is in perfect agreement with the model of ChicK-Watson modified in [5] and formulated as follows : $N/N_0 = A \exp(- KI^nT)$. The abatement A of the bacteria increases according to the accumulated dose of the UV-C for the two power supplies. The abatement of the bacteria is caused by the increase of the irreversible damage which shows the impacts of UV on the bacterial integrity [6]. For the two power supplies, the value K did not show a significant difference. But the value of the abatement A which is 0.459 for low frequency (50 Hz) and 0.026 for high frequency (64kHz) shows the effectiveness of high frequency in the procedure of water treatment by UV radiations. The analysis of the coefficients of reactivation of Pseudomonas aeruginosa for low frequency shows that the bacterial strain is reactivated in light with Cr greater than 1 *(Cr* =1,194). In the darkness, the bacterial strain is not reactivated (Cr = 0.367). At high frequency, Cr for the rest times in light and in darkness is less than 1. So, the strain is not reactivated and can not repair the damage caused by UV.

CONCLUSION

The kinetics of bacterial inactivation is in perfect agreement with the model of Chick-Watson. The bacterial destruction increases with the UV dose as a result of the increasing of the irreversible damage incited by the stress factor and the disturbance of the integrity of genetic material. The coefficient K did not show a significant difference for the two power supplies. On the other hand, the values of A show the effectiveness of the high frequency power supply to inactive the bacteria in more reduced time compared to a traditional power supply at 50 Hz. The study of bacterial repair exposed to the light and the darkness for various rest periods, shows that the capacity of the bacteria to repair the aggressions generated by irradiation UV also constitutes an important parameter in the process of disinfection. For high frequency we did not notice a reactivation of the strain neither in the light nor in the darkness. For low frequency we noticed a reactivation of the strain in the light. Consequently, this bacterial strain could surmount the damage induced by UV by the mechanism of photoreactivation. These results show the importance of high frequency UV irradiation produced by a germicidal lamp to inactivate the bacteria in short time and to prevent their reactivation.

REFERENCE

[1] Jae J. Kim, Sundin G. W., Construction and Analyss of Photolyase Mutants of Pseudomonas aeruginosa and Pseudomonas syringae, Appl. Environ. Microbiol., 67, pp. 1405-1411 (2001).

[2] M.Ben Mustapha, B.Mrabet, A.Chammem, E.Elheni, L.Bouslimi, M. Stambouli, Etude et réalisation d'une alimentation auto-oscillante et régulée pour lampe UV, STA (2006).

[3] Hardalo C., et Edberg S. C., Pseudomonas aeruginosa: assessment of risk from drinking water Crit. Rev. Microbiol., 23, pp. 47-75(1997).

[4] Stover C.K., Pham X.Q., Erwin A.L., Mizoguchi S.D., Warner P. et Olsen M.V., Complete genome sequence of Pseudomonas aeruginosa PAO1, an apportunistic pathogen Nature., 406, pp. 959-964(2000).

[5] Hassen A., and al. UV disinfection of treated waste water in a large-scale pilot plant and inactivation of selected bacteria in a laboratory UV device, *Bioresource Technology*., pp. 1-9(1999).

[6] Marisol D. Labas., Rodollf J Brandi et Alberto E. Cassano., Acontribution to UV dose concept for bacterie disinfection in well mixed photoreactors, 116, pp. 197-202(2005).

Capacitively-coupled alternating-current light-emitting device for 12 V applications

Giuseppe Y. Mak , Edmund Y. Lam , H. W. Choi
Department of Electrical and Electronic Engineering, The University of Hong Kong , Pokfulam Road, Hong Kong,
Contact: yhgmak@eee.hku.hk

ABSTRACT

The fabrication of monolithic GaN-based alternating-current light-emitting device (ac-LED) with Wheatstone bridge circuitry and an externally-coupled capacitor for voltage smoothening is demonstrated. On-chip LEDs are first isolated by laser-micromachined trenches and subsequently interconnected by the deposition of metal lines. By coupling capacitors to the dc path, it is found that the radiant flux increases with capacitance. The maximum wall-plug efficiency (WPE) obtained is 4.0% when the capacitance coupled to the dc path is 330 μF. This device could be applied to 12 V_{rms} lighting applications that conform to the MR16 or MR11 standards.

INTRODUCTION

LEDs are inherently p-n junctions which only lights up when they are forward-biased. In order to make use of both half-cycles of the sinusoidal ac voltage, external rectifiers are usually installed within LED luminaires to provide dc voltage. To eliminate the energy wastage during rectification, there is an increasing interest towards LED chips where ac voltage can be directly applied. Anti-parallel design [1] is one of the earliest solutions, but it suffers from the fact that only half of the on-chip LEDs light up in each half-cycle, resulting in flickering (synchronous with the ac frequency) and low efficiency. As a more efficient design, Wheatstone bridge [2] was proposed where the diode property of LED is utilized to form an on-chip rectifier circuit. The rectified dc voltage is further applied to another series of on-chip LEDs (called dc path thereafter) so that they light up in both half-cycles. However, flickering still exists (twice the ac frequency) as the rectified dc voltage still falls towards 0 V periodically. In this paper, we demonstrate the addition of a capacitor to smoothen the voltage across the dc path for the reduction of flickering. The fabrication process of the bridge-type ac-LED chip is also reported.

EXPERIMENT AND RESULTS

Fig. 1 illustrates the schematic structure of our ac-LED chip. The fabrication process began with a bare GaN blue-light LED wafer ($\lambda_{emission}$ = 470 nm) with indium tin oxide (ITO) layer (~ 200 nm) as the current spreading layer. The active region was formed by standard photolithography, wet etch of ITO by aqua regia (volume ratio of $HCl:H_2O:HNO_3$ = 4:2:1) and reactive ion etch of p-GaN and InGaN/GaN multi-quantum well (MQW) by Ar/CHF_3 plasma. Device isolation was then realized by a direct-write laser micromachining process [3]. This process produces trenches that terminate abruptly at the GaN/sapphire interface. Following that, a SiO_2 layer (100 nm) for passivation was deposited by electron beam evaporation. This layer was annealed in N_2 ambient at 500°C for 5 min to improve adhesion. Photolithographic patterning and Ar/CHF_3 dry etch were then performed to expose the p-contact and n-contact regions. To connect the isolated LEDs, Ti/Al (40 nm/300 nm) interconnects were deposited by standard lift-off process. This metal layer was also annealed in N_2 ambient at 600°C for 30 s in order to reduce the contact resistance. The ac-LED die was finally packaged in a gold-plated TO-header for testing.

Regarding the circuit design (Fig. 2), since the LEDs along the dc path will turn on all the time, more LEDs are placed along the dc path to increase the chip area utilization ratio (defined the ratio of the light-emitting area to the area of the whole chip). While 2 LEDs are connected in parallel along each branch of the rectifier, 8 LEDs are in parallel along the dc path. The area of each LED is 230 μm x 230 μm, which amounts to a chip of about 1 mm x 1 mm in size. Fig. 3 shows the operation of the chip at 12 V_{rms}. A range of electrolytic capacitors (3.3 μF - 3300 μF) were tested to examine the effect of capacitance on the potential difference (p.d.) across the dc path, radiant flux and WPE (defined by (radiant flux)/$I_{rms}V_{rms}$). As shown in Fig. 4, when no capacitor is connected, the p.d. across the dc path is often below the usual turn-on voltage of GaN blue-light LED (~ 3.5 V). The charging and discharging of the capacitors helps to smooth the voltage ripples. By using capacitors of 330 μF or above, the ripples are small enough to enable continuous operation of dc-path LEDs. This result correlates well with the radiant flux increase (~ 2 mW) and WPE (attaining 4.0% for 330 μF) for the mentioned capacitances (Fig. 5).

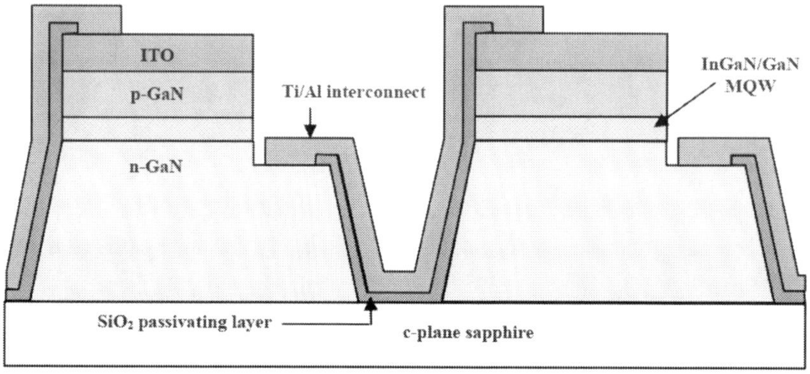

Fig. 1 Schematic structure of two interconnected on-chip LEDs.

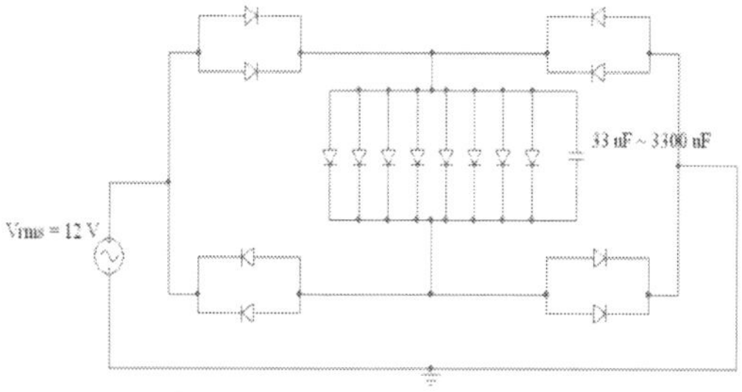

Fig. 2 Circuit diagram of ac-LED chip.

Fig. 3 Microphotograph of ac-LED operating at 12 V_{rms}.

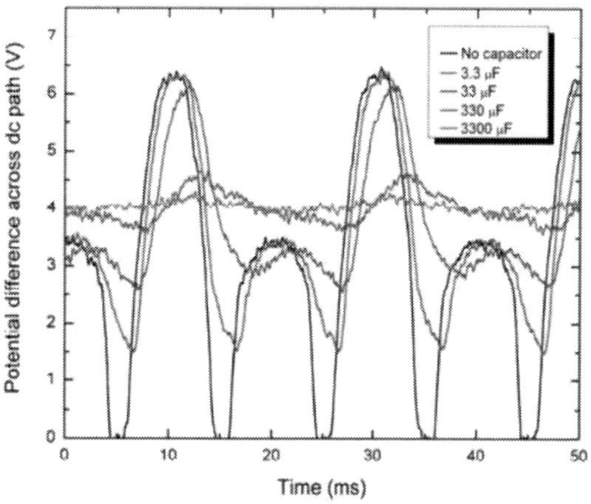

Fig 4 Variation of p.d. across dc path.

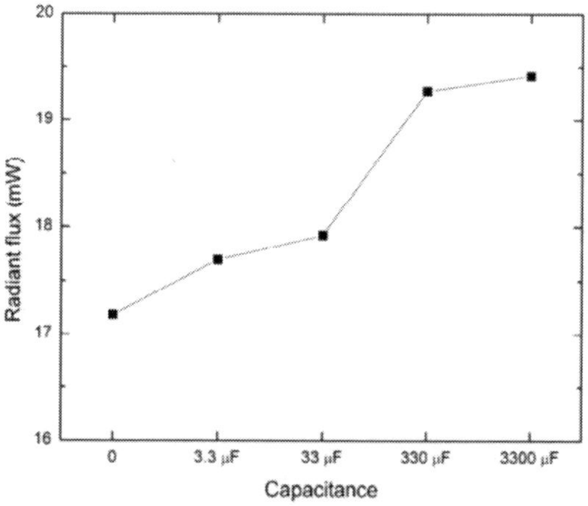

Fig. 5 Trend of radiant flux with increasing capacitance.

[1] J. Ao, H. Sato, T. Mizobuchi, K. Morioka, S. Kawano, Y. Muramoto, Y. Lee, D. Sato, Y. Ohno, and S. Sakai, Phys. Stat. Sol. (a), 194 (2002) 376.
[2] J. Cho, J. Jung, J.H. Chae, H. Kim, J.W. Lee, S. Yoon, C. Sone, T. Jang, Y. Park, and E. Yoon, Jap. J. Appl. Phys., 46 (2007) L1 194.
[3] G. Y. Mak, E. Y. Lam, and H. W. Choi, J. Vac. Sci. Tech. B, 28 (2010) 380.

Site-controlled GaN photonic crystals by nanosphere lithography

K. H. Li, H.W. Choi*
Department of Electrical and Electronic Engineering, The University of Hong Kong, Hong Kong
**Corresponding author: e-mail hwchoi@hku.hk*

ABSTRACT

We report on the fabrication of ordered array of site-controlled nano-pillars on GaN by nanosphere lithography. Utilizing a self-assembled two-dimenional silica nanosphere array as an etch mask, nano-pillars with vertical sidewalls are obtained. Prior to transferring the pattern to the GaN substrate, the diameters of individual nanospheres in the array are tuned using an etching process with a CHF_3-based recipe. With such an approach, not only can the dimensions of the nano-pillars be adjusted, but photonic band gap properties of the self-assembled nano-structures can also be modified.

INTRODUCTION

The phenomenal success of Gallium nitride (GaN) based light-emitting diodes (LEDs) is clearly evident from the rapidly-rising solid-state lighting market. Nevertheless, conventional GaN-based LEDs suffer from limited light extraction as the majority of light generated at the multi-quantum-wells (MQWs) remains trapped within the device. Recently, significant improvements have been demonstrated using dimensional downscaled structures compared with conventional broad-area LEDs, including micro-disks [1] and nano-rings [2]. Moreover, photonic crystals, which are attractive and promising optical structures for controlling spontaneous emission [3], possess a important property for LEDs – the existence of a photonic band gap forbidding the propagation of light within a range of wavelength and channeling photons into directions that are useful for light extraction. Typically, the nano-patterns of photonic crystal structure are fabricated using electron beam lithography and nanoimprint lithography [4], which both provide for highly nanoscale patterns. The main drawbacks of these techniques are high cost and inefficiency, thus being not suitable for mass production. An alternative self-assembly method for large scale production is nanosphere lithography. A severe limitation of this technique is that the spheres naturally form hexagonal closed-packed arrays, whereas photonic crystal structures require a periodic arrangement of two alternating media with finite separations. The sites and dimensions of nano-pillars thus play an important role in determining the light extraction efficiency through the photonic crystal structure, affecting the existence and characteristics of photonic bandgaps. In this report, the diameters of closed-packed silica nanopheres are tuned using a dry etch process, prior to transferring pattern to GaN, so that dimension-controllable of nano-pillars can be achieved.

EXPERIMENTAL DETAILS AND RESULTS

Figure 1 illustrates the fabrication process flow of nano-pillars by nanosphere lithography. The starting material is a GaN-based LED structure containing InGaN/GaN multi-quantum wells (MQWs) with emission wavelength of ~ 470 nm. Silica (SiO2) nanospheres suspended in deionized water, with mean diameter 500 nm, are mixed with Sodium dodecyl sulfate (SDS) in order to lower the surface tension of the mixture. The self-assembled monolayer can be formed on the top surface of the sample by spin-coating 4ul of suspension at 1000 revolutions per minute. Prior to transferring the pattern to GaN, the dimensions of the silica nanospheres are first modified by a reactive ion etching (RIE) process using CHF_3-based plasmas under low RF power. Figure 2 plots the diameters of nanospheres as a function of etch duration, which show that the slope after 20 min is four-times steeper than before, indicating that the diameter of nanosphere shrinks effectively once the etch duration exceeds the twentieth minute. The low power and CHF_3-based etching conditions ensure the silica nanospheres shrink without significant damage. The field-emission scanning electron microscope (FE-SEM) image in Figure 2 shows a nanosphere array with dimensional tuning after etching for 30mins.The shrunken nanosphere pattern acts as an etch mask and it is subsequently transferred the GaN wafer by inductive coupled plasma (ICP) etching using Cl_2 and BCl_3 as etchant gases. The spaces between the closed-packed pillars are exposed, as shown in Figure 3. Photoluminescence (PL) spectra in Figure 4 show both enhancements of PL intensity, as well as a systematic blue shift with decreasing pillar diameter, signifying strain relaxation in nanostructures.

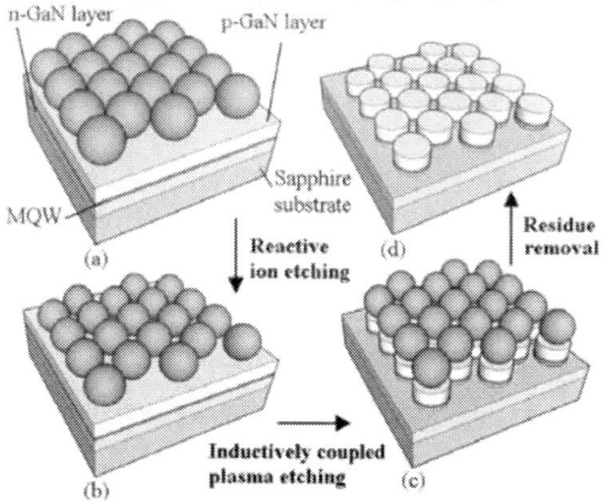

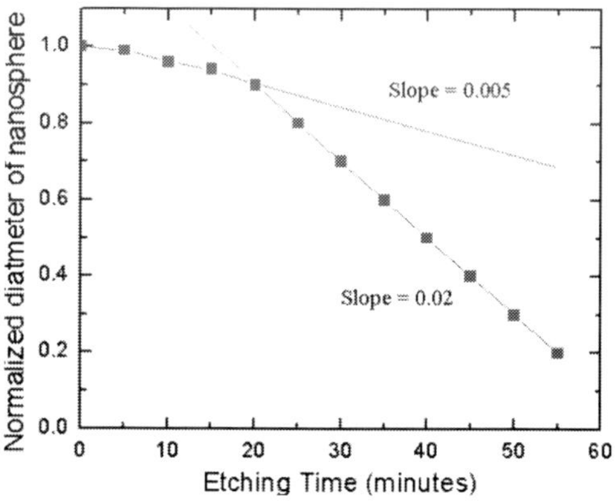

Figure 1(left). Schematic diagrams demonstrating the fabrication flow. (a) Silica microspheresare coated on surface of p-GaN layer. (b) Dimension control of spheres using reaction ion etching. (c) Pattern transfers to GaN using ICP etching. (d) Nanopillar array formed after reside removal.
Figure 2(right). The plot of normalized diameter of nanosphere(500nm) verses etching time.

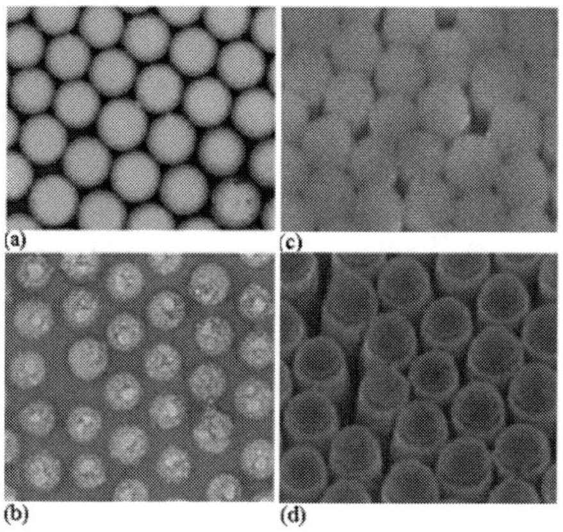

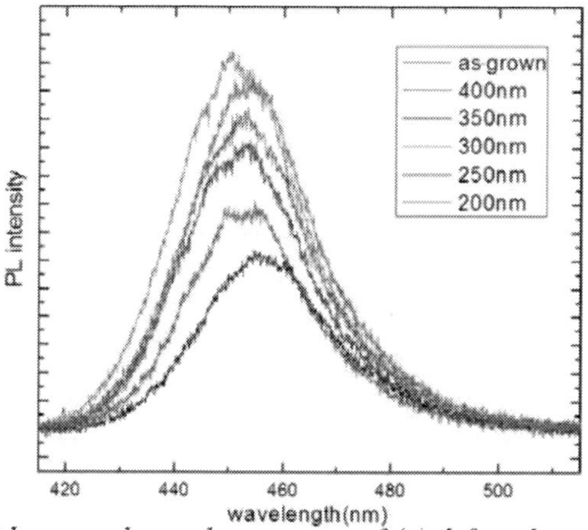

Figure 3. FE-SEM images showing ordered hexagonal monolayer arrays of (a) defore dimensional tuning and (b) after dimensional tuning using reactive ion etching; nanopillar array (c) without dimension-modifying treatment (d) with dimension-modifying treatment before nanosphere removal.
Figure 4. PL spectra of as grown and nanopillar array with dimension-modifying treatment.

REFERENCES

[1] S.X. Jin et al., Appl. Phys. Lett. 77, 3236 (2000)
[2] H.W. Choi et al, Appl. Phys. Lett. 89, 211101 (2006)
[3] S. Noda et al., Nature Photonics 1, 449 (2007)
[4] S.H. Kim et al., Nanotechnology 18, 055306 (2007)

Zinc Oxide Nanorods/Polymer Hybrid Junctions for White Light Emitting Diodes

M. Willander[1], N. Bano[1], S. Hussain[1], I. Hussain[1], O. Nur[1]

[1]*Department of Science and Technology, Campus Norrköping, Linköping University, SE-601 74 Norrköping, Sweden*

Contact: magwi @itn.liu.se

ABSTRACT

We report white-light luminescence from ZnO-organic hybrid junctions grown on glass substrate by low temperature aqueous chemical growth (ACG). The configuration used for the hybrid white light emitting diodes (WLEDs) consists of two-layers of polymers on glass with top ZnO nanorods (NRs). The inorganic/organic heterojunction has been fabricated by spin coating the p-type polymer Poly (3,4-ethylenedioxythiophene) poly(styrenesulfonate) (PEDOT:PSS) for hole injection with an ionization potential of 5.1 eV and Poly(9, 9-dioctylfluorene) (PFO) is used as blue emitting material with a bandgap of 3.3 eV. ZnO nanorods (NRs) are grown on top of the organic layers. The WLEDs were characterized by current-voltage (I–V) characteristics, room temperature photoluminescence (PL) and electroluminescence (EL). The EL spectrum reveals a broad emission band covering the range from 420 to 800 nm, and the emissions causing this white luminescence were identified.

INTRODUCTION

Zinc oxide (ZnO) a II-VI direct wide bandgap (3.4 eV at room temperature) semiconductor possesses excellent physical and chemical properties, among them is relatively large exciton binding energy, the possibility to obtain white light emission due to the existence of many deep levels that emits covering the whole visible spectrum, has recently attracted a renewed global attention [1]. The excellent luminescence property of ZnO, although known for a long time, has not been explored until recently, the main reason is the fact that there is a persistent difficulty in growing stable p-type doped ZnO. Nevertheless, the self organized growth property of ZnO has shown to be very beneficial for the development of different photonic devices based on ZnO [1]. Among these different photonic applications is the development of 'intrinsic' white light emitting diodes using a technology which is environmental friendly. ZnO offers this possibility as it is a bio-safe material and doesn't involve hazards materials during growth. Due to the self organized growth property of ZnO it has become possible to grow ZnO in the nanostructure form on many different p-type substrates, including those with even amorphous nature [2]. One of the interesting possible substrates to be used is polymeric semiconductors. One reason is that these semiconducting materials can be processed on flexile substrates or glass and hence are suitable for large area LEDs development.

In this paper we demonstrate intrinsic white light emission using a hybrid ZnO nanorods/p-polymer light deposited on glass substrate. The ZnO nanorods were grown at 90 °C using an aqueous chemical growth on polymer coated glass substrates. The components and origin of intrinsic white light constituents were identified.

EXPERIMENTAL

A glass substrate was washed using iso-propanol, acetone, and de-ionized water in an ultra-sonic path. Conducting PEDOT: PSS was then spin-coated on the glass substrate and baked at 150 °C to form a uniform film. A small apart of the PEDO: PSS to be used for ohmic contacts was covered and the rest was covered by PFO using spin coating. Then a two step growth procedure was used to grow well aligned ZnO nanorods on top of the PEDOT: PSS/PFO layers. For details of the growth the readers are directed to [1]. The ZnO nanorods/PFO-PEDOT: PSS multi-layer structure was then processed to fabricated LEDs as described in [1]. Electrical and electroluminescence (EL) were measured to characterize the diode rectification and light emission from the device.

RESULTS AND DISCUSSION

Figure 1 below shows the current voltage (I-V) characteristics of the fabricated LED. As can be clearly seen the fabricated device possesses good rectification implying that a reasonable p-n hybrid junction has been achieved. Analyzing this I-V has indicated that the ideality factor of the diode is 2.27-3.7 for all the fabricated diodes. The rather high value of the ideality factor implies that the thermionic emission is not the dominating current process.

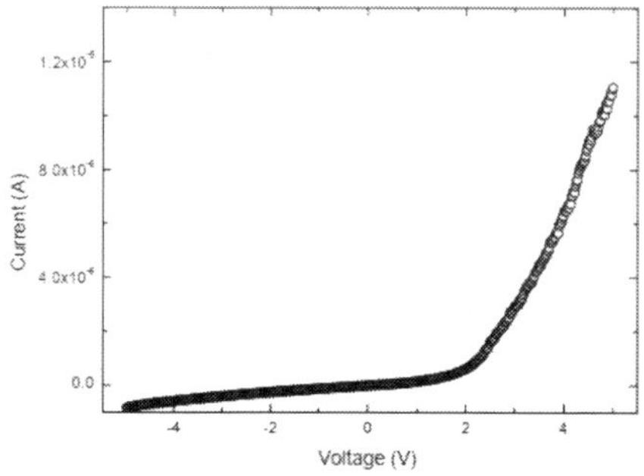

Figure 1: Rectification characteristics obtained from the ZnO nanorods/polymer hybrid pn heterojunction.

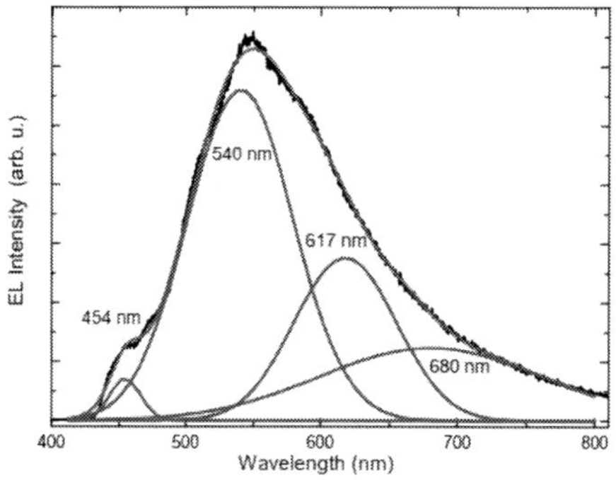

Figure 2: Electroluminescence (EL) spectrum obtained for one of the ZnO NRs/Polymer hybrid junctions revealing a broad spectrum covering the whole visible region.

Figure 2 shows the EL spectrum obtained form the LEDs. As clearly seen white luminescence as can be seen a wide emission covering the range between 420 nm and up to over 750 nm is observed. A Gaussian fitting to the observed white emission indicates that the white light is composed of four different emissions as indicated in the figure. These are a blue emission at 454 nm, a green emission at 540 nm, an orange emission at 617 nm and a red emission at 680 nm.

CONCLUSION

In conclusion using a simple low temperature (100 °C) chemical process, intrinsic white light emission was demonstrated using ZNO nanorods/polymer hybrid junctions. The demonstrated device is suitable for the development of large area white LED on glass substrates and hence can be integrated into existing armature technologies.

REFERENCES

[1] M. Willander, O. Nur, Q. X. Zhao, L. L. Yang, M. Lorenz, B. Q. Cao, J. Zúñiga Pérez, C Czekalla, G. Zimmermann, and M. Grundmann, A. Bakin, A. Behrends, M. Al- Suleiman, A. El-Shaer, A. Che Mofor, B. Postels and A. Waag, N. Boukos and A. Travlos H. S. Kwack, J. Guinard, and D. Le Si Dang, Zinc oxide Nanorod based Photonic Devices: Recent Progress in Growth, Light Emitting Diodes and Lasers. Nanotechnology 20 (2009) 332001.

[2] M. Willander, O. Nur, N. Bano, and K. Sultana. Zinc oxide nanorods based heterostructures on solid and soft substrates for light emitting diodes applications, New Journal of Physics 11 (2009) 125020.

Synthesis and Characterization of EuZn and TbZn complex emitting red and green colors for organic light-emitting diodes

Byoung-Hwa Kwon, Sung Wook Kim, Duk Young Jeon
Department of Materials Science and Engineering, Korea Advanced Institute of Science and Technology, 335 Gwahangno, Yuseong-gu, Daejeon 305-701, Republic of Korea
Contact: dyjeon@kaist.ac.kr

ABSTRACT

EuZn and TbZn complexes were synthesized by a organometallic method. As the reaction temperature increases, organic materials and Zn ions coordinate with Eu or Tb ions, and they form heterodinuclear complexes. They show strong Eu^{3+} and Tb^{3+} characteristic emission due to the energy transfer from organic ligands to lantanide ions, so it has the potential for OLEDs.

GENERAL

Inorganic phosphor with strong and stable luminescence has been researched for light-converting devices[1]. The surface property of phosphor is hydrophobic, so it has the demerit for the application of a organic-based light source such as organic light-emitting diodes (OLEDs)[2] and polymer light-emitting diodes (PLEDs)[3]. On the other hand, lanthanide metal complexes can have various luminescent properties[4] and be well dispersed in organic layer. The research for lanthanide complex synthesized by simple process is significant for efficient OLEDs. Herein we report the synthesis and optical property of EuZn and TbZn dinuclear complex.

EnZn and TbZn complex were synthesized by a organometallic method. Europium acetate hydrate or terbium acetylacetonate powder was mixed with zinc stearate, octadecene (ODE), and tetracosane (TCA). The mixed solution was heated to 310°C under argon ambient atmosphere, and then was cooled down to room temperature. Final products were dispersed in hexane. The optical and structural properties were characterized by photoluminescence (PL), photoluminescence excitation (PLE), and nuclear magnetic resonance (NMR). Figure 1-(a) is PL spectra of Eu and Zn precursor-mixed solution as the temperature increases from 120 °C to 310 °C. In 120 °C, there is weak emission at 594, 615nm. The weak and broad peak around 368nm is related to organic materials such as ODE and TCA. As the reaction temperature becomes higher than 210 °C, Eu^{3+} characteristic emission at 580, 594, 615, 654, 702nm increases largely and broad emission in 350-650nm also appears. The PLE spectra of Eu and EuZn complex is shown in Fig. 1-(b). Sharp excitation at 395nm is characteristic Eu^{3+} excitation ($^5L_6 \rightarrow {}^7F_0$), and broad excitation in 220-420nm is related to organic ligands of the complex. EuZn complex shows stronger excitation of the segmental ligand than Eu complex. It means that EuZn complex has proper structure for the efficient energy transfer from the organic ligand to Eu^{3+} ion. Fig. 1-(c) is the emission spectra of Eu, Zn, and EuZn complex after the purification process. The ionic molecules and non-reacted metal ions removed by centrifugal separator. Metal complex may be formed by coordinate bond between metal ion and organic ligand. Zn complex shows the weak and broad emission in 350-600nm. Eu complex also shows the weak Eu^{3+} emission, and the intensity is similar to that of organic ligand in 350-500nm. However, EuZn complex has strong Eu^{3+} red emission, and it is totally different from both Eu complex and Zn complex. The ^{1}H-NMR spectra of EuZn complex in D-chloroform ($CDCl_3$) is shown in Fig. 2. The raw materials such as TCA, ODE and zinc stearate have resonances for methylene (CH_2) group (δ = 1.26−1.31 ppm) and methyl (CH_3) group (δ = ~0.88 ppm) in a ratio of 1:0.10. In 120 °C temperature, the peak position of EuZn complex is 1.26, 0.88ppm, and the ratio is similar with raw organic materials. Although the resonance position at 310 °C is almost same, the ratio of the peak intensity considerably changes from 1:0.16 to 1:0.68. The hydrocarbon chain of the capping materials and stearate ion may break down and react with Eu and Zn ion to form metal complexes. In figure 3, it shows the PL and PLE spectra of TbZn complex. TbZn complex have Tb^{3+} characteristic emission at 491, 545, 584, 591, 623nm. The excitation spectrum measured at 545nm emission shows two distinct band. The excitation peaks around 314nm and 350nm may be due to the organic ligands and Tb^{3+} ion. Red-emitting EuZn and green-emitting TbZn dinuclear complexes synthesized by a simple pyrolysis process have the potential as a organic-based light source.

Figures and equations

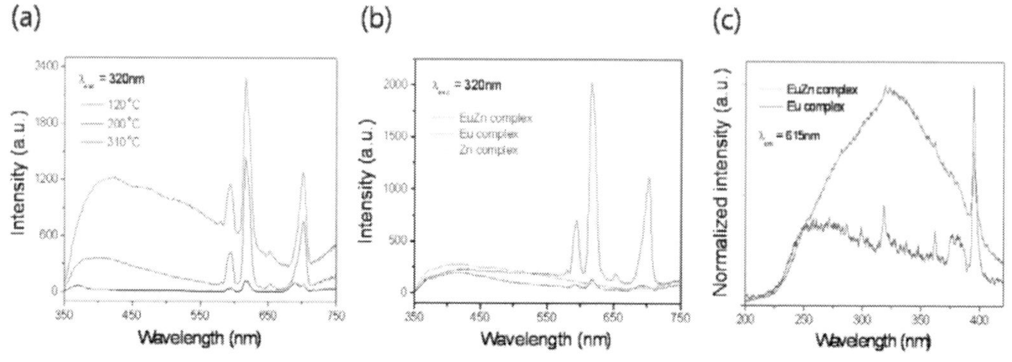

Figure 1. Photoluminescence spectra of (a) EuZn complex at 120, 200, 310°C, (b) Eu, Zn, and EuZn complex after purification process, and (c) photoluminescence excitation spectra of Eu and EuZn complex.

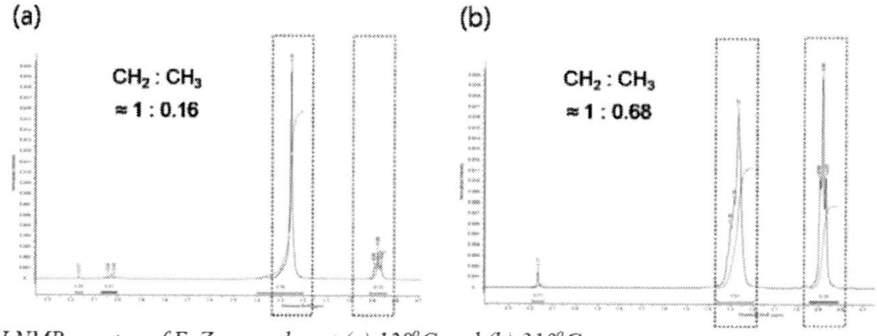

Figure 2. ^1H NMR spectra of EuZn complex at (a) 120°C and (b) 310°C.

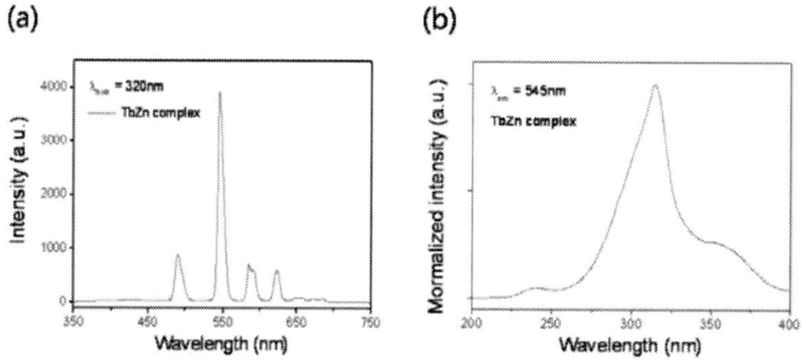

Figure 3. (a)The photoluminescence and (b) photoluminescence excitation spectra of TbZn complex.

REFERENCES

[1] H. S. Jang, B.-H. Kwon, and D. Y. Jeon, et. al., Appl. Phys. Lett. 95, 161901 (2009).
[2] G. He, D. Guo, and C. Duan, et. al., Angew. Chem. Int. Ed. 48, 6132 (2009).
[3] L. Zhang, B. Li, and W. Li, et. al., J. Lumi. 128, 67 (2008).
[4] C. Piguet, J.-C. G. Bunzli, and O. Schaad, et. al., J. Am. Chem. Soc. 118, 6681 (1996).

Near Field UV Output Measurements for Amalgam Lamps

R. Kilgour and D.G. Knight
Trojan Technologies, 3020 Gore Road, London, Ontario, Canada N5V 4T7
Contact: gknight@trojanuv.com

ABSTRACT
The UV output of low pressure amalgam lamps as determined by the near-field technique is shown to be uniform over the entire lamp, except for within 1 cm of the lamp filament. This is shown for both new and 12,000 hour aged lamps, so that accurate lamp maintenance data can be determined. The change in UV output with dimming and water temperature change was demonstrated to be similar for both new and aged lamps in this study, except for the expected general decrease in lamp maintenance noted for the aged lamp.

INTRODUCTION
Low pressure (LP) mercury arc lamps are a widely used source of ultraviolet (UV) light for water disinfection. The near-field measurement technique is a reliable method of obtaining the decrease in lamp output with lamp age (lamp maintenance), effect of water temperature on lamp output, and effect of dimming on lamp output; measured under the actual operating conditions of the lamp [1]. The absolute UV output power of a lamp can be determined using near-field with far-field lamp measurements [2]. However, to provide accurate data for the validation of lamp properties, it would be highly desirable to demonstrate that the near-field data gathered for one point on the lamp is representative of the near-field output for the entire lamp. It would also be desirable to show that this is the case for aged as well as new lamps, so that accurate lamp maintenance can be determined. Measurement of near-field output along the length of amalgam lamps as well as around the lamp is conducted for new and 12,000 hour aged lamps, to demonstrate the uniformity of the output values obtained. The effect of aging on lamp properties other than maintenance will be determined as well, to increase understanding of the properties of aged lamps.

METHOD
The near-field test apparatus previously described [1] is used to determine the output of new 100 hour and 12,000 hour aged GA64T6 amalgam lamps operated at 100%, 80% and 60% power levels. Water temperature is maintained at 20°C, except for lamp output versus water temperature tests. To measure the near-field output along the length of the lamp, the assembly holding the fibre optic probe is moved along the lamp at 5 cm intervals as measured from the filament at the socket end of the lamp. The total arc length of the lamp is 147 cm. The lamp output as a function of angle around the lamp is measured 50 cm from the socket filament, and determined by rotating the lamp within the sleeve in 45° increments with the fibre optic probe in the upper vertical position and the amalgam spot within the lamp initially at the bottom of the lamp. The amalgam temperature for the spot closest to the socket end of the lamp was determined for each measurement point by placing a thermocouple on the exterior of the lamp at the spot location.

RESULTS
The normalized UV output as a function of distance along the lamp arc for a 100 hour lamp at 100, 80 and 60% electrical power is shown in Figure 1. The UV output is very uniform along the length of the lamp with two exceptions; when the output measurement is taken ≤1 cm from a filament of the lamp, and when the UV output is determined adjacent to the socket end amalgam spot. In the first case, a decrease in intensity of the arc next to the filament can account for this observation. In the second case, the Teflon holder of the fibre optic probe must be placed over the amalgam spot, which could change the thermal conditions of the spot. The plots of amalgam temperature versus position confirms this, since an upward spike in amalgam temperature of 2 - 4 °C is noted at all electrical powers when the probe covers the spot. The change in amalgam temperature can therefore change the mercury vapour pressure in the lamp, and alter the UV output as observed. Since the amalgam spot temperature will not change in normal operation, the UV output value above the amalgam spot can be considered an artefact of the measurement method. Considering the data > 1 cm from the filament and excluding the point above the amalgam spot, the average normalized UV output values for 100, 80 and 60% power are 0.96±0.01, 0.71±0.02 and 0.45±0.01 respectively, where the uncertainties are ±1σ in normalized UV output. Similar data for a 12,000 hour aged lamp is shown in Figure 2, where the UV output ≤ 1 cm from the filament is lower than the output at other points along the lamp. Since lowering of the UV output is not noted further along the lamp, lamp end darkening is not expected which is supported by visual inspection of aged lamps. The amalgam spot temperature shows a similar

increase when the UV output is measured above the spot, but there is no corresponding spike in UV output above the spot. Since the average spot temperature at each electrical power has increased, an upward shift in amalgam temperature may not cause a corresponding change in mercury vapour pressure. The average normalized UV output values for 100, 80 and 60% power are 0.97±0.04, 0.74±0.02 and 0.44±0.02 respectively. In comparison with the 100 hour lamp results, there is slightly more variation in lamp output along the lamp, and no significant change in the decrease in normalized lamp output with dimming.

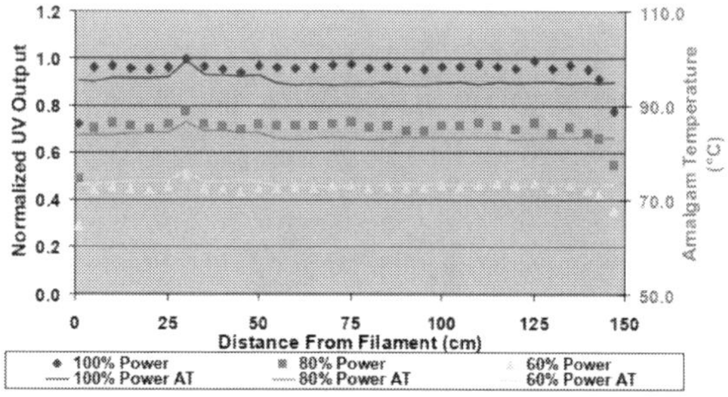

Figure 1: Normalized UV output of 100 hour lamp (data points) and amalgam temperature (line) at 100, 80 and 60% power.

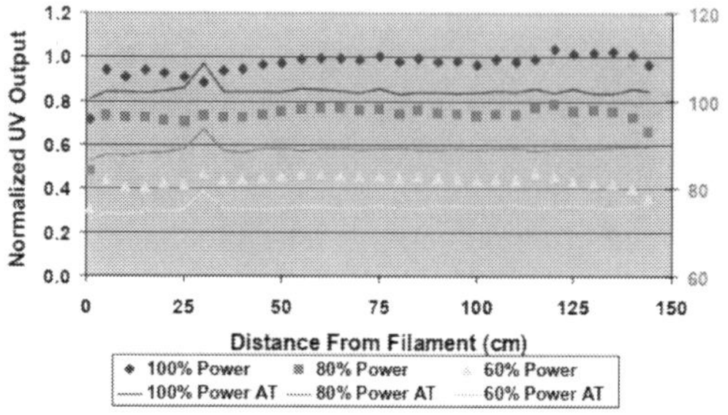

Figure 2: Normalized UV output of 12,000 hour lamp (data points) and amalgam temperature (line) at 100, 80 and 60% power.

For the UV output as a function of angle, Figure 3 shows very uniform output for all angles measured except for 180°, when the amalgam spots are in the upwards position. In this position there is a decrease in UV output which is especially noted when the lamp is dimmed. This is believed to be a result of the location of the amalgam spot which has caused a change in the mercury vapour pressure. The average normalized UV output values for 100, 80 and 60% power are 0.98±0.02, 0.74±0.02 and 0.44±0.01 respectively when the 180° data is omitted, and the data along the lamp is in agreement within the uncertainty limits. Finally, the normalized lamp UV output as a function of water temperature for new and aged lamps is shown in Figure 4. The changes in UV output with temperature are very similar for both lamps. The only significant change is the consistent 3-5% decrease in lamp output noted for the aged lamp at all temperatures measured.

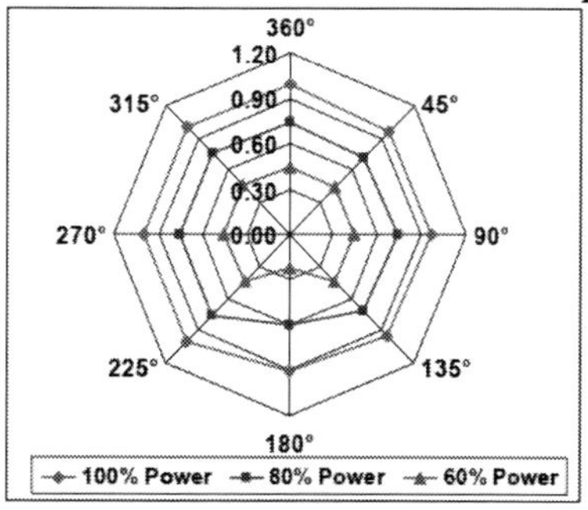

Figure 3: Normalized UV output as a function of angle for 100 hour lamp.

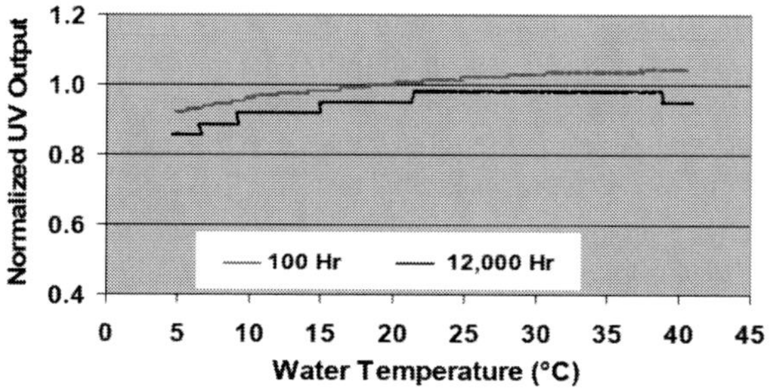

Figure 4: Normalized UV output of new and aged lamps as a function of water temperature

CONCLUSIONS

Excluding near-field data obtained within 1 cm of a lamp filament, over the amalgam spot, or with the amalgam spot in the upwards position, the near-field UV output is very uniform when measured at any point on the lamp at a given electrical input power. This is observed for both new 100 hour lamps, and aged 12,000 hour lamps. This result indicates that accurate lamp maintenance data can be obtained using the near-field technique for validation purposes. For the lamps studied in this report, there are no significant changes in the normalized UV output versus dimming or UV output versus water temperature data when new and aged lamps are compared, except for the expected decrease in lamp maintenance for the aged lamp.

REFERENCES

[1] R. Kilgour and D.G. Knight, Proc. 11th Int. Symp. On the Science and Technology of Light Sources, Shanghai China, FAST-LS press, 2007, p. 167-168.

[2] J. Robinson, Proceedings 2007 World Congress on Ozone and Ultraviolet Technologies, Los Angeles CA, p. 1381-1390.

Coupled Simulation of Phosphor Emission and Light Propagation in white LEDs

Kunio Ishida, Takahiro Sato, Yasushi Hattori, Iwao Mitsuishi, and Shinya Nunoue
Corporate Research and Development Center, Toshiba Corporation, Kawasaki 212-8582, Japan
Contact: ishida@arl.rdc.toshiba.co.jp

ABSTRACT

We calculated the spatial distribution of light intensity in the resin layer of phosphor-based white LEDs with various structures. Using the previously proposed simulation method for the luminescence profiles of white light-emitting diodes (LEDs)[1], we take into account the effect of light scattering by phosphor particles as well as the reabsorption effect by phosphor species. Comparing the calculated results, we show that this method is efficient to design optimum device structures.

Recent advance in white light-emitting diodes (LEDs) has attracted our attention regarding general illumination applications[2-4]. Phosphor-based white LEDs, in particular, are known as a strong candidate for commercial use, and many authors have been studying to find optimized device structure[5].

It has been pointed out that multicolor phosphor blends have several advantages to realize ideal white light sources, i.e., high efficiency and high color rendition. However, it has been experimentally observed that, when multiple phosphor species are simultaneously implemented in the device, the light output cannot be described by a simple superposition of the luminescence spectra of individual phosphor species. In our previous paper[1] we showed that an efficient simulation method for these devices was established by employing a phenomenological model of phosphors describing the absorption/emission processes of photons coupled with the method by Kubelka and Munk[6].

In our method, by combining the stationary solutions of rate equations for phosphor emission/absorption processes with the Kubelka-Munk equations, we obtained coupled equations for the light intensity as[1]

$$\frac{\partial I_F(\lambda,z)}{\partial z} = \sum_i \left\{ \frac{A_i N_2^i L_i(\lambda)}{2} - B_i N_g^i a_i(\lambda) I_F(\lambda,z) \right\} - S(I_F(\lambda,z) - I_B(\lambda,z)) \quad (1)$$

$$\frac{\partial I_B(\lambda,z)}{\partial z} = \sum_i \left\{ -\frac{A_i N_2^i L_i(\lambda)}{2} + B_i N_g^i a_i(\lambda) I_B(\lambda,z) \right\} + S(I_F(\lambda,z) - I_B(\lambda,z)), \quad (2)$$

where $I_F(\lambda, z)$ and $I_B(\lambda, z)$ denote the intensity of forward-scattering light and backward-scattering light for wavelength λ at the depth z, respectively. i is an index to distinguish the phosphor species from each other, and N_g^i, N_1^i, and N_2^i denote the stationary solutions of the rate equations for the ground state, the final state of absorption, and the initial state of emission, respectively. $a_i(\lambda)$ and $L_i(\lambda)$ are the excitation luminescence spectra (EPL) and the photoluminescence spectra (PL) of each phosphor species, by which we take into account the wavelength dependence of absorption-emission profiles of each phosphor species.

We numerically calculated the light distribution in the phosphor resins, where the phosphors for blue (B), bluish-green (BG), yellow (Y), and red (R) are based on: $BaMgAl_{10}O_{17}$:Eu,Mn, (Ba, Ca, Sr)$_{10}(PO_4)_6Cl_2$:Eu, $(Sr,Ba)_2SiO_4$:Eu, and $CaAlSiN_3$:Eu, respectively, and the phosphors are excited by near-UV light (λ=405nm). The values of the parameters for phosphor species were obtained to reproduce experimental data of white LEDs with the above phosphors. As for the resin structures, we chose two cases: (a) a single-layer structure in which all the phosphors are uniformly distributed and (b) a multilayer structure composed of red, yellow, bluish-green, and blue layers from the side irradiated by UV light. Figure 1 shows the pallet-mapped plot of the spatial distribution of $I_F(\lambda,z)$ for (a) and (b). For the single-layer case (a), the light intensity increases with z initially. However, due to the reabsorption effect[1], the blue/bluish-green light takes its maximum value inside the resin and

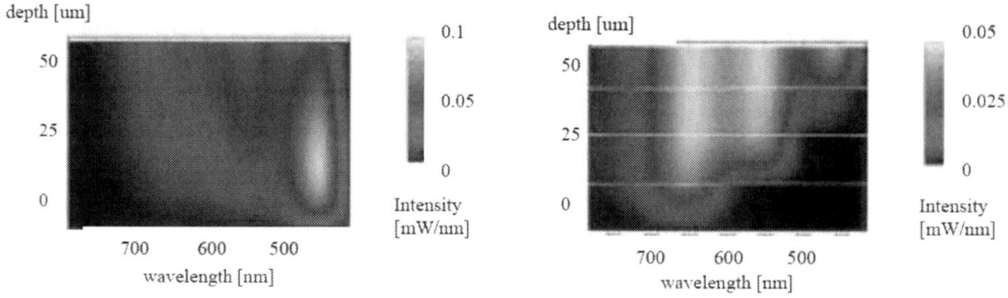

Figure 1. A pallet-mapped plot of the light intensity $I_F(\lambda, z)$ for (a) a single-layer resin with phosphors uniformly distributed and (b) a multilayer resin with red(R), yellow(Y), bluish-green(BG), and blue(B) layers shown in the figure.

decreases afterwards, which causes the energy loss. On the contrary, when the multiple layer structure is taken, Fig. 1-(b) shows that the decay of the light intensity is not observed for any phosphor species, i.e., the reabsorption effect is suppressed in the multiple layer structure. We, however, found that the intensity of blue light output does not increase even when the reabsorption effect does not take place. This feature is understood as follows: when the multilayer structure is taken, the UV light is mainly absorbed by the red and yellow phosphors, and only a portion of the UV light is used to excite the blue phosphors.

Thus, the amount of blue phosphors is relatively large to maintain sufficient intensity of blue light output. In other words, the amount of phosphors to obtain required output spectra is determined by the distribution of injected excitation energy among phosphor species, which is difficult to estimate without numerical simulations.

We stress that the present method is applicable to other resin structures by choosing appropriate boundary conditions, and that the calculation of the light intensity, color rendition, chromaticity of the light propagating in the resins is available.

In particular, by obtaining the spatial distribution of the energy loss in the resin, we will find a clue to the thermal design of the devices, which shows that the present method is an effective and powerful tool to design the device structure of white LEDs.

REFERENCES

[1] K. Ishida, I. Mitsuishi, Y. Hattori, and S. Nunoue, Appl. Phys. Lett., 93 (2008) 241910.
[2] T. Taguchi, J. Light & Vis. Env. 27 (2003) 131.
[3] D. A. Steigerwald, J. C. Bhat, D. Collns, R. M. Fletcher, M. O. Holcomb, M. J. Ludowise, P. S. Martin, and S. L. Rudaz, IEEE J. Sel. Top. Quantum Electron. 8 (2002) 310.
[4] M. S. Shur and A. Zukauskas, Proc. IEEE 93 (2005) 1691.
[5] N. T. Tran and F. G. Shi, Proc. Int'l. Microsystems, Packaging, Assembly Conf. (IMPACT), (2007) 255.
[6] P. Kubelka and F. Munk, Z. Tech. Phys., 12 (1931) 593.

Novel insights in the electron and hole mobility in organic semiconductors used in OLEDs for lighting applications.

R.J. de Vries[1,2], R. Coehoorn[1,2] and R.A.J. Janssen[2]
[1] *Philips Research Laboratories, High Tech Campus 4, 5653 AE Eindhoven, The Netherlands*
[2]*Molecular Materials and Nanosystems, Department of Applied Physics, P.O. Box 513, 5600 MB Eindhoven, The Netherlands*
Contact: rein.de.vries@philips.com

ABSTRACT

The development of organic light-emitting diodes (OLEDs) for lighting applications is hampered by a lack of understanding of the carrier mobility which determines the electric conduction in organic materials. In order to rationally design state-of-the-art OLEDs, which consist of multiple organic layers with different functions, a proper understanding of the carrier mobility is of crucial importance. In earlier work, only a field and temperature dependence of the mobility was assumed. Recent developments show that the effect of carrier concentration on the mobility is very important [1,2]. Furthermore, distinguishing between hopping transport in between molecules with a random Gaussian distributions of energy levels and hopping in between molecules with positionally correlated energy levels is of crucial importance for understanding transport in multilayer devices. We find that current (J) versus voltage (V) characteristics in polymer OLED materials can be described best by a model based on random Gaussian disorder whereas the small-molecule materials studied (hole transport in α-NPD and electron transport in BAlq and Alq_3) are characterized more convincingly by the correlated disorder model.

ROLE OF DISORDER IN POLYMER AND SMALL MOLECULE OLED MATERIALS

Figure 1 shows as an example a state-of-the-art multilayer OLED based on small-molecule materials in which layers with a clear function can be distinguished. Holes are injected from an indium tin oxide (ITO) anode into a hole transport layer which consists of α-NPD (N,N'-bis(1-napthyl)-N,N'-diphenyl-1,1'-biphenyl-4,4'-diamine) whereas electrons are injected from a lithium fluoride cathode into an electron transport layer made of BAlq (bis(2-methyl-8-quinolinolato)(4-phenylphenolato)aluminium).

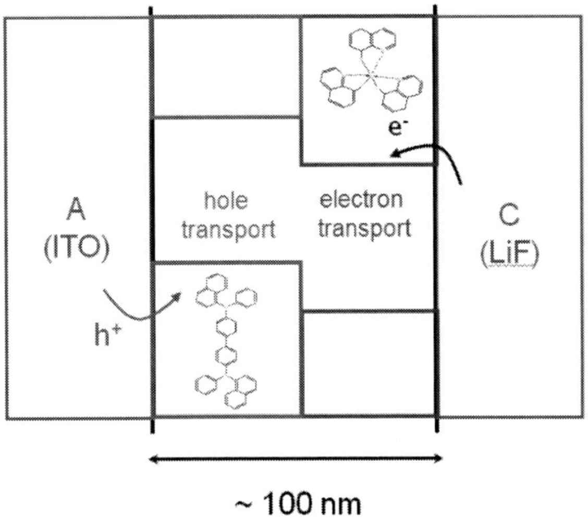

Fig 1. Schematic diagram of a multilayer OLED stack which consists of the small molecule material α-NPD which transports holes and an electron transport layer that consists of BAlq. BAlq has a sky-blue emission colour. These small molecule material layers are positioned in between an ITO anode and a LiF cathode by which holes and electrons are injected, respectively.

Recent studies [3,4] show that the measured $J(V)$ characteristics for different layer thicknesses (L) and temperatures (T) for devices that contain only an α-NPD or BAlq layer in between electrodes, can only be described well using the experimentally found molecular density, if a mobility model based on hopping in a positionally correlated potential landscape is assumed. Figure 2 shows the positionally correlated energy level landscape assumed (bottom part, "extended correlated disorder model", ECDM) as compared to the case of fully random energy levels (top part, "extended Gaussian disorder model", EGDM). In these two models, a very different density and field dependence of the mobility is obtained [1,2]. Contrary to the small molecule materials studied, it was found [5,6] that transport in a blue-emitting polyfluorene based co-polymer should be described using the random Gaussian mobility model (EGDM). Again, distinguishing the models is achieved based on a comparison with the experimental hopping site density.

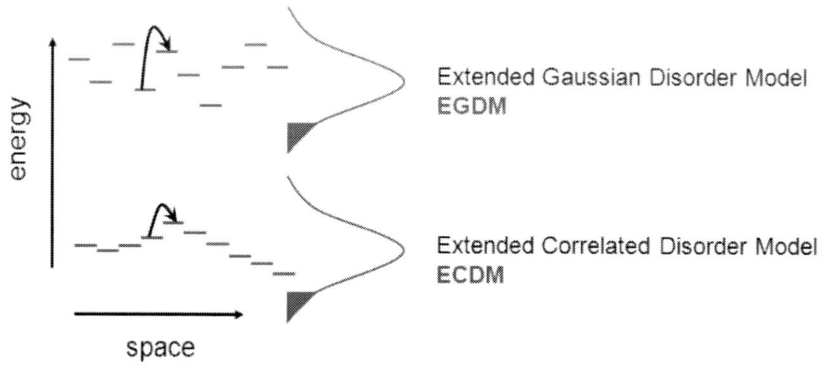

Fig 2. Hopping in a random Gaussian disordered energy landscape (upper figure) versus hopping in a spatially correlated potential landscape (lower figure). Both models describe the total density of states as a Gaussian but lead to very different mobilities at comparable fields and densities. The word "extended" stands for the density dependence of the mobility which is appropriately included in both models

In multilayer OLEDs in which interfaces are present, a locally high density of carriers occurs close to interfaces. Since the predicted mobility at high carrier densities is very different in both models and interfaces strongly determine transport characteristics, it is of crucial importance to distinguish between these mobility models.

APPLICATION TO ALQ$_3$ (TRIS (8-HYDROXYQUINOLINOLATO) ALUMINIUM)

An ongoing investigation on the electron transport in the well studied [7,8] archetypical OLED material Alq$_3$ shows that very similar to the transport in BAlq, the correlated disorder model gives a more appropriate description of the $J(V,L,T)$ characteristics at the molecular Alq$_3$ density. This indicates the presence of correlated disorder. Moreover, it was found that as compared to the results of molecular dynamics calculations on this material [8] a significantly lower width of the density of states was found (about 0.1 eV as compared to ~0.3 eV in ref. 8). More intensive theoretical and modeling studies will be needed to elucidate the origin of this discrepancy.

REFERENCES

[1] W.F. Pasveer *et al.*, Phys. Rev. Lett. **94**, 206601 (2005)
[2] M. Bouhassoune *et al.*, Org. Electr. **10**, 437 (2009)
[3] S.L.M. van Mensfoort *et al.*, (unpublished)
[4] S.L.M. van Mensfoort *et al.*, (unpublished)
[5] S.L.M. van Mensfoort *et al.*, Phys. Rev. B **78**, 085208 (2008)
[6] R.J. de Vries *et al.*, Appl. Phys. Lett. **94**, 163307 (2009)
[7] W. Bruetting *et al.*, Org. Electr. **2**, 1-36 (2001)
[8] Y. Nagata *et al.*, J. Chem. Phys. **129**, 034709 (2008)

Design of an LED Street Lamp with Rectangular Light Spot

H. Shen, X. Zhou, W. Zhang, M. Liu*

Department of Light Sources and Illuminating Engineering, Fudan University, No. 220 Handan Road, Shanghai, 200433, China

Contact: mqliu@fudan.edu.cn

ABSTRACT

The design of an LED street lamp is introduced. The LED street lamp consists of free-curvature-surface lens as the optical elements in front of the LEDs. The lens is design so that the shape of the light spot is rectangular to get the maximum application efficiency. The LED street lamps are installed in Fudan campus and the test results are satisfying.

INTRODUCTION

In China, the LED industry has developed rapidly since 2003. Recently LED street lamps are expected to widely replace the traditional HPS street lamps for energy saving. The LED street lamps have many advantages compared to the HPS. First, the luminous efficacy of LEDs is now high enough to be used in illumination. Second, the LED street lamps can achieve higher luminaire efficiency, due to the fact that the LEDs are commonly semi-spherical directional lighting. Third, the LEDs have longer life so the maintenance cost is low. Moreover, the LEDs are mercury free, and so on.

This paper introduces the design of an LED street lamp with rectangular light spot.

DESIGN

The LED street lamp introduced here is designed that the shape of its light spot is rectangular. Then all the light of the lamps is projected to the useful area. The light spot is controlled by the lens in front of the LEDs. It is called free-curvature-surface lens, which is developed by the free-form optical design theory.

The recent optical design methods based on free-form theory are: Ries's Tailoring method [1, 2], Benitez's SMS (simultaneous multiple surface) method [3], Oliker's Geometric and Variational method [4, 5], Parkyn's Lawnmower method [6, 7]. According to these theories, Ding developed a free-curvature-surface reflector and lens used in the projector to produce uniform illumination [8, 9]. Wang studied the discontinuous free-form lens with precise beam controlling [10]. And we developed the free-curvature-surface lens for LED street lamps.

The surface of the lens is calculated by the energy mapping method and the Snell's Law. The energy mapping method can be described by Figure 1 and Equation 1:

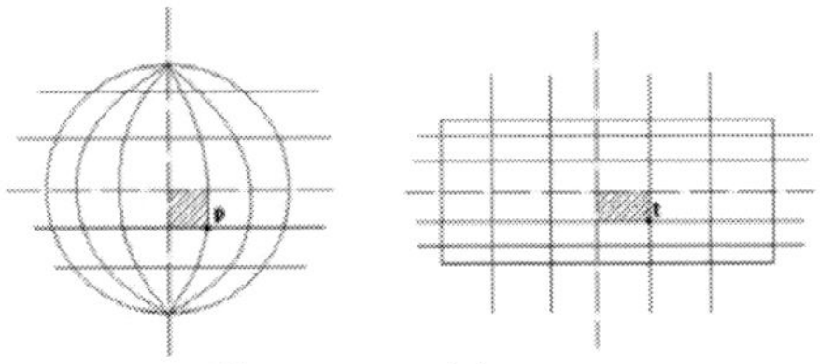

Figure 1. Illustration of the energy mapping

$$\int_{\frac{\pi}{2}}^{\frac{\pi}{2}-\varphi}\int_{\frac{\pi}{2}}^{\frac{\pi}{2}-\theta} I_s(\theta,\varphi)\sin(\theta)\mathrm{d}\theta\mathrm{d}\varphi = \int_{x0}^{x0+\Delta x}\int_{z}^{z+\Delta z} E_t(x_0+\Delta x, z_0+\Delta z)\mathrm{d}x\mathrm{d}z \quad (1)$$

where $I_t(\theta,\varphi)$ is the luminous intensity distribution function of the LED, and $E_t(x_0+\Delta x, z_0+\Delta z)$ is the illuminance distribution function on the target plane. Light beam from the LED with the direction ($\pi/2$, $\pi/2$) is transmitted to the position (x_0, z_0). Details are not given in this paper.

After the calculation of the lens surface data, simulation and practical test are carried out. The results are shown in Figure 2:

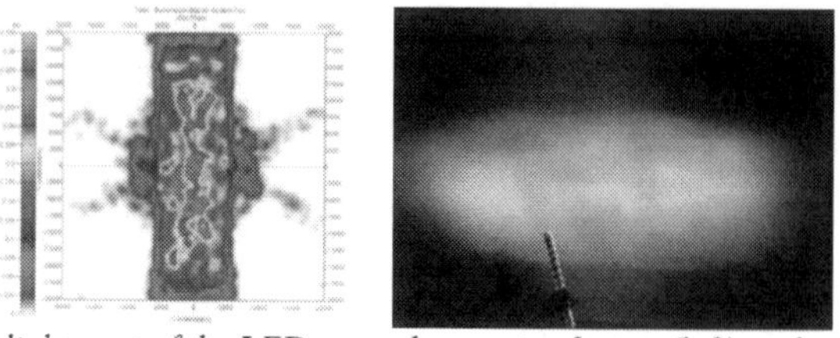

Figure 2. The light spot of the LED street lamp, simulation (left) and practice (right)

INSTALLATION IN FUDAN CAMPUS

Ten LED street lamps are installed on a two-lane road in the campus of Fudan University. The previous HPS lamps are replaced with the poles unchanged. The road is a cement road. Its central width is 7m and the two pedestrian lanes are 3m wide each. The lamps are installed on both sides, crossed, with 7.6m height and 25m interval. A picture of an LED street lamp is given in Figure 3:

***Figure 3**. The LED street lamps in Fudan campus*

The field test results are: average illuminance 32 lx, illuminance uniformity 0.7. The results meet the requirements of the National standard. Most students interview expressed their satisfaction with the visibility and comfort of the white light. However, an observable yellow spot is found in the middle of the road. It is due to the nonuniform spatial distribution of the color of the white LEDs.

REFERENCES

[1] H.Ries, J.Muschaweck. Tailored freeform optical surfaces [J]. J. Opt. Soc. Am. A, 2002, **19**(3): 590~ 595.

[2] H.Ries, J.Muschaweck. Tailoring freeform lenses for illumination [C]. SPIE, 2001, 4442:43~50.

[3] Pablo Benitez , Juan C.Minano ,Jose Blen et al. . Simultaneous multiple surface optical design method in three dimensions [J]. Opt. Eng., 2004, **43**(7): 1489 ~ 1502.

[4] V.Oliker. Geometric and variational. methods in optical design of reflecting surfaces with prescribed irradiance properties [C]. SPIE, 2005,5924:592407.

[5] S.Kochengin. V. Oliker,O. VonTempski. On the design of reflectors with prespecified distribution of virtual sources and intensities [J]. Inverse Problems, 1998, 14:661~678.

[6] William A. Parkyn. Design of illumination lenses via extrinsic differential geometry [C] .SPIE, 1998, 3428:154~162.

[7] Bill Parkyn, David Pelka. Free-form illumination lenses designed by a pasedo-rectangular lawnmower algorithm [C]. SPIE,2006,6338:633803.

[8] Ding Yi , Gu Peifu. Freeform reflector for uniform illumination [J]. Acta Optica Sinica, 2007, **27**(3): 540 ~544 (in Chinese).

[9] Yi Ding, Xu Liu, Zhen-rong Zheng. Freeform LED lens for uniform Illuminating [J]. Optics Express, 2008, **16**(7):12958~12966.

[10] Lin Wang, Keyuan Qian, Yi Luo. Discontinuous free-form lens design for prescribed irradiance [J]. Applied Optics, 2007, **46** (18): 3716 ~ 3723.

Energy distribution of below-gap states in InGaN quantum wells revealed by two-wavelength excited photoluminescence

N. Kamata[1] , T. Yamaguchi[1] , T. Fukuda[1] , Y. Arakawa[2]

[1] *Dept. of Functional Materials Science, Saitama University, 255 Shimo-Ohkubo,Saitama 338-8570, Japan*

[2]*Institute of Industrial Science, the University of Tokyo, 4-6-1 Komaba, Meguro-ku, Tokyo 153-8505, Japan*

Contact: kamata@fms.saitama-u.ac.jp

Since below-gap states originating from lattice defects and residual impurities deteriorate the internal quantum efficiency, their quantitative characterization is indispensable to optimize the growth process. By changing the energy of below-gap excitation light and observing the intensity change of band-edge photoluminescence, we studied an energy distribution of below-gap states in $In_{0.15}Ga_{0.85}N/In_{0.02}Ga_{0.98}N$ quantum wells of NUV (390nm) emission. The energy distribution of states in a sample with Si-doping density of $1x10^{18}$ cm^{-3}, centred at 1.6 eV, is consistent with an interpretation by the combination of V_{Ga}, and others closely located within 0.6eV below the conduction band. Increasing the doping density up to $4x10^{18}$ cm^{-3} is effective for reducing the density of below-gap states and improving the efficiency.

INTRODUCTION

Lattice defects and residual impurities during growth form below-gap states inside the band-gap of a crystal and act as nonradiative recombination (NRR) centers. For lighting applications where ever larger light output is desirable without increasing heat dissipation, the quality of epitaxial layers on lattice-mismatched substrates is becoming more and more crucial [1-3]. In order to improve light output and external quantum efficiency of light emitting diodes (LEDs), it is indispensable to detect and characterize such below-gap states quantitatively.

By combining a below-gap excitation (BGE) light in a photoluminescence (PL) measurement in GaP, below-gap states were detected [4]. We exemplified that a quantitative characterization of the density, electron and hole capture rates of the below-gap state becomes possible by analyzing the BGE density dependence of the PL intensity change based on the trap-filling model [5-7]. Since then, the method of two-wavelength excited PL (TWEPL) became a non-contacting tool for wafers and various device configurations of GaAs/AlGaAs, InGaAs/GaAs, InGaN/GaN and GaN/AlGaN etc.

In this paper, we studied $In_{0.15}Ga_{0.85}N/In_{0.02}Ga_{0.98}N$ quantum well structures with different Si-doping levels by the TWEPL method and determined an energy distribution of below-gap states by using the BGE sources ranging from 1.27eV to 1.95eV.

TWO-WAVELENGTH EXITED PHOTOLUMINESCENCE

Above-gap excitation (AGE) light excites electrons and holes into the conduction and valence bands, respectively, and their radiative recombination produces a band-edge PL of the intensity I_{AGE}. In the TWEPL, another BGE light is superposed on the AGE, and the PL intensity under both AGE and BGE, $I_{AGE+BGE}$, is recorded. When the photon energy of the BGE corresponds to activate a NRR process via below-gap states, the normalized PL intensity, defined as $I_{AGE+BGE}/I_{AGE}$, changes from unity, showing the influence of the below-gap states selected by the BGE energy. In case the value of $I_{AGE+BGE}$ decreases from unity, the NRR process is interpreted by the combination of two levels and the BGE energy corresponds to the energy difference between them [5, 7]

EXPERIMENTAL

Samples studied here are $In_{0.15}Ga_{0.85}N/In_{0.02}Ga_{0.98}N$ quantum wells [8] with Si-doping densities of $1x10^{18}$ cm^{-3} (sample A), $2x10^{18}$ cm^{-3} (B) and $4x10^{18}$ cm^{-3} (C) at the quantum well region. Low-temperature grown GaN buffer (25 nm), GaN buffer (1.7 μ m), $In_{0.02}Ga_{0.98}N$ barrier (5 nm), $In_{0.15}Ga_{0.85}N$ well (3 nm) and $In_{0.02}Ga_{0.98}N$ barrier (18 nm) were grown on sapphire substrate in sequence by MOCVD technique.

The PL was measured by the AGE of a He-Cd laser (3.82 eV) between the temperature range of 12K and 300K. The TWEPL measurement was accomplished at the temperature of 20 K by using BGE of laser diodes ranging from 1.27eV to 1.95eV photon energies.

BGE DENSITY DEPENDENCE

Band-edge PL centred at about 3.2eV (390nm) was observed in each sample at 20K. The AGE density was set at 23.2mW/cm^2, then a BGE light of 1.27eV was superposed on the AGE and the PL intensity change was recorded as a function of the BGE density. As shown in Fig. 1, the normalized PL intensity decreased with increasing the BGE density. The shift from unity corresponds to the enhancement of NRR rate through the combination of two below-gap states. The density of below-gap states was reduced in samples B and C when compared with that in sample A. Saturating tendency of the normalized PL intensity above 500mW/cm^2, slightly observed in the sample C, represents also the lower density of the below-gap state [5].

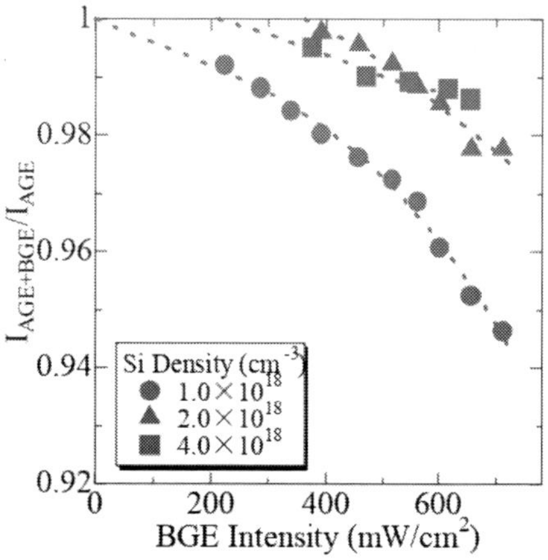

Fig. 1 The normalized PL intensity as a function of BGE density at 20K.

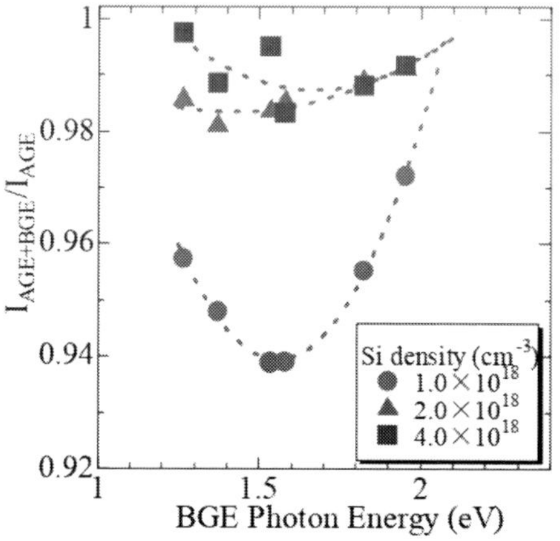

Fig. 2 The normalized PL intensity as a function of BGE energy at 20K. The shift from unity corresponds to the density of below-gap states.

ENERGY DISTRIBUTION

The BGE energy was varied between 1.27eV and 1.95eV by changing laser diodes while keeping the BGE density nearly constant, i. e. 500~600mW/cm^2. Measured normalized PL intensity as a function of BGE energy was shown in Fig. 2. The sample A showed an energy distribution centred at 1.6eV with the half width of c. 0.8eV. The distribution is consistent with an interpretation by the combination of the reported levels of V_{Ga}, 0.9eV above the valence band, and others closely located within 0.6eV below the conduction band [1, 2]. Increasing the doping density from 1x10^{18} cm^{-3} to 4x10^{18} cm^{-3} reduced the peak intensity change at least by a factor of 3. These results imply that Si-doping above 2x10^{18}cm^{-3} reduces the density of below-gap states and effective for the improvement of internal quantum efficiency.

CONCLUSION

We investigated below-gap states in $In_{0.15}Ga_{0.85}N/In_{0.02}Ga_{0.98}N$ quantum wells with different Si-doing levels by the TWEPL method. The sample with 1x10^{18} cm^{-3} Si-doping showed 4% PL intensity decrease in response to the BGE of 1.27eV at 600mW/cm^2. Its energy distribution was centred at 1.6 eV and the distribution is consistent to an interpretation by the combination of reported below-gap levels in GaN-based materials. Increasing the doping density up to 4x10^{18} cm^{-3} reduced the density of below-gap states, improving the internal quantum efficiency.

References

[1] J. I. Pankove, T. D. Moustakas, Gallium Nitride I, Academic Press, London (1998).
[2] F. A. Ponce, GaN and Related Materials (ed. S. J. Pearton) Chap. 5 (1997) 141.
[3] H. Hirayama, N. Noguchi, N. Kamata, Appl. Phys. Exp. 3 (2010) 032102.
[4] H. G. Grimmeis, B. Monemar, Phys. Stat. Sol. (a) 19 (1973) 505.
[5] E. Kanoh et al., J. Lumin. 63 (1995) 235.
[6] K. Hoshino et al., J. Lum. 79 (1998) 39.
[7] N. Kamata et al., Recent research Developments in Quantum Electronics 1 (1999) 123.
[8] N. Kamata et al., Proc. Int. Conf. on White LEDs and Solid State Lighting, Taipei (2009)MA1-6.

CO/Xe 2m-long microwave discharge lamp for VUV light source

A. Hatta, T. Sakai, K. Kawamura
Department of Electronic and Photonic Systems Engineering, Kochi University of Technology, 185 Miyanokuchi, Tosayamada-cho, Kami-city, Kochi 782-8502, Japan
Contact: hatta.akimitsu@kochi-tech.ac.jp

ABSTRACT

Microwave excited CO/Xe mixture gas discharge lamp has been developed for application to VUV light source ranging from 150 to 210 nm. The microwave discharge was excited by travelling microwave through a coaxial line consisting of a central plasma rod as the inner conductor and the outer metal tube. Intensive VUV emission can be extracted through a slit on the outer conductor. By alternating operation of a couple of magnetron power sources from both the ends of the coaxial discharge tube, the plasma light source was extended to 2m long.

INTRODUCTION

It has been known that molecular gases have many kinds of emission band in VUV, UV, Visible, and IR ranges. In the electrical gas discharge, however, the molecular gases are not stable due to their chemical reactivity with metal electrodes. The electrode less microwave discharge will enable us to use a variety of material as light emitting medium. It has been found that microwave plasma of CO/Xe gas mixture emit intensive VUV light in the range from 150 to 210. The most intensive peak in the VUV emission spectrum appears around 172 nm which is close to that of Xe excimer lamp widely used in industry, such as surface cleaning of glass substrates for liquid crystal display (LCD) panel. The Xe excimer lamp is usually operated by dielectric barrier discharge (DBD). It requires that the length of the lamp should be extended with the increase of LCD panel size. The conventional microwave discharge has been limited in size due to its short wavelength. In this paper, a novel microwave discharge lamp of CO/Xe gas mixture is presented. By using travelling microwave for plasma excitation, it has been achieve to extend the discharge length to 2m.

EXPERIMENTAL

Figure 1 shows schematic drawing of the 2m-long microwave discharge lamp. A quartz discharge tube of 15 mm-O.D.,13 mm-I.D., about 195 cm-long was filled with CO(50%)/Xe at 200 Pa. Both the end of the tube was specially shaped to fit into the inner conductor of microwave coaxial line (WX-39). The discharge tube was supported coaxially from both the ends to a Cu tube of the same size as the outer conductor of WX-39 coaxial line. A 3mm-wide slit was opened on the outer tube to extract the emitted light. The discharge tube was directly connected to the antennas of a couple of magnetrons at both the ends. The magnetron power sources used for the discharge were products in consumer field for the conventional 2.45 GHz microwave ovens. The microwave power sources were alternately

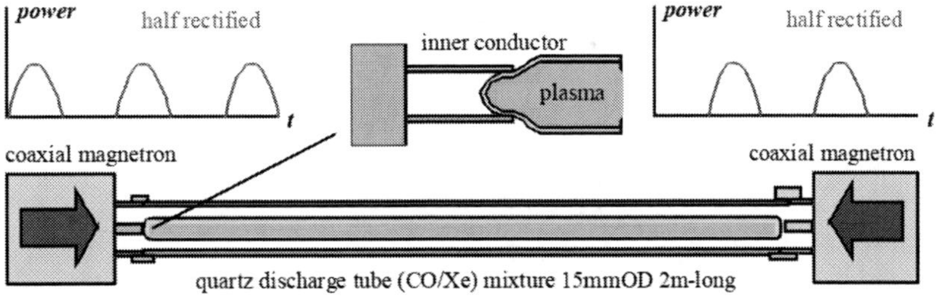

Figure 1. Schematic drawing for the 2m-long CO/Xe M.W. discharge VUV lamp.

operated by half-wave rectified sinusoidal wave at 60 Hz to avoid interference. The time averaged power of each magnetron power source was expected as 600 W from the catalogue specification although the incident and reflected powers were not monitored.

RESULTS AND DISCUSSION

Figure 2. 2m-long microwave plasma VUV light sourace of CO(50%)/Xe at 200 Pa operated by 600 W half rectified sinusoidal wave from each side and both sides.

Figure 2 shows photographs of the CO/Xe microwave discharges excited from each side and from both sides. By operating the single side power source, the microwave plasma was nearly 1.5 m in length. By alternating operation of both the power source, whole the discharge tube was brightly and almost uniformly glowed.

The emission color of the plasma looked almost white due to radiation from Xe recombination process and also a little bit green due to C_2 molecules at both the ends. Due to difficulty in connecting the lamp to a vacuum spectrometer, the VUV spectrum was not investigated. The VUV emission spectra have been investigated in the previous experiments. After a few seconds of operation, it smelled intensive ozone. From the fact, it was expected that the CO/Xe microwave lamp radiated intensive VUV light with resulting in production of large amount of ozone.

SUMMARY

It has been successfully demonstrated that a novel type of long microwave plasma VUV light source was operated using CO/Xe gas mixture. The length of the discharge tube was extended to 2m by alternating operation from both the ends.

Acknowledgements

This work was partly supported by Japan Science and Technology Agency (JST) innovation satellite Kochi from 2005 to 2008. The quartz discharge tubes were fabricated by ORC Manufacturing CO., LTD. The authors greatly appreciate to Mr. K. Yanai, M. Uno, and Dr. J.S. Oh for their experimental assistance.

Study of the 700 nm emitting spectrum using GaInP quantum dots in the AlGaInP-based light emitting diodes

Hwa Sub Oh[1,2], Sang Mook Kim[1], Huyn Haeng Lee[1], Jong Hyeob Baek[1,*], and Joon Seop Kwak[2,*]

[1]*LED Device Team, Korea Photonics Technology Institute (KOPTI), 971-35 Wolchul-dong, Buk-gu, Gwangju 500-460, Korea*

[2]*Department of Printed Electronics Engineering, Sunchon National University, Chonnam 540-742, Korea*

Contact: hsoh@kopti.re.kr

Recently, AlGaInP-based light emitting diodes (LEDs) have experienced an impressive evolution in both device performance and market volume. However, development of new applications is required in order to realize their full potential in areas such as use as a light source for auto focusing in digital cameras, special illumination for particular functions in agriculture, and in full color displays. To enlarge their utility in these applications, it is necessary to fabricate and understand a new structure capable of emitting longer wavelengths of around 700 nm. In particular, AlGaInP heterostructure LEDs are lattice-matched with respect to the GaAs substrate, which limits the emitting spectrum to around 650 nm at the longer peak wavelength side. To fabricate an LED structure capable of emitting at a 700 nm peak wavelength, the composition (x) of $Ga_xIn_{1-x}P$ material in the active layer requires a compressive strain of larger than 1 %. This large lattice mismatch, however, causes significant problems in terms of both growth and device properties due to the formation of defects. To overcome these problems, it is necessary to relieve the well strain via the formation of islands, referred to as a Stranski-Krastanow (S-K) growth mode, in order to prevent the generation of dislocations [1,2,3]. However, in AlGaInP-based LEDs emitting at a 700 nm peak wavelength, the effects of well strain on the epitaxial growth and the realization of device performance has yet to be extensively studied.

In this study, we investigate the behaviors of morphological and optical characteristics on the composition of $Ga_{0.33}In_{0.67}P$ material and demonstrate the performance of a device emitting at around 700 nm using quantum dot (QD)-based LEDs.

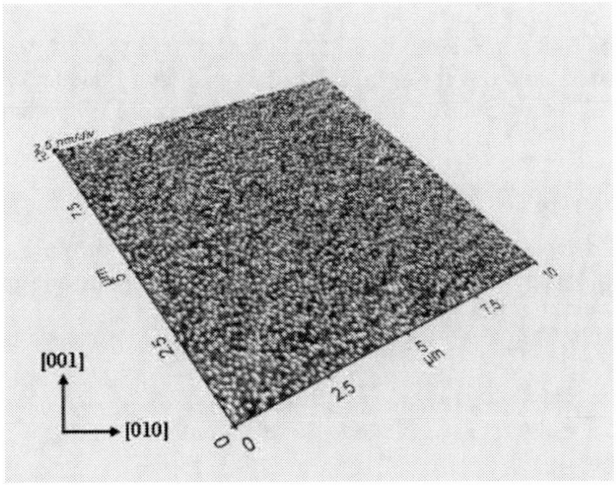

Figure 1. AFM images of 5 ML $Ga_{0.33}In_{0.67}P$ grown on a lattice-matched $(Al_{0.36}Ga_{0.64})_{0.5}In_{0.5}P$ barrier layer.

Figure 1 shows the AFM images of $Ga_{0.33}In_{0.67}P$ quantum dots grown on the lattice-matched $(Al_{0.36}Ga_{0.64})_{0.5}In_{0.5}P$ barrier layer. The $Ga_{0.33}In_{0.67}P$ QDs were self-assembled on lattice-matched barrier layers grown on GaAs substrate offering a lattice mismatch of 1.37 %. The QDs with a density of 3 x 10^9 cm^{-2} had an average height of 5.5 nm and diameter of 160 nm. To further investigate the shape of the dots, scan size was magnified to 1.0 μm x 1.0 μm. From this increased magnification, it could be seen that the size of dots was relatively irregular and large with respect to InP QDs, due to the insufficient strain energy for the high quality QD ensembles. Moreover, elongation of the dots along [0-11] was directly related to steps on the vicinal surfaces, which are oriented along [100] in the present miscut.

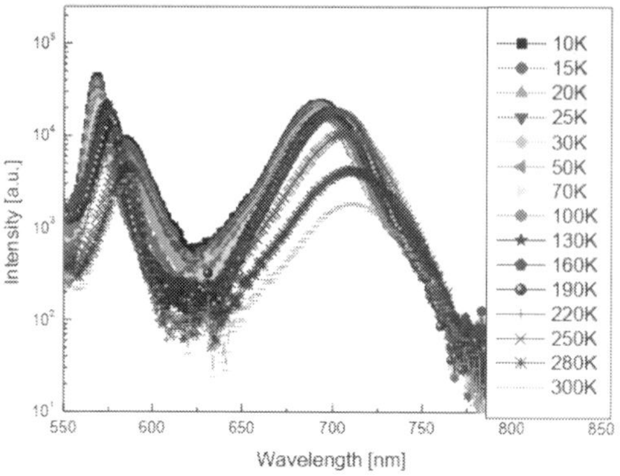

Figure 2. Temperature dependent photoluminescence measurements of the $Ga_{0.33}In_{0.67}P/(Al_{0.36}Ga_{0.64})_{0.5}In_{0.5}P$ QD structure.

To examine the optical behaviors of the $Ga_{0.33}In_{0.67}P$ dots, TD-PL measurements were carried out the temperature was increased from 10 to 300 K. The low temperature PL spectrum at 10 K consists of three major peaks. At this temperature, the relatively broad peak at 691 nm is interpreted as being due to $Ga_{0.33}In_{0.67}P$ QDs and the relatively narrower peak at 568 nm as due to the $(Al_{0.36}Ga_{0.64})_{0.5}In_{0.5}P$ barrier; the inhomogeneous broadening of the spectrum for the $Ga_{0.33}In_{0.67}P$ dots represents the size and compositional distribution of the QDs. The signal at 584 nm can be attributed to the two-dimensional (2-D) wetting layer, as is consistent with the S-K growth mode and not a purely three-dimensional (3-D) Volmer-Weber (V-W) growth mode. As the temperature is increased to 300 K, the spectrum for the $Ga_{0.33}In_{0.67}P$ dots was well maintained. As such, these data indicate that $Ga_{0.33}In_{0.67}P$ QDs can act as an active layer for the 700-nm emitting LEDs.

To compare the PL characteristics of the $Ga_{0.33}In_{0.67}P$ spectrum according to the measuring temperature, the normalized PL intensity is defined as the integrated intensity of the $Ga_{0.33}In_{0.67}P$ spectrum divided by the integrated intensity at 10 K. The normalized intensity slightly decreases as the measuring temperature is increased from 10 to 160 K, owing to the loss of holes located in the wetting layer to the barrier layers. The normalized intensity shows a peak value around 210 K, due to the thermally activated carriers in the aluminum oxide complexes; the Al-containing barrier layers inevitably incorporate oxygen to form AlO_x complexes during epitaxial growth because of the high reactivity of aluminum with oxygen. After activation, these complexes act as a trapping site for carriers, and this lowers the internal quantum efficiency. As the measuring temperature is further increased to 300 K, the normalized intensity rapidly decreased because the carriers escape from the dots to the barrier.

Acknowledgements

This work was supported by the Korea Research Foundation Grant from the Small and Medium Business Administration (S1059702), and was also supported by the WCU program at Sunchon National University.

REFERENCES

[1] P. Blood, Mat. Sci. and Eng. B **66**, 174 (1999).
[2] C. M. Reaves, R. I. Pelzel, G. C. Hsueh, W. H. Weinberg, and S. P. DenBaars, Appl. Phys. Lett. **69**, 3878 (1996).
[3] J. Porsche, M. Ost, T. Riedl, A. Hangleiter, and F. Scholz, Mat. Sci. and Eng. B **74**, 263 (2000).

Thermal modelling of a halogen IRC lamp

G.Van De Poel[1], S.Mekala[1],N.Parisé[2]
[1] *R&D Department, Havells Sylvania, Industriepark 13, 3300 Tienen, Belgium*
[2] *MAYA HTT Ltd., 4999-400 Ste-Catherine W.,Montreal, QC, Canada*
Contact: gunther.vandepoel@havells-sylvania.com

ABSTRACT

Thermal modelling becomes more and more important to reduce the development time of a new product, the number of prototypes to be made and to optimise the product design. In this study numerical simulations of temperature and flow are performed on an IRC (infrared radiation coating) halogen capsule in free burning operation. In a first simplified model we will build a conduction-only model that neglects the gas loss. In this model we will model the temperature of the coating, the pinch temperature (weld connection Molybdenum to electrical connector) and the connector temperature. In a second more advanced model (thermal-coupled flow) we will add the gas properties inside the bulb cavity and the real filament geometry to model the convective streams inside the bulb. Both simulations are performed in vertical base up position of the lamp as this is the worst case scenario. Finally simulated temperatures will be compared with real measurements.

FINITE ELEMENT ANALYSIS PROCESS

In this study we will simulate a Halogen IRC 20W burner with a CFD (Computational Fluid Dynamics) and Heat Transfer coupled solver technology. We first start with explaining the lamp geometry. The lamps consists of a tungsten filament, inert gas, Mo foils, electrical connectors, quartz glass bulb covered with an infrared radiation coating which reflects the IR back to the filament. Figure 1 shows the bulb structure.

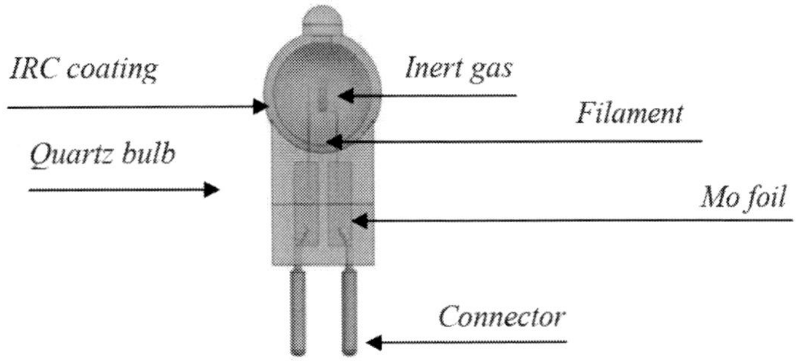

Figure 1: Lamp geometry model

After selecting the solver we idealize the part file (unnecessary blends, chamfers, pockets or holes which have no influence on the thermal model will be removed from the original part file). Once this is done we construct the finite element model by creating the necessary meshes (1D, 2D and 3D meshes) and we apply the material, physical and thermo-optical properties to it. This last step is critical as the thermo-optical properties have a strong influence on the radiative heat transfer calculations made by the software.

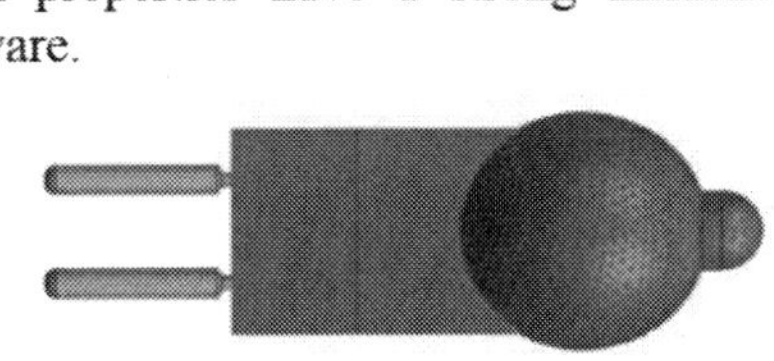

Figure 2: 3D mesh of quartz bulb

In a next step the boundary conditions (loads and constraints) are applied to the finite element model. The model is then solved and the results are reviewed in the post-processing mode of the software.
In a first simplified model we simulate the lamp only with conductive and radiative heat transfer. The

gas loss effects were neglected. The simplified filament (1D beam elements) was simulated as a radiative heat source and a view-factor based algorithm was used to calculate radiative heat transfer within the burner structure and to the environment. Convection to the environment was modelled with a full CFD solve of the external ambient air, at 25°C for this study. The second model used included gas properties inside the bulb cavity and the real filament geometry to model the convective streams inside the bulb.

RESULTS AND DISCUSSIONS

For this model, comparison is made between simulation and experimental data for three monitor points: i) maximum coating temperature, ii) pinch temperature and iii) connector temperature. The first scenario was run without inclusion of gas loss effects inside the burner and simulation results are tabulated in Table 1. A second scenario was run with the inclusion of the gas loss effects and the results are also given in Table 1.

Table 1 – Summary of Results

Monitor Point	Experimental	Simulation #1	Simulation #2
Coating temperature (°C)	283	268	288
Pinch temperature (°C)	195	199	199
Connector temperature (°C)	106	107	107

Figure 3 shows temperature distributions obtained in the burner assembly and the inert gas, for a vertical down orientation.

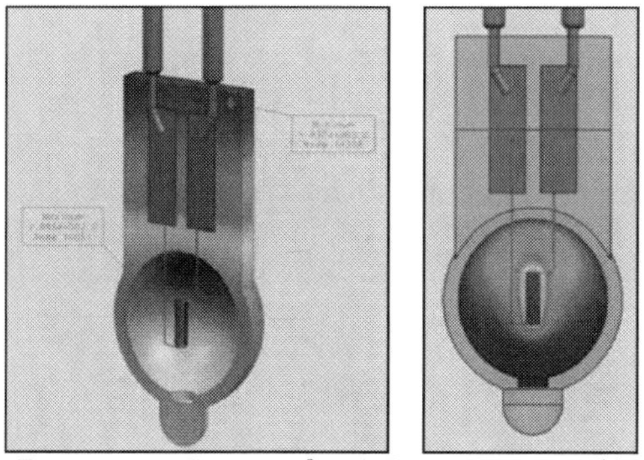

Figure 3 – Temperature Contours for a) Quartz Burner & b) Xenon Gas

RESULTS AND DISCUSSIONS

Study results show a good correlation between experimental testing and numerical simulation. It is also noted that including gas loss effects has a non-negligible effect on result accuracy and should be included for appropriate temperature predictions. Gas loss does not influence our pinch and connector temperature predictions in this work because it was assumed that the pinch temperature was mainly due to conductive effects. Future work could involve refining the representation of the Mo foils to include additional effects, but it is believed that the initial assumptions were correct. A final note can also be made that to obtain good correspondence between testing and simulation, special care must be taken in utilizing accurate material properties, especially to model the radiative heat transfer which is critical in this kind of problem. During the poster sessions we will display more details about other obtained results: operating pressure, gas flow velocity, gas temperature distribution and what will be the influence of increasing the power of the filament on the IR coating. An attempt will be made to include the reflector optics to know the influence of the heat reflected back by the reflector to the quartz burner.

REFERENCES

[1] Milan R. Vukchevich, The science of Incandescence, GEL Nela Press, 1992
[2] Y. S. Touloukin and D. P. DeWitt, Thermal Radiative Properties of metallic Elements and Alloys, Thermophysical Properties of matter, Volume 7, Plenum Publishing, pp. 776-791
[3] www.mayahtt.com

Analysis of Degradation by On-Off Test in Ceramic Metal-Halide Lamp

J.K. Yang[1], W.Y. Kim[1], D.H. Park[1]
[1] *Department of Electrical material, wonkwang university, 344-2, shin yong dong, Iksan city, 570-749, South Korea*
Contact: duck2214@wonkwang.ac.kr

ABSTRACT

Recently, Ceramic metal-halide lamp using PCA arc tube has been developed, and the lamps are being used in various ranges such as street lighting, stage lighting, projector and industrial lighting. Therefore, the study about reliability of lamp and degradation are regarded as an important part.
In this paper, to know the affection of arc tube from Accelerated degradation test in ceramic metal-halide lamp, structural and optical properties of lamp were measured for 2,000 hours with 20 minute On/Off test based on the reliability standard RSC 0085. As results, blackening phenomena have occurred as degradation of the arc tube and electrode. Also, structural distortion has happened as change of vapour pressure inside the arc tube. In addition, the color temperature and color rendering index were changed as loss of gas composition.

INTRODUCTION

In case of ceramic metal-halide lamp, because the ceramic arc tube can operate at high temperature, the new material can be used for high color rendering index. Also, sodium losses can be reduced by the strong resistance to reaction with sodium during the life of the lamp. Therefore, the lamp has high color rendering, color stability, improved luminous maintenance ratio and long life characteristics, compared to conventional quartz lamps. Due to the improved properties, The scope has been expanded to the street lamps and the security light, Sports Complex, factories, gymnasiums, billboards, TV, stage lighting, projectors, and even though car headlights.
While this usage has expanded, reliability of the lamp was being considered, more importantly. Actually, the degradation of arc tube is occurred from the chemical recombination and the loss of internal configuration such as iodine compounds, tungsten and rare earth elements in Ceramic metal halide is occurred.
In this paper, to analyze the reliability of ceramic metal-halide lamp, optical properties changes were analyzed after accelerated degradation of the lamp.

EXPERIMENTS AND RESULTS

Condition of sample for accelerated degradation is On/Off test with Interval of 20 minute for 2,000 hours in accordance with the reliability standard RSC 0085. The experimental temperature was 25 ℃ and magnetic ballasts for 150W metal halide were adopted. Also, integrated sphere (PMS-50, EVERFINE) were used for optical properties, and x-ray NDT system (SMX-225CT Special, Shimadzu Corp), FE-SEM and EDS(S-4800, Hitachi Corp.) is used to analyze the structural properties.

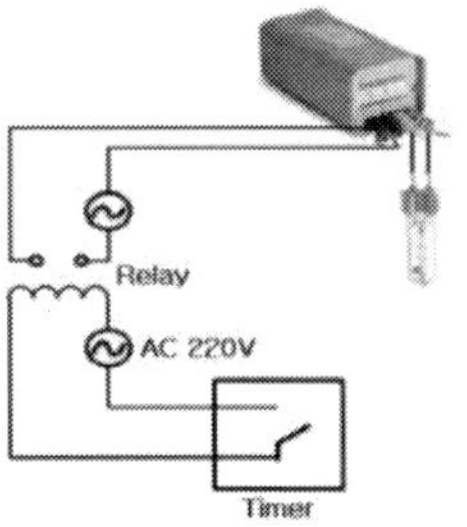

Fig. 1 Experimental setup

As results, blackening and distortion of electrode were confirmed in ceramic arc tube such as fig. 2. Actually, PCB arc tube takes the biggest stress in cool-down state. Therefore, repetition of stress from On/Off test is caused the degradation. In addition,
In Fig 3, the tungsten in electrode surface was reduced from 9.32 to 8.98 percent due to the radiation of electron, whereas sodium, alumina, iodine elements was increased two times, approximately. In arc tube surface, tungsten was infiltrated into the PCA crystal. Figure 5 shows the spectral distribution of the additives in the arc tube. In 380 ~ 520nm, the loss of Dy, I was appeared, and the loss of Na, TI, I showed through the wavelength of 530 ~ 580nm. Therefore the color rendering index is also decreased in R9 region.

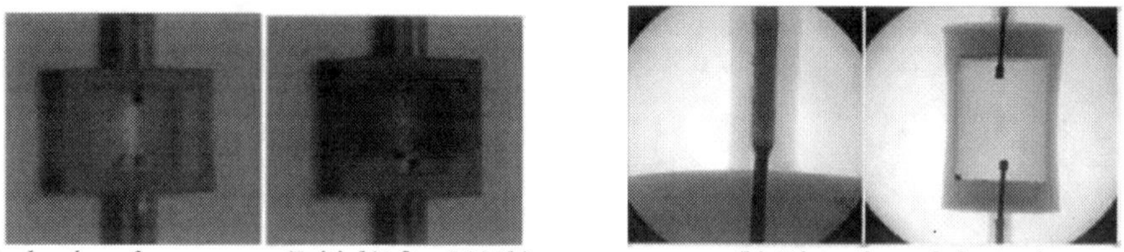

(a) Blackening phenomenon (Initial/After 2000h) (b) Distortion of electrodes

Fig. 2 Change of arc tube by degradation

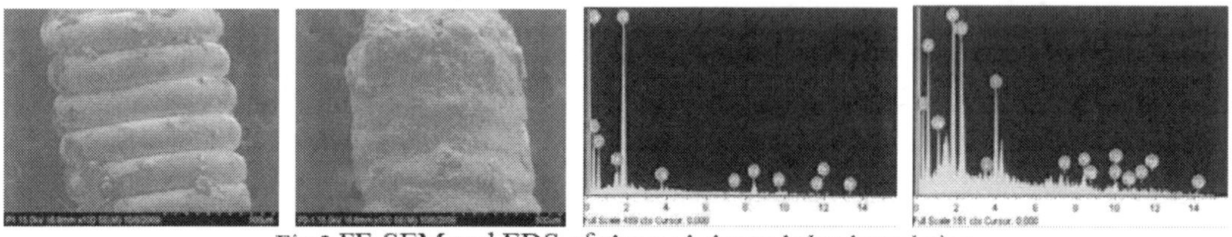

Fig.3 FE-SEM and EDS of changed electrode by degradation

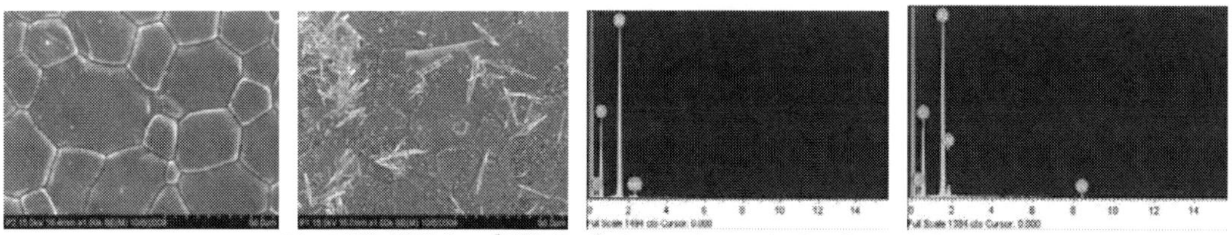

Fig.4 FE-SEM and EDS of changed arc tube by degradation

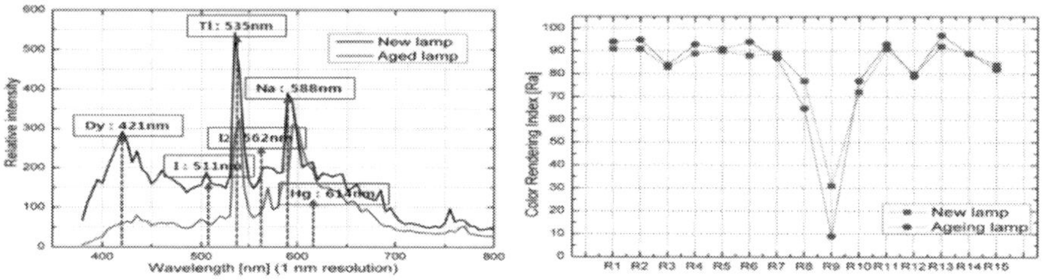

(a) Spectrum of lamp (b) Color rendering index of lamp

Fig.5Optical Properties of ceramic metal halide lamp

CONCLUSION

In this paper, to analyze the reliability of ceramic metal-halide lamp, optical properties changes were analyzed after accelerated degradation of the lamp. Blackening phenomena have occurred as degradation of the arc tube and electrode. Also, structural distortion has happened as change of vapour pressure inside the arc tube. In addition, the color temperature and color rendering index were changed as loss of gas composition.

ACKNOWLEDGEMENT

This work was financially supported by the development project of lighting driving system by using solar cell from Small Business Administration

REFERENCES

[1] Walter P Lapatovich, “Metal-halide lamp design: atomic and molecular data needed", IOP science Physica Scripta, Issue T134, 014024(6pp), 2009.

[2] Yuji SHATA, Toshiro KASUYA, Yasuyuki KIMURA, Shigeo GOTOH, Motoi WADA, “Study on the Wall Blackening of a High Intensity Discharge Lamp” , J. Plasma Fusion Res. SERIES, Vol. 8, 2009.

Visual color matching of LED and tungsten-halogen light sources

K. McGroddy, R. Petluri, P.K. Tseng, J. Yriberri, G. Harbers
Xicato Inc, San Jose, CA 95129 USA
Contact: kelly.mcgroddy@xicato.com

ABSTRACT

Many challenges still exist to achieve mass adoption of Light Emitting Diodes (LEDs) for general illumination. Among these challenges is the problem of color consistency between different LED sources and between LED and halogen sources. In this work visual sensitivity to color differences between test sources with specified controlled color differences has been quantified, specifically for architectural applications. Additionally, we found that color differences are observed between LED sources with different color rendering properties and between halogen and LED sources, even when they are matched to the same (CIE 1931) chromaticity coordinates. This is related to color matching function failures, which we found have a significant effect on the perception of visual color differences. These results help lighting specifiers and manufacturers understand how to avoid observable color variations in LED lighting installations.

INTRODUCTION

An example of a solid state lighting installation is shown in figure 1 where the proximity of the fixtures makes the installation sensitive to color differences between the fixtures. While no color differences were observed during or after the installation in this case, there are many examples where color differences are noticed. It is difficult to achieve color consistency between fixtures in general, but even more so with solid state lighting fixtures. To address this issue, acceptable color differences must be determined and the method of quantifying perceived color should be well understood.

Fig 1: A retail lighting application using solid state lighting fixtures

ACCEPTABLE COLOR DIFFERENCES

How color consistent do LED fixtures need to be? The ground work for determining visual sensitivity to small color differences was done by MacAdam in 1942[1], with many follow-up studies (for an overview see [2]). Our focus is on the chromaticity of Tungsten-Halogen lamps and the region around the 3000K color point at the black body locus (figure 2). To determine color sensitivities in practical lighting installations, we used a test setup comprising of SSL modules manufactured by Xicato that can be built up to achieve any desired color point with very high precision. The modules are mounted in sets of two, with the color points of the sets located at the corners of the CCT and Δuv ranges shown in figure 2, and observers were asked to comment on the color differences. It was tested with hundreds of observers from Europe, US, and Japan under many different conditions.

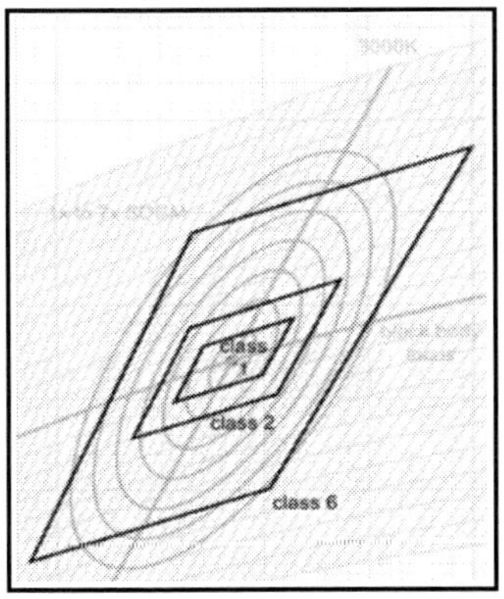

Fig 2: Test samples are located at the corners of these quadrangles, which are here shown overlayed over the 1x – 7x SDCM ellipses in the CIE 1931 diagram. A CCT and Δuv grid is overlayed in blue.

We obtained consistent feedback from virtually all observers, independent of the viewing conditions. The results indicate that a tolerance in CCT of ± 60K and a Δuv of ±0.001 (~2x SDCM, class 1) is adequate for virtually all lighting applications. It also indicates that a range of ± 100K and Δuv of ± 0.002 (~4x SDCM, class 2) limits the applications due to color consistency errors. A range of ± 175K and Δuv of ± 0.006 (~7-8 SDCM, class 6, also ANSI NEMA) is unacceptable for any type of lighting application.

COLOR MATCHING FUNCTION FAILURES

In the experiments described in the previous sections, all the modules had a color rendering index Ra of about 80. In further experiments, we tested visual matching between these modules and modules with a color rendering index of about Ra 98 (similar to Tungsten-Halogen lamps). We found that the colors were very noticeably different, even after precisely matching the CIE 1931 chromaticity coordinates. In addition, we found that the high CRI modules matched the color of Tungsten-Halogen lamps much more closely than the modules with low CRI. After this observation, we adjusted the chromaticity of the low CRI version to achieve a nearly perfect visual match between the low and high CRI versions. Although we see a big difference in CCT (33K), Δuv (0.0021), and x,y coordinates (see Table 1) we got a virtually perfect visual match. Failure of the CIE 1931 color matching functions, particularly in the blue region (450nm - 470nm) have been reported by multiple authors [6,7]. This error is significantly higher than what would be acceptable for a class 1 color consistency, and falls within the realm of a class 2 consistency.

Table 1 Color data for two visual matched samples

	Sample A	Sample B	Δ
CCT	3013	2980	33
Δuv	0.0017	-0.0004	0.0021
Ra	79.4	98.3	-18.8
CIE1931 x	0.4387	0.4379	0.0007
CIE1931 y	0.4093	0.4037	0.0056

DISCUSSION AND CONCLUSIONS

Achieving good visual color consistency between light fixtures is difficult. This is true for not only conventional lamp fixtures, but in particular for solid state lighting fixtures. Although difficult, we have found that good color consistency can be achieved. In architectural lighting applications CCT variations should be limited to ±60K, with Δuv's of ±0.001, corresponding to class 1 color consistency as defined in this paper. Current binning practices used by LED manufacturers are commonly inadequate and result in color differences larger than these results show to be acceptable. Additionally color matching functions are incorrect and color points need to be corrected.

New standards are required for solid state lighting to address these issues and other quality aspects, as failure to do so will delay energy saving and carbon dioxide emission reduction opportunities.

REFERENCES

[1] MacAdam, D.L., Visual sensitivities to color differences in daylight, J.Opt.Soc.Am. 32, 247 (1942)

[2] Wyszecki & Stiles, Color Science, 2nd Ed, Wiley (2000)

[3] Robertson, Alan R. Computation of Correlated Color Temperature and Distribution Temperature". JOSA 58 11, 1528–1535 (1968)

[4] ANSI NEMA ASNLG C78.377-2008, Specifications for the Chromaticity of Solid State Lighting Products (2008)

[5] Rensselaer Polytechnic Institute, Lighting Research Center, Developing Color Tolerance Criteria for White LEDs, Assist Program.

[6] Vos, J.J. Colorimetric and photometric properties of a 2 degree fundamental observer, Color Res. & Appl. 3, 125 (1978)

[7] Stiles, W.S. & Burch, J.M., N.P.L. colour-matching function investigation, Proc. Nat. Acad. Sci. 45, 100 (1959).

Effect of Pulse Voltage Slope on High Pressure Xenon HID Lamps Breakdown Voltage

A. Sato[1] , N. Brates[1] , K. Noro[1] , N. Babaeva[2] and M. Kushner[2]
[1] *Universal Lighting Technologies Inc., 216 West Cummings Park, Woburn, MA 01801 USA*
[2]*EECS Department, University of Michigan, 1301 Beal Ave, Ann Arbor, MI, 4810 USA*
Contact: asato@us.pewg.panasonic.com

ABSTRACT

Generally, the breakdown voltage of high intensity discharge (HID) lamps depends on the slope of the ignition voltage pulse. Experiments had shown[1] that a wider pulse leads to a lower breakdown voltage. However, the mechanism(s) of these findings are not totally understood. In this study, we report the modelling results of the breakdown processes in HID lamps. We determined the breakdown voltage for wide and narrow pulses in HID lamps filled with 8atm xenon. The results show the incremental voltage at the end of the formative time-lag is proportional to the square root of pulse slope. Therefore, a wider pulse leads to a lower breakdown voltage.

INTRODUCTION

One utilization of the high pressure xenon HID lamps is the automotive headlight lamp. In this application, it is important to reduce lamp ignition voltage and thus, reduce the complexity of electronic ballasts. The goals of this study are to understand the mechanisms of breakdown process and to determine the optimum ignition pulse to achieve the lowest breakdown voltage. The breakdown process can be characterized by statistical and formative time lags. The statistical time lag was discussed in Ref. [2]. Our paper details the results of numerical modelling of the formative time lag in a high pressure xenon HID lamp. The numerical model we used is an electric potential solver for an entire 2-d unstructured mesh[3].

RESULTS

The fundamental equations in this model are:

$$-\nabla \cdot \varepsilon \nabla \Phi = \sum_j N_j q_j + \rho_s \qquad (1),$$

$$\frac{\partial N_j}{\partial t} = -\nabla \cdot \vec{\phi}_j + S_j \qquad (2),$$

$$\frac{\partial \rho_s}{\partial t} = \sum_j -\nabla \cdot (q_j \vec{\phi}_j (1+\gamma_j)) - \nabla \cdot (\sigma(-\nabla\Phi)) \qquad (3),$$

where: ε, Φ, ρ_s, N, ϕ, γ, σ, S and q are the permittivity, electric potential, surface charge density, species number density, species flux, electron secondary emission coefficient, conductivity of solid materials, source terms and elementary charge, respectively. The subscript denotes the identity of the species. The Poisson equation (Eq.1) describing the electric potential, the transport equations (Eq.2) for conservation of the charged species and the surface charge balance (Eq.3) are simultaneously solved. A 2-d unstructured mesh is made using a commercial mesh generator, Skymesh2. The 2-d geometry of an automotive headlight HID lamp and the mesh structure of discharge area are shown in Fig. 1. Only the discharge vessel was considered in the numerical model, the outer envelope and the support frame were ignored. The upper electrode is the anode, and the lower electrode is the cathode. An ignition pulse of negative polarity is applied on the lower electrode, whereas, the upper electrode is grounded. The slope of the negative pulse, dV/dt, was varied from 25 to 400 V/ns, with a fixed peak voltage of -30 kV (minus polarity). The arc tube has an inner diameter of 2.7mm, an electrode gap of 4.2 mm and is filled with

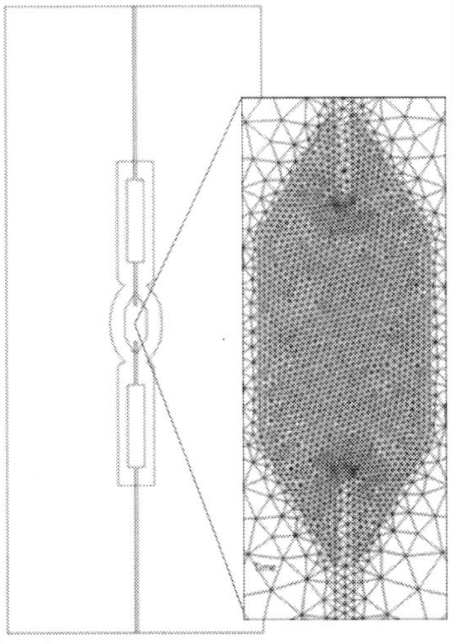

Fig. 1 Geometry of arc tube

8 atm of Xe. In general, the breakdown process is characterized by a statistical and a formative time lag phases. The statistical time lag is defined as the time of appearance of the first electron near the cathode. In our simulations, we initiated the discharge by injecting seed electrons from the negative electrode. The injection starts when the cathode voltage reaches -10kV. The formative time lag is defined as the time required to build up the discharge from the instant the first electrons appear until the discharge reaches equilibrium value[4]. The electron density inside the arc tube during the breakdown process is shown in Fig. 2. Figure 2(a) shows the initial conditions (time=0) of the breakdown process, whereas Fig. 2(b) illustrates the end of the statistical time lag phase and the injection of the seed electrons. The movement of the electron cloud from cathode to anode due to the electric field pulse is shown in Fig. 2(c). Figure 2(d) shows the streamer reaching the cathode that is the end of the breakdown process.

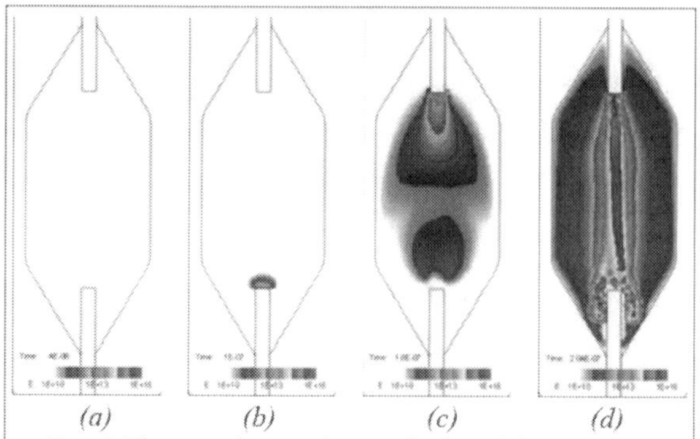

Fig. 2 Electron density during the breakdown process

The breakdown voltage is separated into two components: the voltage increment at the end of the statistical time lag (V_s) and the voltage increment at the end of the formative time lag (V_f). They are shown in Fig. 3. In this study, the V_s was fixed at -10 kV, therefore the magnitude of pulse slope influences V_f only. The correlation between the pulse slope dV/dt and V_f is a square root function and it is shown in Fig. 4 (solid line). This correlation is explained by the square root relationship between the electron velocity and the applied electric field.

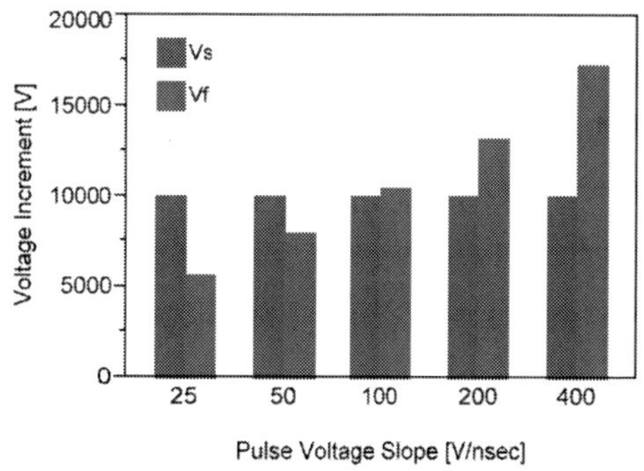

Fig. 3 V_s and V_f in various pulse slopes

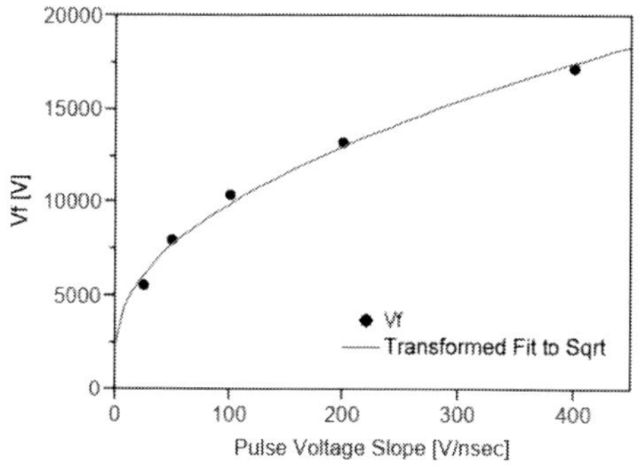

Fig. 4 Effect of pulse voltage slope on V_f

CONCLUSIONS

The effect of the ignition pulse slope dV/dt on the voltage increment at the end of the formative time lag in xenon HID lamps was investigated using a plasma model. We found that a wider ignition pulse leads to a lower breakdown voltage. The statistical time lag of the breakdown process was simplified by cathode injection of seed electrons at the earlier stages of discharge development.

REFERENCES

[1] M. Wendt, S. Peters, D. Loffhagen, A. Kloss, and M. Kettlitz, "Breakdown characteristics of high pressure xenon lamps", J. Phys. D: Appl. Phys. 42, 185208 (2009).

[2] M. Czichy, T. Hartmann, J. Mentel and P. Awakowicz, "Ignition of mercury-free high intensity discharge lamps", J. Phys. D: Appl. Phys. 41, 144027 (2008).

[3] A. Bhoj and M. Kushner, "Avalanche process in an idealized lamp: II. Modeling of breakdown in Ar/Xe electric discharges", J. Phys. D: Appl. Phys. 37, 2510 (2004)

[4] A. Engel, "Ionized gases", AIP Press (1994)

* This work was supported by Panasonic Electric Works Co., Ltd.

Effect of Electrode Structure on Luminous characteristics of Small Diameter Short Xe CCFL

K. Ikeda , H. Kurokawa , H. Motomura, M. Jinno
Department of Electrical and Electronic Engineering, Ehime University
3 Bunkyo-cho, Matsuyama, Ehime 790-8577, Japan
Contact: mjin@mayu.ee.ehime-u.ac.jp

INTRODUCTION

The cold cathode fluorescent lamps (CCFL) have been developed mainly for the backlight for liquid crystal displays (LCD). Recently LEDs are used rapidly for LCS back light and CCFL becomes minor. However UV-LEDs are still under developing and Hg-free CCFL is expected as a cheap and environmental friendly UV light source. The authors have been developing pulsed Xe-FL for lighting with diameter around 30 mm and we found the effect of auxiliary external electrode (AEE) [1-3]. In this paper, we report the effect of the structure of electrodes on small diameter short Xe-CCFL with AEE.

EXPERIMENTS

The schematics of lamp used for the experiments are shown in Figures 1 and 2. As shown in these figures two types of lamps were used with different shape of electrodes. On is Y-shape and the other is sleeve type and both type of inner electrodes are made of Ni. Inner diameter of each lamp is 9.8mm and the distance between electrodes is 50mm. Each lamp has AEE with 3mm width at the point 5mm from the edge of the cathode. Xe gas is filled with the pressures of 60, 70 and 80 Torr. Inner wall of the lamps are coated with 3 colour phosphor.

Figure 3 shows the electrical circuit to operate the lamps. By this ballast circuit the lamps are operated with repetitive pulsed voltage and current.

In this research first we optimized the pulse frequency of the voltage waveform to maximize the luminous flux. Since the luminous characteristics of the lamp also depend on the size of the electrode, the diameter and the length of the electrode were varied and the best combination of them to maximize the luminous flux was investigated. The diameter of the electrode was 1, 3, 5 and 7mm, and the length of the electrodes was 5, 7 and 10mm.

RESULTS & DISCUSSION

Figure 4 shows the luminous flux versus pulse frequency for the Y-shape electrode lamps with different Xe filling pressure. The maximum value of luminance was 10090 cd/m^2 and the luminous flux was calculated as 38.5 lm from this value of the luminance. The frequency at which the maximum value of luminous flux is obtained increases as the Xe filling pressure increases. This can be explained as follows: Since the pulse width is constant as 2μs, as the frequency increases, the non-discharge period (zero voltage period) becomes shorter. As the Xe filling pressure increases, both the electron collision frequency and the ion collision frequency increase and probability of recombination increases. This means as the pressure increases the numbers of both residual ion and electron decrease at the voltage rising phase. If we can assume that the VUV emission intensity is proportional to the number density of electron, the length of non-discharge period must become shorter as the pressure increases. This is the reason why the pulse frequency at which the maximum luminous flux is observed increases as the pressure increases.

The maximum luminous flux is measured with optimizing the value of ballast resistor for the variety of combinations of the diameter f and length of cathode and those of anode. The relation between the maximum luminous flux and cathode diameter is shown in Fig. 5. In this figure, the length of cathode is shown near each symbol. When the cathode and anode have same structure, maximum value of the luminance is 12200 cd/m^2 and from this value, the luminous flux is calculated as 49.6 lm. When the structures of cathode and anode are different, maximum value of the luminance is 12280 cd/m^2 and from this value the luminous flux is calculated as 50.9 lm.

As shown in Figure 5, the luminous flux tends to increase as cathode diameter decreases, cathode length increases, anode diameter increases or anode decreases. This can be explained as follows: To decrease in cathode diameter enhances the electric field of near cathode and electron emission from cathode becomes easy. To increase in cathode length enhances hollow effect and this also increases electron emission from cathode. To increase in anode diameter or to decrease in anode length increases the surface area contacting with positive column and plasma is expanded in radius direction. By expanding plasma in radial direction, the positive column can keep diffusive mode, that is, glow discharge, with much higher current. All these effect reduces plasma impedance and increases plasma density, and finally increases luminous flux.

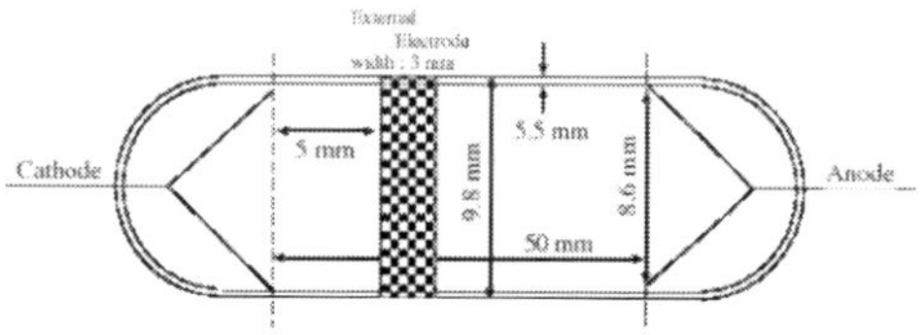

Figure 1. Structure of Xe CCFL with Y-shape Electrodes

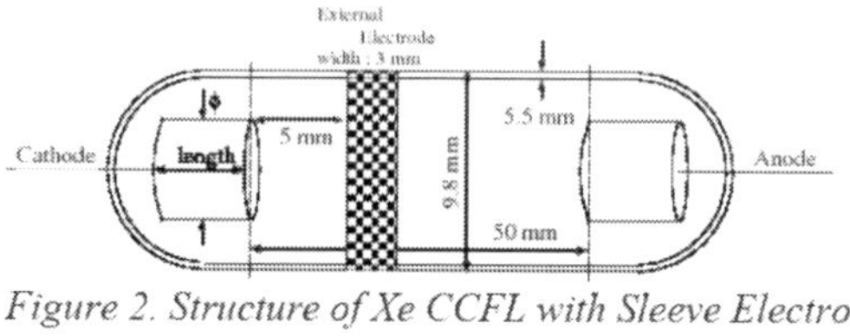

Figure 2. Structure of Xe CCFL with Sleeve Electrodes

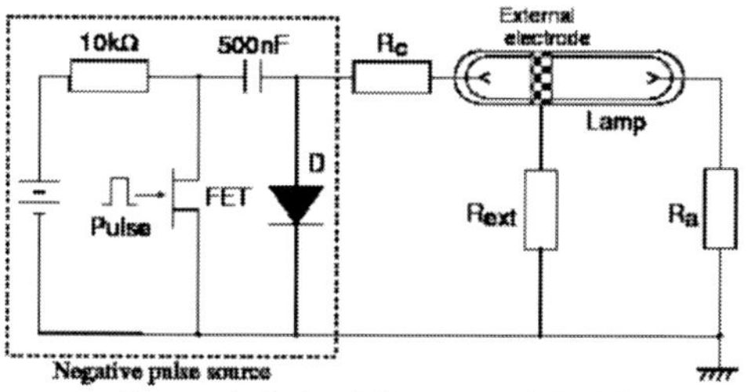

Figure 3. Pulsed Circuit and CCFL

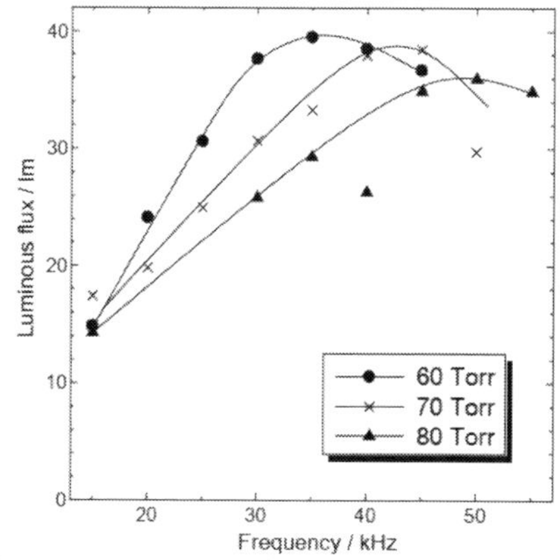

Figure 4. Luminous Flux versus Pulse Frequency for Y-shape Electrode CCFL. (R_c = 1k Ω; R_{ext} = 1M Ω; $R_{ballast}$ = 200k Ω; Pulse width = 2 μs)

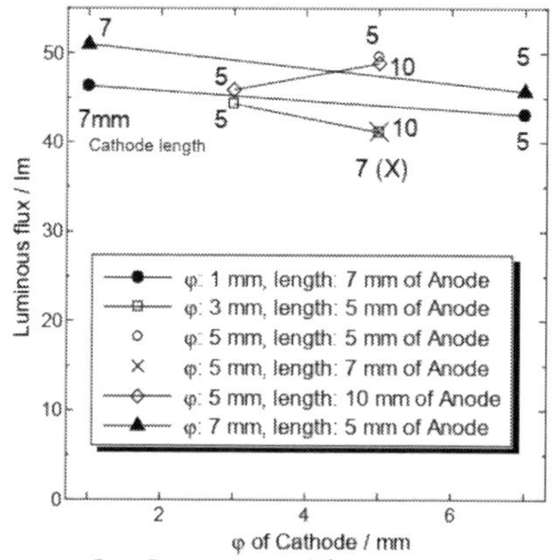

Figure 5. Luminous Flux versus cathode Diameter for Sleeve Electrodes CCFL. (R_c = 10k Ω; R_{ext} = 1M Ω; $R_{ballast}$ = Optimal value; Pulse frequency = 30kHz; Pulse width = 2 μs)

CONCLUSION

In this research, we found that the frequency at which maximum luminous flux is achieved increases as Xe filling pressure increases. Also we found that in order to increase luminous flux cathode should be longer with smaller diameter and anode should be short with larger diameter.

REFERENCES

[1] M. Jinno, M. Okanoto *et al* : "Luminance and efficacy improvement of low-pressure xenon pulsed fluorescent lamps by using an auxiliary external electrode", J. Phys. D: Appl. Phys. **40** pp.3889-3895 art. No. S08 (2007).

[2] M. Jinno and H. Motomura : "Mercury-Free Light Sources: Xenon FL and Molecular Radiators", Proc. of LS-11, Shanghai, pp.351-360 (2007).

[3] M. Hirahaya, M. Takeda *et al* : "Effect of the auxiliary external electrode on xenon fluorescent lamp depending on its diameter", Proceedings of IEEE ICPS 2009, San Diego, art. no. 5227352 (2009).

Session Two

Quantifying Circadian Light

M. S. Rea
Lighting Research Center, Rensselaer Polytechnic Institute, 21 Union Street, Troy, NY 12180, USA
Contact: ream@rpi.edu

ABSTRACT
A model of human circadian phototransduction has been developed to calculate circadian light for any spectral irradiance distribution. Using the model, one can determine how different light sources can stimulate the human circadian system at different levels of irradiance.

GENERAL

Optical radiation incident on the human retina stimulates both the visual and the circadian systems. Each system converts retinal radiation into neural signals that, in the case of the visual system, provides a detailed spatial-temporal representation of the external world and, in the case of the circadian system, orchestrates the timing of all our biological functions, from single cells to overt behavior [1,2].

A model of human circadian phototransduction [3] has been developed to calculate circadian light (CL) for any spectral irradiance distribution [4]. Like illuminance, CL is spectrally weighted irradiance, but for the human circadian system. Additional File #1 in reference [4] gives the complete method for calculating circadian light. The model is based on studies measuring light-induced nocturnal melatonin suppression, an orthodox marker of the circadian system's response to light. Figure 1 shows the spectral sensitivity of the human circadian system based upon nocturnal melatonin suppression from monochromatic light [5,6] together with the modeled spectral sensitivity of the circadian system for both narrowband and polychromatic stimuli. Because the model includes spectral opponency, responses from light stimuli created by combinations of narrowband sources cannot be predicted from the spectral sensitivity derived from narrowband light stimuli alone. Figure 2 shows published data from a wide range of studies measuring light-induced melatonin suppression and how, after transforming optical radiation into units of CL, these data appear to lie along a common transfer function.

Based upon the model then, it is possible determine how effective different light sources would be for stimulating the human circadian system at different levels of irradiance. Since, again, the modeled system uses several non-linear calculations it is necessary to first set an expected criterion response and then determine the spectral irradiance needed to produce that response. Table 1 shows several common light sources and the amount of photopic illuminance needed to produce different criterion responses by the human circadian system.

Figure 1 (left): *The photopic luminous efficiency function (V(λ)) together with the spectral sensitivity of the human circadian system to narrow-band and polychromatic sources based upon the model of circadian phototransduction [3]. Relative sensitivity to different narrow-band spectra based on data from [5] and[6] are also shown.*
Figure 2 (Right): *Light-induced melatonin suppression values (%) from a variety of published studies using narrow-band and polychromatic sources plotted as a function of circadian light (CL).*

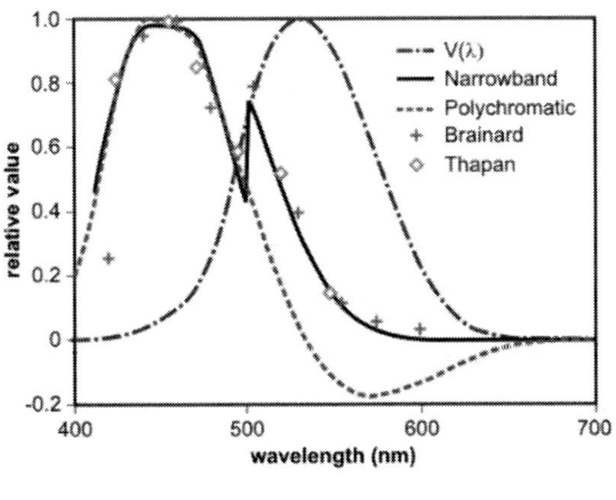

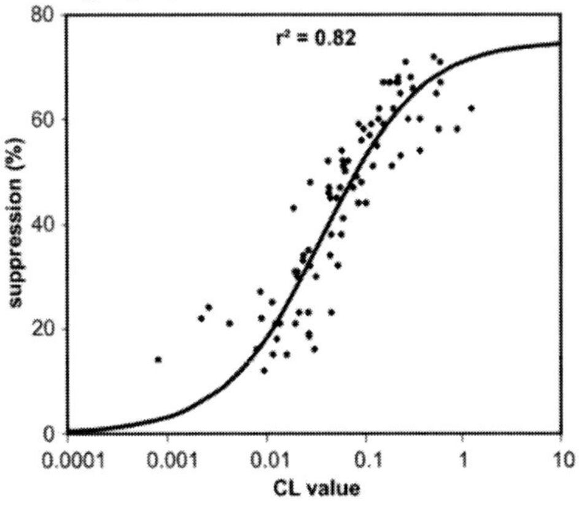

Table 1: *Illuminance and relative electrical power to achieve* ***25%*** *melatonin suppression and* ***50%*** *melatonin suppression for a fixed 2.3 mm diameter pupil*

Light Source	25% melatonin suppression		50% melatonin suppression	
	Illuminance (lux)	Relative Electric Power	Illuminance (lux)	Relative Electric Power
Blue LED (Luxeon Rebel, λ_{peak} = 470 nm)	9	1	30	1
8000 K Lumilux Skywhite fluorescent (OSI)	79	6	266	6
12300 K 3-way CFL (Philips)	60	6	203	6
69000 K White LED (Cree XRE coolwhite)	94	9	318	9
3350 K linear fluorescent (GE F32T8 SP35)	144	10	501	10
6220 K Colortone 75 fluorescent (Philips)	104	12	349	12
4100 K linear fluorescent (GE F32T8 SP41)	214	14	515	14
5200 K LED phosphor white (Luxeon Star)	154	15	708	14
2700 K compact fluorescent (Greenlite15WELS-M)	207	17	722	18
2050 K High Pressure Sodium (400 watt)	357	19	1246	20
2856 K incandescent A lamp	147	72	511	75
Daylight (CIE D65)	80	NA	270	NA

The values in Table 1 provide a comparison among common light sources in terms of their effectiveness, at different levels, for stimulating the human circadian system. Due to the non-linear nature of circadian phototransduction and the reticence of government and industry to use spectral sensitivity functions other than V(λ), it seems unlikely that lighting standards and light sources will be readily changed to reflect this information. Notwithstanding the conservative nature of regulations and manufacturing, much more basic research is in fact needed to understand how light incident on the retina affects the health and well-being of humans through circadian system stimulation. This does not mean, however, that lighting practice cannot change in the absence of supporting standards and specially designed light sources. A recent paper by Figueiro [7] for example lays out a very practical and effective approach for improving lighting for senior housing to improve visibility, reduce falls, and improve sleep efficiency. Such changes could improve the quality of life for these individuals irrespective of supporting standards and absent new light sources. To accomplish an effective design however, information like that presented in Table 1 is critical for maximizing social benefit while minimizing energy use and costs. Indeed, for the foreseeable future, social progress in this area of lighting will have to rely on thoughtful and informed professional designers and engineers who combine science with lighting standards.

ACKNOWLEDGEMENTS

The author thanks Andrew Bierman and Mariana Figueiro for their technical contributions to this paper. Support was provided by the National Science Foundation (NSF) Smart Lighting Engineering Research Center (EEC-0812056).

REFERENCES

[1] E.J. Someren and Riemersma-Van Der Lek R.F. 2007. Live to the rhythm, slave to the rhythm. Sleep Med Rev. 11:465-84.
[2] R.G. Foster and Wulff K. 2005. The rhythm of rest and excess. Nat Rev Neurosci. 6(5): 407-14.
[3] M.S. Rea, Figueiro M, Bullough J, Bierman A. 2005. A model of phototransduction by the human circadian system. Brain Res Rev. 50:213-28.
[4] M.S. Rea, Figueiro M.G., Bierman A, Bullough J.D. 2010. Circadian Light. Journal of Circadian Rhythms. 8(2). Available Online at: http://www.jcircadianrhythms.com/content/8/1/2
[5] G. Brainard, Hanifin J, Greeson J, Byrne B, Glickman G, Gerner E, Rollag M. 2001. Action spectrum for melatonin regulation in humans: Evidence for a novel circadian photoreceptor. J. Neurosci. 21:6405-12.
[6] K. Thapan, Arendt J and Skene DJ. 2001. An action spectrum for melatonin suppression: evidence for a novel non-rod, non-cone photoreceptor system in humans. J Physiol. 535:261-67.
[7] M. G. Figueiro. 2008. A proposed 24 hour lighting scheme for older adults, Lighting Research and Technology. 40: 153-60.

Plasma lamps for biological Rhythms (PLACAR)

R. Kozakov[1], D. Kunz[2], C. Stoll[2], L. Hitzschke[3], J. Meltke[4], A. Deitmann[5], G. Roth[6], H. Rudolph[7], K. Barkowsky[1], H. Schöpp[1]

[1] INP-Greifswald, F-Hausdorff-Str. 2, Greifswald, D-17489, Germany
[2] Charité – University medicine, Sleep Research & Clinical Chronobiology, Gr. Hamburger-Str. 5-11, Berlin, D-10115, Germany
[3] OSRAM GmbH, Werner-von-Siemens-Str. 6, 86159 Augsburg, Germany
[4] NARVA BEL, Erzstr. 22, 09618 Brand-Erbisdorf, Germany
[5] NARVA G.L.E, Herzbergstr. 26, 10365 Berlin, Germany
[6] LITEC-LLL, Brandteichstr. 19, 17489 Greifswald, Germany
[7] TRILUX GmbH, Heidestr. , 59753 Arnsberg, Germany
Contact: schoepp@inp-greifswald.de

ABSTRACT

Known effects of light on melatonin suppression were earlier obtained exclusively under strict laboratory conditions. We investigated lamps and fixtures in naturalistic situations. In different common lighting situations melatonin secretion was controlled by saliva samples. The lamps were chosen for good colour rendering index and high efficacy but with different colour temperatures and illuminance values. It was shown that lamps suppress melatonin in different extent depending on their spectral composition. The melatonin suppression can be sufficiently described by a simple model based on its spectral sensitivity curve.

TARGET AND INVESTIGATION

Investigations in laboratories and usual life situations are often not directly comparable. The project (PLACAR) fills the gap between investigations under strict laboratory conditions and applications in real life situations. It is known that circadian rhythm in humans exists with an increasing melatonin concentration in the evening. It is also known that the melatonin secretion can be influenced by light. Such investigations were carried out with dilated pupils and special lamps. In PLACAR we used commercially available lamps and some modifications of them. We choose fluorescent and HID lamps as well as dielectric barrier discharge lamps. Some of the lamps were modified to achieve lower circadian action.

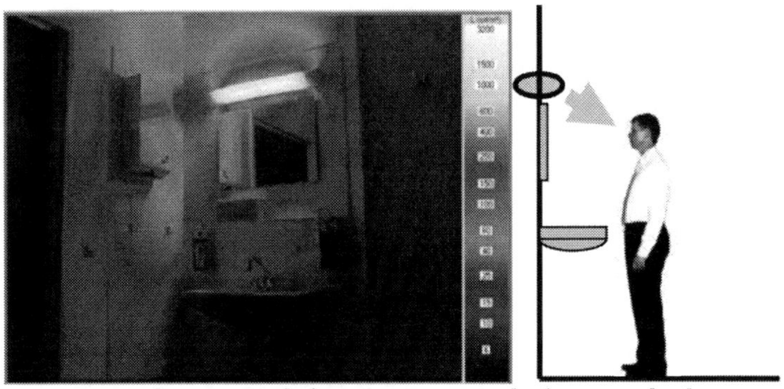

Fig. 1: Luminance picture of an evening lighting situation in a bathroom with a luminance camera (LMK)

The lamps were characterized in an integrating sphere and optical emission spectroscopy set-ups. Measurements for luminance, illuminance and spectral irradiance for the different lighting situations were carried out with a calibrated luminance camera (LMK), illuminance meter and compact spectrometer in the sleep laboratory. The measuring points were equal to the later positions of the test persons eyes. The illuminance varied between 80 and 500 lx. The lamps and fixtures (bathroom, hall or office) were positioned either in front or above the test persons. An example of an evening lighting situation in a bathroom is given in Fig. 1.

Test persons were in age of 25 to 35 years, healthy, of both sexes and lacked sleep disorders. There were no other restrictions for the choice. They kept constant bed-times and remained in the lab only for the evenings of the test period. No other changes of their normal life were necessary. To ensure reproducibility the tests followed a standardized routine. During the winter period test persons arrived the lab at least several hours before their habitual bed-time and stayed in a dim light situation (<1 lx) for that time. Saliva samples were collected every half an hour. One hour before the habitual bed-time they took position in front of the lighting situation which was switched on for half an hour. Within this period as well as ten minutes before and afterwards saliva samples were taken every ten minutes. Thereafter they stayed again in dim light for an hour with sample collection every half an hour. After this procedure they went home. Saliva samples were analyzed by an external lab to quantify melatonin concentration.

RESULTS

The obtained results are given in Fig. 2. The data in figure 2 represent the melatonin suppression values for the tested lighting situations as function of circadian relevant dose. The dose is calculated according to the equation shown in the figure, where A is the pupil area, s_{ms} is the melatonin suppression sensitivity curve [1] and E is the spectral irradiance at cornea. The designations are as follows: FL – fluorescent lamp, DBD – xenon barrier discharge fluorescent lamp, NAT – sodium high pressure lamp, HID – metal halide high pressure lamp. The number designates the correlated colour temperature. Figure 2 shows that there exists statistically significant correlation between circadian dose and melatonin suppression. This confirms that the melatonin suppression measurements in usual lighting conditions can describe the basic features of the non-visual action of light and gives possibility to optimise lamps for appropriate biological action.

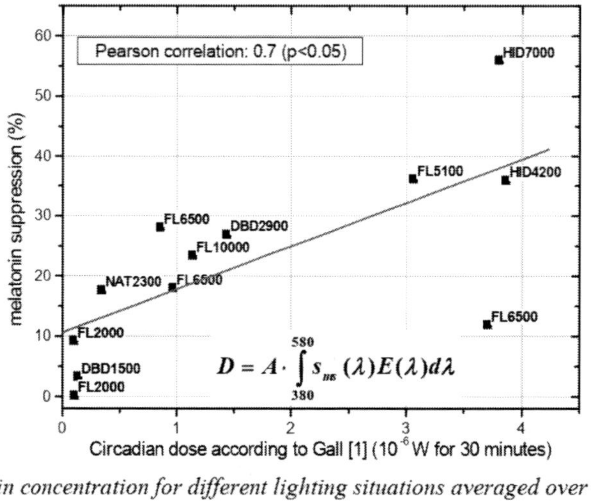

Fig. 2: Melatonin concentration for different lighting situations averaged over 10 test persons

SUMMARY

Investigation in real life situations with different commercially available lamps and fixtures show that melatonin suppression takes place. It depends on properties of the spectrum and can be described by the melatonin suppression sensitivity curve. Nevertheless all significant parameters, like pupil size and illumination time have to be taken into account. The results can be implemented into the lamp manufacturing process to create lamps for morning or evening application.

The project was founded by German BMBF under grant FKZ: 13N 8968-74.

REFERENCES

[1] D.Gall and V. Lapuente, Licht 7-8 (2002) 860-871

Investigation of Visual Perception under Dynamic Color Rendering Modulation for Augmented Energy Savings

Maria Thompson[1], Joseph Laski[1], Makarand Chipalkatti[2]

[1]*OSRAM SYLVANIA CRSL Central Research & Services Laboratory, 71 Cherry Hill Drive Beverly MA 01915 USA*

[2]*OSRAM SYLVANIA Dept of Government Affairs and Public Policy, 100 Endicott Street Danvers MA 01923 USA*

Contact: maria.thompson@sylvania.com

ABSTRACT

Digitally controlled lighting systems equipped with colored LEDs can produce a range of different qualities of light, adjustable to users' requirements. In this context, we explore a novel lighting control concept which involves a strategic control of RYGB LEDs forming white light, employed to increase energy efficiency. This paper reports and discusses the results of an ongoing research project applying the principles of color rendering modulation for energy savings, which encompasses two main domains: (a) a tunable LED light source with accompanying control software; and (b) human factors studies concerning how surfaces and spaces are perceived under a light source with changing color rendering. Results so far have supported the two fundamental hypotheses: (a) when fully developed, color rendering modulation could potentially enable power reductions of at least 25%; (b) the majority of subjects did not detect the color changes.

THE LED TUNABLE PANEL & THE COLOR RENDERING MODULATION STRATEGY

There is a fundamental trade-off between color rendering (CR) and luminous efficacy. Specifically, we can reduce CR and increase efficacy when we redistribute spectral energy intensity in favor of wavelengths where the eye response is greatest, the yellow/green region of the spectrum [1, 2, 3]. The notable point is that while changing CR, the light level and color temperature can stay constant, and hence the appearance of the modified white light remains unchanged. Such a control strategy enables a LED system to dynamically shift from 'energy saving mode' (low CR) to 'quality mode' (high CR) according to occupancy. The strategy is to bring the lighting of unoccupied spaces to minimum color rendering to save energy without altering the appearance or light levels of the illumination.

While dimming is well understood as an effective method to save energy, research shows that standard dimming methods may result in limited savings as they can be unacceptable during normal working (peak energy usage) hours, and in certain large public spaces, when darker areas may appear uninviting or unsafe [4]. A lowering of color rendering where dimming is not acceptable could save energy currently wasted in unoccupied areas which must otherwise remain lit at full power. It is worth noticing that this does not preclude a *combined* dimming and CR modulation approach. Another circumstance that could make use of the CR modulation technique can be found in spaces making use of daylight. In most cases, artificial lighting used in day-lit spaces must stay on the entire business day. If the appropriate color rendition is provided by the daylight component, tunable LED illumination could be set to lowest CR mode with no loss in visual quality, thus augmenting the benefits of daylight harvesting.

In this study, spectra of higher luminous efficacy were obtained by increasing the emission in the yellow/green region of the spectrum where $V(\lambda)$ peaks. Specifically, the approach was to change the ratio of red emission to yellow emission (while making minor adjustments to blue and green). The main objective was to compensate for the reduced red emission with yellow emission to attain higher efficacy sources; hence, obtaining a favorable tradeoff between color rendering and efficacy depends on the availability of higher efficiency yellow LEDs. Presently, all colors of LEDs are increasing in efficiency to the point where this CR modulation strategy is a realistic option. The most recent experimental implementations of the LED tunable panel were used to rebalance from highest to lowest color rendering to reduce the wattage by as much as 12% of the initial wattage with the same lumen output and color temperature. Future optimization of these prototypes is realistically anticipated to achieve at least 25% power savings when applying the CR modulation technique.

THE HUMAN FACTORS EXPERIMENTS

The question of how colored surfaces and spaces will be perceived under light with changing color rendering is critical to the acceptability of the CR modulation concept. We investigate the hypothesis that a significant range of color distortions would be unnoticeable in the periphery of the users' main field of view (surrounding unoccupied areas).

Three incremental series of psychophysical experiments were performed based on subjective assessment of color changes under modulated color rendering produced by tunable LED sources [5, 6]. Results from our early baseline experiments, using color samples viewed within booths, indicate reduction in sensitivity when color changes were not seen side-by-side, and also confirmed that perceived color distortions were mostly on saturated red samples. Subsequent experiments tested the same hypothesis but in a more elaborated and unbiased set-up, which included: (a) simulated 'real life' context involving a full scale mockup of an architectural scenario, instead of the viewing booths; (b) lighting operated with highly automated CR modulation for continuous and seamless transitions among spectra, advancing from the sharp transitions from the previous phase; (c) CR modulation performed on two different positions outside of subjects' main field-of-view, instead of exclusively direct observation. Previous research from other sources strongly indicated that color vision decreases as the observed scene moves away from central vision [7], and that people cannot detect even sizeable visual environmental changes when these are slow progressive changes [8]. The results from our previous studies and from other researchers support that a color rendering modulation strategy can be made viable from the perspective of visual perception.

Our most recent ongoing human factors studies investigate the same hypothesis, but with a larger, more sophisticated experimental set up, again advancing from the previous phase. The visual experiments will take place at a real office space environment, where twelve units of our most recent LED tunable panels will be installed in public and private office areas and the people working under this illumination will be queried periodically and unexpectedly about perception of changes in lighting.

REFERENCES

[1] Zukauskas A., Vaicekauskas R., Ivanauskas F., Gaska R., Shur M., "Optimization of the White Polychromatic Semiconductor Lamps", Appl Physics Letters, volume 80, number 2, 2002.

[2] Schubert E. F., Chhajed S., Xi Y., Li Y., Gessmann, "Influence of Junction temperature on Chromaticity and Color-Rendering Properties of Trichromatic White-Light Sources based on LEDs", Journal of Applied Physics 97, 054506, 2005.

[3] US PATENT # 7,598,859 B2. Laski J., Chipalkatti M. October 6th 2009.

[4] Jennings J., Rubinstein F., DiBartolomeo D., and Blanc S., "Comparison of control options in private offices in an advanced lighting controls test bed," Journal of the Illuminating Engineering Society, Vol. 29, No. 2, pp. 39-60, 2000.

[5] Thompson M., Levin R., O'Reilly U.M. "Psychophysical Evaluations of Various Color Rendering from LED-based Architectural Lighting", Proceeding of the SPIE Seventh International Conference on Solid State Lighting #6669, San Diego CA, August 2007.

[6] Thompson M., "Psychophysical Evaluations of Modulated Color Rendering for Energy Performance of LED-based Architectural Lighting", PhD Thesis, MIT, Cambridge, Massachusetts, February 2007.

[7] Stabell, U. & Stabell, B., 1982, 'Color vision in the peripheral retina under photopic conditions', Vision Research, 22, 839-844.

[8] Simons D. & Ambinder M., 2005, 'Change Blindness Theory and Consequences', American Psychological Society, Vol. 14, no. 1, pp. 44-48.

Increase growth yield in Horticulture by improvements in LED Lighting.

A. Valster[1] , C. Tanase[1] , C.L. Hsu[2] , U. van Slooten[1], J.P Jacobs[3]
[1] *Philips Lighting, Mathildelaan 1, 5611BD Eindhoven, The Netherlands*
[2]*Epistar corporation, 5 Li-hsin 5th Rd, Science-based Industrial Park, Hsinchu, Taiwan 300, R.O.C.*
[3]*Philips Lighting, Philips Campus, Building 2, No.9, Lane 888 Tian Lin Road, Shanghai 200233*
Contact: cristina.tanase@philips.com

ABSTRACT

Light intensity, distribution and spectrum are eminent for boosting year around growth and quality within the Horticulture sector. LED lighting has benefits in flexible positioning the right intensity and spectrum at plant level enabling growers the optimum light at every stage of a crop's growth. This paper describes how LED light applications within Horticulture in combination with a world-record breakthrough in improvements at LED level, can increase the growth yield in real-life greenhouse tests opening the way for commercial exploration of another solid-state lighting revolution.

APPLICATION

One of the main challenges for our growing world population is the production of food hence the world wide interest for advances in the horticulture industry. At the same time, the horticulture sector is highly competitive with strong focus on substantial reduction of energy use, increased production, improved product quality, and possibilities for earlier or year-round production. Light is an important horticultural production tool and a key factor in plant research. As we all know light in photosynthesis is the driving force for plant growth. Less known is that the light intensity and spectrum also influences a great number of physiological reactions in plants, and is used in recurrent cycles in order to control processes as dormancy, flowering, and seed-stalk development. Hence by tuning light we expect to create new advances in plant growth for the horticulture industry.
The plant uses efficiently blue and deep red light for photosynthesis (absorption peaks of chlorophyll A are at 465nm and 665nm) and far-red light (730nm) to set the circadian rhythms. To meet overall efficiency and reliability especially at the red part of the spectrum major LED improvement was needed and will be described in the LED section below.
Taking both advances in LEDs and the application of LED lighting technologies we will show how to save energy consumption and improving plant quality by optimizing the growth process. With LED lighting, the growth light can be tuned on a very detailed and well-controlled level, which makes it possible to apply the optimum light recipe at every stage of a crop's growth. This paper describes a new approach in horticulture by offering a tailored solution with a specific LED light spectrum, light level and positioning, matching the needs of a specific crop and its growing environment.

Figure 1. Growing plants in closed chambers in multilayer application (left) and in a greenhouse (right). In both situations specific combinations of red and blue LED light is used for optimal growth.

To find the optimal growth solution, a lot of knowledge of the influence of lighting in all its aspects, and the relation with other variables influencing growth becomes crucial for success. In developing LED-based solutions, Philips conducts field tests together with growers and breeders as well as experts from the academic community. Evidence of 15% improved yield in the growth season 08/09 will be presented by using inter lighting LED systems place between the crops in a commercial tomato greenhouse. The effects of inter-lighting on light interception, on vertical gradients of leaf photosynthetic characteristics and on crop production and development has been investigated in cucumbers and simulations of light attenuation and canopy photosynthesis show that crop photo synthesis increases by more uniform vertical irradiance within crops [1].

LED Design and Fabrication

Optimal plant growth enhancement is obtained using besides some blue an excess of light at deep red and occasionally far red wavelengths. LEDs using AlGaAs technology may cover both red wavelength regimes, however, at the expensive of low efficiency and limited reliability. This paper describes for the first time the availability in AlGaInP technology at both red wavelengths including world class high efficiency and reliability.

LED structure is grown on a temporary n+-GaAs substrate by low-pressure OMVPE technique. A large challenge for LEDs in the red wavelength regime is the light extraction efficiency, which is limited to approximately 2% due to substrate absorption and total internal reflection. An external quantum efficiency of 55% has been reported for AlGaInP-GaP LEDs based on Transparent Substrate (TS) technology [2]. Although TS-AlInGaP can reach high EQE, due to high cost of GaP substrate and an expensive high temperature bonding process with high precision orientation alignment, makes the TS-AlInGaP technology not very cost effective. Another approach is using the so-called metal bonding (MB) process, which has much less technical constraints for the overall device production process [3] resulting in a efficient and cost effective LED technology for the horticulture application. Due to reliability issues in high power applications using AlGaAs materials, the LED structure consists of a GaInP-AlGaInP multiple-quantum-well active region for both emission at 650 nm as well as 730nm. Following the metal mirror bonding process, the GaAs substrate was removed by chemical etching. Various electrode patterns were deposited onto the n-side cladding layer for ohmic contact. The epitaxial LED layers are then mesa-etched to generate isolated mesas with dimensions of 14x14 mil, 20×20 mil and 42×42 mil. Current crowding leads to non-uniform current density distribution in the active layer, resulting in a non-uniformly pumped active layer with low external quantum efficiency. Current spreading calculations [Figure 2] showing that a major part of the current flow under the electrode contacts. We conclude that current crowding is also a dominant loss mechanism. It also causes reliability issue for long term life testing.

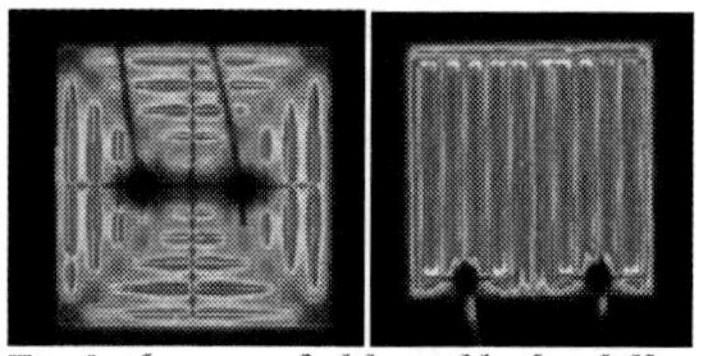

Fig.2, the near field profile for different designs

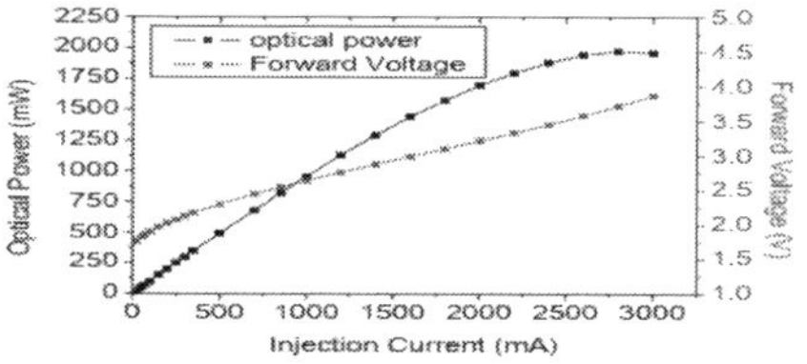

Fig.3, L-I and I-V curves for the 42mil chip at 660nm

A wall plug efficiency of 46% at 660nm is achieved which is the worlds highest reported efficiency for AlGaInP thin-film LEDs [Table 1]. Figure 3 shows the optical power and forward voltage verse injection currents. For the 42mil chip, the saturation current was around 2.8A. The 1000hrs 85% 85℃ reliability test was also performed showing less than 5% power decay.

Product Series	Wd (nm)	IF (mA)	VF (V)	Power (mW)	WPE (%)
ES-SMBRPN14B	656	20	2.18	16.5	38
ES-SMBRPN20B	657	100	2.1	83	39.6
ES-SMBRPN42B	657	350	2.16	347	46
ES-SMSFPN42A	739	350	2.14	180	26.5

Table 1 characterises of 660nm and 730nm chips

[1] G. Trouwborst et, The responses of light interception, photosynthesis and fruit yield of cucumber to LED-lighting within the canopy, Physiologia Plantarum, vol. 138, pp.289-300, 2010

[2] M.R. Krames et, High-power truncated –inverted-pyramid $(Al_xGa_{1-x})_{0.5}In_{0.5}P/GaP$ light emitting diodes exhibiting >50% external quantum efficiency, Appl. Phys. Lett., vol.75, pp.2365-2367, 1999.

[3] Y.J. Lee et, Increasing the extraction efficiency of AlGaInP LEDs via n-side surface roughening, IEEE Photon. Technol. Lett., vol.17, no.11, pp.2289-2291, Nov. 2005

A Picture of the Lighting Future after Edison

T. Noll

CL D-E&L&M, OSRAM GmbH, Berliner Allee 65, Augsburg, 86136

Contact: t.noll@osram.de

Due to massive pressure on governments for realization of promised CO_2 savings (Kyoto, Lisbon, Copenhagen,...), the global phase out of low efficient incandescent lamps is no longer a question of "whether" but just a question of "when" and "to what extent". Many governments are just preparing legislation for "banning the bulb" or have already released a concrete phase out plan like the EU [1]. Open points are still directional lighting and how to handle halogen lamps.

The contribution of the lamp industry has been to develop and bring more and more energy saving retrofits into market which shall replace 1:1 (size, lumen, performance, safety, ...) their less efficient counterparts. Meanwhile retrofits are available in HAL technology (replacement GLS), CFL technology (replacement GLS and HAL) and LED technology (replacement GLS, HAL, FL and CFL). The question whether retrofits make sense in terms of light quality and safety sometimes seems to be of less importance.

However, it is important to emphasize that retrofits for screw based lamps like CFLi or LED cannot utilize the intrinsic efficiency potential of their technology which should be close to 90 lm/W for CFLi and above 100 lm/W for LED. The root causes are too high current density $I_{discharge}$ and cold spot temperature T_{Hg} in applications (CFLi) and too high pn-junction temperature T_j (LEDs). High temperature of integrated electronics additionally sets limits to applications and/or lifetime of the lamp. And anyway, improving efficiency of CFLi is in conflict with some megatrends like (1) miniaturization (realize GLS shape also for 100W GLS equivalents) and (2) use of additional outer bulbs. While (1) needs a switch from T4 to less efficient T3 or T2 tubes, both (1) and (2) make amalgam technology mandatory which - in case of consumer lighting - is in conflict with the requirement of a fast lumen run up.

On the other side luminaires are often designed in a way which is detrimental for the system (=fixture) efficiency: while for incandescent lamps narrow downlights have no negative effect on lamp efficiency, retrofits with integrated electronics suffer even more by still higher temperatures of sensitive parts and electronics. In addition many luminaires face the problem to have a Light Output Ratio (LOR) of much less than 1 and therefore cut off efficiency "at the last meters to the customer".

Therefore, if we consider energy saving and Carbon footprint as a serious problem for the future, we have to provide answers on the following questions:

(1) Should in a future GLS free "Post-Edison-world" new luminaires be still designed for use of Edison lamp holders anyway and, if the answer is "no":
(2) Which lamp holder system should be preferred in future luminaires which shall fulfill new "Generic Eco-Design requirements" [2]?

As above mentioned shortcomings of retrofits and luminaires cannot be resolved in a simple way it is very probable that retrofits will be an interim solution until a better solution is available. While Zhaga [3] and the LED light engine may be a good approach for pushing LED technology in professional lighting, in consumer lighting people have the choice between (1) use of retrofits - with all the problems mentioned above - or (2) buying a new luminaire with built in LEDs ("LED grave"), thus generating a very strong and long lasting negative lock-in effect [2]: consumers have to stick with the LED technology at the time of purchase - up to 50.000 h - and therefore cannot participate on the technical progress of this fast developing technology.

If we could start off with a blank sheet of paper like Edison, we are able to resolve some of the paradigms which partially stick to conventional lamps and luminaires. One of these paradigms is the "shape of the bulb", the "godfather for all retrofit developments", because it invites luminaire industry to design luminaires with negative impact on temperatures and LOR. This paradigm can be resolved by making the lamp flat, which in turn will make it "much more difficult" to realize hot fixtures anyway. Another paradigm is the unfavorable arrangement of electronics above the lamp, especially if the lamp is operated base up in a narrow downlight. This paradigm can be resolved by rearranging the electronics in the lamp plane, but outside the lamp body, which in turn allows to realize real flat lamps or "Power Discs" (PDs) with height down to 12mm.

A sketch of PDs with Ø120mm is shown in Fig. 1 for all existing technologies (HAL, LED, HAL and HID).

Also upcoming technologies like OLED or LED light engine can be hosted in the luminaire, along with what is called an "Electronic Power Disc" (EPD), which will allow to add any useful electronics to the luminaire.

Fig. 2 shows a sketch of 4FL (Future Flat Flexible (Fluorescent) Luminaire) with seven PDs in the version of a pendulum luminaire. One of the major differences to conventional luminaires is that the user can customize the luminaire according to his personal needs and the special illumination tasks in the room (dining, reading, standby light,...). He has the choice between functional and emotional, affordable and luxurious lumens. In a preferred version of 4FL the PDs are operated @ SELV, allowing to ban some of the electronic components which are needed anyway in each of the 230V PD, into the SELV module. The luminaire can be operated via power line or remote control, where the receiver is part of the SELV module ("CPU of the luminaire").

The main advantages of 4FL and PD are seen as follows:

Technics & energy saving	Flat (down to 12mm) and thermally relaxed design of PD with possibility for convective cooling
	Avoids by design hot fixtures and keeps critical temperatures low (T_{Hg}, Tj, electronics)
	Allows to increase efficiency up to 90 lm/W for CFLi and >75 lm/W for LED
	Allows to selectively utilize strengths of individual technologies for different applications (BAT)
	Allows operation of lamps with different technologies (HAL, CFLi, LED,...) in one luminaire
Customer benefit	Allows bidirectional lighting with different colours on floor and ceiling
	Allows to host OLED and Zhaga light engine [3] (depending on system definition)
	Allows adjustment of the beam axis in case of reflector applications (Goniometer principle)
	Allows to increase the intelligence of luminaires (comfort dimming, colour adjustment, light scenes)
	Allows to integrate LMS systems and remote control as part of luminaire (Green Field Approach)
Environment & Safety	Is interesting for consumer as well as for professional lighting
	Allows to improve quality of light (Ra, colour adjustment, circadian function, flicker, HPF, EMI,...)
	Uses simple and extremely safe lamp holder system (no electric shock anymore in case of SELV)
	Fulfils probably all future "Eco design requirements" for luminaires [2] w/o negative lock-in effect
	Is a light tower regarding material effort, Hg dose, packaging & transportation effort and recycling

Fig. 1: Sketch of PD and EPD Design

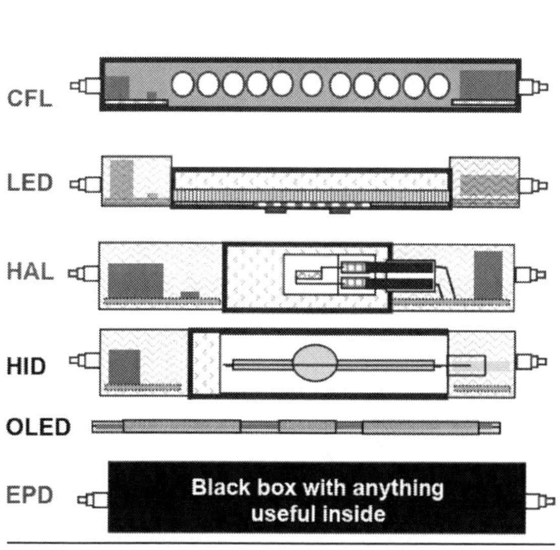

Fig. 2 : Design example "4FL as pendulum luminaire"

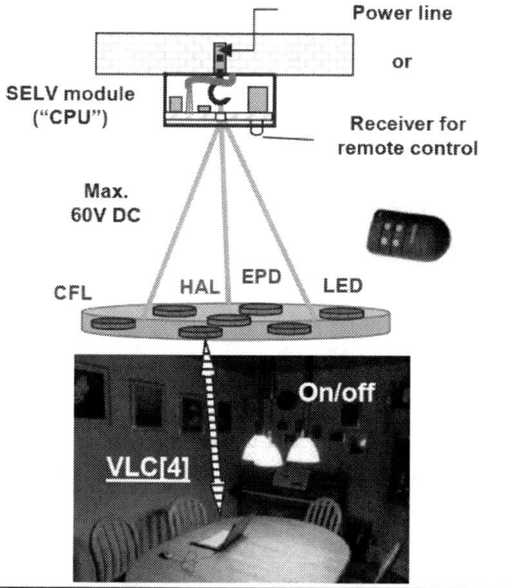

[1] Directive EC No 244/2009 and Amendment 859/2009
[2] VITO Report Lot 19: Domestic lighting (ContractTREN/07/D3/390-2006/S07.72702)
[3] http://idw-online.de/pages/de/news354343
[4] http://www.siemens.com/innovation/en/news_events/ct_pressreleases/e_research_news/2010/e_22_resnews_1002_1.htm

Krypton-85 gas mixture radioactivity comparisons

A. Tom Gallagher[1], B. Jorge Silva[1], C. Tony Almeida[1], D. Michael Baselice[1], E. Mark Yang[2], F. Pritam Chaudhari[3]

[1]LindeGasNorth America LLC, [2]Linde Electronic and Specialty Gas Division of China, [3]BOCIndia Limited, Contact: tomg@spectragases.com

ABSTRACT: To provide radiation activity measurements on krypton-85 gas mixtures shielded by different lead thickness and at different distances for the purpose of serving the lighting industry with data for safety and health reasons.

The use of radioactive krypton-85 gas mixtures in metal halide lamps, high pressure sodium lamps, CFL's and other light giving sources has become more prevalent in recent years. Some of the advantages of using krypton-85 in lamps are improved ionization, decreasing starting voltage, increasing the consistency of the starting voltage, decreasing ballast size and increasing ballast life. With the use of krypton-85 comes the responsibility of the light bulb manufacturer to protect all personnel from exposure of gamma radiation that is emitted by the krypton-85 gas mixture.

EXPERIMENT PARAMETERS:

a. Instrumentation achieved on a Fluke Pressurized Ion Chamber Survey Meter, Model #451P-RYR, calibrated on January 27, 2009 with a sensitivity of +/-10%.
b. Testing achieved on gas mixtures during the month of February, 2009 and October, 2009 by Linde at their radioactivity facility in the USA
c. Test cylinders are carbon steel with CGA 580 brass valve:
   1. Three water liter, Type "A" cylinder for 400 liter cylinders:
      - Height with valve (50.0 cm/20 inches) x Diameter (10.0 cm/4 inches)
   2. Eleven water liter, Type "A" cylinder for 1400 liter cylinder:
      - Height with valve (71.1 cm/28 inches) x Diameter (17.0 cm/6.7 inches)

TEST #1: Radioactivity Measurements on 0.145 mCi/liter Kr-85 + Argon, 400 liter cylinder

	urem/hour measurement from side of carbon steel cylinder			
	0 cm/0'	30.48 cm/1'	60.96 cm/2'	152.4 cm/5'
a. No Lead	4000	600	250	110
b. Lead Thickness: 0.635 cm/1/4"	900	175	120	100
c. Lead Thickness: 1.27 cm/ 1/2"	280	120	70	60
d. Cardboard: 0.3175 cm/ 1/8"	3700	600	240	120

TEST #2: Radioactivity Measurements on 0.08 mCi/liter Kr-85 + Argon, 400 liter cylinder

	urem/hour measurement from side of carbon steel cylinder			
	0 cm/0'	30.48 cm/1'	60.96 cm/2'	152.4 cm/5'
a. No Lead	1800	220	90	30
b. Lead Thickness: 0.635 cm/1/4"	320	70	30	15
c. Lead Thickness: 1.27 cm/ 1/2"	130	24	20	12
d. Cardboard: 0.3175 cm/ 1/8"	1780	200	90	24

TEST #3: Radioactivity Measurements on 0.5 mCi/liter Kr-85 + Argon, 1400 liter cylinder

	urem/hour measurement from side of carbon steel cylinder			
	0 cm/0'	30.48 cm/1'	60.96 cm/2'	152.4 cm/5'

a. No Lead	33,000	6300	2300	600
b. Lead Thickness: 0.635 cm/1/4"	5500	1400	500	200
c, Lead Thickness: 1.27 cm/1/2"	1800	500	260	70
d. Cardboard: 0.3175 cm/1/8"	33,000	6000	2200	550

TEST #4: Radioactivity Measurements on 0.5 mCi/liter Kr-85 + Argon, 400 liter cylinder

	urem/hour measurement from side of carbon steel cylinder			
	0 cm/0'	30.48 cm/1'	60.96 cm/2'	152.4 cm/5'
a. No Lead	17,500	2100	730	250
b. Lead Thickness: 0.635 cm/1/4"	2800	600	250	80
c. Lead Thickness: 1.27 cm/1/2"	880	250	120	46
d. Cardboard: 0.3175 cm/1/8"	16,500	2000	720	180
e. Aluminum: 0.3175 cm/1/8"	8,500	1600	700	150

In relation to radiation exposure limits for humans: according to Title 10, Code of Federal Regulations, sub-part C of the United States Department of Energy in the 10CFR – Occupational Dose, Subpart C, Section 20.1201 (Regulations Manual),
the Occupational Dose Limits for Adults (non pregnant) are:

a. Whole Body: Total Effective Dose Equivalent = 5 rem/year
b. Organ Tissue Deep Dose Equivalent = 50 rem/year
c. Lens of the Eye = 15 rem/year
d. Skin and Extremities – Shallow Dose Equivalent = 50 rem/year

CONCLUSIONS: If a person worked ten hours/day, 250 days/year within a distance of five feet (152.4 cm) of the following kr-85 + argon gas cylinders protected with one half inch (1.275 cm) lead thickness or protected with no lead then the rates of radiation exposure per year would be:

	Krypton-85 + Argon Gas Mixture	Lead Thickness	
		1.275 cm/0.5"	0 cm/0"
a.	1400 liters, 0.5 mCi/liter	0.175 rem/Yr	1.50 rem/Yr
b.	400 liters, 0.5 mCi/liter	0.115 rem/Yr	0.625 rem/Yr

And if a person worked ten hours/day, 250 days/year within a distance of two feet (60.96 cm) of the following kr-85 + argon gas cylinders protected with one half inch (1.275 cm) lead thickness or protected with no lead then the rates of radiation exposure per year would be:

	Krypton-85 + Argon Gas Mixture	Lead Thickness	
		1.275 cm/0.5"	0 cm/0"
a.	1400 liters, 0.5 mCi/liter	0.65 rem/Yr	5.75 rem/Yr
b.	400 liters, 0.5 mCi/liter	0.3 rem/Yr	1.8 rem/Yr

Note: These calculations are worst case scenarios without taking in consideration of the decaying of the Krypton-85 isotope.

Therefore a person, who works ten hours per day, 250 days per year, must remain more then two feet (60.96 cm) from a 1400 liter, 0.5 mCi/liter Kr-85 + Argon gas cylinder with no lead protection to stay below the annual United States of Energy recommended radiation exposure limits for the whole body (5 rem/Year).

Spectroscopic Investigation of Indium Bromide for Lighting Purposes

H.C.J. Mulders[1], G.M.W. Kroesen[1], M. Haverlag[1,2]

[1]*Department of Physics, Eindhoven University of Technology, P.O.Box 513, 5600MB Eindhoven, The Netherlands*
[2]*LightLabs, Koninklijke Philips Electronics N.V., Mathildelaan 1, 5611BD, Eindhoven, The Netherlands*
Contact: h.c.j.mulders@tue.nl

ABSTRACT

Laser Induced Fluorescence was used to study the radiative properties of InBr for lighting purposes. Results include the temperature dependence of the fluorescence decay time, spectroscopic constants and rotational temperature determination from a LIF spectrum.

INTRODUCTION

There are a few different lamp designs that are used to improve on the efficiency of the incandescent lamp. The low pressure discharge lamp is known in the form of the fluorescent tube. This lamp works via a low pressure mercury discharge that produces UV-photons very efficiently that it then converts to visible light somewhat less efficiently. The Stokes Losses that are incurred in this conversion cause the majority of the energy loss in this lamp. The high pressure discharge lamp is often seen in shop lighting. It works by the direct emission of visible light from a high pressure mercury discharge that has some (usually metal halides) additives in it. The high pressure discharge conditions make for a less efficient discharge. However, the light production efficiency of both these discharges is comparable ($\sim$ 100 lm/W).

Ideally, a discharge lamp would combine the best properties of both of these lamp designs: a high discharge efficiency, like the low pressure discharge lamp has, with the direct production of *visible* light [1]. The desired properties of the radiating species for this new lamp should include relatively high vapor pressures at low temperatures and it should radiate at or very near the visible spectrum to reduce or avoid the Stokes Losses altogether. Indium Bromide [2] [3] is suggested as a model species to study, because it radiates between 360 and 390 nm and has reasonable vapor pressures at low temperatures (510 K). The spectrum of Indium Bromide will be studied to find optimal operating conditions as well as fundamental properties of the molecule. To do this, a laser induced fluorescence experiment is designed and built.

SETUP

A cylindrical tube with a diameter of 3 cm and a length of 30 cm is housed inside an oven that provides a homogenous temperature over the volume of the tube. Axially through the tube, a laser can be shone that provides excitation of the studied molecules. The laser used is either the 4th harmonic of a YAG laser or a high repetition rate (1 kHz) tunable dye laser which is pumped by the 3rd harmonic of a YAG pump laser. The high repetition frequency is necessary to prevent saturation effects, while still being able to measure meaningful fluorescence spectra.

The fluorescence is detected in a perpendicular direction. The fluorescence photons are guided through a monochromator and then detected by either an iCCD camera for spectral resolution or a photomultiplier tube connected to a multichannel scaler for time resolution.

RESULTS

DETEX Plots

In spectrally resolved LIF measurements both the excitation wave number and the detection wave number cover a certain wave number range.

For the interpretation of the spectrally resolved LIF data, the so-called DETEX plots were introduced, an example of which is shown in figure 1. A DETEX plot shows the fluorescence emission color-coded in a graph with the **det**ection wave number on the horizontal axis and the **ex**citation wave number on the vertical axis.

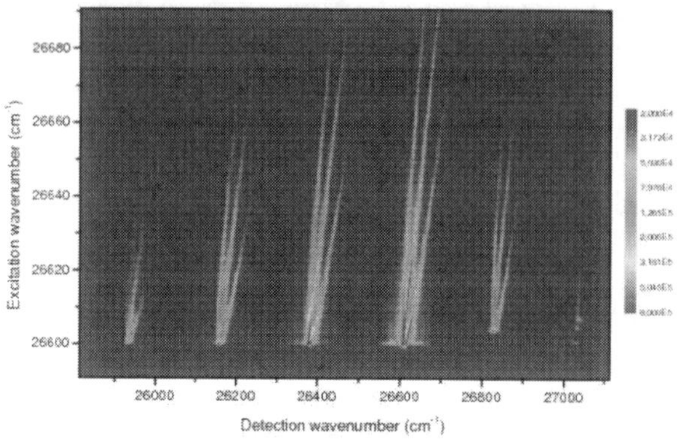

Figure 1. Example of a DETEX *plot. The intensity of the measured spectrum is color coded for each x-y-coördinate.*

These plots are very helpful for the interpretation of rovibronic spectra. The unprecedented detail visible in the DETEX plots enables a better unraveling of spectra than previously possible. This enabled the determination of various spectroscopic constants with greater accuracy.

The DETEX plots helped produce lists of rotational transitions, however the beginning of these lists was generally missing. To still be able to assign the correct rotational quantum number to the observed transition, a method was developed based on the assumption that the unknown rotational constant must be similar to the rotational constant of a level nearby. This method can be useful in the determination of rotational constants for other species too.

Temperature Dependence of LIF

The experiments with the 266 nm excitation showed that collisions between InBr molecules contribute very significantly to the relaxation processes at temperatures typical for the operation of the envisioned lamp. This leads to the conclusion that the molecular component in the radiation produced by an InBr discharge is (strongly) limited by the collisional relaxation processes.

Indium itself is of course also a heavily radiating atom, so the use of Indium (perhaps in the form of InBr) as an additive to a chemical mix for a lamp is not categorically deemed as unwise. However the molecule InBr can be discarded as a candidate to be the primary radiator in an intermediate pressure lamp.

REFERENCES

[1] R. Hilbig, A. Koerber, J. Baier, and R. Scholl. Molecular discharges as light sources. In G. Zissis, editor, *Proceedings of the 10th International Symposium on the Science and Technology of Light Sources*, number 182 in Conference Series, pages 75–84, Bristol and Philadelphia, 2004. IoP.

[2] R. Hilbig, A. Koerber, D. Hayashi, S. Schwan, and R. Scholl. Investigations on low pressure indium mono-halide discharges. In M.Q. Liu and R. Devonshire, editors, *Proceedings of the 11th International Symposium on the Science and Technology of Light Sources*, pages 423–424, Sheffield, 2007. FAST-LS Ltd.

[3] H.C.J. Mulders and W.W. Stoffels. Lif-measurements on a low pressure rf-driven inbr lamp. In M.Q. Liu and R. Devonshire, editors, *Proceedings of the 11th International Symposium on the Science and Technology of Light Sources*, pages 119–120, Sheffield, 2007. FAST-LS Ltd.

High average power VUV and UV excilamps

V. Tarasenko, S. Avdeev, M. Erofeev, M. Lomaev, E. Sosnin, V. Skakun, D. Shitz
High Current Electronics Institute, Akademicheskii ave., 2/3, Tomsk, 634055, Russia
Contact: VFT@loi.hcei.tsc.ru

ABSTRACT

In the present paper, the review of the basic results obtained at the Laboratory of Optical Radiation at High Current Electronics Institute SB RAS during 2008-2010 years is presented. Efficient radiation of Ar_2, Kr_2, Xe_2, KrBr*, KrCl*, XeI*, XeBr*, XeCl*, Cl_2* molecules and I atoms was obtained in rare gas or in rare gas - Br_2 (Cl_2, I_2) mixtures.

GENERAL

Study of radiation parameters and lifetime period of the manufactured barrier discharge excilamp has been performed. Average power of radiation for the just sealed off KrCl ($\lambda \sim 222$ nm) excilamp exceeded 100 W was obtained. The emitter with length of 1200 mm at external diameter 65 mm was tested. Length of a radiating surface has made 1100 mm. The tests were carried out with air cooling system. Maximal radiation power density was of 50 mW/cm^2, that corresponds to excilamp radiant power more than 100 W. UV output power of ~75 W and efficiency up to 10 %, respectively, at $\lambda \sim$ 308 nm (XeCl* excilamp) were obtained under excitation by pulses with frequency of 100 kHz. Application of water cooling allows increasing the radiant emittance of DBD coaxial excilamps. It was noted that in the coaxial quartz tube KrCl-excilamps excited by a barrier discharge, the maximum emission efficiencies are observed for a discharge in the form of diffuse cones. VUV output power of ~120 W at $\lambda \sim 172$ nm was obtained under excitation by pulses with frequency of ~100 kHz.
The lifetime of gas mixtures in small XeCl and KrCl barrier discharge excilamps over 12000 and 8000 hours was demonstrated. There were analyzed the changes in the state of the quartz surface of KrCl-emitters after long life time. At the end of the test, the internal surface of the bulbs was covered with a matt coating comprised by the polymer products of chlorsyloxanes, which is an inevitable process in the excilamps containing molecular chlorine as the halogene carrier. The analysis of the optical characteristics of the modified quartz has shown the following. The transparency of the modified quartz in the UV-range spectrum has dropped. The transparency loss difference between the external and internal tubes indicates that the wall temperature influences significantly on the working mixture degradation. This complies with the assumption that the reaction of the chlorsyloxanes formation is more active in the area with a higher temperature: the temperature of the internal wall of the heated excilamp ~ 150 ^{0}C, while that of the external one is ~ 100 ^{0}C. Therefore, it was drawn the conclusion that in order to operate the excilamp emitters subjected to higher energy contributions, it is necessary to use more intensive internal cooling, e.g. water cooling.
Dynamics of discharge formation in KrCl excilamp has been studied. It has been shown that transition to stationary stage of discharge (in the form of separate microdischarges consisting from cones) takes place within one second as four stages with different discharge forms. Before formation of the stationary fourth stage of discharge, which has the highest power of radiation, the spark stage of discharge is registered during which the bright branchy channels (sparks) are observed. It is supposed that discharge transition from spark to diffuse discharge is determined by formation of runaway electrons and X-ray in the gap.
An efficient low-pressure sealed-off cylindrical excilamp with capacitive discharge excitation is under discussion. Investigation was carried out on the characteristics of XeCl ($\lambda \sim 308$ nm), XeBr ($\lambda \sim 282$ nm), KrCl ($\lambda \sim 222$ nm) and XeI ($\lambda \sim 253$ nm) capacitive discharge excilamps. High efficiency of exciplex molecules and simple design were obtained under capacitive HF discharge excitation. The lifetime of gas mixture in KrCl*, XeBr*, XeCl*, XeI* capacitive discharge excilamps over 3000 hours was demonstrated.
The excilamps of various wavelengths with a flat exit window having an area of >200 cm^2 was development. The photo of the exilamps are shown in Fig.1(in the operational condition) and Fig.2

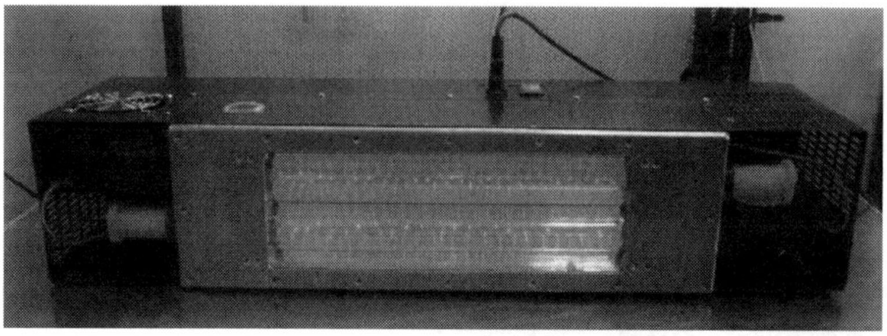

Fig.1. The photo of the KrCl exilamp No.1 in the operational condition.

Fig.2. The photo of the KrCl exilamp No. 2 with seven radiators.

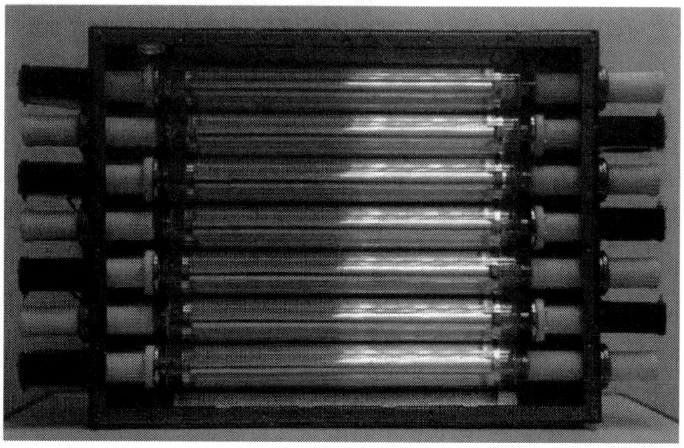

The excilamp No. 1 consists of two cylindrical radiators equipped with reflectors, an exciting generator, and an air cooling system. All the components are placed in 750 × 150 × 210 mm housing. The radiator is made of two GE214 quartz tubes (wall thickness is 2 mm) of different diameters, which are placed coaxially and soldered at their ends. The length of the radiating surface is 315 mm, the diameter of the outer quartz tube is 43 mm, and the diameter of the inner tube is 23 mm. The outer electrode connected to the earth is made of a 150-μm-diameter steel wire wound over the outer quartz tube with a ~1 mm step. 50-μm-thick AlMg-foil reflector is placed at one end of the radiator. While optimizing the operating conditions of the excilamp, we also studied radiators without reflectors. An AlMg-foil high-voltage electrode, to which high-voltage pulses were applied, was inserted into the inner quartz tube. A fan pumping the air through the excilamp's inner tube was used for cooling, and additional fans cooling the excilamp's outer surface from the side of the reflector were also used. The radiators of the excilamps were excited by ±4.2 kV pulses arriving with a 40-kHz repetition rate from an opposite polarity pulse generator. As working molecules, we selected KrCl* (λ ~ 222 nm), Cl_2 (λ ~ 258 nm), and KrBr (λ ~ 207 nm). The obtained average radiation powers in the UV spectrum region behind a flat window are 10.2, 1.55, and 4.15 W, respectively, see Table. The excilamps are intended to uniformly irradiate various objects with areas of up to 200 cm^2 in air and various gaseous media. They are also equipped with a system preventing the ozone generated

Molecule	Power density, mW/cm2	Average power, W	Total efficiency, %
KrCl*	24	10.2	3
Cl2 *	3.6	1.55	0.42
KrBr*	9.7	4.15	1

The excilamp No. 1 consists of seven cylindrical radiators equipped with reflectors, an exciting seven generators, and an air cooling system. As working molecule, we selected KrCl* (λ ~ 222 nm). The obtained average radiation power in the UV spectrum region behind a flat window is ~ 20 W.

This study was supported by the International Science and Technology Center, project no. 3583p.

SEM Investigation of Al_2O_3 Transport Phenomena in Elliptical Ceramic Metal Halide Lamps.

Joseph Gielen[1], Rudy Geens[1]
1 Central Analysis Lab, Havells Sylvania, Industriepark 13, 3300 Tienen, Belgium
Contact: jos.gielen@havells-sylvania.com

1. INTRODUCTION

Ceramic Metal Halide lamps (CMH) contain a diversity of chemical substances. The arc tube itself is made of Poly Crystalline Alumina (PCA), the electrodes are tungsten and the chemical fill in these lamps generally contains sodium iodide, thallium iodide and a mixture of rare earth iodides. At the high lamp operation temperature these substances can form complex interactions, both physical and chemical. For the development of CMH lamps it is necessary to understand these interactions. When a system is at a constant temperature most interactions will tend to equilibrium and a steady state is reached. In real lamps however, there are always temperature gradients due to gravity, geometry and convection. Most of the interactions between different materials are temperature dependant and it is this dependency that will drive material transports between regions with different temperatures.
The material transport processes in Metal Halide lamps are well known and several mechanisms have been proposed to describe the chemical and physical phenomena that occur in these lamps [1, 2]. This paper reports new experimental evidence for the occurrence of these phenomena and provides a better understanding of the physical details.

2. GAS PHASE TRANSPORT

The chemical reaction between Rare earth (Re) iodides and PCA occurs at the hottest part of CMH bulbs and forms AlI_3. This AlI_3 may react with Al_2O_3 at the high temperature side of the bulb.

$$ReI_3 + Al_2O_3 \rightarrow ReAlO_3 + AlI_3 \qquad (1)$$

$$AlI_3 + Al_2O_3 \leftrightarrow 3AlOI \qquad (2)$$

Both AlI_3 and the formed AlOI are in the gas phase at the temperatures inside the arc tube under operational conditions. At the colder part of the bulb the reaction is reversed and Al_2O_3 is deposited there while AlI_3 is reformed. The shift in the equilibrium position of reaction 2 due to the temperature gradient inside the arc tube drives an Al_2O_3 material transport from the hot side of the bulb (upper half) to the colder side (lower half).
The results of reaction (1) were indeed found in the elliptical CMH arc tubes. Both in vertically and horizontally burned lamps there are locations in the upper part of the bulb where the rare earth elements did react with the PCA. Picture 1 shows the location of the formed $ReAlO_3$ crystals in a vertically burned lamp and pictures 2 and 3 show a SEM image and an EDX spectrum of these crystals.

Pict. 1: Location of $ReAlO_3$ crystals.

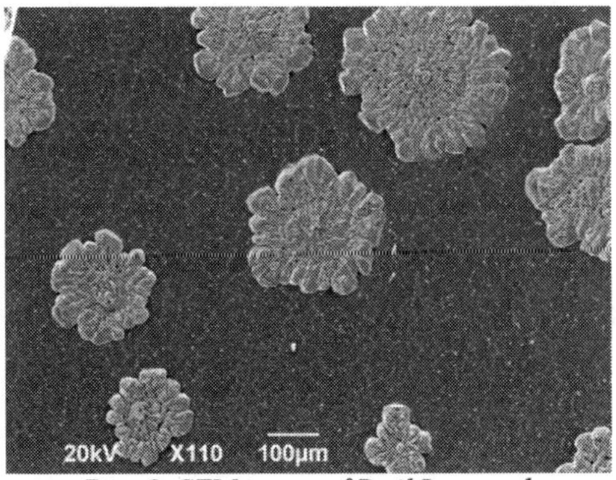

Pict. 2: SEM image of $ReAlO_3$ crystals

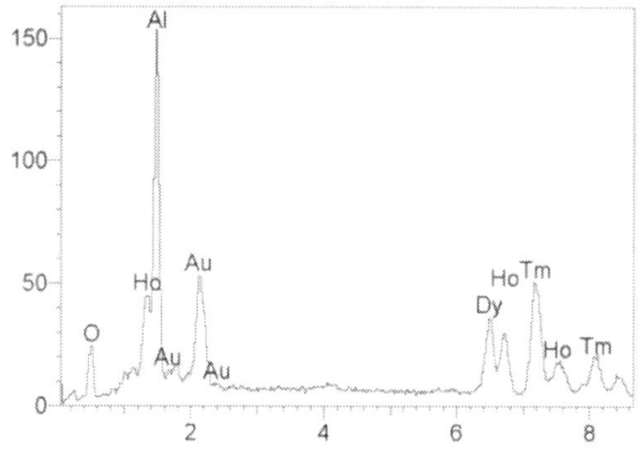

Pict.3: EDX spectrum of the $ReAlO_3$ crystals

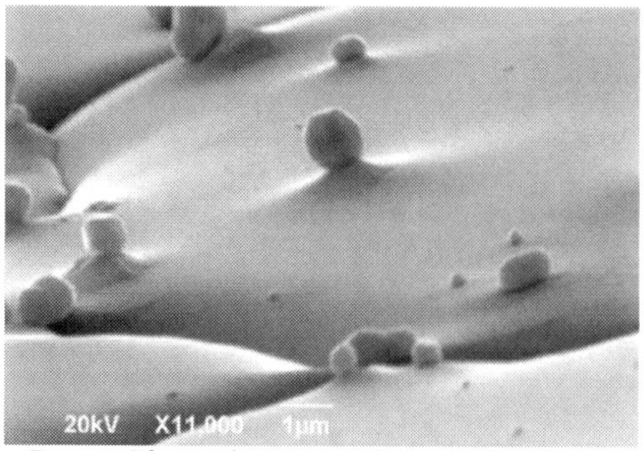

Pict. 4: Elevated W grains due to etching of the Al_2O_3 surface.

The Al_2O_3 transport from the hot to the colder part of the bulb was also observed. At the hot end the PCA corrosion that occurs when Al_2O_3 reacts with AlI_3 causes an etching of the PCA grains. This corrosion does not occur below W grains - originating from the electrodes - that are adhering to the PCA and therefore these W grains are slowly elevated when the rest of the Al_2O_3 is etched away. Picture 4 shows such elevated W grains in a lamp that burned horizontally for 1100 hrs.

In the colder part of the bulb, the deposition of Al_2O_3 on the PCA grains causes some crystal planes to grow faster than others. This leads to overgrowth of some top planes by faster growing side planes. Picture 5 shows an example of Al_2O_3 deposition in a CMH lamp.

This gas phase Al_2O_3 transport was found in all lamps even after only 500 hrs life. This is strange because not all lamps have formed the $ReAlO_3$ crystals. This may indicate that there are other Al_2O_3 transport mechanisms or that the AlI_3 may be formed by other reactions that do not produce $ReAlO_3$ crystals. The amount of Al_2O_3 transport in the gas phase is very low and does not influence the lamp operation.

3. TRANSPORT BY MOLTEN CHEMICAL FILL

A well known Al_2O_3 transport mechanism that can limit the life of CMH lamps occurs due to the temperature dependence of the solubility of Al_2O_3 in the molten chemistry. Droplets of salts in areas with a large wall temperature gradient can produce pin holes in the PCA walls in a time of only a few hundred hours. The transported Al_2O_3 is deposited next to the hole under the colder end of the droplet. Pictures 6 and 7 show the PCA corrosion by molten chemistry in a vertically burned lamp.

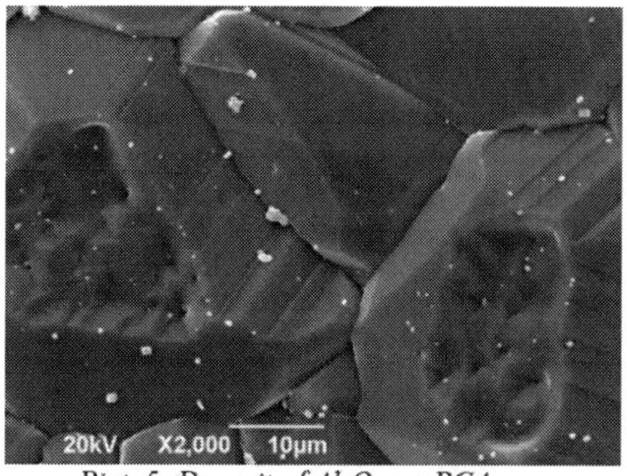

Pict. 5: Deposit of Al_2O_3 on PCA.

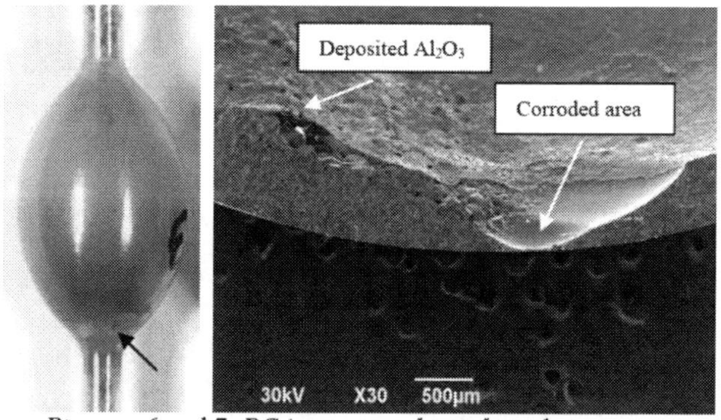

Pictures 6 and 7: PCA corrosion by molten chemistry

4. REFERENCES

[1]: Theo G M M Kappen, "Status quo of ceramic material for metal halide discharge lamps", J. Phys. D: Appl. Phys. 38 (2005) 3033-3039.

[2]: W. van Erk, "Transport processes in metal halide gas discharge lamps", Pure Appl. Chem., Vol. 72, No11, pp. 2159-2166, 2000.

Research on characteristics of the LED lighting using an human body electrical signal

Woo-Young Cheon[1], Gi-Hoon Kim[1], Jin-Hong Kim[1], Young-Woo Kim[1]
[1] *LED/SSL R&DB Sector Solid State Lighting & Display Center, Korea Photonics Technology Institute*
Gwangju 500-460, Korea
wychoen@kopti.re.kr

ABSTRACT

Recently a variety of efforts are conducted to replace the illumination light source with LED. In areas ranging from the interior lighting to the outdoor lighting has been proceeds with various fields. By increasing the efficiency of LED, replacement for traditional light sources that are further acceleration is expected. However, LED lights for the assessment of human responsiveness is not actively being made. In this study, LED lighting, so the impact on the human body to the experiment by measuring the electrical signals to study the difference was between traditional lighting. The experiment to measure the electrical signals of human body using BioPAC's system. the 20-30 male participated in the experiment.

GENERAL

In this study, using LED lighting products for each of the human body in order to evaluate the degree of response to electrical signals were measured.
The measured electrical signals include an electrocardiogram, oxygen saturation, skin temperature, and EEG were measured.
Measured by the 1-2 people each product by selecting the subject human was in progress, along with the measurement of electrical signals by the questionnaire survey was also conducted.
LED light source and the human response to the existing degree of comparison, conventional light sources to measure the electrical signals were measured on the body.
By measuring the frequency analysis of electrocardiogram primarily a result of the frequency of 10Hz or less was part account for the majority.
The rapid changes on an electrocardiogram that indicates there was not visible.
Pulse of the oxygenation of the few visible changes in skin temperature between the two was the total time for a change was gradually reduced.
If the voltage on the eye the entire time was kept in the same potential.
With an existing light on the impact of light EEG electroencephalogram were measured to determine.
EEG is a measure of the frontal and parietal areas were measured by focusing on parts. Positioned in accordance with international conventions on Fp1, Fp2, F3, F4, C3, C4, P3, P4 portion of the EEG was measured.
All measures for EEG time-frequency analysis was carried on.
60Hz noise on the power of a few parts, but overall show the frequency of EEG frequencies in most areas is less than 50Hz. In EEG frequency range divided 4 alpha, beta, theta, delta, gamma and other areas are divided by two is.
Examination of the external disturbance can be excluded as possible Shild Room was in the process.
BioPAC MP150 Data acqusition system's use was an experiment in progress.

[Fig 1. Measurement System]

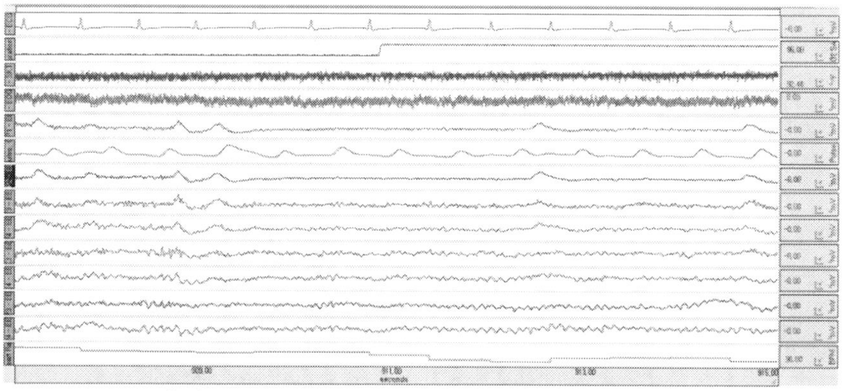

[Fig 2. Measured human body electrical signal]

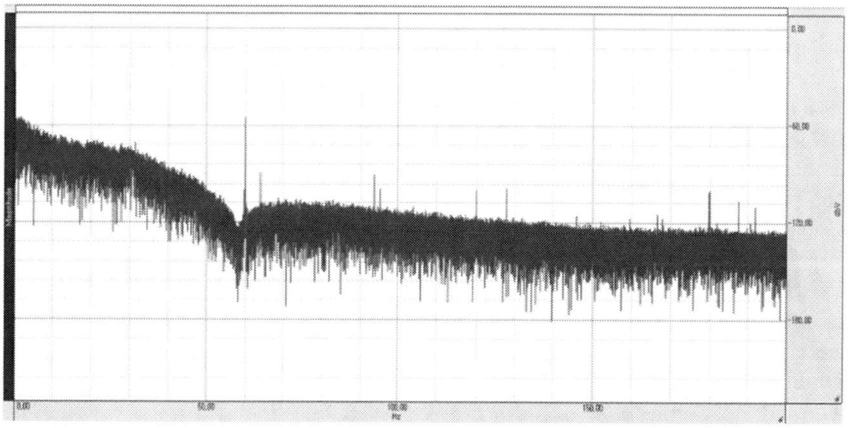

[Fig 3. FFT analysis of measured EEG Signal]

REFERENCES

[1] Ian T. Ferguson, Yoon-Soo Park, Nadarajah Narendran, Steven P.DenBaars "Solid State Lighting and Displays", 31 July-1 August 2001

[2] CVCE 2005 Secretariat, "CVCE 2005 Digest of Technical Papers", December 2005.

White LED Characterization under Variable Current Operation

J. M. Alonso, D. Gacio, L. Campa

Electrical & Electronics Eng. Dept., Universidad de Oviedo, Campus de Viesques, Gijón, 33204, Spain

marcos@uniovi.es

ABSTRACT

The change in the light emitted by white LEDs against current variations is one of the main concerns when designing LED power supplies. In this paper the electrical, thermal and photometrical behaviour of several white LEDs under variable current operation is investigated. It is concluded that LED current variations can be admitted for many cost-effective applications without an excessive penalty in the LED light quality.

INTRODUCTION

The use of white LEDs represents a highly interesting application in general daylight illumination for home and offices. LEDs are inherently monochromatic emitters, but there are several ways to generate white light employing LED technology and different phosphor coatings [1]. One of the main concerns when using LEDs for designing new and retrofit luminaires is the possible change of the LED light characteristics against LED current variations [1][2]. LEDs can be supplied from the input source in different ways. The most typical one is to provide low-ripple dc current to the LED. This makes necessary the use of regulation circuitry to compensate for variations in the input voltage, in LED resistance and threshold voltage, in temperature, etc. In some cost-effective applications, the use of other simpler solutions could be of interest. This usually means to supply the LEDs with a dc current with high ripple, obtained from rectifiers or other low-cost electronic converters. On the other hand, in some applications the dimming of the LED light is requested. It is recommended by some manufacturers to use pulse width modulation (PWM) maintaining a constant LED peak current in order to assure that the LED will behave as stated in the datasheet [3]. However, in cost-effective applications the inclusion of PWM feature increases cost and reduces efficiency, especially for high power applications. In this cases, the use of variable LED current would also simplify the dimming process. This is the motivation for this work in which the behaviour of white LEDs under variable current control is investigated.

LABORATORY TEST BENCH

Fig. 1 illustrates the laboratory test bench used in the experiments. The LED can be supplied with different current levels using a dc voltage source and a series resistance. The LED spectrum and photometric characteristics are measured using an integrating sphere and a spectrometer connected to a personal computer (PC). The software running in the PC calculates all the important parameters. The heatsink temperature is also measured so that the LED junction temperature can be calculated using the LED junction-ambient thermal resistance given by the manufacturer. In the experiments the current through the LED is incremented from zero up to the maximum current in 100 mA steps. The parameters that have been measured are: power (PD, W), junction temperature (TJ, °C), forward voltage (VF, V), correlated colour temperature (CCT, K), colour rendering index (Ra), CIE 1931 chromaticity coordinates (x, y, z), relative lumen output (Flux) and relative luminous efficacy given as relative lumen output over LED power (EF).

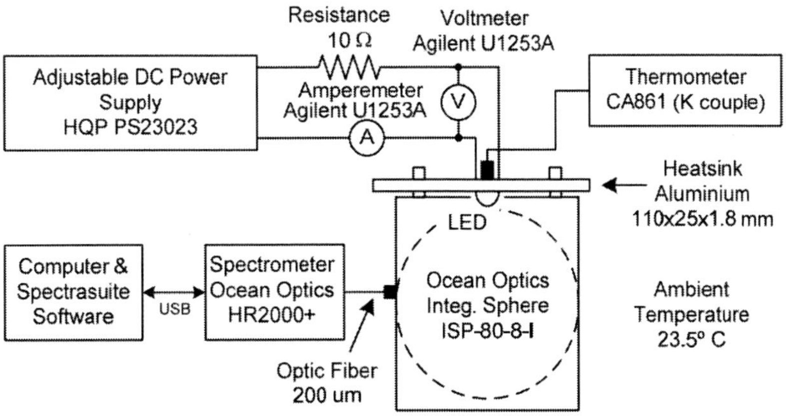

Fig. 1. Laboratory test bench.

RESULTS

Several white LEDs were tested. The results obtained from two of them are shown in Fig. 2.

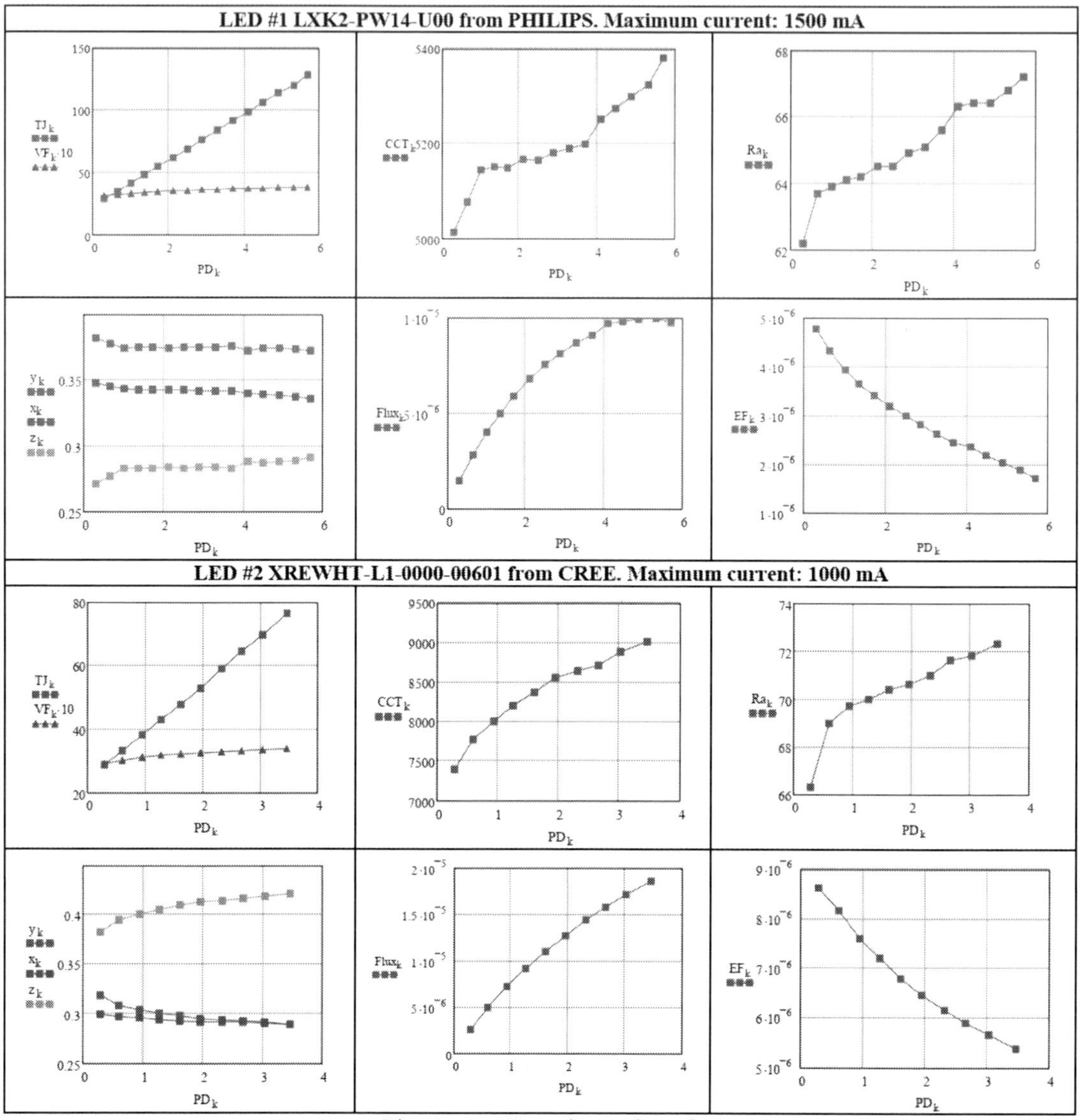

Fig. 2. Experimental results.

CONCLUSIONS

The following conclusions are extracted from the experiments:

- The CCT increases with the LED power; an 8% for LED #1 and 25% for LED #2. Still, in both cases the CCT is high enough to be inside the CIE white light region.
- The CRI Ra increases with the LED power; an 8% for LED #1 and 9% for LED #2. For each diode a similar colour reproduction capability is maintained in the whole range.
- The chromaticity coordinates remains nearly constant for the whole range and especially for the higher power range. The coordinates are well inside the CIE white light region for the whole range.
- The luminous efficacy highly decreases as the power increases. This is also an effect of the higher junction temperature. The variation is -65% for LED #1 and -38% for LED #2.

As final conclusion it can be said that white LEDs can be used under variable current operation without excessive penalty in light performance. Further experiments should be carried out to investigate the aging effects on the LED light under variable current operation.

REFERENCES

[1] E. F. Schubert; "Light Emitting Diodes," Second Edition. Cambridge University Press, 2006.
[2] "LED Luminaire Design Guide," App. Note CLD-AP15.000., Cree, 2007.
[3] "Luxeon K2 Emitter," Technical Datasheet DS51, Philips Lumiled Lighting Company, 2006.
[4] "Cree® XLamp® XR-E LED Data Sheet," Cree Inc., 2007.

The influence of charge effects on the DC breakdown properties of HID lamps

P. Tant, J. Driesen, G. Deconinck
Department of Electrical Engineering ESAT-ELECTA, K.U.Leuven
Kasteelpark Arenberg 10, 3001 Leuven, Belgium
Contact: peter.tant@esat.kuleuven.be

ABSTRACT

This paper describes DC breakdown voltage measurements on a metal halide lamp. The lamp (with short-circuited electrodes) is temporarily elevated to a high potential with respect to ground in between two consecutive experiments. This bias voltage has an influence on the statistical spread and the minimum breakdown voltage, caused by charges deposited on the arc tube wall.

INTRODUCTION

The ignition of high intensity discharge (HID) lamps requires high voltages (often in the kV range), especially when restarting warm lamps. Reducing these ignition voltages is of great interest and widely investigated [1]. Most of this research focuses on pulse-like voltage waveforms of only a few ns or µs long [1-3]. Other research is based on longer-duration AC and DC waveforms, which lead to different breakdown properties and other effects [3]. The breakdown experiments discussed in this paper make use of the test setup described in [4] to produce slowly rising DC voltages, i.e. DC ramps or pulses with a certain ramp rate and polarity. As Fig. 1 shows, breakdown occurs at a certain point in time (t_{BD}). The breakdown voltage V_{BD} is defined as the DC voltage across the lamp just before breakdown takes place, and is statistically distributed with a typical spread of several tens or hundreds of volts. Therefore, 75 samples for V_{BD} are collected in each measurement series, with an interval of Ts = 10 s. As Fig. 1 shows, during the breakdown experiments, one electrode of the lamp is connected to ground, while the other is connected to the DC voltage pulse v(t) with a positive or negative polarity. In between the samples, the entire lamp is elevated to a high common-mode potential with respect to earth: the bias voltage v(t) = V_{BIAS}. Both electrodes are short-circuited during this phase, breakdown does not occur. V_{BIAS} is removed 500 ms before the next sample begins.

MEASUREMENT RESULTS AND DISCUSSION

The presented experiments are performed on a Philips MHN-TD 70 W metal halide lamp, with a ramp rate of 600 V/s, and a variable bias voltage V_{BIAS}. Fig. 2 shows the measured V_{BD} distributions as a function of V_{BIAS} (positive and negative). Each distribution contains 75 measurement results. The boxplots represent the interquartile range (IQR) and the 5% and 95% quantiles (thin line). Clearly, the

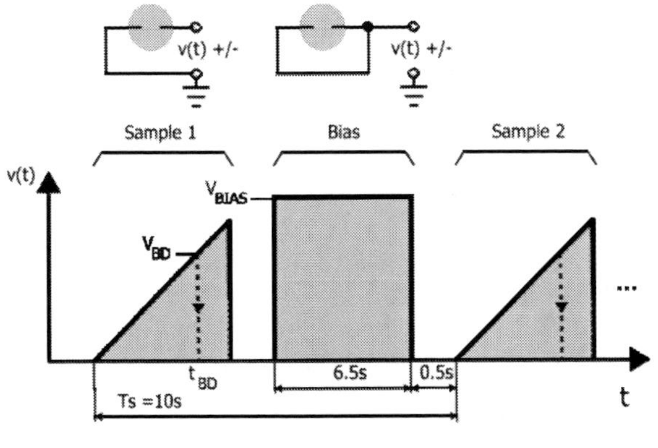

Fig. 1: Description of the breakdown experiments.

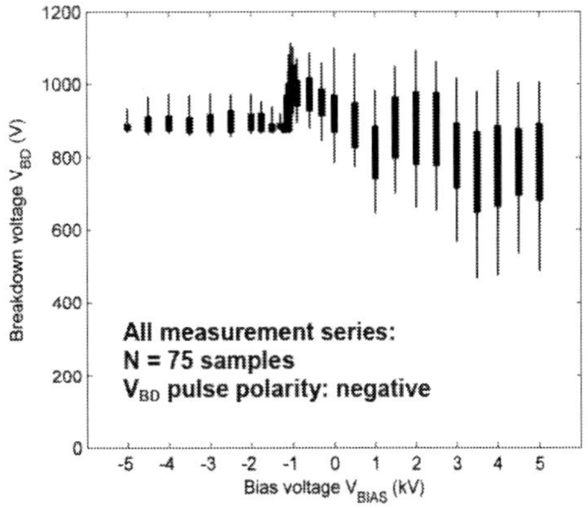

Fig. 2: Distribution of the breakdown voltage as a function of bias voltage.

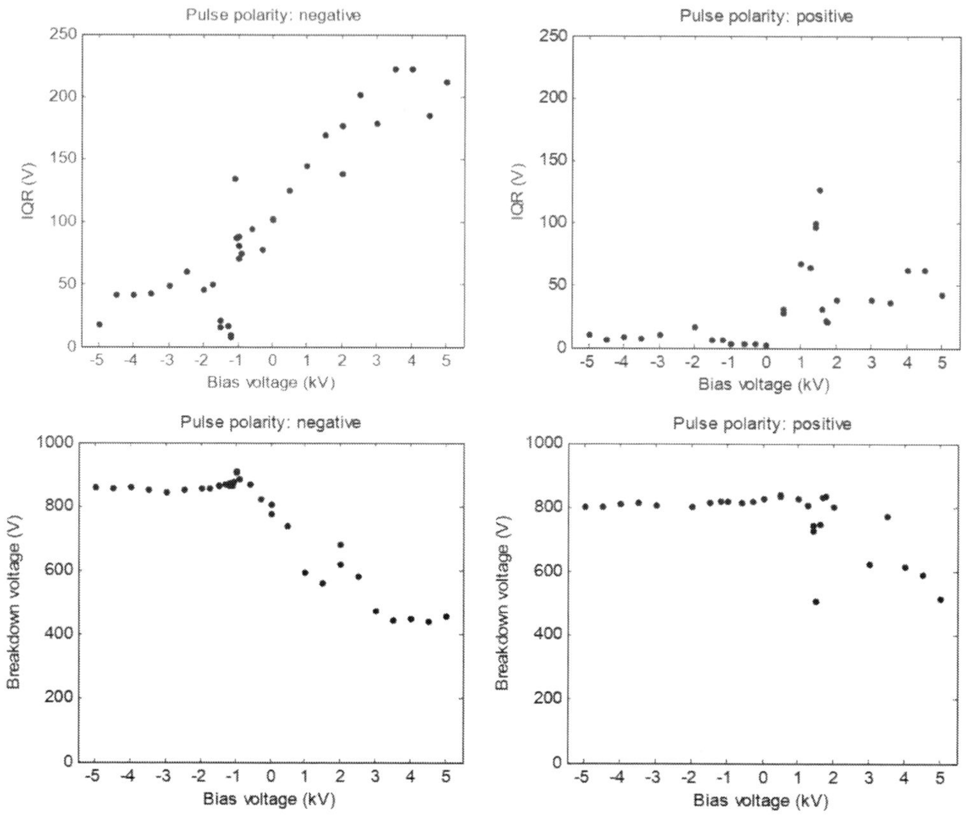

Fig. 3: Interquartile range (IQR) and minimum breakdown voltage as a function of V_{BIAS}.

polarity and the magnitude of V_{BIAS} have a strong influence on the range and the spread (IQR) of the distributions. Fig. 3 shows the IQR and the minimum breakdown voltage as a function of V_{BIAS}, for both negative (left) and positive (right) pulse polarities. Generally, a higher or more positive bias voltage leads to a higher IQR, and a lower minimum breakdown voltage. For both pulse polarities, a negative bias voltage results in a low and constant IQR, and a high and constant minimum V_{BD}. Discontinuities appear in the IQR when V_{BIAS} is around +1 kV and –1 kV. The minimum breakdown voltage then starts decreasing. It is believed that charge carriers are deposited on the arc tube wall during the bias phase, and later influence the electric field at the cathode during the measurement phase. A positive charge ($V_{BIAS} > 0$) enhances the field at the cathode, leading to a higher spread and a reduced minimum breakdown voltage. The effects are less pronounced for positive pulse voltages (the right figures), since the cathode is then grounded instead of connected to v(t) (increasing magnitude).

ACKNOWLEDGEMENTS

The research performed at the K.U.Leuven is financially supported by the Research Foundation - Flanders (FWO). Peter Tant has a PhD fellowship of the Research Foundation - Flanders (FWO).

REFERENCES

[1] B. Lay, R.S. Moss, S. Rauf, M.J. Kushner, "Breakdown processes in metal halide lamps," *Plasma Sources Sci. Technol.* **12** (2003), pp.: 8–21

[2] M. Wendt, S. Peters, D. Loffhagen, A. Kloss, M. Kettlitz, "Breakdown characteristics of high pressure xenon lamps," *J. Phys. D: Appl. Phys.* **42** (2009), 12p.

[3] J. Beckers, F. Manders, P.C.H. Aben, W.W. Stoffels, M. Haverlag, "Pulse, dc and ac breakdown in high pressure gas discharge lamps," *J. Phys. D: Appl. Phys.* **41** (2008), 6p.

[4] P. Tant, B. Vanbrabant, J. Driesen, G. Deconinck, "A variable frequency high-voltage power supply for hot-restrike modelling of HID lamps," *12 Eur. Conf. Pow. Electr. and App.*, 2007, 8p.

Study of atomic state distribution functions in non-thermal high-frequency electrodeless discharge plasmas

N. Denisova[1] , G. Revalde[2] , A. Skudra[2]
[1] *Institute of Theoretical and Applied Mechanics, Novosibirsk, Russia*
[2] *Institute of Atomic Physics and Spectroscopy, University of Latvia, Riga, Latvia*
Contact:denisova@itam.nsc.ru

ABSTRACT

This work presents theoretical and experimental study of the atomic state distribution function (ASDF) in low-pressure non-thermal plasma sources using the quasi steady collision-radiative model. The detailed analysis of the ASDF in high-frequency (HF) electrodeless discharges has been performed using self-consistent numerical model. Calculations have been made for Ar, Kr, He and Ar+Hg plasmas. The intensities of the Ar, Kr, He and Hg spectral lines are measured at a wide range of gas pressures and HF generator current. The detailed comparative analysis of the experimental and theoretical curves has been performed.

THEORY

The atomic state distribution function (ASTF) contains very important information on plasma processes.
Elementary processes which govern the excitation kinetics are considered:

1. Ionization and excitation/de-excitation by electron collisions
2. Ionization by heavy particles collisions
3. Penning ionization
4. Radiative transitions

The population density balance equations describing ASTF in a steady state approximation are written in the general form as follows:

$$n_m\{\sum_{k\neq m} W_{mk}n_e + WI_m n_e + \sum_{k<m} A_{mk}\theta_{mk}\} + W_{heavy}n^* n^* + W_{Penning}n^* n - \\ -\{\sum_{k\neq m} W_{km}n_k n_e + \sum_{k>m} A_{km}\theta_{km}n_k\} = 0 \quad , \qquad (1)$$

The Einstein coefficients A_{mk}, and electron collision rate coefficients for excitation/de-excitation and ionization processes, W_{mk} and WI_m, were calculated. θ_{mk} is the escape factor which takes into account the effect of radiation imprisonment and is mainly dependent on the optical thickness. Heavy particles collision ionization $W_{heavy}n^* n^*$ and Penning ionization $W_{Penning}n^* n$ are included into the population density balance equations for Ar+Hg plasma.
The problem to limit the number of atomic levels that should be taken into consideration is very important in calculation of ASDF. The simple physical criterion we tested was:

$$Wn_e > A\,. \qquad (2)$$

Collisional depopulation processes should be predominate if compare to radiation processes.
Theoretical estimations have been predicted that the densities of the upper excited levels should be at first increase, reach maximum and then decrease with pressures.

EXPERIMENTAL

The geometry and design of high frequency electrodeless lamps used in our research was described in details in [1]. The lamps were produced by Laboratory of high-resolution spectroscopy and light source technology at the Institute of Atomic Physics and Spectroscopy by the special technology developed during quite a long time period. The lamps were made from quartz of a cylindrical and

spherical form of 20 mm diameter. For the Ar+Hg lamps the side arm, serving as a "cold spot", was thermostated to control the pressure of the mercury vapor.
The lamp bulb was inserted into the HF generator coil to induce an inductively coupled discharge. The HF generator of approximately 100 MHz frequency was used in experiments for excitation of the discharge. The generator power was changed by varying the current from 40 to 200 mA. For measurements of the spectral line intensities, the radiation, emitted from the lamp, was passed through a monochromator to a photomultiplier and registered by a photon counter.

NUMERICAL MODELING

The developed HF electrodeless discharge model combines the module of plasma parameters, the module of the electromagnetic field and the module of the emitting spectral lines intensities [2].
In numerical calculation the line intensities are presented in the following form:

$$I(\vec{r}) = \int_{\nu} \varepsilon_{\nu} d\nu \frac{\hbar \nu A}{4\pi} \int_{l} n^{*}(l) dl , \qquad (3)$$

where I is the line intensity, A the Einstein coefficient for the radiation transition, n^{*} the value of the upper level population, ε_{ν} spectral line profile. Integral is taken along the line of observation l.

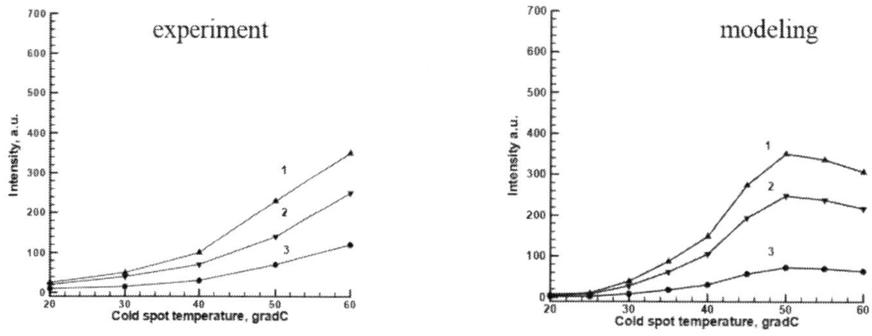

Figure 1 *Dependence of spectral lines intensities in Ar+Hg HF electrodeless discharge versus mercury pressure. Argon pressure 2Torr.*
Curve 1 – Hg line 435.8nm, 2- 546.1nm, 3 – 404.7nm.

We can see from the pictures, that good agreement is observed between experimental data and model in the interval of cold spot temperatures 20° C- 50° C. Then, numerical model demonstrates slow decreasing of intensity. This decrease is in good agreement with theory, but, most likely, it should be at more high value of cold spot temperature.

ACKNOWLEDGMENT: The work was supported by ESF project Nr. 2009/0210/1DP/1.1.1.2.0/09/APIA/VIAA/100.

REFERENCES

[1] Revalde G, Skudra A 1998 *J.Phys. D:Appl. Phys.* **31** 3343-3348.
[2] N.Denisova, G.Revalde, A.Skudra, G.Zissis, N.Zorina *J.Phys.D: Appl.Phys.* 2005, **38** 3275-3284.

The Energy Balance of High Intensity Discharge Lamps

A.J. Rijke[1], S.Nijdam[1], M. Haverlag[1,2], J.J.A.M. van der Mullen[1]

[1]*Department of Physics, Eindhoven University of Technology, P.O.Box 513, 5600MB Eindhoven, The Netherlands*

[2]*LightLabs, Philips Lighting, P.O. Box 80020, 5600 JM Eindhoven, The Netherlands*

Contact: A.J.Rijke@tue.nl

ABSTRACT

We want to determine the energy balance of an HID lamp, by mapping the input energy and the energy flows to the total output energy of the lamp. For such a mapping, we need to determine the total spectral radiance. Therefore we will analyze absolute calibrated integrating sphere measurements of the complete spectrum. The current focus of the project is the infrared range of the spectrum. New in this work is that absolute calibrated integrating sphere measurements will be employed in the infrared part of the spectrum. Calibrating an integrating sphere in this wavelength range will be discussed.

INTRODUCTION

Currently lighting consumes about 20% of the worldwide electricity use. High intensity discharge lamps (HID lamps) are one of the most important light sources worldwide. They are used when a small lamp is needed with a high efficacy.

All energy flows inside the lamp are mapped to determine the energy balance. A big loss factor in HID lamps is infrared radiation. An integral measurement over the infrared spectrum can help to understand the loss processes. From the details of the spectrum the contribution of atomic, molecular, continuum and blackbody radiation can be separated. With this knowledge lamp improvements can be made.

ENERGY BALANCE

The energy balance is a mapping of all energy flows inside the lamp. From the total spectral radiance the energy balance will be reconstructed. To support the division of the radiated power amongst the different processes and aid in understanding the lamp, simulations, and electrode models and also side-on measurements will be used. Absolute calibrated integrating sphere measurement of the complete spectrum will be analyzed.

INFRARED MEASUREMENTS

The current focus of the project is on the infrared range of the spectrum. This part of the spectrum is interesting because a lot of the loss processes appear here. Also the largest losses are found in the IR.

As far as we know, other authors studying the infrared radiation of HID lamps have based their work on side-on measurements, and/or have calibrated their spectra assuming that the burner wall produces black body radiation and/or do not measure the radiation beyond 2.5 μm spectrally resolved. [1, 2, 3].

New in this work is that absolute calibrated integrating sphere measurements will be used in the infrared part of the spectrum up to 10 μm. This is done to include the spectral range where zero or unity transmission of the burner wall can not be assumed. We will extrapolate the emission of the burner wall at higher wavelengths, under the assumption that it follows the blackbody curve.

SETUP

The integrating sphere setup used for this experiment has a roughly spherical wall with an inner diameter of about 28 cm. The wall is made of a 5 mm thick copper plate, with water cooling on the outside to achieve temperature stability. The inside of the sphere is pearl blasted and subsequently coated with aluminium to obtain a diffuse reflection in a broad spectral range. For infrared measurements gold provides a higher reflectivity, but the aluminium coating is chosen so that also measurements in the visible part of the spectrum can also be performed. Light leaves the sphere though a hole in the wall and is then send into a Fourier transform spectrometer (a Bruker Vertex80 vacuum system) by using two parabolic mirrors. A baffle is placed between the exit hole and the lamp to prevent direct light from the lamp entering the spectrometer.

CALIBRATION OF THE INTEGRATING SPHERE

A good calibration is a key element for the success of the measurements and therefore for the accuracy of the energy balance. Particularly in the IR this is a challenge, because nearly all standard calibration sources are either not designed for the use in the mid-IR, or are only suitable for side-on measurements.
To perform the calibration of the integrating sphere in the IR a calibration procedure is under development. Key element in the procedure is resistive heating a metallic strip inside the integrating sphere. If the temperature profile of the strip and its emissivity as a function of temperature, wavelength and emission angle are known, the strip can be used to calibrate the integrating sphere.
Important properties of the source are stability and reproducibility. However, for a lot of materials the emissivity degrades over time due to oxidation. To prevent this, for standard lamps such as the tungsten ribbon lamp [4] the strip is put inside a glass envelope free of oxidizing gasses. For an integrating sphere at the wavelengths important in this research, putting the ribbon inside an envelope is not an option. For example tungsten oxidizes too rapidly, for the vacuum circumstances practically obtainable in an integrating sphere setup, and is therefore not a suitable material.

MATERIALS UNDER INVESTIGATION

Two types of materials have been investigated. First we have searched for materials that have a protective oxide layer, such as a layer of aluminum oxide. A material currently under attention is Kanthal-A, an alloy that consist of iron, chromium and aluminium. An advantage of this type of material is that its protective oxide layer prevents further oxidation; therefore a good stability can be expected. A disadvantage is that the oxide layer may get damaged after a heating cycle [5]. This has a negative effect on reproducibility.
The second type of materials under investigation is noble metals. Currently we are investigating the properties of platinum. The corrosion resistance of noble metals prevents surface changes due to oxidation. Compared to other noble metals platinum has a good combination of a relatively high electrical resistivity with a relatively low thermal conductivity which aids in heating the strip with electric current. The results obtained so far show that platinum has a good stability.
A disadvantage of platinum is that not enough emissivity values applicable for our situation, in the spectral and temperature range of interest, are given in literature (see e.g. [6] for an overview) to have an accurate determination of the total spectral emission of the platinum strip. To tackle this issue, the emissivity of platinum will be measured for more temperatures and wavelengths ourselves.

REFERENCES

[1] Program on Technology Innovation: Advanced Light Source Research: ALITE-II: Losses in High Intensity Discharge Light Sources. EPRI, Palo Alto, CA, Philips Lighting Company, Somerset, NJ, General Electric Company, Niskayuna, NY, Osram Sylvania. Beverly, MA: 2006. 1012994. page 4-9.

[2] D.J. Smith, G A Bonvallet and J E Lawler. Infrared losses from a Na/Sc metal-halide high intensity discharge arc lamp. Journal of Physics D: Applied Physics 36 (2003), pages 1519–1528

[3] D.O. Wharmby. Energy balance of high-pressure sodium discharges under controlled vapour conditions. Journal of Physics D: Applied Physics 17 (1984) pages 367-378.

[4] J.C. de Vos. The emissivity of tungsten ribbon : the tungsten striplamp as a standard source of radiation. Vrije Universiteit Amsterdam, Amsterdam, 1953

[5] V.K. Tolpygo, D.R. Clarke. Spalling failure of a-alumina films grown by oxidation: I. Dependence on cooling rate and metal thickness. Materials Science and Engineering A278 (2000) pages 142–150.

[6] Y.S. Touloukian, David P. DeWitt. Thermophysical properties of matter part 7: Thermal radiative properties : metallic elements and alloys. S.l. : IFI/Plenum, New York, 1970.

Measurement of 254 nm radiant flux in high loaded T5 low pressure mercury discharge lamps with Keitz method

Yuan Zhang[1, 2] and Shanduan Zhang[1]

[1] *Institute for Electric Light Sources, Fudan University, 220 Handan Road, Shanghai 200433, People's R China*

[2] *LAPLACE (Laboratoire Plasma et Conversion d'Energie), Universite Paul Sabatier - Toulouse III, Build. 3R2, 118 rte de Narbonne, F-31062 Toulouse Cedex 9, FRANCE*

Contact: sdzhang@fudan.edu.cn

ABSTRACT

The 254 nm irradiance was measured at some certain distances from a high loaded T5 low pressure mercury discharge lamp for cold temperatures 20-80 °C and currents 0.6-2.0 A. Keitz formula was used to calculate the radiant flux, which has only small deviation from gonio-radiometry. The difference among formulas for linear lamp, cylinder lamp and Keitz is discussed. The results show the optimum radiant efficiency of 254 nm line can be achieved within cold spot temperature 40-50 °C for high loaded T5 lamps.

INTRODUCTION

Low pressure mercury lamp (LPM) is a well-known efficient UV light source at wavelength 254 nm, which is widely used in disinfection and sterilization [1]. The 254 nm radiant flux of LPM is an important parameter for various applications. Nowadays two equipments are usually applied to radiant flux measurement: (i) integrating sphere and (ii) gonio-radiometer. The former is convenient but its accuracy is affected by the quality of its inner surface coating and lamp's absorption to reflecting UV. It is not recommended when the linear lamp is long. The latter is accurate but more complicated.

Keitz method [2] is convenient and accurate enough for the linear tubular LPM. It is proposed by the main supplier of the world [3]. In this paper Keitz method is used to achieve the 254 nm radiant flux of LPM (T5 with length 0.8 m and power 125 W) and compared with gonio-radiometry to estimate the accuracy. Furthermore, the 254nm radiant flux at different cold spot temperature and current is also measured and discussed.

COMPARISION BETWEEN KEITZ FORMULAR AND GONIO-RADIOMETRY

The sphere surface is divided into n belts with equal area in gonio-radiometry. Given the radiation of linear tube is cylindrical symmetry, the total radiant flux Φ should be:

$$\Phi = \frac{4\pi r^2}{n}\sum_{i=1}^{n} E_i \qquad (1)$$

where E_i is the irradiance at the middle of each sphere belt, r is the radius of imaginary sphere and also the distance of measurement. In the experiment n = 30. If the linear tube is Lambert source, the radiant flux can be calculated with Keitz formula [2]

$$\Phi = \frac{2\pi^2 DL}{2\alpha + \sin 2\alpha} E , \qquad (2)$$

where E is the irradiance, L is the length of the lighting parts of the tube, D is the distance of measurement, α is half of the spread angle from the detector to the lighting part in radian. The detector, whose normal is vertical to the tube, should exactly face the middle of the lighting parts.

Other two simplified formulas are also compared with gonio-radiometry: Line light source formula $\Phi = \pi^2 D^2 E$, and cylindrical light source formula $\Phi = 2\pi DLE$.

Tab. 1 shows that when the measuring distance varies from 1.5 m to 3.0 m ($2L \le D \le 4L$) the differences between results of Keitz formula and gonio-radiometry are lower than 1%. As a comparison, it has a poor accuracy in short D for linear light source formula and in long D for cylinder light source formula. If the distance is larger than $4L$, the linear light source formula would also be acceptable, because the diameter of the tube could be ignorable compared with the distance.

Tab.1 Irradiance, radiant flux and their deviation to gonio-radiometry for an amalgam T5 LPM at power 125 W

D (m)	E_n (μW cm^{-2})	Radiant flux (W)				Deviation from gonio-radiometry		
		gonio-radiometry	Keitz	Linear	Cylinder	Keitz	Linear	Cylinder
0.50	1202	39.29	40.47	29.66	29.68	3.0%	-24.5%	-24.5%
1.00	380.0	40.74	41.24	37.50	18.77	1.9%	-7.3%	-53.6%
1.50	189.9	43.69	44.07	42.17	14.07	0.9%	-3.5%	-67.8%
2.00	115.5	46.14	46.56	45.40	11.36	0.9%	-1.6%	-75.4%
2.50	75.87	47.86	47.57	46.80	9.370	-0.6%	-2.2%	-80.4%
3.00	52.43	46.72	47.10	46.57	7.770	0.8%	-0.3%	-83.4%

COLD SPOT TEMPERATURE DEPENDENCE OF RADIANT EFFICIENCY

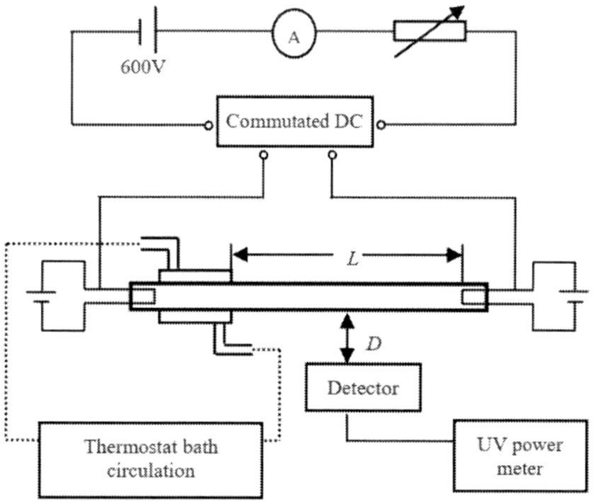

Fig. 1 Schematic configuration of experiment setup.

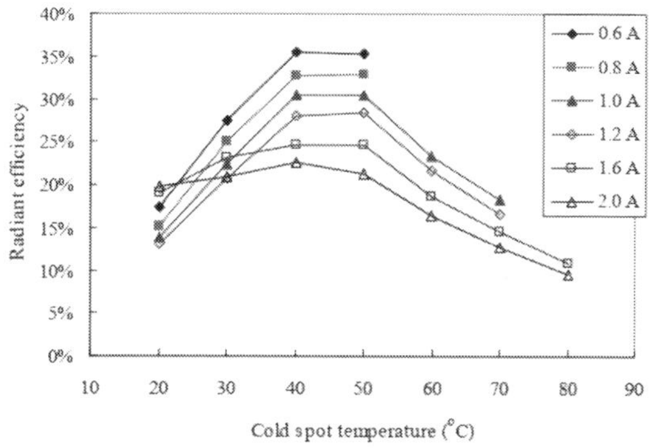

Fig. 2 cold spot temperature dependence of 254 nm radiant efficiency at various current.

Figure 1 shows the experiment setup. A LPM with liquid mercury dosing is driven by commutated DC, which is inverted from DC. An adjustable resistance (0-300 ohm) works as a ballast to control the lamp current within 0.6-2.0 A. The filaments are heated by two DC power supply. A water jacket covering one end of the tube controls the cold spot temperatures of the lamp. A UV power meter (Hamamatsu C8026) and a UV detector (Hamamatsu H8025-254) are used to measure the irradiance of 254 nm line. Keitz formula is used to calculate the radiant flux with $D = 3L$.

The results are shown in Figure 2. As cold spot temperature increases, the 254 nm radiant efficiency has a maximum within 40-50 °C, resulting from the competition between the parallel increase in density of radiative states and radiation trapping.

Figure 2 also shows the 254 nm radiant efficiency decrease with current. The reason is the losses of elastic collision and non-UV radiation increase, while the 254 nm radiation has a flat trend and lamp power has almost a linear relationship with current as the current increases. It means increasing the current only is not an efficient method for high loaded LPM. The filling gas species and pressure should be considered carefully.

REFERENCES

[1] Matsunaga T, Hieda K and Nikaid S 1991 *Photochem. Photobiol.* **54** 403

[2] Keitz H 1971 Light Calculations and Measurements (London: Macmillan)

[3] Lawal O *et al* 2008 *IUVA News* **10** 14

A Cost-effective Design for Better Performance of CFLi in the Downlight

Jingzhou Wang[1]

[1] *GBU CFLi, Philips Lighting, #9 Lane 888 Tianlin Road, Shanghai,200233, P.R.C*

Contact: jingzhou.wang@philips.com)

INTRODUCTION

Thanks to much less energy consumption, and longer life, nowadays more and more consumers replace traditional incandescent bulbs with energy saving lamps, such as integrated compact fluorescent lamps (CFLi). It is not uncommon to see that CFLi is widely used in the downlight and enclosed fixtures, although this application is not suggested by most CFLi manufacturers. The extra temperature loads on the electronic components inevitably reduce the lifetime. The field conditions are preferably taken into account during the design phase, and thus a proper design is expected to lower failure rate.

SELF-EXCITED CIRCUIT ANALYSIS

In investigation, the failed components caused by high temperature are the main failure modes. So for a more robust design, the temperature of the components in the circuit must be well controlled. To limit the temperature rise, the primary choice is to decrease the input power in high temperature environments and consequently decreasing the burner power.

The most widely used topology in CFLi is voltage feedback self-excited half bridge using bipolar transistors, driven by a toroid with three coupled windings. The behaviour of this topology as a relay control system is well known [1]. However, taking the effects of saturated core and the storage time of the BJT into account, analysis will become difficult. According to the reference paper [2], as shown in Fig. 1, the switching period can be divided into two main time intervals: (a) the time interval of the core operated in non-saturation region, and (b) the time interval during the storage time of a power BJT. To simplify the analysis, the inductor current is substituted with a triangle waveform as illustrated in Fig.2, and the equations are rewritten as below:

$$\frac{I_{sat}}{N_p \times dI} + \frac{I_{pk}}{dI} = t_1 \qquad (1)$$

$$\tau_e \times \ln\left(1 + \frac{N_p \times N_s \times L_m \times dI}{V_{be(sat)}}\right) = T_s \qquad (2)$$

Where I_{sat} is the saturation current of the toroid, N_p and N_s are the primary and secondary turns of the toroid, L_m is the inductance factor of the toroid, $V_{be(sat)}$ is the base-emitter saturation voltage of the transistor, τ_e is the minority carrier life time in the base, T_s is the storage time of the BJT, dI is the slope of the triangle current, I_{pk} is the triangle peak current.

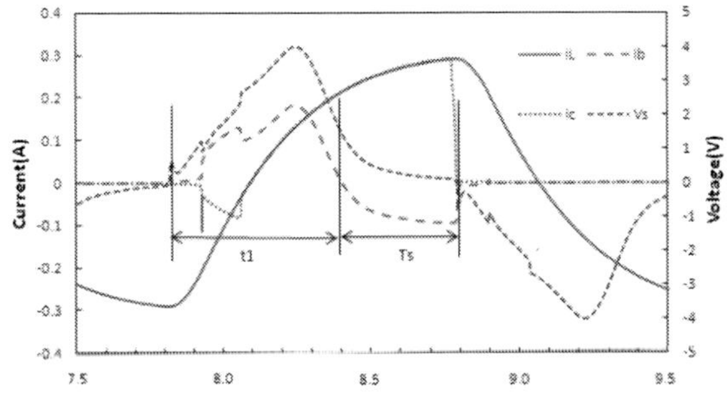

Fig.1 Practical waveforms

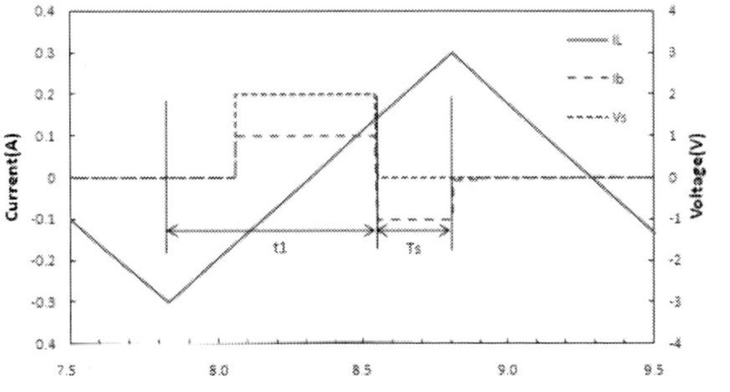

Fig.2 Substitute triangle current for inductor current

TEMPERATURE CHARACTERISTIC OF FERRITE MATERIAL

To reach the primary goal of decreasing the power in a high temperature environment, utilizing the temperature characteristics of the toroid, i.e. the ferrite material, is a cost-effective approach. Considering the operation mode of the toroid, which normally work in deep saturation, a combination of parameters of the ferrite material affects the frequency and consequently the input power. Based on the analysis above, two parameters in a supplier's specification are critical, namely amplitude permeability μ_a and saturation magnetic flux density B_s. Two types of Mn-Zn soft ferrite materials, PC40 and HS52 (TDK) are in common use for CFLi circuits.

The initial permeability of PC40 is low while the Curie temperature is high. As shown in Fig.3, when excited with a sine wave current, the induced voltage drops and saturation current stays nearly the same with rising temperature, shorting the storage time of the BJT according to the analysis above. However, the frequency does not change too much, as the storage time will increase when the temperature of the BJT goes up. On the other hand, the initial permeability of HS52 is high and Curie temperature is low. As illustrated in Fig.3, the induced voltage stays nearly the same but saturation current drops, thus the driving pulse width will short, and the frequency will increase much more. Table.1 shows the components temperature comparison resulting for the two different materials when used in a real application.

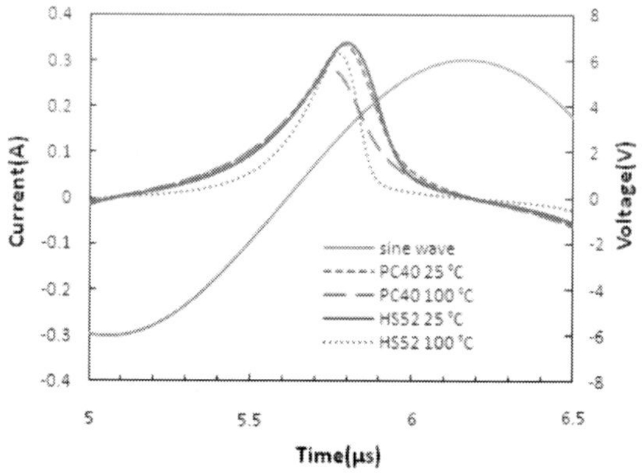

Fig.3 Temperature characteristics of PC40 and HS52

To well control the temperature of the components, obviously it is better to use the high initial permeability material, but the Curie temperature for this kind of material is usually lower than 150°C. To avoid the lamp extinguishing at high temperature, two toroids in series with different materials can be used as illustrated in Fig.4. T1 is using PC40 type material whereas HS52 is used for T2. If temperature rises above the Curie point of T2, T1 is still functional to keep the lamp burning under lower wattages.

A new material combining the advantage of the two different materials is now available, having the temperature character of HS52 but much higher Curie temperature. It has been demonstrated to be suitable for the design for the downlight application.

Tornador 23W	T1	T2	TOR	L	C+	P_in(W)	Fop(kHz)
PC40							
250V/25C	106.1	102.1	93.3	106.0	71.1	25.4	50.3
Down light	**158.9**	**160.5**	**149.1**	**160.1**	**112.3**	**23.7**	**50.1**
HS52							
250V/25C	104.0	100.3	92.6	103.8	79.7	24.6	56.7
Down light	**141.4**	**140.7**	**132.9**	**140.0**	**102.6**	**19.8**	59.4

Table.1 Temperature comparison

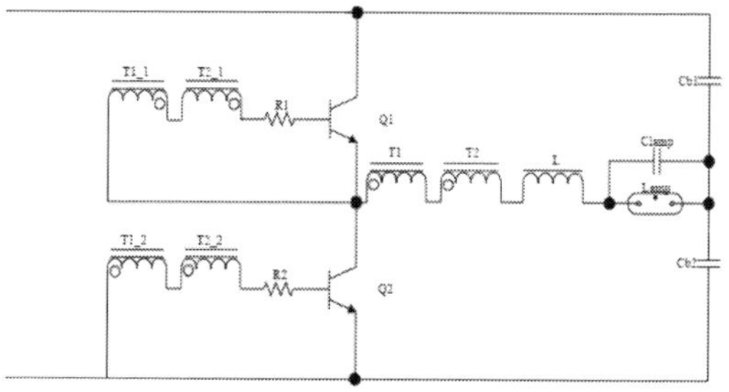

Fig.4 Combination of using both of the ferrite material

REFERENCES

[1] R. N. do Prado, A. R. Seidel, F. E. Bisogno and R. K. Pavão, "Self-Oscillating Electronic Ballast Design Based on Point of View of Control System", IAS2001, record.

[2] T.-H. Yu, H.-M. Huang, T.-F. Wu, "Self Excited Half-Bridge Series Resonant Parallel Loaded Fluorescent Lamp Electronic Ballasts". APEC '95. Conference Proceedings 1995. Tenth Annual, 5-9 Mar 1995 Page(s):657 - 664 vol.2

Design of FR4 PCB with via-holes for improvement thermal resistance and thermal flow in High power LED

S.I. Lee[1], I.T. Shin[1], J.K. Yang[1], D.H.Park[1]
[1] *Department of Information and communication, wonkwang university, 344-2, shin yong dong, Iksan city, 570-749, South Korea*
Contact: player231@naver.com

ABSTRACT

In this paper, we confirmed the thermal & optical properties for improving the heat transfer coefficient by changing the via hole Size in FR4 PCB with the same area.
First of all, to measure the thermal resistance from PCB configuration, the K-factor was measured , which is relative constant between LED junction temperature and forward bias. And then, while 1mA of detecting current applied, K-factor of the LED Package was measured in the constant temperature chamber by changing the temperature from 25 to 85 [℃] with 15 [℃] interval. Also junction temperature by the time was confirmed from measuring △TSP.
As results, thermal resistance in metal PCB came out to the lowest as 26 [℃/W] and thermal resistance in FR4 PCB without via-holes emerged as the highest as 69 [℃/W].
Moreover, when comparing to the initial value, luminous flux in metal PCB was reduced to 5 [%] and luminous flux in FR4 PCB was reduced to 13 [%] similar to thermal resistance results.

INTRODUCTION

Recently, the interest about the loss of luminous efficiency and degradation of chip is increasing according to the increment of driving current for LED lighting.
Therefore, to solve the heat problem, high price metal PCB with high thermal conductivity used and the heatsink is using for smooth heat emission. However, merit of LED is fading from increment of price and huge dimension.
In this paper, the thermal and optical properties of LED was analyzed with formation of via-holes in FR4 PCB and Metal PCB for decrement of LED lighting and improvement of thermal flow.

EXPERIMENTS AND RESULTS

First of all, we have make the metal and the FR4 PCB of 50×50[mm] size with 0.3mm, 0.6mm 0.9mm, 1.2mm via-holes, and copper of the same quantity was used in via-hole of PCB.
Sensing current of 1mA was applied in LED package for measuring the K-factor in constant temperature chamber from 25 [℃] to 85 [℃] with the interval 15 [℃], and then we measured the forward voltage of LED. Also, △TSP, which is voltage gap according to the temperature in sensing current was measured with Keithley 2430 such as Fig. 1, while the driving current of 350 [mA] and sensing current of 1 [mA] was applying.

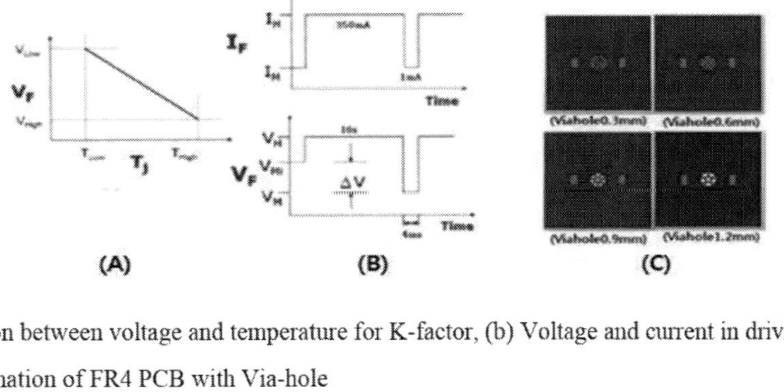

Fig.1 (a) Relation between voltage and temperature for K-factor, (b) Voltage and current in driving and sensing current, (c) Formation of FR4 PCB with Via-hole

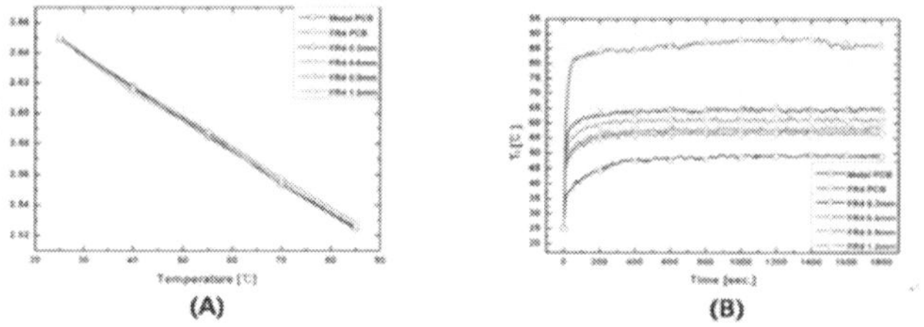

Fig 2. (A) K-factor of LED PKG (B) Variation of junction temperature with time increasing.

Fig.2 shows the relation between sensing current and junction temperature, when the driving current was applied. The sensing current is proportioned to junction temperature like (A) of Fig.2.
In this graph (A) of Fig.2, the gradient called K-factor.
As results from FR4 PCB formation, the temperature of FR4 PCB is more higher than that of Metal PCB as 40 [℃], and the temperature decreased with via-hole as the gap of 10 [℃].

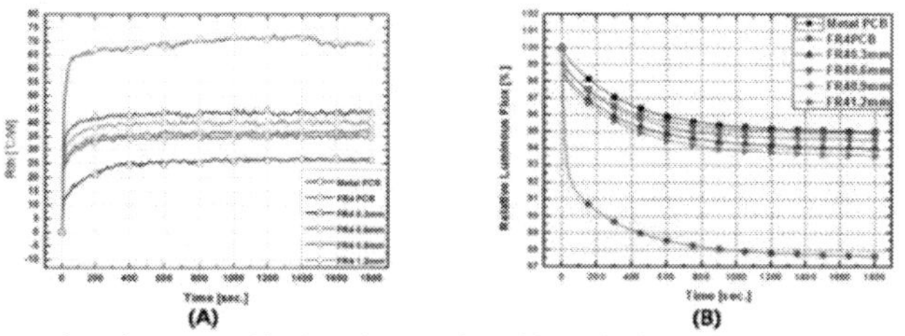

Fig. 3 (a)Variation of thermal resistance with time increasing (b)Variation of luminous flux with time increasing.

the thermal resistance was calculated with equation (1),(2),(3), where, K is K-factor, ΔTSP is temperature sensitivity parameter, Tj is junction temperature, and RθJX is thermal resistance between junction temperature and ambient temperature.

$$\Delta Tj = K \times \Delta TSP \quad (1) \qquad Tj = Tj0 + \Delta Tj \quad (2) \qquad R\theta JX = \frac{Tj - Tx}{PH}$$

Thermal resistance in Metal PCB show 26.2[℃/W] and Thermal resistance in FR4 PCB show 69.2[℃/W]. However, the thermal resistance could be decreased by using the via-holes, and the thermal resistance of FR4 PCB with via-hole of 0.6mm show about 34.6[℃/W] like (B) of Fig.3. Moreover, luminous flux in FR4 PCB was more decreased than other PCB as 12.4%.

CONCLUSION

In this paper, the thermal and optical properties of LED was analyzed with formation of via-holes in FR4 PCB and Metal PCB for decrement of LED lighting and improvement of thermal flow.
Thermal resistance could be decreased by using the via-holes and luminous flux is also improved.

ACKNOWLEDGEMENT

This work was financially supported by the program for the Training of Graduate Student in Regional Innovation which was conducted by the ministry of Commerce Industry and Energy of the Korean Government and also supported by the project for development of energy and resource technology from Korea Energy Management Corporation.

REFERENCES

[1] J. Hu, L. Yang, W. J. Hwang, and M. W. Shin, "Thermal and mechanical analysis of delamination in GaN-based light-emitting diode package", J.Cryst. Growth, Vol. 288, No. 1, PP. 157-161, Feb. 2006

[2] J. Hu, L. Yang, and M. W. Shin, " Mechanism and thermal effect of delaminstion in light-emitting diode package", Microelectron. J. Vol. 38, No. 2, PP. 157 -163, Feb. 2007.

Wall blacking of a fluorescent lamp by a short duration turn-on/turn-off operation

S. Inoue[1], T. Kasuya[1], M. Wada[1], S. Gotoh[2]
[1]*Graduate School of Engineering, Doshisha University, Kyotanabe, Kyoto 610-0321, Japan*
[2]*Panasonic Electric Works Co., Ltd., Kadoma, Osaka 571-8686, Japan*
Contact:dtj0108@mail4.doshisha.ac.jp

ABSTRACT

Wall blackening of a fluorescent lamp was observed to see the effect due to a short duration turn-on/turn-off operation. The lamp had shown a rapid wall blackening at the lamp end, when it was operated with a turn-on/turn-off procedure from the as-received condition. Any continuous operation at the rated power with the time longer than 6.25 hours prior to the turn-on/turn-off operation had indicated moderate progress of wall blackening. The electrode showing deeper blackening tended to produce reduced electron emission corresponding to depletion of emitter from the electrode.

INTRODUCTION

Electrodes of a fluorescent lamp are made of tungsten and coated with alkaline earth oxide materials called emitter. Emitter coating enhances electron emission from the electrode when the electrode works as the cathode for discharge. Emitter on the electrode can be lost by both evaporation and sputtering, while the latter process is believed only effective during lamp ignition phase [1]. For this reason, electrode accelerated testing is considered possible by operating the lamp with short duration turn-on/turn-off procedure. In Japanese Industrial Standards, a procedure to evaluate degradation of a fluorescent lamp is specified with the initial conditioning of the electrode by continuous operation at the rated power for 100 hr. The length of this continuous operation time for initial conditioning procedure is examined to see if it affects the wall blackening of the lamp tube, as wall blackening observed at the lamp end is considered related with release of materials from the electrodes.

EXPERIMENTAL METHOD

A fluorescent lamp (Panasonic FHF32-EX-N-H) coupled to an electronic ballast (Panasonic HEX32HF11/24HK-3EXH) was used as a test sample. A pair of photodiodes, one installed above the electrode and the other installed at the center of the lamp, is contained in a housing of the electronic ballast. Light signals through 10 mm diameter holes are detected by these diodes and the wall blackening at the lamp tube end has been quantified from the ratio of the photodiode signals. The unit was driven by an AC 100 V line through a timer-relay circuit to realize 5 sec turn-on, 5 sec turn-off operation.

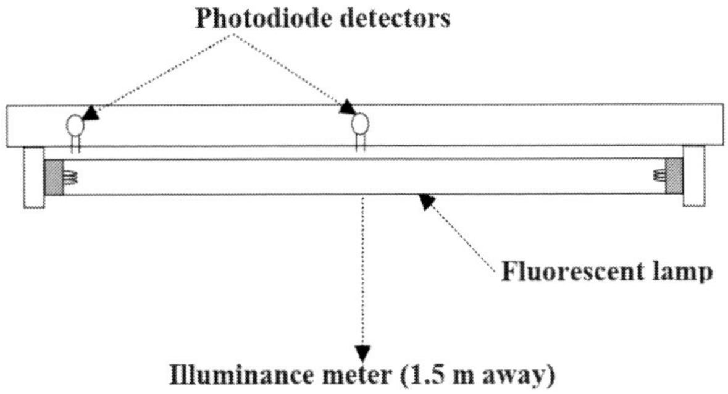

Fig. 1. The diagram of experimental apparatus.

After every 5,000 times of operation, the entire unit was mounted in a dark box to measure its illuminance. Temperature on the wall of the lamp was measured at the same time for several different locations with thermocouples. The changes in temperature, illuminance and output from the photodiodes were constantly monitored for the 30 min after turning on the test lamp. An image of the entire lamp tube was taken by a digital camera toward the end of the steady state operation. The picture image had been converted to a set of digital data to produce a light intensity distribution along the lamp.

Four kinds of lamp initial conditioning were prepared. The first lamp was continuously operated for 100 hr at the rated power as specified in Japanese Industrial Standard (JIS C 7617-2) [2]. The second one was conditioned with 25 hr, or one quarter of the conditioning time of the JIS standard. The third lamp was operated for only 6.25 hr, or one quarter of 25 hr. Finally, the last lamp was put into short duration turn-on/turn-off operation directly from the as-received condition.

EXPERIMENTAL RESULTS

Figure 2 shows the ratio of output from the photodiode mounted at the lamp tube end and that at the lamp center. The ordinate indicates how the wall blackening at the lamp end proceeds. Among the four initial conditionings, the 0 hr sample, or the sample put into operation without any period of continuous operation at the rated power, exhibited the steepest slope. Other three lamps showed similar characteristics up to 25,000 cycles, and the one with the 100 hr preconditioning time deviated from the other two lamps with smaller conditioning time. The light intensity profile obtained from the image of a digital camera picture shows that the effective length of the positive column of the lamp had reduced as shown in Fig. 3. It also shows that the luminous intensity distribution exhibits a gradual decreasing trend from the center to the ends of the lamp.

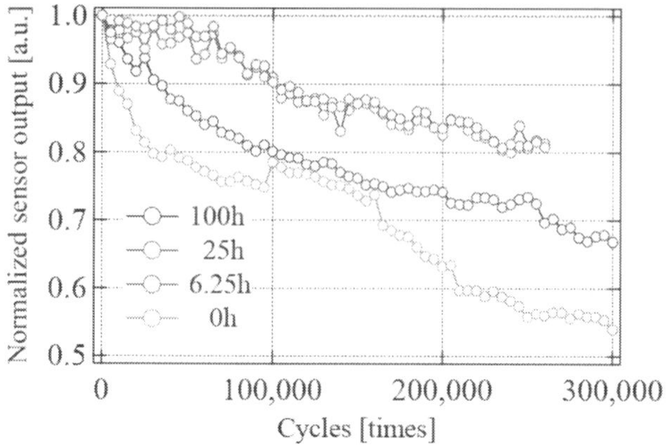

Fig. 2. Ratios of light flux at the lamp end to light flux at the lamp center plotted as functions of number of short time turn-on/turn-off cycles.

Current-voltage characteristics were measured for lamps of different duration of preconditioning, and as-received lamps. Figure 4 shows an example of an oscillograph trace. As it has been pointed out by Lester [3], an asymmetry in current emission from two cathodes was observed in the trace. the correlation between the advancement of blackening at the tube ends and the current asymmetry was only observed for the lamp with the severest blackening at the lamp end.

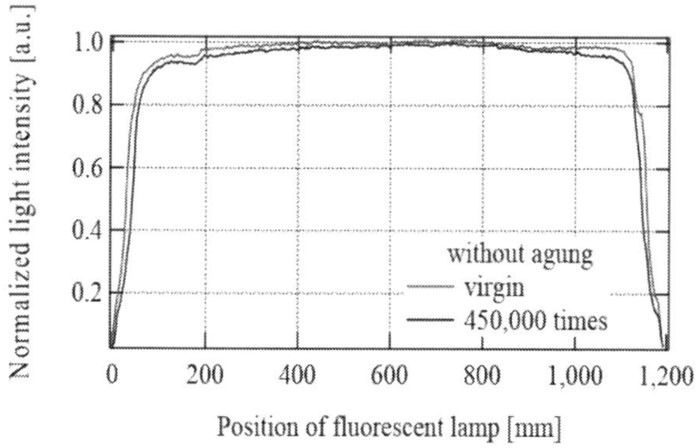

Fig. 3. Light intensity distribution obtained from data reduced from a digital camera picture.

SUMMARY

Blackening at the ends of a fluorescent lamp is accelerated by short duration turn-on/turn-off operation. Continuous operation prior to the turn-on/turn-off cycles may stabilize adhesion of emitter on the lamp electrodes. Only a duration longer than 6 hr seems enough to prepare a sample which is not too sensitive to this pretreatment procedure. Conversely, a sample should be immediately operated without the pretreatment of continuous operation, if a sample lamp with a damaged electrode is immediately necessary.

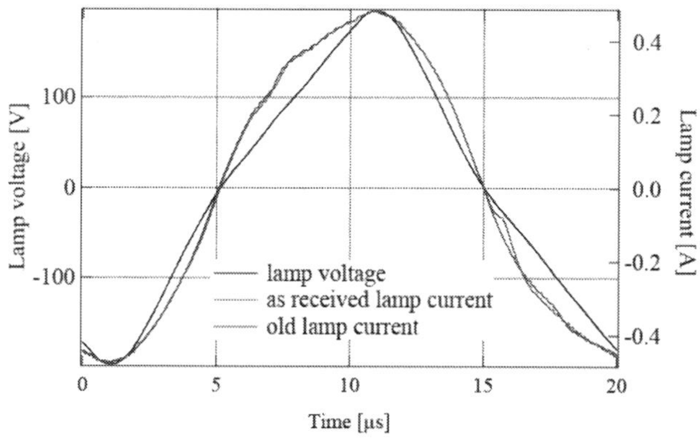

Fig. 4. An oscillograph showing lamp voltage and lamp current.

REFERENCES

[1] J. F. Waymouth, "Electric Discharge Lamps", The MIT press, (1971).
[2] Japanese Industrial Standard No. C 7617-2, (2003), (in Japanese).
[3] J. N. Lester, J. Illum. Engg. Soc. 31, 78 (2002).

A method to evaluate the acceptability of lightsources as replacements for incandescent lamps

Peter Raynham[1], S.A.Mucklejohn[2] and B.Preston[2]
[1] *Bartlett School of Graduate Studies, University College London, 1-19 Torrington Place, London WC1E 7HB, UK*
[2] *Ceravision Limited, The Mansion, Bletchley Park, Milton Keynes MK3 6EB, UK*
Contact: p.raynham@ucl.ac.uk

ABSTRACT

A method to asses the acceptability of lightsources as replacements for incandescent lamps based on the Farnsworth-Munsell 100 hue test and an assessment of run-up times is described.

INTRODUCTION

To understand the colour and run-up time requirements of lightsources a series of tests have been carried out to study colour discrimination and subjective brightness under controlled conditions. This study has been conducted partly in response to criticism from end users about the suitability of some lightsources as replacements for incandescent lamps.

The colour discrimination tests are based on the Fansworth-Munsell 100 hue test [1] which involves placing a series of coloured tiles in the correct order. The 100 hue test has been widely employed for measuring the colour vision of individuals. In this study, the test has been used to monitor the individual's colour vision performance under different lightsources. The 100 hue test quantifies colour vision abnormalities and aptitude by inviting subjects to place coloured tiles in order of hue. The tiles used are based on colours from the Munsell Colour system they have fixed value (lightness) and chroma (colour purity) and vary only in hue.

The tests are carried out in a colour comparison booth which is illuminated with a series of different lightsources each giving similar levels of photopic illuminance but with different spectra. Table 1 gives a summary of the standard colour properties of the light sources measured at a point just above the coloured tiles in the colour booth.

Lamp	CCT	Ra8	Ra14
Incandescent	2,430	96.5	95.3
Fluorescent 840	4,155	77.9	66.8
LED	4,300	76.6	69.3
CFL (Cool White)	5,500	83.8	76.2
Fluorescent Graphica	6,600	98.0	97.0

Table 1 Properties of the illumination used

The colour discrimination ability varies widely from person to person and thus a large number of test subjects are needed to ensure that the findings of the experiment are valid for the population as a whole. Moreover, there is a clear trend that as subjects age their colour discrimination become poorer and thus for the experiment it is important to ensure that subjects have a wide range of ages. Subjects are given an initial training run and then have to perform the 100 hue test with each of the light sources in turn, with a short break between each test. The results of the tests are analysed using the Farnsworth scoring system and the results used to rank the colour discrimination of the various light sources.

The test of run-up time test uses a booth containing a number of normal domestic objects such as fabrics, magazines and fruit; illuminated by a LED source. The subjects are asked to switch on the light in the booth and observe the booth for a minute. At the end of that time they fill in a questionnaire relating to the subjective brightness of the scene. The process is repeated a number of times with different run-up profiles for the light output. The order of scene presentation is random with each run up profile being used twice in the series of tests. Also included in the test series are some scenes where the light comes on instantly but at a level lower than the final level used in the other scenes. The run up profiles vary from the virtually instant light from LEDs through the quick run up of incandescent lamps to the slower run of linear and compact fluorescent. The analysis of the results permits an assessment of the impact of run up time on the perceived brightness of an illuminated environment.

A more detailed description of the experiments together with analyses of the results and the spectral output of the lightsources used will be presented at the conference.

REFERENCES

[1] Dean Farnsworth, *J. Opt. Soc. Am.*, 33, 568 (1943).

Ultraviolet Light Emitting Diodes for Water Treatment

D.G. Knight, F. Daynouri and S. Jolly
Trojan Technologies, 3020 Gore Road, London, Ontario, Canada N5V 4T7
Contact: gknight@trojanuv.com

ABSTRACT
Light emitting diodes offer the potential for a high irradiance instant-on UV source for water disinfection and environmental contamination treatment. We demonstrate an LED based lamp where UVA is developed << 1 second, is linearly dimmable to 5% of the full power, and achieves an irradiance of > 20 mW/cm^2. Destruction of N-nitrosamines using direct photolysis is demonstrated. Increases in external quantum efficiency > 100% is achieved for UVC LEDs by directly coupling the optical output of the semiconductor device to water.

INTRODUCTION
Water disinfection and environmental contamination treatment (ECT) using UV light is normally carried out with mercury arc lamps; where the radiation in the UVC region of the electromagnetic spectrum is used to inactivate harmful organisms through damage to nucleic acids in the case of disinfection, or decompose contaminant molecules through direct photolysis or with the aid of advanced oxidation agents in the case of ECT. The recent use of AlGaN based UV light emitting diodes (LEDs) has provided an alternate UV source, but at a power of only ~1mW for individual LED devices at 280 nm. Disinfection using UVC LEDs has been demonstrated [1]. A high power UVA LED lamp will be used to demonstrate destruction of chemical contaminants, as well as the potential advantages of UV LEDs for water treatment compared to a conventional mercury lamp. A method of increasing the power of UVC LEDs so that they would become a more practical UV source for water treatment will be shown as well.

METHOD
A high power UVA LED lamp was constructed using a 4 x 4 array of 16 Nichia NCU033 365 nm LEDs with 2.0 cm spacing between diodes. Irradiance from the LED lamp is measured with a solar blind radiometer with a peak response at 360 nm. A comparison mercury amalgam lamp emits UV at 254 nm, where the lamp radiation is detected using a radiometer with a 254 nm band pass filter. Water containing a mixture of 200 ppb each of four nitrosamines was placed in a dish to a depth of 2.0 cm, where the distance between the LED lamp and liquid surface is 1.5 cm. The measured irradiance was 11.3 mW/cm^2 at a distance of 2.5 cm from the lamp. The concentration of each of the n-nitrosamines was determined as a function of UV exposure time using high pressure liquid chromatography. The total UV flux was measured for bare prototype UVC LEDs mounted on a TO-8 base, using an integrating sphere and spectrometer. The external quantum efficiency was augmented by placing an approximately hemispherical water droplet of 2.5 mm height on the sub mount, covering the LED chip, and measuring the UV flux.

RESULTS
To determine the output of the LED lamp as a function of time and electrical input power, the radiometer head was placed at the center of the lamp at a distance of 2.5 cm from the lamp face. For measurement of the UV irradiance as a function of time the current was set at 240 mA, and a constant value of 11.3 mW/cm^2 was obtained within 1 second. The UV output for the mercury lamp was obtained by setting the lamp driver at 100% power and observing the radiometer output. The normalized lamp irradiance versus time is shown in Figure 1 for both lamps. The mercury lamp requires 2.2 minutes to reach 80% of the steady state output. This result shows that the LED lamp reaches maximum power much more quickly than a mercury lamp, and can be regarded as an instant-on UV source since the UVA LEDs used are rated at a rise time of 28 ns [2]. Figure 2 shows the normalized irradiance as a function of the input electrical power for the LED and mercury lamp. The LED lamp has an extremely linear UV output down to 5% of the electrical input power, and can be driven at 1% electrical power and still emit UV light. In contrast, the mercury lamp arc will extinguish at electrical input powers < 30%. The maximum irradiance for the LED lamp at a distance of 2.5 cm is 20.7 mW/cm^2 at 466 mA. The destruction of N-nitrosamines as a function of UV exposure time is shown in figure 3, where an excellent linear relationship between the base 10 log of the nitrosamine concentration and exposure time is obtained. For all nitrosamines except NDMA, the concentration is reduced by a factor of ten within 30 minutes. Since the absorption coefficient of N-nitrosamines is ~100 times higher in the UVC compared to the UVA at 365 nm, the development of UVC LEDs of equivalent

irradiance should result in more rapid destruction of these harmful compounds.

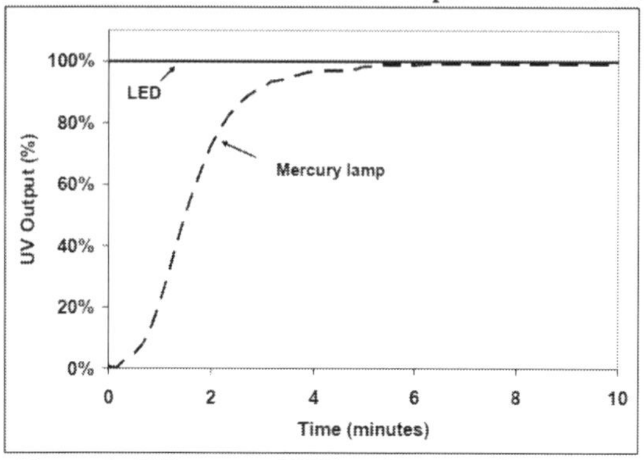

Figure 1: UV output of LED and mercury lamp as a function of time after lamp start.

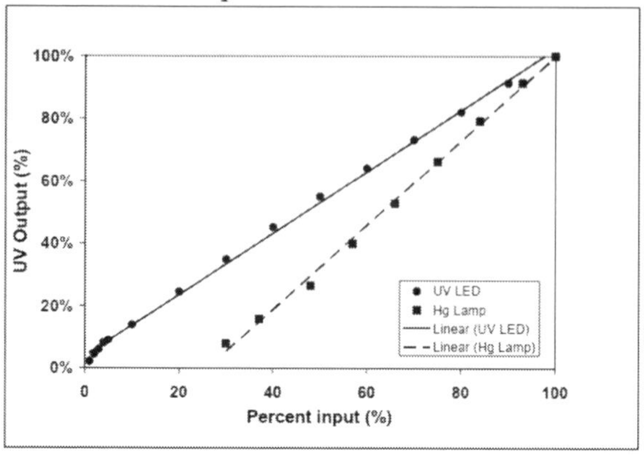

Figure 2: UV output of LED and mercury lamp as a function of electrical power input.

The UV flux emitted from UVC LED chips that are either bare, or covered with a water droplet to enhance the UV output of the LED chip, is shown in figure 4. Duplicate runs were conducted, showing a doubling of the UV output over the entire region of the UV spectrum when water is used to couple the UV light from the chip. The average increase in UV output of UVC LED chips is 130±30% with water coupling for duplicate runs conducted for three chips. A logical explanation for these observations is the increase in escape angle available to a device with a semiconductor/water versus a semiconductor/air interface [3]. For water treatment the LED structure can have optical coupling directly to water as proposed in the patent application of Knight et al. [4], and the claim of increased UV output for such devices is supported here.

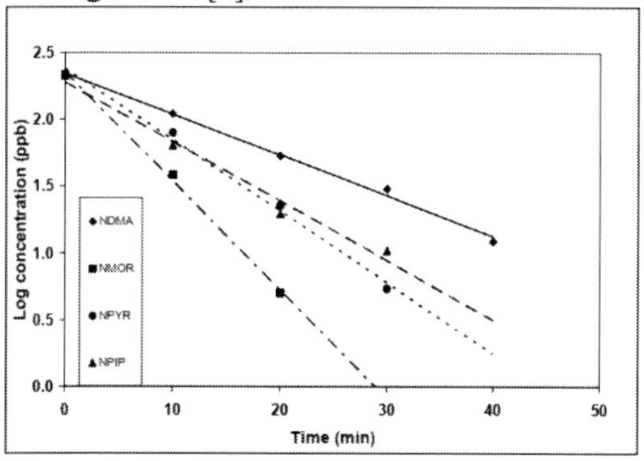

Figure 3: Destruction of N-nitrosamines with UVA LED lamp

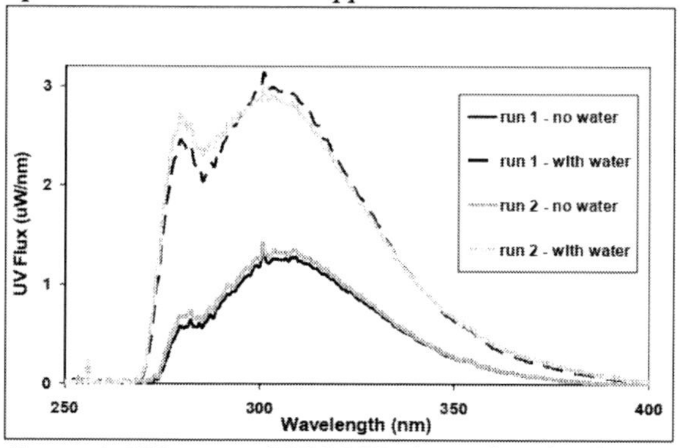

Figure 4: UV output of UVC LED chip with and without water coupling

CONCLUSIONS

A high irradiance UVA LED lamp is demonstrated that is instant-on, has a UV output that varies linearly with the applied electrical power down to 5% of the maximum electrical power used, and is capable of destroying harmful chemicals. LED lamp technology will be highly advantageous for water treatment systems requiring many on/off cycles and dimming of lamps. A method to boost the external quantum efficiency of UVC LEDs by > 100% to obtain higher UV output from these devices is demonstrated here. The optical output of the LED chip is directly coupled to the water being treated, allowing more UV from the chip to escape the semiconductor structure and into the water.

REFERENCES

[1] C. Chatterley and K. Linden, Disinfection 2009 Conference, Atlanta GA, February 28 – March 3, 2009.
[2] Nichia Corporation, NCCU033 Technical Information LA-KSE4150, June 2004.
[3] A. Zukauskas, M. Shur and R. Gaska, Introduction to Solid-State Lighting, John Wiley and Sons, 2002, p. 83-115.
[4] D.G. Knight, J. Fraser and M. Sasges, U.S. Patent application 20090026385, January 29, 2009.

Modeling Performance Limits for Tungsten Halogen Energy Saving Lamps using Infra-red Reflecting Films

Subba Rao Mekala[1], Gunther Van de Poel[1]

[1]*Research and Development, Technology and Competence Center, Havells-Sylvania Lighting Belgium NV, Industriepark 13, Soldatenplein Z2, B-3300 Tienen, BELGIUM*
Contact: subbarao.mekala@havells-sylvania.com

INTRODUCTION

European Union (EU) ban on the sale of incandescent light bulbs by September 2012 as part of a plan to cut down on emissions of greenhouse gases, as these light sources are very inefficient, only a small portion of the electrical power they draw is converted into visible light. Numerous improvements in lamp efficacy by means of infra-red reflecting (IR) films have been described in scientific publications and patents over the last few decades. The infra-red (IR) film is an interference multi-layer coating composed of alternate transparent dielectric thin layers of high and low refractive indices, and it constitutes an interference band-pass filter[1, 2]. Its spectral reflectance to infra-red radiation is determined mainly by the design of its multi-layer structure, such as a choice of coating materials, number of layers and thickness of individual layers[3]. Good adhesion and layer integrity are also important if the optical coating is to be environmentally durable. The IR film coating process currently employs chemical vapour deposition or dipping technique[4, 5]; sputtering [6] and more recently Micro-Dyn Sputtering® technique [7]. The authors applied a indigeniously developed lamp model to 12V lamp types using different IR filter characteristics data [7] and made quantitative theoretical predictions regarding lamp performance. Also, these predictions are compared to literature values and to the new European Commission (EC) regulations for non-directional household lamps [8].

LAMP MODELING

In the present study, the lamp model has a tungsten C8 filament as a light source with mandrel to wire diameter ratios ranging between 6.8-8. The ratio of filament length L to its diameter D is in the range of 1.9-3.8. The lamps are modelled with xenon fill gas having an ellipse outer bulb shape made of Quartz glass with a ratio of 1.05 for major to minor axis. The spectral power distribution of a tungsten filament at $3000K$, typical of these lamps is shown in Fig. 1; also shown is the reflectance in the near IR band for 47-layer *conventional* and 63-layer *hybrid* IR filters. A raytracing program which is integrated in the lamp model, was applied with these IR filter characteristics to predict the efficacy gains and lamp life. The simulated coiling parameters for the uncoated lamps are shown in Table. 1. The 63-layer *hybrid* IR coating results in $\sim$ 80% efficacy gain, where as the 47-layer IR coating results in $\sim$ 40% efficacy gain. It is evident from Fig. 2 that the lamp life is reduced to half with 63-layer IR coating compared to 47-layer IR coating because of the higher efficacies with the former coating. However, both coatings when applied to the bulb wall will satisfy the recent EU stage 6 ecodesign requirements for non-directional household lamps as shown in Fig. 3. The modeled results are in good agreement with the previously reported results.

Type	uncoated			47-layer IR coating					new **Hybrid** 63-layer IR coating				
12V	Watt.	lumens	LPW	Watt.	lumens	LPW	Life h	η gain %	Watt.	lumens	LPW	Life h	η gain %
20W	21.2	330	15.6	20.2	447	22.1	5100	42.2	19.6	548	28	2400	79.6
35W	38.2	641	16.8	36.4	875	24	6200	43.3	35.4	1060	29.9	3000	78.4
50W	56.6	916	16.2	53.9	1256	23.3	7000	44	52.5	1509	28.7	4000	77.6

Table 1. Modeled luminous efficay gain, η, and life after coating with 47-layer and 63-layer IR coatings for 12V lamps. LPW is an acronym for lumens per watt.

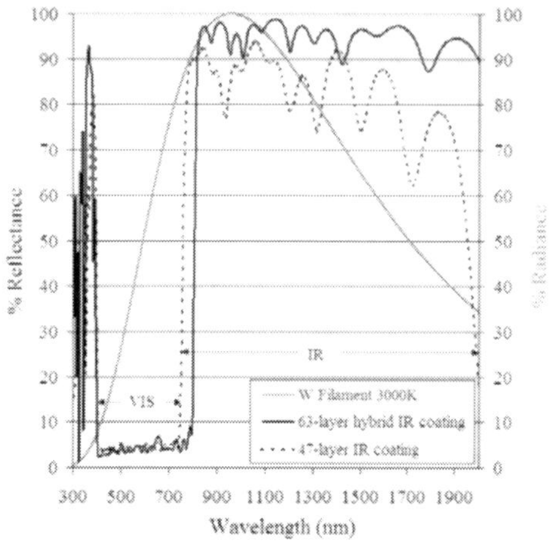

Figure 1. Reflectance spectra of new Hybrid 63-layer IR coating compared to 47-layer TiO_2-SiO_2 dielectric film [7]. Spectral emittance of 3000K tungsten filament is also shown.

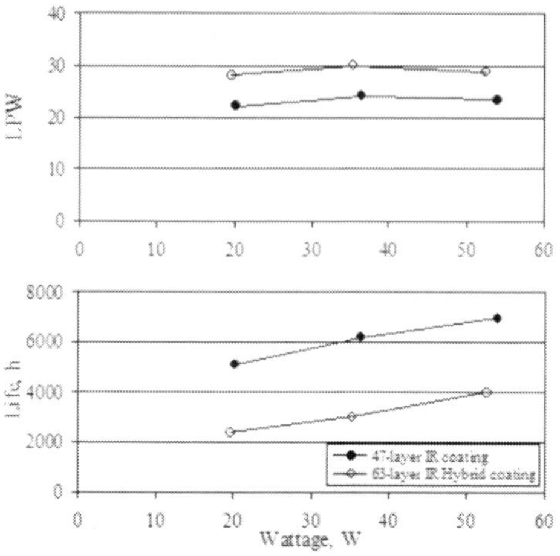

Figure 2. Modeled efficacy gain and life for 12V lamps as a function of wattage with 47-layer and 63-layer IR coatings.

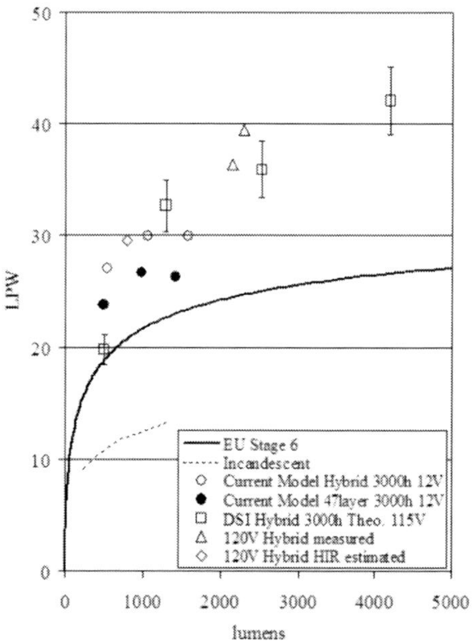

Figure 3. Luminous efficacy as a function of luminous flux for various lamp types [7], using different IR coatings compared to current lamp model. Lower limit for efficacy, $\frac{Luminous\ Flux,\ \phi}{Maximum\ Power,\ P_{max}} = \frac{\phi}{0.6\cdot(0.88\cdot\sqrt{\phi}+0.049\cdot\phi)}$, *derived from EU stage 6 curve [8] for non-directional household lamps (applicable from September 2016 onwards) is also shown as a function of ϕ.*

CONCLUSION

The current model confirms that the new 63-layer coating offers an increased efficacy compared to traditional 47-layer IR coating but with some sacrificed life. With new upcoming energy regulations in force for developing countries, the future looks bright for low voltage tungsten halogen lamps using IR films resulting in higher efficacy, for household applications, accent lighting etc.

REFERENCES

[1] H. Bach, D. Krause, Thin films on glass. Heidelberg Springer, 1997: Chapter 6.

[2] H. Pulker, Coating on glass. Amsterdam Elsevier, 1984.

[3] R. S. Bergman, Application of thin film reflecting coating technology to lamps. Proc. LS-6 1992, 173-190.

[4] L. Goldstein, The design, construction and performance of an incandescent light source with a transparent heat mirror. Lighting Res. Tech. 1986, 18, 93-98.

[5] A. Kawakatsu, Tungsten halogen lamps with infra-red reflective films. J. Illum. Engng. Inst. Jpn. 1985, 69, 537-542.

[6] Y. Omata, N. Hashimoto, S. Kawagoe, T. Suemitsu and M. Yokoyama, Sputtering deposition of infra-red reflecting films on ellipsoidal bulbs of energy saving lamps. Lighting. Res. Technol. 2002, 34, 111-119.

[7] L. Bartolomei, B. Gray and S. Pearson, Deposition Sciences, Inc. Santa Rosa, CA, USA, private communication, 2010.

[8] Commission Regulation (EC) No. 244/2009 of 18 March 2009, Official Journal of the European Union dated 24.3.2009.

Low-Pressure Gas Discharge UV Source Using a Thermo-Field Emission Carbon Nanotube Array Cathode

M. Dionne, S. Coulombe, J.-L. Meunier
[1] *Department of Chemical Engineering, McGill University, 3610 University, Montréal, H3A 2B2, Canada*
Contact: martin.dionne@mail.mcgill.ca

ABSTRACT

We produced composite electrodes comprising carbon nanotubes (CNT) arrays embedded in anodic aluminium oxide (AAO) templates grown directly on commercial aluminum by thermal chemical vapour deposition (thermal CVD). We then used our composite electrodes as cathodes of low-pressure gas discharges (~1-2 Torr, ~20-30 mA) in mixtures of Ar, N_2 and H_20. The measured discharge parameters are different from those found in the literature and show a new domain of operation in thermo-field emission at low temperature and voltage. Significantly reduced discharge voltages are observed with the composite cathodes (up to 84%). Strong emission in the near UV range is observed with N_2-rich discharge. The composite cathodes have sustained gas discharges for durations up to 500 hrs. The observed damages to the composite cathodes suggest a temperature limit of 45-60 °C for the electrode's bulk temperature. The presence of a low concentration of water vapour is also necessary to maintain the enhanced electron emission conditions from the composite cathodes.

1. INTRODUCTION

Field emission from carbon nanotubes has been intensively studied for potential applications such as electron sources in flat panel displays [1], X-ray sources [2], gas discharge tubes [3] and low pressure lamps [4]. Our previous work [5, 6] showed that very dense arrays of short CNT tips can withstand large surface-averaged current densities J_{eq}. With anodic aluminum oxide (AAO) templates, the very high pore density matches the desired surface density of emitters [7, 8]. These templates also show a good compatibility with CNTs [9, 10], which are used as electron emitters. In this paper, we report on a simple one step anodization procedure and a straightforward CNT growth process for our as-anodized aluminium substrates. We then report on low-voltage, non-destructive thermo-field (T-F) emission-sustained DC gas discharges produced in various gas mixtures of Ar, N_2 and H_20. The near UV emission spectra are presented and an attempt is made to elucidate the role of water.

2. EXPERIMENTAL PROCEDURE

Mirror-like 0.8 mm (0.032") thick aluminum 6061 sheets were cut into 50 x 20 mm coupons and degreased in acetone for 60 min before the anodization was performed. The later was carried out in 0.2 M oxalic acid at 20-25 °C. The anodization voltage was set to 40 V and maintained for 4 hrs. The anodized and rinsed samples were dipped for 5-10 min into a 48 hrs old alcohol-based 50 mMol/l ferric nitrate solution [11]. Thermal CVD experiments were performed in a Lindberg Blue M furnace. CNT synthesis was performed at 580 °C by injecting C_2H_2 for 5-15 min in a H_2 atmosphere. DC gas discharges were produced using the as-produced composite cathodes. A stainless steel ring located 3 cm away from the cathode face was used as the anode (also grounded to the vacuum chamber). We used Ar, N_2 and water vapour as feed gases and the total pressure was maintained at 2 Torr. H_2O and its carrier gas were used simultaneously as optional additives in a total proportion of 10-15 v/v%. The composite cathode bulk temperature was measured with a thermocouple. We used an OceanOptics USB S2000 spectrometer to monitor the UV-VIS emission from 1-500 hours long discharges.

3. RESULTS AND DISCUSSION

Fig. 1a) shows the time evolution of the electrical discharge voltage (ΔV) during four experiments which lasted approximately 24 hrs each, and which made use of different composite cathode samples. In each experiment, ΔV first decreased to a minimal value within the first hour and then, either remained stable or increased due to CNT contamination or water desorption. Figs. 1b) shows the UV-visible spectrum of each discharge. At 2 Torr, the minimum ΔV value observed was 84 V for a N_2-water discharge, while this number was 38 V for an Ar discharge. A N_2-water discharge operated for

500 hrs maintained a ΔV in the 87-98 V range. N_2-water discharges produced a large amount of UV. Our study of the possible damages revealed that exposed metal surfaces next to the AAO templates were vulnerable and caused significant contamination. The CNTs survived unless the electrode bulk temperature exceeded 60 °C for several hours. The presence of adsorbed water molecules on the CNT surface was identified as the cause of the observed field emission currents in vacuum when high voltages are applied and no discharge takes place [12]. Our previous modelling work [see 5, 13] provided an effective work function of 0.9 eV for the water vapor-CNT system and we predicted that self-sustained T-F emission was possible for CNT arrays below 1 atm of chamber pressure. We maintained the discharge parameters over 100-500 hrs periods by renewing the moisture content of the chamber and cooling the sample below 45 °C. These low voltage discharges producing strong UV photon fluxes as well as some visible light without relying on the standard Ar/Hg fill are extremely promising for future applications in gas discharge lighting.

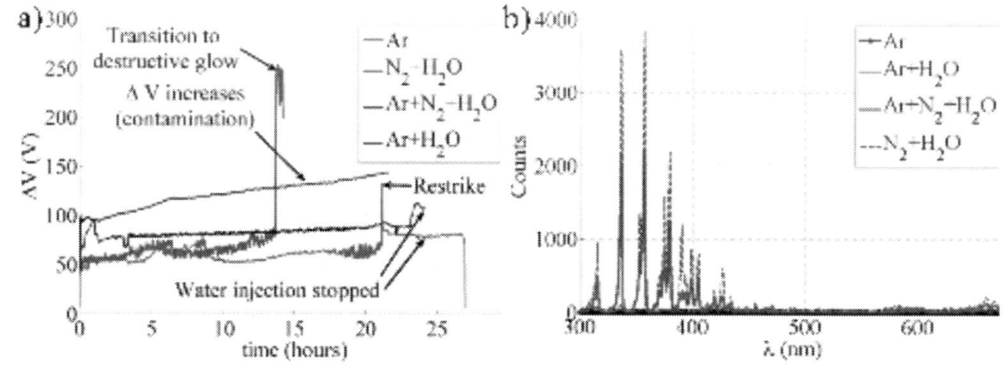

Figure 1: a) ΔV as a function of time for 16-26 hours discharges. b) UV-visible spectra for each discharge.

4. CONCLUSION

We produced CNT array electrodes using anodized sheets of commercial grade aluminium and a standard thermal CVD protocol for the growth of CNTs. The resulting CNT array electrodes were used as cathodes of low-pressure DC gas discharges in various gas compositions. We observed a significant reduction of the operating voltage with the CNT-covered aluminum cathodes. Based on our observations and modeling work, we suggest that thermo-field emission is enabled at low cathode temperatures and applied electric fields due to the extremely low work function of the water vapor-CNT system and the fine elongated geometry of the CNTs which enhances the surface electric field at their tips. We are now updating our theoretical models for plasma-enhanced T-F emission from vertically-aligned CNT arrays and will report the results at the conference and in a forthcoming paper.

5. REFERENCES

[1] Bonard J-M, Kind H, Stockli, Nilsson L-O, 2001, *Solid-State Electronics*, **45**, 893-914.
[2] Zhang J, Yang G, Cheng Y et al, 2005, *Appl. Phys. Lett.*, 86, 184104.
[3] Rosen R, Simendinger W, Debbault C, Shimoda H, Fleming L, Stoner B and Zhou O, 2000, *Appl. Phys. Lett.*, **76** (13) 1668-70.
[4] Sun *et al*, 2005, US Patent no 6949877 B2.
[5] Dionne M, Coulombe S and Meunier J-L, 2008, *IEEE Trans. Electron. Devices*, **55** (6), 1298-1305.
[6] Dionne M, Coulombe S and Meunier J-L, 2009, *Phys. Rev B15*, **80**, 085429.
[7] Lee W, Ji R and Nielsch K, 2006, *Nature*, **5**, 741-7.
[8] Lee W, Ji R, Gösele U and Nielsch K, 2006, *Nature Materials*, **5**, 741-7.
[9] Sklar G P, Paramguru K, Misra M and LaCombe J C, 2005, *Nanotechnology*, **16**, 1265-71.
[10] Sang Suh J, Seok Jeong K and Lee J S, 2002, *Appl. Phys. Lett.*, **80** (13), 2392-4.
[11] Nerushev O A, Sveningsson M, Falk L K L and Rohmund F, 2001, *J. Mater. Chem.*, **11**, 1122-32.
[12] Dean K A, Chalamala B R, 2000, *Appl.. Phys.Lett.*, **76** (3), 375-7.
[13] Dionne M, Coulombe S and Meunier J.-L, *Proceedings of the19th International Symposium on Plasma Chemistry*, August 26th-31st, 2009, Bochum, Germany.

Determination of $HgBr_2$ and $SiBr_4$ reaction products in electrodeless metal halide discharge lamps

T. Emilsson[1], D.A. Doughty[2] and T.R. Brumleve[1]
[1] *APL Engineered Materials, Inc., Urbana, IL 61802 USA*
[2] *Topanga Technologies, Inc., 2178 Paragon Drive, San Jose, CA 95131*
Contact: timothy_brumleve@adlt.com

ABSTRACT

Ultraviolet spectrophotometric molecular absorption measurements in the 190-250 nm range are described which can provide useful diagnostics for $HgBr_2$ and $SiBr_4$ impurities in quartz metal halide lamps. These impurities are important for understanding lamp performance and life. The technique is illustrated with applications to microwave-powered electrodeless lamps containing InBr and rare earth bromide doses.

INTRODUCTION

Reactions between the quartz wall and the metal halide dose can ultimately limit useful lifetimes of metal halide lamps. Chemical reactions between the dose and the quartz wall remove halide components from the dose and can result in significant changes in color and luminous flux. These effects are important for both electrodeless lamps and lamps with tungsten electrodes. A measure of the rates and extents of gaseous wall reactions products can provide a diagnostic of lamp performance and life. [1,2]. Gas phase iodides derived from wall reactions have previously been measured non-destructively in lamps by absorption spectroscopy [3]. In this study we illustrate the determination of $HgBr_2$ and $SiBr_4$ reaction products in the presence of InBr, using ultraviolet (UV) absorption spectroscopy.

EXPERIMENTAL

The electrodeless lamps used here consisted of cylindrical quartz envelopes ~15 mm long, ~8 mm outer diameter, and are dosed with mercury, indium bromide and rare earth bromides. Excitation of the lamps is accomplished with a microwave coupler that inputs ~150 watts to the arc tube. Lamps with three different burn times were studied: TT3 (5 minute burn), TT2 (with about one hour total burning and multiple starts), and TT1 (1000 hours of continuous burning). UV absorption spectra were recorded using a Hewlett-Packard Model 8453 spectrophotometer over a wavelength range of 190 to 250 nm. A small heated aluminum block was constructed to hold the ~8 mm electrodeless arc tubes and empty quartz blanks in the optical path of the spectrophotometer. The aluminum block was fitted with an inlet for cold N_2 gas to allow cooling below ambient temperature. The calibration spectra for $HgBr_2$ were measured in a larger furnace as described in [3]. Calibrations for both gaseous $HgBr_2$ and $SiBr_4$ species were performed by adding sufficient amounts of each compound to a calibration arc tube and using the literature values [4] of vapor pressure to establish absorbance/path length versus pressure correlations.

RESULTS AND CONCLUSIONS

Molecular absorbances for the saturated $SiBr_4$ and $HgBr_2$ standards were observed in the 190 to 250 nm range as a function of temperature, and were used to construct calibration curves (absorbance/path length, A/ℓ, vs. pressure). The slopes of the calibration curves (which pass through the origin at (0,0)) for $SiBr_4$ and $HgBr_2$ are 34.9 $(cm\text{-}atm)^{-1}$ and 234 $(cm\text{-}atm)^{-1}$, respectively. $SiBr_4$ shows a large absorbance tail in the 190 to 220 nm range (see Fig. 2). $HgBr_2$, as seen in Fig. 1, shows a distinct peak at 198 nm, in good agreement with literature spectra [5,6]. The spectrum of the 1000 hour lamp shown in Fig 2 also shows high absorbances at temperatures above ~250 C, attributable to the vaporization of the InBr component in the lamp dose. The recommended $HgBr_2$ analysis temperature for the actual lamp samples is thus ~200 C, high enough to vaporize all of the $HgBr_2$ but low enough to limit the vapor pressure of the large amounts of InBr in the dose. The $HgBr_2$ spectra were blank corrected using a blank at ambient temperature. Similarly, the recommended temperature for determination of $SiBr_4$ is ~3 C, high enough to vaporize all of the $SiBr_4$ but low enough to limit the vapor pressure of any $HgBr_2$ present to negligible values. $SiBr_4$ spectra were blank corrected using a blank at approximately -40 C.

The potential presence of HBr (a possible hydrolysis product of H_2O contamination and rare earth bromide doses) has not been evaluated here. Literature spectral data [6] indicate that HBr has an absorbance spectrum similar to that of $SiBr_4$ shown in Fig. 1. However, HBr does not appear to be a significant interference in the samples examined here. Results for the partial pressures of $HgBr_2$ and $SiBr_4$, calculated from the absorption values at 198 and 192 nm, respectively, are shown in Table I for the 3 lamp samples. Detection limits for both gaseous $HgBr_2$ and $SiBr_4$ are estimated to be on the order of 0.0005 atm and 0.002 atm, respectively. All 3 lamps examined showed $SiBr_4$ levels below detection limit, and only the 1000 hr lamp showed a level of $HgBr_2$ (0.0023 atm) greater than the detection limit. The UV spectrophotometric measurements described here thus provide useful diagnostics for $HgBr_2$ and $SiBr_4$ impurities in quartz metal halide lamps.

Table I. $HgBr_2$ Results at 198 nm and $SiBr_4$ Results at 192 nm

Sample ID	Hours	T measured	$HgBr_2$ Pressure / atm	T measured	$SiBr_4$ Pressure / atm
TT1	1000	204 C	0.0023	4 C	<0.002
TT2	1	201 C	<0.0005	3 C	<0.002
TT3	0.08	200 C	<0.0005	-9 C	<0.002

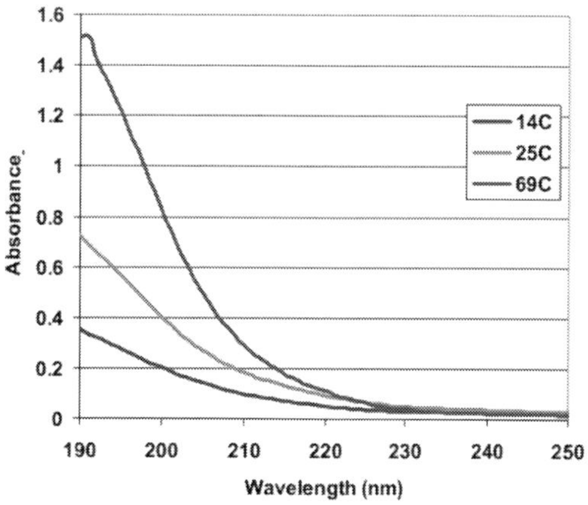

Figure 1. Absorbance spectra of saturated $SiBr_4$ standards as a function of temperature.

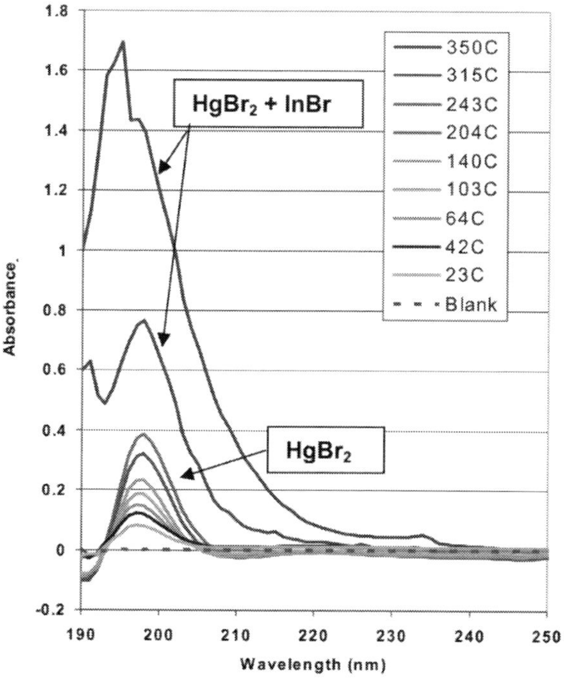

Figure 2. Absorbance spectra of 1000 hour lamp sample TT1 as a function of temperature.

REFERENCES

[1] W. van Erk and T. Rietveld, Philips J. Res., 42(1), 1987, 102.

[2] C.V. Varanasi, *et al.*, Proc. Ninth International Symposium on the Science and Technology of Light Sources (LS:9), Aug. 2001, p. 383.

[3] T. Emilsson *et al.*, Proc. Ninth International Symposium on the Science and Technology of Light Sources (LS:9), Aug. 2001, p. 135.

[4] S.C. Hansen, *et al.*, "Vapor Pressure of Metal Bromides and Iodides", 2nd Edition, Sept. 2006, available from APL Engineered Materials.

[5] B.E. Wilcomb, *et al.*, Chem. Phys. Lett., 25(2), 1980, 239.

[6] C. Roxio and A. Mandl, J. Appl. Phys., 51(6), 1980, 2969.

Perceived Brightness in LEDs

Tsung-Hsun Yang[1], Shih-Fang Liao[1], Cheng-Chung Lee[1], Kirk Chang[2]

[1] *Department of Optics and Photonics, National Central University, Jhongli 32001, Taiwan*
[2] *Display Solution BU, Delta Electronics, Inc., Jhongli 32063, Taiwan*
Contact: thyang@dop.ncu.edu.tw.

ABSTRACT

It is commonly observed that the light-emitting diode (LED) light sources can provide with more perceived brightness by less stimulus luminance. In this work, the Helmholtz-Kohlrausch effect (H-K effect) in human vision has been studied for the enhanced perceived brightness from LEDs. In fact, the variable-chromatic-color method developed by Nayatani's group has been applied for evaluating the enhanced quantity in the H-K effect. A real visual matching experiment by 24 observers with normal color vision is carried out to figure out the specific luminance of equal perceived brightness between LED and the reference light sources in dark room. Finally, the result from the data analysis proposes the additive property in H-K effect.

BACKGROUND

Considering the white light is made of red, green, and blue primaries in both LED and the traditional lamp projectors, the corresponding chromaticities and lightness of each primary are characterized and summarized in Table 1 respectively. These two projectors are almost the same except for the light sources only. The stimulus luminance from the lamp projector (645 cd/m^2) is more brighter than that from the LED projector (364 cd/m^2). However, as watching by human eyes, every observer feels the white made by LED projector is much brighter than the traditional lamp projector. No doubts, the LED projector provides with higher perceptive brightness than the lamp projector does.

Table 1. Characteristics of the LED and the lamp projector.

	Lamp			LED		
	R	G	B	R	G	B
x	0.638	0.301	0.151	0.704	0.184	0.146
y	0.330	0.595	0.069	0.294	0.719	0.040
White(D_{65})	645 lm/m^2			364 lm/m^2		

The perceived brightness of LED projector is commonly considered as the result of the H-K effect[1, 2, 3, 4]. For any two color stimuli with the same luminance, the perceived brightness induced by the color stimuli of higher purity will be higher than that of lower purity. According to Nayatani's variable-chromatic-color (VCC) method[5], the effective enhanced ratio (γ_{vcc}) of the perceived brightness for LED projector is $\gamma_{vcc}^{LED} = 2.12$ in dark room environment, which is higher than $\gamma_{vcc}^{Lamp} = 1.58$ for lamp projector. The $\gamma_{vcc}s$ show that the white light from LED source introduces more perceived brightness than lamp projector. Qualitatively, it matches to the practical observation experience.

In the following, a human eyes matching experiment is performed for further quantitatively studying on the perceived brightness of mixing colors in the projectors. Because the color gamut of the LED projector is larger than that of the lamp projector, each chromatic light made by the lamp projector can be reproduced by the LED projector. Meanwhile, even the same chromaticities in the both projectors, both the spectral compositions are completely different to each other. Then, does the VCC method still work to evaluate the chromatic light by mixing with the red, green, blue primaries? It is an interesting and urgent subject to many applications in lighting and display technologies.

EXPERIMENT

The chromatic light producing by LED projector and the reference project are chosen for the six primaries based on the REC709 color gamut standard, that is red, green, blue, cyan, yellow, and magenta. These two color gamuts are similar to each other, and it is approached that the chromaticities of the chosen chromatic primaries in LED projector are the same as those in the lamp projector.

There are 24 observers joining the experiment. Every observer has been tested for normal color vision by the FM-100 Hue test[6]. The observers shall adapt to the dark room environment at least for 5 minutes before the experiment. There are two patterns showing the same color but with different luminance level, one from the lamp projector and the other from the LED projector. Each observer was requested to recognize which pattern being brighter. In the meantime, each frame holds for 3 seconds.

RESULTS

The experimental data for the most possible equal perceived luminance levels between two projectors are compared with the prediction by extending the VCC method. Actually, the prediction of the perceived LED brightness for any color is assumed to the summation of the individual enhancement of its red, green and blue composing primaries, rather than the enhancement of the mixing color itself. As shown in Fig. 1, the additive property of VCC method for H-K effect holds for all the tested colors.

ACKNOWLEDGEMENTS

This wok was partly sponsored by the National Science Council, Taiwan, under the grant number of NSC 95-2221-E-008-113-MY3.

REFERENCES

[1] G. Wyszecki and W. S. Stiles, Color Science, 2nd Edition, Wiley, New York (1982).

[2] Mark D. Fairchild, Color Appearance Models, Adddison Wesley Longman (1998).

[3] Miyoshi Ayama and Mitsuo Ikeda, Brightness-to-Luminance Ratio of Colored Light in the Entire Chromaticity Diagram, Color Research and Application, 23(1998) 274.

[4] Eiichiro Toriumi, Sosuke Morii, Miyosi Ayama and Takeshi Kumagai, The Illumination Engineering Institute of Japan, 24 (2000) 28.

[5] Yoshinobu Nayatani, Hiroaki Sobagaki, and Kenjiro Hashimoto, Color Research and Application, 19 (1994) 246;*ibid* 19 (1994) 262.

[6] Dean Farnsworth, The Farnsworth-Munsell 100-Hue Test for the examination of Color Discrimination, Maryland: Munsell Company, Inc., (1957) 1.

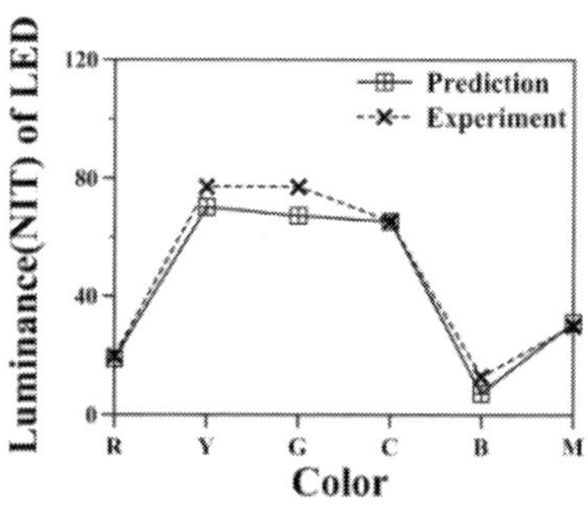

Figure 1. Comparison between the experimental data and the prediction values.

Implementation of optimal control in optically induced plasma region with non-uniform for threshold breakdown

Xue. Wang[1], J.H. Yun[1], G.B. Hong[1], Y.K. Kim[1]
[1]*Department of Information and Communication, Wonkwang University*
344-2 Shinyong-dong, Iksan Chonbuk 570-749, South Korea
Contact: asdfx3105@naver.com

ABSTRACT

We study on the implementation of thermal analysis for LED chip breakdown in optically controlled non-uniform plasma region. The peak response of the reference pulses in different plasma layers has been evaluated by means of the reflection function of dielectric lines. The lines were used with LED-chip region containing an optically induced plasma region, which was illuminated by a light source. The variation of optimal response in breakdown point in plasma region with differentially localized pulses has been evaluated analytically in a LED-chip response.

1. INTRODUCTION

There has been increasing interest in the application of light-wave technology for the control, generation and measurement. The dielectric constant of the optically illuminated LED chip region takes a complex form and this can be useful in the design of optically controlled plasma region with non-uniform. The considerable interest in optical control of LED chip, which has illuminated with light source. A high-intensity illumination system made of transient plasma structures can be utilized to circumvent the ubiquitous optical damage problem since the structures are already fully ionized. In this study, we analyze the behaviour of LED chip thermal analysis in plasma region by calculating the dependence of the reflection function of the line on the plasma properties. We have also modelled the dielectric plasma region using a more accurate non-uniform dielectric profile characterizing the plasma created by an absorbed optical absorption.[1-3]

2. EFFECTS OF NON-UNIFORM LED-PLASMA REGION

When a LED chip material is illuminated with light, which form a thin layer of plasma near the surface of the material. The induced plasma density decreases approximately as $\exp(-\alpha_1|x|)$ and x is the position along the penetration direction of the light beam. The illumination is also considered to be non-uniform such that the variation along y(width) is of the Gaussian type. The dielectric constant in the plasma induced region can be analyzed from the equation of motion of charge carriers in the chip considering the classical electron-hole plasma theory as predicted by the Drude-Lorentz equation.[4]

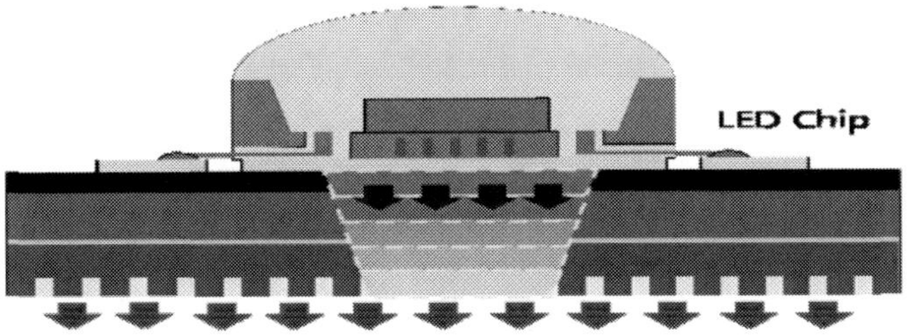

Fig. 1 : Configuration of LED-chip plasma region with optically controlled light source

Fig. 1 shows the configuration of LED-chip plasma region with optically controlled light source. The plasma region ΔZ is assumed to have a uniform density of free carriers. The relative permittivity of the plasma induced in the chip region is given by [5]

$$\varepsilon_p = \varepsilon_s - \sum_{i=e,h} \frac{\omega_{pi}^2}{\omega^2 + \gamma_i^2}\left(1 + j\frac{\gamma_i}{\omega}\right)$$

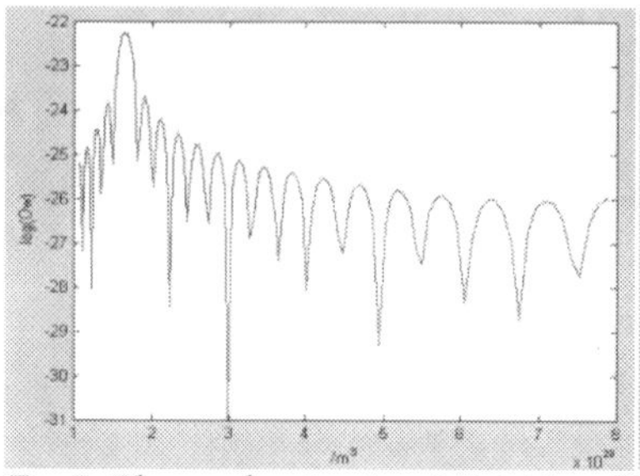

Fig 2. Plasma density in $\alpha_l LD<<1$.

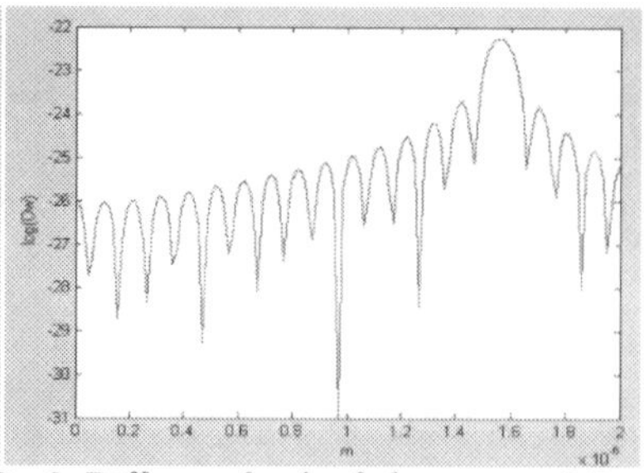

Fig 3. Different depth of plasma in $\alpha_l LD<<1$.

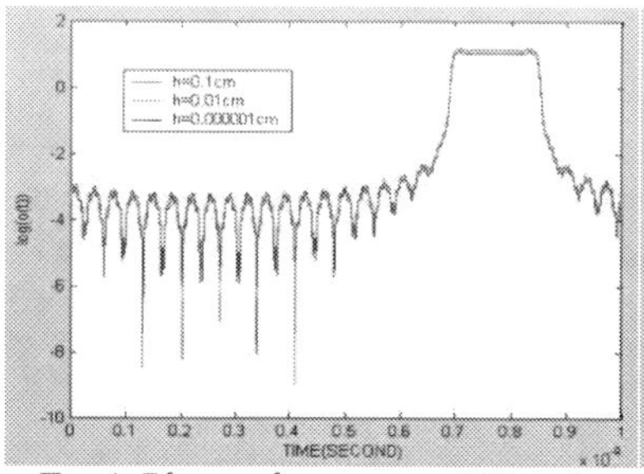

Fig 4. Plasma density in $\alpha_l LD>>1$.

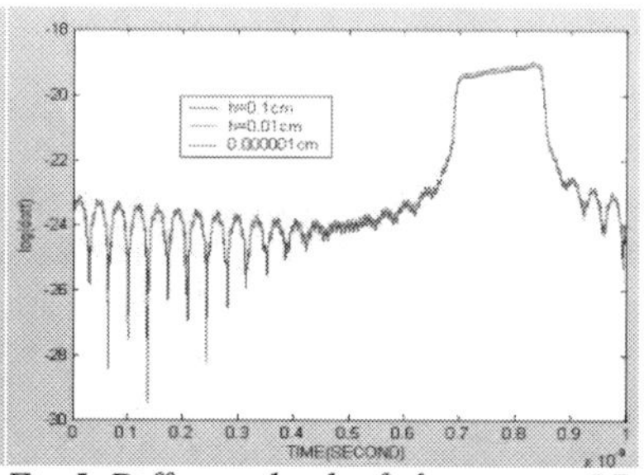

Fig 5. Different depth of plasma in $\alpha_l LD>>1$.

The LED thermal response of the variation in the plasma region, which has the form of a pulse-modulated sinusoid signal, is shown in Figure 2 and Figure 3. In Fig. 4 we depict the response of the fourier transform of the signal at depths of 0.1 cm, through 0.01cm in the plasma depth and the characteristics of the corresponding density response. The magnitude is largest at a depth of 0.1 cm and the least in a depth of 0.01 cm. In Fig. 5 the magnitudes differ only slightly at depths of 0.1cm, through 0.01cm in the plasma layer and the largest magnitude is about a few dB.

3. CONCLUSION

In this paper, we have implementation the system LED chip threshold breakdown in optically controlled non-uniform plasma region. The peak response of the reference pulses in different plasma layers has been evaluated function of plasma density and depth. The LED-chip region containing an optically induced plasma region, which was illuminated by a light source, has shifted as a few breakdown threshold region. The variation of thermal response in breakdown point in plasma region with differentially localized LED-chip materials.

4. ACKNOWLEDGEMENT

This work was financially supported by Small Business Administration through Industry-Academic Joint Technology Development Project in 2009.

5. REPERENCE

[1] W. K. Lee, I. H. Choi, K. Y. Shin, K. C. Hwang and S. W. Han, "Accelerating Aging of Transmission Line Porcelain Suspension Insulators by Autoclaving" KIEEME Trans. Electronics, Vol. 9, No. 4, pp. 147~155, Sep (2008).

[2] T. Ueda, M. Tsutsumi, "Left-Handed Transmission Characteristics of Ferrite Microstrip Lines without Series Capacitive Loading", IEICE Trans. Electronics, Vol. E-89C, No. 9, pp. 1318~1323, Sep (2006).

[3] A. Tip, "Linear dispersive dielectrics as limits of Drude Lorentz systems", PHYSICAL REVIEW, Vol. E69, pp. 166 Jan (2004).

[4] S. K. Dana, H. Shimasaki, and M. Tsutsumi, " Efficient Optical Control of Millimeter Waves in a Slot Line on Semiconductor Plasma Substrate", IEEE Trans. on Microwave theory and Tech, Vol. 50, No. 1, pp. 207~210 Jan (2002)

[5] Matthew N. O. Sadiku, "Elements of electromagnetic (3/E)", (Oxford University Press, 2000) pp. 240

Effect of the color shift on human perception enhancement by repetitive pulsed operation of LEDs

Hideki Motomura, Mitsuhide Abe, Shodai Miyake, Masafumi Jinno
Department of Electrical and Electronic Engineering, Ehime University
3 Bunkyo-cho, Matsuyama 790-8577, Japan
Contact: hmoto@mayu.ee.ehime-u.ac.jp

ABSTRACT

The authors reported that the human perception is enhanced for the repetitive pulsed light even over the critical fusion frequency (CFF) by using fast turn-on and turn-off light such as LEDs [1–3]. However, the color shift of the pulsed LED may affect the observer's decision. The authors tried separating the effect of the color shift from statistically derived value of human perception enhancement degree. It is found that the effect of the pulsed operation itself exists in some cases.

INTRODUCTION

The authors reported that the human perception is enhanced for the repetitive pulsed light even over the critical fusion frequency (CFF) by using fast turn-on and turn-off light such as LEDs [1–3]. We have proved this effect from the statistical analysis from experiments using 20 observers and concluded that the human perception was enhanced up to 2 times by pulsed operation of LEDs at the frequency of 60 Hz and the duty ratio of 5%. However, the pulsed operation of LEDs with small duty ratio means very large injection current during limited on-period; this results in the shift of the emission wavelength, and then the color shift of the pulsed LED may affect the observer's decision. In this study, the authors tried separating the effect of the color shift from statistically derived value of human perception enhancement degree mentioned above.

EXPERIMENTAL AND ANALYSIS PROCEDURE

The procedure of experiments using observer's decision and statistical analysis is same as previous works [1–3], thus it is briefly described here. The LEDs used in this study was Nichia, NSPB500S (blue, 464 nm), NSPG510S (green, 520 nm), NSPR510CS (red, 633 nm) and NSPW500DS (blue-YAG type quasi-white, 464 and 560 nm). For the direct light observation two same LEDs were set on the same plate with 17 mm of distance between the centers of each LED. The distance from the LEDs to the observer was 1 m. For the indirect light observation two same miniature cars were used as targets of illumination. They were placed in a black box which has two compartments and they were separated by a black wall. The LEDs were set on the ceiling of the each compartment just above each car. The distance from the car to the observer is 30 cm. In both cases one LED was operated by DC current and the other was operated by pulsed current with frequency of 60 Hz. The duty ratio was changed from 2 to 60%. Under the fixed duty ratio condition, two LEDs were operated 50 times with random input power and an observer repeated judgment of specifying brighter LED (or same brightness) 50 times. Figure 1 shows an example of an observer's 50 times judgments arranged on the luminous flux of DC and pulsed LED. Pairs of the fluxes of DC and pulsed LED when the observer felt that the pulsed LED was brighter are shown as blue cross marks. The marks of green triangle and red circle correspond to the judgments of same brightness and brighter DC operation respectively. Here, the slope K_{Φ} of the line

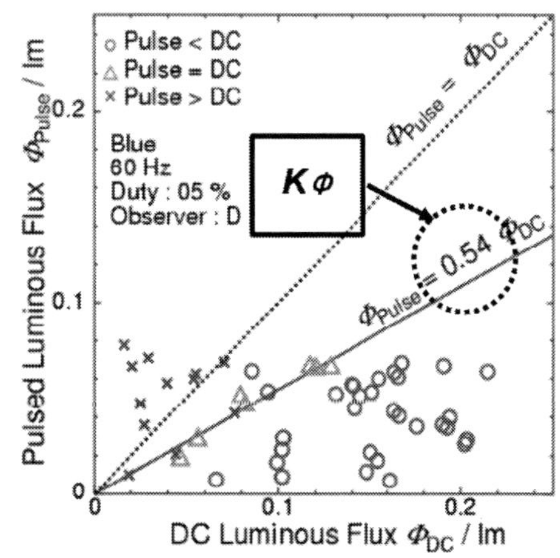

Fig. 1: An example of an observer's 50 times judgments arranged on the luminous flux of each LED.

obtained from the least square fitting on the points corresponding to the judgments of same brightness was calculated as 0.54. We defined the human perception enhancement degree as inverse of this slope; in this case $1/K_{\Phi}=1.85$ thus the pulsed operation enhanced the human perception by 1.85 times.

In order to separate the effect of the color shift from the human perception enhancement degree defined above, for all the operating condition, distances on the CIELAB color space between two points corresponding to the DC and pulsed operation were calculated as,

$$\Delta E = \sqrt{(L^*_{DC} - L^*_{pulse})^2 + (a^*_{DC} - a^*_{pulse})^2 + (b^*_{DC} - b^*_{pulse})^2} \ . \qquad (1)$$

After that, the human perception enhancement degree $1/K_{\Phi}$ was arranged as a function of the distance on the CIELAB ΔE.

RESULTS

Figure 2 shows an example of relationship between the human perception enhancement degree $1/K_{\Phi}$ and the distance on the CIELAB ΔE for (a) direct and (b) indirect observation of the blue LED. In both cases $1/K_{\Phi}$ increases when ΔE increases. It means the human perception is enhanced by color shift by pulsed operation. It is seems that the data points are going to the intercept of $1/K_{\Phi}=1$ linearly in the case of the direct observation, whereas the data points have an asymptote and have the intercept of $1/K_{\Phi}\approx 1.3$ in the case of indirect observation. It means that the effect of the pulsed operation itself exists in the case of indirect observation of the blue LED. Similar analysis for the other color LEDs with more observers is under progress.

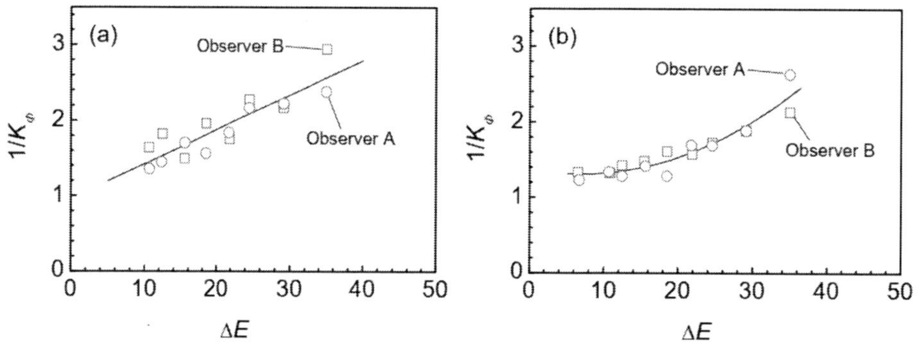

Fig. 2: An example of relationship between the human perception enhancement degree $1/K_{\Phi}$ and the distance on the CIELAB ΔE for (a) direct and (b) indirect observation of the blue LED.

CONCLUSION

The authors tried separating the effect of the color shift from statistically derived value of human perception enhancement. It is found that the effect of the pulsed operation itself exists in some cases. Analysis of the condition where the effect of the pulsed operation itself appears and investigation of its mechanism is under progress.

ACKNOWLEGMENT

This work was partly supported by NEDO through Industrial Technology Research Development Project (09E52013d).

REFERENCES

[1] M. Jinno *et al.*, J. Light & Vis. Env. **32** (2008) 161
[2] M. Jinno *et al.*, J. Light & Vis. Env. **32** (2008) 170
[3] M. Jinno *et al.*, Proc. 2nd Int. Conf. White LEDs and Solid State Lighting, Taipei, TA3-7 (2009)

Broad excitation band Barium silicate phosphors for LEDs, $Ba_4Si_6O_{16}:Eu^{2+}$ and $Ba_5Si_8O_{21}:Eu^{2+}$

T. Ishigaki[1] , K. Sato[2] , K. Uematsu[3] , K. Toda[1,2], M. Sato[1,3]

[1] *Center for Transdiciplinary Research, Niigata University, 8050 Ikarashi 2-no-cho, Nishi-ku, Niigata 950-2181, JAPAN*

[2] *Graduate School of Science and Technology, Niigata University, 8050 Ikarashi 2-no-cho, Niigata 950-2181, JAPAN*

[3] *Department of Chemistry and Chemical Engineering, Niigata University, 8050 Ikarashi 2-no-cho, Niigata 950-2181, JAPAN*

Contact:tishigaki@eng.niigata-u.ac.jp

ABSTRACT

Blue green emitted barium silicate phosphors, $Ba_4Si_6O_{16}:Eu^{2+}$, $Ba_5Si_8O_{21}:Eu^{2+}$ were prepared by a conventional solid state reaction. These silicate phosphors can be efficiently excited by n-UV light, yielding an intense blue green emission. Under the excitation of near UV light, the emission bands are broader than that of $Ba_2SiO_4:Eu^{2+}$ phosphor.

GENERAL

Since the investigation of the blue and n-UV LEDs, there is an interest in developing white LEDs for lighting proposes. The Eu^{2+} activated silicates seem suitable for the lighting application. In 1968, Barry and Blasse et al. first reported divalent europium in barium orthosilicate [1, 2]. Thereafter, Poort investigated the luminescence properties of $Ba_2SiO_4:Eu^{2+}$ at different cation sites [3]. According to these reports, barium-silicate is suitable for photoluminescent host compound. The advantage of the application for LEDs are novel broad excitation bands based on alkaline earth silicate host matrix $Ba_4Si_6O_{16}$ doped with Eu^{2+} for n-UV excitable white light emitting diodes have successfully investigated. The phosphors were synthesised by solid state reaction.

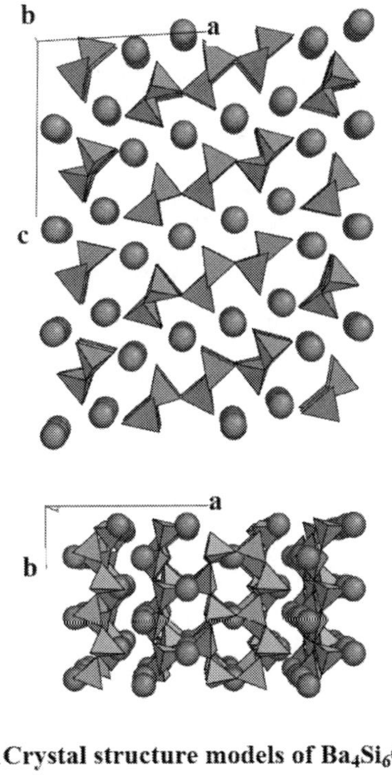

Fig. 1Crystal structure models of $Ba_4Si_6O_{16}$[4]

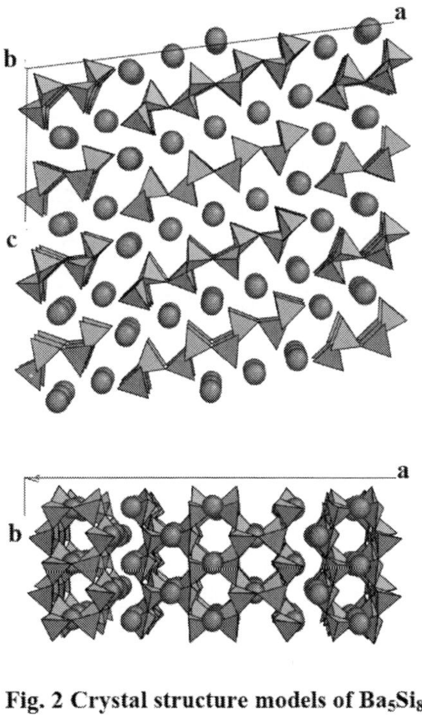

Fig. 2 Crystal structure models of $Ba_5Si_8O_{21}$[5]

The Ba^{2+} ions were located 4 kinds of asymmetric position in the $Ba_4Si_6O_{16}$ lattice. On the other hand, $Ba_5Si_8O_{21}$ has 3 asymmetric barium ions positions.

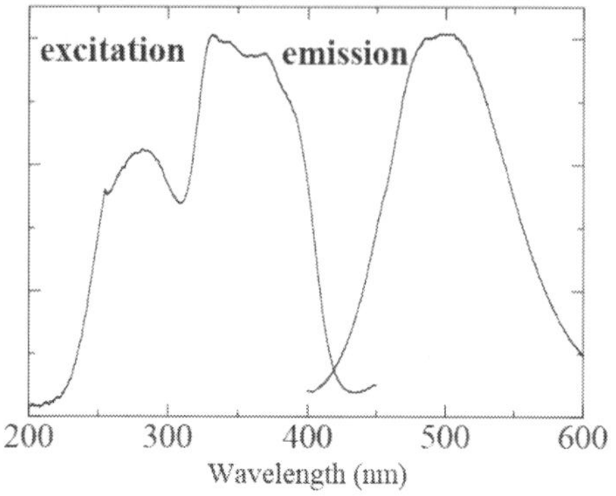

Fig. 3 Excitation and emission spectra of $Ba_4Si_6O_{16}:Eu^{2+}$

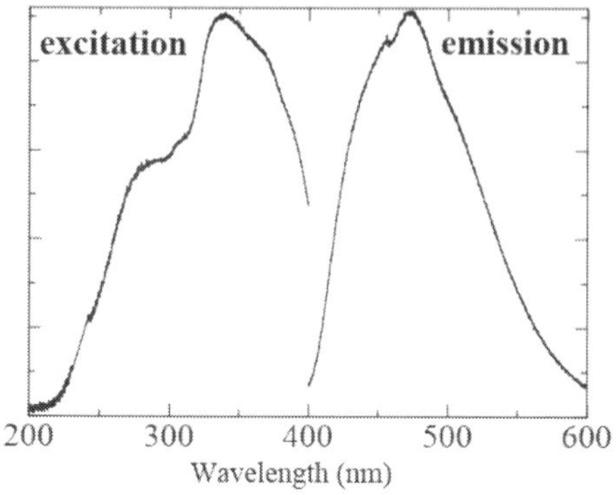

Fig. 4 Excitation and emission spectra of $Ba_5Si_8O_{21}:Eu^{2+}$

Photoluminescence spectra of the $Ba_4Si_6O_{16}:Eu^{2+}$ sample was shown in figure 3. This compound shows a wide band excitation and emission. It is because, the emission centre, Eu^{2+} was substituted asymmetric atomic position of Ba^{2+}. It is said that the coordination of emission centre was strongly related with photoluminescence properties. As same as the $Ba_4Si_6O_{16}:Eu^{2+}$ sample, $Ba_5Si_8O_{21}:Eu^{2+}$ sample was shown wide band excitation and emission, too.

As a result, these barium silicate phosphor, $Ba_4Si_6O_{16}:Eu^{2+}$ and $Ba_5Si_8O_{21}:Eu^{2+}$ are expected as green phosphor for tricolored white LED excited by n-UV LED.

REFERENCES

[1] T.L. Barry, J. Electrochem. Soc. 115 (1968), 1181
[2] G. Blasse, W.L. Wanmaker, J.W. Tervrugt and A. Bril, Philips Res. Rep. 23 (1968), 189
[3] S.H.M. Poort, W. Janssen and G. Blasse, J. Alloys Compd. 260 (1997), 93
[4] O S. Filipenko, Doklady Akademii Nauk SSSR, 196 (1971) 137
[5] K. F. Hasse, Zeitschrift fuer Kristallographie, 153 (1980) 3

Surface analytical investigation of the Tungsten build-up in Ceramic Metal Halide lamps

Z. Tóth[1], T. Russell[2], T; Z. Kolozsi[3], Z, V. K. Josepovits[3]

[1] *GE Lighting, GE Hungary, Vaci ut 77, Budapest, H-1044, Hungary*

[2] *GE Lighting, 1975 Noble Road, East Cleveland, 44112-6300, USA*

[3]*Dep.t of Atomic Physics, Budapest University of Technology and Economics, Budafoki út 8., H-1111, Budapest*

contact: zoltan2.toth@GE.com

ABSTRACT

A technique has been set up to study the very early stage of the lumen depreciation in ceramic metal halide lamps. The capability of measuring sub-monolayer coverage of tungsten on an alumina surface is demonstrated. The method can be used to make lumen maintenance predictions before the long term ageing, as well as to study the layer build up as a function of different surface conditions.

INTRODUCTION

Decrease of the lumen output of the quartz or ceramic metal halide lamps (QMH or CMH) over their life is a well-known phenomenon. The reason for the lumen loss is the blackening of the arc tube wall caused by the transport and condensation of tungsten from the hot electrode tip to the wall. The tungsten particles formed on the arc tube surface block the light of the arc [1]. Ensuring an appropriate oxygen pressure exists in the arc, the tungsten can be transported back to the electrode enabling superior lumen maintenance during lifetime [2], [3].

EXPERIMENTAL

The aim of this investigation was to study the build up of the tungsten layer in the very early stage of the lamp ageing (a few hours to a few hundred hours). At this time the probability of formation of a new tungsten particle on a clean alumina surface is higher, so the rate of the lumen loss is higher also. (It is assumed, that once a tungsten particle is formed it blocks the light completely, so a tungsten particle formed on the top of another has no or minimal influence on lumen loss.) The surface at this stage is covered by sub-micron sized tungsten particles, so to be able to measure the amount of the actual tungsten coverage via traditional techniques – like scanning electron microscopy (SEM) – cannot be used since they would detect the alumina layer below the tungsten particles as well, i.e. one has to use a technique which detects only the topmost few atomic layers. Since the surface coverage is small – below 1% - a very sensitive technique must be used. Based on these, Secondary Ion Mass Spectroscopy (SIMS) with very good depth sensitivity (2-5 monolayers) and elemental sensitivity (up to 1 ppm) [4] has been chosen. To minimize the charge-up of the insulating ceramic surface during the measurement Fast Atom Beam (FAB) has been used to sputter the surface instead of the usual ion gun. To avoid the effect of the curved surface, high wattage lamps with large internal diameter are preferred during the investigation.

RESULTS AND DISCUSSION

As a baseline, a 100 hours aged experimental ceramic metal halide lamp with NaI-SrI_2-TlI-CeI_3 dose with oxygen source has been measured. SEM picture in Figure 1 reveals only few spots on the arc tube surface. Figure 2a and 2b shows the SIMS spectra taken on the wall next to the electrode and next to the middle of the arc gap, respectively. In both cases the tungsten as well as tungsten-oxide signal can be nicely observed. As expected, the intensity of the tungsten is higher next to the electrode (source of the tungsten). Dose elements (Sr and Ce) can also be observed together with their oxides. The presence of the dose oxides is the effect of the reaction of the dose with the alumina wall as well with the oxygen source. Figure 3 shows the thermodynamical calculation of the amount of the cerium-aluminate formed as a function of the added oxygen source. Even with no extra oxygen added cerium-aluminate formation is predicted, and adding the oxygen increases the amount further. Although the intensity of the oxides seems to be in the same order of magnitude of the pure metal, it has to taken into account that oxygen enhances the probability of the formation of a positive ion, so the actual concentration is less, than one would calculate from the pure intensities.

Figure 1. SEM picture of the internal surface of the ceramic metal halide arc tube after 100 hrs of ageing

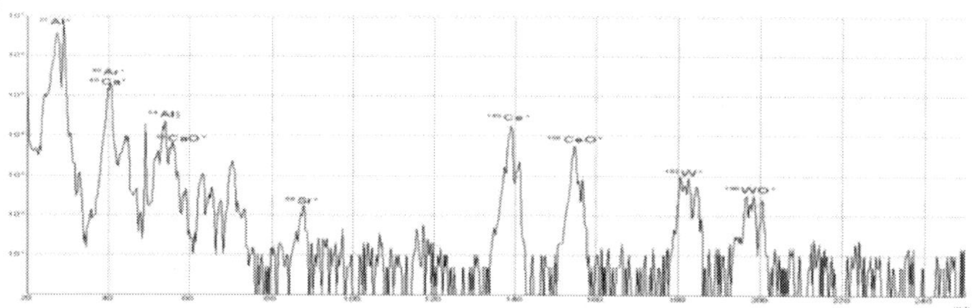

Figure 2a. SIMS spectrum of the internal arc tube surface next to the electrode

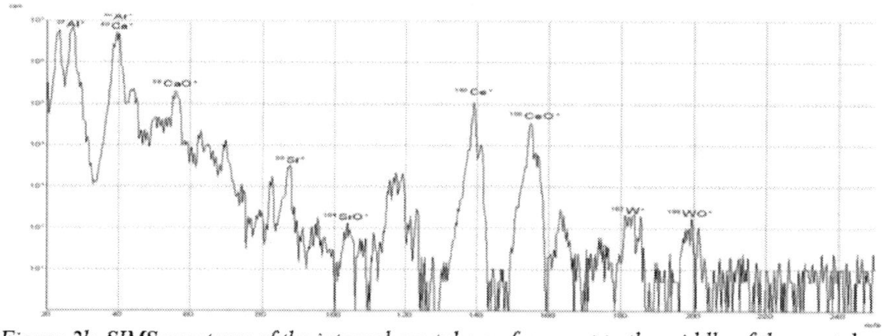

Figure 2b. SIMS spectrum of the internal arc tube surface next to the middle of the arc tube

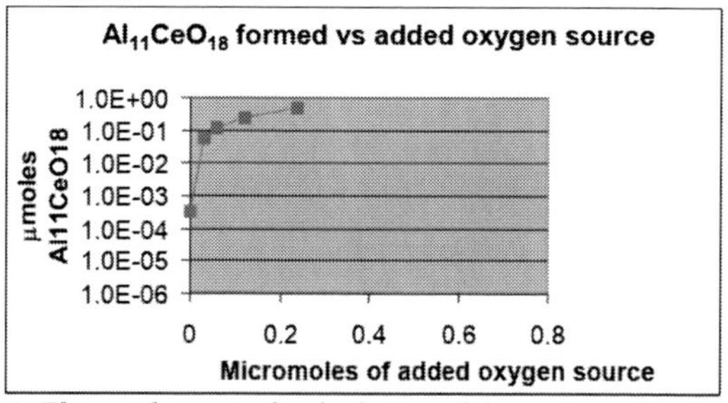

Figure 3. Thermodynamical calculation of the cerium-aluminate formation.

REFERENCES

[1] W. van Erk, Pure Appl. Chem, vol. 22, no. 11 (2000), 2159
[2] J. Tu et al, Proc. of 11th Int. Symp. of the Sci. and Techn. of Light Sources, Shanghai, 2007, 35
[3] J. He, Z. Toth, T. Russell, US Patent Application 2009/0146570 A1
[4] A. Benninghoven et al, Secondary Ion Mass spectrometry, Wiley-Interscience, New York, 1987

Session Three

Roll-to-Roll OLED Lighting Technology

Anil R. Duggal
GE Global Research Center, 1 Research Circle, Niskayuna, NY 12309,
Contact: duggal@research.ge.com

ABSTRACT

This paper describes progress in developing the key technology requirements that can enable a low cost organic light-emitting device (OLED) source for lighting. The results indicate that roll-to-roll fabrication is possible and that solution-processed devices compatible with such processing have reached the 50-60 LPW efficacy range of interest for general lighting applications.

Organic light-emitting device (OLED) technology has been actively pursued since the mid-1980's for displays and such products are now commercially available. Efforts aimed at developing OLEDs for lighting applications started ~10 years ago and are now maturing rapidly.[1-3] Due to the inherent capability of the technology, the focus has been on developing OLEDs as a diffuse light source. At present, most of the key attributes necessary for a commercial product have been demonstrated. In particular, it has been shown that large area OLEDs can be made with acceptable color and color-rendering with efficacy in the 40-60 LPW range and with lifetimes with a light output of ~3000 lumens/m2 in the range between incandescent and fluorescent technology. The key missing attribute to date is a demonstration that low cost manufacture is possible. Simple arguments suggest that to be competitive with the best fluorescent technology for general lighting, manufacturing costs of ~\$20/m2 must be attained.[4] In this paper, we describe GE's efforts at developing low cost OLED lighting.

In their essence, OLEDs are simple devices. They consist of thin organic layers (~100 nm) sandwiched between two electrodes, one of which is transparent. When a voltage is applied, electrons and holes are injected from the electrodes and travel through the organic material until they meet to create an excited state which generates light. Over the years, there have been two basic OLED development paradigms which differ primarily in how the organic layers are deposited during device fabrication. One paradigm – dry coated OLEDs – utilizes physical vapor deposition in high vacuum while the other – wet coated OLEDs – utilizes solution-based coating processes at atmospheric pressure. Most of the important innovations leading to increased device performance have first been demonstrated with dry-coated OLEDs and then subsequently repeated with wet-coated OLEDs. This is because there are extra constraints in building multi-organic-layer devices in wet-coating relative to dry coating such as ensuring solvent orthogonality and material purity.

At the research scale, it is relatively easy to pursue both dry and wet coated OLEDs. However, for manufacturing, a choice needs to be made. Either approach should be able to hit the manufacturing cost target for lighting if the manufacturing scale is made large enough that the cost per unit is dominated by material costs. However, for business viability, where return on investment is important, the key metric that must be considered is the maximum product throughput that can be achieved for a given capital investment. Here, wet-coating of the organic layers has a major advantage since it can be extremely fast and, in contrast to dry-coating, is not limited by a maximum evaporation rate due to the low damage threshold of organic materials.

The goal of GE's effort has been to develop high performance wet-coated OLEDs that can be manufactured with high throughput roll-to-roll processing. Accomplishing this has required efforts in substrate development, roll-to-roll process development, and materials and device designs that yield high performance. Substrate development is required because traditional glass substrates are not compatible with roll-to-roll processing. It is straightforward to replace glass with transparent plastic film as a substrate but the water and oxygen permeability of known plastics is approximately 5 orders of magnitude too high relative to current OLED material sensitivities. Simple transparent inorganic coatings on the plastic film do not solve the permeability problem due to the inevitable pinholes and other defects that occur during the coating process. Various more complex coatings have been demonstrated that appear to be compatible with OLED requirements. GE's solution consists of a coating that is graded between inorganic and organic zones in order to mitigate the effect of coating defects.[5] This coating is deposited using plasma-enhanced chemical vapor deposition technology and is currently being scaled for manufacturing.

In order to prove the feasibility of roll-to-roll wet-coated OLED manufacturing, a research fabrication line was developed. The goal was to utilize the highest throughput coating and patterning methods possible. For the organic layers, a blanket coating method was chosen followed by a simple solvent-assisted wiping technique for low resolution subtractive patterning. For the metal cathode, vacuum-based evaporative deposition was utilized since it is known from the metal coating industry that this can be scaled to high throughput. A key innovation in the line is in the integration of the modules together enabled by an atmospheric pressure to vacuum module. This allows the web to move continuously from the organic coating modules to the vacuum coating modules at high speed. The research line was fully operational in 2007.

The first OLEDs made on GE's research roll-to-toll line utilized simple two-layer polymer-based device architectures. However, to date, polymer devices have been limited in performance relative to small-molecule based devices where the organic layers are vacuum-deposited at low rate. This is likely due to the fact that the highest performance polymer materials have a triplet energy that is not compatible with the full-range of phosphorescent dye materials currently utilized in small-molecule devices.[6]

In order to obtain both the manufacturing advantages of wet-coated polymer devices and the performance advantages of dry-coated small-molecule devices, GE has teamed with Konica Minolta (KM) to develop an OLED based on small-molecule materials where all of the organic layers can be wet-coated using standard roll-to-roll techniques. The key challenges in this endeavor have been in the design and synthesis of molecules with the right combination of electronic, optical, and film-forming properties and the design of devices that minimize the number of organic layers and account for process incompatibilities between the different materials.

Figure 1 shows the most successful white device design achieved by the GE/KM partnership to date. It is a down-emitting device consisting of at least 4 organic layers each of which is successively wet-coated on top of the other. Also shown is an outcoupling layer which enhances light extraction from the substrate and is applied after the device is made. The appropriate white light spectrum is provided by one or more emission layers made up of multiple small-molecules including host materials and various phosphorescent dyes. At present, our highest performing devices are designed to have a color temperature of 2900K and a color-rendering index of 80. The performance of these devices is 56 LPW and 6000 hrs life at a brightness of 1000 cd/m2 (3141 Lumens/m2).

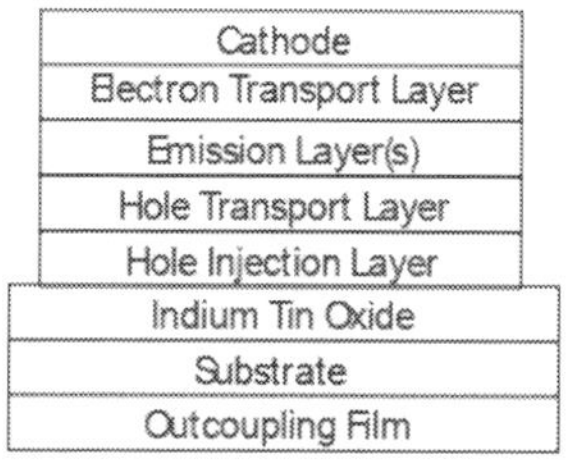

Figure 1: *Schematic of solution-processed small-molecule OLED device structure.*

The performance of these latest GE/KM devices is on par with the best lighting-quality OLEDs made to date by any technique and demonstrates that it is possible to make OLEDs that utilize high performance small-molecules using standard, high speed, wet-coating techniques. This suggests that it will be possible to make OLEDs for lighting at the low cost required for lighting with acceptable capital equipment expenditures. Work is now in progress to replicate these results at large scale using roll-to-roll processing and to further increase the intrinsic device performance.

ACKNOWLEDGEMENTS

This work is the result of nearly a decade long effort by a dedicated team of scientists and engineers at GE. In addition, it is a pleasure to acknowledge the fruitful collaboration with the research team of Konica Minolta in developing high performance solution-coatable small molecule devices.

REFERENCES

[1] A. Bergh, G. Craford, A. Duggal, and R. Haitz, *Physics Today*, **54**, 42, (2001).
[2] B. W. D'Andrade and S. R. Forrest, *Adv. Mat.,* **16**, 1585 (2004).
[3] A. R. Duggal, C. M. Heller, J. J. Shiang, J. Liu, and L. N. Lewis, *J. Display Tech.,* **3**, 184 (2007).
[4] A. R. Duggal, in Organic Electroluminescence, Z. H. Kafafi, Ed. Taylor & Francis, Boca Raton, Fl, pp. 437-466 (2005).
[5] M. Yan, T. W. Kim, A. G. Erlat, M. Pellow, D. F. Foust, J. Liu, M. Schaepkens, C. M. Heller, P. A. McConnelee, T. P. Feist, and A. R. Duggal, *Proc. IEEE*, **93**, 1468 (2005).
[6] S. Madhusoodhanan, P. I. Djurovich, T. E. Hogen-Esch, and M. E. Thompson, *JACS*, **125**, 7796 (2003).

OLED Technical challenges and the chemistry involved

A. Monkman

OEM Research Group, Dept of Physics, University of Durham, UK

Contact: a.p.monkman@durham.ac.uk

ABSTRACT

It is clear from many recent reports of ultra efficient OLEDs that the major challenge for efficient white light generation and the realisation of OLED solid state lighting is an efficient stable long lifetime blue 1. Obviously, this could be fluorescent but phosphorescent would be much better. One major problem for the latter is the need for a high triplet energy host to accommodate the blue phosphor. One major problem with the former is the lack of singlet excited states generated on recombination.

During this presentation I shall briefly show why blue fluorescent, red and green phosphorescent is not a potentially good strategy, due to the contributions of triplet triplet annihilation to the overall yield of known blue emitters. I will focus thereafter on the blue emitter host problem.

Key developments on blue phosphors will be discussed and the potential amount of blue light in the 440nm to 480 nm band required for typical lighting. The stringent requirements for host materials will be given along with problems that can arise through host morphology which limit the usefulness of certain design motifs. A range of blue hosts will then be described both for small molecule and polymer devices. Various strategies for controlling singlet and triplet energies of a host independently will be shown. This is a key requirement, especially as the number of layers within a device must be kept to a minimum to help improve production yield. This is true for both OLED and PLED.

Finally, novel techniques for producing multilayer PLEDs will be shown to give an overview of how chemical properties can be used to aid complex device fabrication. This work is clearly aimed at solution processing of large area lighting panels. As an example, I will report on work in Durham which has led to the demonstration of a green precursor multilayer PLED which has achieved a photonic efficiency of 40 cd/A. Strategies for full RGB phosphorescent white following on from these results will be discussed.

REFERENCES

[1] Recent Advances in White Organic Light-Emitting Materials and Devices (WOLEDs), Kamtekar KT, Monkman AP, Bryce MR., Advanced Materials **22,** 2010, 572-582

Organic Light-Emitting Transistors with an Efficiency that Out-performs the Equivalent OLEDs

R. Capelli[1], S. Toffanin[1], G. Generali[1], H. Usta,[2] A. Facchetti[2] and M. Muccini[1,3]

[1] *Consiglio Nazionale delle Ricerche (CNR), Istituto per lo Studio dei Materiali Nanostrutturati (ISMN), via P. Gobetti 101, I-40129 Bologna, Italy*
[2] *Polyera Corporation, 8045 Lamon Avenue, Skokie, IL 60077 (USA)*
[3] *E.T.C. srl, via P. Gobetti 101, I-40129 Bologna, Italy*
Contact: M.Muccini@bo.ismn.cnr.it

ABSTRACT

The potential of organic semiconductor-based devices for light generation is demonstrated by the commercialization of display technologies based on organic light emitting diodes (OLEDs). Nonetheless, exciton quenching and photon losses processes still limit OLED efficiency and brightness. Organic light-emitting transistors (OLETs) are alternative light sources combining, in the same architecture, the switching mechanism of a thin-film transistor and an electroluminescent device. Thus, OLETs could open a new era in organic optoelectronics and serve as testbeds to address fundamental optoelectronic and photonic general issues. Here, we introduce the concept of using a p-channel/emitter/n-channel tri-layer semiconducting heterostructure in OLETs providing a novel approach to dramatically improve OLET performance and address these open questions. Exciton-charge annihilation and electrode photon losses are prevented. Our devices are >100× more efficient than the equivalent OLED, >2× more efficient than the optimized OLED with the same emitting layer and >10× more efficient than any other reported OLETs.

GENERAL

The organic light-emitting transistor (OLET) is an optoelectronic device having the structure of a thin-film transistor and the capability of light generation [1]. Bright/multicolor OLETs may allow electroluminescent display fabrication with simpler driving circuits. Furthermore, the most advanced OLETs encompass a huge technological potential for the realization of intense nanoscale light sources and highly integrated optoelectronic systems, including the long time searched electrically pumped organic laser. In terms of performance and reliability, OLED technology is by far the most developed and active matrix OLED displays have been introduced into the market. However, detrimental device-related processes affecting OLED operation under high injection conditions are the exciton-charge interactions and the photon losses at the electrodes. The proximity (within tens of nanometer) of the contacts to the OLED light generation region induces losses due to absorption of the emitted photons. Moreover, the highly dense electron and hole currents converge to the light emitting layer, where they form but spatially coexist with the excitons, and lead to significant exciton-charge quenching. Indeed, this mechanism is predicted to be greater than any other quenching effects and should be controlled to enhance OLED efficiency, brightness, and stability even further. Thus, at the base of the OLET development there is the possibility to enable new display/light source technologies, and exploit a transport geometry to suppress deleterious photon losses and exciton quenching mechanisms inherent in the OLED architecture. To date, only the exciton-metal interaction has been successfully addressed in OLETs based on ambipolar single layers. Under proper bias conditions, the spatial location of the light-emitting area is far from the metal electrodes preventing exciton-metal quenching. However, in these single layer devices the charge carrier accumulation and the exciton formation zones largely coincide, leading to severe exciton-charge quenching. Indeed, even in the most impressive demonstration to date, the external quantum efficiency (EQE) does not exceed 0.5-0.6%. Consistently, single layer unipolar OLETs, in which only one type of charge carrier is effectively transported across the channel, reached remarkable results in terms of brightness, but their EQE is only 0.2%, mainly because excitons are subjected to both metal and charge quenching, and electrode induced photon losses. A horizontal PN-heterojunction OLET has also been reported, in which light emission is confined far from the contacts, however, exciton-charge quenching is not avoided. A bi-layer approach has been used to improve OLET brightness, or achieve higher and more balanced charge transport. In the first case a highly efficient luminescent layer is superimposed over a unipolar conducting layer, while in the second case p-type and n-type transport films are directly in contact with each other. However, in both cases this device architecture does not offer any control of the exciton quenching and photon losses as the light emission area is in contact with the minority carrier injecting electrode. Excitons interact with accumulated charges and the metal electrode, while photons are absorbed by the contacts. In this work, we report the first tri-layer heterostructure approach for OLETs enabling simultaneous control of electrode induced photon losses, exciton-metal and exciton-charge interactions. OLET devices with EQE of 5% are demonstrated [2], which exceeds the best OLEDs based on the same emitting layer and optimized transport layers (2,2%).

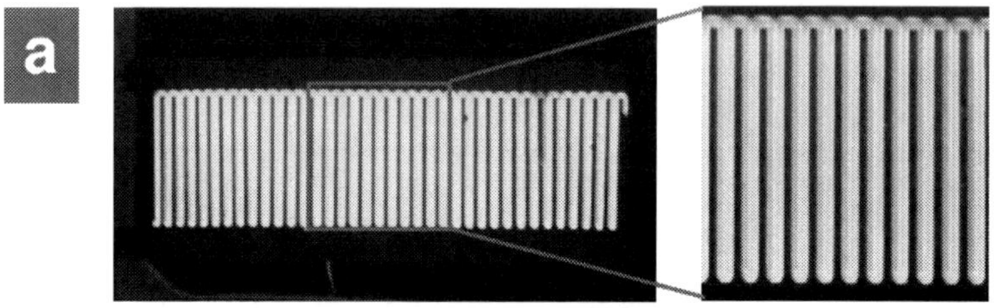

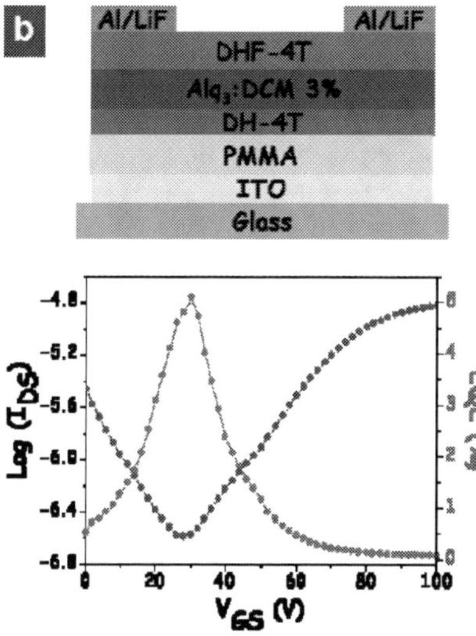

Figure 1. a) Image of a tri-layer interdigitated OLET device, and b) schematic of the device structure with the electrical transfer and EQE profile as a function of the applied V_{GS} potential.

REFERENCES

[1] M. Muccini, A bright future for organic field-effect transistors. *Nature Mater.* **5**, (2006) 605-613.

[2] R. Capelli, S. Toffanin, G. Generali, H. Husta, A. Facchetti and M. Muccini, Organic light-emitting transistors with an efficiency that out-performs the equivalent light-emitting diodes. *Nature Mater.* (2010) DOI: 10.1038/NMAT2751.

Methodology of LED's total luminous flux using a compound parabolic concentrator

Xiaoli Zhou, Haiping Shen, Wanlu Zhang, Muqing Liu

Institute for Electric Light Sources, Fudan University, 220 Handan Road, Shanghai 200433, P. R. China

Contact: mqliu@fudan.edu.cn

ABSTRACT

As the applications of Light Emitting Diodes (LEDs) expand, measurement of LED's total luminous flux is becoming increasingly important. LED is quite different to conventional light sources in terms of spectral and spatial characteristics, which brings difficulties for measuring LED's total luminous flux. This paper describes a fast and feasible method for accurately measuring the total luminous flux of high power LED. A Compound Parabolic Concentrator (CPC) is introduced to collect the light of an LED in conjunction with a detector which in turn measures the luminous flux.

1. INTRODUCTION

LED technology is developing very quickly and is penetrating the market of general lighting [1] due to increasing of LED's efficacy. It is required that the total luminous flux of LEDs should be measured as accurately as possible for general lighting applications.

Generally, the total luminous flux of light sources can be measured by using either a goniophotometer or an integrating sphere [2]. In comparison with the latter, the former is more accurate and needs no standard lamp of total luminous flux but at the expense of a great deal of time. The method by integrating sphere allows fast measurement, and hence is widely adopted for measurements of various conventional light sources [3]. However, LED is quite different from conventional light sources in terms of physical size, flux level, spectrum and spatial distribution, errors occur when the traditional integrating sphere measurement is used to achieve LED's total luminous flux.

This paper presents a new method and portable device using a Compound Parabolic Concentrator (CPC) to measure LED's total luminous flux.

2 FUNDAMENTAL DESIGN OF THE METHOD FOR LED'S LUMINOUS FLUX MEASUREMENT

A Compound Parabolic Concentrator is introduced in this paper to collect the light from a test LED in conjunction with a V(λ)-corrected and cosine-crrrected detector, which carries out the measurement. The system is calibrated with a total luminous flux standard.The total luminous flux of a test LED can be obtained by comparison to the standard LED having smilar spectral distribution. The device using this method is small size and portable for production line.

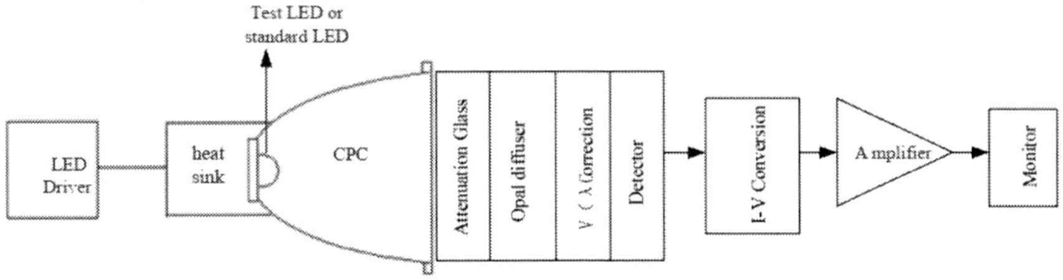

Figure 1. Schematic diagram of LED's luminous flux measurement system.

Figure 1 shows the system schematic diagram of LED's total luminous flux measurement. In this system different heat sinks (with area in proportion to LED's power) are used for test LEDs. The LED's optical center is adjusted at the focus plane of CPC which is acting as a flux concentrator. The CPC's inner surface is coated with a silver paint of 98% reflectance, and its exit port is mounted with a V(λ)-corrected detector. The detector has an opal diffuser for consine correction.An attenuation glass of 90% absorptance is equipped between the detector and CPC to reduce the influence of reflected lights from the detector.

The focal of CPC is: $f = a(1+\sin\theta_{max})$

θ_{max} denotes the maximum incident angle, whereas a denotes the radius of the exit port.

$$c = \left(\frac{d_1}{d_2}\right)^2 = \sin^2\theta_{max} \ , \qquad d_2 = 2a$$

Where d_1, d_2 denote diameters of entrance and exit port respectively, as shown in Figure 2.

CPC is often used in tandem with primary solar concentrators for augmenting the solar flux intensity incident onto solar receivers. In this paper, CPC is used for concentrate light. The test LED is positioned at the CPC's exit aperture; rays from the LED will reach the CPC's entrance with at most one reflection and their angles will be within θ_{max}.

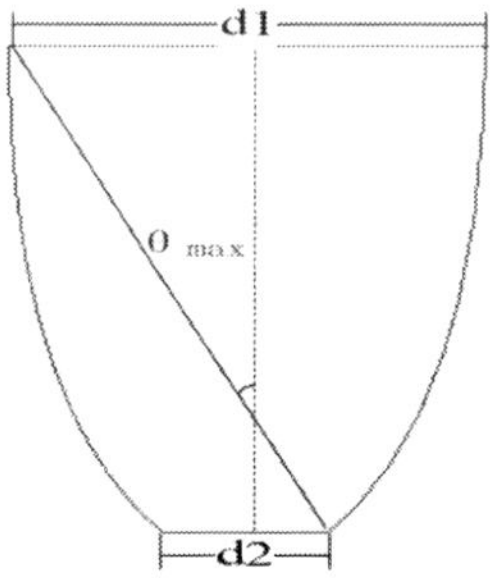

Figure 2. Profile of Compound Parabolic Concentrator.

3 EXPERIMENTAL ANALYSIS

An experimental system was built as shown in Figure 1. Seven LED samples were selected to be measured by this system, of them samples 1 and 2 are in good lambertian spatial distribution, samples 3 and 4 are in similar lambertian spatial distribution, samples 5, 6 and 7 are in quite narrow light distribution-far away from lambertian distribution. Each LED sample was also measured with a goniophotometer(GO-R3000, Everfine, Hanzhou China) and results from the CPC system and the goniophotometer were compared in table 1. Another lambertian white LED was chosen to be the standard lamp to calibrate the CPC system after its flux was measured by the goniophotometer. In all measurements, the whole installation was placed in a dark room; LEDs were driven by a constant DC current of 350mA. The measurements were started 30 minutes after LED was driven for getting a thermal equilibrium. The deviations of this method, as compared with the goniophotometer, are found to be within ±5% for all tested LEDs.

Table 1 Comparison of results using the CPC method with those from using the goniophotometer method.

NO.	Result of goniophotometer method(lm)	Result of CPC method (lm)	Relative error
1	61.83	63.821	3.22%
2	40.36	39.216	-2.83%
3	61.84	61.232	-0.99%
4	70.17	72.46	3.26%
5	39.05	37.287	-4.73%
6	41.67	39.893	-4.45%
7	50.54	51.903	2.69%

4. CONCLUSIONS

A new approach, based on a Compound Parabolic Concentrator was proposed, to measure LED's total luminous flux. The differences of the CPC method, as compared with the goniophotometric results, for seven different types of LEDs are found to be within ±5%, which is applicable for industry.

REFERENCES

[1] Liu M, Zhou D, and Mei Y. The comparison of light efficacy between LED and traditional light sources .Zhaoming Gongcheng Xuebao 17(2006) 41-5

[2] Kathleen Muray. New international recommendations on LED measurements. Proc. SPIE 3140(1997)12-8

[3]CIE 2007. Measurement of LEDs. CIE Technical Report 127-2007 (Vienna: CIE CB)

Fabrication of High-Brightness Large-Area White OLED with Auxiliary Electrodes by Meniscus Printing Process

Y.Shinjo, H.Oh-oka, T.Sawabe, T.Sugizaki, A.Amano, T.Ono, K.Sugi, Y.Mizuno, I.Takasu, J.Yoshida, S.Enomoto, A.Hirao, and I.Amemiya

Electronic Imaging Laboratory, Corporate Research & Development Center, Toshiba Corporation, 1, Komukai-Toshiba-cho, Saiwai-ku, Kawasaki 212-8582, Japan

Contact: *yasushi.shinjo@toshiba.co.jp*

Abstract: *We demonstrated high-brightness large-area white OLED (organic light-emitting diode) consisting of printed organic semiconductor layers. Meniscus printing process was applied to the substrate with 2 μm-high stripe-shape auxiliary electrodes. The OLED panel showed white emission all over the whole emitting area of 58 mm × 52 mm, high average luminance of 10,000 cd/m^2, luminance uniformity of 40 %, and high luminous flux of 95 lm.*

INTRODUCTION

Polymer OLEDs have been the subject of intense research because of their potential for low-cost fabrication by solution processes. To achieve uniform large-area light emission, it is necessary to reduce electric potential drops resulting from limiting conductivity of ITO (indium tin oxide). One of the solutions is to use auxiliary electrodes, which can lower the effective ITO resistance [2]. We employed the meniscus printing process [1] shown in Figure 1 and applied this process to the substrate with stripe-shape auxiliary electrodes.

DEVICE STRUCTURE AND PROCESS

We calculated the electric potential distribution based on the stripe-shape auxiliary electrodes. In the case that the width of the auxiliary electrode is 150 μm and the pitch is 1 mm, the maximum electric potential drop of the anode is about 4 %. To prevent the emission on the auxiliary electrodes, they were covered (200 μm width) with insulating material (polyimide) and then the aperture ratio of the panel was 80 %. As shown in Figure 2, we printed all of the organic semiconductor layers by the meniscus printing process on the substrate with the auxiliary electrodes as mentioned above. After printing organic semiconductor layers, the electron injection layer and Al layer as the cathode electrode were deposited by the vacuum sublimation process and a top encapsulating cover glass was sealed to the device in an N_2 atmosphere. Finally, a light-outcoupling film was attached to the device. As shown in Figure 1, the total height of the auxiliary electrode with insulating layer is about 2 μm. It is key technology how to deposit uniform thin layers of organic semiconductors of several tens of nm thick on the substrate with such a relief structure. The meniscus printing process is a robust process with a relatively large printing gap (100 μm - 800 μm) compared to other printing or coating methods and can be applied to a substrate with irregular surface. Thus, we have chosen the appropriate combination of the auxiliary electrode design and the printing process.

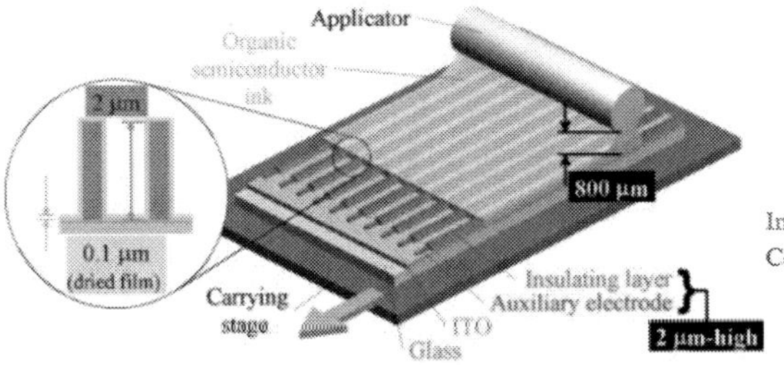

Figure 1. Schematic illustration of the meniscus printing method applied to the ITO glass substrate with 2 μm-high stripe-shape auxiliary electrodes.

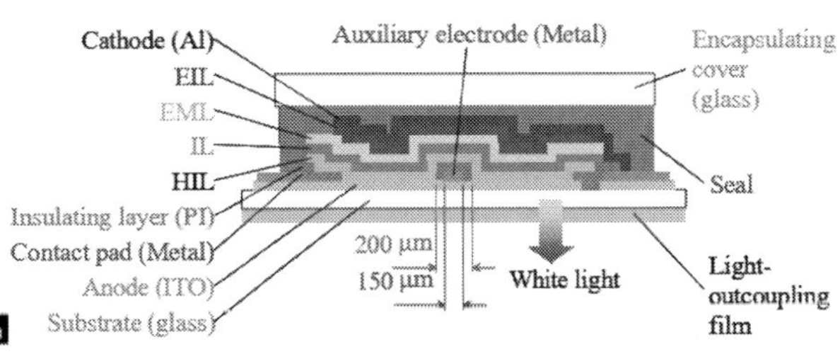

Figure 2. Schematic illustration of sectional structure of the fabricated OLED panel.

Metal : Mo-Nb/Al-Nd/Mo-Nb, ITO : Indium tin oxide, Al : Aluminum, PI : Polyimide, EIL : Electron injection layer, EML : Emitting layer (white), IL : Interlayer, HIL : Hole injection layer

RESULT

Figure 3 shows the typical cross-sectional profiles of the organic semiconductor layers between the banks of the insulating material on the auxiliary electrodes. As shown in the figure, the total thickness accuracy of the three organic semiconductor layers between the banks was ± 5 %. The thickness accuracy is given by

{ (Max. - Min.) / 2 } / { (Max. + Min.) / 2 } × 100 (1)

The elements of Max. and Min. on the right side in Eq.(1) are the maximum and minimum values of the film thickness, respectively. This result was obtained in the case that every layer was printed along the stripe direction of the auxiliary electrodes (Figure 1). Figure 4 shows a demonstration of white OLED panel fabricated by the meniscus printing process. The panel showed white emission all over the whole emitting area of 58 mm × 52 mm, high average luminance of 10,000 cd/m^2, luminance uniformity of 40 %, and high luminous flux of 95 lm. The luminance uniformity is given by

{ (Max. - Min.) / Max. } × 100 (2)

The elements of Max. and Min. in Eq. (2) are the maximum and minimum values of the luminance, respectively. In the calculation of the luminous flux, it was assumed that the emission follows Lambertian distribution. Since ensuring sufficient heat radiation was not considered in the design of the fabricated OLED panel, the value of luminance uniformity was much higher than our expectation.

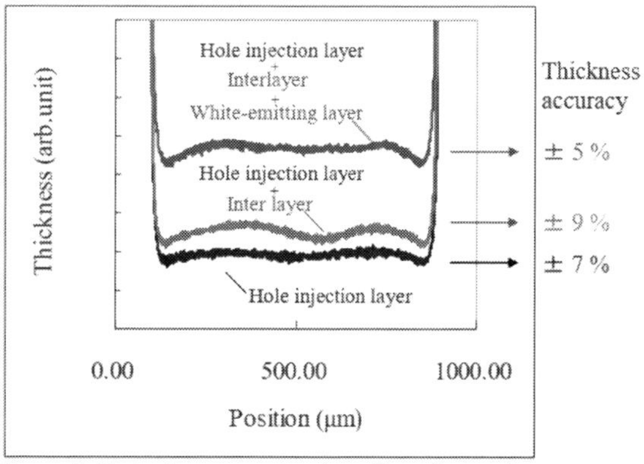

Figure 3. Typical cross-sectional profiles of printed organic semiconductor layers between banks.

Figure 4. Demonstration of white OLED panel fabricated by the meniscus printing process.

Emitting area: 58 mm × 52 mm, Average luminance: 10,000 cd/m^2, Luminance uniformity : 40 %, Luminous flux: 95 lm

CONCLUSION

High-brightness large-area white OLED fabricated by the meniscus printing process was demonstrated. The organic semiconductors were printed onto ITO glass substrate with 2 μm-high stripe-shape auxiliary electrodes. Each organic semiconductor layer was several tens of nm thick and the total thickness accuracy was ± 5 %. The OLED panel showed white emission all over the whole emitting area of 58 mm × 52 mm, high average luminance of 10,000 cd/m^2, luminance uniformity of 40 %, and high luminous flux of 95 lm.

Acknowledgements

We thank Sumitomo Chemical Co., Ltd. and Sumation, Inc. for providing organic semiconductors including white-emitting polymer and interlayer polymer.

REFERENCES

[1] Byoungchoo Park, Mi-Young Han, Mina Kim, Chan Hyuk Park and Young In Lee, IDW '08 DIGEST 1037-1040 (2008).

[2] Joerg Amelung, SPIE Newsroom, 21, April (2008).

On the standard evaluation procedure for biological action of light

R. Kozakov[1] , H. Schöpp[1] , K. Barkowsky[1] , C. Stoll[2], D. Kunz[2]
[1] *Leibnitz Institute for Plasma Science and Technology, F-Hausdorff-Str. 2, Greifswald, D-17489, Germany*
[2]*Charité – University medicine, Sleep Research & Clinical Chronobiology, Gr. Hamburger-Str. 5-11, Berlin, D-10115, Germany*
Contact: kozakov@inp-greifswald.de

ABSTRACT

Recent findings in chronobiology show that light has a major synchronizing role for the inner master clock of a human. This feature of light is transmitted by non-visual light sensitive receptors on retina. A standardized evaluation procedure is required to describe this new discovered property of light. The present paper shows that the current knowledge on non-visual action of light and especially on dose-response dependence is sufficient for the definition of such procedure. The correct incorporation of main dose-influencing factors such as pupil diameter and illuminance and choice of proper illumination time give adequate prediction of the light induced melatonin suppression which can be considered as robust indicator of biological action of light.

INTRODUCTION

The spectral sensitivity curves of visual light receptors in the retina define the colorimetric and photometric properties of light such as Luminous flux, Correlated colour temperature (CCT) and Colour rendering index. The novel light-sensitive receptor brings in the new quality of light – the non-visual action. The quantification of this action should be based on the corresponding spectral sensitivity curve. This was first measured in the works of Brainard[1] and Thapan[2]. With the help of this curve one can calculate the characteristic values describing the strength of non-visual action of given irradiance spectrum. Following this Ansatz Gall has introduced the so called circadian metrics for the light [3]. There are some attempts to define a standard procedure for quantifying the non-visual action of light, for example, DIN has issued a pre-standard [4] were definition of terms of non-visual action of light is summarized.
Present paper discusses some features of non-visual action which have to be considered during definition of its standard evaluation procedure. Also an example of measurement of non-visual action under realistic lighting situations is given and the correlation with the calculated non-visual stimuli is discussed.

MODEL OF BIOLOGICAL ACTION

Up to the moment only semi-empiric models of non-visual action are introduces. Such a model consists of two main parts – the expression for calculation of stimulus and the empirically defined stimulus-response dependence. The stimulus (S) calculation is based on the measured spectral sensitivity curve and can be defined as

$$\boldsymbol{S = A \cdot \Delta t \cdot \int_{380}^{580} s_{ms} E_{\lambda} d\lambda}, \qquad (1)$$

where A – is the pupil area, Δt – is exposure duration, s_{ms} – is melatonin suppression spectral sensitivity curve [1,2,4] and E_{λ} – is the spectral irradiance at cornea. This expression gives us the total stimulus in Joules obtained by the subject during illumination. The stimulus can be used then to predict the strength of non-visual action, for example melatonin suppression. For this purposes one needs the empirical relation between the radiation dose (in Joules) and the melatonin suppression level (in percents). Due to complexity of non-visual action the experimental conditions should coincide to those used during determination of empirical dose-response dependence. This scenario would allow using a single evaluation procedure for evaluation of and comparison between non-visual actions of different light sources.

EXPERIMENT AND RESULTS

The measurements of the melatonin suppression in usual residential lighting situations were performed. Groups of healthy subjects (N=7 to 10) were asked to come to laboratory in the evening. One hour before the habitual bed-time the subjects were exposed to 30 minutes of light. The commercially available luminaries were equipped with the commercially available light sources and used in usual lighting situations like for example bathroom illumination, hall illumination, office illumination. The intensity and spectral composition of radiation were measured by a compact spectrometer and an illuminance meter. The melatonin concentration was evaluated by radioimmunoassay procedure. The suppression (U) of melatonin was determined as

$$U = \left(M_{off} - M_{on}\right) / M_{on}^{\dim} - \left(M_{off}^{\dim} - M_{on}^{\dim}\right) / M_{on}^{\dim} \quad (2)$$

M equals the melatonin concentration in saliva and $M^{\dim}$ equals the melatonin concentration in the control dim-light situation (<1 lx). Five light sources were used – fluorescent lamp (FL) with CCT=2000 K (FL2000) and CCT=6000 K FL6000 in bathroom arrangement, high intensity discharge lamps with CCT=4200 K (HID4200) and CCT=7000 K (HID7000) in hall arrangement and xenon filled dielectric barrier discharge lamp with CCT=2900 K in bathroom arrangement (DBD2900). Illuminance levels at cornea were laying in the range 130 – 500 lx. Figure 1 shows the dependence of measured melatonin suppression (eq. (2)) as a function of stimulus, calculated according to eq. (1) except for multiplication with exposure time Δt.

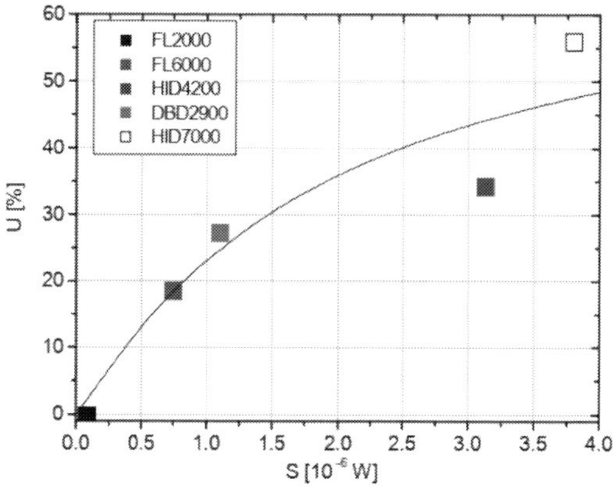

Figure 1: Measured melatonin suppression versus stimulus according to eq(1). Solid curve represents fit of logistic function.

DISCUSSION

The data in figure 1 show a correlation between stimulus and melatonin suppression. This dependence can be fitted by logistic function $U(S) = U_{\max} + \left(U_{\min} - U_{\max}\right) / \left[1 + \left(S / S_0\right)^b\right]$, which is usual for dose-response processes [1,2]. The parameters U_{min}, U_{max}, S_0 and b represent minimal and maximal suppression, half-max dose and rise exponent respectively. The minimal and maximal suppression can be kept constant at 0 and 70 %. The fit results in $S_0=1.9 \cdot 10^{-6}$ W and b=1.1, which is comparable with data from [2] – $2.12 \cdot 10^{-6}$ and 1.5 respectively. These results show that melatonin suppression is a robust marker of non-visual action – model parameters from laboratory investigations with monochromatic light coincide well with the application near investigations with polychromatic lights. Hence it can be used for definition of standard evaluation procedure

This work is supported by BMBF under grant FKZ 13N9779

REFERENCES

[1] G. Brainard et al, Journal of Neuroscience, 21 (2001) 6405-6412.
[2] K. Thapan, J. Arendt and D Skene, Journal of Physiology – London , 535 (2001) 261-267.
[3] D. Gall and V Lapuente, Licht 7-8 (2002) 860-871
[4] DIN V 5053-100, Beuth Verlag, Berlin, 2009

Next generation of automotive HID bulbs: Low power Xenon

T. Anger, M. Haacke, L. Küpper, G. Lüttgens, J.S. Straetmans

Philips Automotive Lighting, BCA Aachen, Philipsstr. 8, D-52068 Aachen
Contact: gunnar.luettgens@philips.com

ABSTRACT

A 25 W low-power HID light source has been developed for automotive headlighting applications to enable significant reductions of CO_2-emissions while maintaining a high level of traffic safety and comfort. This contribution describes the technical properties of 25 W HID light sources and how they translate into headlamp performance.

INTRODUCTION

Automotive headlighting systems need to provide high performance leading to high safety at reduced power consumption and complexity. While 35 Watt HID light sources are the state-of-the-art light sources for high-end automotive headlighting systems, the small and medium vehicle segments are still dominated by the simpler, less efficient halogen headlamps.

Recent developments of the HID technology opened up unprecedented opportunities to realize HID light sources with significantly reduced power for headlighting applications. "Xenon 12V" technology with 25 W power consumption enables power-saving, compact, cost-efficient headlamps which can replace today's Halogen headlamps in many vehicle segments. Compared to traditional Halogen headlamps (2*55 W per car), about 0.65 g (CO_2-emissions)/km can be saved (Fig. 1). The improved performance compared to Halogen headlamps will enhance traffic safety at night significantly.

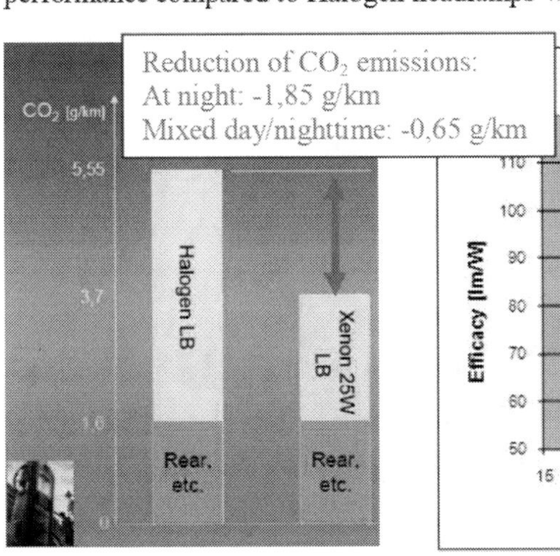

Figure 1: Low beam energy consumption

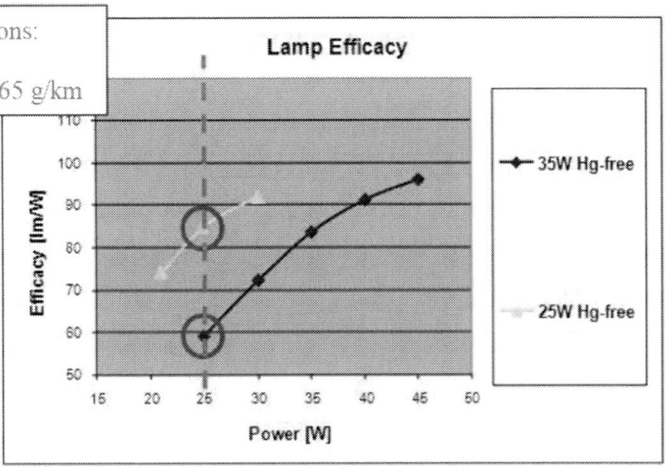

Figure 2: Lamp efficacy 35 W vs. 25 W HID light source

LIGHT SOURCE SPECIFICATIONS

Light sources are designed to balance efficacy and lifetime. Dimming of existing 35 W Xe-HID light sources reduces the efficacy drastically (Fig 2). Hence, for 25 W operation a dedicated HID light source with adapted electrode design, bulb shape and salt filling had to be developed.

The luminance of 25 W HID bulbs is only slightly reduced compared to 35 W HID bulbs, exceeding the luminance of halogen bulbs by a factor of 2 or more (Fig. 3). High luminance allows the headlamp designer to create both a far-reaching high beam as well as broad low beam with the same light source. Table 1 compares the specifications of 25 W HID with Halogen H7 LL and HID 35 W. Electrode distance and arc diffusity differ slightly. The colour temperature of 25 W HID (a major characteristic of Xe-light) is close to 35 W HID bulbs, but emission is shifted towards the black body line.

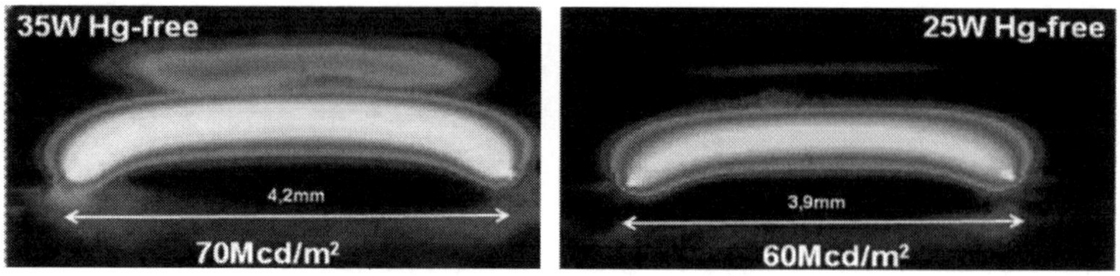

Figure 3: Average luminance of 35 W D3S and 25W HID light sources

Table 1: 25 W HID light source specifications, compared to Halogen and 35 W D3S

Bulb	Power [W]	Power Reduction [%]	Lum. Flux [lm]	Efficacy [lm/W]	U [V]	electr. distance [mm]	Arc diffusity [mm]	T_c [K]	Avg. Luminance [MCd/m²]	Lifetime B3/T_c [hrs]	Current in Steady State [A]
H7 LL	55	-	1400	26	-	4,1 (fila-ment)		3200	29	700/1100	4
35 W D3S	35	30	3200	91	42±8	4,2	0,9	4200	70	1500/2500	3
25 W	25	50	2000	85	42±8	3,9	0,7	4500	60	2000/3000	2

BEAM PERFORMANCE

Table 2 and fig. 4 show the optical performance of the new 25 W light source measured in a projection headlamp (without front screen) and compared to a 35 W D3S and a H7LL in a bi-Halogen system. The 25 W HID system performance lies well between Halogen- and 35 W HID performance, see e.g. low beam flux in the beam (column 6), low beam maximum (column 7) or high beam flux (column 8). In these measurements high beam maximum is comparable to the Halogen values since the projection unit was not optimised for 25 W HID light sources. This can be improved easily.

Table 2: Headlamp performance (FOM=Figure of Merit)

System	Power [W]	Light flux [lm]	T_c [K]	avg. Luminance [Mcd/m²]	LB [lm]	E_{max} LB [lx]	FOM	HB [lm]	E_{max} HB [lx]
Bi-Hal (H7LL)	55	1400	3200	20	350	50	185	600	85
35 W D3S	35	3200	4200	70	1050	80	240	1550	120
25 W (D3S headlamp)	25	2000	4500	60	700	70	225	1100	85

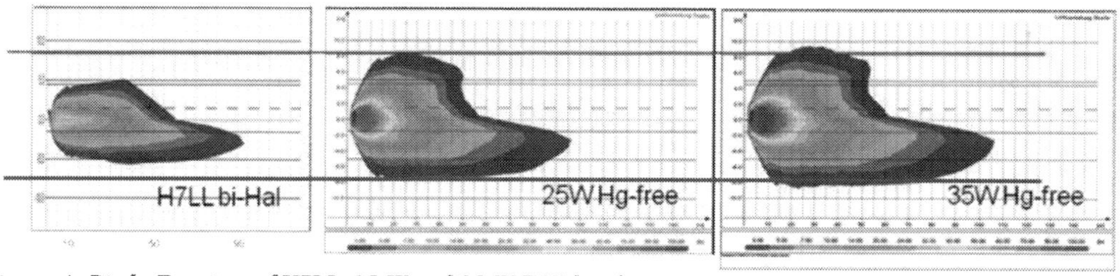

Figure 4: Birds-Eye view of H7LL, 25 W and 35 W D3S low beams

Both HID beams are clearly superior in terms of beam width and vision distance.

SUMMARY

It has been shown that the new 25 W HID technology enables significant energy savings compared to traditional Halogen technology at a superior performance.

REFERENCES

[1] T. Anger, M. Haacke, L. Küpper, N. Lesch, G. Lüttgens, ISAL Proceedings 2009

Predicting Long Lifetimes using Accelerated Reliability Techniques

R. Gibson

HID Lamp Development lab, Philips Lighting, 7265 Route 54, Bath, New York, USA

Contact: ray.gibson@philips.com

ABSTRACT

A continual challenge in lamp development is the need to rapidly test products that have life-time ratings in the tens of thousands of hours. Traditional life testing, under controlled conditions, remains a mainstay test procedure. However, highly reliable statistical analysis methods allow lamp developers to estimate long life times at a fraction of the testing time. These techniques usually involve the well known Weibull distribution together with historic data, multi-level accelerated stress testing, and virtual failure analysis to reduce the necessary testing times for establishing product life.

THE WEIBULL DISTRIBUTION AND HISTORIC SHAPE FACTORS

One rating of life is based on the 50% failure point of a population of lamps. Conventional life estimates can be made once a portion of the population has failures. The distribution of failures might be described by several different functions, but a good choice is the 2-parameter Weibull distribution, equation (1), which is used in a variety of life and strength evaluations [1]. When plotted in log-log form as shown in Figure 1, β (shape factor) corresponds to the slope, and α (scale factor) corresponds to the lateral position with lifetime (t) = α at the 63.2 percentile. Figure 1 is an example of the Weibull distribution for medium power ceramic metal halide discharge (CDM) lamps.

$$F_t = 1 - e^{-(t/\alpha)^{\beta}} \tag{1}$$

When no failures are present in a test, a value of β can be assumed based on historic data of other similar lamps types. Typically, medium power HID lamps (150 – 400 W) have a shape factor between 1 and 4; β = 3 being a good conservative approximation. Without lamp failures to establish the actual shape factor early in testing, the question becomes: what can be said about lamp life without failures? The use of β = 3 can help with this question but the quantity of lamps on test will also influence the answer. A special case of the Weibull distribution when no failures are present gives the time of 50% failures ($t_{50\%}$) based on the actual test time (T), the number of lamps on test (n), and the shape factor (β) as:

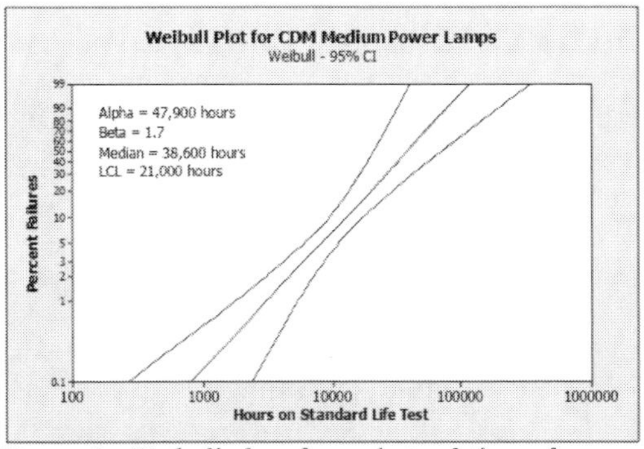

Figure 1: Weibull plot of cumulative failures for standard life test.

$$t_{50\%} = n^{1/\beta} T \tag{2}$$

When no failures are present in a test of 40 lamps, a median life of 20,500 hours can be predicted based on only 6000 hours of actual test time by using β = 3. This type of analysis is known as Bayesian analysis [2]. More sophisticated modeling is achievable in commercial statistical and reliability based software packages.

MULTI-LEVEL ACCELERATED TESTING

Multi-level accelerated testing is another method for reducing overall test times [3]. The use of multiple levels allows estimates of life under extrapolated "at-use" conditions. For example, the 330W CDM lamps were evaluated at 20% to 70% over power. Failures typical of standard life test experience were recorded at times much lower than found under normal conditions. These results are in the relation plot shown in Figure 2. The shape factor of 2.4 is based on the actual failure data and applied to all levels in this analysis.

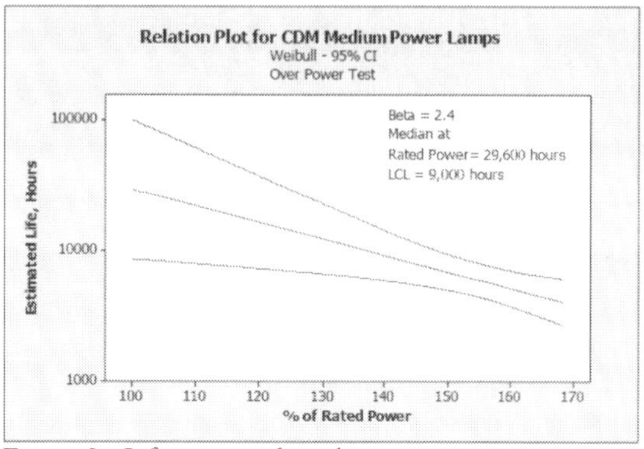

Figure 2: Life estimate based on over power test.

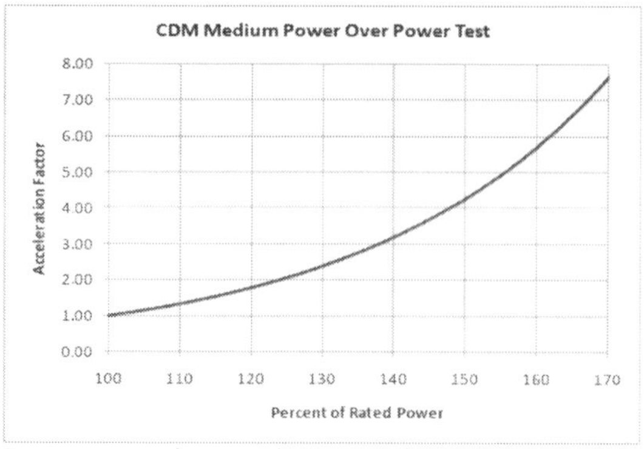

Figure 3: Acceleration factor as a function of over power.

Acceleration factors can be extracted from such data. In this example, the acceleration factor is given as:

$$AF = \alpha_0 / \alpha_{stress} \tag{3}$$

Where α_0 = the Weibull scale factor at zero stress ("at-use" conditions),
α_{stress} = the scale factor at the stressed level.

Modest increases in lamp power can yield significant reduction in test time until failure as can be seen in Figure 3. For example 50% over power should result in ¼ of the testing time. The acceleration factors, however, may be different for different product types and care must be exercised not to force failures that are inconsistent with actual use. Accelerated tests have the advantage of speeding the time to failure. This is important for identifying the failure modes, and making more rapid decisions regarding design modifications prior to product release.

VIRTUAL FAILURE ANALYSIS

The same techniques for standard and accelerated testing analysis can also be used on other *performance* indicators such as color shift, lumen maintenance, temperature degradation, etc. by defining *virtual failures* and applying failure statistics to the population. Analysis of the virtual failures in the context of an accelerated test can then be used to predict or confirm the actual performance at standard use conditions for that performance indicator. For example, a common performance metric in lamps is voltage rise. One form of failure can be when the voltage exceeds the available sustaining voltage (magnetic systems) or the bus voltage (electronic drivers). In one example, the virtual failure was chosen as an (arbitrary) voltage rise greater than 5 V. The predicted results at rated power show that a 5 volt rise can be expected, on average, after 5700 hours of operation. Standard life data shows that the average voltage rise of 5 volts occurred at 5500 hours, in close agreement.

CONCLUSIONS

Predicting operational life times in long life lamps, such as the ceramic metal halide lamps, can be significantly shorter than waiting for 50% of the products to fail. Especially useful is the Weibull distribution and the use of historic based shape factor when no failures have occurred in the early stages of testing. This information can be augmented through the use of multi-level stress testing such as over power, where extrapolation to at-use conditions compares favorably to standard life testing. The use of virtual failures using both standard and accelerated test results can further enhance the estimation of product performance over life, well in advance of the tens of thousands of hours of actual lamp operation.

REFERENCES:

[1] Z. Tóth, Al Juhász, B. Nyíri, *Evaluation of the Mechanical Properties of HID Sealing during Ageing*, Proc. 9th Inte. Symp. Sci. Technol. of Light Sources LS9 (2001), p. 381.
[2] N. Singpurwalla, *Reliability and Risk; A Bayesian Perspective*, John Wiley & Sons, NY. (2006).
[3] W. Nelson, *Accelerated Testing; Statistical Models, Test Plans, and Data Analyses*, John Wiley & Sons, NY. (1990).

The Experimental Study of Alternative Mode with Micro Shut down Power Supply Used for Alimentation of High Pressure Discharge Lamp (HP 400W)

A. Chammam[1] , W. Nsibi[1] , M. Stamboli[1] , G. Zissis[2]

[1] *SEPE, ESSTT, 5 avenue Taha Hussein 1008 Tunis Tunisie*

[2] *Université de toulouse, LAPLACE, CNRS-INP-UPS, 118 route de Narbonne 31062 Toulouse cedex*

Contact: a.chammam@voila.fr

ABSTRACT

The objective of this work consists in studying the influence of the alimentation of the gas-discharge lamp high pressure with power supply, alternative current with micro shut down, and stuffy of the dynamic behaviour of a 400W HP lamp by causing short power shut downs at various moments of the period (each half-period). The acquisition of the current and voltage arc enables us to evaluate the behaviour of the conductance of the lamp.

1- INTRODUCTION

The power supply of the discharge lamps in the low frequency square wave with micro shut down can be very effective to the optimisation of the alimentations of this lamps, indeed it is a power supply at frequency, cyclic report and duration of shut down all variables allowing us to explore the dynamic behaviour of the discharge lamp such as the voltage behaviour, the current of arc and conductance following one or more micro shut downs. In this paper we discuss the design, realisation and use of the power supply with micro shut down to study an example of discharge lamps: the mercury discharge lamp high pressure 400W.

2- MODEL USED FOR THE SIMULATION

For the modulation of the discharge lamp we used the description of the conductance G (I) by a simple equation:

$$\frac{dG}{dt} = a_2 \times I^2 - (b_1 G + b_2 G^2 + b_3 G^3) \qquad (1)$$

The values of parameters have been identified by using a recursive method from the experimental characteristic G (I) of 400W HP lamp [1]: $a_2 = 2.8\ \Omega^{-1}S^{-1}A^{-2}$; $b_1 = 141.2\ S^{-1}$; $b_2 = 40300\ \Omega S$; $b_3 = 753000\ \Omega^3 S^2$

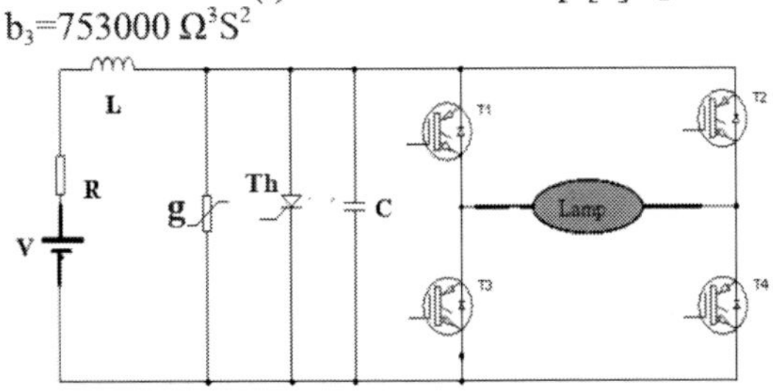

Fig1: Structure of power supply with micro shut down

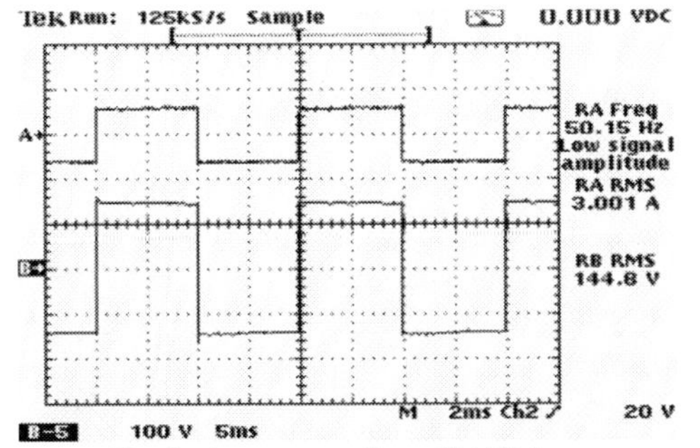

Fig.2: A: current of arc (5A/div); B: voltage of arc (100V/div); 5ms/div

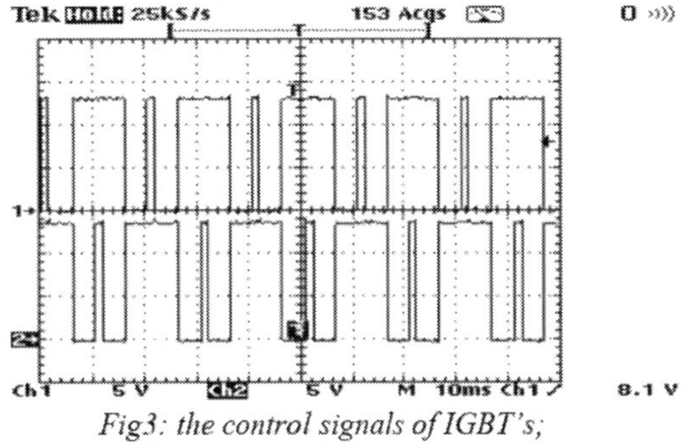

Fig3: the control signals of IGBT's;
1: control signal of IGBT's T1, T4
2: control signal of IGBT's T2, T3; 5ms/div

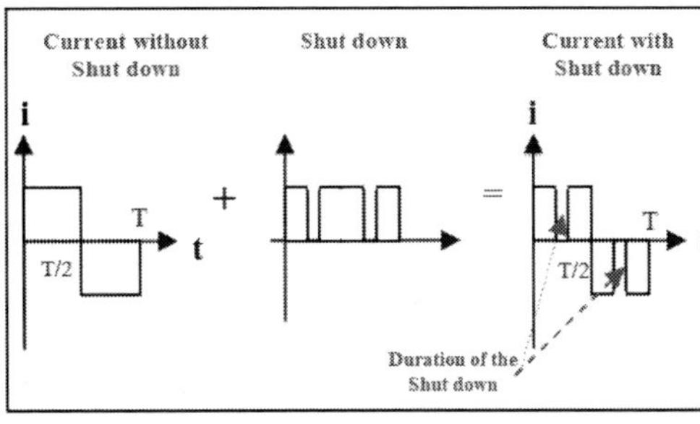

Fig4: Principle of power supply with micro shut down

3- EXPERIMENTAL RESULTS AND DISCUSSION

The principle of the power supply with micro shut down is to make short current shut downs on each half period of the current arc, the number and duration of shut downs are programmable with microcontroller PIC (fig 4). The structure of the power supply with micro shut down is represented on figure 1. It is made up of three parties: the current source (V, L), the protection circuit (g, Th) and the full wave bridge (4 IGBT). The control signals of the four IGBT's are digitally synthesised, programmed in Pascal language, and then stocked in a microcontroller (fig 3). The main characteristics of the discharge lamp used in this investigation are; 72 mm for the inter electrode length, 18,2 mm for the external diameter, 70 mg for the total mercury mass. In our experiment, the discharge lamp is supplied by square wave former of current of average power equal to 400W , the voltage arc 145V and the current equal to 3A (figure 2). In figure 7 and figure 8 test we make one shut down, three shut downs of variable duration on each half period of current arc. On figure 5 we compare the conductance of the lamp measured and calculated as well. During the shut down, the arc current decreases but the conductance of the discharge lamp can not decrease instantaneously, we have the reverse phenomenon just after the application of the shut down, which explains the peak of arc voltage.

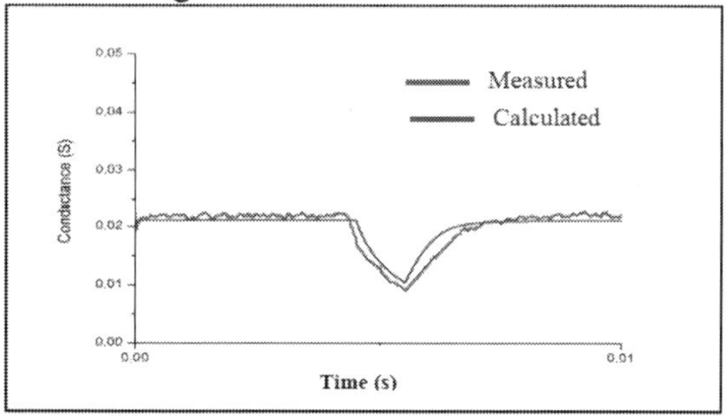

Fig 5: measured and calculated conductance of the lamp

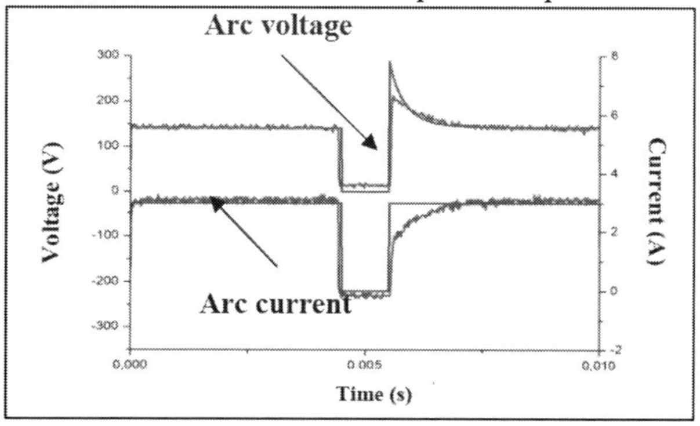

Fig 6: measured and calculated arc voltage and current

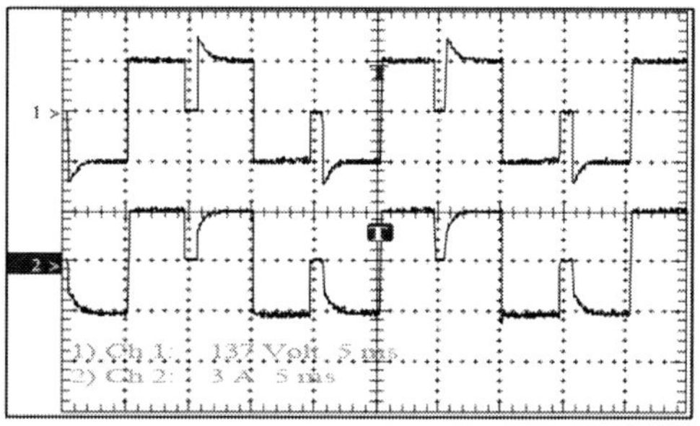

Fig7: voltage and current arc
ch1: arc Voltage (137V/div); ch2: arc Current (3A/div); 5ms/div

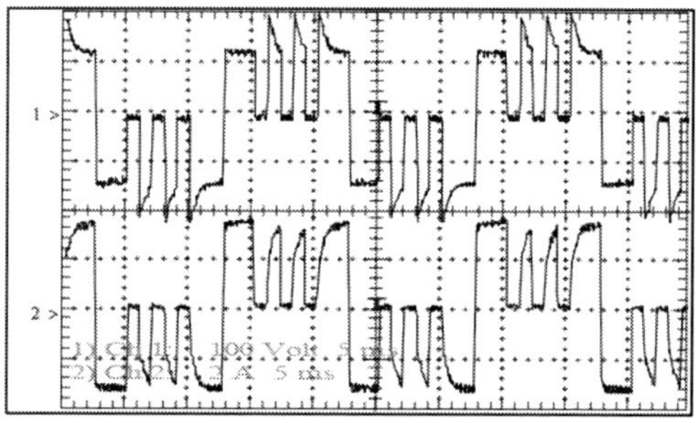

Fig8: Voltage and current arc
ch1: arc Voltage (100V/div); ch2: arc Current (2A/div); 5ms/div

4- CONCLUSION

In this work we made the study, design and the realisation of a power supply with micro shut down and also we validated this power supply in simulation and experimental. Photometric measurements solved in time of the significant lines of a mercury lamp and the use of the metal halides discharge lamps enabled us to analyse the modification of the electric and photometric characteristics of plasma after the addition of metal components constitute the continuation of our work.

REFERENCES

[1] M.Stambouli "Modélisation d'une décharge de mercure haute pression: application à l'analyse de circuit électriques comportant des lampes à décharge" Thèse, (ENSET) université de Tunis (1984).
[2] J.C.Anton, C.Blanco, F.Ferro, G.Zissis "an equivalent conductance model for high intensity discharge lamps," *in Proc.IMTC-IAS, 2002, PP, 1494-1498.*
[3] A.Chammam, H.Elloumi, B.Mrabet, K.Charrada, M.Stambouli, J.J.Damelincourt "Effect of a pulsed power supply on the spectral and electrical characteristics of HID lamps" J.Phys.D:Appl.Phys

Molecular constants and standard enthalpies of formation for the lanthanide monobromides and monoiodides, LnBr<g> and LnI<g>

S.A.Mucklejohn[1]

[1] *Ceravision Limited, The Mansion, Bletchley Park, Milton Keynes MK3 6EB, UK*

Contact: stuart.mucklejohn@ceravision.com

ABSTRACT

Dissociation energies reported in the literature and estimated molecular constants have been used to derive the heat capacities, entropies and standard enthalpies of formation for the lanthanide monohalides, LnBr<g> and LnI<g>, Ln = Ba, La to Lu.

INTRODUCTION

Although a complete set of dissociation energies for the lanthanide monohalides, LnX<g>, has been published [1] there is very little information available about the molecular constants for the monohalides and monoiodides. Cao and Dolg [2] report fundamental frequencies and interatomic separations for LaF<g>, LaCl<g>, LaBr<g>, LaI<g> derived from a series of quantum chemistry calculations. This paper outlines the determination of the standard molar enthalpies of formation for the series LnX<g>, Ln = Ba, La to Lu and X = Br, I, from dissociation energies.

Promethium, Pm, is an unstable element and does not occur naturally but it is included in all theoretical considerations of lanthanide compounds. Although Ba is not a lanthanide element it is often included in studies of the lanthanide series because of its position in the Periodic Table.

Scheme 1. Derivation of standard enthalpy of formation from dissociation energy

Ln<r> + $1/2X_2$<r> = LnX<g>	Standard enthalpy of formation of LnX<g>	{1}
LnX<g> = Ln<g> + X<g>	Dissociation of LnX<g>	{2}
Ln<r> = Ln<g>	Standard enthalpy of formation of Ln<g>	{3}
$1/2X_2$<r> = X<g>	Standard enthalpy of formation of X<g>	{4}

{1} = {3} + {4} - {2}

where <r> is the standard state of the element at the reference temperature

The dissociation energies at 0 K for LnBr<g> and LnI<g> were taken from reference [1], the standard enthalpies of formation for Ln<g>, Br<g> and I<g> were taken from the SGTE Substance Database [3].

The heat capacities of the LnBr<g> and LnI<g> series were calculated from the molecular constants listed in table 1. These were estimated from the ratios of the corresponding parameters for LaCl<g>:LaBr<g> and LaCl<g>:LaI<g> taken from reference [2]. The energy levels used for these calculations were those for the corresponding Ln(II) species [4]. These calculations also provide the molar entropies at 298.15 K. In view of the need to estimate the molecular constants, heat capacities and entropies were calculated without anharmonicity corrections. The uncertainties introduced by this approximation can be estimated by comparing results for LnCl<g> species with and without anharmonicity corrections.

The standard molar enthalpies of formation for LnX<g> were derived from scheme 1. Molar heat capacities for Ln<g> and X<g> were calculated using energy levels for Ln(I) and X(I) as listed in [4].

$$\Delta Cp(T) = C_p(Ln,g,T) + C_p(X,g,T) - C_p(LnX,g,T) \quad \{5\}$$

$$\Delta H(T) = \Delta H(298.15\ K) + \int \Delta C_p(T)dT \quad \{6\}$$

Enthalpies of dissociation at 298.15 K were determined from equations {2}, {5} and {6}.

Table 1. Molecular constants for LnBr<g> and LnI<g>used in this work

	For LnBr<g>			For LnI<g>		
Ln	Molar mass	ω_e	R_e	Molar mass	ω_e	R_e
	$g.mol^{-1}$	cm^{-1}	cm	$g.mol^{-1}$	cm^{-1}	cm
Ba	217.244	194.4	2.840E-08	264.244	153.2	3.082E-08
La	218.814	237.8	2.644E-08	265.814	187.4	2.869E-08
Ce	220.024	239.7	2.631E-08	267.024	188.9	2.855E-08
Pr	220.811	241.5	2.620E-08	267.811	190.3	2.843E-08
Nd	224.144	243.0	2.608E-08	271.144	191.5	2.830E-08
Pm	224.904	244.4	2.596E-08	271.904	192.6	2.817E-08
Sm	230.254	245.5	2.585E-08	277.254	193.5	2.805E-08
Eu	231.864	201.0	2.721E-08	278.864	158.4	2.953E-08
Gd	237.154	247.1	2.565E-08	284.154	194.7	2.783E-08
Tb	238.828	247.6	2.554E-08	285.828	195.1	2.772E-08
Dy	242.404	247.8	2.544E-08	289.404	195.3	2.761E-08
Ho	244.834	247.9	2.536E-08	291.834	195.4	2.752E-08
Er	247.164	247.8	2.527E-08	294.164	195.3	2.743E-08
Tm	248.838	247.3	2.518E-08	295.838	194.9	2.732E-08
Yb	252.944	202.7	2.633E-08	299.944	159.7	2.858E-08
Lu	254.874	246.0	2.504E-08	301.874	193.9	2.718E-08

Table 2. Enthalpies of dissociation ($\Delta_{dis}H_m$) for LnBr<g> & LnI<g> at 298.15 K and standard molar enthalpies of formation ($\Delta_f H_m$) for LnBr<g> & LnI<g> at 298.15 K

	$\Delta_{dis}H_m$	$\Delta_f H_m$	$\Delta_{dis}H_m$	$\Delta_f H_m$
	$J.mol^{-1}$	$J.mol^{-1}$	$J.mol^{-1}$	$J.mol^{-1}$
T	298.15 K	298.15 K	298.15 K	298.15 K
Ln				
Ba	359918	-63048	321028	-29265
La	448585	89135	414888	117725
Ce	375196	153748	335516	188321
Pr	346283	121400	307799	154777
Nd	341771	97228	303285	130607
Pm	337605	33428	299119	66807
Sm	334277	-16810	295791	16569
Eu	328835	-40778	290374	-7424
Gd	374473	135908	335986	169288
Tb	386447	111479	339584	153235
Dy	315705	81417	277217	114798
Ho	323857	88204	276994	129960
Er	363224	62473	317559	103031
Tm	300931	43698	262444	77078
Yb	297742	-33386	259280	-31
Lu	303285	236447	264799	269826

REFERENCES

[1] L.A. Kaledin, M.C. Heaven & R.W. Field, *J. Mol. Spec.*, 193, 285 (1999).
[2] X. Cao & M. Dolg, *J. Theoretical & Computational Chem.*, 4, 583 (2005).
[3] I. Hurtado, D. Neuschutz, 'Thermodynamic Properties of Inorganic Materials', compiled by SGTE, Landolt Börnstein, 19, Springer-Verlag, Berlin (1999–2005).
[4] http://physics.nist.gov/PhysRefData/Handbook/index.html

Determination of Junction Temperature in GaN and AlGaInP Light-Emitting Diodes

Chia-Jung Lee, Chih-Hao Chen, and Ya-Ju Lee

Institute of Electro-Optical Science and Technology, National Taiwan Normal University, 88, Sec.4, Ting-Chou Road, Taipei 116, Taiwan

Contact: yajulee@ntnu.edu.tw

ABSTRACT

The junction temperature of a light-emitting diode (LED) directly and greatly affects its performances. Therefore, the reliable measurement and accurate estimation of the junction temperature of an LED is extremely important. This work proposes an approach for directly determining the dependence of junction temperature on injected currents in GaN and AlGaInP LEDs. Various important physical parameters that affect the junction temperature of an LED are also considered.

INTRODUCTION

III-V-based light-emitting diodes (LEDs) are of current technological interest owing to their potential to improve upon the efficiency of traditional lamp-based lighting systems.[1] Typically, the amount of heat energy that is generated inside LED strongly depends on its construction and epitaxial material, and defines its junction temperature. In this study, a series of temperature-dependent electrical and optical measurements of GaN and AlGaInP LEDs are made to estimate their junction temperatures. The theoretical derivation and possible determinants of the physical dependence of the junction temperature of an LED on the injected current are also analyzed and discussed.

THEORETICAL MODEL

Figure 1 shows the peak wavelength of GaN and AlGaInP LEDs are λ=470nm and λ=630nm, respectively. In practice, the parasitical effects on LEDs are inevitable and introduce additional series (R_S) and parallel (R_P) resistances into the LEDs. Accordingly, the Shockley equation[2] is modified to,

$$I = I_J + I_P = I_S\left(\exp\left(\frac{e(V_f - IR_S)}{n_{ideal}kT}\right) - 1\right) + \frac{(V_f - IR_S)}{R_P} \tag{1}$$

Figure 2(a) presents the effective electronic circuit of an LED with such a parasitical effect. Figures 2 (b) and (c) plot the typical *I-V* characteristics of GaN and AlGaInP LEDs on a semi-log scale at junction temperatures of T=300K and T=400K.

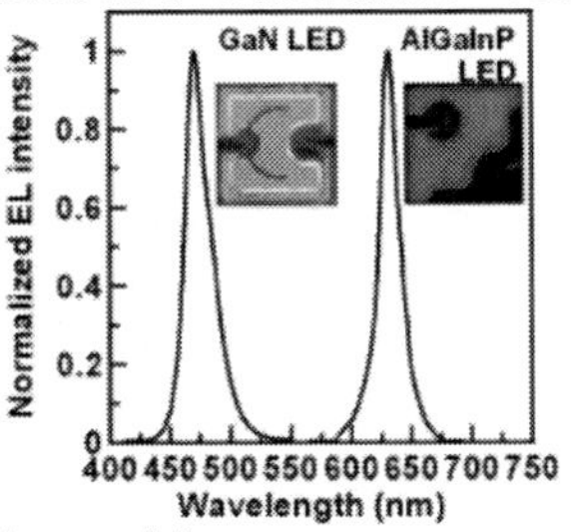

FIG. 1 Electrical luminescence spectra of GaN and AlGaInP LEDs. Insert: photographs of GaN and AlGaInP LEDs obtained at I =20 mA

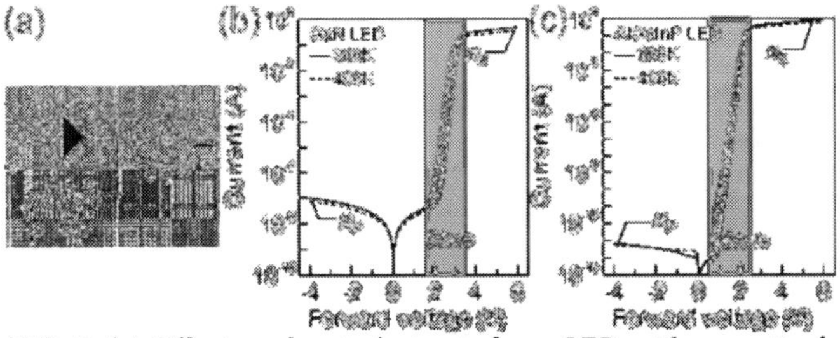

FIG. 2 (a) Effective electrical circuit for a LED with parasitical effects. Dependence of ln|I| vs Vf at 300K and 400K for (b) GaN and (c) AlGaInP LEDs.

Accordingly, $\partial V_f / \partial T = \gamma$ and $\partial V_f / \partial I = R_S$, and both γ and R_S are equal constants. Consequently, V_f can essentially be described as a linear function of I and T:

$$V_f(I,T) = R_S \cdot I + \gamma \cdot T + C \tag{2}$$

In the steady state, the total electrical input-power can be reasonably treated as the summation of the light output-power and the dissipated power. Hence, the dissipated power that is converted to heat, $P_{heat} = P_{in}\text{-}P_{out}$ can be rewritten as,

$$IV_f - \eta I(V_f - IR_S) = \frac{T - T_a}{R_{th}} \tag{3}$$

Substituting V_f from Eq. (2) into Eq. (3) yields a very useful expression for the dependence of the junction temperature of the LED on the injected current:

$$T = \frac{T_a + CR_{th}(1-\eta)I + R_S R_{th} I^2}{1 - \gamma R_{th}(1-\eta)I} \tag{4}$$

For η=1, R_S=0, and R_{th}=0, in the limit of Eq. (4), as T approaches to T_a, the junction temperature is no longer related to the injected current, and is approximately equal to the ambient temperature.

RESULTS AND DISCUSSION

In Fig, 3 (a), $\partial V_f/\partial T$ (γ) remains constant, independently of the injected current. From Fig. 3 (b), γ is estimated to be -3.02 mV/K and -2.23 mV/K for GaN and AlGaInP LEDs, respectively. Clearly, for both samples, V_f increased linearly with injected current as shown in Fig, 3 (c). Despite the difference between junction temperatures, $\partial V_f/\partial I$ (R_S) barely differed. From Fig.3 (d), R_S is estimated to be 12.5Ω and 7Ω for GaN and AlGaInP LEDs, respectively. Substituting η, R_{th}, γ and R_S measured or calculated above into Eq. (4) yields the curve of junction temperature as a function of injected current for GaN and AlGaInP LEDs in Fig. 4. The related physical values for substituting into Eq. (4) are summarized in Table *I*.

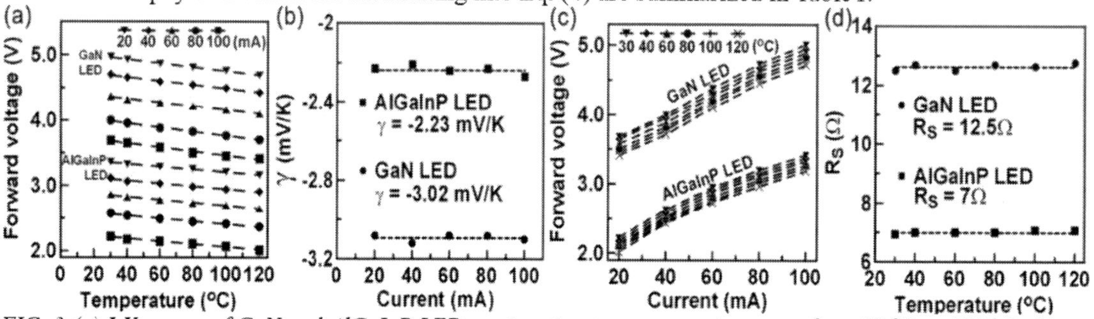

FIG. 3 (a) I-V curves of GaN and AlGaInP LEDs as junction temperature increases from 30 °C to 120 °C; (b) γ vs. injected current for GaN (filled circles) and AlGaInP (filled squares) LEDs; (c) V_f vs. junction temperature for GaN and AlGaInP LEDs with various injected pulse currents; (d) R_S vs. junction temperature for GaN (filled circles) and AlGaInP (filled squares) LEDs.

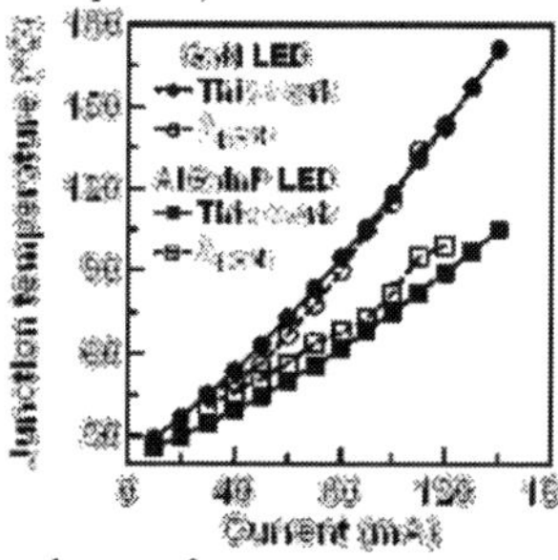

FIG. 4 Dependence of junction temperature on injected current for GaN and AlGaInP LEDs. For comparison, the junction temperature obtained by using the emission peak shift method (λ_{peak}) was also plotted in the figure.

	GaN	AlGaInP
R_s	12.5 Ω	7 Ω
R_p	4x10^7 Ω	6x10^{13} Ω
n_{ideal}	2.24	1.81
γ	-3.02 mV/K	-2.23 mV/K
η	9%	13%
R_{th}	300K/W	265 K/W
C	3.45 V	2.27 V

TABLE I. Physical values for substituting into Eq. (4) to obtain the junction temperature of GaN and AlGaInP LEDs in Figure 4.

SUMMARY

In summary, an approach for estimating the junction temperature of an LED was presented. The partial derivatives the forward voltage with respect to injected current ($\partial V_f/\partial I$) and junction temperature ($\partial V_f/\partial T$) are mutually independent, indicating that the forward voltage of the LED is indeed a linear function of injected current and junction temperature. Further separating the light-output power from the generated heat energy yields the physical dependence of junction temperature on injected current.

REFERENCES

[1] M. R. Krames, O. B. Shchekin, M. M. Regina, G. O. Mueller, L. Zhou, G. Harbers and M. G. Craford, IEEE Journal of Display Technology, vol.3, pp.160-175, 2007.

[2] W. Shockley, Electrons and holes in semiconductors, with applications to transistor electronics, D. Van Nostrand Company Inc., New York, 1950.

Modelling the radiation of diatomic molecules in low pressure discharges

U. Fantz[1,2] , S. Briefi[1] , R. Friedl[1,2]

[1] *Lehrstuhl für Experimentelle Plasmaphysik, Universität Augsburg, Universitätsstr. 1, D-86135 Augsburg, Germany*
[2] *Max-Planck-Institut für Plasmaphysik, EURATOM Association, Boltzmannstr. 2, D-85748 Garching, Germany*
Contact: fantz@physik.uni-augsburg.de

ABSTRACT

The emission spectra of diatomic molecules are simulated by using the molecular constants, rotational and vibrational populations. Examples of metal halide spectra such as InBr and GaI are shown as well as spectra of N_2 and compared with measured spectra. The absolute emission of nitrogen spectra is modelled on the basis of a collisional radiative model providing insight in the dominant population channels for optimising the nitrogen radiation in the VIS and UV range.

INTRODUCTION

Diatomic molecules are considered as candidates for replacing mercury in low pressure discharges [1]. One promising category are the metal halide molecules which radiate in the near UV range (330 – 420 nm) and make the conversion to the visible light by phosphors more efficient compared to mercury in fluorescent lamps due to the lower Stokes shift. Another promising candidate is nitrogen [2] which can be added easily as gas resulting in bulb temperature independent emission. Nitrogen molecules have two prominent emission systems, the 2nd pos. system in the near UV range and the 1st pos. system in the visible range with maximum intensity at 300 – 400 nm and 550 – 700 nm, respectively.

Experimental work on optimisation of the radiation of molecules in low pressure discharges [3,4] can be strongly supported by modelling. Simulation of emission spectra and comparison with measured spectra gives access to parameters which determine the intensity and the shape of the ro-vibrational band emission, namely the rotational, vibrational and electronic population of the excited states. Important population channels can be identified by using a collisional radiative model which allows also to predict molecular radiation for a given set of plasma parameters and to perform systematic dependence studies. The purpose of the paper is to introduce these powerful tools and to show their applications to metal halides such as InBr and to nitrogen.

SIMULATION OF DIATOMIC SPECTRA

The simulation of emission spectra of diatomic molecules is straightforward. By using the molecular constants wavelengths of allowed transitions are calculated rotationally and vibrationally resolved. The line profile is determined by the apparatus profile of the spectroscopic system. The relative intensity of rotational lines is given by the Hönl-London factors and the rotational population whereas the intensities of vibrational transitions depend on the vibrationally resolved transition probabilities and on the vibrational population. Emission of an electronic transition needs the electronic population as input. Due to the lack of a complete set of vibrationally resolved transition probabilities for metal halides these data have been calculated from potential curves solving the Schrödinger equation in the Born-Oppenheimer approximation using the program package `TraDiMo` [5].

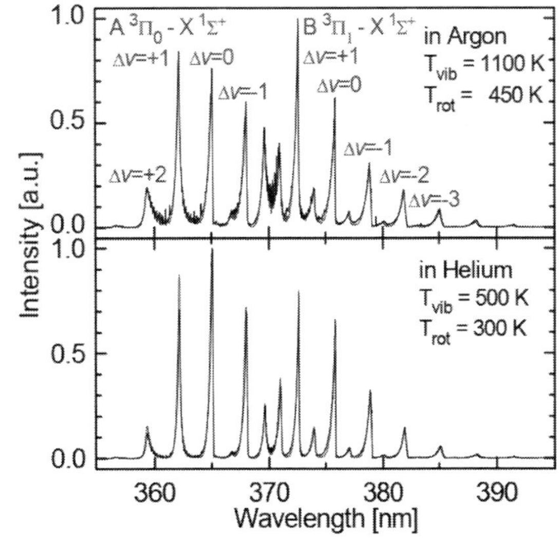

Fig. 1: InBr spectra obtained in RF-discharges with Ar and He compared to simulations (red). The different electronic transitions and vibrational sequences are indicated.

Figure 1 shows simulated InBr spectra which were fitted to measured spectra using the rotational and vibrational temperature as fitting parameters. The band structure depends clearly on these two parameters, in which a higher rotational population broadens the vibrational bands. Details of the simulations are given in [6] and can be carried out for any other diatomic molecule such as the other metal halides (InCl, GaI) or nitrogen which is another prominent example.

COLLISIONAL RADIATIVE MODELLING

A collisional radiative (CR) model balances the dominant population and de-population channels for an excited state. For the nitrogen molecule such a CR model has been established taking into account four electronic states and the coupling to the molecular ion. Electron impact collisions and heavy particle collisions with the ground and metastable state of nitrogen molecules are considered as well as energy transfer with metastable argon atoms for calculating the emission of nitrogen in argon background gas. The set of equations is being solved using the flexible YACORA solver [7]. The model is benchmarked with measurements of nitrogen radiation in an argon arc discharge with plasma parameters known from measurements. Figure 2 shows molecular band emissions of the two prominent nitrogen transitions ($C^3\Pi_u \rightarrow B^3\Pi_g$ and $B^3\Pi_g \rightarrow A^3\Sigma^+_u$) in comparison with results from the CR model and a simple corona model. The corona model takes into account only electron impact excitation from the ground state balanced with spontaneous emission. Obviously the corona model does not reflect the measurements. With the CR model the C → B transition is reproduced very well whereas for the B → A transition further investigations are needed. Heavy particle collisions such as energy pooling and energy transfer with metastable states as well as quenching processes by the background gas are identified as important population mechanisms in the pressure range under investigation.

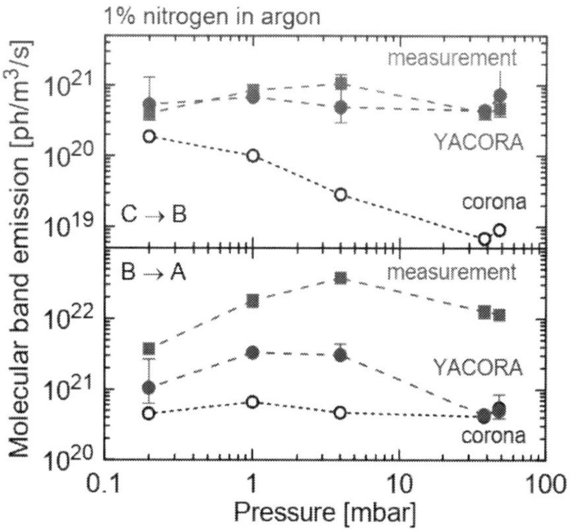

Fig. 2: Measured radiation of the 2nd pos. system (C → B transition) and 1st pos. system (B → A transition) of N_2 [4] compared with results from the simplifying corona model and the collisional radiative model YACORA.

CONCLUSION

Simulation of spectra is a useful tool to determine the ro-vibrational population of a diatomic molecule in the excited state and can be carried out for any molecule provided the molecular constants and transition probabilities are known. Collisional radiative models provide insight in population and de-population channels and can predict the radiation on the basis of plasma parameters. Both models support the efforts to optimise the radiation of diatomic molecules for lighting purposes.

REFERENCES

[1] S. Kitsinelis, G. Zissis, E. Fokitis, J. Phys. D: Appl. Phys. 42 (2009) 045209.
[2] M. Jinno, S. Takubo, Y. Hazata, et al., J. Phys. D: Appl. Phys. 38 (2005) 3312.
[3] S. Briefi, U. Fantz, contribution to this conference.
[4] R. Friedl, U. Fantz, contribution to this conference.
[5] U. Fantz, D. Wünderlich, Atomic Data and Nuclear Data Tables 92 (2006) 853.
[6] S. Briefi, U. Fantz, "Simulation of the emission spectrum of the A-X and B-X transitions of the InBr Molecule for diagnostics in low pressure plasmas" to be submitted.
[7] D. Wünderlich, S. Dietrich, U. Fantz, J. Quant. Spec. Radiat. Transfer 110 (2009) 62.

Atomic transition probabilities of Ce I from Fourier transform spectra

J. E. Lawler[1], J. Chisholm[2], D. E. Nitz[3], M. P. Wood[1], J. Sobeck[4], and E. A. Den Hartog[1]

[1]*Department of Physics, University of Wisconsin, 1150 University Ave., Madison, WI 53706*
[2]*Department of Physics, Boston College, 140 Commonwealth Ave., Chestnut Hill, MA 02467*
[3]*Department of Physics, St. Olaf College, 1520 St. Olaf Ave., Northfield, MN 55057*
[4]*Department of Astronomy and Astrophysics, University of Chicago, 5640 Ellis Ave., Chicago, IL 60637*
Contact: jelawler@wisc.edu

ABSTRACT

Atomic transition probabilities for 2874 lines of the first spectrum of cerium (Ce I) are reported. These data are from new branching fraction measurements on Fourier transform spectra normalized with previously reported radiative lifetimes from time-resolved laser-induced fluorescence measurements. The use of 14 spectra enabled us to correct for optical depth errors on strong lines, make reliable measurements on weak lines, and cover a wavelength range from 360 - 1500nm. Comparisons are made to previous investigations which are less extensive. Accurate Ce I transition probabilities are needed for lighting research and development on metal-halide lamps.

RARE EARTH ELEMENTS IN LIGHTING

For higher power lighting applications requiring tens-of-thousands of lumen and a high CRI it is likely that Metal Halide High Intensity Discharge (MH-HID) lamps will continue to be the technology of choice. The introduction of Ce into MH-HID lamps was a significant milestone in the development of this technology [1]. Efficacies of 177 lumen per Watt (lpw) have been achieved in conventional electroded MH-HID lamps with Ce doses [2]. Currently in development, electrodeless MH-HID lamps are likely to achieve luminous efficacies of ~200 lpw and lifetimes of ~100,000 hours.

Improved basic spectroscopic data is needed for both diagnostic and modelling efforts on advanced MH-HID lamps. The Wisconsin Atomic Transition Probability (WATP) program has recently completed a decade long effort on singly ionized rare earth elements which was motivated primarily by needs of astrophysicists studying the origins of the chemical elements [3]. We are now extending this work to neutral rare earth species. The WATP program technique of combining branching fractions measured using a Fourier transform spectrometer (FTS) with radiative lifetimes measured using time-resolved laser-induced fluorescence (TRLIF) is a well-recognized and accurate method for determining reliable transition probabilities. We report new atomic transition probability measurements for 2874 lines of Ce I with uncertainties in the 5 to 25% range. Radiative lifetimes from Den Hartog et al. [4] provide a foundation for this study of branching fractions and the transition probabilities of Ce I. Branching fraction measurements were attempted on lines from all 153 levels of the lifetime experiment, and were completed for lines from 134 even-parity and 2 odd-parity upper levels. As in earlier work on other rare earth spectra, we used the 1.0 meter FTS at the National Solar Observatory (NSO) on Kitt Peak for Ce I branching fraction measurements. The Kitt Peak FTS has a large etendue (like all interferometric spectrometers), a limit of resolution as small as 0.01 cm^{-1}, wavenumber accuracy to 1 part in 10^8, broad spectral coverage from the UV to IR, and the capability of recording a million point spectrum in 10 minutes [5]. The line density in Ce I is so high that a very high performance FTS is essential. Grating spectrographs are not adequate.

The determination of an accurate relative radiometric calibration is critical to a branching fraction experiment. We made use of both standard lamp calibrations and Ar I and Ar II line calibrations in this Ce I study. All possible transition wavenumbers between known energy levels [6] of Ce I satisfying both the parity change and ΔJ (= -1, 0, or 1) selection rules were computed and used during analysis of FTS data. Systematic errors from missing branches to known lower levels are negligible in our work, because we were able to make measurements on branching fractions as weak as 0.0001. Most even-parity upper levels, depending on their J value, have on the order of 50 to 100 possible transitions to known lower levels. Typically half or more of these possible transitions have some detectable intensity in the FTS spectra. More than 100,000 possible spectral line observations were studied during the analysis of lines from 153 upper levels in 14 different Ce spectra. We set integration limits and occasionally nonzero baselines "interactively" during analysis of the FTS spectra. Figures 1 and 2 are samples of the FTS data for two, the strongest and weakest, of 85 possible lines from even-parity upper level at 19661.076 cm^{-1}. These samples illustrate the excellent spectral resolution and signal-to-noise of our FTS data.

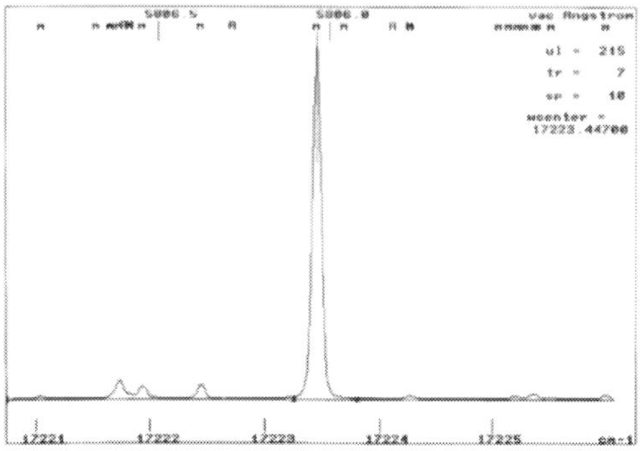

Fig. 1 Strongest line from the selected upper level (branching fraction = 0.479).

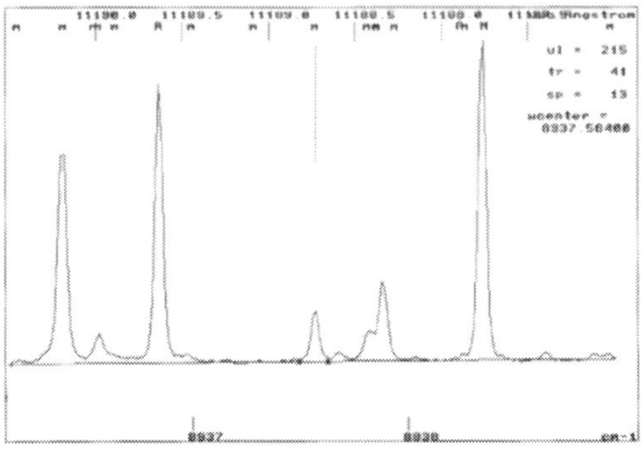

Fig. 2 Weakest line from the selected upper level (branching fraction = 0.00081).

Figures 3 and 4 are comparisons of our results to earlier work by Bisson et al. [7] and by Curry [8]. More complete discussions of our FTS data, relative radiometric calibration, analysis method, the comparisons to earlier work, and a complete table of our final results are in press [9-10].

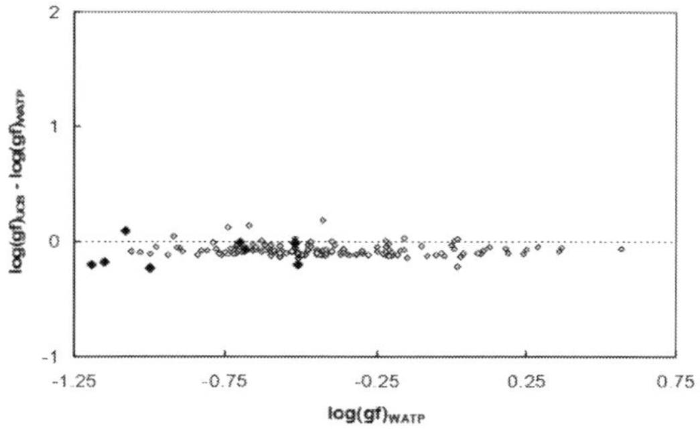

Fig. 3 Comparison of log(gf) values from Bission et al. [7] at the Univ. of California-Berkeley (UCB) to our (WATP) results.

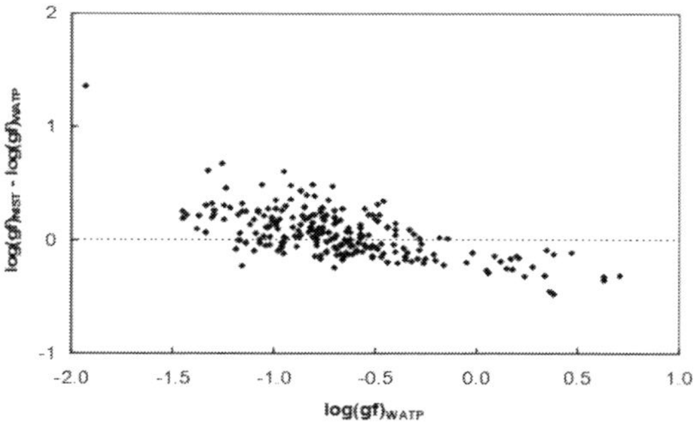

Fig. 4 Comparison of log(gf) values from Curry [8] at NIST to our (WATP) results.

REFERENCES

[1] J. A. J. M. Van Vliet and J. J. F. Geijtenbeek, U.S. Patent 5,973,453 (1999).

[2] W. P. Lapatovich, Physica Scripta T134, 014024 (2009).

[3] J. E. Lawler, C. Sneden, J. J. Cowan, I. I. Ivans, and E. A. Den Hartog, Astrophys. J. Suppl. 182, 51 (2009).

[4] E. A. Den Hartog, K. P. Buettner, and J. E. Lawler, J. Phys. B 42, 085006 (2009).

[5] J. W. Brault, J. Opt. Soc. Am. 66, 1081 (1976).

[6] W. C. Martin, R. Zalubas, and L. Hagan, *Atomic Energy Levels – The Rare Earth Elements, NSRDS NBS 60* (Washington: U. S. G. P. O.) p. 70 (1978).

[7] S. E. Bisson, E. F. Worden, J. G. Conway, B. Comaskey, J. A. D. Stockdale, and F. Nehring, J. Opt. Soc. Am B 8, 1545 (1991).

[8] J. J. Curry, J. Phys. D 42, 135205 (2009).

[9] J. E. Lawler, J. Chisholm, D. E. Nitz, M. P. Wood, J. Sobeck, and E. A. Den Hartog, J. Phys. B in press (2010).

[10] This research was supported at participating institutions by the United States NSF (Grants AST-0453442, CTS-0613277, PHY 02-16783), DoD (ASSURE Program), and DoE (Grant B523820).

Modelling of the dynamic behaviour of thin mercury lines luminance for HID lamps

L.Bouslimi[1], M.Stambouli[1], J.P.Cambronne[2], G.Zissis[2]
[1]*ESSTT, 05Avenue Taha Hussein, Tunis, 1008 Tunisia*
[2] *Université de Toulouse,UPS,INPT, LAPLACE(Laboratoire Plasma et Conversion d'Energie), 118 route de Narbonne, F - 31062 Toulouse cedex 9 , France*
E-mail: lotfi.bouslimi@univ-tlse3.fr

ABSTRACT

A new mathematical model representing the time dependence of high-pressure mercury discharge luminance of the optically thin lines has been determined. The model was conducted under Local Thermodynamic Equilibrium (LTE) assumption and some simplifying assumptions of a channel model. The model consists of a nonlinear differential equation. It makes it possible to calculate easily time-dependent plasma luminance mainly of visible thin lines such as 577nm for any arbitrary supply current waveforms.

INTRODUCTION

The mathematical models for High Intensity Discharge (HID) lamps were widely used in the literature to describe physical phenomena of the plasma inside the lamps with more approximate methods [1]-[2]. Following Elenbaas, most models of high pressure arc lamps have accepted the concept of LTE as a useful approximation, which greatly simplifies the theoretical treatment. The channel model arc was used also by Elenbaas and other authors to simplify the characterization of the high pressure mercury lamp by determining the discharge temperature profile [2]. The temperature and species composition of lamp arcs have been determined primarily by traditional methods of optical spectroscopy. The entire temperature profile and axis temperature have often been obtained from an absolute side-on intensity measurement of an optically thin mercury line such as the 577 nm line, and have been subsequently Abel-inverted or by other optimised methods. The mercury pressure and the atomic transition probability of the atomic line must be also known.

On this matter, a new differential equation representing the time dependence of the variation of high pressure arc thin lines luminance for any given lamp parameter is determined using some channel model assumptions and based on energy balance equation. This is easily useful in order to predict the luminance behaviour of the visible lines inside the discharge lamp for various operating conditions. Some experimental results are also displayed to show the effectiveness of the proposed model and to be compared with results of simulation.

MODEL EQUATIONS

The present approach was conducted under LTE assumption and some simplifying assumptions of a channel model. The description of the discharge lamp state is based on the energy balance equation:

$$\rho C_p \frac{dT_c}{dt} = P_{el} - \left(P_{rad} + P_{cond}\right) \qquad (1)$$

Where, ρ is the density of mercury, T_c is the channel gas temperature, C_p is the specific heat capacity, P_{el} is the electrical input power , $P_{rad} + P_{cond}$ is the power removed from the lamp (radiation plus heat conduction) and dt is the time increments.

Using the simplifying assumptions of the channel model that the actual discharge is replaced by a channel zone [$-\beta R$, βR] which corresponds to an effective temperature T_{eff}, that's taken constant inside this zone, where $\beta \approx 0{,}5$ is a constant and R is the radius of the lamp tube [3]. For the outer zone, the current density and radiation intensity are negligible since the temperature is assumed to be very low. As these assumptions, the total luminance of the thin lines is given then by the following expression:

$$L = \beta R \frac{hc}{2\pi\lambda_0} \frac{A_{ea} g_e}{U} \frac{P}{kT_c} \exp\left(\frac{-eV_e}{kT_c}\right) \qquad (2)$$

Where h is Plank's constant, c is the velocity of light, λ_0 is the wavelength of optically thin line, g_e is the statistical weight of the upper level, e is the charge on an electron, k is Boltzmann's constant, V_e is the excitation potential of the upper level, A_{ea} is the transition probability, U is the partition function and P is the pressure of the gaz.

The mathematical approach of this model is relatively long. It gives finally, a fundamental nonlinear differential equation which describes the arc lamp luminance state as a function of time and of the current input i.

$$\frac{dL(t)}{dt} = A_2 i^2(t) - \left(B_1 L(t) + B_2 L^2(t)\right) \tag{3}$$

Where A_2, $B_1$and B_2, are parameters which depend of the temperature and the parameters of any given lamp. The values of theses parameters have been identified, by simulation using SIMULINK.

RESULTS AND DISCUSSION

The lamp operates vertically and all the measurements have been done in a steady state. The experimental (figure1 (a)) current, voltage, and luminance waveforms of a 400 W high pressure mercury lamp have been compared with the simulated ac characteristics (figure1 (b)). The lamp was operated with a line frequency of 50 Hz and sinusoidal source voltage of 230 Vrms in series with commercial ballast. The characteristics data of which were measured to be L= 0.14H and R=3Ω. Figure1 (c) shows the measured and simulated luminance as a function of time. Explanations on the experimental measurement set-up of the luminance and the lamp parameters are given in [4].

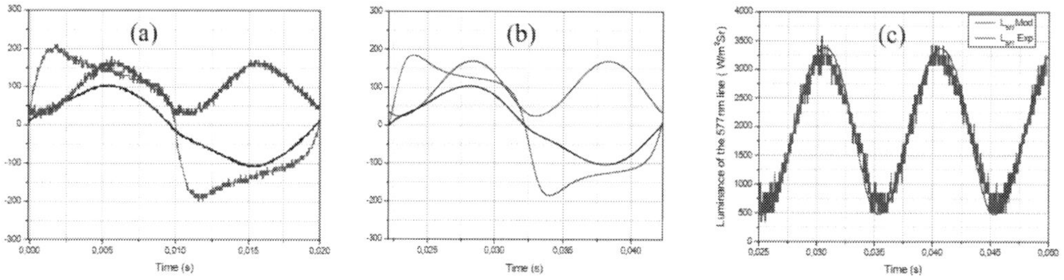

Fig.1: Experimental (a) and simulated (b) Curves of: (—) lamp current (5A/div, 2.5 ms/div); (—) lamp voltage (100 V/div); (—) Evolution of the 577nm line (2000W/m².Sr /div). (c): Curves of the 577nm line luminance (500W/m².Sr, 10 ms/div); (—): experimental; (—): from model

The light difference between the measures and simulated values is due to the choices of the characteristic data R and L of the electromagnetic ballast. According to this model the absolute values of the pressure and the axis temperature inside the lamp are equal to P=3.5atm and T=5600k. We also find that the model imitates faithfully the responses of the lamp.

CONCLUSION

In this paper a new model describing the optically thin lines as a function of time is presented and verified against experimental results. The results found confirm the fair agreement between the model and experiment. It should be also pointed out that this method of analysis enables the prediction of the luminance for various operating conditions and for any given lamp parameters. We can also predict the temperature inside the arc and the pressure for a simple experimental measurement of the luminance of the lamp.

REFERENCES

[1] Stormberg, H.P., and Schafer, R. Time-dependent behaviour of high-pressure mercury discharges, J. Appl. Phys., 1983, 54, (8), pp.4338-4347

[2] Elenbaas, W. the High Pressure Mercury Vapour discharge, W.H.Pub.Company, Amsterdam 1951

[3] M. Stambouli, K.Char., C. Cos., and J-J Dam. Thermalization of the high pressure mercury lamp positive column during the warm-up phase, IEE T.P. SC., vol.23.Issue 2, April 1995- pp.138-144

[4] A. Chammam, H. Ell., B. Mra., K. Char., M. Stam. and J-J. Dam. Effect of a pulsed power supply on the spectral and electrical characteristics of HID lamp, J. Phys.D:App.Phys. 38 1170 (2005).

Phase resolved Dy-density and plasma temperature measurements by absorption- and emission-spectroscopy of Dy spectral lines

M. Westermeier, C. Ruhrmann, J. Reinelt, J. Mentel and P. Awakowicz
Ruhr-University Bochum, Electrical Engineering and Plasma Technology, D-44780 Bochum, Germany

Contact: westermeier@aept.ruhr-uni-bochum.de

ABSTRACT

The density of Dy-atoms is measured in a YAG HID-lamp by means of a broadband absorption technique using a UHP-lamp as a quasi-continuous back-light source in combination with emission spectroscopy. The population density of the atomic ground state of Dy is determined by an analysis of the absorption profile of a Dy-resonance-line and of an excited state by a measurement of the emission coefficient of an optical thin Dy-line. By a combination of both the plasma temperature is deduced. Therefore, measurements at Hg-lines become unnecessary which will be important for diagnostics at Hg-free HID-lamps. The phase resolved densities are used to analyse the emitter effect of Dy on the electrodes.

INTRODUCTION

The lifetime of HID-lamps can be effectively extended by means of the so called "emitter-effect" which reduces the temperature of the tungsten electrodes. The emitter-effect is produced by a monolayer of emitter-materials like Dy or other rare earth elements on the tungsten surface which reduces the effective work-function $e\varphi$ of the electrode [1].

Hence, to characterise and optimise the emitter-effect, a measurement of the Dy-density n_{Dy} and of the plasma temperature T_{Pl} around the electrodes of an HID-lamp is needed. Up to now, the densities of Dy-atoms and Dy-Ions in front of the electrode were determined by measurements of Dy emission spectral lines. The plasma temperature was usually determined by measuring the emission coefficient of a mercury double-line while assuming a constant pressure of Hg within the HID-lamp [1].

THE BROADBAND ABSORPTION MEASUREMENT

Within this work, a new broadband absorption measuring method for Dy-densities is presented. Instead of a complex scanning laser system, a broadband UHP (ultra-high-pressure) lamp is used as backlight source for this measurement [2]. The quasi-continuous radiation of the UHP backlight lamp offers the advantage of a direct absorption-profile determination $I_\nu^{abs,UHP}(\lambda,y)$ of the Dy resonance-line at λ= 626 nm in dependence on λ and the lateral coordinate y with a spectrograph by a single measurement. The included undisturbed UHP-backlight signal $I_\nu(\lambda,0)$ passing through the measured lamp can be determined by a polynomial fit between the wings of the UHP-light profile. The light which is emitted by the lamp $I_\nu^{em}(\lambda,y)$ itself is determined by a reference measurement without the UHP-light source and subtracted. As a result, the optical depth $\tau(\nu,y)$ and the line-absorption coefficient $\alpha_L(r)$ can be calculated which finally yields the radial distribution of the Dy-atom ground state density $n_{0,Dy}(r)$ by means of an inverse Abel transformation of

$$\int_{\Delta\nu}\tau(\nu,y)d\nu=\int_{\Delta\lambda}\ln\left(\frac{I_\nu(0)}{I_\nu^{abs,UHP}}\right)\frac{c_0}{\lambda^2}d\lambda=2\int_{\eta=0}^{\sqrt{R^2-y^2}}\alpha_L(\eta)d\eta=\frac{e^2}{4\pi\varepsilon_0}\frac{\pi}{m_e c_0}f_{nm}2\int_{\eta=0}^{\sqrt{R^2-y^2}}n_{0,Dy}(\eta)d\eta \qquad (1)$$

where R is the radius of the discharge. The measurement has a high accuracy as it does not need an absolute calibration and is not influenced by changes of the optical path e.g. by wall-blackening [2].

Additionally a measurement of the Dy emission-line at λ= 695 nm is performed to determine the line-emission coefficient $\varepsilon_L(r)$. The emission measurement is described in detail in [1] and yields the Dy-atom density of the upper level $n_{u,Dy}(r)$ of this line. In an LTE-plasma, which can be assumed within the measured HID-lamp, the ratio between the population of the atomic ground state and of an upper energy level is given by the temperature dependence of the Boltzmann factor and the statistical weight of the states. Accordingly, the plasma temperature T_{Pl} within the HID-lamp can be calculated by the ratio of the ground state density from the line-absorption coefficient α_L and the density of an upper level from the line-emission coefficient ε_L.

The correlation of an absorption- and an emission measurement of Dy lines leads to the plasma temperature T_{Pl} independently of the Hg filling in the HID-lamp. The method might become very important for future measurements of the plasma temperature in Hg-free HID-lamps.

EXPERIMENTAL SETUP

The experimental setup consists mainly of the UHP-backlight system, the investigated YAG-lamp (yttrium aluminium-garnet) and an imaging path with a spectrograph equipped with a 1200 l/mm grating and a CCD-camera which results in a spectral resolution better than $\Delta\lambda$= 0.15 nm FWHM.

The UHP-lamp within the backlight system is a Philips 120W-type with an operating gas pressure of p= 150 bar. The UHP-backlight is filtered by a 10nm-interference filter to avoid an impact on the measured YAG-lamp. The investigated YAG-lamp (NTD1) consists of an yttrium-aluminium-garnet burner and tungsten electrodes with a diameter of d_E= 360 µm. It is filled with Hg as buffer gas and a mixture of NaTlDy-iodides which is described in detail in [1]. The imaging optics consisting of two achromats and a Dove prism produces a rotatable 1:1 image on the entrance slit of the spectrograph.

RESULTS AND DISCUSSION

Figure 1 a) shows the Dy-atom density in a NTD1 YAG-lamp 125 µm in front of the electrode and in the middle between the electrodes. The radial resolved results are averaged over the area of the electrode diameter. The lamp was operated with an i= 800 mA switched-dc current at f= 100Hz. The total Dy-atom densities deduced from emission and absorption measurements are quite similar which confirms the good quality of the measurements. As usual for 100Hz operation, the density of Dy-atoms is reduced during the anodic phase and enhanced during the cathodic phase by an ion current. In the middle between the electrodes the density is almost constant and slightly lower than the average value at the electrodes.

Figure 1 b) shows the plasma temperature for the same lamp operating parameters as in fig. 1 a). The plasma temperature is determined using the absolute intensity of Hg-lines and the presented emission/absorption-measurement. The results of both measuring methods agree very well for the plasma temperature. It proves the reliability of the Dy-em/abs measurements. The plasma temperature increases in the anodic phase and decreases in the cathodic phase. In the middle of the discharge, the plasma temperature is almost constant and its value is always lower than that in front of the electrode.

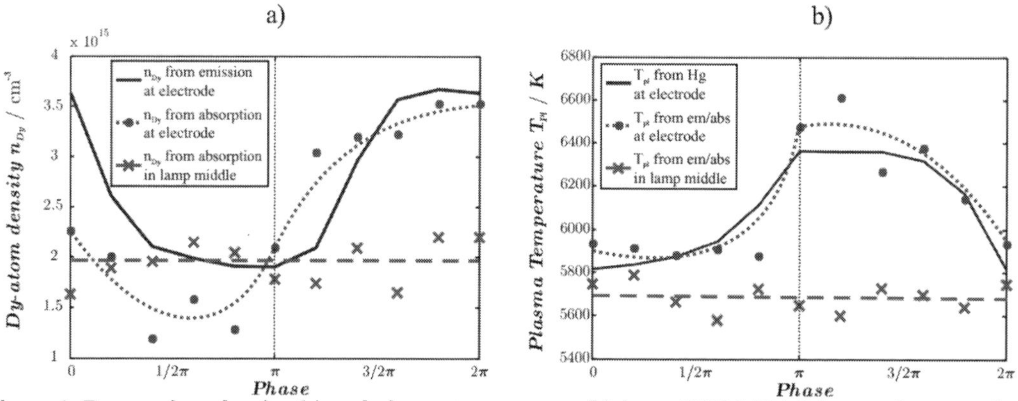

Figure 1: Dysprosium density (a) and plasma temperature (b) in an NTD1-YAG-lamp, phase resolved in front of the electrode and in the discharge middle; i= 800 mA, switched-dc, f= 100 Hz

REFERENCES

[1] O. Langenscheidt *et al*, J. Phys. D: Appl. Phys, **41** (2008) 144005

[2] M. Westermeier *et al*, J. Phys. D: Appl. Phys, **43** (2010) 124015

This work was supported by Philips Lighting, the DFG (GRK 1051) and the RUB-Research School.

Electronic Ballasts and Sodium Electrolysis in Automotive Headlight Lamps

N. Brates[1] , M. Seki[2] , S. Ukegawa[3], J. Connolly[1]

[1] *Universal Lighting Technologies, 216 W Cummings Pk, Woburn, MA 01801, USA*

[2]*Panasonic Electric Works, Ltd., 1115 Shimahata Aza Ohkawazu, Tsubame, Niigata, 9590192, Japan*

[3]*Panasonic Electric Works, Ltd., 1048, Kadoma City, Osaka 571-8686, Japan*

Contact: nbrates@us.pewg.panasonic.com

Abstract – *Modern automotive headlights incorporate High Intensity Discharge (HID) lamps powered by electronic ballasts. Circuit topology of the ballast results in the build-up of a DC potential between lamp and the headlight fixture (i.e., reflector, light screen, etc.). Under certain conditions, the DC bias could cause electrolysis of Sodium contained in the lamp chemical fill, resulting in Sodium loss, thus, short lamp life. Results of our investigations, together with design rules and recommendations for ballast- fixture systems are presented.*

INTRODUCTION

In an automotive HID lamp an arc discharge is created between two electrodes located inside a quartz arc tube, which is filled with xenon and metal halide salts [1]. The HID lamps offer a number of advantages, such as higher luminous efficacy, reduced heat output and much longer lamp life. Typically, an automotive HID lamp has twice the light output of a halogen lamp and still requires less input power. Another benefit of the automotive HID lamps is an average lifetime of 3000 hours, which normally covers the typical lifetime of a car. Igniting, heating and stabilizing the arc of an HID automotive lamp requires electronic control gear. The system consists of an igniter and an electronic ballast unit (inverter). In typical HID automotive ballasts, car battery voltage is stepped-up by a DC-DC converter, followed then by a DC-AC inverter configured as an H-bridge, Figure 1. The role of the inverter is to change the stepped-up DC voltage into an alternating square-wave for lamp operation. Topology requirements for the ballast control and the DC-AC inverter circuitry necessitate that the "LO" output lead (Fig.1) is not grounded. Accordingly, a DC potential, or "floating potential", ΔV_{DC} builds up between the lamp body and the ground (i.e., headlight housing-reflector). The polarity of this DC bias in respect to reflector – either positive or negative - depends on the ballast circuit topology. Typically, the metal halide chemistries used in HID Automotive lamps include NaI [1]. The presence of the Na^+ ions in the arc discharge column (Fig 2.) makes the polarity of this external DC bias critical in avoiding the loss of Na via electrolysis and thus, a short lamp life [2], [3]. The goal of the experimental work described in this paper is to determine limits for bias voltage magnitude, bias polarity and spacing between HID lamp and headlight housing/reflector.

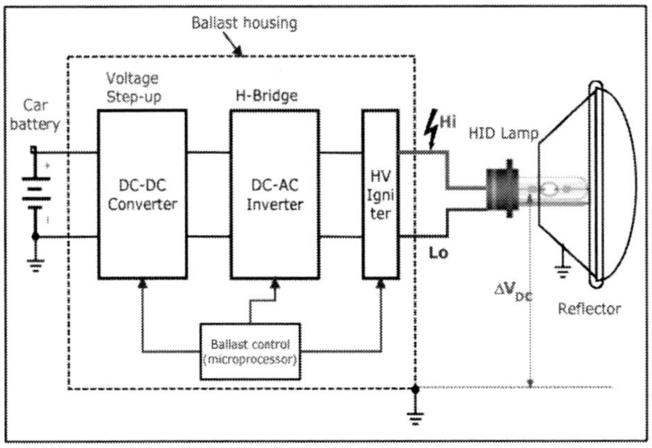

Figure 1. Typical topology of an HID automotive ballast

EXPERIMENTAL

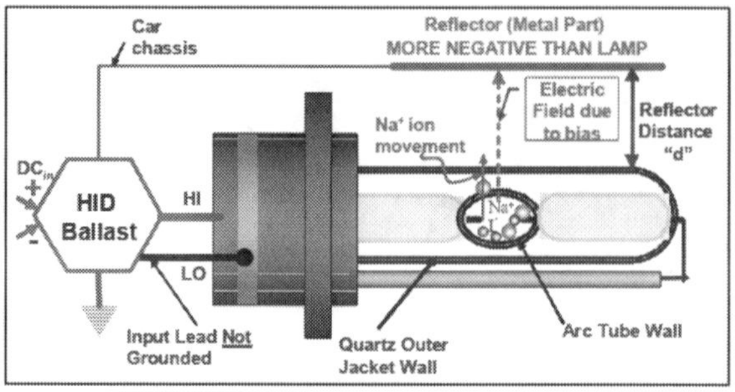

Figure 2. Sodium electrolysis in a HID Lamp

Design of Experiments ("DOE") technique was utilized to determine the limits for bias voltage magnitude, polarity and the "d" spacing between HID lamp and headlight housing (Fig. 2). Currently, there are two types of Automotive HID Lamps: "D2" (mercury containing) and "D4" ("mercury-free"). Experiments included both types, as the lamps construction and chemical fill composition are substantially different, hence, the Na kinetics. The DOE predictors are shown in the Table 1. Experiments responses were: the rate of rise of lamp RMS voltage and lamp life. Lamp life (EOL) is based on the maximum lamp RMS voltage, specified by lamp Manufacturers (e.g., $V_{EOL}\cong 67V_{rms}$ for D4 and $V_{EOL}\cong 117V_{rms}$ for D2). Measurements were carried out at specific time intervals. To accelerate the rate of sodium loss under DC bias conditions, a group of lamps was operated at elevated arc tube wall temperature (about 200°C over nominal). To minimize the number of experiments, we selected a "fractional factorial" design, as opposed to a "fully crossed" one (3x2x3x2).

Predictors (Factors)	Levels		
	High	Medium	Low
Bias Voltage (negative)	500V	225V	45V
Arc Tube Wall Temp. (Maximum)	1098°C		907°C (nominal)
Reflector Distance "d" [Fig.2] (Outer Jacket – Metal parts)	15mm	2mm	0mm touching

Table 1. Experimental Parameters

CONCLUSIONS

Sodium loss due to electrolysis causes rapid increase in lamp voltage and thus, short lamp life. Higher DC bias voltage (negative), higher wall temperature, and closer proximity of metal parts to lamp, result in faster voltage rise. Experimental results are summarized in Fig. 3. The response, rate of rise of lamp RMS voltage [V/hr] is plotted against the three factors: (1) Arc Tube Wall temperature, (2) DC bias voltage magnitude (negative), and (3) reflector distance to lamp (Proximity) "d". The effect of lamp chemistry, "D2" (with Mercury) versus "D4" (Mercury-free) is also included. In the absence of Mercury (i.e., D4 type), the sodium loss occurs at a faster rate, hence, the rate of voltage rise is higher. As expected, when negative biased metal parts are touching the lamp envelope, d=0mm, the rate of lamp voltage rise is the highest. By using the Design of Experiment methodology, it is possible to optimize the ballast topology (i.e., bias, polarity), the lamp fixture design (i.e., reflector parts proximity, maximum lamp temperature).

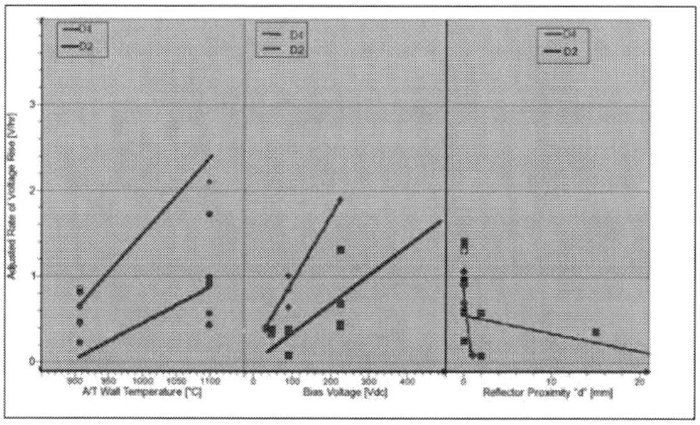

Figure 3. Summary of DOE results.

REFERENCES

[1] P. Flesch, "*Light and Light Sources*" Springer-Verlag, Berlin, Heidelberg 2006
[2] J. F. Waymouth, *"Electric Discharge Lamps"* MIT Press, Cambridge, MA, 1971
[3] F. Nagel and O. Kaposi, Proc. 3rd Symposium on Science and Technology of High Temperature Light Sources, Electrochemical Society, Proceedings **85-2**, Toronto, 1985, pp 72-77.

Dimming Characteristics of Medium and Small Wattage Ceramic Metal Halide Lamps

Shuijun Shi, Wei Li, Ling Chen, Jie Liu and Shanduan Zhang
Institute for Electric Light Sources, Fudan University, 220 Handan Road, Shanghai 200433, People's R China
Contact: sdzhang@fudan.edu.cn

ABSTRACT

Photometric and colour parameters (luminous flux, efficacy and correlated colour temperature et al) of 70 W and 150 W ceramic metal halide (CMH) lamps are measured under dimming conditions. The result proves acceptance of such medium and small wattage CMH lamps operated at reduced power because the key parameters are not affected by dimming.

INTRODUCTION

Ceramic metal halide (CMH) lamps play a key role in the occasions where high and medium wattage light source is needed because their high efficacy and colour rendering index enable the complete lighting system more energy efficient and comfortable. In addition, dimming can further save energy while meeting the requirements for green lighting. Research has been made on 150 W lamps for street lighting under various dimming conditions [1]. The results show that no negative effects result from dimming on lifetime, lumen maintenance and voltage of the lamps. The increasingly mature techniques have brought into the market medium and small wattage CMH lamps, which began to play important roles in indoor lighting. Usually customers need to dim the lamps in indoor applications (such as previous dimmable incandescent lamps) and have higher expectation for photometric and colour performance of light source. Dimming characteristics of ceramic metal halide lamps have been reported mainly about high and medium wattage lamps [2, 3], but seldom is about medium and small wattage lamps and none is about those under 100 watts.

In this paper, parameters of 70, 150 W CMH lamps are measured under dimming conditions (at 210 Hz) to justify dimming for medium and small wattages lamps and discover potential problems.

EXPERIMENT

In our experiment, three types of CMH lamps are measured: 70 W, 4000 K; 150 W, 3000 K; and 150 W, 4000 K (to be numbered as S1, S2 and S3 hereafter). Specific dimmable electronic ballasts are designed to operate the lamps at 60%~100% rated power (70 W lamps) and 70%~100% rated power (150 W lamps).

The output of the e-ballast is responsible for driving the lamp in the integrating sphere. The spectrometer and the integrating sphere are used to obtain photometric, colour and spectral parameter. Electric parameters can be gotten by a four-channel digital storage oscilloscope (100 MHz, 1 Gs/s) coupling with a voltage probe and a current probe.

RESULTS AND DISCUSSIONS

(1) Efficacy, colour rendering index (CRI) and correlated colour temperature (CCT)

Table 1 shows the parameters of the lamps at dimming conditions. For lamp S1, efficacy at lowest power (60% rated power) accounts for 85.8% of that at rated power, CRI reduction is 11.8 and CCT rise is 697 K. For lamp S2 and S3, the three parameters are 95%, 13.3, 697 K and 89.4%, 13.6, 796 K, respectively.

(2) Spectral power distribution (SPD)

Figure 1 illustrates the SPDs of S1 and S3 (520~620 nm). Three trends upon dimming are obvious, which can explain the changes of photometric and colour parameters:

Supported by Key Technologies R & D Program of Shanghai Science and Technology Committee (Grant No. 08DZ1140802)

(i) The peak between 530 nm and 540 nm is enhanced when lamps are dimmed. That partly explains why efficacy can be maintained relatively high at reduced power;
(ii) The continuum near 555 nm is weakened greatly at reduced power, which mainly leads to the loss of luminous flux and colour CRI;
(iii) The peaks at about 595 nm shift to the shorter wavelength and weaken. The two factors will remain the luminous flux in this region while increase CRI.

(3) Chromaticity coordinates and deviation from the black body line (BBL)

Figure 2 shows (u, v) lotus of the three lamps under dimming conditions and figure 3 shows the deviation from the BBL. It can be concluded that CMH lamps with CCT 4000 K has less deviation than that with 3000 K. That means the former has more natural lights and advantages in indoor dimming application.

In conclusion, it is practical to dim medium and small wattage CMH lamps in indoor applications. Lamps with CCT 4000 K are more advantageous thanks to its less colour deviation. Lamps of suitable power can be selected to be operated under dimming conditions according to specific cases.

Table1 Lamp parameters under dimming conditions

	Luminous efficacy (lm/W)				CRI				CCT (K)			
	100%	90%	80%	70%	100%	90%	80%	70%	100%	90%	80%	70%
S1	90.3	89.8	88.2	84.3	91.7	89.7	87.0	84.1	3988	4075	4217	4393
S2	100.9	99.0	98.1	96.0	81.8	76.4	73.0	68.5	3352	3505	3685	3958
S3	97.7	95.9	91.8	87.4	91.1	87.2	82.9	77.5	4268	4566	4840	5064

S1: 70 W, 4000 K; S2: 150 W, 3000 K; S3: 150 W, 4000 K.

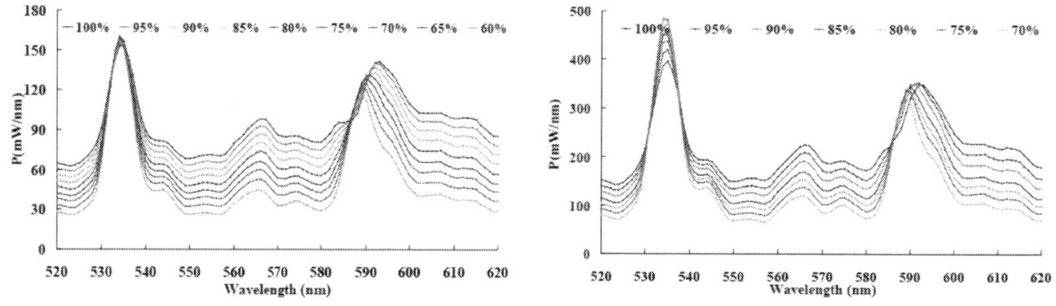

(a) (b)

Fig. 1 Spectral Power Distribution of S1 and S3 (520~620 nm)

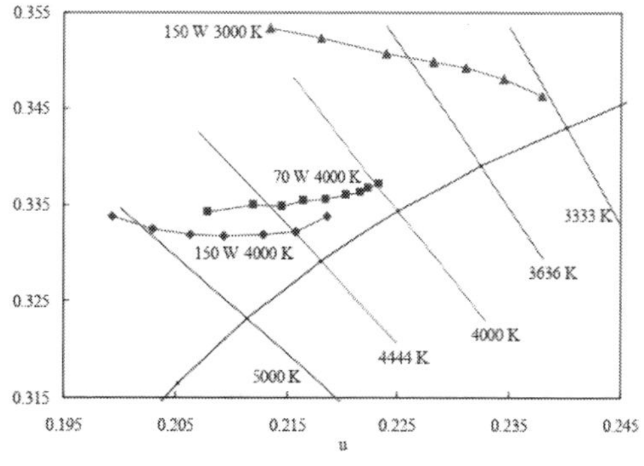

Fig. 2 (u,v) lotus under dimming conditions

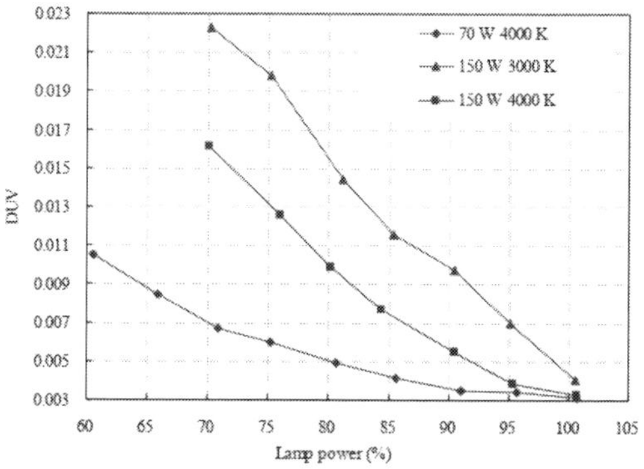

Fig. 3 DUV under dimming conditions

REFERENCES

[1] Guest E C, Girach M H, Mucklejohn S A, *et al.* 2008 *Lighting Res. Technol.* **40** 333
[2] Zhu H L, Lambrechts S and Maya J 2004 *Institute of Physics Conference Series* **182** 377
[3] Kettlitz M, Wendt M, Peters S, *et al* 2007 *Proc. 11th Int. Symp. on the Science and Technology of Light Sources (LS-11, Shanghai, China)* (Sheffield, UK: FAST-LS) p 33-34.

Electrodeless HID Lamps
Excited by Circularly-Polarized Microwave Discharges

Jin Joong Kim, Kyoung Shin Kim, Dong Ho Won, and Hyun Sung Yoon

Lighting Technology Laboratory, Taewon Lighting Co.
Hansin IT Tower (Rm 411), 235 Guro-dong, Guro-gu, Seoul 152-050, Republic of Korea
Contact: kimjj@taewon.co.kr

Abstract: We have developed, to our knowledge, the first commercial-class electrodeless HID (ELS-HID) lamp system using a nonrotating bulb excited by circularly-polarized microwave discharges of InBr vapor at atmospheric pressure. Unlike conventional high pressure "electrodeless HID lamps"—a more generic term for the well-known electrodeless sulfur lamp—the bulb in our lamp system is stationary. The microwaves at 2.46 GHz are circularly polarized in TE_{11} mode by a specially designed polarizer, making the gaseous discharges themselves rotate in the stationary bulb that is placed in a circular cavity. The ELS-HID SolaRay lamp system generates white light of an unoptimized luminous efficacy of 65 $lm/W_{[wall\text{-}plug]}$ with an input power of 1.1 kW and of a color rendering index in excess of 96 Ra. This ELS-HID lamp system has a number of other distinctive features, including a significant change in the spectrum as the input power is varied.

INTRODUCTION

The electrodeless HID lamps are known as one of the most efficient high intensity discharge (HID) lamps [1-4]. The class of electrodeless HID (ELS-HID) lamps—a generic term for the well-known electrodeless sulfur lamp—has a number of attributes desirable for the next generation ELS-HID lamps, including a long system lifetime (>20,000 hours), a lumen maintenance, an excellent color rendering index (CRI ranging 80~98 Ra, depending on the radiators) with a continuous spectrum in the visible range, and the absence of hazardous substances like mercury, to name just a few. Since the advent of the sulfur lamp in the early 1990's [1, 2], there have been limited but continuing efforts to study the physics of white light generation by ELS-HID sulfur lamps and to develop commercial lamp systems [3-9].

One of the drawbacks of "conventional" ELS-HID sulfur lamps is that the bulb itself has to be rotated to keep it from being punctured by hot spots that are formed on the lamp surface by the linearly polarized electric fields in the circular cavity that contains the bulb. At the Tenth Light Source Symposium [Light Sources 2004] held in Toulouse, France [8], we reported the demonstration of rotating discharges in a stationary ELS-HID lamp, using circularly polarized microwaves (CPMs) [7]. Since then, we have developed a commercial-class of ELS-HID lamp systems that uses a *nonrotating* bulb. This paper presents the characteristics of this uniquely distinguishing ELS-HID lamp system.

TECHNICAL DETAILS

The commercial product of ELS-HID lamp system -- so called the SolaRay ELS-HID lamp-- was constructed as a stand-alone set, including an inverter power supply and a kW-class magnetron of 2.46 GHz, using a specially designed CPM generator [7, 8]. The general principle of the CPM generation is described elsewhere [7]. The rotating plasma discharges were confirmed in a separate experiment and were published in Ref. 7. The spectrum was measured by using a spectrometer. The total lumen flux, the correlated color temperature, and the CRI were measured by using a goniometer. The bulb made of quartz (a few cm in diam) is filled with a few mg of InBr powder and a buffer gas. It is placed in the middle of a circular cavity that is coupled to a polarizing waveguide and is excited by circularly polarized microwave discharges in TE_{11} mode. The gaseous discharges themselves rotate in the bulb as was demonstrated in Refs. [7, 8].

RESULTS

In full power operations with the input power of 1.1 kW, a brilliant fireball formed by microwave discharges of InBr or other radiator vapour fills up the bulb and generates approximately 68,000 lm, in an unoptimized condition. Typical emission spectra are shown for two different RF input powers in FIG. 1. The efficiency is substantially increased as the input power is increased. The yellow lines are presumed to be due to the Na impurity emissions. Other typical photometric parameters and features of the lamp systems are summarized in Table. 1. In the conference presentation, we will show other performance features, including the scene of a field site where commercial products are installed.

CONCLUSION

We have developed a commercial-class ELSHID lamp system, using a nonrotating bulb that is excited by circularly polarized microwaves. We expect that the performance of the lamp system will further be improved under optimal conditions of the system. The ELS-HID lamp systems can also be operated with a variety of other radiators, opening up the possibility of utilizing these lamps for a variety of applications in lighting and displays.

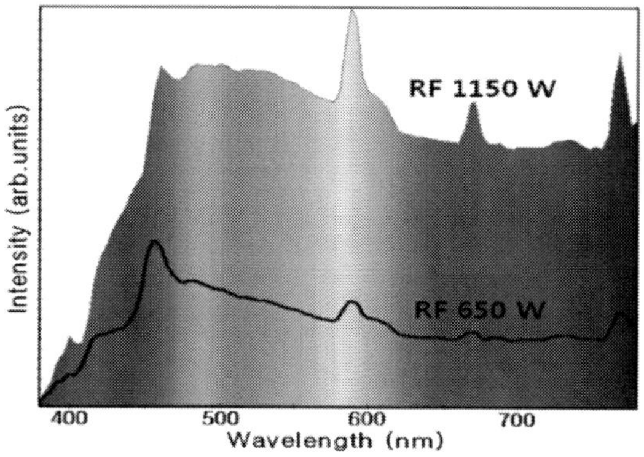

FIG. 1. Spectra of a prototype at two input power levels.

Input power (W)	1,100~1,800
Lumen efficacy [lm/W(wall-plug)]	62~69
Correlated color temperature (K) (depending on input power)	5200~7000
Color rendering index (Ra)	>96
Lumen maintenance (%) [est.]	>80
System lifetime (hrs) [est.]	>20,000
Starting time (s)	<20
Mercury content	0
Bulb rotation	Nonrotating

Table 1. **Specifications of typical ELS-HID Ray lamps.**

REFERENCES

[1] J. T. Dolan, *et al.*, "*A novel high efficiency, microwave powered light source*," Proc. the Sixth International Symposium on the Science and Technology of Light Sources," Budapest, September, 1992, L. Bartha and F. J. Kedves (eds.) (Technical University of Budapest, 1992); U.S. Patent No. 5,404,076 (1995).

[2] B. P. Turner, M. G. Ury, Y. Leng, and W. G. Love, "*Sulfur Lamps – Progress in Their Development*," J. IES **26** (1995) 10.

[3] C. W. Johnston, H. W. P. van der Heijden, G. M. Janssen, J. van Dijk, and J. J. A. M. van der Mullen, J. Phys. D: Appl. Phys. **35** (2002) 342.

[4] C. W. Johnston, J. Jonkers, and J. J. A. M. van der Mullen, J. Phys. D: Appl. Phys. **35** (2002) 1.

[5] Jin J. Kim, *et al.*, Proc. The 9th Int. Symposium on the Science and Technology of Light Sources, 12-16 August 2001, Cornell University, Ithaca, NY, USA (Cornell University Press, Ithaca, NY, 2001) p. 123.

[6] Jin J. Kim, IEEE International Conference on Plasma Science, May 26-30, 2002, Banff, Alberta, Canada (IEEE, Piscataway, NJ, USA, 2002) Conference Record p. 318.

[7] Jin J. Kim, J. T. Ko, Dong H. Won, S. S. Kim, and H-Y Chang, Appl. Phys. Lett. **84** (2004) 2769.

[8] Jin Joong Kim, "*Light Sources Excited by Microwave High Intensity Discharges*," Proc. The 10th International Symposium on the Science and Technology of Light Sources, Toulouse, France, 18-22 July 2004 (Institute of Physics Conference Series Number 182) (Bristol, UK), pp. 85.

[9] LG Electronics, private correspondence and its website information.

Superradiance in the sodium D lines of a MHD lamp

D. Karabourniotis[1], E. Drakakis[2]
[1] *Institute of Plasma Physics, Department of Physics, University of Crete, 71003 Heraklion, Greece*
[2] *Technological Educational Institute, Department of Electrical Engineering, Heraklion, Greece*
Contact: dk@plasma.physics.uoc.gr

ABSTRACT

The superradiance in the D resonance lines of sodium emitted by a high-pressure arc lamp containing mercury and sodium iodide is reproduced by increasing the relative density of the resonant sodium atoms in the plasma periphery. The release of the binding energy of sodium iodide near the lamp wall might be the cause of increase of the excited atom density. The obtained emission results are satisfactory, but they cannot explain the experimentally observed reduction of the optical depth.

INTRODUCTION

A pronounced radiance enhancement is usually observed at the central wavelength of self-reversed lines emitted from absorbing plasmas of metal halide discharge (MHD) lamps [1, 2].
This kind of superradiance takes place within the line center near the transition frequency of self-reversed lines. It was explained [2] as the result of non-radiative transfer of excitation by resonant collisions from the plasma core to the arc periphery involving a strong decrease of the apparent absorption coefficient in the vicinity of the line center followed by an analogous decrease of the apparent optical depth within the same frequency region. It is this apparent optical depth that is measured applying absorption techniques. In emission measurements this strong decrease of the apparent absorption coefficient results in a rapid increase of the radiance within the line center. The diagnostic procedure proposed in [2] assumes monotonically decreasing radial distributions of the emitting and absorbing atoms, contrary to the radial course of sodium atoms in mercury arcs. Therefore, it cannot be directly applied to sodium lines.
Recently, the phenomenon of superradiance was also observed in the central part of aluminium self-reversed resonant lines emitted through the plasma core in the early phase of laser-induced plasmas [3, 4]. This phenomenon was interpreted as being the result of the interaction of the plasma edge with the background air excited by the plasma shockwave leading to an additional excitation of the aluminium vapour that increases the emission from the plasma edge [4].
Similarly, disregarding the fact of the strong reduction in the absorption coefficient in the line center, the superradiance of self-reversed lines emitted from arc lamp plasmas might be assumed related to the presence of an additional source of radiation in the periphery of the plasma. This source could be either the radiative emission of the non-radiative transfer of excitation that occurs in the periphery of the plasma or the release of the binding energy coming from the recombination of the metal halide molecules near the lamp wall or the release of the electronic recombination energy or, the most probable, a combination of these effects. All these effects lead to a local increase in the radial distribution of the emitting atoms near the lamp wall.
This is the key used in this work to reproduce the superradiance in the D resonant lines of sodium emitted by a high-pressure arc lamp containing mercury and sodium iodide (figure 1).

EXPERIMENTAL CONDITIONS

The lamp constructed by PHILIPS, operates at ac, at an input power of 150 W, with an estimated pressure of 2.9 atm, *without external envelope*. The confining discharge tube made by transparent quartz has an internal diameter of 12 mm and the distance between electrode tips is 18 mm. It is filled with 5 mg Hg, 1.85 mg NaI, and 100 mbar Ar/Kr as starting gas. The resolving power of the monochromator was 3.5×10^5; and the measurements were carried out with a band-pass of 15 mÅ at 6000 Å. The spectral measurements were made on a side-to-phase resolved basis. Figure 1 shows the sodium lines D_1 (5896 Å) and D_2 (5890 Å) at maximum current phase (5 ms). The electron temperature was determined from the pair of the mercury lines ending to the metastable levels. The excitation temperature of the sodium resonance levels is assumed to be very close to the gas temperature [5], which in turn is close to the minimum electron temperature that occurs at the voltage zero-crossing moment at 0 ms [6].

THEORETICAL CALCULATIONS AND RESULTS

The calculations were carried out according to the theory of self-absorbed emission lines previously described [1]. The broadening constants for the D_1 and D_2 lines are those derived in [1]. A parabolic temperature profile is assumed. The radial distribution of the relative density of the absorbing sodium atoms is given by $L=(1-r^2)\times exp\{F[1-(1-2.2*r^2)^4]\}$ This profile (L) is derived from physical considerations concerning the depletion of the sodium atoms near the high-temperature arc axis by ionization and the lacking of the free sodium atoms near the discharge wall due to molecular recombination of NaI. The value of F was determined from the condition for line-reversal, which was applied to the blue maximum of the D_2 line, and the measured radiance at this line maximum. The radial course of the ratio $\Lambda = U/L$, where U is the radial distribution of the relative density of the emitting atoms, is shown in figure 2. The reconstructed self-reversed lines along with the experimental one are depicted in figure 1.

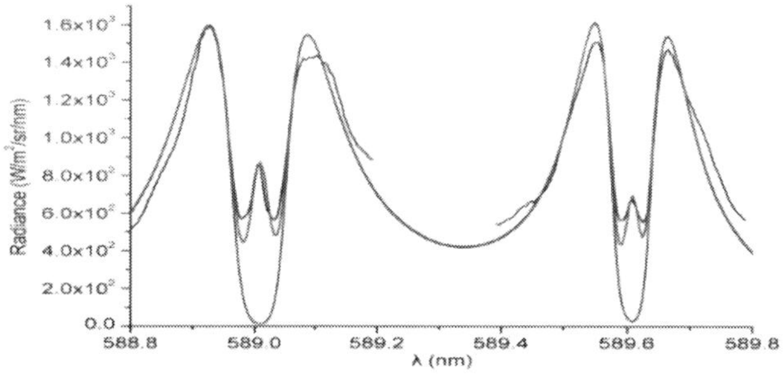

Fig. 1. Sodium D lines; experimental (black); reconstructed with radiation source near the wall (red); reconstructed without additional radiation source (blue).

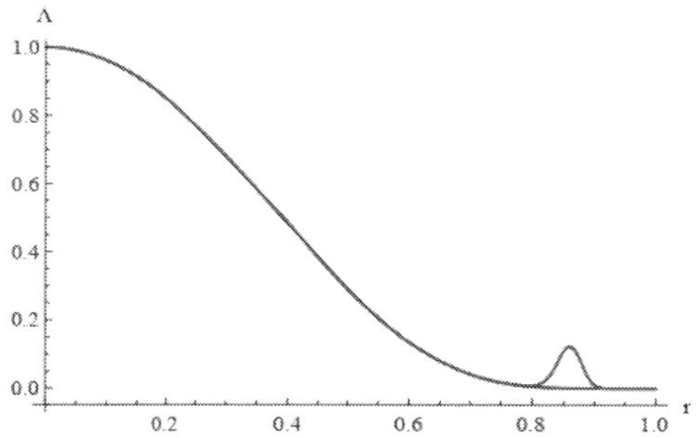

Fig. 2. Radial course of the ratio Λ=U/L; with radiation source near the wall (red); without additional radiation source (blue).

Acknowledgment

Thanks to Dr. H. Schneidenbach (INP- Greifswald) for fruitful and encouraging discussions.

REFERENCES

[1] D. Karabourniotis, J. Phys. D: Appl. Phys., 40 (2007) 6608.
[2] D. Karabourniotis, E. Drakakis, Plasma Phys. Control. Fusion, 50 (2008) 124009.
[3] D. Karabourniotis, M. Ribiere, B. G. Cheron, Appl. Phys. Lett., 93 (2008) 041501.
[4] M. Ribiere, D. Karabourniotis, B. G. Cheron, J. Appl. Phys., 105 (2009) 083309.
[5] D. Karabourniotis, Appl. Phys. Lett., 85 (2003) 5395.
[6] D. Karabourniotis, E. Drakakis, Appl. Phys. Lett., 81 (2002) 1972.

Accelerated testing of an HID lamp by radiative cooling

D. Itoh[1] , T. Kasuya[1] ,M. Wada[1] , S. Gotoh[2]
[1] *Graduate School of Engineering, Doshisha University, Kyotanabe, Kyoto,610-0321, Japan*
[2] *Panasonic Electric Works Co. Ltd., Kadoma, Osaka, 571-8686, Japan*
Contact:dtj0110@mail4.doshisha.ac.jp

ABSTRACT

High intensity discharge lamps were operated in an enclosure cooled down to liquid nitrogen temperature. Substantially longer duration of time was necessary to make glow to arc transition under this cooled condition. When the lamp was operated repetitively with a 2 min turn-on 3 min turn-off procedure, the lamp operated in the cooled condition had shown the transmittance of the discharge tube to decrease down to 20% of the original value. This compares to the reduction in transmittance of the tube down to 60% of the original value for a lamp operated at room temperature with the same repetitive procedure.

INTRODUCTION

High voltage pulses start plasma of a high intensity discharge (HID) lamp under a low pressure condition. During this start-up period, energetic ions strike the electrode surface to sputter out electrode materials [1], while evaporation predominates the electrode damage during steady state arc discharge [2]. Thus, short duration turn-on/turn-off procedure can accelerate electrode degradation of an HID lamp by sputtering. However, the lamp temperature largely affects the lamp ignition time and corresponding electrode damage during lamp start-up period. Once the lamp establishes a steady state discharge, it advances to arc discharge mode from glow discharge mode as the lamp tube temperature and pressure increase due to heating by the plasma. Right after extinguishing the plasma, the lamp temperature is too high to restart a discharge. When the electric circuit turns on the plasma at elevated temperature, the lamp pressure quickly reaches a value sufficient for an arc discharge transition. This will reduce the duration of glow discharge or the period that an electronic ballast induces trains of high voltage pulses to the HID lamp.

LAMP STARTUP

We have measured the above mentioned correlation between the time duration that lamp operates in the glow discharge mode, and the wall temperature of the sample lamp. The experimental setup consists of a 150 W rated power alumina ceramic metal halide lamp (Panasonic, Model MT150CE-W-D A), and an electronic ballast (Panasonic Electric Works, Model NDCX09156BY9). An infrared radiation thermometer (Lutron, Model TM-969) was used to measure the temperature of the ceramic tube before starting a discharge. The duration of glow discharge, or the period that the ballast had produced the trains of high voltage pulses exhibits a nearly straight line on a graph with the abscissa corresponding to the logarithm of the wall temperature of the discharge tube. For comparison, the characteristic for an as received lamp is compared with the one taken for an old lamp, which had been operated for 180 hours total with 800 times of turn-on procedure from room temperature. The old lamp clearly requires longer time to accomplish transition from glow mode to arc mode, indicating the degradation of electrodes.

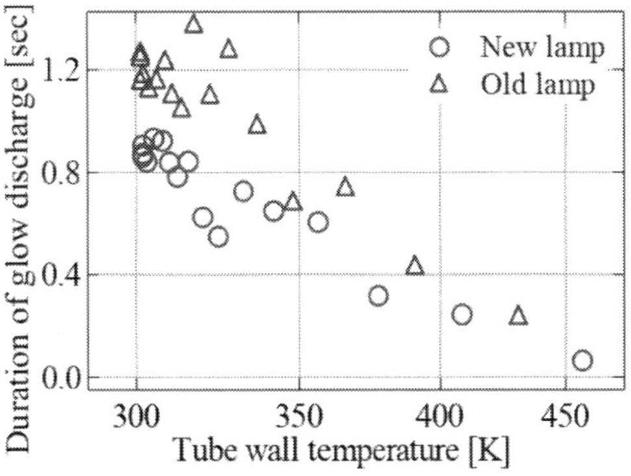

Fig. 1. Durations of glow discharge for new and old lamps plotted as functions of temperatures of discharge tubes.

The sputtering erosion of electrode appears in the optical emission spectrum of Fig. 2 (a) taken at a time about 800 ms after the commencement of a discharge in the test HID lamp. For comparison, a typical optical emission spectrum taken at a time 3000 s after the start of the discharge is shown in Fig. 1 (b). In these figures, the ordinates are plotted in arbitrary unit, and their relative intensities are computed from integration over time. The actual optical emission from W at the time of discharge start up is bright, which has been confirmed by a high speed camera (Red Lake, Model MotionPro HS-3). Thus, the plasma damages the electrode during the initial glow discharge by sputtering, and the cooling of the lamp temperature elongates this period to accelerate lamp electrode degradation.

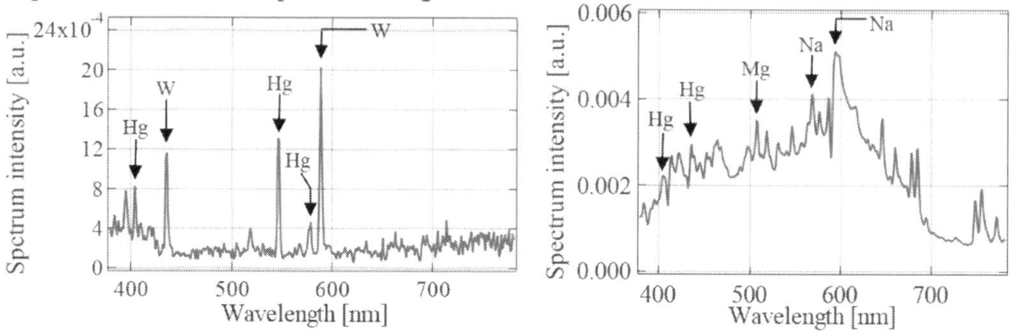

(a) After 800 ms *(b) After 3,000 s*

Fig. 2. Time variations of spectrum distribution from an HID lamp. (a) About 800 msec, and (b) 3000 sec after turning on the discharge.

LAMP OPERATION IN COOLED ENVIRONMENT

The lamp wall temperature had shown a semi-logarithmic dependence against the time after extinguishing the discharge. The efficiency for lamp cooling by photon radiation increases as the temperature of the surrounding wall decreases. Thus an HID lamp was operated in an liquid nitrogen cooled environment as shown in Fig. 3. A cylindrical aluminum bottle with the surface black-anodized contained a test HID lamp at the center. The bottle was dipped into liquid nitrogen kept in a Dewar. A mirror reflected the light from the lamp to record the image of the plasma during operation. This cold ambient temperature operation had successfully elongated the glow discharge period from about 1 second at 300 K to about 2 second at 77 K.

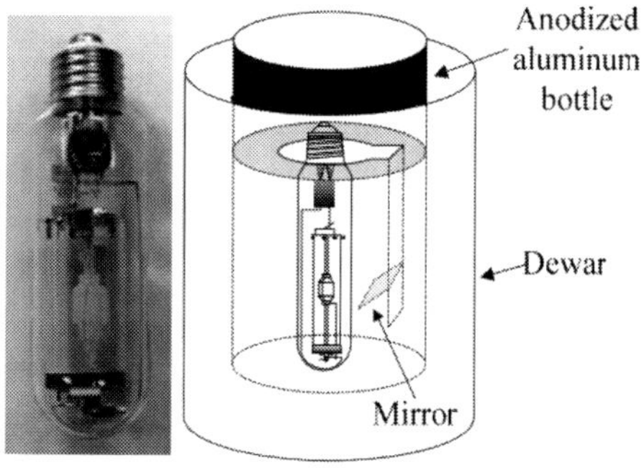

Fig. 3. Picture of the HID lamp and a schematic diagram of the experimental setup.

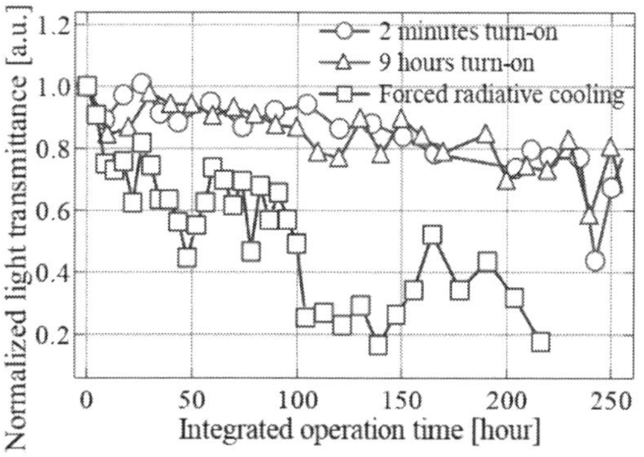

Fig. 4. Change of light transmittance of alumina ceramic discharge tube.

LAMP DEGRADATION BY COOLING

An electronic circuit repetitively turned on an HID lamp for 2 minutes after 3 minutes turn off period to elucidate the effect due to elongated time duration of glow discharge. As Fig. 4 shows, the light transmittance of the lamp tube decreased rapidly when the lamp was operated in a liquid nitrogen cooled environment. In the present condition that the cumulative operation time does not exceed 250 hours, short duration operation accompanying many occasions of discharge start-up exhibits lamp light transmittance reduction similar to the one observed for a lamp operated continuously for 9 hours. Thus, operation of an HID lamp in a liquid nitrogen temperature can largely accelerate electrode degradation of an HID lamp.

REFERENCES

[1] A.N. Bhoj, M.J. Kushner, IEEE Transactions on Plasma Science 33 (2 I), 518 (2005).
[2] P.G.C. Almeida, M. S. Benilov, M.D. Cunha, J. Phys. D. Appl. Phys. 41, art no. 144004 (2008).

The Scalability Challenge in a Large Wireless controlled Lighting System

Cheng Guo[1], Henk van Zeijl[1], Alessandro Baiano[1], Guoqi Zhang[2] and Ignas Niemegeers[3]

1Delft Institute of Microelectronics and Submicrontechnology, Delft University of Technology, Mekelweg 4, Delft, The Netherlands

2Department of Telecommunications, Delft University of Technology, Mekelweg 4, Delft, The Netherlands

3Philips Lighting, Mathildelaan 1, Eindhoven, The Netherlands

Contact: c.guo@tudelft.nl

ABSTRACT

Wireless communications have been facilitated in many domain of our life. Lighting is one among the areas where wireless communications can show its power. Compared to a lighting system connected by wires, advantages of the wireless system include better user experiences, ease of installation, self-organization and ease of update new services. However, there are also challenges in such a system. Specifically we focus on the challenge of scalability when a system has more traffic or grows large. We propose two protocols. The first one organizes a network into clusters and the second one adjusts transmission power of cluster members to only reach their cluster head. Thus contention and collisions can be alleviated. The first-step performance results are also presented.

INTRODUCTION

Air is a good medium to carry signals. Wireless links, in the ISM band, can be realized between any pair of lighting device. Then the devices are connected to each other wirelessly. Each device is equipped with a wireless transceiver and a Micro-Controller Unit (MCU). Certainly more components may be added to the device to form a more complex module. In the left of Figure 1 we show a simple lighting device which only has a transceiver, a MCU, a driver and a LED. However, the middle of the figure shows a complex device which has, except for the components in the simple module, memory, sensors and power management. All the communications in the networked wireless system are carried out through air as shown in the right of Figure 1. In the system every device can communicate with any other devices. The system is controlled by controllers shown as the black keypad in Figure 3. In a large lighting system, if the device to be controlled is not in the communication range of the controller, other devices act as routers to relay a command or a service update package to the intended module.

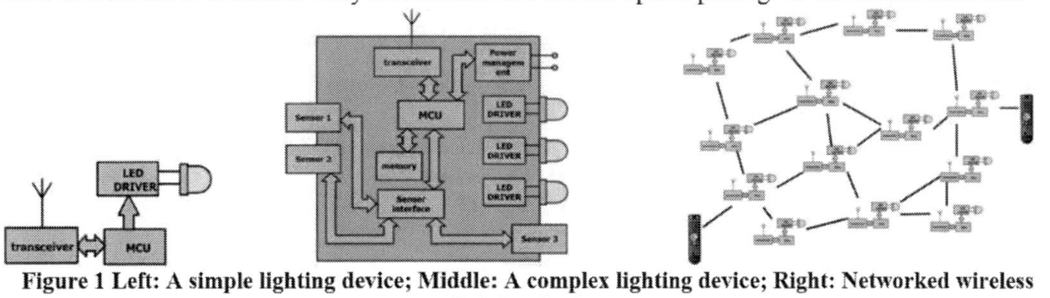

Figure 1 Left: A simple lighting device; Middle: A complex lighting device; Right: Networked wireless lighting devices

Such a system has many advantages. A user has a better experience since control is not restricted in one position any more. Moreover, installation of such a system is much easier due to the absence of wiring. Thanks to the embedded MCU, the system is enabled to have functions such as self-organization, self-healing and system diagnose. Furthermore, because the power consumption of LEDs are so low, a fully standalone lighting device, which is powered by batteries, is possible.

CHALLENGE ON SCALABILITY

Although we have walked the first step to control a lighting system with networked wireless devices, there are still many challenges. In this paper, we especially throw light on the challenge on scalability in large lighting systems. Scalability refers to the ability of a system to be readily enlarged or to handle the growing amount of traffics. Since a lighting system may contain hundreds or thousands of devices, concurrent transmissions from different devices may contend or even collide. To solve the problem, we propose to organize a network into clusters. Each cluster contains one Cluster Head (CH) and several Cluster Members (CMs). All CMs are one hop away from the respective CHs. A CM links only with its own CH to transmit/receive packets. They communicate in cyclic superframes [1] where the CH coordinates intra-cluster communications. Only CHs participate in network control and management. The clustered topology can increase a network's scalability due to the following two reasons. First, a device tends to associate with the CH closest to it, for the sake of better channel condition and shorter transmission range. Assuming devices having transmission power control, CMs can transmit with a lower power. Thus contentions and collisions are reduced among clusters. Second, CHs participate in the control and management of the network and CMs only communicate with respective CHs. For instance, only CHs will rebroadcast route request in a route discovery process, therefore a large amount of redundant traffic is prevented.

CHs actually form a Connected Dominating Set (CDS) [2] in a network. A device v is selected as a CH if it fulfills one of the following two criteria. 1) v has the smallest ID among its neighbors, and 2) v is a MPR selected by its smallest ID neighbor. A device's MPRs are its direct neighbors who cover all its two-hop neighbors (A device's two-hop neighbor is a neighbor in the device's two-hop range neighborhood but not a direct neighbor.). A device's MPR is selected in two steps. 1) Add a device u to v's MPR, if there is a two-hop neighbor of v only connected to u and mark all two-hop neighbors connected to u as covered. 2) Add a device w to v's MPR if w has highest score among its neighbors, device ID is used to break a tie, mark all two-hop neighbors connected to w as covered. Repeat this step until all v's two-hop neighbors are marked as covered. The score used above can be number of uncovered two-hop neighbors or a mixed of desired characteristics of a CH. In [3], we used a mix of remaining energy and number of uncovered neighbors.

Figure 2 shows the result of running our cluster protocol in a network. The network is effectively divided into five clusters and devices choose the closest CH to associate with. We can see that the network is still well connected and all devices are either CHs or have CHs to connect with. Thus the connectivity of the network is not influenced by the clustering protocol. TPC protocol adjusts the transmission power to be just enough to maintain a link between respective CHs and CMs.

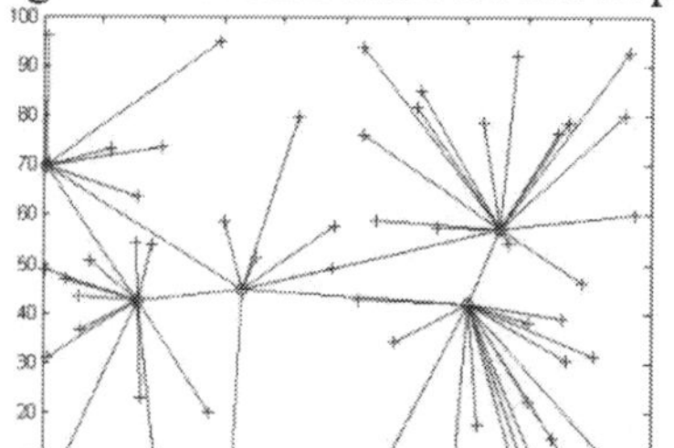

Figure 2 A network organized in clusters

REFERENCES

[1] J. A. Gutierrez, M. Naeve, E. Callaway, M. Bourgeois, V. Mitter, and B. Heile, "IEEE 802.15.4: a developing standard for low-power low cost wireless personal area networks," IEEE/ACM Trans. Networking, vol. 15, pp. 12–19, Sept. 2001.

[2] C. Adjih, P. Jacquet, and L. Viennot, "Computing connected dominant sets with multipoint relays," in Technical Report 4597, INRIA-Rapport de recherche, Oct. 2002.

[3] Cheng Guo, Ramin Hekmat and Przemyslaw Pawelczak, "Analysis and Optimization of Energy Efficient Cluster Forming for Wireless Sensor Networks", Vehicular Technology Conference (VTC) 2007 fall, Baltimore, MD, USA, September 2007.

Efficacy of optical radiation

B.Preston[1]

[1] *Ceravision Limited, The Mansion, Bletchley Park, Milton Keynes MK3 6EB, UK*

Contact: barry.preston@ceravision.com

ABSTRACT

The efficacy of any lamp is determined by the product of two numbers, the efficiency of the source in producing optical power, and the efficacy of the optical power. By constructing flexible artificial spectra maps estimating the maximum efficacy of optical power as a function of chromaticity and Colour Rendering Index (CRI) can be made

INTRODUCTION

The efficiency of the source we define as, the radiated power emitted in the wavelength range 400 – 800nm, divided by the total power supplied to the lamp system. The efficacy of the radiation is the radiated power weighted by the wavelength dependent sensitivity of the human eye divided by the optical power, equation (1).

$$\eta = \frac{680 \int_{400nm}^{800nm} v_{\lambda} \varepsilon_{\lambda} d\lambda}{\int_{400nm}^{800nm} \varepsilon_{\lambda} d\lambda}$$

where

η is the efficacy of optical radiation in lumens per watt (lm/W)

ε_{λ} is the relative spectral energy at wavelength λ

v_{λ} is the eye response under photopic conditions at wavelength λ

In order to maximise the efficacy of a light source both the efficiency and efficacy should be maximized subject to the light source having acceptable chromaticity and colour rendering properties. A question arises as to what is the maximum efficacy of optical power? To make estimates of the maximum efficacy of the optical radiation as a function of both chromaticity and colour rendering index, calculations employ artificial spectra constructed from combinations of Gaussian functions. The maximum efficacy is calculated from equation (1) by permitting the wavelength spread, wavelength position and the relative energy represented by the Gaussian functions to vary while the chromaticity and CRI are constrained to lie within narrow ranges.

CALCULATIONS

The relative energy at wavelength λ is given by

$$\varepsilon_{\lambda} = \sum_{i=1}^{n} \frac{k_i}{\sigma_i} e^{\frac{-(\lambda-\mu_i)^2}{2\sigma_i^2}}$$

where

k_i is a relative energy for the ith gaussian

σ_i is related to the Full Width at Half Maximum (FWHM) of the ith gaussian by $FWHM = 2\sqrt{2\ln 2}c$

μ_i is the centre wavelenth of the ith gaussian

An excel spreadsheet was constructed in which equation (1) is maximized using the excel solver[1] by allowing k_i, σ_i and μ_i to vary subject to the CRI being fixed, the correlated colour temperature (CCT) being confined to a narrow band ±25K and the chromaticity constrained to be close to the black body locus. The CRI is calculated using the CIE method of Measuring and Specifying Colour Rendering Properties of Light Sources[2]

RESULTS

The results of a calculation for three Gaussian spectra whose CRI's lie between 60 and 100 with CCT's varying between 3000K and 6000K are shown in Figure 1. The chromaticities for the spectra are located within the boxes shown in Figure 2.

Figure 1

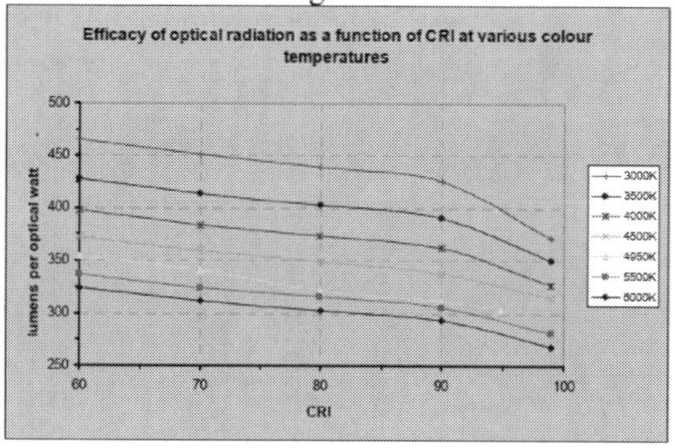

Figure 2

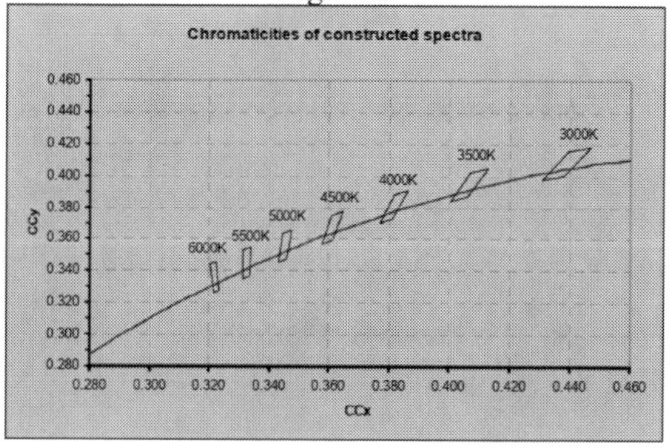

CONCLUSION

The efficacy of optical radiation decreases surprisingly slowly below CRI = 90. Also surprisingly the highest optical efficiencies are found at lower colour temperatures.

REFERENCES

[1] Microsoft® Office Excel 2003 solver add-in
[2] Publication CIE 13.3-1995

Compact Surgical Luminaire with Single PhlatLight® LED

E. Goldfain, R. Tamburrino
Welch Allyn, Inc., 4619 Jordan Rd., Skaneateles Falls, NY 13153 USA
Contact: goldfaine@welchallyn.com

ABSTRACT

We describe the design and construction of a surgical headlight luminaire using a single white LED based on the PhlatLight® technology [1]. Output requirements, design tradeoffs and implementation challenges are briefly reviewed. The paper concludes with an analysis of actual performance results.

INTRODUCTION

Surgical headlights are used by physicians as auxiliary illumination during surgical procedures. The headlight provides three benefits to the user: additional light for improved visualization, shadow free illumination and illumination into deep recesses. Shadow free illumination is achieved by positioning the luminaire in the plane of the surgeon's line of sight (between the eyes). The narrow focused beam of the luminaire allows illumination into deep cavities without illuminating surrounding tissue.

Surgical headlights require significant amount of light to achieve good visualization of tissue. Typically, about 400 lumens minimum within a 120mm spot is required. Typically, a 300W Xenon light source provides light to the luminaire via a fiber optic cable to achieve desired illumination levels. The disadvantage of Xenon lamp technology is that the surgeon is tethered to the light source and cannot move freely around the operating room. In addition, lamp replacement and fiber optic cable expense, light source size, and power consumption are other disadvantages.

LED technology has developed to the point where it can provide Xenon lamp equivalent illumination levels with superior collection optics at a lower cost with the benefits of portability, battery operation, low power consumption, and small footprint.

REQUIREMENTS

A successful surgical LED-based luminaire must balance several mutually exclusive requirements:

1) Ability to provide high collection efficiency from a large LED die over a wide beam angle.
2) Compact and lightweight construction.
3) Maximum simplicity of the optics system (minimal number of components including an adjustable iris diaphragm and a folding mirror).
4) Highest spot quality (light output, uniformity and edge definition) at the prescribed working distance of 16" and for the full range of prescribed working distances (10" to 25").
5) Projected spot must be free of artifacts and any "memory" of either square LED die or its surface structure.

LUMINAIRE CONSTRUCTION

The luminaire design is based on the principle of Koehler illumination and consists of two groups of lenses: a condensing lens (C) and an objective lens (O) with the iris control and folding mirror placed in between. To maintain a compact and cost effective format, C consists of a doublet of elements in high index glass (Schott LaSFN 31 or optical equivalent). To maximize collection efficiency, the first element of C is placed in close proximity of the LED dome and is shaped as a meniscus lens. The function of C is to efficiently collect light from the LED and back-illuminate the iris with a beam that has a uniform light distribution and is properly sized to the iris opening. To comply with the same requirements, O is also designed as a split doublet in high index glass. Particular attention is paid to the selection of glasses and lens curvatures that yield a minimal amount of lateral chromatic aberration at the edge of the spot. The function of O is to project the image of the iris opening at the prescribed spot diameter, with a high degree of light uniformity, no objectionable artifacts inside or outside the spot, good edge definition across the entire range of working distances and iris openings.

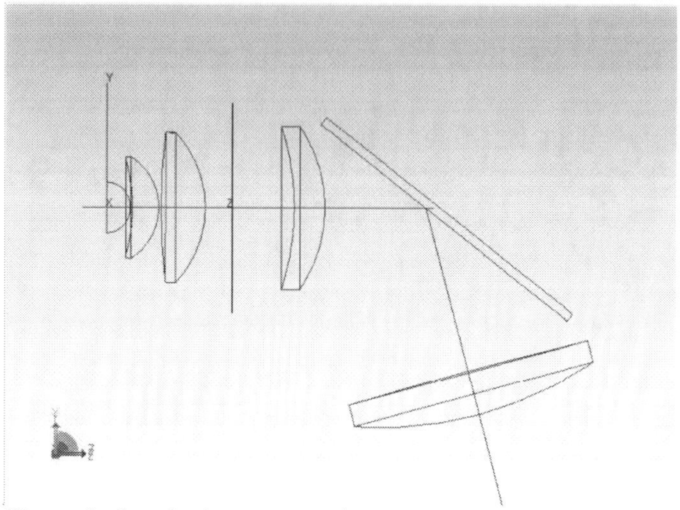

Figure 1. Luminaire construction.

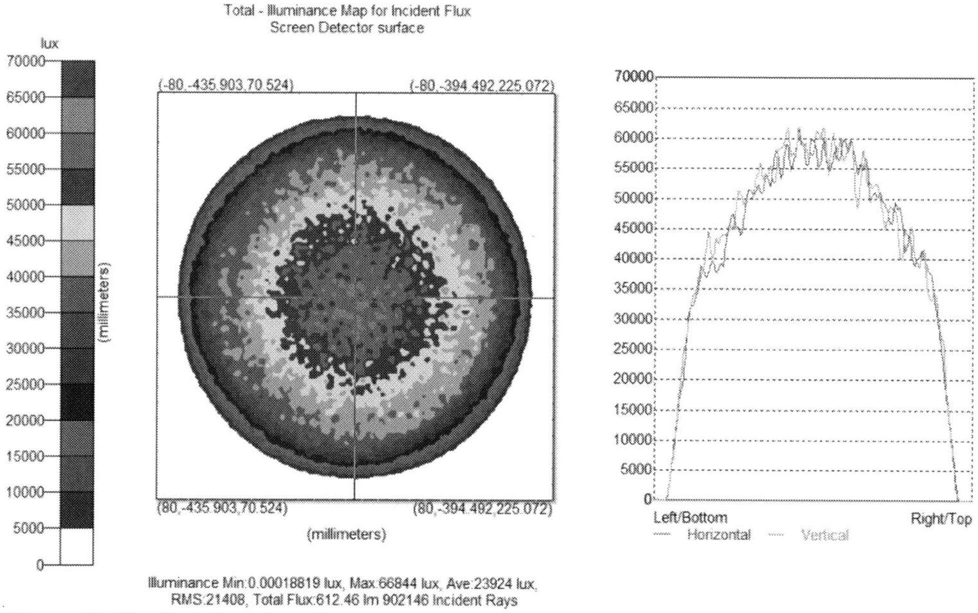

Figure 2. Predicted spot performance.

PERFORMANCE

The headlight luminaire achieved the following performance: Spot diameter: 130 mm at 400 mm object distance; Total lumen output: 500 lumens; Color temperature: 5000 K; Center lux: 50,000 lux; Edge Definition: Superior

REFERENCES

[1] Luminus Devices information available at http://www.luminus.com/content1034

[2] Radiant Imaging/ProSource information available at http://www.radiantimaging.com/?q=products/prosource

[3] TracePro (Illumination and Stray Light Analysis Software) at http://www.lambdares.com/software_products/tracepro/tracepro/

HID lamp acoustic emission signatures and arc distortion

W. Kaiser[1], A.F. Correa[1], R. P. Marques[1]
[1] *Escola Politécnica da Universidade de Sao Paulo*
Av. Prof. Luciano Gualberto trav.3 n. 158, 05508-900 Sao Paulo, SP, BRAZIL
Contact: kaiser@lac.usp.br

ABSTRACT

This paper presents data correlating acoustic emissions in HID lamps measured at acoustic resonance with visual arc distortion and lamp electric quantities such as operation power and impedance.

INTRODUCTION

In spite of the advantages of electronic ballasts, the operation of HID lamps in high frequency remains limited by an unstable and harmful phenomenon called "acoustic resonance" (AR). Theoretically, it originates from pressure wave propagation in the discharge enclosure with reflexes on the walls from which a standing wave will result that causes arc deformation and light fluctuations. The AR resonance frequencies are numerous and vary according to the power injected, the geometry and pressure of the discharge tube, the composition of the gas, the lamp age and model, and even the manufacturing batch. AR exhibits a hysteresis that changes the resonance windows systematically when the supply frequency initially increases and then decreases [1]. There are basically three AR detection mechanisms: acoustic emissions measurement [1], electrical measurements (lamp voltage, current or impedance) [2] and optical detection of arc displacements or disturbances [3].

EXPERIMENTAL APPARATUS AND RESULTS

The arc deformation is measured with the lamp in a quadrangular chamber with two webcams on perpendicular planes to obtain a three-dimensional rendering (see Fig. 1). Light is filtered by a neutral density optic filter with a gain of 1/1024. A class D converter operating at 300 kHz provides current. A programmable waveform generator synthesizes the lamp current waveform and a DSP (Digital Signal Processor) provides three level PWM command signals for class D converter MOSFETs. A LC filter at converter output recomposes the desired lamp current waveform. A manual ignitor starts the lamp and is bypassed at normal operation. The sound emissions were measured employing an Electret-Condenser microphone (Panasonic Model PM 329) with a pre-amplifier supplied by a battery built inside a brass casing placed close to the lamp bulb. Lamp current and voltage are measured to determine the lamp power and ensure operation at nominal power conditions. The measuring system set up is fully automated by a PC, using the software package Matlab.

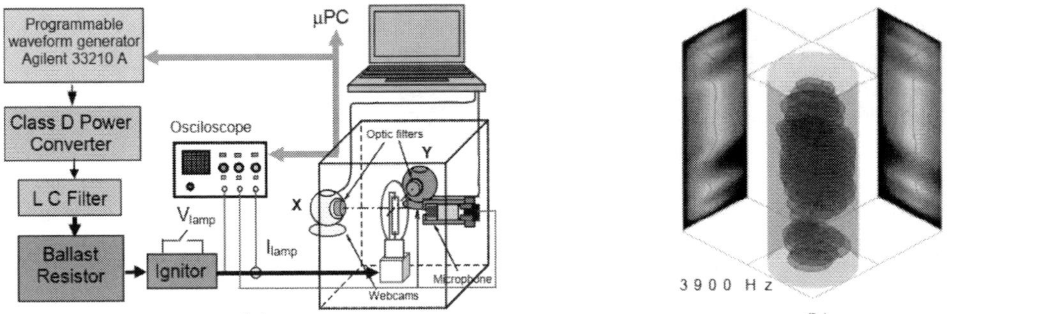

(a) (b)

Figure: 1 - (a) Experimental apparatus and (b) Reconstituted arc image of a HPS 70 W lamp with AR.

Lamp arc models are generated based on the arc images, consisting of a main curve following the center of the arc and a series of ellipses representing the visual width of the arc along the lamp discharge tube (Fig. 1b). Figure 2 presents experimental data from a HPS 70W lamp (Osram NAV). Operation frequency varies from 3800 to 4400 Hz. Measured power and impedance are presented in the first plot, while the sound intensity (normalized) is presented in the second plot. For selected frequencies, the collected arc data are presented below. Intense arc distortion can be noticed in the 3900-4100 Hz range, the same range in which measured impedance is maximal. Sound emission is maximal in the range 4200-4350 Hz, where the impedance curve becomes flat again with respect to frequency (and the lamp arc becomes stable).

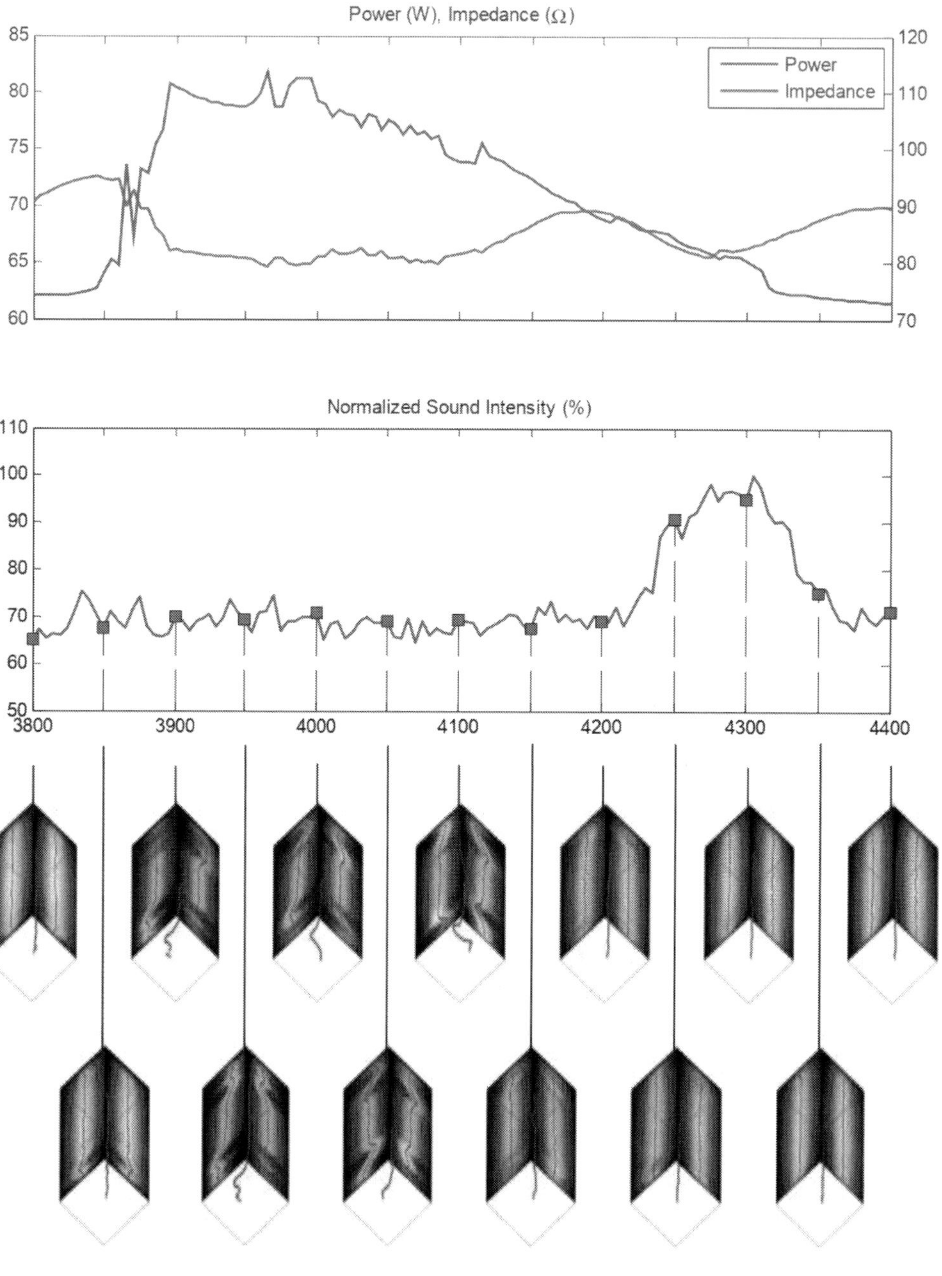

Figure: 2 - Experimental data for a HPS 70W lamp (Osram NAV).

REFERENCES

[1] H. P. Stormberg and, R. Schafer, Lighting Research and Technology, v.15, n.3, pp. 127-132, 1983.
[2] B. Siessegger, K. Guenther, H. Gueldner, G. Hirschmann, 39th IEEE PESC, pp. 1898 – 1904, 2008
[3] W. Kaiser, R.P. Marques, A.F. Correa, 43th IEEE IAS Annual Meeting Conference, pp 1-6, 2009.

Waveform effect on the efficiency of coaxial $KrCl^*$ excilamps

Xiaobo Zhuang[1], Xiangfen Feng[2], Markus Roth[3], Oliver Rosier[3], Shaolong Zhu[1], and Shanduan Zhang[1]

[1] *Institute for Electric Light Sources, Fudan University, 220 Handan Road, Shanghai 200433, People's R China*
[2] *Corporate Technology, Siemens Ltd China, No. 1 Pudong Avenue, Shanghai 200120, People's R China*
[3] *Business Unit Specialty Lighting, OSRAM GmbH, Dr. Eugen Kersting Str. 6, 51688 Wipperfürth, Germany*
Contact: sdzhang@fudan.edu.cn

ABSTRACT

The 222 nm radiant efficiency were measured for coaxial $KrCl^*$ excilamps based on the dielectric barrier discharge. The lamps were driven by sinusoidal, rectangular and pulsed electronic control gear (ECG). The 222 nm radiant power is derived from irradiance, according to the Keitz formula. The results show the maximum efficiency of 222 nm radiation is 9.2% for a $KrCl^*$ excilamp filled with 2 mbar Cl_2 and 198 mbar Kr, driven by a pulsed ECG. The effects of waveform are discussed.

INTRODUCTION

Conventional UV lamps mismatch the absorption cross section of chlorinated substrates at the wavelength in the range of 210 to 230 nm. $KrCl^*$ excimer sources, emitting an intense narrow band UV radiation at 222 nm are eligible for selective irradiation of chemical objects. 222 nm radiation is proving to make a considerable contribution to the treatment of water purification and air deodorization etc [1]. Normally, the geometries of a $KrCl^*$ lamp are coaxial and pumped by dielectric barrier discharge (DBD), but the dimensions, optimal gas species as well as power supplies are different. Owing to difficulties in accurate measurements of 222 nm radiant power, optimal parameters and efficiency values differ dramatically [1].

EXPERIMENTS

A coaxial geometry is adopted for $KrCl^*$ excilamps. The lamp dimensions are: the outer quartz tube has outer diameter (OD) 40 mm, wall thickness 1.2 mm, and length 400 mm; the inner quartz tube has inner/outer diameter (ID/OD) 14/16 mm, and length 500 mm. Two pieces of stainless steel foils, attached firmly to the inner surface of 14/16 mm tube, are used as a high voltage electrode. A ring metal mesh is mounted on the outer tube as a grounded electrode. The annular discharge cavity, whose gap is 10.8 mm, is filled with 2 mbar chlorine and 198 mbar krypton.

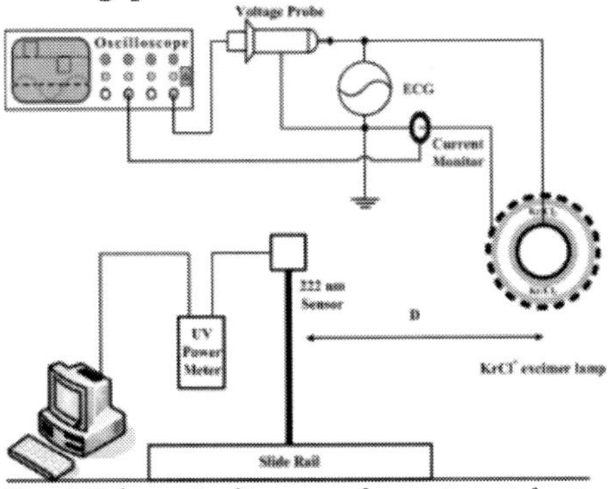

Fig. 1 Schematic diagram of experimental setup

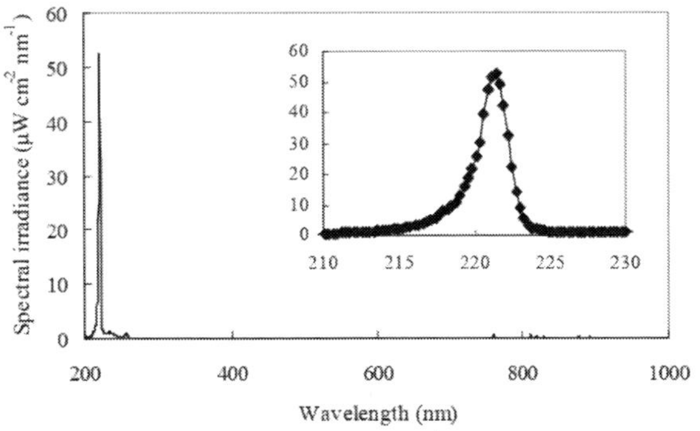

Fig. 2 Spectrum of $KrCl^*$ excilamp

Figure 1 shows the experimental setup. The excilamp, whose orientation is horizontal, is fixed with two Teflon rods, driven by 3 electronic control gears (ECGs): a rectangular ECG with frequency 40-70 kHz; a pulsed ECG with pulsed width about 0.75 μs and repeating frequency 90 kHz; and a sinusoidal ECG with frequency 35.8-48.4 kHz, which depends on the impedance of the lamp at different power levels. A high voltage probe and a current probe are connected to a digital storage oscilloscope to measure the voltage, current and transient power waveform of the excilamp. The sensor (Hamamatsu H8025-222) of the UV power meter (Hamamatsu C8026) is used to detect the

222 nm band irradiance of the excimer radiation. Since the excilamp can be considered as a linear light source, i.e. the diameter is much less than the length, the radiant power of 222 nm can be derived from the Keitz formula [2]: $P_{rad} = 2\pi^2 DLE/(2\alpha + \sin 2\alpha)$, where E is the measured irradiance, D the distance from lamp centre to the sensor of UV power meter, L the discharge length of the lamp covered with metal mesh, α the half angle, $\tan\alpha = L/(2D)$.

RESULTS AND DISCUSSION

The spectrum of KrCl excilamp is measured by a calibrated CCD spectrometer (Ocean Optics HR4000 CG-UV-NIR, FWHM 0.7 nm). Figure 2 shows the spectrum of the excilamp, driven by the rectangular ECG at input power 400 W. It is obvious that the only strong band in the range of 200-1000 nm is at 222 nm, with full width at half maximum (FWHM) 2.4 nm, which is emitted by $KrCl^*$ excimer. There are some weak atom lines of krypton in range of 750-900 nm.

Figure 3 shows the effect of the waveform on radiant efficiency. The 222 nm efficiency decreases remarkably when the power setting is over 200 W, especially for the sinusoidal ECG. This trend is in good agreement with the results obtained by Panchenko *et al* [3]. It is clear that the pulsed ECG has the highest efficiency (9.2% at 200 W) perhaps because the rapid rise time, leads to more excitation from the excimer combination collision, and less heating collisions. The rectangular ECG has a relative smooth efficiency trend for various powers, while the efficiency of sinusoidal ECG is the lowest, especially at high power.

This phenomena can be explained by the reaction kinetics for forming the $KrCl^*$ excimer molecules. As the power setting increases, the discharge current and electron density rise, leading to the increase of stepwise ionization rate of Kr^* and the decrease of chlorine molecules. At the relatively low pressure of 200 mbar, the dominant excimer forming mechanism is the harpooning reaction [4]: $Kr^* + Cl_2 \rightarrow KrCl^* + Cl$. Due to the loss of Kr^* by stepwise ionization and the loss of chlorine molecules due to their dissociation by high-energy electrons, the 222 nm radiation will weaken and hence the efficiency will fall with the increasing power.

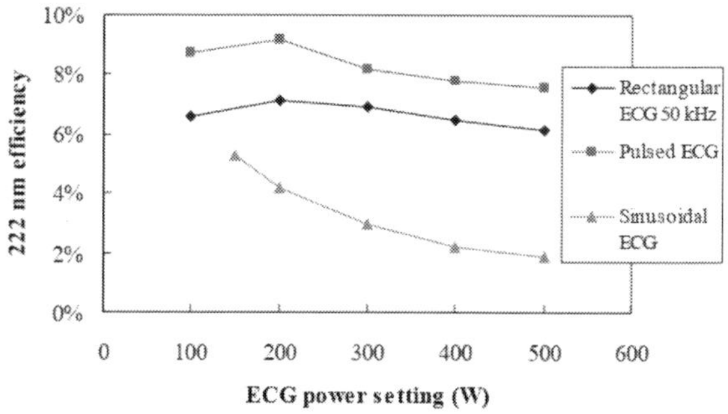

Fig. 3 Waveform effect on 222 nm radiant efficiency

The reaction kinetics for forming $KrCl^*$ excimer begin with collisional excitation, ionization or dissociative attachment of energetic electrons to krypton or chlorine. The key steps in the formation of $KrCl^*$ are described as follows.

The high-energy electrons excite and ionize the krypton or chlorine atoms, or dissociate the chlorine molecules: $e + Kr \rightarrow Kr^* + e$, $e + Kr \rightarrow Kr^+ + 2e$, $e + Cl_2 \rightarrow Cl + Cl^-$. $KrCl^*$ excimer are then created by a three-body recombination of the positive krypton ions and the negative chlorine ions ($Kr^+ + Cl^- + M \rightarrow KrCl^* + M$, where M is a collisional third partner, which in many cases can be an atom or molecule of the active species, or otherwise) or the Harpooning reactions in which the excited Kr^* species transfers its loosely bound electron to the chlorine molecule, which lead to the formation of an electronically excited state of $KrCl^*$.

The excimers produced are not stable and will rapidly decompose, giving up their excitation energy in the form of a UV photon. When the $KrCl^*$ excimers relax to their ground states they emit photons at 222 nm: $KrCl^* \rightarrow Kr + Cl + $ UV radiation (222 nm).

REFERENCES

[1] Sosnin E A, Oppenlander T and Tarasenko V F 2006 *J. Photochem. Photobiol. C* **7** 145

[2] Lawal O *et al* 2008 *IUVA News* **10** 14

[3] Panchenko A N and Tarasenko V F 2008 *Quantum Electron.* **38** 88

[4] Neiger M 1992 *Proc. 6th Int. Symp. on the Science and Technology of Light Sources* (LS-6, Budapest, Hungary) p 75

Absolute determination of the quantum efficiency of phosphor powders

S. Leyre[1], L. Pandey[2], J. Hofkens[2], W. Fyen[3], P. Hanselaer[1,4]

[1] *Light & Lighting Laboratory, Catholic University College St. Lieven, Gebroeders Desmetstraat 1, Gent, 9000 Belgium*
[2]*Department of Chemistry, K.U. Leuven, Celestijnenlaan 200F, Heverlee, 3001 Belgium*
[3]*R&D, K.U. Leuven, Minderbroedersstraat 8a, Leuven, 3000 Belgium*
[4]*Dept. ESAT/ELECTA, K.U. Leuven, Kasteelpark Arenberg 10 - bus 2445, Leuven, 3001 Belgium*
Contact: Sven. Leyre@kahosl.be

ABSTRACT

An absolute method to obtain the quantum efficiency of phosphor powders is reported. With this method, only a small (<1g) is required and sample preparation is significantly simplified since the amount of phosphor doesn't require accurate control. The results are not always consistent with results obtained using a relative measurement procedure.

INTRODUCTION

One of the most important properties of phosphors is the quantum efficiency (QE). It is defined as the ratio of the number of emitted photons to the number of absorbed photons (Eq. 1):

$$QE = \frac{\#\ emitted\ photons}{\#\ absorbed\ photons} \tag{1}$$

At this moment, QE measurements of phosphors have not been standardized. Essentially, there are two approaches to measure QE. In the photo-acoustical approach, an acoustic wave is induced by a temperature rise of an illuminated sample. This wave is proportional to the non-radiative transitions in the phosphor [1]. When using optical determination methods, absolute or relative measurements can be performed. Wrighton *et al* described an absolute measurement procedure for phosphor powders based on reflectance spectra [2]. This method requires a correction for self-absorption [3]. A relative measurement procedure was proposed by Bril *et al.* comparing reflectance spectra of a sample and a reference phosphor [4]. With this procedure, reliable results can only be expected when sample and reference have identical properties, such as excitation wavelength and absorption cross-section [5]. This requires a very controlled preparation of the sample. In this paper, a new absolute method to obtain QE of phosphor powders is described. With this method, the sample preparation procedure is not critical and there is no need for a self-absorption correction.

MEASUREMENT SETUP

The setup is inspired by the setup proposed by de Mello *et al.* [6]. The measurement system consists of an excitation source, an integrating sphere with sample holder, a spectrometer with CCD which is connected to the sphere with an optical fibre, and a PC (Fig. 1). The excitation source is a 300 W Xenon light source with an interference filter (10 nm bandwidth) and secondary optics to direct a beam of radiation onto the sample. The excitation wavelength can be selected using the appropriate interference filter. In order to obtain an irradiance measuring head with a good cosine response, a diffuser was positioned in front of the fibre entrance [7]. The phosphor powder is put in a quartz cuvette which is positioned in the centre of the sphere. The quartz cuvette has diffuse windows which scatter the incident beam. In this way, the optical paths for the excitation and emission radiation are more or less identical, which is a condition for performing accurate integrating sphere measurements. The relative spectral response of the setup was obtained by measuring the spectrum of a calibrated halogen light source. Since the halogen source has little output below 400 nm, even low stray light levels within the spectrometer can cause significant errors for this wavelength region. For this reason, a calibrated deuterium source was used to obtain the relative spectral response below 400 nm.

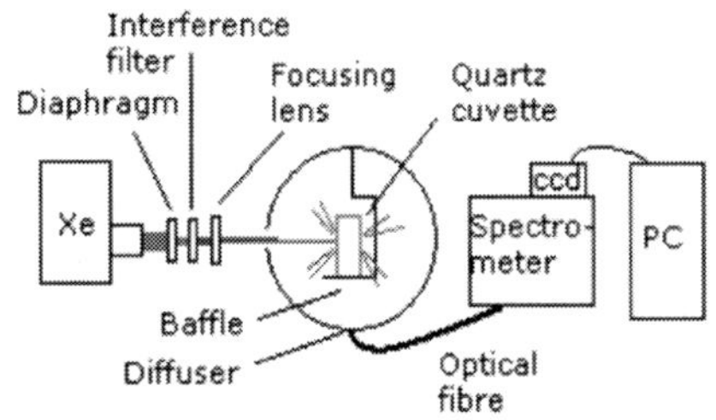

Fig. 1: Schematic representation of the measurement setup with an integrating sphere

MEASUREMENT PROCEDURE

Two spectral measurements are required to obtain the QE of the phosphor. With the first measurement, a blank sample (an empty quartz cuvette) is positioned in the centre of the sphere. From the spectrum, the flux of photons entering the sphere at the excitation wavelength (L_a) and the "dark" signal at the emission wavelengths (P_a) can be obtained. A second measurement is done after the phosphor has been put in the quartz cuvette. Only a small amount of phosphor powder (~ 100-500 mg) is put into the cuvette, resulting in a negligible substitution error. Some powder sticks to the walls and some lies on the bottom of the cuvette. With this second measurement, the flux of photons which are not absorbed by the phosphor at the excitation wavelength (L_b) and the flux of photons emitted by the phosphor at the emission wavelengths (P_b) can be obtained. The QE of the phosphor can thus be calculated according to Eq. 2:

$$QE = \frac{P_b - P_a}{L_a - L_b} \quad (2)$$

MEASUREMENT RESULTS

The QE of the UV phosphors $BaMgAl_{10}O_{17}:EU^{2+}$ (which emits blue light) and $BaAl_{12}O_{19}:Mn^{2+}$ (which emits green light) have been determined. Furthermore, $Y_3Al_5O_{12}:Ce^{3+}$ phosphors used in white phosphor LEDs and provided by two manufacturers have been compared. The values of the QE for these phosphors are gathered in Table 1, together with the QE values specified by the manufacturers. The UV-excitable phosphors and the YAG-phosphors were characterized at an excitation wavelength of 380 nm and 460 nm respectively. The results for the UV phosphors are in good agreement with the specifications but the values obtained for the YAG phosphors are much lower than those expected. At this moment we cannot explain the inconsistency between our absolute measurements and the specified values which have been determined with the relative measurement procedure.

Phosphor	*Specified QE*	*Measured QE*
$BaMgAl_{10}O_{17}: Eu^{2+}$	0.96 ± 0.02	0.95 ± 0.04
$BaAl_{12}O_{19}: Mn^{2+}$	0.83 ± 0.02	0.85 ± 0.04
$Y_3Al_5O_{12}: Ce^{3+}$	0.93 ± 0.02	0.75 ± 0.04
$Y_3Al_5O_{12}: Ce^{3+}$	0.93 ± 0.01	0.73 ± 0.04

Table 1: Specified and measured QE values.

VALIDATION

The influence of the amount of phosphor in the cuvette was investigated by changing the amount from 100 to 500 mg. This induced a variation of the relative absorption ($(L_a - L_b)/L_a$) from 0.2 to 0.7. Within a standard deviation of 0.04, the observed QE was independent on the amount of phosphor in the cuvette. However, from an absorption of 0.2 onwards, the QE values became less reproducible. These results indicate that the self-absorption within the sample is negligible. If the self-absorption was significant, the observed QE would decrease with increasing absorption [3]. The influence of non-identical optical paths for emission and excitation was investigated by performing QE measurements in a cuvette with smooth side walls. This led to an average decrease of the observed QE with 0.04. Light originating from the centre of the sphere seems to be detected less efficiently than light reflected from the sphere wall. This effect is probably caused by the baffle, which is positioned between the sample and the detector in order to avoid direct irradiation of the detector.

CONCLUSION

An absolute method to obtain the QE of phosphor powders was proposed. In the integrating sphere, a similar optical path for the excitation and the emission was maintained. Since only small amounts of phosphor are used, self-absorption by the phosphor can be neglected. For the YAG phosphor, inconsistencies were found between our results and those obtained with a commonly used relative measurement method. A more elaborate investigation of both measurement methods needs to be performed.

REFERENCES

[1] W. Lahmann, H. Ludewig, Chem. Phys. Lett. 45 (1977) 177.

[2] M. Wrighton, D. Ginley, D. Morse, J. Phys. Chem. 78 (1974) 2229.

[3] T. Ahn, R.O. Al-Kaysi, A.M. Müller, K.M. Wentz, C.J. Bardeen, Rev. Sci. Instrum. 78 (2007) 086105.

[4] A. Bril, A.W. de Jager-Veenis, J. Electrochem. Soc., 123 (1976) 396.

[5] O.L. Malta, H.F. Brito, J.F.S. Menezes, F.R. Conçalves e Silva, C. de Mello Donéga, S. Alves Jr., Chem. Phys. Lett. 282 (1998) 233.

[6] J.C. de Mello, H.F. Wittmann, R.H. Friend, Adv. Mater., 9 (1997) 230.

[7] P. Hanselaer, A. Keppens, S. Forment, W.R. Ryckaert, G. Deconinck, Meas. Sci. Technol., 20 (2009) 095111.

Non-destructive estimation of Xe filling pressure in mercury-free automotive metal halide lamp

Kyosuke Enoki, Hideki Motomura, Masafumi Jinno
Department of Electrical and Electronic Engineering, Ehime University
3 Bunkyo-cho, Matsuyama 790-8577, Japan
Contact: hmoto@mayu.ee.ehime-u.ac.jp

ABSTRACT

Mercury-free metal halide lamps for automotive purpose have been developing in the market. When mercury is not used, the electric and emission characteristics of the lamp strongly depend on the xenon filling pressure. Therefore a simple and non-destructive gas pressure estimation technique used in production lines is required to obtain stable performance of the lamps as commercial products. The authors have developed an estimation method by which the gas pressure is estimated from the current peak value at the initial stage of ignition under pulsed operation. It is shown that the accuracy of the order of ±(0.1–0.3) atm is obtained using an empirical formula.

INTRODUCTION

Nowadays the market demands for non-toxic and environmentally friendly products. For automotive headlamps, mercury-free metal halide lamps (MHLs) are already available in the market. Since xenon has the important role of storing thermal energy in mercury-free MHLs, the lifetime of the lamp and the optical properties such as spectra, color temperature, color rendering index (CRI) will have large deviation, if the deviation of the filling pressure of xenon is significant. Therefore a non-destructive and simple method to estimate the xenon filling pressure in an automotive mercury-free MHL which is applicable in the manufacturing process is required. As a non-destructive method, optical methods that use line shifts or broadening of atomic spectra are popular. However optical methods require monochromators and optical detectors, and complicated data analysis with curve fitting. Therefore the optical method is suitable for the scientific measurements in a laboratory but not suitable for measurements in a factory that require simplicity and low-cost. In this paper the authors propose a new and simple method of xenon filling pressure estimation by measuring the peak current in the ignition stage of mercury-free MHLs.

EXPERIMENTAL SETUP

D4 lamps for automotive headlight with different xenon filling pressure are used for the measurements. In these lamps ScI_3, NaI, ZnI_2 and InBr are introduced as the light emitting materials and xenon is also introduced as the buffer gas. The xenon filling pressures are 10 atm, 12.1 atm, 13.5 atm and 15.4 atm. For each pressure four lamps are prepared. All the lamps have passed the manufacturer's inspection about the electrical characteristics such as voltage, current, power consumption, and optical characteristics such as total flux, color coordinate, etc. These 16 D4 mercury-free MHLs are operated by a commercial ballast designed for these lamps and the lamp currents are measured. The lamp current is picked up at the point between the ballast and the lamp by a current probe (Tektronix, TCP202) and an oscilloscope (Tektronix, TDS3054). The outer bulb temperature is measured by a pyrometer. The accuracy of the pyrometer is ±1°C and the range of measurement is 0–220°C.

RESULTS AND DISCUSSION

Figure 1 shows an example of the lamp current waveform up to 2 s and 100 s just after ignition. It is found that the current increases up to 1 s, maintains a constant peak value of 3 A up to 2 s. After that the current decreases after slight increase, and finally reaches steady state working (300 Hz rectangular waveform). In this study, average value of the current peak from 1.2 to 1.8 s was calculated and used for estimation of the filling pressure.

Figure 2 shows above mentioned current values averaged over 40 times measurements for each filling pressure and the outer bulb temperature condition as a function of inverse of xenon filling pressure. The outer bulb temperature was set to these values by turning off and cooling the lamp at room temperature after steady state burning. The error bars show confidence interval with confidence level of 0.95, which is represented as $\pm 1.960\sigma/\sqrt{N}$, where σ is standard deviation and N is number of measurement (40). Here, the current is represented as $I=en_eSv_e$, where e is the charge of an electron, n_e is the electron density, S is the arc cross-section area and v_d is electron drift velocity. Here, the drift velocity is shown as $v_d=eE/(m_e\nu_e)$, where E is the electric field, m_e is the mass of an electron, ν_e is the electron collision frequency with neutral molecules. It is shown as $\nu_e=n\sigma_e\langle v_e\rangle=P\sigma_e\langle v_e\rangle/(kT)$, where n is the neutral molecule density, σ_e is the electron collision cross-section with neutral molecules, $\langle v_e\rangle$ is the electron thermal velocity, P is the working pressure, k is the Boltzmann factor and T is working gas temperature. Finally, the current is represented as,

$$I = en_eSv_d = \frac{e^2 n_e kSET}{m_e P\sigma_e\langle v_e\rangle}. \quad (1)$$

Here, by assuming that the electron density coming from ionization of xenon atoms is proportional to the pressure (αP) and that from ionization of metals depends only on the gas temperature (β), Eq (1) is written as,

$$I(P,T) = A(T) + \frac{B(T)}{P}, \quad (2)$$

where

$$A(T) = \frac{e^2 kSET}{m_e\sigma_e\langle v_e\rangle}\alpha,\ B(T) = \frac{e^2 kSET}{m_e\sigma_e\langle v_e\rangle}\beta. \quad (3)$$

Since measurements were done at the ignition stage (1.2–1.8 s after ignition), by assuming that the increases in the gas temperature and pressure are small and substituting them to the outer bulb temperature and the filling pressure, we can fit the measured data by Eq. (2). We determined the four sets of values of $A(T)$ and $B(T)$ for each outer bulb temperature, by deriving the intercept and the slope values for each outer bulb temperature by the least square fitting as shown in Fig. 2, and then fitting them as a function of T by the least square method again. Figure 3 shows measured data at the outer bulb temperature of 100°C with the empirical formula curve of Eq. (2). It is found that the empirical formula agrees well with experimental result. The confidence interval of the current corresponds to the pressure range of the order of ±(0.1–0.3) atm according to Eq. (2). It means the filling pressure will be estimated by this order of the accuracy.

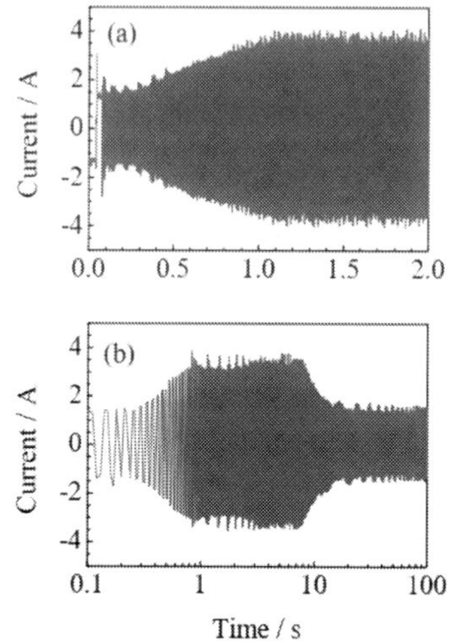

Fig. 1: An example of the lamp current waveform up to (a) 2 s and (b) 100 s just after ignition.

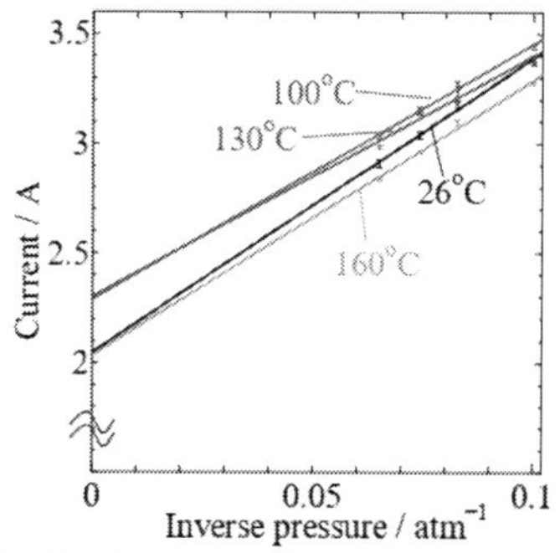

Fig. 2: Current values averaged over 40 times measurements for each condition as a function of inverse of xenon filling pressure.

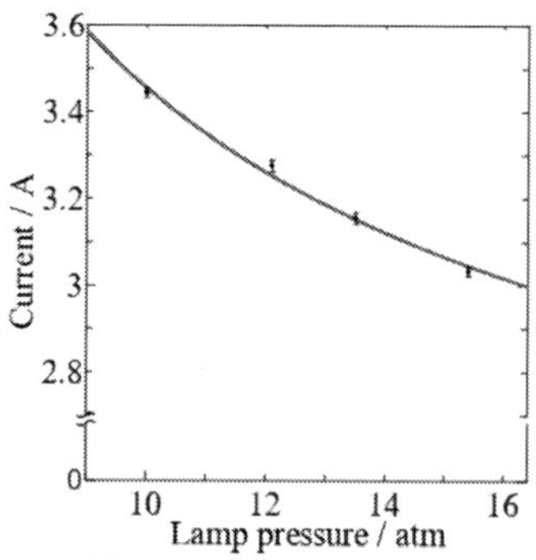

Fig. 3: Measured data at the outer bulb temperature of 100°C with the empirical formula curve of Eq. (2)

CONCLUSION

The authors propose an estimation method of xenon filling pressure of automotive mercury-free MHLs by measuring the average lamp current at ignition stage. The estimation accuracy was order of ±(0.1–0.3) atm according to derived empirical formula.

Thermal management of LED Lamp

A. Motoya , M. Kai , Y. Manabe, S.Shida
Lightnig company, Panasonic corporation
Contact: motoya.atsushi@jp.panasonic.com

ABSTRACT

The luminous efficacy of the LED depends on the temperature of P-N junction (Tj). If Tj is increased excessively, the luminous efficacy of lamp is decreased. Therefore, we need the thermal management of LED module to keep the Tj properly. In this paper, to investigate the proper thermal management structure for LED lamp, following three steps has reported by means of CFD model.

First, the relation between the thermal resistance against the surroundings and lamp body volume was studied. Second, the limitation was cleared in case of putting additional fin on body as heat sink. Third, the possibility to adopt any surface treatment of the body for increasing the heat flux dissipated was considered. As the results, we confirmed that the surface treatment to improve the emissivity is quite efficient way for thermal management of LED lamp.

INTRODUCTION

Recently to protect the global environment, Many people begin to use the high efficiently lamps instead of standard lamps all over the world. For example, in Australia they announced to prohibit selling the standard lamps by 2010. Moreover, they announced to prohibit using them by 2010 in EU.

LED lamp is one of high efficient lamps. Recently they developed many commodities of LED lamps in many companies, and when we use the LED lamps, we can use the same lighting system of standard lamps, so LED lamps are remarkably spread all over the world.

CHARACTERISTICS OF THE LED LAMPS

The LED lamp is prepared according to the following procedure. The LED chip is mounted on the board, and next it is sealed with the resin that mixed phosphor. We usually call it LED module. When we put the electric power, LED module emits the 30% of it as white light and the 70% of it as heat [1]. By the way, the luminous efficacy of the LED depends on the temperature of P-N junction (Tj). If Tj is increased, the luminous efficacy of lamp is decreased. It is necessary that we put much electric power to obtain more luminous flux. However, when we put much electric power, the heat capacity is increased. As the results, Tj is increased, and then the luminous efficacy is decreased. Therefore, the thermal management of LED module is very important to increase the luminous efficacy.

We simulated the relation between the putting electric power and the Tj. We showed the simulation model in Fig 1. When we put much electric power, Tj was increased. Especially, when we put 3.1W into the LED module, Tj was increased to 380[K]. And then, luminous efficacy of it was decreased more than the case of 373[K] by 5%. Therefore, thermal management was necessary to decrease Tj to 373 [K].

APPROACH OF THERMAL MANAGEMENT CHARACTERISTICS OF THE LED LAMPS

First we researched the energy balance of LED lamp. As the results, we confirmed that the most important element of LED lamp was the lamp body for thermal management. It was necessary to decrease thermal resistance of lamp body to 25[K/W] to decrease Tj to 373[K].

Thermal resistance of it depended on the lamp body volume[2]. We simulated the relation between thermal resistance and lamp body volume. We showed the result of simulation in Fig.2. When lamp body volume was the 25700mm3, thermal resistance of it was 25[K/W]. However, we must not make LED lamp of this size, because it was larger than standard lamp and the consumer could not use the same lighting system of standard lamp. Therefore, we must use another approach for thermal management.

One of the approaches of improving the heat capability is putting additional fin on body as heat sink. In the case of putting additional fin on body, we simulated the relation between the thermal resistance and volume. We changed the depth and thickness and width of the fin. When we put it on body, thermal resistance was decreased little than the case of without fin. But, it was not efficiently to decrease the thermal resistance.

One of the approaches of improving the heat capability was increasing the emissivity body surface. We simulated the relation between the thermal resistance and the emissivity of body surface when the body volume was 18604mm3. When the emissivity of it was 0.8, the thermal resistance of it was 25[K/W]. And then, Tj was 373[K]. Therefore when the emissivity of it was 0.8, we could decrease Tj to our target.

One of the ways of improving the emissivity was treating almite treatment. Therefore, we treated it the almite treatment. As the results, we could make it emissivity of our target. And we could decrease Tj to our target. As the results, we developed the very small size and more efficient LED lamp.

CONCLUSION

We investigated the proper thermal management structure for LED lamp, following three steps has reported by means of CFD model.

First, the relation between the thermal resistance against the surroundings and lamp body volume was studied.

Second, the limitation was cleared in case of putting additional fin on body as heat sink.

Third, the possibility to adopt any surface treatment of the body for increasing the heat flux dissipated was considered.

As the results, we confirmed that the surface treatment to improve the emissivity is quite efficient way for thermal management of LED lamp.

As the results, we developed very small size and more efficient LED lamp.

Figure1. The CFD model of LED lamp

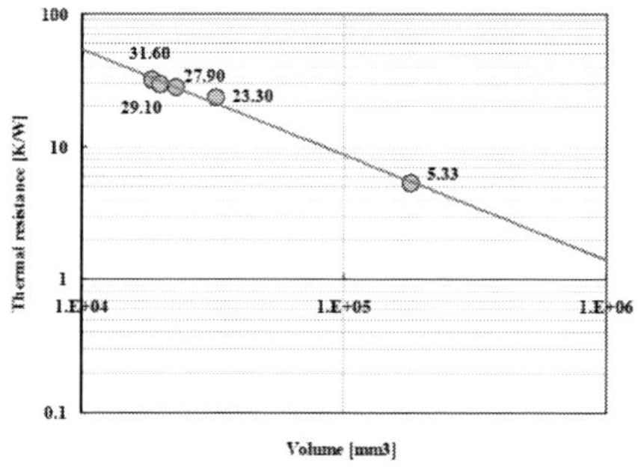

Figure2. Relation between thermal resistance and enveloping volume

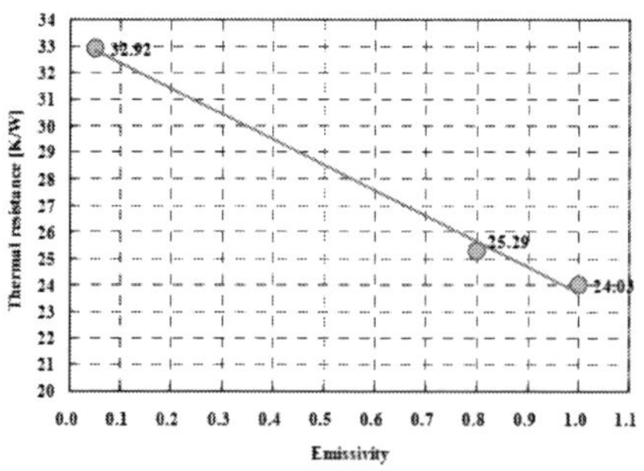

Figure3. Relation between emissivity and thermal resistance

REFERENCE

[1] Lighting hand book , pp. 102- 146 (2003)

[2] Naoki Kunimine , *Heat design complete introduction for electronics*, pp.170 (1997)

Session Four

Construction and experimental validation of models for HID lamps

Joost van der Mullen

Faculty of Applied Physics; Eindhoven University of Technology
Den Dolech 2; 5600 MB Eindhoven, The Netherlands
j.j.a.m.v.d.mullen@tue.nl

ABSTRACT

This contribution gives a description of numerical models constructed with the PLASIMO toolkit and aiming to describe the plasmas in HID lamps. Apart for the simple HID lamps based on pure Hg, we have also constructed models for Metal Halide lamps that apart form Hg contain small additives of NaI or DyI_3. The models combine fluid descriptions with modules that deal with the generation and transport of radiation. The input of a model is the set of control parameters, such as the fill chemistry, driving power and vessel shape. The output is formed by plasma properties; such as the voltage over the lamp and the production of heat and radiation. The models were validated by comparing the results with those of experiments on earth, in the international space station and in a centrifuge. Several experimental techniques have been applied.

I. INTRODUCTION

Metal-halide (MH) lamps form an important subclass of high intensity discharge (HID) lamps. They are based on a Hg vapour at which a small amount of metal-halides such as DyI_3, TlI, and/or NaI are added. Even though the additive concentration is much less than that of the buffer gas mercury, most light of MH lamps is emitted by the added metals in the visible region. This results in a high power efficiency (up to 40%) and a much better colour rendering.

However, despite the clear advantages the growth of the metal-halide lamp market has been hampered by a number of limitations. One of these is the *segregation of colours* caused by the non-uniform distribution of the additives over the lamp that originates form the competition between three transport mechanisms: diffusion, convection and radiation.

In order to study the additive distribution in vertically burning metal halide lamps the plasma simulation platform PLAsma SImulation MOdel (PLASIMO) has been used to develop a numerical model. The model assumes axial symmetry, a steady laminar flow and the presence of local thermodynamic equilibrium (LTE). Two different transport modes were used; *the fluid* and the *radiation* transport modes. The first mode deals with the solution of the balances of mass, momentum and energy using the control volume approach, the radiation transport is treated by means of the ray-tracing method.

The input of this model is the set of control parameters, such as the fill chemistry, driving power and vessel shape. The output is formed by plasma properties; such as the voltage over the lamp electrodes, the production of heat and the emission of radiation. Along the way many more plasma properties are calculated: the distribution of the elements over the lamp and the fields of the temperature, the velocities and the pressure. The model attempts to calculate these parameters in a self consistent manner from first principles. There is no place for fudge factors.

Using this model numerical simulations have been performed of lamps containing mercury with sodium iodide and dysprosium iodide (Na and Dy lamps). Initial studies were carried with Na as the simulation of the Na lamps is relatively easy. The chemistry is rather simple and only a few Na lines have to be taken into account in the ray-tracing procedure. The modelling of the Dy lamp is more complicated; a proper chemical description needs to include more species and an enormous number of radiative transitions has to be dealt with. However, there was no choice, we had to construct a model for the Dy lamp since the experimental validation of a Na lamp is not easy.

Apart from describing the structure of the models, we also pay attention to the experimental validation. This article can be seen as the overview of the work done at the Eindhoven University of Technology as condensed in the theses of Janssen [1], Van Dijk [2], Van der Heijden [3], Colin Johnston [4], Zhu [5], Nimalasuriya [6], Beks [7] and Flikweert [8] . The URLs to these works is given in the list of references providing access to further details and giving references to publications on which these theses are based.

Demixing

When the metal-halide lamp is operated, the Hg is entirely vaporised forming the buffer gas. However, the few milligrams of DyI_3 additive do not evaporate completely, leaving a liquid salt pool at the coldest spot at the burner wall. The additive molecules diffuse from the relatively cool wall (~ 1200 K) toward the hot region (~ 6000 K) of the arc where they dissociate. At the centre the metal atoms are excited and ionised. As Dy atoms diffuse back to the wall they encounter I atoms in the cooler gas near the walls and recombine back to molecules.
Three principal regions can be identified within the radial density distribution of elemental Dy (cf figure 1):

1) the *near-wall* region where DyI_3 molecules are predominant,
2) the *mantle* region between the wall and the arc core where Dy atoms dominate, and
3) the *core* region where the Dy is almost completely ionized in the form of Dy^+ .

The molecules DyI_2 and DyI are relatively unstable and do not have a dominant presence in any region of the discharge. When burned vertically, the discharge emits light that is non-uniform in colour along it's axis.

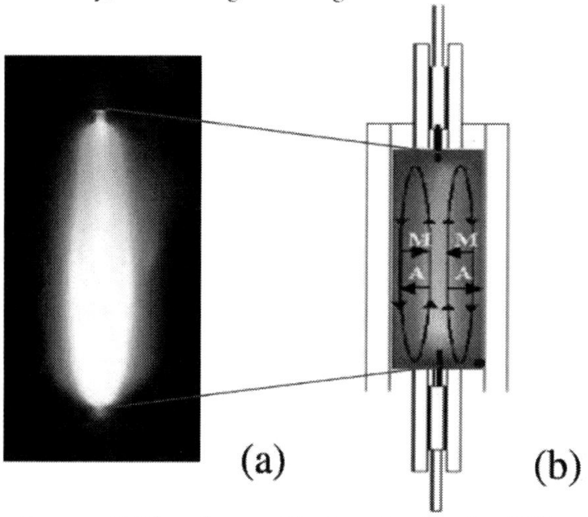

Figure 1. (a) Colour separation in an MH lamp burner; (b) schematic view of an MH lamp; diffusion and convection of atoms (A) and molecules (M) are indicated by arrows.

This colour segregation is mainly caused by the interplay between convection and diffusion which determines the distribution of the plasma species. The radiation transport has also some influence, albeit smaller. Radiation can lead to local cooling (emission) or heating (absorption). This will change the local temperature balance and thus the species distribution (gradually).
First we consider the *radial* distribution which is mainly ruled by the *temperature gradient* and *diffusion*. The temperature profile that is high in the centre and steeply decreases towards the wall, leads to a hollow profile of the mass density due to the ideal gas law. Moreover with the temperature change the chemical composition of the mixture changes as well (cf figure 2). This generates diffusion processes of the different constituents. However, the atoms and molecules have different diffusion velocities. The smaller and lighter Dy atoms diffuse faster outward than the larger and heavier molecules (DyI, DyI_2 , DyI_3) diffuse inward (cf figure 1). Ambipolar diffusion, in particular, causes the ions to diffuse out of the core faster than neutral atoms or molecules diffuse inwards.
These difference in diffusion velocities results in an even more hollow profile of the *elemental density* of Dy; this effect is called ***radial segregation***.

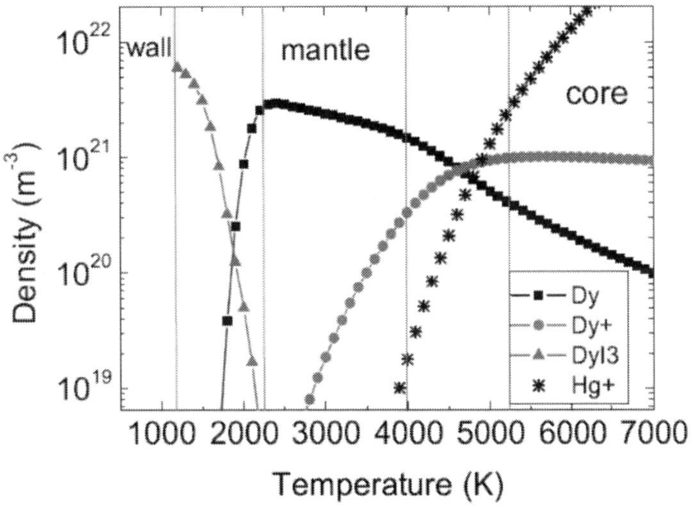

Figure 2: Theoretical density distribution of Dy atoms, ions and molecules, and Hg ions as a function of temperature at the mid-plane of a metal-halide lamp. For clarity, DyI_2 and DyI have been omitted.

Next we study the *axial* direction that is mainly governed by *convection*. Due to the high density in the *mantle and wall* region this plasma part will be pull down by gravity. Due to continuity equation the hot inner gas in the centre of the arc has to move upwards. This movement of the bulk gas drags the high concentration of elemental Dy near the wall downwards, whereas the lower concentration of Dy in the centre, caused by the radial segregation, is dragged upwards. As a consequence, a high density of Dy accumulates at the bottom, a phenomenon which is known as ***axial segregation***.
The latter obviously only occurs in presence of convection. Since there is no convection under *micro- gravity* conditions, axial segregation does not occur, but radial segregation does. By mean of experiments in the international space station (ISS) we can attain a situation in which only the combination of diffusion and radiation is active. Under *hyper-gravity* conditions, as attained in a centrifuge, the convection will be important and tuneable, just by increasing the rotational velocity the centrifugal force will increase and herewith the convection [8].
In order to quantify the degree of segregation as found by models and experiments it is useful to introduce the concept of *segregation parameter* k_s and *segregation depth* τ_s defined as

$$k_S(\alpha) = p_\alpha^{-1}\, \partial p_\alpha / \partial z \qquad \text{and} \qquad \tau(\alpha) = L\, k_s(\alpha) \tag{1}$$

where z is the axial coordinate, L the axial length of the discharge and p_α the elemental pressure, a quantity that will be defined below. So, if $\tau(\alpha)$ is large the length of the discharge is large compared to the e-folding distance of the segregation of element α.

II. THE MODEL

As stated before the model has been constructed using the PLASIMO platform [1,2]. This can be seen as a toolkit with ingredients ready to assemble models for the description of non-LTE or LTE, flowing or non-flowing plasmas driven by electro-magnetic sources ranging in frequency from DC to several GHz. The model constructed for HID lamps is an LTE model that is able to describe the competition between convection, diffusion and radiation transport. In [3,4] PLASIMO was employed to construct a model for the Sulphur lamp. In [3] a method was developed to handle the generation and transport of radiation; this will be used here.

The model constructed applies to a lamp operated at a constant power of 130 W using a mixture of Hg and

DyI_3. The distance between the electrodes is 18 mm the inner length of the burner is 20 mm while the diameter is 8 mm. We assume that the discharge is axially symmetric. The model thus uses a 2D rotationally symmetric grid. All equations are discretized on a structured finite volume mesh (cf figure 3).

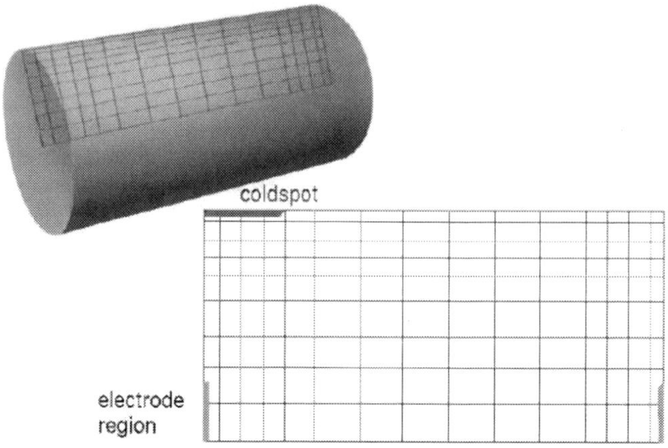

Figure 3. Schematic view of the grid and geometry used in the model. The actual model is with 34 axial and 18 radial positions not as coarse as the depicted grid.

One of the abilities of PLASIMO is to play with the dimensions of and the power coupled to the burner. The features given here are selected such that model-results can be compared with experimental results performed on the COST reference lamp [5,6,8].

2.1 Basis equations

A short overview of the basic equations solved in the model is given below. More details can be found in [1,2,3] and [7].

Energy balance

A central role is played by the energy balance that aims to calculate the plasma temperature. The temperature obtained in turn, strongly influences the transport coefficients, the plasma composition, the flow and the radiation generation and transport. The temperature is determined by

$$\nabla \cdot (C_p \mathbf{u} T) - \nabla \cdot (\lambda_c \nabla T) = P + Q_{rad}, \qquad (2)$$

where C_p is the heat capacity at constant pressure, **u** the bulk velocity, λ_c the thermal conductivity, $P = \sigma_{el} E^2$ the Ohmic dissipation, and Q_{rad} a term related to radiation generation and transport. This term is negative if emission dominates and positive if local absorption exceeds local emission. The latter takes place in the mantle of the discharge. Viscous energy terms are neglected. For the boundary conditions we assume that the electrodes have a surface temperature of 2900 K, while the rest of the wall has a temperature of 1200 K.

Radiation generation and transport

The term Q_{rad} is obtained by the solution of the radiation transport equation [3]

$$dI_v(v)/ds = j_v(v) - k(v)I_v(v), \qquad (3)$$

where $I_v(v)$ is the spectral intensity (in $Wm^{-1}\ sr^{-1}\ Hz^{-1}$), $j_v(v)$ the local emission coefficient ($Wm^{-3}Hz^{-1}$) and $k(v)$ the local absorption coefficient (m^{-1}). Equation (2) is solved by means of the ray-trace method. This consists of a network of rays that cross the plasma in various directions. Each ray (or beam) starts outside the plasma with a $I_v(v)$ =0 for any v-value; after entering the plasma the beam is filled or drained due to emission and absorption processes. This is done for many values of the frequency, v. After tracing several beams in different directions the $I_v(\nu)$ can be determined on each plasma grid location and the radiation term can be calculated by [3,7]:

$$Q_{rad} = \int_v \{-4\pi j_v(\nu) + \int_{4\pi} k(\nu) I_v(\nu)\, d\Omega\}\, d\nu. \qquad (4)$$

For the precise form of Q_{rad} for a DyI_3 -Hg mixture we refer to [7] where a method is given to separate the optically thin from optically thick radiation.

Particle transport

Since we assume LTE, the particle densities are described by the local temperature, pressure and elemental composition. To introduce the concepts and nomenclature used in this study we take the example of a water plasma (cf [3,7]). The hydrogen nucleus (in fact the protons) in such a plasma can be carried as part of the H atoms, excited hydrogen atoms H* or as part of OH/OH^- radicals, H_2O molecules etc.. In the latter case the displacements of one H_2O molecule implies the movement of two protons. It is therefore useful to define the *elemental density* of hydrogen as

$$n_{\{H\}} = 2n_{H2O} + n_{OH-} + 2n_{H2} + n_H + n_{H*} \ldots. \qquad (5)$$

Note the difference between the notation $n_{\{H\}}$ and n_H. The former represents the *elemental* density of H, i.e. the presence of H nuclei per volume irrespective their state of binding or excitation whereas n_H refers to the concentration of unbound atomic H in the ground state. Equation (5) thus shows that the density of a particular element is given by the density of the species multiplied with the number of nuclei of the element in that species. In general, we will use Greek subscripts for elements and Latin subscripts for species (also named 'compounds'). To distinguish the elemental density of a specific element from the density of the free (unexcited) atoms we will use braces (e.g. $n_{\{H\}}$ versus n_H).

In general the elemental density is given by: $n_\alpha = \Sigma_i R_{i\alpha}$, with $R_{i\alpha}$ the stoichiometric coefficient that gives the number of elements α in species i. The index runs over all species in the plasma. If a species j does not contain a nucleus of α it is simply $R_{j\alpha} = 0$. In the same way we define the elemental fluxes $\mathbf{\Gamma}_\alpha = \Sigma_i R_{i\alpha} n_i \mathbf{u}_i$ with n_i and $\mathbf{u}_i$ the density and velocity of compound (species) i. Since elements are neither created nor destroyed, the continuity equation for elements is source-less and reads:

$$\nabla . \mathbf{\Gamma}_\alpha = 0. \qquad (6)$$

Similarly, we have for the bulk density $\rho = \Sigma_i \rho_i$ the continuity equation

$$\nabla . \rho\mathbf{u} = 0 \qquad (7)$$

with the **u** the bulk velocity.

The elemental pressure p_α is defined as the pressure that is build up by all the molecules, atoms, ions etc. that carry (nuclei of) the element α. It can thus be written as

$$p_\alpha = \Sigma_i R_{i\alpha} p_i \qquad (8)$$

with p_i the partial pressure of species i. The value of p_α is determined by eqn 6 which can be written as

$$\nabla . \{(D_\alpha / kT) \nabla p_\alpha + (p_\alpha / kT)\, \mathbf{c}_\alpha\} = 0 \qquad (9)$$

by employing a 'pseudo-convective velocity' $\mathbf{c}_\alpha$ [3,7] and effective diffusion coefficient

$$D_\alpha = p_\alpha^{-1} \Sigma_i R_{i\alpha} D_i p_i. \qquad (10)$$

The diffusion coefficient D_i is calculated from the binary diffusion coefficients D_{ij}. The binary diffusion coefficients D_{ij} for the diffusion of species i through species j depends on the differential cross-section of the interacting species.

The *bulk velocity* is obtained by solving the Navier Stokes equation

$$\rho \mathbf{u}.\nabla \mathbf{u} = -\nabla p + \nabla.\mu\nabla \mathbf{u} + \rho \mathbf{a}_g \quad (11)$$

in conjunction with the continuity equation (equation (7)) using the SIMPLE [7] algorithm. Here p is the total pressure, μ the dynamic viscosity and $\mathbf{a}_g$ the acceleration.

Ohmic heating

The power to the plasma is supplied by Ohmic heating. By solving the Poisson equation:

$$\nabla.\ \sigma_{el}\nabla\phi\ = 0; \quad (12)$$

we get local values of the potential ϕ from which the electric field $\mathbf{E} = -\nabla\phi$ and the current density $J = \sigma_{el} E = \sigma_{el} \nabla\phi$ can be derived. The following boundary conditions are imposed:

1. There is no current through the walls, resulting in a homogeneous Neumann boundary condition.
2. One electrode is kept at zero potential, which leads to a Dirichlet condition $\phi = 0$ at that electrode.
3. The potential of the other electrode is initially put at 100 V. This value is adjusted during the iteration process and determined by the fact that the power dissipated in the discharge equals 110 W. This is equivalent to the actual lamp power of 130 W of which 20 W is consumed by electrode losses and 110 W by ohmic dissipation of the discharge.

The selection of cross-sections

In the basic equations summarized above, important roles are played by transport coefficients, such as the diffusion coefficients D_i, the thermal conductivity λ_c and the electrical conductivity σ_{el}. These coefficients are calculated from collision integrals that are based on differential cross sections [7].

Ideally, one would like to have cross-sections for each collision-type. In practice such data is difficult to gather or to calculate since for only a few interactions energy-dependent integral cross-sections σ_{ij} are available. When available these are used in the model. For collisions for which such data could not be found, approximations were made. The collisions can be classified along the following categories:

1. Charged-charged (c-c) collisions; in this case the shielded Coulomb cross-sections is a good and generally applicable approach.
2. Neutral-neutral (n-n) collisions; in cases where dedicated cross-section values are not available these interactions are described as hard sphere collisions.
3. Charged-neutral (c-n) collisions; this category is the most complicated. In principle, a full quantum mechanical (QM) treatment is needed in which e.g. the Ramsauer effect has to be taken into account. For each pair of interacting particles the QM aspects are different. If results of experiments or QM calculation are not available the use of the Langevin cross-section is the only option [7].

In the study of the influence of the values of the cross-sections on the transport coefficients D_i, λ_c and σ_{el} it is important to note that Hg atoms form the most dominant species in the discharge; all other species are present in small concentrations. This simplifies the model.

Due to the low elemental concentration of Dy and I we can neglect the heat generation that is liberated in the formation of DyI_x molecules. The reactive part of heat conductivity can therefore be neglected and only the frozen part remains which is determined by the collisions of Hg atoms mutually. These Hg-Hg collisions are of the n-n category and experimental results confirm that the hard sphere approach is indeed valid here. This implies that there is not much doubts on the thermal heat conductivity λ_c. The choice of D_i and σ_{el} is less trivial and will be guided by comparing the model results with those of experiments.

III. EXPERIMENTAL VALIDATION

Various methods have been employed to get insight in the lamp parameters. The experiments verify the model results, whereas the model aids the interpretation of the experimental results. The experiments are part of a poly-diagnostic study [5, 6] meaning that in general different experimental methods are employed. The following experimental methods were used:

- *Voltage measurement* (VM), to determine the plasma resistance.
- *Absolute line intensity measurements* (ALI); for the temperatures in the Hg, Dy and Dy^+ systems.
- *X-ray absorption* (XRA), to find the gas temperature T_g field
- *X-ray fluorescence* (XRF), to establish the element concentration as function of plasma position.
- *Thomson scattering* (TS) to measure the electron density and temperature fields.

The lamps were measured under three gravity conditions

a) The *normal gravity* condition of $a_g = 1$ g
b) *microgravity* with $a_g = 0$ as obtained in the ISS [7] and
c) *hyper-gravity* of an acceleration between $1 < a_g < 10$ g, realized in a centrifuge [8].

All methods listed above can be and were applied to the normal 1 g situation. For micro- and hypergravity only the two methods VM and ALI are feasible.

The function of the micro-gravity experiments in the ISS is that the complex interaction between convection and diffusion can be unravelled by "switching off" the gravity. This is advantageous since the metal-halide lamp is easier to model if convection is absent.
Below we give some results of the comparison of experimental with model results for the micro-gravity case, so without convection. The lamp is operated at 130 W (of which 110 W dissipated in the discharge and 20 W spend on electrode losses) and contains 10 mg Hg and 4 mg DyI_3.
We start with voltage measurements. Comparing the results of this relative simple VM method with model calculations can give insight into electrical conductivity σ_{el}, a quantity that strongly depends on the value of the elastic interactions between electrons with Hg atoms. This e-Hg interaction is of the c-n category but fortunately the corresponding cross-section as delivered by Rockwood is well known [7]. This was verified by comparing the model results based the Rockwood cross-sections with those based on the Langevin cross-section. The voltage drop over the lamp was calculated for an Ohmic dissipation of 110 W. Using the Rockwood cross section the model predicts a voltage drop of 99 V, which is close to the measured value of 100 V. The calculations based on the Langevin cross-section, however, produces a voltage of 71 V, which is much too low. We may therefore conclude that the model gives a good description of the potential distribution over the lamp and the corresponding Ohmic heating provided the Rockwood values are used.

The ALI method was applied on the systems of Hg, Dy and Dy^+. In this way it could be determined whether the spectral power as predicted by the model agrees with experimental results. Moreover the ALI method was used to determine the densities of several emitting atomic (or ionic) levels [7] from which the atomic state distribution function (ASDF) of Hg, Dy and Dy^+ could be determined. The ASDF of the Hg-system was used to determine the temperature field and the ASDF of Dy and Dy+ were constructed to find the spatial concentration of Dy and Dy^+ and the segregation of elemental Dy. Apart from that we used the slopes in the Dy and Dy^+ systems for a cross validation of the T-value obtained from the Hg-ASDF.
The temperature profile as deduced from the Hg system is given in figure 3 where its is compared with two model results that differ from each other in the value of the cross section of the interaction between Hg and Dy+. Namely for the case that the cross section follows the Langevin cross section, i.e. σ(Hg, Dy+) = σ_{LV} and the case for which σ(Hg, Dy+) = 1.2 σ_{LV}. The comparison shows that the model based on the Langevin cross-section is in good agreement with the experiment. For the case that σ(Hg, Dy+) = 1.2 σ_{LV} we get a broader temperature profile with a lower axis temperature.
The reason why the higher σ(Hg, Dy+) value reduces the central temperature is that by increasing σ (Hg, Dy+) the ambipolar diffusion in the central region is hampered which causes the Dy particles to be accumulated in the centre. The temperature can then decrease slightly as the discharge will burn more on Dy

that has a lower ionisation potential than Hg (5.93 eV versus 10.43 eV). However, the change in the temperature distribution remains in the error margin of 10% so that no firm conclusions can be made about the optimum value for σ (Hg, Dy+).
Several aspects of the model were investigated by using the ALI, XRA, and XRF methods. The details can be found in [6], [7] and [8]. The main conclusion is that the model predictions fairly agree with the experimental results. In this comparison we should keep in mind that also the experimental results are not perfect and limited in accuracy. One of the big uncertainty is that the cold spot temperature of the lamps is not easy to determine and that a slight variation in this temperature can lead to an drastic change in the vapour pressure of the additives.

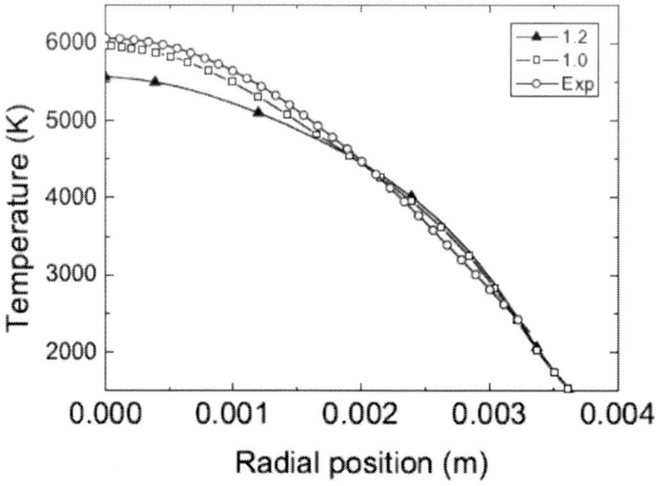

Figure 4: Radial temperature profile for a lamp containing 10 mg Hg and 4 mg DyI_3 at the midplane of the lamp. The numbers 1.0 and 1.2 refer to the two cross-sections that were used, the normal Langevin cross-section σ_{LV} and 1.2 σ_{LV}. Exp denotes the experimental results.

Trends
By running the model for various power settings and burner dimensions insight was obtained in the main trends. It was found that these can be put in one simple scaling law that predicts the conditions for which the lamp is most strongly de-mixed. This simple scaling law relates the average segregation depth τ_S (cf equation (1)) with the Peclet number. The Peclet number Pe is given by the ratio of the axial convection and radial diffusion velocities. In terms of the effective diffusion coefficient of the element D_α and the axial convection speed V_z it reads:

$$\mathrm{Pe} = R^2 V_z/(LD_\alpha) \tag{13}$$

with L the length and R the radius of the lamp. By using simple considerations a simple model based on the axial velocity was developed [7] to estimate V_z. This means that for most conditions the Pe can be determined and from this the average segregation depth τ_S.
This simple relation can be used as an engineering tool.

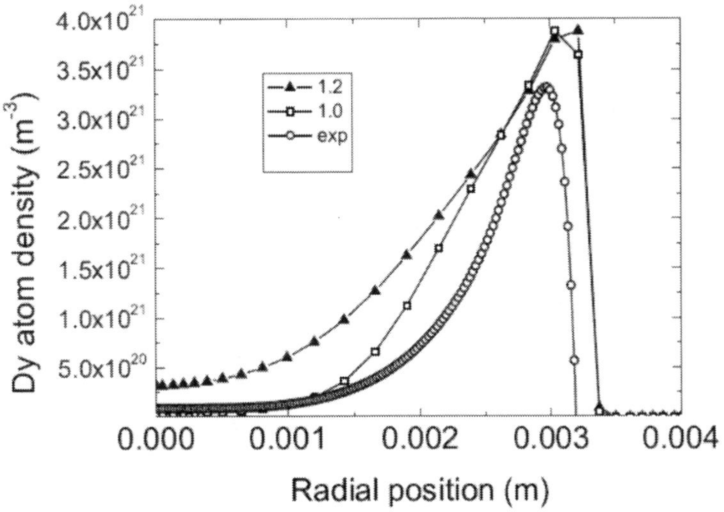

Figure 5: Radial Dy atom system density profile for a lamp containing 10 mg Hg and 4 mg DyI_3 at the mid-plane of the lamp. The numbers 1.0 and 1.2 refer to the two σ(Hg, Dy+) values σ(Hg, Dy+) = σ_{LV} and σ(Hg, Dy+) = 1.2 σ_{LV}

The validity of the LTE assumption.

The models developed with PLASIMO to model HID lamps are based on the assumption of the presence of LTE. More specifically this implies among others that

1) the electrons are ruled by Maxwellian electron energy distribution functions EEDF,
2) the electron temperature equals the heavy particles temperature, $T_e = T_g$,
3) the excited states in the atomic or ionic systems are populated according to Boltzmann,
4) the ionization degree obeys Saha's law,
5) the dissociation degree follows the law of Guldberg-Waage.

Since the model describes the plasma as a whole, from cathode to anode and from wall to central axis we should realize that the conditions given above must be valid for any plasma position. This is a very strong demand that is certainly not fulfilled. The subsequent question is whether this has serious consequences for the predictive value of the model w.r.t. the light technical properties of the lamps.

It is known that the presence of a Maxwellian EEDF demands for a sufficiently high degree of ionization. As the electron density goes down whereas the heavy particle density goes up for increasing radial distances we can expect deviations from Maxwell above certain r-values.

On the conference we will report results of the comparison of Thomson scattering (TS) with x-ray absorption experiments performed on pure Hg HID lamps. The former gives the electron temperature T_e and the latter the gas temperature T_g. It is found that in the outer regions the equality $T_e = T_g$ does not hold.

In the section formed by pages 90-93 of [6] it was found that the Boltzmann balance in Dy+ does not hold in the outer regions, that is beyond $r = 0.5R$ (R is the radius of the burner). In fact for $r>0.5R$ the Boltzmann is replaced by the corona balance.

The experimental values for n_e obtained with TS show that the ionization degree does not follow Saha's laws, especially not in the outer plasma region.

CONCLUDING

We may state that the models constructed with the PLASIMO platform give valuable descriptions of the plasmas in HID lamps. These Grand models give insight in the competition between convection, diffusion and radiation and show how this leads to the main plasma features. On bases of these results a good engineering model could be made that predicts the segregation parameter as function of the Peclet number. However, for a more detailed description of the burner plasma we have to abandon the LTE assumption.

REFERENCES

[1] Ger Janssen (2000); *Design of a General Plasma Simulation Model, Fundamental Aspects and Applications*; http://alexandria.tue.nl/extra2/200013063.pdf
[2] Jan van Dijk (2001); *Modelling of Plasma Light Sources; an object –oriented approach;* http://alexandria.tue.nl/extra2/200210525.pdf
[3] Harm van der Heijden (2003); *Modelling of Radiative Transfer in Light Sources;* http://alexandria.tue.nl/extra2/200310729.pdf
[4] Colin Johnston, (2003); *Transport and equilibrium in molecular plasmas: the sulfur lamp;* http://alexandria.tue.nl/extra2/200310819.pdf
[5] Xiaoyan Zhu (2005); *Active spectroscopy on HID lamps, Exploration of various methods;* http://alexandria.tue.nl/extra2/200510706.pdf
[6] Tanya Nimalasuriya (2007); *Transport Phenomena in Metal-Halide Lamps: a poly-diagnostic study;* http://alexandria.tue.nl/extra2/200712152.pdf
[7] Mark Beks (April 2008); *Modelling Additive Transport in Metal Halide Lamps;* http://alexandria.tue.nl/extra2/200810199.pdf
[8]Arjan Flikweert (2008); *Spectroscopy on Metal-Halide Lamps under Varying Gravity Conditions*; http://alexandria.tue.nl/extra2/200811537.pdf

Plasma Diagnostics in Discharge-Type Light Sources by Using Electromagnetic Waves

Kunihide Tachibana
Promotion Laboratory for Plasma and Photonic Science Researches,
Graduate School of Science and Engineering, Ehime University,
3 Bunkyo-cho, Matsuyama, Ehime 790-8544, Japan
Contact: tachibana.kunihide@eng.ehime-u.ac.jp

ABSTRACT

Several new methods have been applied to the diagnostics of plasma parameters in discharge-type light sources by using electromagnetic waves in the infrared to far-infrared ranges. A mm-wave transmission method was successfully used for the estimation of electron density n_e in a CCFL with its two-dimensional array. A THz time-domain spectroscopic method was used for the measurement of n_e and the elastic (momentum transfer) collision frequency ν_{me} of electrons in an rf-driven Hg-Ar filled lamp, while a heterodyne interferometric method with a CO_2-laser operating at 10.6 μm was used for the spatially resolved measurement of n_e in a mercury-free Xe-lamp.

1. INTRODUCTION

We frequently encounter the situations in which it is difficult to access to the target plasmas in various discharge-type light sources for performing the diagnostics of plasma parameters by electric or spectroscopic methods due to finite sizes or opaque envelopes. In some cases, however, the difficulties are solved by using electromagnetic waves of infrared to far-infrared ranges, which can penetrate the glass or ceramic envelopes. Although the spatial resolution of the methods may not be good, we can also find the solution with a combination of appropriate theoretical model in the simulation of transmission characteristics of electromagnetic waves.

For an example, one can imagine a cold-cathode fluorescent lamp (CCFL) of which inner diameter is 1 to 2 mm. It is usually difficult to insert an electrical probe into the positive column for the plasma diagnostics. The use of lasers with visible to infrared wavelengths is also impossible due to the glass envelope and the inner-coated fluorescent material. We have tried to use a transmission method of electromagnetic waves of 10 ~ 100 GHz range through an array of CCFLs, and analyzed the transmittance characteristics in a similar manner to the photonic crystals in the visible range of optics. The results can provide us with the plasma parameters with a certain satisfaction.

In this talk, I will explain several new methods, including the above mentioned method, for the plasma diagnostics in various discharge-type light sources by using electromagnetic waves in the infrared to far-infrared ranges.

2. MM-WAVE TRANSMISSION METHOD

The dielectric constant in uniform plasma for the frequency ω of incident electromagnetic waves is expressed in the Drude model as

$$\varepsilon_p = 1 - \left(\frac{\omega_{pe}}{\omega}\right)\frac{1}{1 - i(\nu_{me}/\omega)}, \tag{1}$$

where ω_{pe} is the electron plasma frequency given by $(\varepsilon_0 k_B T_e / e^2 n_e)^{1/2}$ with the dielectric constant in vacuum ε_0, Boltzmann constant k_B, electron charge e and electron temperature T_e. Now, let's consider the situation shown in **Fig. 1(a)**, where many plasma columns are arranged in a square lattice structure with a lattice constant a. In this situation, the dielectric constant depends on the propagation direction such as like the electron wave in a crystal lattice of solid and can be expressed by the Fourier expansion form using the reciprocal lattice vector [1]. The result of a numerical calculation for the

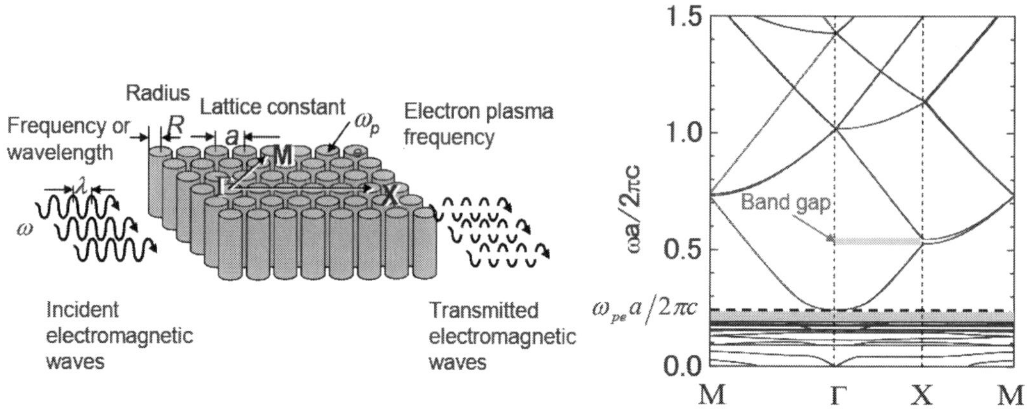

(a) Columnar plasma array (b) Photonic band for TE mode

Fig. 1 (a) Square lattice array of cylindrical plasma columns and (b) the dispersion relation (photonic band diagram) for propagation of electromagnetic waves in the TE mode with electrical vector perpendicular to the columns.

dispersion relation between ω and the wave vector k is shown in **Fig. 1(b)**. It is seen that band gaps appear in the frequency range of $\omega > \omega_{pe}$, and the propagation of electromagnetic waves is forbidden in those band gaps. Below ω_{pe} the dielectric constant becomes imaginary and the electromagnetic waves are attenuated except for the sharp resonance like frequencies due to the surface wave propagation modes.

When we scan the frequency and see the transmittance as a function of ω, we obtain the result as shown in **Fig. 2(a)** for the propagation along Γ-X direction [2,3]. In this example, we used a columnar plasma array composed of micro hollow cathode discharges with a lattice constant of 2.5 mm. A band gap appears at around $\omega a/2\pi c = 0.5$, corresponding to the frequency range of 60 – 63 GHz. Although the band gap frequency is mostly determined by the lattice structure, the peak attenuation and the band width depend on the plasma density n_e and the number of rows of plasma columns. In the case of our model experiment, the transmittance became about 0.7 with 17 rows and 0.3 with 30 rows as shown in **Fig. 2(b)**. The comparison with a simulation of the transmittance as a function of n_e, the experimental value of n_e can be estimated as about $1\text{x}10^{13}$ cm^{-3} [3].

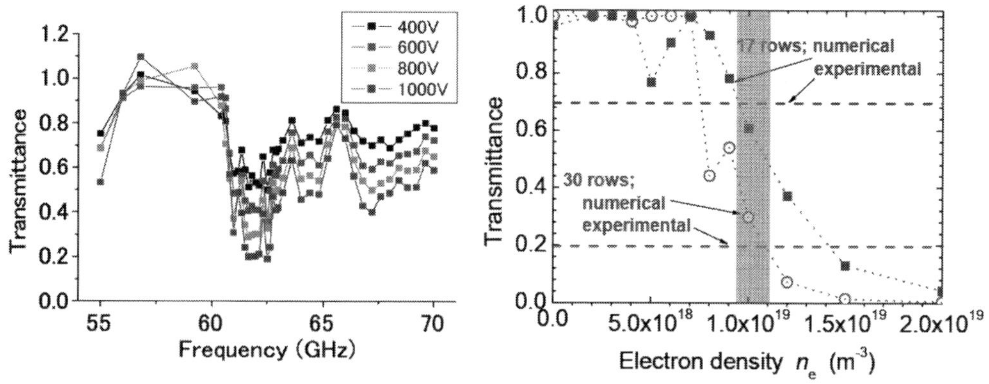

(a) (b)

Fig. 2 (a) Frequency dependent transmittance of electromagnetic waves through plasma columns arranged in a square lattice (see Fig. 1) and the calculated transmittance as a function of electron density in two cases with 17 and 30 rows in comparison with experimental data.

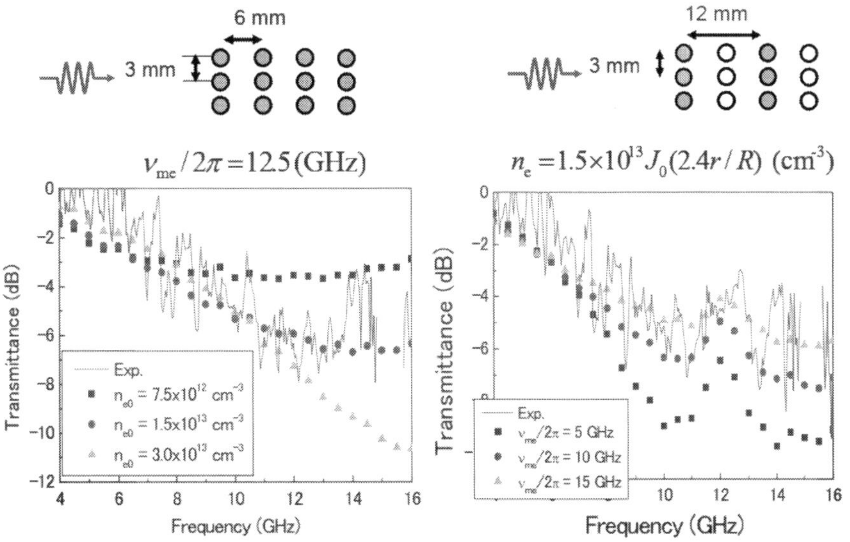

Fig. 3 Two types of square lattice arrays of CCFLs and the calculated transmittance characteristics as functions of frequency by using n_{e0} and ν_{me} as adjustable parameters in comparison with experimental data.

Alternatively, if we use the frequency range below ω_{pe}, we may observe the frequency dependent attenuation due to the complex dielectric constant. There also appear sharp resonance-like transmissions due to the flat bands. In this situation, we tested the transmittance using a vertical array of CCFLs arranged with a pitch of 3mm in columns and 6 mm or 12 mm in rows as illustrated in the upper part of **Fig. 3**. Measured results of the transmittance with these two arrangements are shown in the lower part of the same figure, where each CCDFL was operated at a current of 10 mA. We also simulated the transmittance as a function of ω using n_e and ν_{me} as adjustable parameters. From the best fit condition for both the arrangements, the experimental values of n_{e0} and $\nu_{me}/2\pi$ are estimated to be 1.5x10^{13} cm^{-3} and 12.5 GHz, respectively. Here, we assumed that the radial distribution of n_e in the positive column is given by the Bessel function of the zero-th order as $n_e = n_{e0}J_0(2.4r/R)$. From this result, the plasma in a CCFL is known in a highly collisional condition.

3. CO_2-LASER HETERODYNE INTREFEROMETRY

Heterodyne interferometry with a CO_2 laser is often used in plasma diagnostics due to its high special resolution [4], although the detectivity is limited by its wavelength of 10.6 μm and infrared windows such as ZnSe crystals are required for the transmittance in the optical path. A typical setup for the heterodyne interferometry measurement is shown in **Fig. 4(a)** [5]. The CO_2-laser beam is split into two parts by a Mach-Zehnder interferometer composed of two beam splitters and two mirrors. In one arm of the interferometer a test plasma is set and in the other beam an acoust-optic modulator (AOM) is inserted to up-shift the laser frequency by the modulation frequency $\Delta\omega/2\pi$ (40 MHz in this case). The output beams are merged together and the beat signal at 40 MHz is detected by a lock-in amplifier and the phase shift caused by the presence of the test plasma is derived with respect to the reference provided from the modulation signal.

The phase shift due to the presence of the test plasma is expressed as

$$\Delta\Phi = \frac{2\pi}{\lambda}(N-1)l\,. \tag{2}$$

Here λ is the wavelength of the probe laser, l is the optical length in the test plasma, and (N-1) is the change of the refractive index given by

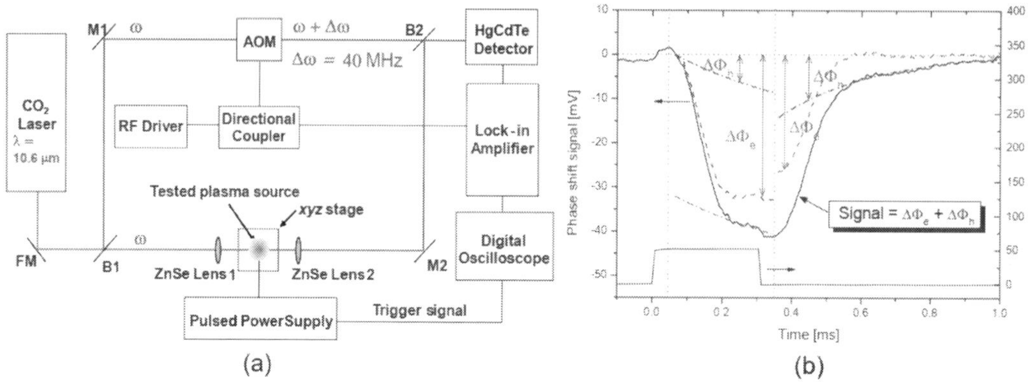

Fig. 4 (a) Setup for CO_2-laser heterodyne spectrometry and (b) example of phase shift signal measured in a micro hollow cathode discharge operated in pulsed mode.

$$N-1=-\frac{e^2}{2\left(c^2 m_e \varepsilon_0 4\pi^2\right)}\lambda^2 n_e+\left(A+\frac{B}{\lambda^2}\right)\frac{n_h}{n_{h0}}\ , \tag{3}$$

where m_e is the electron mass and A and B are constants. The first term stands for the change due to the generation of plasma with the electron density n_e and the second term is due to the change of the gas (heavy particle) density by Joule heating from n_{h0} to n_h.

Since $n_h < n_{h0}$, both terms contribute to the decrease in N that makes the separation difficult. However, if we operate the discharge in a pulsed mode, we can perform the separation according to their different behaviours. An example of the phase shift signal is shown in **Fig. 4(b)** where the measurement was done in a pulsed hollow cathode discharge with 2 mm bore operated in He at 9.3 kPa with a discharge current of 50 mA [5]. It is noted that after the rise of the current a rapid change occur and a slower one follows, corresponding to the two terms of eq.(3), respectively. A similar tendency is also seen in the fall period.

We applied this method to the measurement of a Xe filled lamp, which we are developing for a Hg-free fluorescent lamp [6]. The experimental configuration and the results of estimated values of n_e

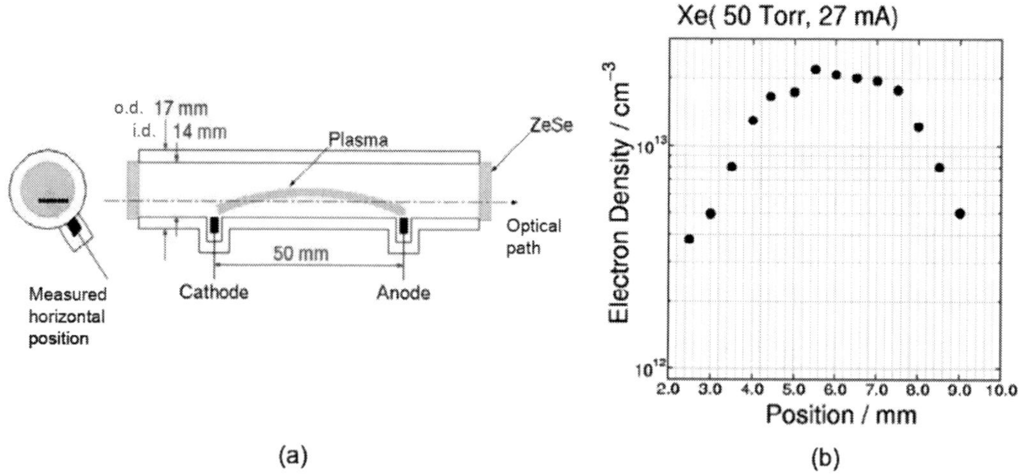

Fig. 5 Structure of Xe-lamp and experimental configuration for CO_2-laser heterodyne spectrometry and (b) the measured results of n_e with scanning the horizontal position.

measured by changing the horizontal (radial) position are shown in **Fig. 5**. The inner diameter of the discharge tube was 14 mm and the discharge length was 50 mm. Xe gas was filled at 6.7 kPa and the square-wave modulated pulsed discharge was operated at 27 mA. It is seen that the peak value of n_e becomes about 2×10^{13} cm^{-3} at the central position. In this example the separation of the two terms are not completely done due to the large contribution of the gas temperature change, so that there remains a possibility of over-estimation of n_e values. It is also noted that the plasma column was rather constricted and bowed along the tube axis at a relatively higher operating pressure. We will try to refine the data with improving the operating conditions.

4. THz TIME DOMAIN SPECTROSCOPY

The frequency range from 0.1 to 10 THz is also interesting for plasma diagnostics because the detectivity can be increased in a comparison with the CO_2-laser heterodyne method although there is a drawback in the spatial resolution [7,8]. The usual setup for THz time domain spectroscopy (TDS) with a fs-laser as the light source is shown in **Fig. 6(a)**. The fs-laser beam is separated by a beam splitter and the first beam excites the antenna to pump the THz electromagnetic waves, which is detected by the other antenna by the second probing beam with a scanned time delay generated by a moving corner cube. The time resolved signal corresponding to the electric field strength of THz waves is acquired by a lock-in amplifier at the frequency of AC bias voltage applied to the transmitting antenna. The signal is quite similar to the interferogram in FT-IR spectroscopy. It is processed by a fast Fourier-transformation to produce the transmittance (real part) and phase shift (imaginary part) spectra as functions of the frequency ω. We tried to apply this method for the measurements of n_e and ν_{me} in an RF driven discharge in a glass tube filled with Ar at 400 Pa and a small amount of Hg as shown in **Fig. 6(b)**. The beam of THz waves collimated by a off-axis parabolic mirror is transmitted through apertures of 6 mm bore placed on both sides of the tube for realizing the

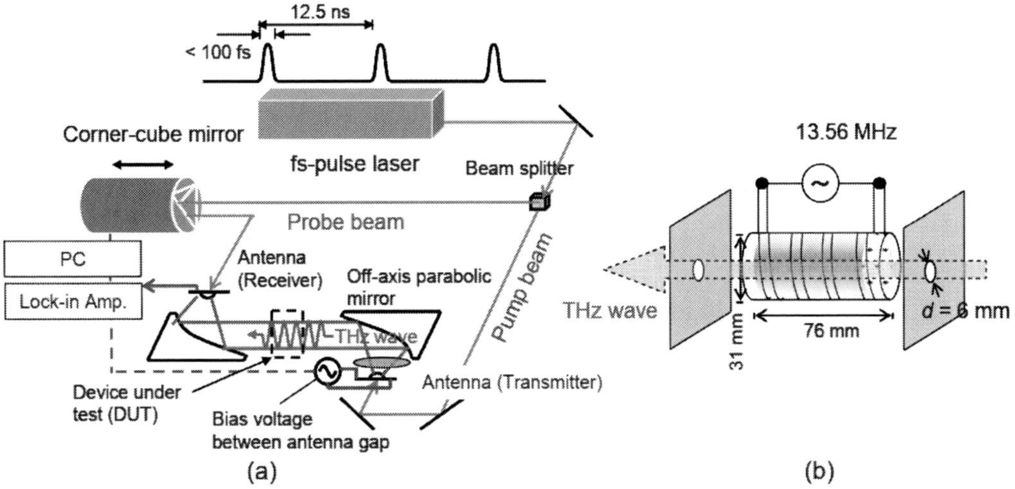

Fig. 6 (a) Schematic arrangement for THz-TDS measurement and (b) the experimental configuration with a tested discharge lamp.

spatial resolution.

Examples of the transmittance and the phase shift spectra are shown in **Fig. 7**, which were measured along the central axis. There are two adjustable parameters for the data fitting, i.e., n_e and ν_{me}; n_e (or plasma frequency $f_{pe} = \omega_{pe}/2\pi$) is more sensitive for the variation of the transmittance spectrum while ν_{me} is for the phase shift spectrum. With the best fits to both data we estimated that f_{pe} = 34 GHz ($n_{e0} = 1.4 \times 10^{13}$ cm^{-3}) and ν_{me} = 5 GHz. The radial distribution obtained by scanning the transmitting position is shown in **Fig. 8**. From the result it looks that $n_e(r)$ takes a parabolic or zero-th order Bessel function-like profile. However, it is seen in Fig. 7 that the fitting becomes poor at the frequency range below 0.1 THz. One of the possible reasons for that disagreement may be due to the refraction of the THz beam transmitting through a non-uniform media, of which effect becomes larger

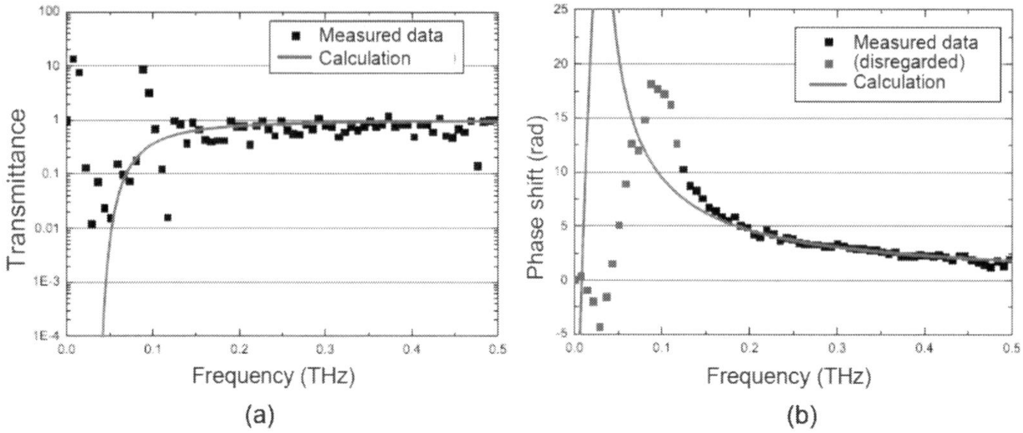

Fig. 7 Examples of experimental data for (a) transmittance and (b) phase shift measured in a lamp shown in Fig. 6(a) along the central axis and the fitted curves with adjusted f_{pe} and ν_{me}.

as the frequency decreases. We need more detailed analyses toward the final goal for deriving the results with better accuracy.

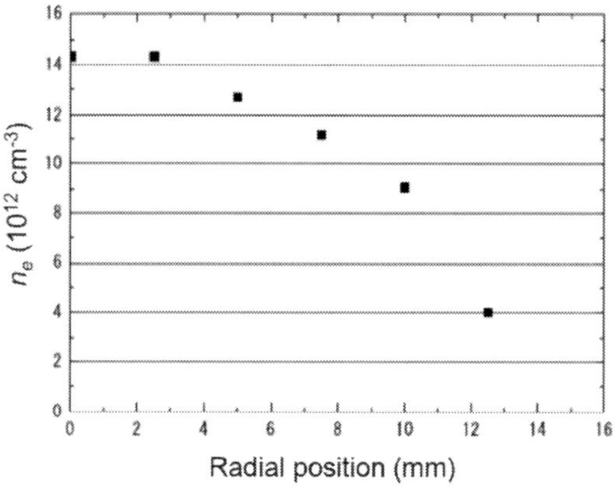

Fig. 8 Radial distribution of n_e measured in a discharge lamp shown in Fig. 6(b).

5. CONCLUDING REMARKS

For the measurement of plasma parameters in discharge-type lamps, laser Thomson scattering and other spectroscopic methods using line profiles or spectral line intensities have been used commonly. In this work, however, we tried to use electromagnetic waves in the infrared to far-infrared ranges in order to avoid the difficulties caused by the opaque envelopes of the lamps in the UV to visible ranges. We showed an example to avoid the inherent poor spatial resolution of our methods by using an array of columnar plasmas and an appropriate model for simulating the effective transmittance through the array to derive the plasma parameters of each component. The THz-TDS method is proved to have high potential in plasma diagnostics and will be used popularly in near future.

[1] O. Sakai, T. Sakaguchi, K. Tachibana, J. Appl. Phys. **101** (2007) 073304.

[2] T. Sakaguchi, O. Sakai, K. Tachibana, J. Appl. Phys. **101** (2007) 073305.

[3] O. Sakai, T. Sakaguchi, T. Naito, D.-S. Lee, K. Tachibana, Plasma Phys. Control. Fusion, **49** (2007) B453.

[4] F. Leipold, R. H. Stark, A. El-Habachi, K. H. Schoenbach, J. Phys. D: Appl. Phys. **33** (2000) 2268.

[5] J.-Y. Choi, N. Takano, K. Urabe, K. Tachibana, Plasma Sources Sci. Technol. **18** (2009) 035013.

[6] N. D. Ahmad, A. Kondo, H. Motomura, M. Jinno, J. Phys. D: Appl. Phys. **42** (2009) 095202.

[7] M. C. Nuss, J. Orenstein, *Millimeter and Submillimeter Wave Spectroscopy of Solids*, ed. G. Gruner (Springer, Berlin, 1998).

[8] S. P. Jamison, J. Shen, D. R. Jones, R. C. Issac, B. E. Ersfeld, D. Clark, D. A. Jaroszynski, J. Appl. Phys. **93** (2003) 4334.

Modeling of Emitter Loss in Fluorescent Lamps

I. Kobayashi[1], R. Bruckshaw[1], M. Kai[2], Y. Manabe[2], Y. Yamagata[3] and R. Devonshire[4]

[1] Cavendish Instruments Ltd., Aizlewood's Mill, Nursery Street, Sheffield S3 8GG, UK
[2] Panasonic Corporation, Lighting Company, 1-1 Saiwai-cho, Takatsuki City, Osaka 569-1193, Japan
[3]Department of Engineering Sciences and Materials, Kyushu University, 6-1 Kasugakoen, Kasuga, Fukuoka 816-8560, Japan. [4]Department of Chemistry, University of Sheffield, Sheffield S3 7HF, UK
Contact: r.devonshire@sheffield.ac.uk

ABSTRACT

Laser-induced-fluorescence (LIF) measurements of barium (Ba(I)) in the near-electrode regions of AC and DC operated fluorescent lamps have revealed interesting features [A. Samir, *et al.*, *IEEJ Trans. FM*, 127 (2007) 543]. The observations, which relate indirectly to electrode degradation, are interpreted here using the electrode sub-model of a system model of FL and CFL lamps. A thermal evaporation model of barium loss is developed that uses model-predicted, evolving, electrode surface temperature profiles, and a material- and temperature-dependent rate equation. The results are compared with those of a gaseous diffusion model of barium loss where the barium vapour density at the electrode's surface is assumed to be the equilibrium vapour pressure. The electrode model is used to predict lamp life for both design and accelerated life test lamp operating conditions, and also the electrode's electrical resistance for use in the lamp system electrical sub-model.

INTRODUCTION

The operating lives of fluorescent lamps, FL and CFL, are determined principally by the lifetimes of their electrode(s). Lamp failure occurs when the low work function, triple-oxide (BaO-SrO-CaO), emitter material coating the electrodes is exhausted as a result of evaporation at the prevailing high electrode surface temperatures, and to other loss mechanisms occurring mainly at ignition. The loss processes can be followed by observing *in situ* barium-containing species in the near-electrode regions using spectroscopic techniques such as spontaneous emission, resonant absorption, and laser-induced-fluorescence (LIF) [1], by Raman studies of the surfaces of electrodes *in situ*, or by Raman imaging of cross-sections of extracted electrodes [2]. Models of operating FL electrodes have been reported previously [3][4][5]. Here an electrode model is developed and used to interpret recent barium LIF measurements [1], to predict lamp lifetime and the effect on lifetime of changes in electrode design and operating conditions, and to predict the electrode electrical resistance during life.

METHOD

The electrode model used here is a component of a new system model of low pressure discharges comprising ballast, electrode, electrode-plasma interface, cathode region, positive column and anode region sub-models. Example cathode and anode half-cycle electrode temperature distributions predicted by the electrode model for "stick" electrodes in a DC operated lamp are shown in Fig. 1.

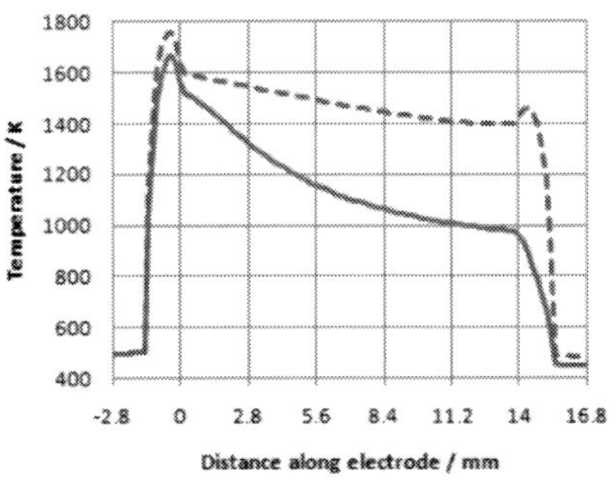

Figure 1. Anode (*dashed line*) and cathode (*continuous line*) temperature profiles predicted by the electrode model for a DC operated lamp

The rate of emitter evaporation was modeled as a first-order kinetic process whose rate coefficient is material and temperature dependent. Combining the kinetic model with model-predicted electrode temperature profiles leads to a computationally efficient method of calculating local and net barium loss rates. These rates are compared with those predicted by a gaseous diffusion flux model [6]. The rate equation's parameters are derived from the LIF data [1] and/or from lamp lifetime data.

RESULTS AND CONCLUSIONS

Fig. 2 a) shows the temperature profile along the axis of a linear "stick" electrode at two stages of electrode life, namely, at the start of life; and at the point where 50% of the emitter material remains. Combination of such model-predicted temperature profiles with the kinetic model identifies the set of kinetic parameters that gives agreement between predicted (assumed to be 100% barium loss) and measured electrode lifetimes. Fig. 2 b) is an example output from such a procedure which shows the fractional loss of the initial total barium as a function of lamp burn time using kinetic parameters adjusted to give an electrode (lamp) lifetime of approximately 10,000 h. An independent route to the kinetic parameters is provided by LIF intensity *vs* heater current data given in Ref. 1.

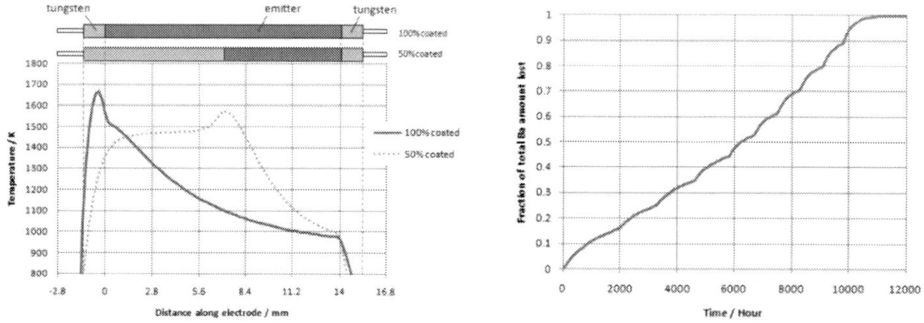

Figure 2. a) Electrode temperature profiles at different stages of life;

Figure 2. b) Fractional loss of total Ba as a function of burn time.

Once a set of kinetic parameters has been identified they can be used then to predict the effect on life of changes in the design of the electrode or the electrode region and/or of changes in lamp operating conditions, e.g., lamp current. Of particular interest is the application of the model in accelerated lamp life testing which provides a critical test of the choice of the kinetic parameters. The model has been used successfully to describe both DC and mains and high frequency AC operated lamps. The electrode sub-model provides a range of useful information such as a detailed electrode energy balance, which is an important element of the lamp's overall efficacy and a sensitive function of electrode design. The model provides also values for the electrode's electrical resistance during lamp life for inclusion in the lamp driver circuit sub-model [7] as well as values of the hot-to-cold resistance ratio, an important and useful measurable quantity in lamp manufacture.

REFERENCES

[1] A. Samir, G. Yamashita, Y. Egashira, S. Shimada, Y. Yamagata, K. Uchino, Y. Manabe and T. Ueda, *IEEJ Trans. FM*, 127 (2007) 543.

[2] S. Bott, R. Forrest, R. Devonshire, E. Arnold and G. Desinger, *Procceedings 9th International Symposium on the Scienec and Technology of Light Sources*, Ithaca, N.Y., 2001, p221.

[3] T. F. Soules, J. H. Ingold, A. K. Bhattacharya and R. H. Springer, *J. Illum. Eng. Soc*, 18 (1989) 81

[4] M. Haverlag, A. Kraus, J. Sormani, J. Heuvelmans, A. Geven, L. Kaldenhoven and G. Heijne, *J. Phys. D: Appl. Phys.*, 35 (2002) 1695.

[5] Yu. B. Golubovskii, S. Hadrath, H. Lange, I. A. Porokhova, F. Sigeneger and J. Ehlbeck, *J. Phys. D: Appl. Phys.*, 39 (2006) 4601.

[6] J. F. Waymouth, "*Electric Discharge Lamps*," M.I.T. Press, Cambridge MA, 1971, p. 107.

[7] K. H. Loo, D.A. Stone, R.C. Tozer and R. Devonshire, *IEEE Trans.Power Electron.*, 20(5) (2005) 1178-1185.

Measured Vapor Pressures in Metal-Halide Systems with and without Complexing Agents

J. J. Curry[1], E. Estupiñán[2], W. P. Lapatovich[3], A. Henins[1], and S.D. Shastri[4]

[1]*National Institute of Standards and Technology, Gaithersburg, MD 20899 USA*
[2]*CRSL, OSRAM SYLVANIA, 71 Cherry Hill Drive, Beverly, MA, 01915 USA*
[3]*CRSL, OSRAM SYLVANIA (retired)*
[4]*Advanced Photon Source, Argonne National Laboratory, Argonne, IL, 60439 USA*
Contact: jjcurry@nist.gov

ABSTRACT

We have measured, using x-ray induced fluorescence, elemental vapor pressures in equilibrium with a few metal-halide salt systems relevant to ceramic metal-halide high-intensity discharge lamps. Vapor pressures of both metals and halides in the systems DyI_3, DyI_3 - InI, TmI_3, and TmI_3 - TlI have been obtained for temperatures up to and exceeding 1300 K. The rare-earth vapor pressures in the pure DyI_3 and TmI_3 systems show good agreement with existing data. The rare-earth vapor pressures in the mixed systems DyI_3 - InI and TmI_3 - TlI show strong enhancement, with the enhancement factor decreasing with temperature. Iodine vapor pressures show evidence of a small amount of water vapor contamination.

BACKGROUND

The measurements presented here were undertaken to obtain a better understanding of the vapor pressures of metal-halide salt systems in the range of cold spot temperatures relevant to ceramic metal-halide high-intensity discharge lamps. Experimental quantification of the effects of hetero-complex formation (e.g., $Dy_mIn_nI_{3m+n}$ in the DyI_3-InI system) on the rare-earth vapor pressures are of particular interest[1]. The use of x-ray induced fluorescence (XRIF) for these measurements was based on the speed with which data can be obtained relative to Knudsen cell techniques, the applicability of XRIF to higher vapor pressures, and the immunity of XRIF to certain systematic effects. XRIF measures elemental densities, N_i, from which elemental vapor pressures, p_i, are defined as

$$p_i = N_i kT = kT \sum_j c_{ji} n_j \tag{1}$$

where the summation is over densities, n_j, of all species, j, containing element i and c_{ji} are the stoichiometric coefficients of element i in molecules of species j. k is Boltzmann's constant and T the measured temperature.

TECHNIQUE

The vapor pressure measurements were made on the Sector 1-ID Beamline at the Advanced Photon Source. Without focusing, the cross section of the monochromatic beam is approximately 1 mm × 1 mm and the beam flux is on the order of 10^{12} ph/s. The experimental arrangement is shown from above in Figure 1. The x-ray beam enters from the left and passes through a set of beam-defining slits, S, an ionization chamber, IC_1, an oven containing the vapor cell, and a second ionization chamber, IC_2. The nitrogen-filled ionization chambers monitor the flux entering and exiting the oven. X-ray fluorescence induced by the synchrotron beam is observed at right angles to the incident beam by a solid-state Ge detector.

The metal-halide salt system whose vapor pressure is to be measured is contained in a cylindrical poly-crystalline alumina (PCA) cell. Each of the cells was dosed with a few mg of the desired salt and 5 torr of Ar. The vapor cell is held in the oven by a two-piece tantalum slug that fits snugly around the cell. Tantalum provides control of the x rays through entrance and exit holes. A third small hole in the tantalum slug and an aperture, A, define the field of view of the detector. The tantalum also ensures a more uniform distribution of temperature in the vapor cell. Measurements were taken with beam energies of 61 keV and 86 keV. The lower beam energy provided larger fluorescence signals because of the larger absorption cross sections, but the higher beam energy allowed us to check for the presence of heavy contaminants. None were found.

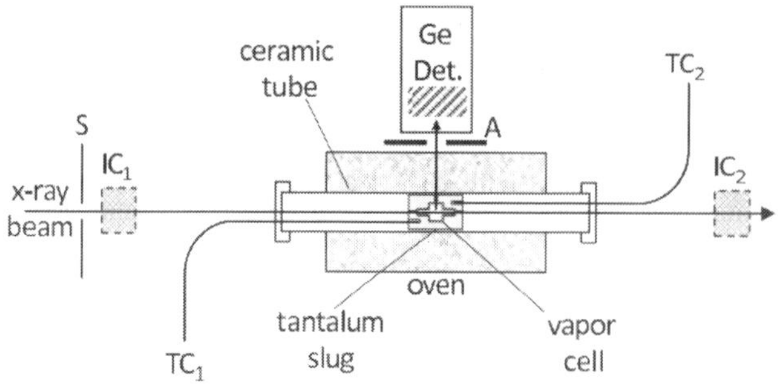

Figure 1. Apparatus for measuring elemental vapor pressures.

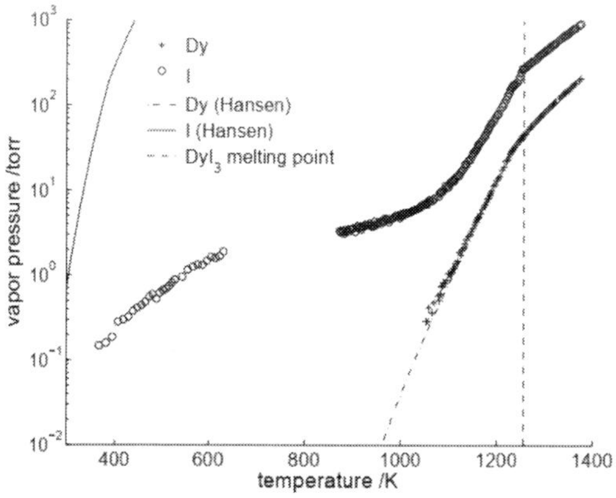

Figure 2. Preliminary data for the pure DyI_3 system compared against data from Hansen et al.[2]

RESULTS

The elemental vapor pressures for Dy, Tm, Tl, In, and I were obtained for the salt systems DyI_3, DyI_3 - InI, TmI_3, and TmI_3 - TlI. Measurements were made over the range of 300 K to 1350 K although vapor pressures below about 100 mtorr generally did not have sufficient signal-to-noise. Figure 2 shows preliminary data from the DyI_3 salt along with comparisons to data from the compilation of Hansen et al.[2] The Dy vapor pressure is in good agreement with previous measurements, but the iodine pressure, particulary at lower temperatures is anomalously high.

ACKNOWLEDGMENTS

We are grateful for the expert technical assistance of Jeff Neil and Michael Quilici in the fabrication and evaluation of these test cells.

REFERENCES

[1] W. P. Lapatovich and J. A. Baglio, Chemical Complexing and Effects on Metal Halide Lamp Performance, Paper O26:I, 9th Intl. Conf. on Sci. and Tech. of Light Sources, Cornell Uni., Ithaca, NY 12-16 Aug. 2001.

[2] S. Hansen, J. Getchius, and T. R. Brumleve, Vapor Pressures of Metal Bromides and Iodides, APL Engineered Materials, Inc., Urbana, IL (1998).

Study on a new material to replace the thoriated-tugsten electrode

Tadao Uetsuki[1], Akira Matsuo[2], Katsuyuki Morii[3], Hisao Hatanaka[3]
[1]*Tsuyama National College of Technology, 624-1 Numa Tsuyama Okayama, 708-8509 Japan*
[2]*Nippon Tungsten Co.,Ltd.,3173-2 Sonobe Kiyama-cho Miyaki-gun Saga, 841-0203 Japan*
[3]*HIMEJI RIKA CO.,LTD.,17 Suginouchi Yumesaki-cho Himeji Hyogo, 671-2111 Japan*
Contact: uetsuki@tsuyama-ct.ac.jp

ABSTRACT

Thoriated-tungsten (ThO_2-W) is popularly used for the electrodes of a super-high pressure mercury lamp, or for arc-welding, because of its excellent performance as to the electron emission. Because it is radioactive, however, a new material which can replace it would be desirable. We have attempted to find a material to replace ThO_2-W as the electrode. We have found that neodymium oxide or samarium oxide made under certain conditions produces encouraging results.

INTRODUCTION

ThO_2-W is popularly used for the electrodes of a super-high pressure mercury lamp used in semi-conductor processing, or for arc-welding. This is because the work function of ThO_2-W is much lower than that of the tungsten, therefore its electron emission performance is excellent compared with tungsten. This lower work function leads to lower cathode temperatures for the electron emission, thus improving the energy consumption and the life of the lamp. However, ThO_2-W is not desirable for the global environment because of its radioactivity. Therefore it would be very useful to make a new non-radioactive material to replace ThO_2-W as the electrode, and many research efforts have been attempted for this purpose. However, the materials which were researched till now could not improve upon ThO_2-W lifetime when they were used as the electrode in the lamp. This means these materials have higher work function than ThO_2-W. The higher work function causes higher electrode temperatures, leading to faster evaporation of the electrode material and shorter lamp life.

The purpose of our research is to make a new electrode material having better performance than ThO_2-W. We have studied which oxide is best by controlling the particle radius or the volume of metal oxides mixed into the tungsten. In this paper we report the results obtained by studying neodymium-oxide-tungsten and samarium-oxide-tungsten as the electrodes of a super-high pressure mercury lamp used for the semi-conductor processing.

EXPERIMENTAL RESULTS AND DISCUSSION

Fig.1 shows the dependence of the lamp voltage on the operating time, and Fig.2 shows the dependence of the UV output on the operating time. The increasing rate of the lamp voltage indicates electrode evaporation, and the decreasing rate of the UV output indicates the transparency of the lamp wall by a deposit of the vaporized electrode material. The data obtained from Fig.1 and Fig.2 reveal the following three points. The first is that the pure tungsten (pure-W) electrode and the samarium oxide tungsten (5%Sm_2O_3-W) electrode evaporate more heavily than the ThO_2-W electrode. The second point is that the neodymium oxide tungsten (5%Nd_2O_3-W) electrode evaporates at the same rate as the ThO_2-W electrode. The third point is that the samarium oxide tungsten (2%Sm_2O_3-W) electrode and the neodymium oxide tungsten (2%Nd_2O_3-W) electrode evaporate less than the ThO_2-W electrode.

Fig.3 shows photographs of the dependence of the electrode appearances on the operating time. It can be seen that there is no discrepancy between these results and the discussion results above.

Fig.4 shows the temperature distributions on the electrodes. These data show the following points. The temperature gradient is steep and the tip temperature is high for the electrodes whose evaporating rates are high, such as the pure-W electrode and the 5%Sm_2O_3-W electrode. On the other hand, the temperature gradient is gentle and the tip temperature is low for the electrodes whose evaporating rates are low, such as the 2%Sm_2O_3-W electrode. It can be considered from these results that the electron emission area of the pure-W electrode and the 5%Sm_2O_3-W electrode is small, and that he electron emission area of the 2%Sm_2O_3-W electrode is large.

CONCLUSION

We have searched for a new material which can replace the ThO_2-W as the electrode. We found that the 2%Nd_2O_3-W electrode and the 2%Sm_2O_3-W electrode bring about encouraging results. We have found that the volume and size of the oxide mixed into tungsten is very important as well as the kind of oxide used.

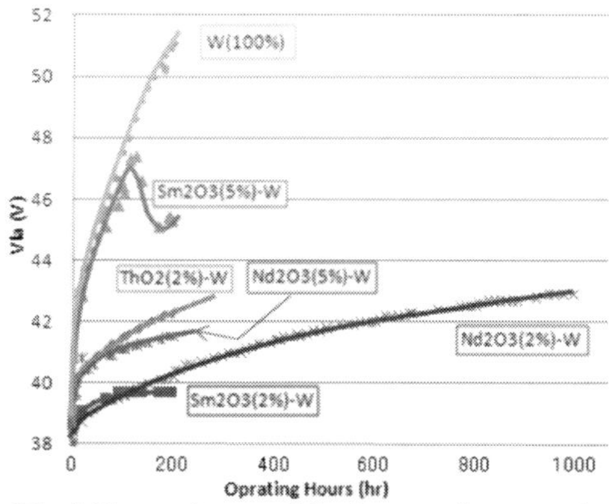

Fig.1 Dependence of the lamp voltage on the operating time

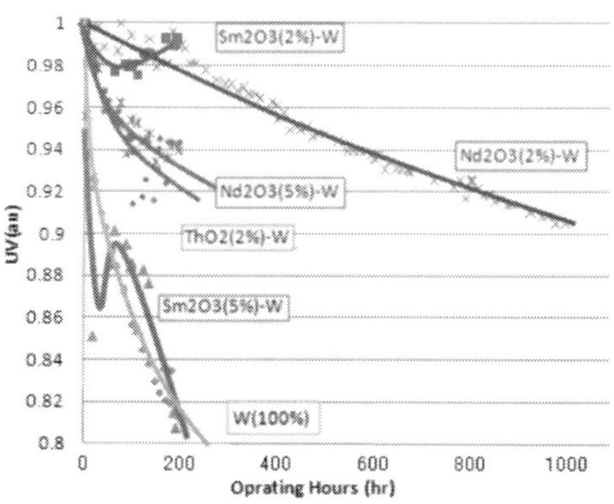

Fig.2 Dependence of the UV output on the operating time

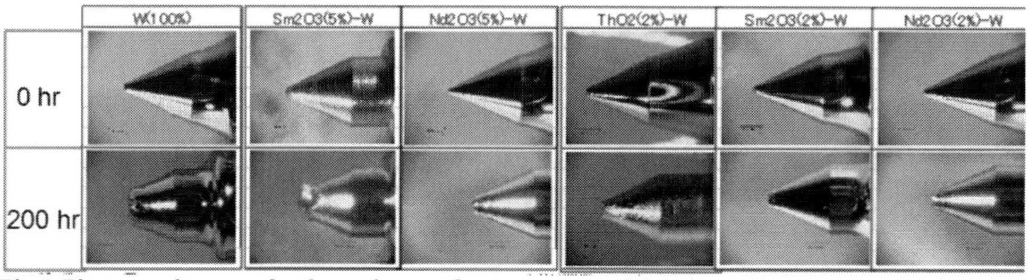

Fig.3 Photographs as to the dependence of appearances of the electrodes on the operating time

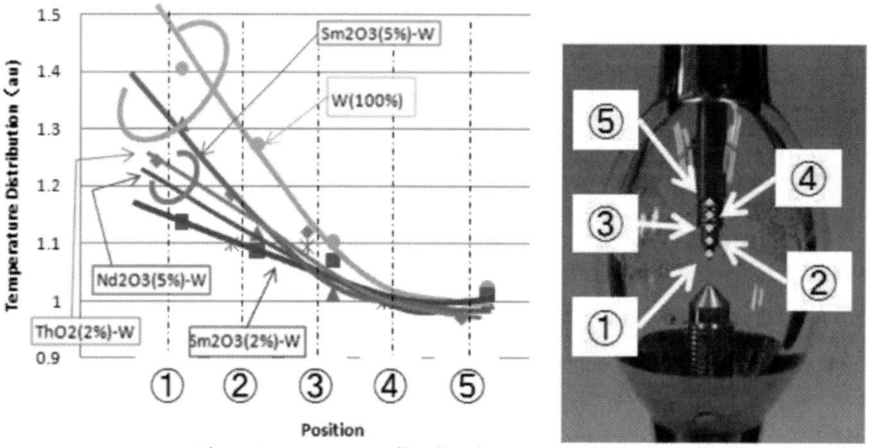

Fig.4 Temperature distributions on the electrode

Modelling mercury-free HID lamps: Breakdown characteristics and thermodynamics

N. Y. Babaeva[1], A. Sato[2] , N. Brates[2], K. Noro[2], and M. J. Kushner[1]
[1]*University of Michigan, EECS Department, 1301 Beal Ave, Ann Arbor, MI 48109 USA*
[2]*Universal Lighting Technologies Inc., Woburn, MA 01801 USA*
Contact: mjkush@umich.edu

ABSTRACT

The modelling of mercury free HID lamps is challenging due to the complexity of gas mixtures and sensitivity of breakdown characteristics to the initial physical conditions. In this paper we report on results from a computational investigation of breakdown characteristics and the plasma properties of mercury-free HID lamps having metal halide fills. We discuss breakdown time as a function of voltage rise times, polarities and plasma composition (e.g., hot and cold restart). We also discuss the thermodynamic properties for doses containing $Xe/NaI/ScI_3/ZnI_2$. The simulator used in this work, *nonPDPSIM*, is a plasma hydrodynamics model in which continuity, momentum and energy equations are solved for charged species with solution of Poisson's equation. The model is coupled with a thermodynamics module providing local thermodynamic equilibrium (LTE) properties. Algorithms were developed to represent the transition of the lamp from a kinetics-Poisson regime during breakdown to an LTE-ambipolar regime as the arc begins to form.

BREAKDOWN CHARACTERISTICS

The breakdown process of HID lamps can be characterized by a statistical and a formative time lag. One mode of breakdown requires a high ignition voltage and is characterized by initial plasma formation in the volume of the lamp between the electrode tips. Another ignition mode is characterized by streamer propagating along the wall. In this case, the lamp breaks down at lower voltages.

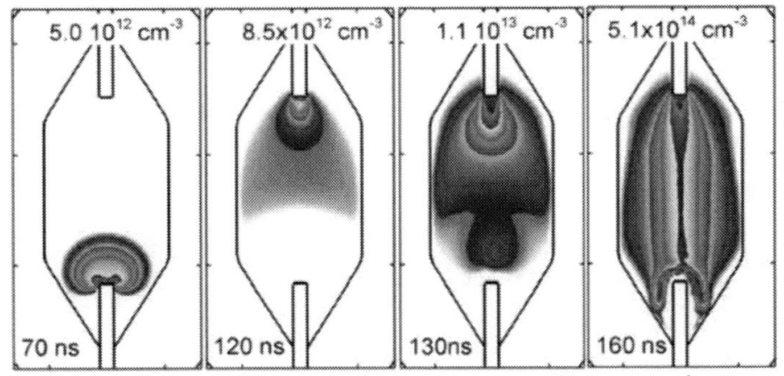

Figure 1. Electron density (3 dec) at 4 times. Negative polarity is applied to bottom electrode. The voltage rise time dV/dt=150 V/ns. Breakdown time lag is mainly determined by the time-of-flight of seed electrons.

Dosed HID lamps often have a condensed and mildly conductive salt layer on the interior of one side of the lamp. In the absence of the salt layer, the formative breakdown time is determined by the time-of-flight of seed electrons produced in the vicinity of the cathode. We studied the formative breakdown time-lag in a lamp of D4 geometry having a xenon gas fill of 8 atm by injecting electrons into the gap resulting from roughness enhanced field emission from the negative (relative to plasma) electrode. The D4 lamp is 2.7 mm in diameter with an electrode spacing of 4.2\ mm. After emission from the cathode, the electron cloud drifts towards the opposite electrode. Plasma parameters for a negative pulse with dV/dt=150 V/ns applied to the bottom electrode are shown in Figure 1. The total voltage span is 30 kV. Upon reaching the positive upper electrode, the cloud initiates a positive streamer toward the negative electrode. The positive streamer propagates downward towards the cathode where it meets the opposite negative streamer and breakdown occurs. For these conditions the cloud drifts in rapidly increasing electric field during the first 50 ns of flight (first and second frames). The breakdown then rapidly develops during the following 20-30 ns (second and third frames).

The existence of condensed salt layers on the inner wall of the HID lamp and on the electrodes influences the ignition behaviour of the lamp. For example, we found that streamers tend to propagate along the inner wall with the salt deposits. This is due to the conductive nature of the salt layers that focuses the electric fields produced by the electrodes. For example, the effect of a conductive salt layer on breakdown time is shown in Fig. 2. The operating conditions are as in Fig. 1. The salt layer has a thickness of a few microns with conductivity $10^{-3}\ \Omega^{-1}\ cm^{-1}$. The breakdown generally occurs more rapidly and asymmetrically. The effect is more pronounced when the salts are deposited as separate islands (as opposed to a uniform film) due to local field enhancement near the islands' edges.

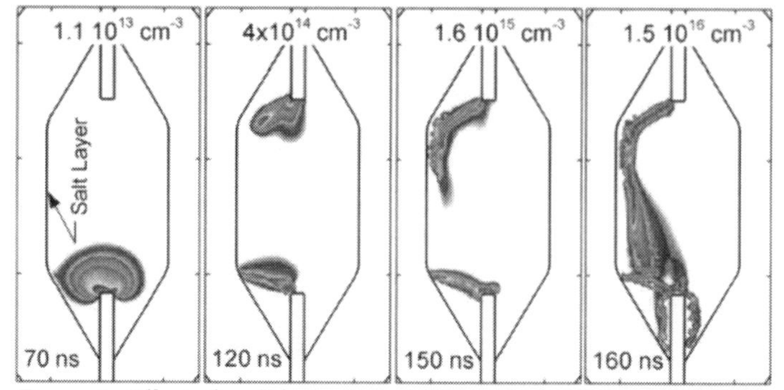

Figure 2. Effect of salt layer on breakdown time. Electron density (2 dec) is shown at 4 successive time moments Conditions as in figure1. Breakdown time lag is determined by time-of-flight of seed electrons as well as by faster development of streamer due to enhancement of electric filed by the salt layer.

THERMODYNAMICS

In mercury-free HID lamps, the mercury is often replaced by ZnI_2. This substitution, along with the use of conventional metal halides such as NaI and ScI_3, add both complexity and potential variability to the system. For example, it has been experimentally shown that the lamp operating voltage can be controlled by varying the amount of ZnI_2. To develop models of such lamps assuming LTE, knowledge of the thermodynamic properties of both dose substances and plasma composition are required. Calculation of the plasma composition does not necessarily use reaction rate constants when detailed balancing between direct and reverse reactions is used. Detailed energy balance modelling, however, requires data for enthalpy, entropy and heat capacitance of species [1]. For example, the LTE plasma composition for a D4 lamp is shown in Fig. 3 as a function of temperature. The fill of the lamp in relative mole fractions (gas phase) is $Xe/NaI/ScI_3/ZnI_2 = 1/3.16\times10^{-4}/4.63\times10^{-5}/4.48\times10^{-5}$. The effects of mixing, segregation and ionization of light and heavy additives on thermal, electrical conductivity and I-V characteristics will be discussed. The thermodynamic database constructed for these doses will be also discussed [1].

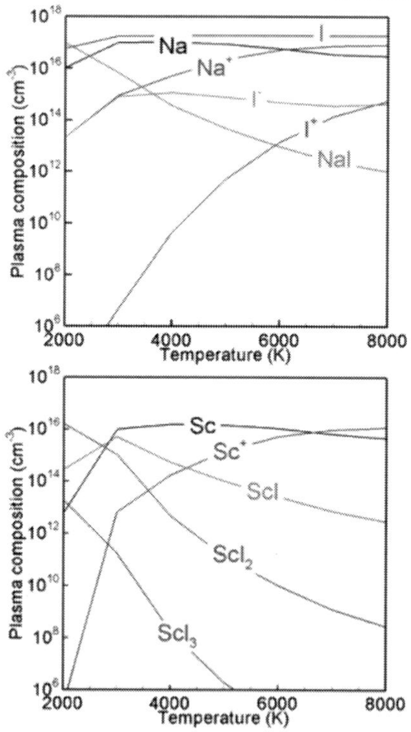

Figure 3. D4 ($Xe/NaI/ScI_3/ZnI_2$ fill) plasma composition vs temperature (a) NaI molecule and its dissociation fractions , (b) ScI_3 molecule and its dissociation fractions. Na is almost totally ionized at high temperatures, ionization of Sc occurs at higher temperatures.

REFERENCES

[1] M.W. Chase et. al., JANAF Thermochemical Tables. Fourth Edition. Part I and II. (1992).

* Work supported by Universal Lighting Technologies, Inc.

Sputtering yields of alkaline earth oxides on W electrodes of a fluorescent lamp

S. Gotoh[1] , T. Kenmotsu[2], M. Wada[3]
[1]*Panasonic Electric Works, Kadoma, Osaka 571-8686, Japan*
[2]*Faculty of Life and Medical Sciences, Doshisha University, Kyoto 610-0321, Japan*
[3]*Graduate School of Engineering, Doshisha University, Kyoto 610-0321, Japan*
Contact: mwada@mail.doshisha.ac.jp

ABSTRACT

Effects due to surface structure upon sputtering yields of materials coated on electrodes of a fluorescent lamp have been evaluated with a sputtering simulation code ACAT (Atomic Collisions in Amorphous Target). The calculated sputtering yields largely differ depending upon whether alkaline earth oxides reside on tungsten surface forming a thin layer, or they constitute a volume of powders. They also show a strong dependence upon the projectile mass, particularly at very low incident energy. Possible mechanisms leading to loss of oxides from a fluorescent lamp electrode is discussed.

ELECTRODE GEOMETRY OF A FLUORESCENT LAMP

Alkaline earth oxides coat the surface of electrodes of low pressure Hg-rare gas discharge lamps to enhance electron emission from the electrodes. Admixture of alkaline earth powders used for electrodes of low pressure discharge lamps is called emitter, and the elements in the powder are to be sputtered by ion bombardments in plasma. The sputtering yields for any projectile-target pair can be computed by means of a simple formula with tabulated constants [1]. However, as emitter materials are in the form of powder, they fill the region in between the coiled tungsten (W) filament rather than coating the surface of metallic electrodes. A typical condition of a W filament electrode impregnated with emitter is shown in Fig. 1 (a). In Fig. 1 (b) is shown the magnified view of the filament electrode indicating two regions are present on the electrode. In one region, the tungsten filament is exposed to the plasma. In the other region, emitter powder confronts the plasma, without any metallic components present beneath the powder layer. The difference between these two structures may cause different loss rates of emitter by charged particle incidence to these parts of electrodes. For this kind of problem, the exact trajectory of incident ion and the knocked-off particles forming the collision cascade should be computed to obtain resultant sputtering yield. Thus, sputtering yields for some typical projectile-target pair with two different conditions of emitter materials were computed by running the ACAT code based on the Monte-Carlo method [2].

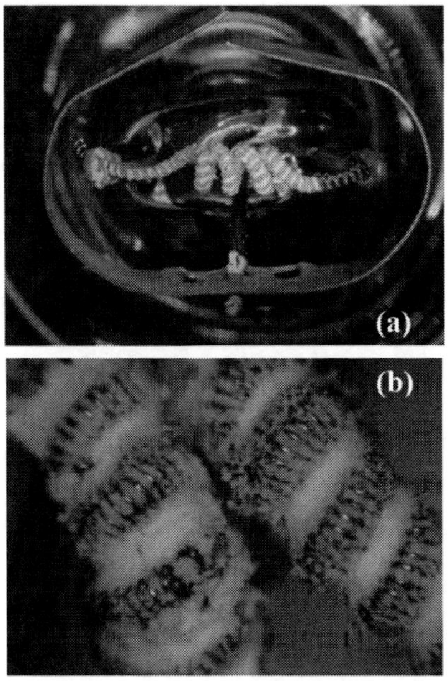

Fig. 1. Pictures showing emitter in a W electrode. (a) Overall view of an electrode impregnating alkaline earth oxide cathode materials. (b) Magnified image of the emitter impregnated tungsten discharge cathode of a fluorescent lamp.

ACAT CALCULATIONS

Two different geometries are employed for ACAT calculation. In one geometry, the surface of W is covered by a layer of oxygen, and then covered by a layer of Ba. Namely, BaO forms a double layer on top of W. This surface condition will be expressed by the symbol (Ba+O+W). A homogeneous region of stacked BaO powder is the other geometry. This surface geometry is specified with the symbol (BaO). Letting Ar^+ and Hg^+ as the projectiles, sputtering yields are calculated for these two geometries. In addition to the geometrical difference, the surface binding energy, or the energy necessary to leave the surface, had been assumed to be that of pure Ba for the case of (Ba+O+W), and that of binding energy of BaO for the case of (BaO). The incoming projectiles are assumed to strikes the surface of the target at the right angle. Namely, the incident ions are injected toward the direction of surface normal. The results are summarized in Fig. 2.

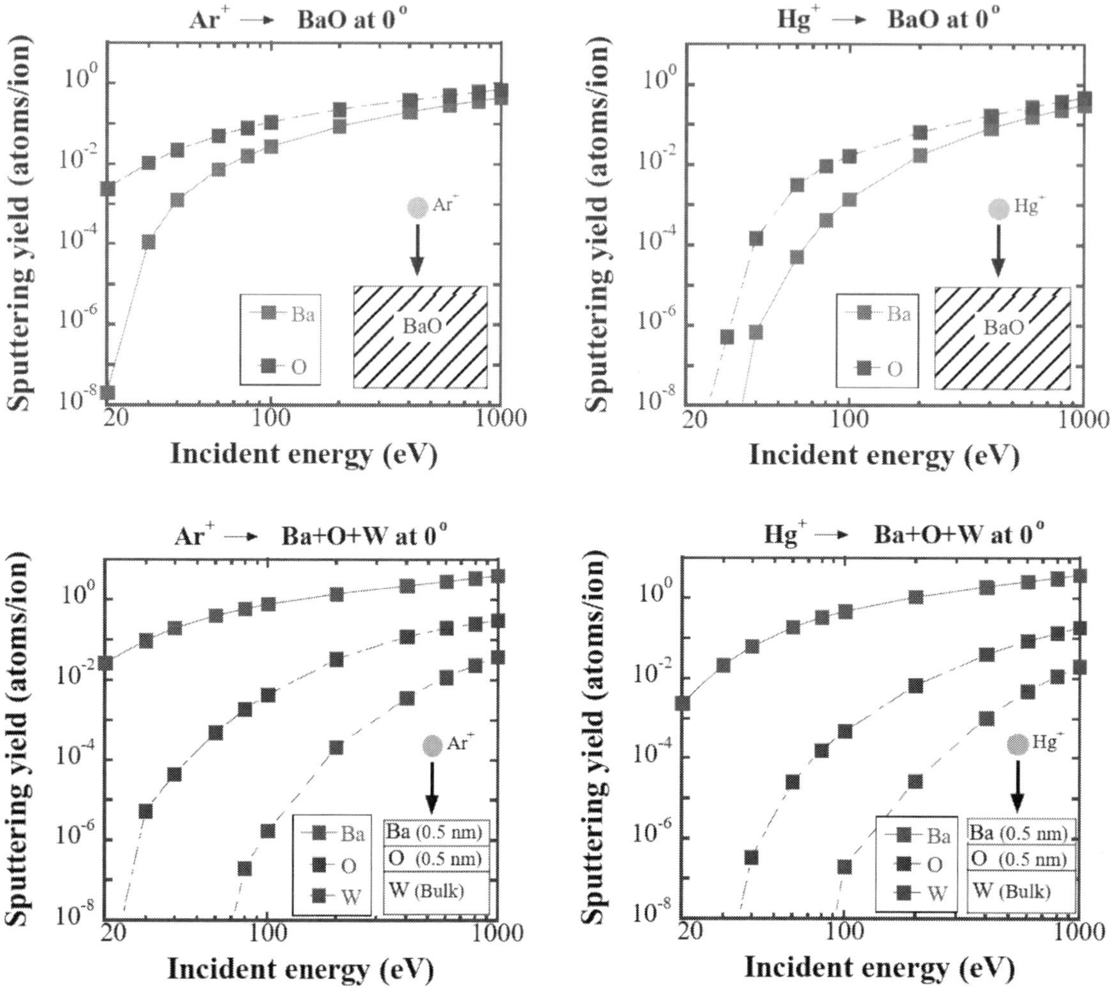

Fig. 2. Sputtering yields calculated by the ACAT code for two different geometry of (Ba+O+W), or the layered target formed on the W electrode, and (BaO), or the bulk BaO volume.

SUMMARY

The sputtering yields of emitter materials for the actual electrode conditions in a fluorescent lamp had been shown to differ, depending upon the geometry of the emitter coverage on the W filament. For the condition that thin layer of BaO adheres on W, Ba will be lost rapidly. As the oxygen rich condition reduces the binding energy of the residual oxide [3], the Ba depleted layer will be lost rapidly as Ba atoms are sputtered out of the outer most surface of the layered electrode. On the contrary, the bulk BaO system will lose O by Ar^+ bombardment. The residual atomic Ba will be lost by vaporization, and again the loss rate can be greater than the value assumed from the pure element system. The validity of the present model should be carefully re-examined based upon experimental data, but the importance of the surface condition, which had been overlooked in the past study, should be taken into account in order to properly estimate the emitter loss due to sputtering.

REFERENCES

[1] Y. Yamamura, H. Tawara, Nuclear Data Tables, **62**, 149, (1996).
[2] Y. Yamamura, Y. Mizuno, IIPJ-AM-40, Inst. Plasma Physics, Nagoya Univ. (1985).
[3] V. Boutou, M.A. Lebeault-Dorget, A.R. Allouche, C. Bordas, J. Chevaleyre, Z. Phys. D 40, 448 (1997).

An Appropriate Arrangement of Multiple LEDs for Optimal Power Need

Guan-Chyun Hsieh[1] , Cheng-Chih Chu[2]
[1] *Department of Electrical Engineering, Chung Yuan Christian University, Chung-Li, Taiwan 32023, R.O.C.*
[2] *Grenergy Opto, Inc., Hsin-Chu, Taiwan 300, R.O.C.*
Contact: gchsieh@dec.ee.cycu.edu.tw

ABSTRACT

The arrangement of multiple LEDs (multi-LEDs) for large-scale LED panels or lighting apparatuses has been very anxious to study the issue. Means nothing more than commonly used in series, parallel, and/or both co-existing modes, but they are based on individual thoughts to another. No conclusion that what is most appropriate. In this paper, by mean-value theorem an optimal operation point is derived to find appropriate combinations for multiple LEDs (multi-LEDs) as well as proper power need for design. The combination modes of multi-LEDs can be easily estimated by only considering the number of LEDs. Experiments show satisfactory results with reasonable power design; all apparent in this method is correct and practical.

INTRODUCTION

In large-scale LCD panels or lighting apparatuses, LEDs displays for backlighting or lighting may need a power to drive hundreds of LEDs in modes of series and/or parallel strings [2]. Such as the use of all in series mode, a string of higher number of LEDs requires very high-voltage supply with very small operation current. On the contrary, if taken all in parallel mode, huge number strings of LEDs connected in parallel will need a power supply with large current and very low operating voltage. However, the mentioned methods may cause difficulties in the design of power supply and impracticability. Literature on the arrangement of multi-LEDs is scarce and scattered. Most literatures are only to promote the driving strategies but few for considering the arrangement for multi-LEDs [1-3]. The main purpose of this paper is to find optimal combinations of multi-LEDs and power need for design rationalizing for stabilizing the LED luminance.

MODELING AND ANALYSIS OF OPTIMAL LEDS ARRANGEMENT

Generally, the power consideration of LED can only refer to its rated power for placing its appropriate operating point not over safe-operating area (SOA). For multiple LEDs (multi-LEDs), the basic reference mentioned above is still valid. Interestingly, if multi-LEDs, such as N LEDs, are used simply in the form of all in series or all in parallel connection, a power dissipation curve of multi-LEDs can then be built and easily described by

$$P_D = NV_D \cdot I_D \equiv V_{\max} I_{\min} = V_D \cdot NI_D \equiv V_{\min} I_{\max} \quad (1)$$

Where P_D is the power dissipation by N LEDs, V_D and I_D respectively the forward voltage and operation current of a single LED; N the number of LEDs; define $V_{max}=NV_D$, $V_{min}=V_D$, $I_{max}=NI_D$, and $I_{min}=I_D$. As shown in Figure 1, we can easily plot the power dissipation (PD) curve of multi-LEDs in the I_D-V_D plane to describe their power behaviours according to (1). Then, we can build lots of possible combinations of multi-LEDs along the profile of PD curve in any kind of combinations such as series and parallel strings. Accordingly, the physical arrangement models are predicted in Figure 2, in which the number of parallel strings is always the same as that of LEDs in each string. In order to gain an appropriate operation mode for multi-LEDs and find proper power need for design, a strategy to derive average conductance of multi-LEDs for searching an optimal operation point is proposed. Mean-value theorem is then used to derive our theme along thc profile of PD curve [4]. Based on P_D curve, the multi-LEDs current i_D can then be represented in terms of voltage v_D and given by

$$i_D = f(v_D) = P_D/v_D \quad (2)$$

From (2), since $f(v_D)$ is continuous in $[V_{min}, V_{max}]$ and differentiable in (V_{min}, V_{max}), there exists some point $v_{Dc} \in (V_{\min}, V_{\max})$ such that [4]

$$f(V_{\max}) - f(V_{\min}) = f'(v_{Dc})(V_{\max} - V_{\min}) \quad (3)$$

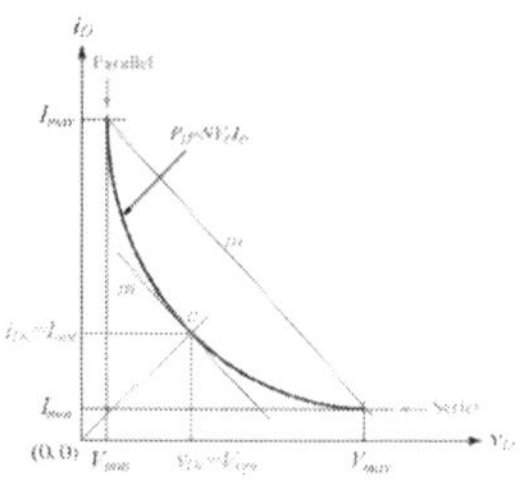

Figure 1 Power dissipation curve of multiple LEDs for finding series-parallel combination in LED arrangement

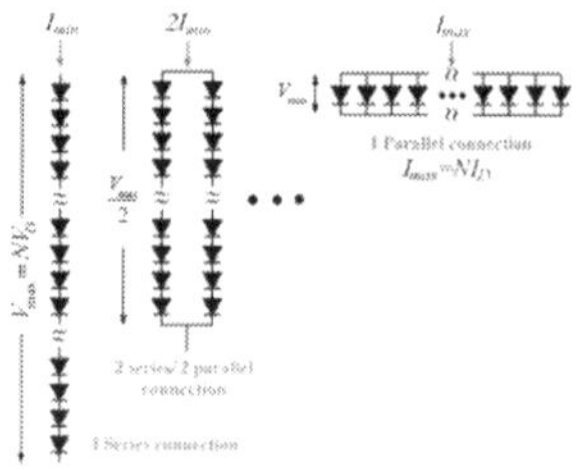

Figure 2 Predicted series/parallel combinations for optimal LEDs arrangement

From (2) and (3), we have

$$f'(v_{Dc}) = -P_D / v_{Dc}^2 \quad (4)$$

$$V_{opt} = \sqrt{V_{\max} V_{\min}} \,. \quad (5)$$

$$I_{opt} = \sqrt{I_{\max} I_{\min}} \,. \quad (6)$$

where $v_{Dc} \equiv V_{opt}$. We specify the point c on PD curve as the operating point of multi-LEDs given by (V_{opt}, I_{opt}), where is the reference to determine the power need. If the appropriate combination consisting of q parallel strings and p LEDs in each string, the power need is then given by $V_{opt} = \sqrt{V_{\max} V_{\min}} = pV_D$ and $I_{opt} = \sqrt{I_{\max} I_{\min}} = qI_D$. A simple expression is obtained as $p = q = \sqrt{N}$, which interestingly shows that the appropriate arrangement can be easily estimated from the number of LEDs, in which the number of parallel strings and that of LEDs in each string are actually the same.

REALIZATION AND EXPERIMENTS

A design example of LED display with area 41cm×31cm = 1271cm^2 is to be fulfilled about 600 LEDs, according to the size of the specified white LED (67-21UWC/S400-X9/TR10). After calculation in real area, the possible quantity of LEDs is 588. The practical photos are shown in Figure 3.

(a) (b) (c)

Figure 5 Practical design of arrangement of multi-LEDs by proposed strategy: (a) front panel, (b) rear panel for layout strategy, and (c) power supply for drives

CONCLUSION

In this study, the LED optimum operating point is determined by mean-value theorem, which gives the appropriate combination of LEDs and the power need for design. The estimation of the LED combinations modes can be easily obtained by accounting the number of LEDs.

ACKNOWLEDEGMENT

The authors would like to thank the National Science Council for his sponsor in this project, NSC98-2221-E-033-071-MY2

REFERENCES

[1] M. Nishikawa, Y. Ishizuka, H. Matsuo, and K. Shigematsu, "An LED drive circuit with constant-output-current control and constant-luminance control," in *Proc. IEEE Int. Telecommun. Energy Conf.*, 2006, pp. 1–6.

[2] C. C. Chen, C. Y. Wu, Y. M. Chen, and T. F. Wu, "Sequential color LED backlight driving system for LCD panels," *IEEE Trans. Power Electron.*, vol. 22, no. 3, pp. 919–925, May 2007.

[3] W. K. Lun, *et al*, "Bilevel Current Driving Technique for LEDs." *IEEE Trans. Power Electron.*, vol. 24, no. 12, pp. 2920-2932, 2009.

[4] G. B. Thomas, M. D. Weir, J. Hass, and F. R. Giordano, *Thomas' Calculus*, 11th ed., 2008.

COST Action MP0602: High Temperature Lead-Free Solders (HISOLD)

S.A. Mucklejohn[1], Z. Tóth[2] and N.Hoo[3]

[1] *Ceravision Limited, The Mansion, Bletchley Park, Buckinghamshire MK3 6EB, UK*

[2] *GE Hungary, Vaci ut 77, Budapest, H-1044, Hungary*

[3] *ITRI, Unit 3, Curo Park, Frogmore, St.Albans, Hertfordshire AL2 2DD, UK*

Contact: stuart.mucklejohn@ceravision.com

ABSTRACT

The paper describes the COST Action MP0602, which was established to seek lead-free replacements for high melting temperature solders (T_m >230°C) used in High Intensity Discharge lamps and printed circuit boards for various types of light sources.

INTRODUCTION

EU directives 2002/95/EC (Waste from Electrical and Electronic Equipment, the WEEE directive) and 2002/95/EC (Restriction of the use of certain Hazardous Substances in electrical and electronic equipment, the RoHS directive) restrict the use of lead-containing solders in electrical and electronic equipment to a few well-defined exemptions. One such an example is lead in high melting temperature type solders (i.e. Sn-Pb solder alloys containing more than 85% Pb). These exemptions are subject to review, usually every 4 years. Lighting equipment is specifically included in both the WEEE and RoHS directives. A typical example is High Intensity Discharge (HID) lamps. Solders are used both in the base as well as side of the lamp cap to connect the lamps' electrical feedthroughs to the base (Figure 1.). The maximum allowed temperatures in these points are in the range of 190-250°C, depending on the cap type (specified in IEC 60662). Printed circuit boards (PCB) in the electronics for energy saving fluorescent or HID lamps as well as LEDs also employ high lead-content solders. Other applications benefiting from the use of lead-free solders besides the lighting industry will include the electronics industry, power circuits and the automotive industry.

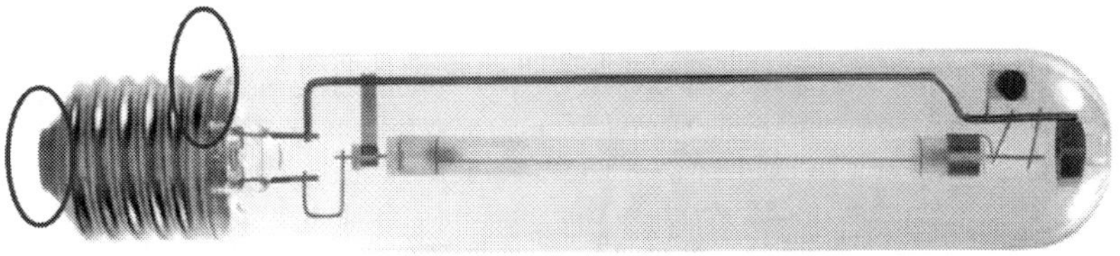

Figure 1. Photograph of a typical HID lamp showing the base and side connections with solder

AIM AND OPERATION OF THE ACTION

COST (European Cooperation in Science and Technology) is a European funding program supporting cooperation among scientists and researchers [1]. Action MP0602 has been established to seek lead-free replacements for high melting temperature solders (T_m >230°C). The full title of the Action 'Advanced Solder Materials for High Temperature Application – their nature, design, process and control in a multi-scale domain' reflects the wide range of topics covered. Groups involved in the Action are undertaking: the development of a database of materials properties (the meso-scale); the study of the properties of solder joints (the macro-scale); the investigation of the characteristics of the solder - substrate interface (the micro- and nano-scale). The main objective of the Action is, through investigations on the meso-, macro- and micro-scale, to increase the fundamental knowledge of the crucial properties of alloys that might be suitable for use as environmentally friendly lead-free alternatives to existing high melting temperature solders. The aim is to identify promising materials with a set of suitable properties, such as melting temperature, wettability and surface tension, which will allow them to be used successfully in a variety of industrial applications. Furthermore, health and economic issues will be taken into account during the evaluation in addition to physical, chemical and mechanical behaviour. An outline of the work to be conducted during this COST Action, taken from the Memorandum of Understanding, is given below.

'On the meso-scale: The application of thermodynamics and kinetics of multicomponent systems and the establishment of materials property databases for Pb-free alloy systems suitable for high-temperature solder applications. The extent of existing information about key Pb-free systems up to now deemed suitable for high-temperature soldering is limited and in many cases inconsistent. The aim is to compile a set of databases containing information on phase diagrams, thermodynamic, structural, physical, chemical, electrical, mechanical and process related properties of possible solders and corresponding joint materials.

On the macro-scale: The creation of a phenomenological description and models for the prediction of corrosion behaviour, deformation processes, failure modes and other mechanically related problems occurring in the soldered structure during fabrication and service at high temperatures. In particular, this COST Action will develop processing structure-property relations, an understanding of thermo-mechanical fatigue, scale and constraining effects of the thermo-mechanical response, the durability of interfaces and intermetallics and identify optimum process conditions.

On the micro- and submicro-scale: Reactive phase formation remains the most mysterious and least understood problem in solid-state science. This involves the formation of intermetallic compounds at solder-substrate interfaces, the development of texture in the reaction products in concentration gradients and the development of defect structures in the vicinity of the reaction interface. The existence of sharp concentration gradients and huge diffusion fluxes in the proximity of the interphase interface (in the region of nucleation) changes the thermodynamics as well as the very notion of critical nucleus and incubation time.'

During 2008 and 2009 it became clear that some alloys had been identified with suitable melting temperatures to be considered as replacements for conventional lead-containing high melting temperature solders. A team was therefore formed with the task of defining a protocol for the quantitative evaluation of alloys to be used as replacements for high temperature lead-containing solders in products and industrial processes.

The main objective of this protocol is to help industry, especially small & medium sized enterprises (SMEs), appreciate the risks and the effort needed to validate material substitutions into existing products and processes. The protocol is based upon the need to meet the expectations of customers and end users so that any material substitution does not result in a lowering of product performance or reliability. These customer expectations may be summarised by: Unchanged performance; Unchanged reliability; Unchanged product availability; Unchanged end of life mechanisms; Unchanged failure modes; No detriment to visual appearance; Small cost difference. After the any material substitution, the product must be completely compatible with current fittings & uses.

The material substitution must be able to withstand all stage of the product lifecycle from manufacture through quality assurance testing, packaging, shipping, storage, distribution, installation, operation, maintenance, replacement to disposal, recycling and reuse. The protocol has 3 three sections: Design verification; Production verification; Product compliance with EU directives and international standards.

The Chairman of COST Action MP0602 is Ales Kroupa from the Academy of Sciences, Brno, Czech Republic, the Vice-chairman is Andy Watson from the University of Leeds, UK. More details about this COST Action can be found on the website [2].

REFERENCES

[1] http://www.cost.esf.org/
[2] http://cost602.ipm.cz

Towards a System Level Reliability Approach for Solid State Lighting

F.E. Evertz[1], W.D. van Driel[1,2] , J. Kloosterman[1], G. Vanlier[1], G.Q. Zhang[1,2]
[1]*Philips Lighting, Mathildelaan 1, 5611BD Eindhoven, The Netherlands*
[2]*Delft University of Technology, Mekelweg , 2628CD Delft, The Netherlands*
Contact: willem.van.driel@philips.com

ABSTRACT

Reliability is an important aspect of LED-based systems and modules, as long lifetime is an important selling argument. Besides, LED modules are specific solutions for specific applications. As a result of the strong customization, reliability estimations for LED-based modules start practically from scratch. Therefore there is a need for a generic system level reliability approach for lifetime and reliability estimations of LED systems. This paper describes our approach to determine system reliability based on the known failure distributions of the sub-components.

INTRODUCTION

In engineering, reliability is the ability of a system or component to perform its required functions under stated conditions for a specified period of time. It is often reported in terms of a probability [1, 2]. It is already very challenging to understand and predict the reliability of a macro-level system, because reliability is always a multi-disciplinary issue, and strongly associated with materials, design, manufacturing process, testing and application conditions. System reliability, herein, mainly addresses the reliability of a number of micro/nano components to form a system, is even more challenging. Given the 20.000 hours lifetime specification for Solid State Lighting (SSL) systems, design for system reliability is necessary in order to comply with the high demands and severe requirements of LED systems. Typical LED failure modes for the sub-components are:

- On Level 0: Bare LED
  - Degradation of active region
  - Electro-migration causing dislocations
  - Diffusion of metal atoms to the active region
  - Current crowding (uneven current distribution)
- On Level 1: Packaged LED
  - Yellowing of packaging materials (degradation/aging)
  - Electrostatic discharge (ESD)
  - Interconnect failure (solder or die-attach)
  - Cracks (f.e. vertical die crack)
  - Delamination (at any interface)
  - Wire bond failure
- On Level 2: LED on substrate
  - Cracks (f.e. in the ceramic)
  - Solder fatigue
  - PCB metallization problem (mis-alignment, crack, fatigue)
- On Level 3: SSL system / module
  - Casing cracks
  - Driver failures
  - Optic degradation (browning, cracks, reflection change)

In practice, a malfunction of the system might be induced by the failure and/or degradation of the subsystems/interfaces. Extra costs, in terms of exceed efforts/designs/parts, has been applied to the system in order to secure the guaranteed reliability performance of SSL system. The most SSL system designs allow few failures of the subsystem/interface during the application period. Hence, a significant cost reduction can be achieved when the system level reliability is well understood by proper experimental and/or simulation techniques.

SYSTEM LEVEL APPRAOCH

The major challenge to generate a generic system level approach for solid state lighting is the broad range of products and applications [3]. The user conditions and requirement for a retrofit bulb are different than those for a streetlight luminary or a freezer module. For a SSL system reliability prediction the knowledge of the application conditions, the system architecture and the failure modes of the sub-components should be combined into a system reliability model. To create such a model we have developed and successfully applied the following approach, see Figure 1:

1. **Define use conditions**

Environmental conditions like temperature, humidity, vibrations. But also application conditions like switching behaviour.

2. **Define the required function**

Retrofit lamps or modules require a minimal lumen output; further requirements on colour consistency can be defined. Specific modules can have additional requirements.

3. **Decompose the system**
    - Identify component failure modes
    - Translate stated system conditions to component conditions
4. **Gather data on components failures**

By studying literature, information of suppliers and tests on component and sub-system level, failure statistics and degradation models of the different components are created.

5. **Combine component failures to system failures**

To predict system failures the failure distributions and degradation models should be combined. The challenge is to properly combine the so called catastrophic failures, which provide the distribution of the time of failure of a component with the degradation models, which provide the distribution of parameter (e.g. flux or reflectivity) at a given time [4, 5, 6].

6. **Lifetime statement**

Once the system reliability has been created it can be used for lifetime statements. For example B50L70 is 35,000 hours: after 35,000 hours 50% of the systems have a lumen maintenance of 70%.
We have applied this approach to a number of SSL systems, among which are retrofit bulbs and office lighting modules. In future we plan to apply it to an even more challenging application: automotive.

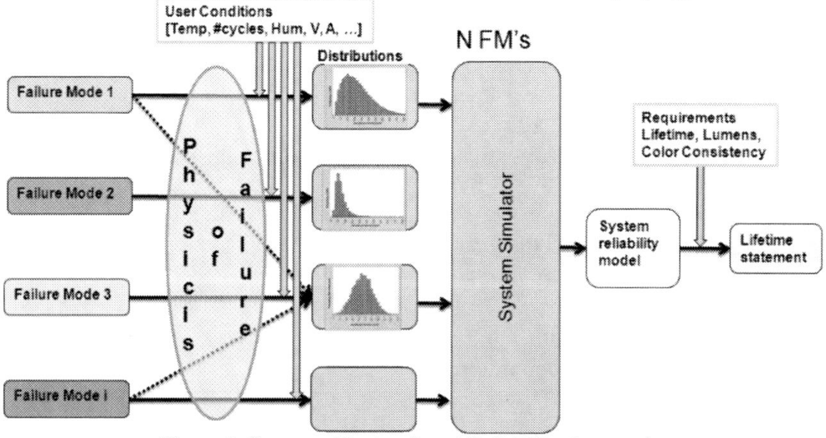

Figure 1: Process of System Level Reliability Approach

REFERENCES

[1] http://en.wikipedia.org/wiki/Reliability

[2] G.Q.Zhang and A.J. van Roosmalen: Reliability Challenges in the nanoelectronics era, Journal of Microelectronics & Reliability 46 (2006), pp. 1403-1414.

[3] LED applications, US Department of Energy (DOE), 2009 www.ssl.energy.gov

[4] Mehmet Arik,,Physics of Failure based on hierarchical models, presented at the DOE R&D workshop, February 2010

[5] LED Reliability and Lumen Maintenance, www.philipslumileds.com

[6] R. Hechfellner and S. Landau, Understanding LED performance, Led Lighting 2009.

Effect of Design Parameters on the UV Radiation Characteristics of Low Pressure Mercury-Rare Gas Discharge Lamp for LCD Backlight

Katsuhide Misono

Department of Electrical and Electronic Engineering, Tsuruoka National College of Technology, 104 Aza Sawada, Inooka, Tsuruoka-shi, Yamagata-ken, 997-8511 Japan

Contact: katsuhide_misono@yahoo.co.jp

ABSTRACT

We investigated the effect of design parameters on the UV (254(nm)) output and its efficiency by using the model we developed previously. The use of the orthogonal array reduced the number of calculation before we reached the best design parameters. We found the possible design parameters that improve both UV output and its efficiency. The efficiency obtained is as high as that of the fluorescent lamp for general lighting.

1. INTRODUCTION

We previously developed a practical model to analyze the characteristics of low pressure mercury-rare gas discharge lamp with the accuracy of 10% or better, and applied this model to design the possible energy saving CCFL for LCD backlight [1]. This paper discusses the effect of design parameters such as gas type, gas pressure, mercury vapour pressure, tube diameter, and wall load on the radiation characteristics, i.e. UV (254(nm)) radiation output and its efficiency, by using the model.

2. SIMULATION PROCEDURES

The number of calculation is very large if we execute all the possible combinations of design parameters. Thus, we adopted the orthogonal arrays L18 and L9 [2] to reduce the number of calculation and yet, find the main effect to the radiation characteristics. The distance between the electrodes was fixed to 1(m), and only the discharge in the positive column was considered. The wall load was varied from 50 to 200 (mW/cm^2) from the view point of practical interest. The design parameters were allocated inside the orthogonal array, and the wall loads were allocated outside the orthogonal array. We carried out the following two simulations.

Simulation1:

Gas type, gas pressure, mercury vapour pressure, and tube diameter were taken as the design parameters, and the radiation characteristics were calculated for the wall load given above. L18 orthogonal array in which the design parameters were allocated is shown in Table 1.

Simulation 2:

After Simulation 1, the design parameters that give better values of the radiation characteristics were obtained. We paid attention to these design parameters, and carried out the calculation again for tuning the best design parameters by changing around the design parameters obtained in Simulation 1.

Table 1. L18 Orthogonal Array for Simulation 1

No.	gas type	Pgas	PHg	dia	e	e	e
1	Ne/Ar=100/0	10	6	2.5	-	-	-
2	Ne/Ar=100/0	20	13	3.5	-	-	-
3	Ne/Ar=100/0	50	25	4.5	-	-	-
4	Ne/Ar=80/20	10	6	3.5	-	-	-
5	Ne/Ar=80/20	20	13	4.5	-	-	-
6	Ne/Ar=80/20	50	25	2.5	-	-	-
7	Ne/Ar=60/40	10	13	2.5	-	-	-
8	Ne/Ar=60/40	20	25	3.5	-	-	-
9	Ne/Ar=60/40	50	6	4.5	-	-	-
10	Ne/Ar=40/60	10	25	4.5	-	-	-
11	Ne/Ar=40/60	20	6	2.5	-	-	-
12	Ne/Ar=40/60	50	13	3.5	-	-	-
13	Ne/Ar=20/80	10	13	4.5	-	-	-
14	Ne/Ar=20/80	20	25	2.5	-	-	-
15	Ne/Ar=20/80	50	6	3.5	-	-	-
16	Ne/Ar=0/100	10	25	3.5	-	-	-
17	Ne/Ar=0/100	20	6	4.5	-	-	-
18	Ne/Ar=0/100	50	13	2.5	-	-	-

3. RESULTS

Simulation 1:

The relationship between UV radiation power and its efficiency by taking the wall load as parameter is shown in Figure 1. From this graph, we find that the condition L18-13 gives good radiation characteristics for wide range of wall load.

Simulation 2:

By considering the results of Simulation 1, we carried out the next calculation by changing the design parameters around the condition L18-13. L9 orthogonal array in which the design parameters were allocated is shown in Table 2. The relationship between UV radiation power and its efficiency by taking the wall load as parameter is shown in Figure 2. From this graph, we find that the condition L9-4 gives both high UV radiation power and efficiency for wide range of wall load. It is interesting that the condition L9-7 also gives the similar UV output and efficiency for low wall load region.

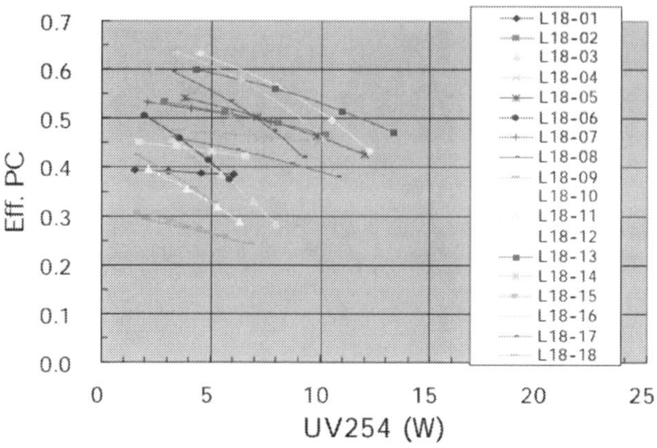

Figure 1. Relationship between UV output and Efficiency

Table 2. L9 Orthogonal Array for Simulation 2

No.	Ne/Ar	Pgas	PHg	e
1	Ne/Ar=40/60	5	9	3.5
2	Ne/Ar=40/60	10	13	4.5
3	Ne/Ar=40/60	15	18	5.5
4	Ne/Ar=20/80	5	13	5.5
5	Ne/Ar=20/80	10	18	3.5
6	Ne/Ar=20/80	15	9	4.5
7	Ne/Ar=0/100	5	18	4.5
8	Ne/Ar=0/100	10	9	5.5
9	Ne/Ar=0/100	15	13	3.5

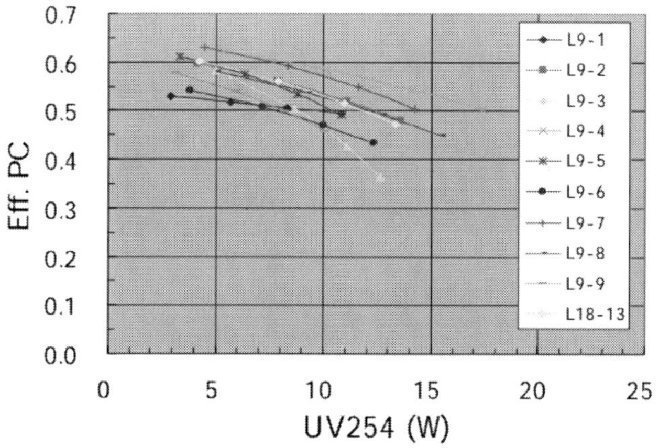

Figure 2. Relationship between UV output and Efficiency

4. DISCUSSION

From Figure 2, we can say that the conditions L9-4 is the reasonable candidate to make the radiation characteristics best. Notice that these characteristics do not change much even if the parameters are varied in the given ranges.

This implies that this design is robust to the radiation characteristics, and this is important to the design of the practical fluorescent lamp for backlight. Moreover, L9-4 shows a little reduction of the efficiency and linear increase of UV output as the wall load increases. This suggests that this design can reduce the number of fluorescent lamps installed in the backlight, which means cost down.

The orthogonal array is a useful technique to find the vital few design parameters from a large number of possible design parameters with less calculation.

5. CONCLUSION

We investigated the effect of design parameters on the UV output and its efficiency by using the model we previously developed. The design parameters that possibly make the radiation characteristics better are obtained. The UV radiation efficiency is expected to be as high as that of the fluorescent lamp for general lighting. The use of orthogonal array reduces the number of calculation for determining the vital design parameters.

REFERENCES

[1] K. Misono: Practical Model to Analyze the Characteristics of CCFL for Backlighting, #FMCp-15, IDW2009 (Miyazaki, Japan, 2009)

[2] G. Taguchi and Y. Yokoyama: Introduction to Quality Engineering, (Japanese Standards Association, 2007) (in Japanese)

Potential health impacts of LED lighting

D. Attia[1], A. Barlier-Salsi[2], F. Behar-Cohen[3], JP. Césarini[1], O. Enouf[4], M. Garcia[5], C. Martinsons[5], O. Merckel[1], S. Picaud[6], F. Viénot[7], G. Zissis[8]

[1] *French agency for environmental and occupationnal health safety (AFSSET), 253, avenue du Général Leclerc, 94701 Maisons-Alfort, France*

[2] *Institut National de Recherche et de Sécurité (INRS), Rue du Morvan CS 60027, 54519 VANDOEUVRE Cedex*

[3] *INSERM U86, Centre des Cordeliers, 15 rue de l'Ecole de Médecine, 75270, Paris, Cedex 06 and, Department of Ophthalmology, Hôpital Hôtel-Dieu, 1 Place du Parvis Notre-Dame, 75004, Paris, France*

[4] *Laboratoire National de Métrologie et d'Essais (LNE), 29, Avenue Roger Hennequin, 78197 Trappes, France*

[5] *Centre Scientifique et Technique du Bâtiment (CSTB), 24 rue Joseph Fourier - 38400 St Martin d'Hères – France and 11 rue Henri Picherit, BP 82341, 44323 Nantes Cedex 3*

[6] *INSERM, U592, Institut de la Vision, Paris, France*

[7] *Centre de Recherches sur la Conservation des Documents Graphiques, CNRS FRE 2743 - Muséum National d'Histoire Naturelle, Paris, France*

[8] *LAPLACE, Université Toulouse III, 118 Route de Narbonne 31062, Toulouse cedex 9, France*

Contact: dina.attia@afsset.fr

ABSTRACT

The French Agency for Environmental and Occupational Health Safety (AFSSET) is a public body reporting to the French Ministers for ecology, for health and for employment. AFSSET established a working group in order to conduct a health risk assessment on LED (Light Emitting Diode) lamp systems. We present here the framework of this work and the first conclusions of the working group. Blue light hazard and circadian rhythms disorders are of a particular concern with white LEDs. Direct viewing of LEDs should be avoided. Vulnerable and photosensitive population should be adequately protected.

GENERAL

Working method

AFSSET established a working group of independent experts in order to conduct a health risk assessment on LED (Light Emitting Diode) lamp systems. The working group gathers skills in both lighting physics and ophthalmology. The goal of this work is to give a general overview of light hazards associated with LEDs. Visible light emitting diodes potentially pose 5 types of hazards: damage to ocular tissue (essentially related to blue light hazard), glare, circadian rhythm disorders, dermatological effects and flashing light induced seizures.

Blue Light Hazard

The most popular technique to produce high intensity white LEDs is to couple a blue LED with down-conversion phosphors. Hence, blue light hazard is of particular concern with white LEDs. Retinal damage from blue LEDs has been demonstrated in laboratory in animals [1], but not observed in humans. Figure 1 shows the blue light hazard function (light phototoxicity function) [2] and an example of a white LED spectrum. Some measurements have been performed to classify some LED lamp systems according to CIE S 009/IEC 62471 standard ("Photobiological safety of lamps and lamp systems"). This standard defines 4 risk groups (group 0: exempt group, group 1: weak risk, group 2: moderate risk, group 3: high risk). According to this standard, LED lamps systems can be assessed at a distance which produced an illuminance of 500 lx (like for common use lamps) or at 20 cm from the lamp (like particular lamps)(see Table 1). Most classical lamp systems are classified in 0 or 1 group, whereas tested LED lamp systems are classified in 0, 1 or 2 group.

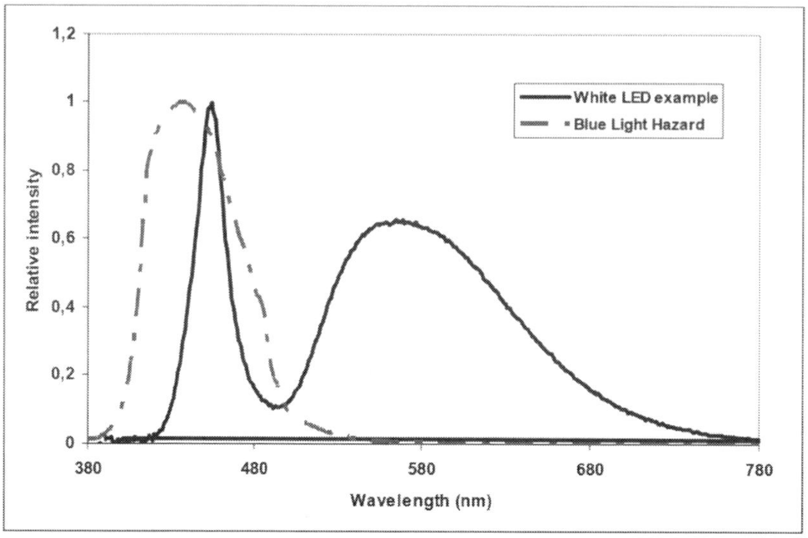

Figure 1 : Blue Light Hazard Function and spectrum of a white LED

Risque class	At 500 lx	At 20 cm
Lamp (dichroic lamp shape) - 1 LED 3 W	**1**	**1**
Lamp system - 1 LED 10 W	**1**	**2**
Lamp system with 6 LEDs – total power 15 W	**0**	**1**

Table 1 : Risk assessment for different LEDs lamps according to CIE S 009/IEC 62471 standard

Circadian rhythms and pupillary reflex

A newly discovered photopigment found in the human retina (melanopsin) regulates non-visual functions of light such as circadian rythms and pupillary reflex [3][4]. The melanopsin response to light is maximal at roughly 484 nm [5]. White LEDs spectra present a dip between 460 and 480 nm. The potential effect of this dip on health still needs to be quantified and assessed.

REFERENCES

[1] W. Dawson, T. Nakanishi-Ueda, D. Armstrong, D. Reitze, D. Samuelson, M. Hope, S. Fukuda, M. Matsuishi, T. Ozawa, T. Ueda and R. Koide. Local Fundus Response to Blue (LED and Laser) and Infrared Sources (LED and Laser). Academic Press. (2001).

[2] International Commission on Non Ionizing Radiation Protection, (ICNIRP). Guidelines on Limits of Exposure to Broad-Band Incoherent Optical Radiation. (1997).

[3] D.M. Berson, F.A. Dunn, M. Takao. Phototransduction by retinal ganglion cells that set the circadian clock. Science (New York, NY. 2002 Feb 8;295(5557):1070-3.

[4] S. Hattar, R.J. Lucas, N. Mrosovsky, S. Thompson, R.H. Douglas, M.W. Hankins, et al. Melanopsin and rod-cone photoreceptive systems account for all major accessory visual functions in mice. Nature. 2003 Jul 3;424(6944):76-81.

[5] L.S. Mure, PL. Cornut, C. Rieux, E. Drouyer, P. Denis, C. Gronfier, H.M. Cooper. Melanopsin Bistability: A Fly's Eye Technology in the Human Retina. Plos One, (2009). 4:e5991.

Power Balance Studies in Ar-Hg and Kr-Hg T2 Discharges

Graeme Lister[1] , Qiuyi Han[2] , Hao Jang[2] , Shanduan Zhang[2] and Tzvetelina Petrova[3]
[1] *CRSL, OSRAM SYLVANIA, 71 Cherry Hill Drive, Beverly, MA, 01915, USA*
[2] *Institute for Electric Light Sources, Fudan University, 220 Handan Rd, Shanghai 200433, People's R of China*
[3] *Berkeley Scholars Inc., 6551 Mid Cities Avenue, Beltsville, MD 20705, USA*
Contact: graeme.lister@sylvania.com

INTRODUCTION

This paper describes an experimental and numerical study of the positive column of T2 (*OD*=7 mm) lamps, which is an extension of work previously reported [1]. We have measured the electric field and the exitance of 254 nm radiation, and all the principal near UV and visible lines emitted, and compared the results with those of a numerical model.

T2 lamps belong to a class of *FL* referred to as "highly loaded" discharges, so-called because of the very high current density during operation. T2 *CFL* operate at 100 mA, resulting in a current density ~0.5 A cm^{-2}, compared to standard T8 operation (200 mA) ~0.05 A cm^{-2}. Such highly loaded lamps have traditionally provided a challenge to modellers, due to both the *cataphoresis* (mercury depletion at the centre of the discharge) and the inadequate data available for ionization processes at high electron densities obtained in these discharges.

EXPERIMENTAL OVERVIEW

The experimental set up to measure the electric field and absolute radiation from T2 discharges has been described in detail in [1]. The essential steps are:

(1) The positive column of the lamps is produced in a long (25-40 cm) quartz tube, with inner diameter (*ID*) 5 mm and outer diameter (*OD*) 7 mm.
(2) The discharges are driven by a commutated dc 50 Hz power supply.
(3) Four electrostatic probes at the centre of the tube, each separated by a distance of 5 cm, and sealed in the tube wall, are used to measure the electric field in the positive column.
(4) The emitted radiation is guided to a spectrometer, calibrated using a deuterium tungsten halogen standard lamp, to obtain the relative irradiance of each spectral line.
(5) The discharge tubes are covered with black paper, except for a 1 mm aperture in the centre, mounted with a molybdenum diaphragm normal to the vertical plane of the lamp axis. Absolute radiance of each line is obtained using a D lamp calibrated by NIST .
(6) The angular dependence of the light emitted from the aperture (the "Koedam factor") is then measured, in order to calculate the absolute radiation power emitted from the lamp.

ONE DIMENSIONAL FLUID MODEL

The total electrical power per unit length W dissipated in the positive column of a fluorescent lamp is the product of the discharge current I, and the electric field E_z, which can be written as

$$W = IE_z = W_{rad} + W_{heat} + W_{wall} \qquad (1)$$

where, W_{rad}, W_{heat} and W_{wall} are the power dissipation per unit length due to radiation, gas heating, and electronic wall losses respectively. The numerical model used to calculate the various contributions to the power balance in T2 lamps is described in [1]. The main features of the model are:

(i) Diffusion equations for the ground and 7 excited states of Hg, including cataphoresis, and 3 excited states of Ar, are solved to compute the total particle flux to the walls.
(ii) Electron and ion densities, together with the radial ambipolar field, are solved self consistently, from Poisson's equation and the mobility equations for Hg and Ar ions.
(iii) Rate coefficients for all electron collision processes are computed from the radially dependent EEDF, obtained from a self-consistent solution of the Boltzmann equation.
(iv) Radiation processes are evaluated through frequency dependent cell-to-cell coupling constants, evaluated from the detailed emission and absorption profiles [1].

RESULTS

The main difficulty in accurately modelling "highly loaded" FL is to compute the electric field, and thus the electric power in the discharge. This is important if the power balance in equation (1), is to be quantitatively determined. Figure 1 a) shows that the model reproduces the electric field in an Ar-Hg discharge, at the standard operating cold spot temperature of 50^0C quite well, while at 30^0C (where cataphoresis is important) and 70^0C (where radiation trapping is higher) calculations are of order 10% below the measured values. This is a considerable improvement over previous calculations. The power balance in Figure 1 b) shows that the model and experiment are in reasonable agreement for the efficiency of 254 nm radiation, although the experiment shows a larger pressure dependence.
In the case of Kr-Hg discharges, the calculated values of electric field are somewhat greater than the measured values, and the discrepancy is larger than for the Ar-Hg case. The Kr transport data is currently being re-evaluated, to address this discrepancy. Once again, the fraction of electrical energy converted to 254 nm radiation is in good agreement with the measured values. In both cases, the discharge is shown to be an excellent emitter, with 70-80% of electrical energy converted to radiation.

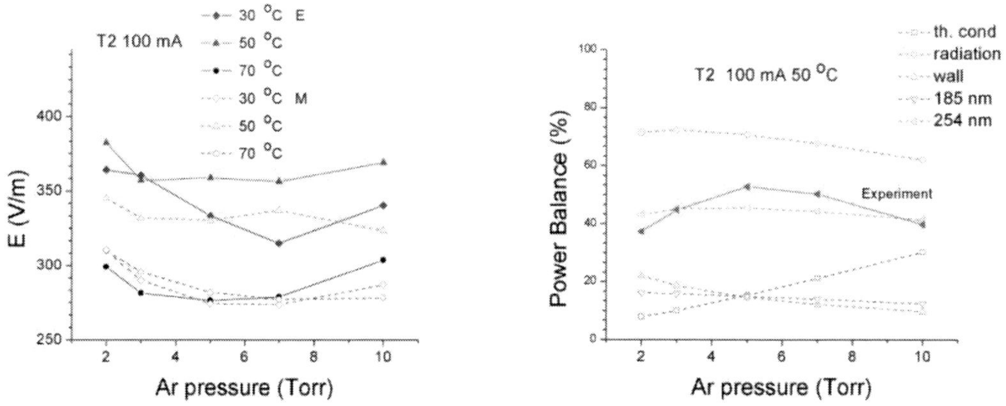

(a) (b)

Figure 1: (a) Measurements (E) *and calculations* (M) *of electric field in T2 positive column Ar-Hg discharges (b) Computed power balance and comparison with measured 254 nm radiation.*

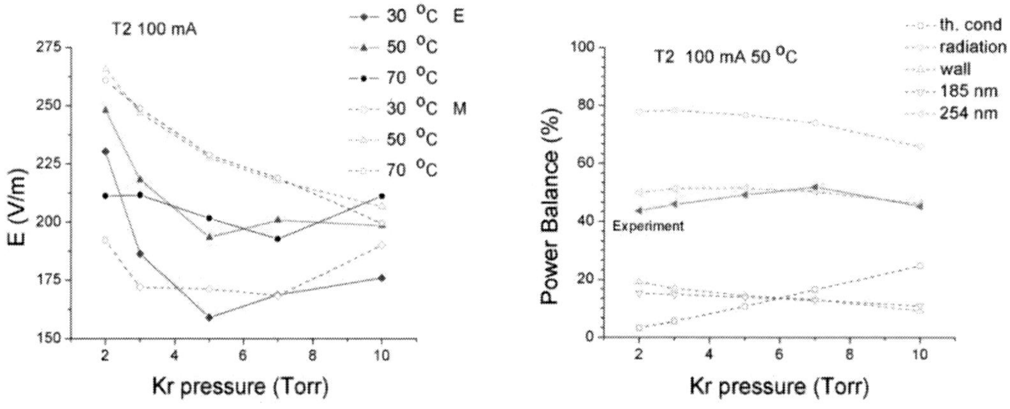

(a) (b)

Figure 2: As for Figure 1, Kr-Hg discharges

REFERENCES

[1] Q Han, H Jiang, S Zhu, S Zhang, G Lister and T Petrova 2008 *J. Phys. D: Appl. Phys.* **41** 144008

[2] G M Petrov and J L Giuliani 2003 *J. Appl. Phys.* **94** 62

Radiative Transistion Probabilities for Neutral Cerium

J. J. Curry[1] and D. E. Nitz[2]

[1]National Institute of Standards and Technology, Gaithersburg, MD 20899 USA
[2]Department of Physics, St. Olaf College, Northfield, MN 55057 USA
Contact: jjcurry@nist.gov

ABSTRACT

Radiative transition probabilities for neutral cerium have been obtained by analysis of existing line intensity data under the assumption that the intensities can be described by a Boltzmann equilibrium with an empirical temperature. Values for 600 transitions obtained from such an analysis[1] of intensities measured by Meggers et al.[2] have shown ± 55 % agreement with values measured more directly by Lawler et al.[3] We are currently applying the same analysis to intensities measured by Bisson et al.[4] The goal of this project is to obtain transition probabilities for 10,000 or more of the 20,000 significant transitions of neutral cerium. It is hoped the new data will be suitable for modeling of the broad spectral distributions emitted from metal-halide high-intensity discharge lamps containing CeI_3.

BACKGROUND

Cerium, because of its rich emission spectrum, has recently enabled substantial increases in the luminous efficacy of metal-halide high-intensity discharge (HID) lamps[5]. An unpublished line list for neutral cerium by W. C. Martin contains approximately 20,000 classified lines over the range of 338.5 nm to 1000 nm. Transition probabilities have been published for less than 4000 of these transitions[1, 3, 4]. The goal of this effort is to obtain experimental values from existing data for more than half of the 20,000 classified lines, with an accuracy sufficient for modeling the broad spectral distribution emitted from an HID lamp containing CeI_3.

METHOD

The basis of a Boltzmann analysis is the assumption that the intensity values, I_{ul}, in a data set are related to each other by a Boltzmann factor, i.e.,

$$I_{ul} = \beta \frac{g_u A_{ul}}{\lambda_{ul}} \exp^{-E_u/kT} \tag{1}$$

where the subscript ul signifies a transition between upper level u and lower level l, g_u is the upper level degeneracy, A is the transition probability, λ is the transition wavelength, E_u is the upper level energy, k is Boltzmann's constant, β is a scale factor, and T is a temperature. If A_{ul} is known for a sufficiently large subset of the transitions for which intensity values are known, β and T can be determined from a least-squares fit to $\ln(\frac{I\lambda}{gA})$ versus E_u (see Figure 1). The empirical values of β and T can then be used to obtain transition probabilities corresponding to the remaining intensity values.

RESULTS

Using a Boltzmann analysis, one of us (Curry) has obtained transition probabilities for nearly 600 strong transitions in neutral cerium[1]. These values were obtained from intensities measured by Meggers et al.[2] and a smaller set of transition probabilities measured by Bisson et al.[4] Based on the statistical fit of the Boltzmann model, an uncertainty of ± 35 % was assigned to most of the resulting transition probabilities. More recently, Lawler et al.[3] have measured values for nearly 3000 transitions obtained from an independent technique that is considered to be much more accurate. The Lawler data set overlaps with Curry's for 229 transitions. With 4 outliers excluded, Curry's values average 21 % greater than Lawler's with a $1 - \sigma$ standard deviation from the mean of 54 %. (Figure 2). Lawler et al. note that their values are systematically larger than Bisson's values where the two data sets overlap. Since Curry's values were normalized by the Bisson data, this is a probable reason for the systematic difference with Lawler et al.

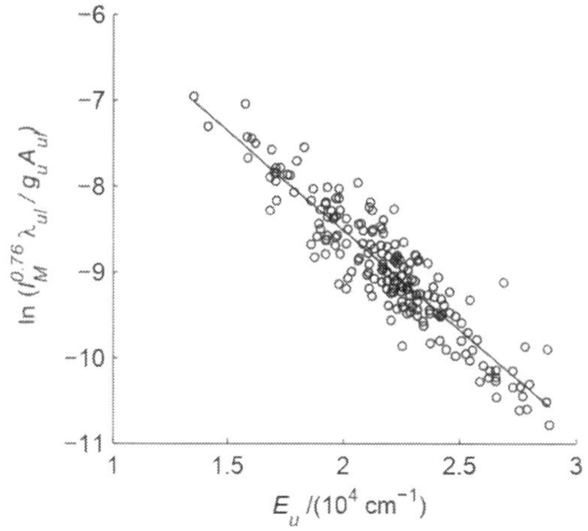

Figure 1. Boltzmann plot for neutral cerium transitions with intensities from Meggers et al.[2] and gA-values from Bisson et al.[4]. Compressing the intensity scale by the power 0.76 improved the fit. The resulting Boltzmann parameters are $T = 6200 \pm 200$ K and $\beta = 0.021 \pm 0.003$ nm· s.

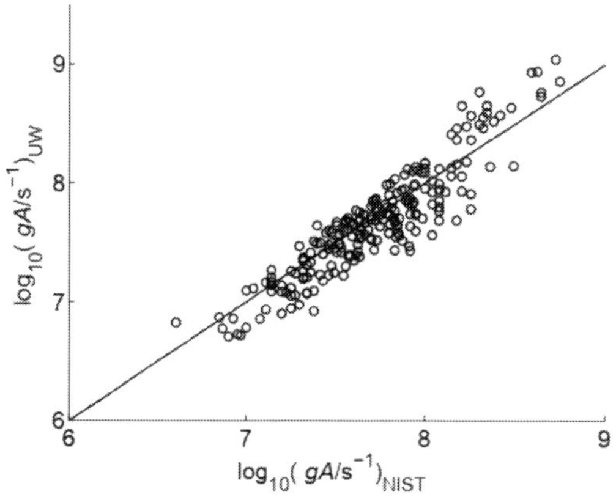

Figure 2. Comparison of gA values from Curry (denoted by the subscript NIST[1]) with those of Lawler et al.[3] (denoted by the subscript UW). The straight line indicates perfect agreement.

We are currently applying a Boltzmann analysis to a more comprehensive set of intensities obtained by Bisson et al.[4], who used a substantially different method than that used by Meggers et al.

REFERENCES

[1] J. J. Curry, J. Phys. D: Appl. Phys. **42**, 135205 (2009).

[2] W. F. Meggers, C. H. Corliss, and B. F. Scribner, *Tables of Spectral Line Intensities*, National Bureau of Standards Monograph 32 (U.S. Department of Commerce, 1961).

[3] J. E. Lawler, J. Chisolm, D. E. Nitz, M. P. Wood, J. Sobeck, and E. A. Den Hartog, J. Phys. B: At. Mol. Opt. Phys. **43** (2010) 085701.

[4] S. E. Bisson, E. F. Worden, J. G. Conway, B. Comaskey, J. A. D. Stockdale, and F. Nehring, J. Opt. Soc. Am. B **8**, 1545 (1991).

[5] W. P. Lapatovich, Phys. Scr. **T134** 014024 (2009).

Spectral intensity of the N_2 emission in rare gas low-pressure arc discharges

R. Friedl[1,2], U. Fantz[1,2], F. Vogel[1]

[1]*Lehrstuhl für Experimentelle Plasmaphysik, Universität Augsburg, Universitätsstr. 1, 86135 Augsburg, Germany*
[2]*Max-Planck-Institut für Plasmaphysik, EURATOM Association, Boltzmannstr. 2, 85748 Garching, Germany*
Contact: roland.friedl@physik.uni-augsburg.de

ABSTRACT

As an alternative to hazardous mercury in lamps for general lighting nitrogen is being discussed. Molecular nitrogen bands emit both in the near UV and in the visible spectral range which reduces the conversion losses. For analysing the potential of nitrogen low-pressure arc discharges in rare gas background were characterised by means of optical emission spectroscopy. The dependence of the spectral intensity of the molecular nitrogen emission on gas mixture, pressure and discharge current is investigated. With regard to the application as a light source, the efficiencies are determined.

INTRODUCTION

Conventional fluorescent lamps use mercury as UV emitter at 254 nm and convert this emission via phosphors into the visible spectral range. The disadvantages of this method to produce visible light are the harmfulness of the mercury and the great energy loss due to the conversion. In contrast nitrogen is environment-friendly, chemical inert and in addition almost unlimited available. Furthermore its prominent emission systems, the 1st positive system ($B^3\Pi_g \rightarrow A^3\Sigma_u^+$) and the 2nd positive system ($C^3\Pi_u \rightarrow B^3\Pi_g$), are very useful for lighting applications [1]. While the B-A-transition emits from the visible to the near infrared ($500 - 2500$ nm), the C-B-transition lies in the near ultraviolet spectral range ($260 - 430$ nm). Therefore the great advantages of using nitrogen as emitting element are the lower conversion losses and the possibility to use visible light right away. The aim of the work is to characterize nitrogen low-pressure arc discharges.

EXPERIMENTAL SETUP

The principal setup is designed similar to conventional fluorescent lamps, i.e. the discharge vessel has a tube geometry (26 mm in diameter) and electrical power is provided to the discharge via internal electrodes (electrode gap 27 cm). The vessel is integrated in a vacuum system to provide a wide accessible parameter space of pressure and nitrogen concentration in different background rare gases. The electrodes consist of a tungsten wire which is 3-fold twisted, coated with emitter material and can be heated separately. The power supply generates AC voltages up to 650 V at frequencies from 5 to 50 kHz (so far driven at 20 kHz). For diagnostics the electrical parameters discharge current I, voltage U and power P_{el} are recorded by a power meter. Optical emission spectroscopy is used with a line of sight which is centered radially and axially. A survey spectrometer and a high resolution spectrometer are applied ($\Delta\lambda_{FWHM} = 1.3$ nm and 0.05 nm).

RESULTS

The dependence of the emission intensity on the concentration of nitrogen in the background gas ($0.1 - 10$ %), on the composition of the background rare gases (Ar, Ne, He), on the absolute pressure ($10^{-1} - 10^{2}$ mbar), and on the discharge current ($100 - 500$ mA) was investigated.

Figure 1 shows an example of an emission spectrum of a nitrogen argon arc discharge, indicating both the nitrogen systems and the different argon transitions. In contrast to nitrogen RF plasmas or glow discharges, such an arc discharge has an intense emission from the 1st positive system of nitrogen in its spectra, which is very useful in applications as a light source.

To explain the dependence of the emission with varying nitrogen content and pressure, the electron density and temperature were determined. The electron density was calculated from the discharge current using the known electron drift velocities [2]. The electron temperature could then be determined by using effective emission rate coefficients calculated with a collisional radiative model for the background gas.

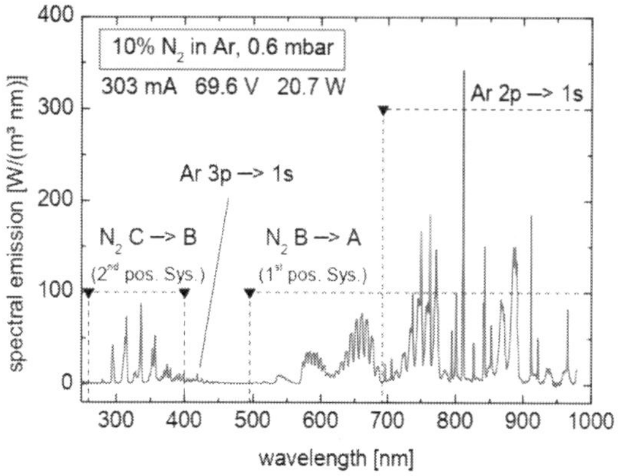

Figure 1. Typical spectrum of nitrogen argon arc discharges with denoted parts of the emission.

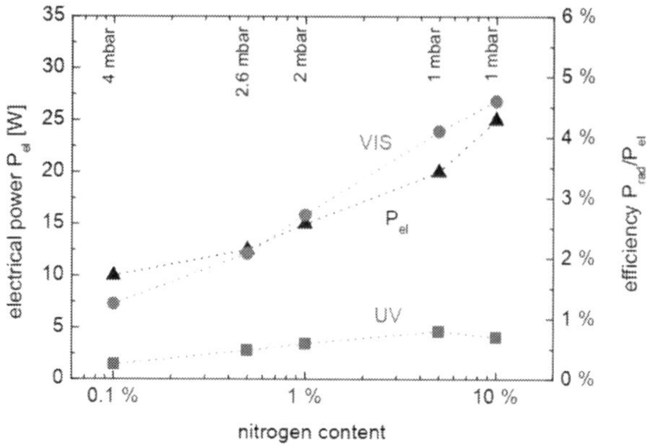

Figure 2. Power consumption and radiation efficiencies at optimal absolute pressures for an argon background at 300 mA discharge current.

With increasing nitrogen content the emission from the background gas decreases due to a decreasing electron density, whereas the nitrogen emission increases due to the higher molecule density (slightly compensated by the lower n_e). When increasing the pressure a maximum in the nitrogen emission in the mid-pressure area develops. This is because of increasing n_e while T_e decreases (as expected due to the ionization balance). The background emission in contrast is mainly determined by the lower electron temperature and so decreases drastically when increasing the pressure.

By establishing a collisional radiative model for the electronic states X, A, B and C of nitrogen [3], the heavy particle processes like energy pooling reactions and energy transfer from the metastable states of the background gas together with quenching processes by the background gas could be identified to have a great influence on the density of the excited nitrogen states and therefore on the emission.

The gas composition largely influences the electrical power consumption which is needed to sustain a constant discharge current. This can be taken into account by using efficiencies which are defined as the ratio of incoupled electrical power P_{el} to emitted radiation power P_{rad}. Figure 2 shows these values for the optimal absolute pressure (most efficient emission) of each nitrogen concentration in an argon background separated in the ultraviolet and in the visible spectral range. One can clearly see an increase in the efficiency for producing UV and for visible radiation with higher content of nitrogen and pressures around 1 mbar.

CONCLUSIONS AND OUTLOOK

The 1st positive system of nitrogen in the VIS range is very prominent in the spectra of nitrogen arc discharges and can be used together with the UV emission in applications as a light source. The intensity and efficiency of the nitrogen emission strongly depends on the nitrogen content and the discharge pressure due to varying plasma parameters. Heavy particle collisions have a significant influence over the investigated parameter space. The most efficient radiation until now is obtained at a gas filling with high concentrations of nitrogen in an argon background and a pressure of 1 mbar: 5 % of visible and 1 % of UV efficiency can be reached.

Investigations on different background gases including mixtures of rare gases will be presented on the poster.

REFERENCES

[1] M. Jinno, S. Takubo, Y. Hazata, et al., J. Phys. D: Appl. Phys. 38 (2005) 3312.

[2] J. V. Jovanovic et al., J. Phys. D: Appl. Phys. 42 (2009) 045202.

[3] U. Fantz, S. Briefi, R. Friedl, contribution to this conference.

Research on a Tube Element LED

Z.L. Chen[1], Z.L. Chen[2]
[1] *Yancheng Haomai Lighting Ltd., Yancheng Shi, Jiangshu Sheng, 224761 P.R. China*
[2] *Voxlearning Inc., Shangdi Zhongguancun Chuangyedasha, Beijing, 100085 P.R.China*
Contact: zongliechen@yahoo.com.cn

ABSTRACT

We have conducted a research on one type of tube element LED which is made of micro- chip. In this type of LED's manufacturing process, first of all, we will paste some chips with the size of no more than 25μm×25μm on the substrate with good light transmission and heat-conducting. We put them into a glass tube with good light transmission after conducting electric connection with the comb structure of reflector board which can conduct electricity and heat, and conduct air-tight sealing, then charge it with over one atm of dry nitrogen or some other inert gases. We conduct comparative testing of the LED called as the tube element after powering it on, its luminous efficiency is unexpectedly increased by over 100% compared with that of the same passel of chips with the size of 1mm×1mm.

1. Introduction

By the reason of the current technology, no matter what we choose the chips with mm or μm class for sealing the LED, the external quantum efficiency is very low, its problem on the beam given off from the active layer of the LED. Secondly, the heat radiation of the junction of the LED is hard to be solved. Thirdly, the compatibility and interchangeability of the LED with those traditional fluorescent lamps is not to be taken into consideration.

As we know, there is almost no loss when the photons pass through the obstacles with the thickness of no more than 0.38μm, so the efficiency of the external quantum led out by the beam in the zone of light side surface of LED chips shall be 100% in theory. The length that the chips with size of 1mm×1mm can reach the zone with 100% of external quantum efficiency is only 4mm. But if we carve the chips above into ones with size of 25μm×25μm, then the total length will reach 160mm, so as to increase the area of the zone with 100% efficiency of external quantum led out by the beam by 40 times.

2. Technical Specifications

2.1 Design of Chip Size

It is that the smaller the size of chip is designed, the higher efficiency of external quantum is obtained.

2.2 Components and Structure of Tube Element LED

2.21 Components

a. Element Glass Tube

The element glass tube is the outer covering for sealing the tube element LED, and it consists of common glass tube with caliber of ϕ 0.5~26.5mm. It is characterized by transparency, inner-coated white and vapor deposited reflecting layer.

b. Light-emitting Baseplate

The light-emitting baseplate is used for sticking to the chips, and it is composed of good light transmission and heat-conducting materials, such as semi-transparent polycrystalline aluminium oxide PCA board.

c. Reflecting Body

It is made of aluminium in the shape of comb structure inserted mutually and sticked tightly to the inner wall of the half-element glass tube. Apart from efficiently reflecting the visible light which comes from the chips of the light-emitting baseplate. The two comb parts are respectively connected with P or N junction near them through metal wire to conduct electricity and heat. It can make most of heat generated by the chips conducted, then it will give off the heat to the air space through the closely sticking and heat exchange between the reflecting body and the tube wall.

d. Concave-convex Terminal of Power Source
The concave-convex terminals are located at the two ends of the element glass tube. The lead-out wire of the concave-convex terminals conducts electrical connection with the two comb of aluminium plate of the reflecting body, which not only is convenient to switch on the input power, but also allows the same size of tube element LED by concave-convex or convex-concave terminals at both ends to freely butt into any length of or geometrical luminous body.

e. Inert Gas
We charge the element glass tube with over one atm of dry nitrogen or other inert gases where the air within such it is pumped down, which not only can make electronic photonic devices under the state of "Dew Point" on the long-term basis, but also can generate convection current along with the heat generated from the operating of the chips, and it will quickly radiate the heat into the inner wall of it, then give off the heat to the air space through heat conduction of tube wall.

2.22 Structure

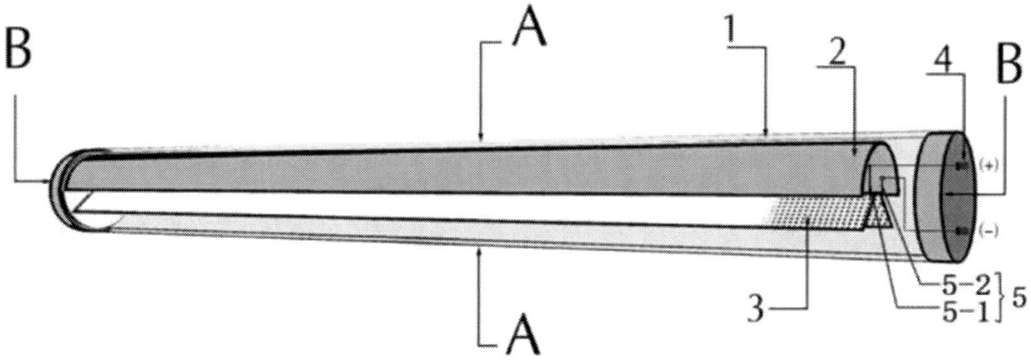

1.the element glass ; 2.the refraction body; 3. the light-emitting base plate; 4. the socket of concave-convex; 5(5-1,5-2).the wire between 2 and 3 or between 2 and 4

Figure 1

The structure of the tube element LED is shown in the Figure 1. The shape of the structure shown in the Figure 1 is a straight-tube column, and it also can be designed into other suitable geometrics. Two pieces of light-emitting baseplate are set up in the Figure 1, and one piece or more pieces of the baseplate can also be set up according to actual needs.

3. Conclusion and Discussion
Such as above-mentioned, which will undoubtedly bring about a revolutionary solution to the two main difficulties on the light lead-out and heat radiation of the large-area chips adopted by the LED with high power. In addition, such sealing form not only can immediately be used for common white-light illumination, but also synchronistically for the back lighting of LCD.

4. Acknowledgement
The author extends thanks to professor Mr. Chen Dahua of Fudan university for his concern about this research, and sincerely dedicates this document to the deceased professor Mr. Cai Zuquan, who is the father of Chinese new light source.

5. References
1. A. Zukauskas, Michael S. Shur, Remis Gaska:"Introduction to Solid-State Lightion" (2002)
2. AReportof the Development in China' Semiconductor Illumination (2005).
3. M.Q.Liu and R.Devonshire :"LIGHT SOURCES 2007"

Modeling of the discharge plasma parameters on noble gases with water vapor admixture

A. General, V. Kelman, Yu. Zhmenyak, Yu. Shpenik
Department of Quantum Electronics, Institute of Electronic Physics, National Academy of Sciences of Ukraine, Universitetska 21, Uzhgorod, 88017, Ukraine
Contact: heneral-andrij@rambler.ru

ABSTRACT

Water – rare gas discharge plasma parameters were calculated for hydroxyl 306 – 315 nm UV source. Obtained results demonstrate the best parameters for Ar – H_2O mixture.

RESULTS AND DISCUSSION

A large number of practical applications [1] of high-power sources of spontaneous UV radiation favors the investigation of the ways of improvement their parameters and especially the development of environmentally secure sources, for instance, based on water vapor. The main output power of water vapor sources is concentrated at the UV-B spectral region peaked at 309 nm (A→X band of OH hydroxyl, $\Delta\lambda$ = 306 - 315 nm). The typical UV spectra of water vapor discharge can be found, for example at [2]. These emitters may be used further as alternative sources instead of hazardous and expensive gas medium on XeCl molecules (B→X band, $\Delta\lambda$ = 307 - 310 nm).

This paper presents the calculated data on electron energies distribution functions (EEDF) and average electron energies in discharges containing of noble gases (Ar, Xe, Kr) and water vapor (up to 1%) mixtures, as well as constant rates for electron impact dissociative excitation of hydroxyl radical $A^2\Sigma$ state.

The EEDF of discharge plasma was calculated for R - H_2O (99% - 1%) mixtures where R is noble gas atom (Ar, Kr, Xe), using the calculation technique [3]. The plasma parameters were calculated for real experimental conditions: plasma ionization rate 10^{-3} - 10^{-2} (the EEDF is considerably non-equilibrium), electron concentration 10^{14} - 10^{15} cm^{-3} and E/N = 100 - 600 Td (1 Td = 10^{-21} V·m^2). The effective cross-sections for elementary electron processes for the calculation model were taken from [4, 5].

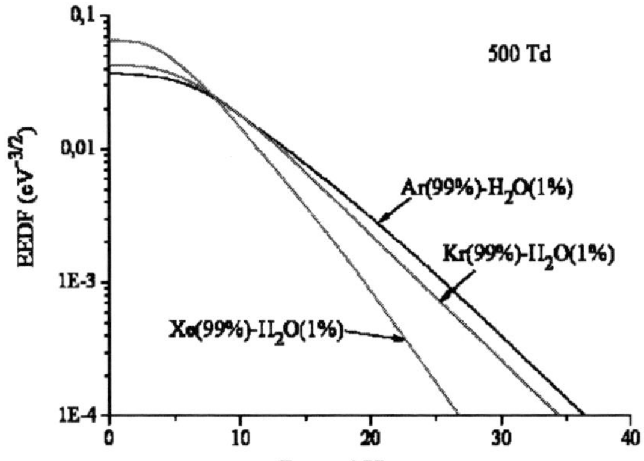

Figure 1. Calculated electron energies distribution functions

The calculated EEDF in the noble-gas and water vapor mixture discharges for different mixtures are presented at Fig.1 and the E/N parameter dependences of average electron energy – at Fig.2. The average electron energy is maximal for the Ar-H_2O mixture and essentially exceeds its value for other mixtures (Fig.2). The excitation rate constants of emitting $OH^*(A^2\Sigma)$ state as a function of specific electric field E/N were calculated aiming to increase the radiation efficiency of experimental radiation source on OH* radical bands. Rate constants were calculated by averaging of proper reaction cross-sections on EEDF. The results are presented at Fig.3. for electron impact dissociative excitation of $OH^*(A^2\Sigma)$ radicals.

The OH^* radical $A^2\Sigma$ state in the water vapor discharge mainly forms due to the processes of electron impact

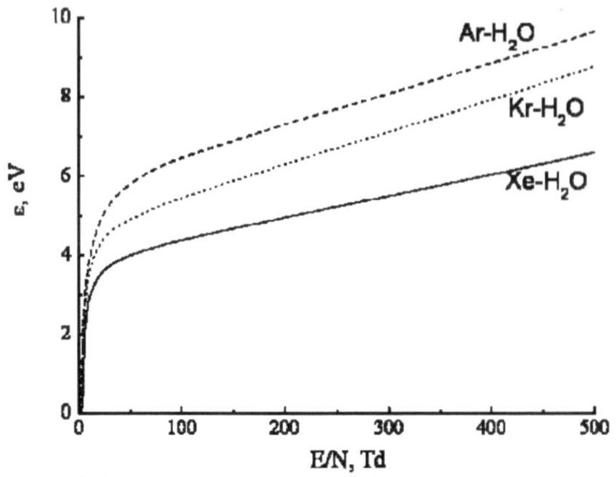

Figure 2. Electric field E/N dependences of calculated average electron energies

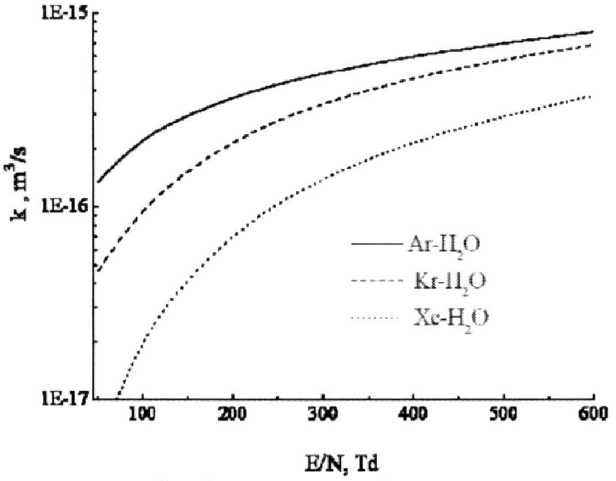

Figure 3. Calculated constant rates for electron impact dissociative excitation of $OH^*(A^2\Sigma)$ radicals

dissociative excitation of water molecules. It is obvious (see Fig.3) that the excitation rate constants for OH*(A) state for the Ar-H_2O mixture greater than in the case of Kr-H_2O and Xe-H_2O mixtures, i.e. the excitation of the emitting state of the OH* radical is much more effective in the Ar-H_2O mixture. With the growth of E/N value the constant rates of dissociative excitation increases due to higher average energy of electrons (see Fig. 2).

Moreover, the OH* $A^2\Sigma$ state radicals may appear additionally [6] as a result of Ar* atoms interaction with H_2O molecules which relaxing through the highly excited $\tilde{B}(OH^*(A)+H)$ state producing the $A^2\Sigma$ state of OH* radical. On the other hand, the excitation process similar to the N_2-laser case where Ar atoms effectively excites the C $^3\Pi_u$ nitrogen state (excitation energy 10.9 eV). So, the Ar atoms can excites the C $^2\Sigma^+$ state of hydroxyl molecules with the same excitation energy of 10.9 eV, as in our case. So, more adequate model including such second kind collisions with energy transfer must be developed in future.

The calculated discharge plasma parameters of R-H_2O mixtures further will be used for determination of the output parameters of UV sources, understanding of the main processes contributing to the UV emission excitation, raising of power and efficiency of the sources on high pressure barrier discharge.

REFERENCES

[1] V.S. Dmitruck, E.A. Sosnin, I.A. Obgol'tz. Proc. of SPIE. 6263 (2006) 316.
[2] A.K. Shuaibov, A.A. General, V.A. Kelman, I.V. Shevera. Tech. Phys. Lett. 34 (2008) 588
[3] G.J.M. Hagelaar, L.C. Pytchford. Plasma Sources Sci. Technol. 14 (2005) 722.
[4] Y. Itikawa, N. Mason. J. Phys. Chem. Ref. Data. 34 (2005) 1.
[5] J.D. Gorfinkiel, L.A. Morgan and J. Tennyson. J. Phys. B: At. Mol. Opt. Phys. 35 (2002) 543.
[6] A. Kanaev, L. Museur, F. Edery, T. Laarmann, T. Möller. Eur. Phys. J. D. 20 (2002) 261.

Influence of Quartz Glass Properties on VUV-Performance of LP-Amalgam-Lamps

E. Arnold, F.-J. Schilling, H.-D. Witzke
Heraeus Noblelight GmbH, R & D Department,
63450 Hanau, Heraeusstr. 12-14, Germany
Contact: *erich.arnold@heraeus.com*

ABSTRACT

OH-containing quartz glass is favoured for the production of low pressure amalgam lamps as high intensity 185nm photon sources due to its ageing characteristic under high VUV-photon fluxes. For longer exposure times a recovery effect is observed which guarantees higher transmittance in the 185nm spectral region due to recombination of radiation induced defect centres.

INTRODUCTION

The VUV-photon flux at 185nm from low pressure mercury lamps are used since decades for the production of ultra pure water in the semiconductor industry. These lamps are mainly made from quartz glass tubes based on pegmatite sand. New applications like surface activation, odour abatement and grease destruction in kitchen hoods require higher photon fluxes at 185nm which can only be achieved by high power Amalgam lamps (specific power per arc length ~4W/cm). The useful operational time of these lamps is limited by the generation of colour centres inside the fused quartz bulk material affecting the transmission properties in the VUV. There are two typical colour centres absorbing photons: the E'-centre (at 215nm SiH+hv=>Si*+H) and the non-bridging oxygen hole centre NBOH (at 265nm SiOH+hv=>SiO*+H).

Experiment and Results

Ageing experiments near room temperature of different quartz glass types under VUV irradiation are reported. Electrical molten quartz glass samples produced from pegmatite sand are compared to the results obtained using synthetic fused silica. The samples were thermally annealed to relax the internal stress produced during the production process and then irradiated with VUV-photons (~30mW/cm²) in an inert gas atmosphere. Transmission measurements were carried out on each of these samples after well defined irradiation times. Fig. 1 shows the ageing effect of natural fused quartz with low OH-content typically used for LP-mercury lamps. The transmission loss is dominated from the beginning by a monotonic increasing absorption band centred at 215nm. This absorption band is related to the build-up of E'-defects from SiH precursors shown in Fig. 2. NBOH-defects are not produced due to the absence of SiOH groups in this OH-free material with oxygen deficient structure (ODC II absorbing at 242nm).

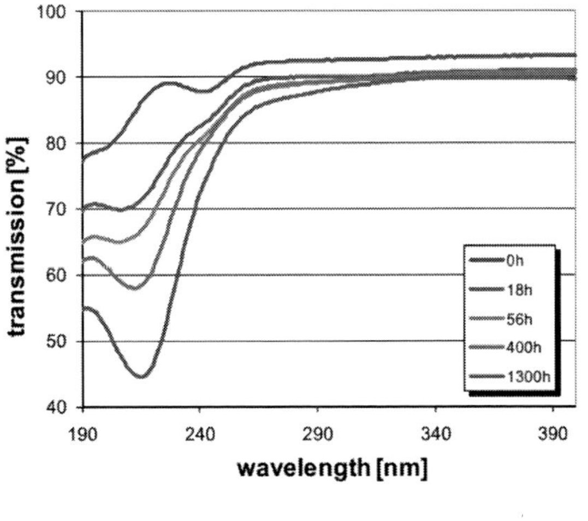

Fig. 1: Transmission of natural fused silica ([OH]<1ppm d=1mm) at different VUV-exposure times

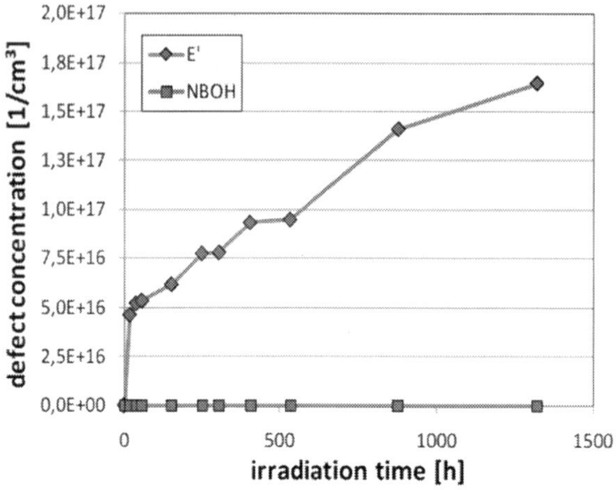

Fig. 2: Defect concentrations of natural fused silica ([OH]<1ppm) under VUV-radiation

Synthetic fused silica with a lower impurity level and a certain content of stable OH-groups (~250ppm) shows a quite different pattern (Fig. 3). Transmission measurements of 1mm thick samples down to 190nm show only weak absorption bands at 215nm (E') and 265nm (NBOH). A recovery of transmission is observed for longer exposure times (Fig. 3) due to the recombination reaction Si* + SiO* => Si-O-Si (Fig. 4) under irradiation.

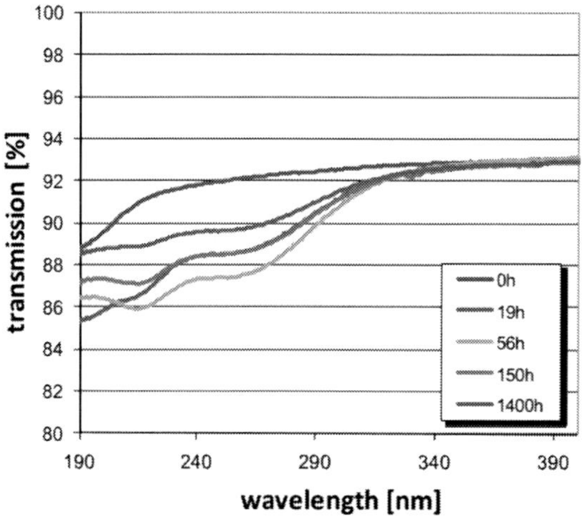

Fig. 3: Transmission of synthetic fused silica ([OH]~250ppm d=1mm) at different VUV-exposure times

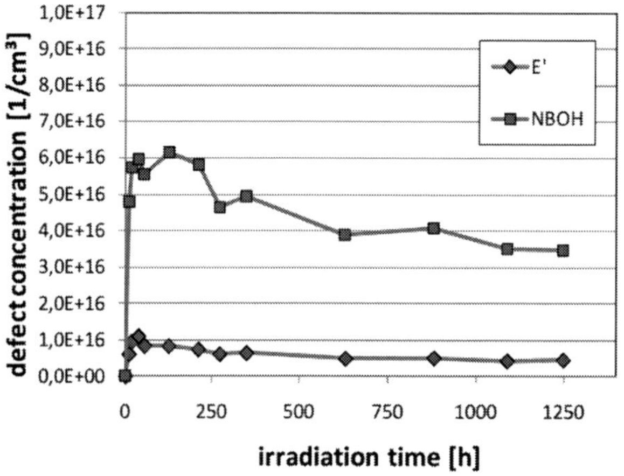

Fig. 4: Defect concentrations of synthetic fused silica ([OH]~250ppm) under VUV-radiation

This material perfectly matches the requirements for long life high power VUV-lamps due to its weak radiation damage and recovery feature caused by the balance of low NBOH and E'-defect formation.

Luminous efficacy and uniformity studies of mercury free flat lamp

Bruno Caillier[1], Philippe Guillot[1], Laurent Therese[1] and Philippe Belenguer[2]

[1] *DPHE, Université Jean François Champollion, place de Verdun, 81012 Albi cedex 9, France*
[2] *LAPLACE, Université Paul Sabatier, 118 route de Narbonne, 31062 Toulouse cedex 4, France*
Contact: bruno.caillier@univ-jfc.fr

ABSTRACT

In this paper, we will study the spatial emission of neon-xenon dielectric barrier discharge lamp for plane to plane and complementary coplanar electrode geometries. Based on optical and electrical measurements, the luminous efficacy will be obtained. We will discuss the influences, on homogeneity and efficiency, of the applied voltage (waveforms, amplitude) and the electrode geometry.

GENERAL

Xenon excimer dielectric barrier discharge is a good candidate for mercury free lightning [1]. The studied device (Fig.1) is consisting of two glass plates separated by a constant gas gap. The glass thickness is 4 mm and the gap 2 mm. A transparent conducting material (electrode) has been deposited on both external sides and phosphors on both internal sides of the dielectric. The lamp is filled with a neon-50%xenon gas mixture and operates at 250 mbar. In a previous work, for sinusoidal excitation, we have shown the influence of the applied voltage (amplitude and frequency) on the consuming power, the light emission and mostly on the non-homogeneity of the discharge [2,3].

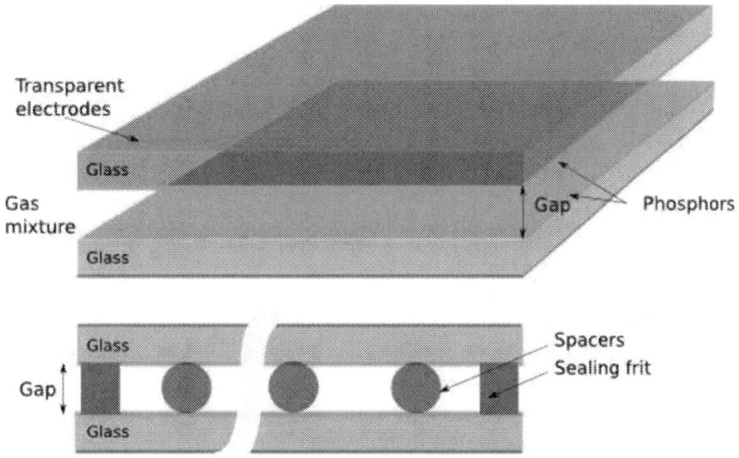

Figure 1 : standard plane to plane dielectric barrier discharge lamp.

In this work, electrodes are first deposited all over the external dielectric surface corresponding to the standard configuration, plane to plane. Another configuration is also proposed : several parallel strips (a few centimeter width, separated by the same distance) covering half of the surface on one side and covering the complementary surface on the other side (Fig.2). This configuration is called complementary coplanar geometry [4,5]. In these last conditions, the discharge starts at the edge of the electrodes and spreads along the electrodes when the voltage increases.

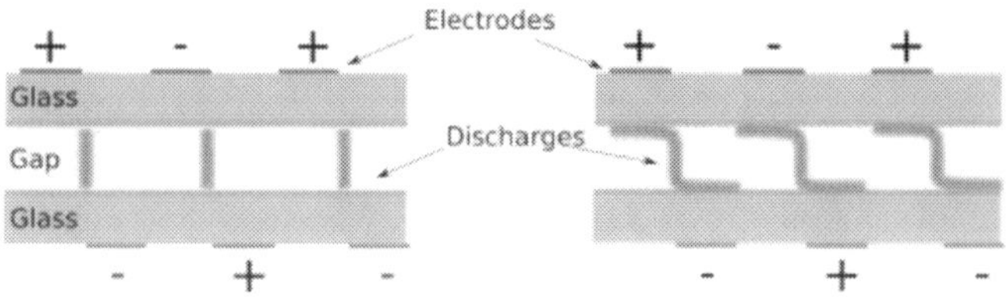

Figure 2 : complementary coplanar dielectric barrier discharge lamp.

We will compare the spatial light emission (250mbar, Ne-Xe 50%, 20kHz) in plane to plane configuration (sinusoidal and pulsed voltages) and in complementary coplanar configuration (sinusoidal and pulsed voltage).

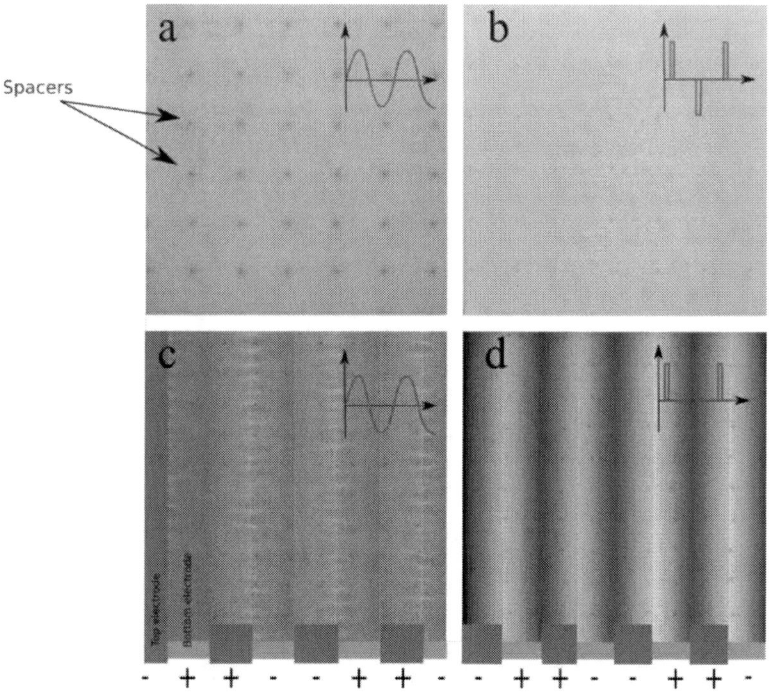

Figure 3 : 250mbar, 20kHz, Ne-Xe 50%, image : (a) plan-plan sinusoidal voltage (b) plan-plan pulse voltage (c) coplanar configuration sinusoidal voltage (d) coplanar configuration pulse voltage

In plane to plane configuration, from sinusoidal to pulsed excitation, the signal shape modifications involve a uniform discharge due to the pulsed voltage characteristics (low rise time). From plane to plane to complementary coplanar configuration, an increase of the efficiency has been observed.
We will present quantitative results for these different conditions and discuss on the influence of the electrode geometry and applied voltage on the emission efficacy.

REFERENCES

[1] U. Kogelschatz, B. Eliasson and W. Egli 1997 J. Physique IV C4 7 47
[2] T. Beaudette et al., "Experimental characterization of dielectric barrier discharges for mercury free flat lamps", 59th Annual Gaseous Electronics Conference, 10-13/10, 2006, Columbus, USA
[3] T. Beaudette et al "Influence of heating process on phosphor efficacy for mercury free flat lamp", LS11, Shangai, Chine, May 2007.
[4] G. Auday et al. Patent FR2915311 (a1) October 2008
[5] G. Auday et al Patent FR2890232 (a1) March 2007

Feed-through Solutions for HID Lamps with Ceramic Arc Tubes

V. Baier[1], T. Eckardt[1], N. Gübler[1], J. Schielke[1], B. Spaniol[1]
[1] *Engineered Materials Division, W. C. Heraeus GmbH, Heraeusstrasse 12-14, 63450 Hanau, Germany*
Contact: tanja.eckardt@heraeus.com

ABSTRACT

Feed-through components are decisive elements in advanced high intensity discharge lamps in terms of performance, lifetime and processability. It is well established to apply materials with compatible thermal expansion behaviour compared to polycrystalline alumina (PCA). Possible material solutions are niobium alloys (NbZr1), ceramic metal composites (Mo-Al_2O_3 cermet) or micro-alloyed iridium. NbZr1 is offered as tubes or caps made by metal extrusion, cylindrical cermet pins are processed by a powder-metallurgical route and iridium wire is manufactured by hot-forming processes. Further development has lead to electrode assemblies combining tungsten electrodes and feed-through elements by reliable welding technologies maintaining close dimensional specifications.

INTRODUCTION

Materials used in feed-throughs in high intensity discharge (HID) lamps have to fulfil various requirements. Besides electrical conductivity, thermal and chemical resistance they have to ensure a gas-tight sealing of the discharge vessel for several hundred or even thousands of hours. In case of quartz glass lamps, thin molybdenum foils reduce thermal stresses due to their ductility and elliptical shape. For lamps with ceramic discharge tubes, similar approaches have been suggested by co-sintering molybdenum caps with wall thicknesses of 0.1 mm between two ceramic bodies [1]. The mismatch of thermal expansion, however, between molybdenum and alumina is significant and limits lamp lifetime due to leakage after thermal cycling. Therefore current designs of HID lamps with ceramic tubes usually apply materials with similar thermal expansion in order to reduce tensile stresses within the seal. In this presentation a review of feed-through solutions for ceramic arc tubes is given with special emphasis on their properties and manufacturing technologies.

COMPARISON OF FEEDTHROUGH SOLUTIONS

Niobium Alloy Components

Translucent, polycrystalline alumina (PCA), which is the standard material in HID and sodium discharge lamps, has a coefficient of thermal expansion (CTE) of $8.1*10^{-6}$ K^{-1} (40-800°C) [1]. Niobium metal shows similar expansion behaviour. By alloying pure niobium metal with 1 wt.-% of zirconium several benefits can be achieved: The melting point is still above 2400°C, the thermal expansion coefficient increases to $7.5*10^{-6}$ K^{-1} (20-100°C) and its tensile strength is improved by stabilising a fine-grain microstructure. Its chemical stability against metal vapours makes the NbZr1 alloy an ideal candidate for feed-through parts in lamps with a ceramic vessel. An extremely wide range of geometries can be manufactured by extrusion. These components may be combined to assemblies by conventional or laser welding and sealed to ceramic tubes by using glass frits [2].

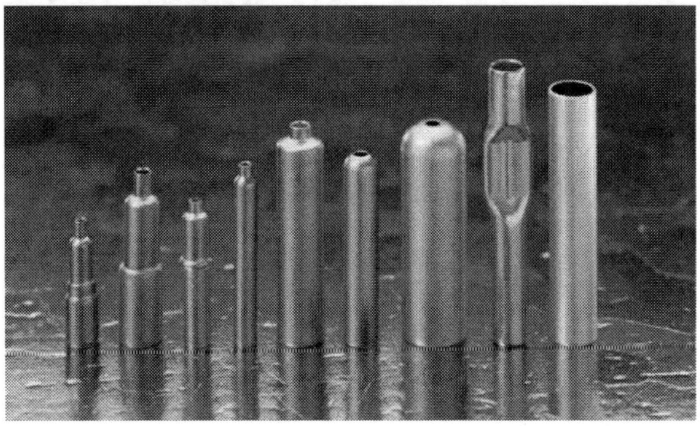

Figure 1: NbZr1 parts manufactured by extrusion

Molybdenum-Alumina Cermet

Composites made from ceramic (Al_2O_3) and metal (Mo) also offer similar thermal expansion behaviour compared to PCA and a good electrical conductivity due to the percolation of the metallic phase. It is therefore essential to achieve a homogenous microstructure and optimal phase distribution. These cermets with an amount of 55 wt.-% molybdenum exhibit good chemical stability against corrosive metal halides that are constituents of lamp fillings. These composites also demonstrate a moderate thermal conductivity that prevents temperature increase in the socket zone of discharge lamps.

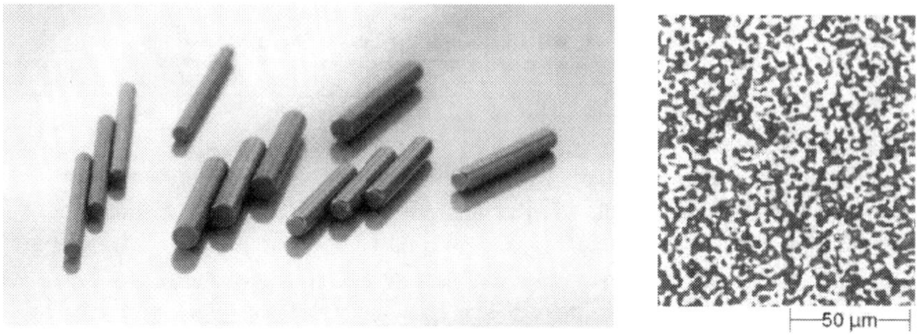

Figure 2: Feed-through pins made of molybdenum-alumina cermet and their microstructure (right)

Iridium Wire

Iridium metal with a melting point of 2454°C has a thermal expansion behaviour that is compatible with PCA; a CTE of 7.1 10^{-6} K^{-1} (20-300°C) was determined. Manufacturing iridium products like wire, pins or spheres requires hot forming facilities. Starting with a melted and cast ingot a profile is achieved by forging and rolling, finally it can be hot-drawn down to diameters smaller than 0.7 mm. Subsequent thermal annealing steps enable different microstructures and states of strength.

Additionally, micro-alloying with refractory metals in the range of 50 to 100 ppm results in a fine-grain stabilised micro-structure which has a positive effect on long-term stability.

Iridium as a precious metal demonstrates outstanding chemical stability and low vapour pressure at elevated temperatures. Due to their higher costs iridium components are considered for special lamp designs with exceptional performance specifications.

Figure 3: Rolled profile of iridium with diameter of 3.8 mm

Electrode Assemblies

Assemblies of tungsten electrodes and feed-through elements combine different materials, e.g. refractory metals (W, Mo, Nb) and cermets ($Mo-Al_2O_3$) which have to ensure specific functions in their individual position. Most important is the joining technology, whereby laser and friction welding demonstrate reliable tools with regard to tight tolerances of straightness, bending and processability at the lamp manufacturer's site.

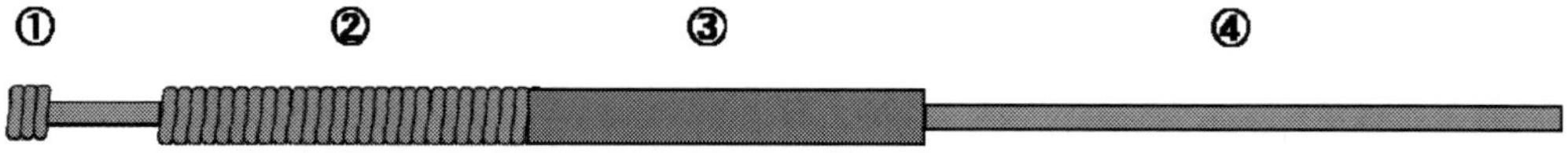

Figure 4: Electrode system for ceramic metal halide lamps assembled from tungsten electrode (1), molybdenum coil rod (2), cermet (3), niobium lead (4)

REFERENCES

[1] K. Watanabe, Proc. of 11th Int. Sym. on the Sci. and Tech. Of Light Sources, 20th – 24th May, Shanghai, LS 11 (2007) 89-98.

[2] G. Dobos, K. V. Josepovits, Z. Toth, I. Csanyi, L. Kocsanyi, Proc. of 11th Int. Sym. on the Sci. and Tech. Of Light Sources, 20th – 24th May, Shanghai, LS 11 (2007) 81-82.

Optical design for bike forward lighting based on high power LEDs

Yi-Chien Lo, Chin-Wei Chen , Ching-Cherng Sun
Optics of Solid-state lighting Laboratory,
Institute of Lighting & Display Sciences/Department of Optics and Photonics
National Central University, Taiwan (R.O.C.)
Phone: +886-3-4276240, Fax: +886-3-4252897, E-Mail: yjlo@dop.ncu.edu.tw

ABSTRACT

In our paper, we design a compact optical module with a high-contrast light pattern based on a high-power LED. According to the mid-field model, we establish the LED model precisely. The optical module is applicable to K-mark regulation.

INTRODUCTION

LEDs have been regarded as the most important and promising solid-state light source for the next generation lighting due to the advantages in efficiency, life time, color performance, environmental protection and reliability [1-2]. The technique with use of blue chip with yellow phosphors enables white-light LED more efficient and cost-effective. Recently, the efficacy of white light LED has achieved a certain leading level. In Feb. 2010, Cree reported 208 lm/W power LED at 350mA [3], which is a milestone in LED lighting. More applications with use of LEDs will be proposed and the date for replacing the traditional illumination lamp is approached. In this paper, we design a compact optical module with a high-contrast light pattern based on a high power LED. According to the mid-field model, we establish the LED model precisely. The optical module is applicable to K-mark regulation.

OPTICAL MODELING FOR A HIGH-POWER LED

The light source is made by a Taiwan company as shown in Fig. 1. The measured output flux is around 100 lm at 350mA. In the process of optical modelling, we starts at the measurement of the geometrical parameters of the package. Then, we build up the effective emitting surface at right location and right optical performance. The emitted lights are traced by using Monte Carlo Ray tracing program such as ASAP [4-6]. In the mid-field model, the light pattern will vary from one distance to another once the light source is not an ideal point source. [7] Accordingly, several light patterns at different mid-field distances are simulated and are compared with the corresponding experimental measurements..

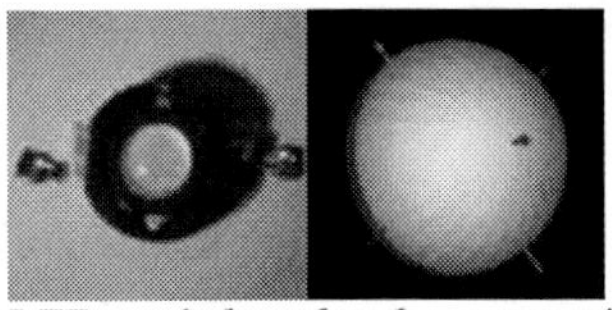

Fig. 1 1W LED made by a local company in Taiwan.

The light patterns in different distances in the mid-field range are shown in Fig. 2. In order to determine the similarity between the simulated pattern and the measured pattern, we apply the normalized cross correlation (NCC) [7]. From our experience in designing LED lighting systems for industrial applications, the NCC value suggested higher than 99% gives enough accuracy for practical applications [7,8].

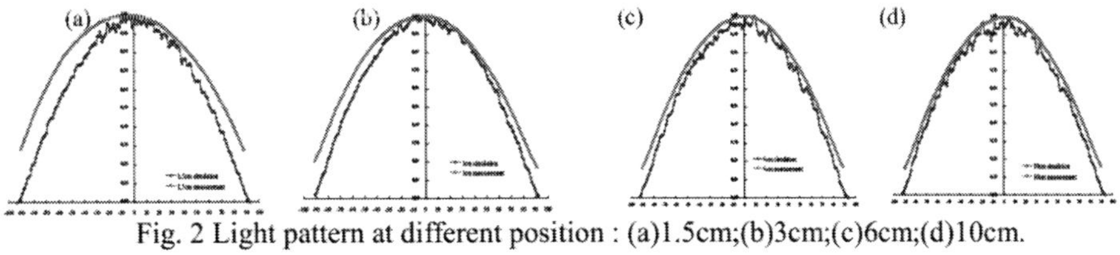

Fig. 2 Light pattern at different position : (a)1.5cm;(b)3cm;(c)6cm;(d)10cm.

OPTICAL DESIGN FOR BIKE FORWARD LIGHTING

After completing the optical model of the light source, we turn to design a compact optical module for the bike lamp. The size of the geometrical structure is about 30×20×30 mm^3 shown as Fig. 3. The light pattern is shown as Fig. 4. In order to reduce the ring-like noise, we modify the curvature of the mirror. The modified light pattern is shown as Fig. 5. According to the K-mark regulation for bike lamp in Germany, the contrast of the cut-off line should be higher than 5. In the design, the contrast of the pattern reaches 15 to avoid contrast failure after fabrication. The illuminance of the hot spot is about 10.5(lx) at 10 m. However, the distribution of the illuminance in the central region at the regulation is like a Gaussian distribution. Therefore, its light pattern is more acceptable. The roadway illumination is also designed to have a high performance, as shown in Fig. 6. The optical efficiency is around 55%.

Fig. 3 Geometrical structure with a small size.

Fig. 4 Light pattern at the regulation plane.

Fig. 5 Optimized pattern at the regulation plane.

Fig. 6 Light pattern on the roadway.

SUMMARY

In the paper, we propose a design of a compact optical module of bike K-mark regulation based on LEDs. The optical model of the LED has been constructed successfully with the mid-field model. The NCC between the simulation and the experimental measurement is always higher than 99%. Based on the precise model, we design the optical module with a high contrast and uniform pattern in the central region at the regulation pattern. In the design, the contrast of the pattern is higher than 15. The size of the geometrical structure is about 30×20×30 mm^3 and the optical efficiency is about 55%.

ACKNOWLEDGMENT

This study is sponsored by the National Science Council of the Republic of China with the contract no. 97-2221-E-008-025-MY3. The authors would like to thank Breault Research Organization for the support of ASAP simulation program.

REFERENCES

[1] A. Zukauskas, M. S. Shur, and R. Caska, *Introduction to Solid-state Lighting,* (John Wiley & Sons, New York, 2002).

[2] D. A. Steigerwald, J. C. Bhat, D. Collins, R. M. Fletcher, M. O. Holcomb, M. J. Ludowise, P. S. Martin, and S. L. Rudaz, "Illumination with Solid State Lighting Technology," IEEE J. Selected Topics in Quantum Electron **8**, 310-320 (2002).

[3] http://www.cree.com/

[4] S. J. Lee, "Analysis of light-emitting diode by Monte Carlo photo simulation," Appl. Opt. **40**, 1427 (2001).

[5] C. C. Sun, C. Y. Lin, T. X. Lee, and T. H. Yang, "Enhancement of light extraction of GaN-based LED with introducing micro-structure array, " Opt. Eng. **43**, 1700 (2004).

[6] T. X. Lee, C. Y. Lin, S. H. Ma, and C. C. Sun," Analysis of position-dependent light extraction of GaN-based LEDs," Opt. Express **13**, 4175 (2005).

[7] C. C. Sun, T. X. Lee, S. H. Ma, Y. L. Lee, and S. M. Huang, "Precise optical modeling for LED lighting verified by cross correlation in the midfield region," Opt. Lett. **31**, 2193-2195 (2006).

[8] Wei-Ting Chien, Ching-Cherng Sun , and Ivan Moreno, "Precise optical model of multi-chip white LEDs," Opt. Express **15**, 7572 (2007).

EMI from high-pressure Xe lamps – a clean experimental study of the underlying near-anode plasma instability

U. Hechtfischer[1] and T. Vos[2]

[1] *Philips Lighting, GBU Automotive Lamps, Philipsstraße 8, 52068 Aachen, Germany*
[2] *Philips Lighting, GBU HID, P.O. Box 80020, 5600 JM Eindhoven, The Netherlands*
Contact: ulrich.hechtfischer@philips.com

ABSTRACT

EMI from high-pressure Xe lamps is linked to a periodic lamp-voltage instability, probably in the near-anode plasma. We have measured the dependence of the pulse rate on lamp current and anode temperature independently and found an unexpected clear effect of anode temperature, not lamp current.

INTRODUCTION

High-pressure xenon (Xe) discharge lamps [1], both the classic types filled with only Xe and the Xe-metal-halide types occasionally cause strong electromagnetic interference (EMI). The lamps then emit a broad (GHz) RF spectrum that can disturb radio transmission. The physical origin of this phenomenon is unclear and the little existing knowledge mainly heuristic: The EMI normally occurs below a certain critical lamp current [2] and has to do with the design of the anode [3]. This points to an effect in the near-anode plasma.

Understanding the true origin is very important, for both practical reasons (design of EMI-free lamps) and fundamental reasons (plasma physics). The first step, before the development of a theory, is a clean experimental characterization. The very few existing descriptions [4, 5, 3] show a fragmentary, inconsistent picture: There are doubts as to the origin at the anode [4], and the observed dependencies on lamp current I, current density j, anode shape, anode temperature T_a, *etc.* are not really clear. The onset of the phenomenon is generally believed to be I-controlled. Here, we present a new experimental study that decouples the main EMI-influencing factors anode temperature and current (density) in a much cleaner way than before, with surprising results.

EXPERIMENT

The high-pressure Xe experimental lamps were in principle normal Xe mercury-free metal-halide quartz discharge lamps for automotive headlighting with a rated power of 35 W, but without the metal-halide fill. The lamps had an arc gap of 3.6 mm between electrode rods of 0.3 mm diameter and about 2 mm free length. The Xe pressure (cold) was about 1.5 MPa. The lamps were operated on a laboratory current source with freely programmable current $I(t)$ between 0.2 A and 1.2 A. In order to measure electrode temperature, monitor electrode geometry, and observe the near-electrode plasma, the lamps were imaged with a CCD camera.

At low current, the test lamps typically showed EMI, which was monitored not directly by detecting the radiation with an antenna [5, 3], but indirectly by measuring the common-mode current I_{cm} with a high-frequency current probe around both lamp wires together. The I_{cm} signal consisted of periodic RF bursts, with a repetition frequency ϵ in the range 1–10 MHz. The bursts coincided with sharp lamp-voltage spikes like those described by Anders *et al.* [4]. During a spike, U rose from about 31 V to 38–39 V during only a few nanoseconds. This voltage instability is the reason for the EMI, the sharpness of the spikes being responsible for the broad spectrum.

Our aim was to measure the occurrence and pulse repetition frequency ϵ of the instability as a function of I and T_a; varying these *independently* is difficult because the anode heat load is roughly proportional to I [6, 7]. In the earlier work [4, 5, 3], this correlation has not been taken fully into account. Here, an independent variation of I and T_a in the same lamp was achieved by applying a stepped-DC pattern $I(t)$ and measuring T_a and ϵ [Fig. 1 (a)]. The T_a values (dots) follow $I(t)$ with a clear lag (time constant $\approx$ 60 ms) due to the heat capacity of the electrode; this creates pairs of a given I value with unusual (i.e. transient) T_a values. By carefully designing the $I(t)$ pattern (length $\approx$ 20 s), we could map out the I–T_a parameter space in Fig. 1 (b) (I = 0.2–1.2 A, T_a = 1700–3400 K) in a single measurement.

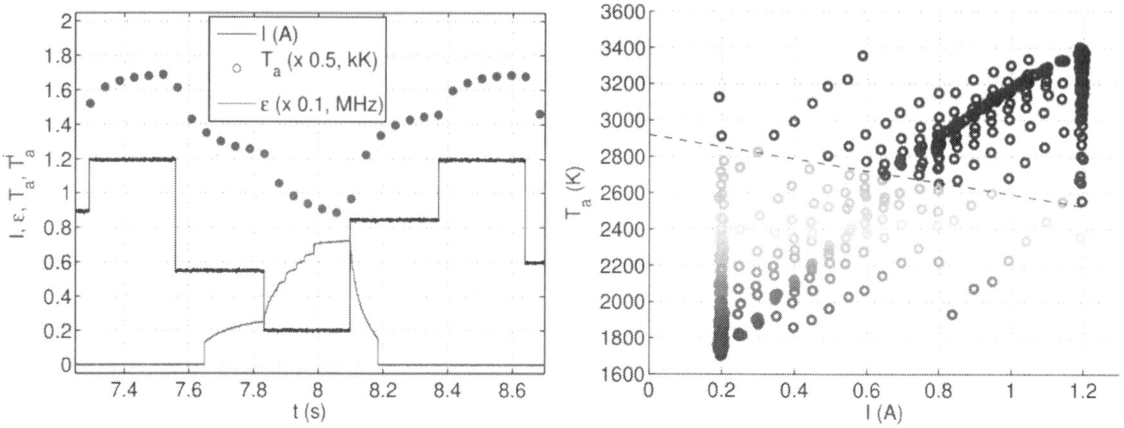

Figure 1. (a) Measured $I(t)$, T_a, and pulse-repetition frequency ϵ during the stepped-DC experiment; (b) resulting I, T_a parameter space (blue dots: stable; colored dots: instable, false color indicating the frequency ϵ).

RESULTS

Surprisingly, the boundary between the stable region [blue dots in Fig. 1 (b)]) and the instable region (colored dots) is almost horizontal, i.e., the existence of the instability depends mainly on T_a and hardly on I. The linear boundary is approximately $2920\,\text{K} - 330\,\text{K/A} \times I$ (dashed line). Even the small apparent influence of I on the boundary is most probably only an experimental artifact: T_a could not be measured at the true electrode tip, but only on the shaft, near the tip. An estimate for the true tip temperature is $T_a^{\text{tip}} = T_a + R_{\text{tip}}P_{\text{in}}$, with R_{tip} the thermal resistance between the tip and the position where T_a was measured, $P_{\text{in}} = U_{\text{H,a}}I$ the anode heat load, and $U_{\text{H,a}} = 5.5\,\text{V}$ the anode heating voltage [7]. Based on thermal simulations of the electrode, we estimated $R_{\text{tip}} \approx 50\,\text{K/W}$, slightly depending on the assumed size of the arc attachment. This leads to $R_{\text{tip}}U_{\text{H,a}} = 275\,\text{K/A}$, similar to the slope of the boundary in Fig. 1 (b). Thus, if T_a^{tip} rather than T_a was used to plot the existence region, the onset of the instability would depend *only* on anode (tip) temperature. Similarly, even the frequency ϵ [shown as color in Fig. 1 (b)] depends mainly on temperature. The experiment was repeated for different anode front shapes (in the same lamp, via melting) and with other parameters varied, but the results (boundary, ϵ characteristic) changed little.

These results contradict earlier conclusions [4, 3], but the actual earlier data turn out to be compatible with the new premise that anode temperature is most important for controlling the voltage instability in the near-anode plasma. No theoretical explanation for this influence nor for the instability itself will be tried at this point, but the consistent experimental picture now at hand will be both a valuable help in practical lamp engineering, and a tight constraint for any future theoretical treatment of the effect.

REFERENCES

[1] M. Rehmet, IEE Proc. A **127**, 190 (1980).

[2] Perkin Elmer Optoelectronics, *Cermax Lamp Engineering Guide*, available from *http://optoelectronics.perkinelmer.com/content/manuals/gde_cermax.pdf.*

[3] O. B. Minayeva and D. A. Doughty, Abstracts of the 60th Gaseous Electronics Conference 2007 (Arlington, VA, USA), Bull. Am. Phys. Soc. **52**, No. 9.

[4] A. Anders, H. Schneidenbach, and D. Sünder, IEEE Trans. Plasma. Sci. **19**, 324 (1991).

[5] U. Krönert, M. Stucky, and U. Müller, *Abschlussbericht BMBF Projekt 13N7110/2* (Fachhochschule Kaiserslautern, Germany, 2000).

[6] J. Mentel and J. Heberlein, J. Phys. D: Appl. Phys. **43** 023002 (2010).

[7] N. A. Almeida, M. S. Benilov, U. Hechtfischer, G. V. Naidis, J. Phys. D: Appl. Phys. **42**, 045210 (2009).

System Integration Options for LED Module

A. Baiano[1], H.w.v. Zeijl[1], G. Cheng[1], S. Tarashioon[1], P. Sarro[1], G.Q. Zang[2]
[1]*Delft Institute of Microsystems and Nanoelectronics (DIMES) of Delft University of Technology, Feldmaanweg 17, Delft, 2628CT, The Netherlands*
[2]*Faculty Mechanical, Maritime and Materials Engineering of Delft University of Technology, Mekelweg 2, Delft, 2628CD, The Netherlands*
Contact: a.baiano@tudelft.nl

ABSTRACT

An overview of a multi-functional, compact and cost-effective light emitting diode (LED) module is given. The most compact LED module proposed is made by a driver supplying LED, which is working as light emitting and light detector. Furthermore, measurement result of AN4240 high-brightness LED of Seoul Semiconductor shows that current ratio of 10^3 for on/off ambient light is obtained, making it possible the use of LED as photodetector.

INTRODUCTION

Solid state lighting (SSL) is a novel lighting technology that allows the introduction of new lighting concepts and functionalities. Furthermore, advances in light emitting diode (LED) technology like high lumen for watt (lum/W), improved reliability and cost reduction allows the manufacturing of light sources and luminaries with more functionality novel form factors etc. compared with conventional lighting systems. One of the key components in SSL is the LED driver circuit. Consequently, research on driver topology for LEDs has been extensively enhanced, not only to optimize the driver design with respect to efficiency, also a high level of integration of the driver in the SSL system allows the co-integration of logic and sensor technology to enhance the performance and reliability of the SSL system. In this article, several integration strategies will be reviewed.

LED MODULE INTEGRATION OPTIONS

The main requirement for a LED driver is to provide a stable flicker free LED light output. To obtain this, AC/DC or DC/DC converter (depending on power source) must be used. The converter topologies are basically classified in three categories: linear, switching [1]-[4] and charge-pump mode [5]-[6]. This choice is based on the application which LED is used for. The linear mode is the simplest in designing and the most compact because of it is made without any inductor or transformer, which are bulky and non-scalable components. However, in applications where output voltage is substantially lower than the input voltage, switching mode is more suitable than the linear mode because of its high efficient power conversion. Furthermore, linear topologies cannot provide an output voltage higher than input voltage (step-up conversion). In such a case, switching mode is essential. Recently, an important attention is also given to the charge-pump mode [5]-[6], where high efficiency can be obtained without the use of inductors. Only capacitors are used as energy storage elements to create either a step-down or step-up DC/DC converter. Formerly, charge-pump driver was only for low-voltage application. Nowadays, there are charge pump topologies that are suitable for high voltage applications [6]. For a high level integration of the driver circuit with logic and sensor circuits a monolithic or heterogeneous integration on IC level is preferred. Consequently the driver topology must be suitable for IC integration.

Several driver-sensor integration schemes are given in figure 1, ranging from the most basic LED module (Fig. 1a) to advanced SSL systems equipped with multiple sensors and logic (Fig 1e). The sensor integration has consequences for the driver design. In general 3 types of integration can be categorized: (I) sensor integration at the driver output side, (II) sensor integration at the driver input side and (III) advanced multiple sensor integration with logic. Obviously cost and functionality are important boundary conditions and the integration scheme must match with the application economics. In integration scenario (I) either the LED is used as a photo diode (Fig. 1b) or a sensor with an identical form factor as the LED is connected parallel to the LED (Fig 1c). This scenario allows a cost effective integration of the sensor in the luminaries. The IV characteristics of two AN4240 high-brightness LED's (Seoul Semiconductor) are given in figure 2. Despite the different forward and dark reverse characteristics, the reverse bias current under illumination is similar for both devices indicating the potential to use LED as photo detectors for ambient light sensing. Integration scenario (II) requires an integration of sensor control electronics at the input side of the driver but the driver output can be single mode (Fig 1d). Finally, in figure 1e an integration scenario is given where multiple sensors are integrated in the SSL module. This however requires high level logic to process the different sensor inputs.

As a conclusion, experimental result shows that LED modules can be used as photodiodes. This, together with an optimized driver design can yield a cost effective integration of a light sensor into medium level SSL systems. Furthermore, several sensor-driver integration scenarios are discussed, ranging from basic to high level multi sensor SSL systems. These scenarios can be used as guideline for optimal sensor-driver designing.

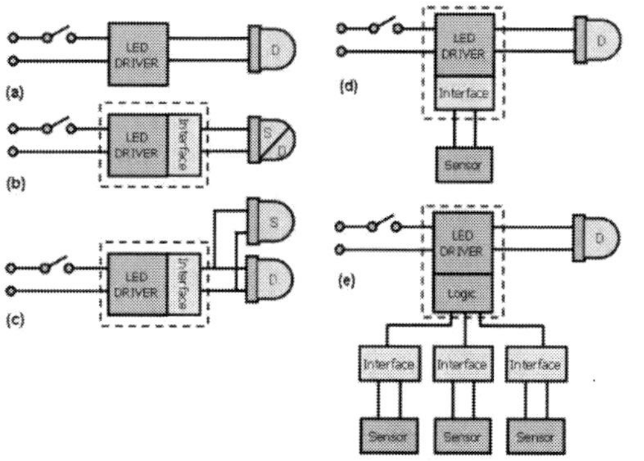

Fig. 1. Different driver-sensor integration scenario of LED module.

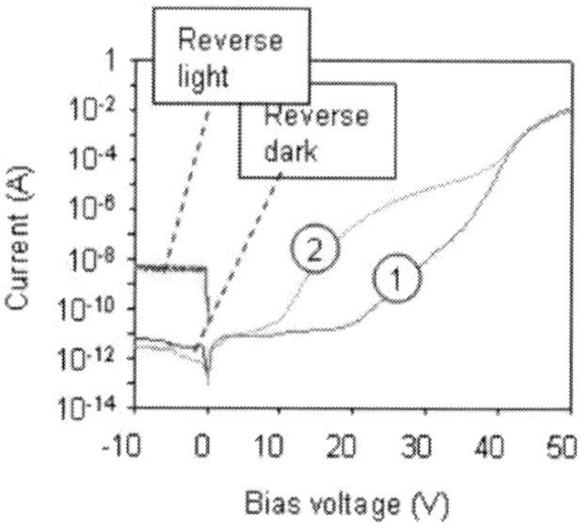

Fig. 2. Forward and reverse bias measurement result of LED under light and dark light conditions. The device under test is the high brightness AN4240 Acriche of Seoul Semiconductor.

REFERENCES

[1] Brown, Marty. 2001. Power Supply Cookbook. Woburn MA: Newnes.

[2] Pressman, Abraham I. 1998. Switching Power Supply Design. New York: McGraw-Hill.

[3] Billings, Keith. 1999. Switch-Mode Power Supply Handbook. New York: McGraw-Hill.

[4] Harrison, Linden T. 2005. Current Sources & Voltage References. Burlington MA: Newnes.

[5] Pelliconi, R.; Iezzi, D.; Baroni, A.; Pasotti, M.; Rolandi, P.L.; , "Power efficient charge pump in deep submicron standard CMOS technology," *Solid-State Circuits, IEEE Journal of*, vol.38, no.6, pp. 1068- 1071, June 2003.

[6] Richard, J.-F.; Savaria, Y.; , "High voltage charge pump using standard CMOS technology," *Circuits and Systems, 2004. NEWCAS 2004. The 2nd Annual IEEE Northeast Workshop on* , vol., no., pp. 317- 320, 20-23 June 2004.

Modification of the blackbody radiation using a selective emitter

T. Matsumoto[1], S. Omori[1], M. Tomita[2]
[1] *Research and Development Center, Stanley Electric Corporation, 5-9-5 Tokodai, Tsukuba 300-2635, Japan*
[2]*Department of Physics, Shizuoka University, 836 Ohya, Suruga 422-8529, Japan*
Contact: takahiro_matsumoto@stanley.co.jp

ABSTRACT

We reduced the infrared contribution to the blackbody radiation spectrum and shifted the peak emission to shorter wavelengths by using a selective emitter with high emissivity in the visible wavelength region and low emissivity in the infrared wavelength region. The conversion efficiency from input electric power to the shorter wavelength radiation was quantitatively evaluated and the efficiencies in excess of 80% were produced.

INTRODUCTION

Incandescent lamps based on the principle of blackbody radiation have many advantages compared with other types of modern electric light sources: they require no external regulating equipment, and are inexpensive to manufacture. As a result, the incandescent lamp has been used worldwide for room lighting and car headlights for more than 100 years. However, the luminous efficiency of the incandescent lamp is low, on the order of 15-20 lm/W. Therefore, the incandescent lamp is gradually being replaced by other types of electric light sources such as fluorescent lamps, high-intensity discharge lamps, and light-emitting diodes. These newer technologies seem to provide more visible light and less heat for a given electrical energy input; thus, some jurisdictions are in the process of eliminating incandescent light bulbs in favor of more energy-efficient lighting.

Instead of turning to new light-source technologies, an incandescent lamp of higher luminous efficiency could be obtained by modifying the blackbody radiation of the incandescent bulb itself, that is, by reducing the infrared radiation contribution to the blackbody spectrum without reducing the visible wavelength radiation. Many studies have reported the modification of blackbody radiation using one- [1,2], two- [3,4], or three-dimensional [5,6] tailored structures. All of these studies were based on sophisticated and time-consuming technologies. In addition, these studies have not provided clear evidence for the modification of the blackbody radiation spectrum.

Here we demonstrate that using a selective emitter with high emissivity in the shorter wavelength region and low emissivity in the longer wavelength region, we obtained the modification of the blackbody radiation spectrum whose shorter infrared radiation output was enhanced and whose longer infrared contribution was reduced. Precise measurements of the emitter's radiation spectrum at several temperatures and the conversion efficiencies from input electric power to shorter wavelength radiation were quantitatively evaluated with high accuracy. We found that the efficiency of the selective emitter exceeded 80 %.

EXPERIMENTAL AND RESULTS

Figure 1 shows the quantitative model that explains this modification of the blackbody radiation. Based on Kirchhoff's law for a system in thermal equilibrium in which transmittance is negligibly small, the emissivity $\varepsilon(\lambda)$ can be related to the reflectance $R(\lambda)$, as $\varepsilon(\lambda)=1-R(\lambda)$. Therefore, a material with high reflectance in the infrared wavelength region and low reflectance in the visible wavelength region (the brown line) will display a reverse trend in emissivity (the blue line). The thermal radiation spectrum $\Phi(\lambda)$ becomes the product of the blackbody radiation spectrum $B(\lambda)$ (green line) and the emissivity $\varepsilon(\lambda)$ as $\Phi(\lambda)=\varepsilon(\lambda)\,B(\lambda)$ as shown in red. Thus, we can reduce infrared radiation by imposing a step-function-like reflectivity onto the surface of the heating material. The step-function-like reflectivity was obtained by the thin film deposition performed by co-sputtering Cr metal and Cr_2O_3 oxide onto a copper substrate. The thickness of the Cr and Cr_2O_3 co-deposited film was 100 nm.

Figure 2 shows the thermal radiation spectrum of the emissivity-modulated layer obtained using a

Fourier transform infrared spectrometer at 580 K (red circles). For comparison, the thermal radiation spectrum from a copper plate was included (open squares). To protect the sample from oxidation and energy losses due to natural air convection, the sample was placed in a vacuum chamber that was evacuated to about 10^{-3} Pa. The temperatures of the sample and the chamber were measured using type-K thermocouples (0.1Φ). The thermal radiation spectrum of the copper plate clearly obeys Planck's law of blackbody radiation.

On the other hand, the peak position of the thermal radiation spectrum of the emitter (solid circles) is shifted toward shorter wavelengths, the longer-wavelength emission is reduced, and the shorter-wavelength emission is enhanced compared to the thermal radiation spectrum of the copper plate. The reductions in longer-wavelength emission, along with the enhancements of shorter-wavelength emission at each temperature, clearly demonstrate the wavelength dependence of the emissivity.

CONCLUSION

If we can construct an emitter with a reflectance cutoff wavelength in the visible region, and the emitter does not degrade at high temperatures, the radiation spectrum of the incandescent light source will have no infrared component. This complete reduction of the infrared radiation will ensure a highly efficient incandescent light source.

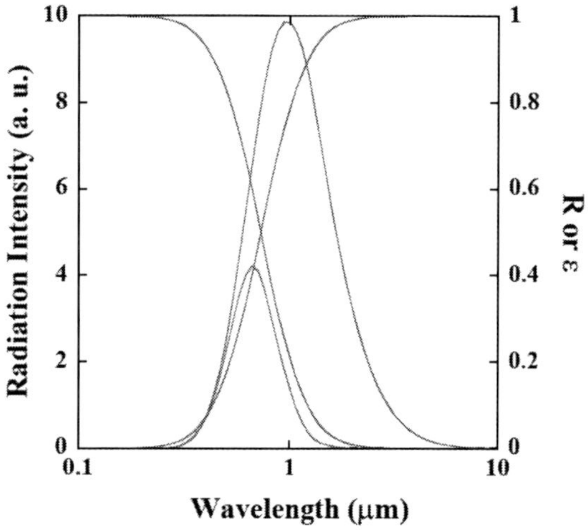

Figure 1 Quantitative model for the modification of the blackbody radiation. The black line represents the reflectance spectrum, the blue line is the emissivity spectrum described by ε(λ) = 1 - R(λ), the green line is the blackbody radiation spectrum, and the red line is the modulated thermal radiation created by the product of the blackbody radiation spectrum and the emissivity ε(λ).

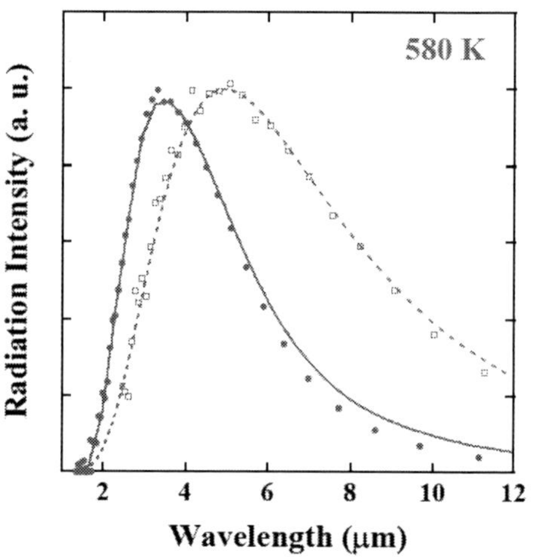

Figure 2 Thermal radiation spectrum from the emitter obtained by a FTIR spectrometer at 580 K (red circles). For comparison, the thermal radiation spectrum from a copper plate is shown (open squares).

REFERENCES

[1] J. J. Greffet, R. Carminati, K. Joulain, J. P. Mulet, S. Malnguy, Y. Chen, Nature 416 (2002) 61.
[2] P. J. Hesketh, J. N. Zemel, B. Gebhart, Nature 324 (1986) 549.
[3] F. Kusunoki, T. Kohama, T. Hiroshima, S. Fukumoto, J. Takahara, T. Kobayashi, Jpn. J. Appl. Phys. 43 (2004) 5253.
[4] H. Sai, Y. Kanamori, H. Yugami, Appl. Phys. Lett. 82 (2003) 1685.
[5] J. G. Fleming, S. Y. Lin, I. El-Kady, R. Biswas, K. M. Ho, Nature 417 (2002) 52.
[6] S. Y. Li, J. G. Fleming, I. El-Kady, Opt. Lett. 28 (2003) 1909.

Multi-wavelength DBD-driven exciplex lamp operated with the mercury bromide/rare gases mixtures

M. M. Guivan[1], A. A. Malinina[1], A. Brablec[2]
[1]*Department of Quantum Electronics, Uzhgorod National University, Pidgirna 46, Uzhgorod 88000, Ukraine*
[2]*Department of Physical Electronics, Masaryk University, Kotlářska 2, Brno 61137, Czech Republic*

Contact: m_guivan@rambler.ru

ABSTRACT

The emission spectra from atmospheric-pressure dielectric barrier discharge (DBD) with $HgBr_2$/He or $HgBr_2$/Xe/Kr mixtures, as well as the electrical characteristics have been investigated at the repetition frequency of sinusoidal voltage pulses up to 125 kHz. In the spectra, the research has revealed radiation from HgBr(*B–X, C–X*) exciplex molecules, atomic lines of mercury and rare gases, and, in mixtures with xenon, radiation of XeBr(*B–X, B–A*) exciplex molecules. Regularities in the spectral characteristics of the radiation from the gas-discharge plasma are discussed. The high-frequency atmospheric-pressure barrier discharge in mixtures of mercury dibromide with gases can be used in multi-wavelength exciplex lamps, operating in the UV and visible regions.

INTRODUCTION

Mercury monohalides HgX* (X=Cl, Br, I) are known to be the intense radiation sources in the visible spectral range [1]. Up to now no data are available in the literature on excitation processes of mercury bromide at high repetition rates ($f \sim 100$ kHz) under conditions of high-pressure gas discharge plasma, which is necessary for the creation of efficient and powerful sources of visible light. In this study, the results of our investigation of the spectral and electric characteristics of atmospheric-pressure DBD-driven small-size lamp ($f = 125$ kHz) based on mixtures of mercury dibromide vapour with helium or xenon and krypton have been presented.

RESULTS AND DISCUSSION

A small-size cylindrical DBD-driven radiator made of a quartz tube with an outer diameter of 6 mm and having one dielectric barrier with the capacitance of $C_d = 6$ pF was used. A molybdenum electrode was placed inside the radiator along its axis. The discharge gap was 1.5 mm. The external electrode, placed on the tube surface, was made of a mesh with the transmission of 90%. To excite a working mixture a generator of high voltage AC sinusoidal pulses was used; peak-to-peak voltage up to 10 kV, $f = 1 - 125$ kHz. The voltage U applied to the lamp was measured using a high voltage probe (Tektronix P6015A). The waveforms of current I and charge Q were registered by means of a 50 Ω resistor or low-inductance 8 nF capacitor placed in series with the lamp and monitored by digital oscilloscope LeCroy WaveRunner 6100A (1 GHz, 10 GS/s). The radiation emitted by the discharge was registered and analyzed in the spectral range of 200–740 nm. The spectra were recorded by the Jobin Yvon FHR 1000 monochromator (grating 2400 grooves/mm, quartz optical fiber, the spectral resolution of about 8 pm) with the high-speed CCD detector.

The characteristics of the lamp were studied in the $HgBr_2$/He and $HgBr_2$/Xe/Kr mixtures. At the initial stage (the first 20–30 s), the colour of the discharge was dependent on the component composition of a mixture. As the mixture self-heated, the discharge colour became green-blue.

In each half-cycle of the applied voltage U_0, the current waveform consisted of a displacement current and conductivity current peaks, with characteristics depending on the operating frequency and applied voltage. The peak current did not exceed 52 mA and duration was in the range 0.2–0.5 μs.

The dynamics in the emission spectra of HgBr* exciplex lamp operated with the $HgBr_2$/Xe/Kr mixture (8% Xe + 92% Kr) at total atmospheric pressure and $f = 120$ kHz with changing of temperature is shown in figure 1. At low temperatures ($T \sim 40$°C) the emission bands of the exciplex XeBr($B \rightarrow X$) and XeBr($B \rightarrow A$) molecules peaked at 281 and 320 nm together with the atomic mercury lines and OH* bands have been observed (figure 1(a)). In the spectra obtained, a wide band

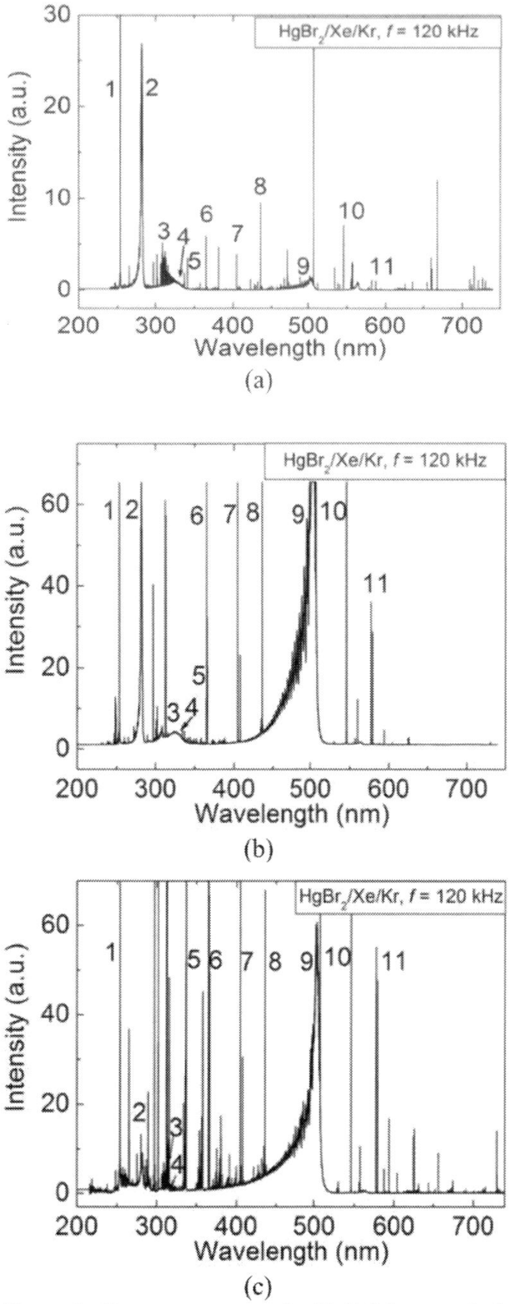

Figure 1: *Emission spectra of a HgBr* lamp with the $HgBr_2/Xe/Kr$ mixture: (a) T = 40°C, (b) T = 80°C, (c) T = 160°C. 1 – 254nm Hg*, 2 – 281nm XeBr*, 3 – 306nm OH*, 4 – 320nm XeBr*, 5 – 337nm N_2*, 6 – 365nm Hg*, 7 – 404nm Hg*, 8 – 436nm Hg*, 9 – 502nm HgBr*, 10 – 546nm Hg*, 11 – 578nm Hg*.*

system peaking at λ = 502 nm stands out, which intensity increased with the increasing of temperature (figure 1 (b, c)). This system corresponds to the electronic–vibrational $B\,^2\Sigma^+ \rightarrow X\,^2\Sigma^+$ transition of the exciplex HgBr* molecules [2]. The HgBr($B \rightarrow X$) radiation manifests itself starting from $\lambda \sim$ 340 nm. In addition, the weak HgBr($C \rightarrow X$) bands were observed in the range of 285–295 nm.

Besides the HgBr($B \rightarrow X$) radiation and mercury lines, the emission spectra of DBD in the $HgBr_2/He$ mixture included atomic lines of the buffer gas helium at 589 and 706 nm, as well as the $N_2(C^3\Pi_u \rightarrow B^3\Pi_g)$ bands appearing due to a presence of nitrogen as a residual gas (p<10 Pa). The intensity of the helium lines decreased considerably as the working medium warmed up.

It can be assumed that the emission at the observed spectral bands and lines probably occurs through the following reactions:

$HgBr_2 + e \rightarrow HgBr^* + Br + e$ or (Br^-)

$HgBr_2 + Xe^* \rightarrow HgBr^* + Br + Xe$,

$HgBr^* \rightarrow HgBr + h\nu$ (λ_{max} = 502 nm),

$HgBr_2 + e \rightarrow Hg^* + 2Br\ (Br_2^*, Br_2, Br_2^-) + (e)$,

$HgBr_2 + e \rightarrow Hg^* + Br_2^+ + 2e$,

$HgBr_2 + e \rightarrow Hg^* + Br^+ + Br + 2e$,

$HgBr_2 + e \rightarrow Hg^* + Br^* + Br + e$,

$Xe^*(Xe_2^*), + Br_2 \rightarrow XeBr^* + (Xe) + Br$,

$Xe^+\ (Xe_2^+, Xe_3^+) + Br_2^- + M \rightarrow XeBr^* + Br +$ (Xe, 2Xe) + M,

$Xe^+ + Br^- + M \rightarrow XeBr^* + M$,

$XeBr^* \rightarrow XeBr + h\nu$ (λ_{max} = 281 nm),

where M is the third particle.

SUMMARY

The intense broadband radiation of HgBr($B \rightarrow X$) exciplex with a peak at λ = 502 nm has been revealed from the dielectric barrier discharge (f = 120 kHz) in a self-heating mode. The multi-wavelength operation mode of the DBD-driven atmospheric pressure high-frequency HgBr* excilamp has been realized. It has been shown the possibility to regulate the emission spectrum of the excilamp by means of changing the working temperature. An atmospheric-pressure barrier discharge with the $HgBr_2/Xe/Kr$ mixture can be used for the creating a self-heated excilamp emitting in the UV and green-blue spectral regions.

REFERENCES

[1] Ch. K. Rhodes, Excimer Lasers (Berlin: Springer) 1979 &1984.

[2] R.W. Pearse, A.G Gaydon, The Identification of Molecular Spectra (New York: Wiley) 1963.

ElectroCandoLuminescence - A Novel Light Source Technology

S. Kitsinelis, H. Motomura and M. Jinno

Department of Electrical and Electronic Engineering, Ehime University, Japan, 3 Bunkyo-cho, Matsuyama, Ehime 790-8577, JAPAN

Contact: skitsinelis@ath.forthnet.gr

ABSTRACT

We describe a new technology that fuses the effect of candoluminescence with the technology of electrical incandescence. It is known that under heating certain materials emit light of shorter wavelengths than one would expect from a black body under the same temperature. This effect of candoluminescence has been described in the past with the use of various materials under flames and also utilised in gas lamps. In this paper we demonstrate that a common material such as silica that exhibits the candoluminescent effect strongly under a flame emitting very bright white light can be added to a typical incandescent lamp in order to change the lamp's emission characteristics. The heating of the filament which is at temperatures similar to most flames can induce the candoluminescent effect and the end result could be an evolved but still simple incandescent lamp of much higher efficacy, colour temperature and flux.

GENERAL

Candoluminescence is the term used to describe the light given off by certain materials which have been heated to incandescence and emit light at shorter wavelengths than would be expected for a typical blackbody radiator. The phenomenon is noted in certain transition metal and rare earth metal oxide materials (ceramics) such as zinc oxide and cerium oxide or thorium dioxide, where some of the light from incandescence causes fluorescence of the material. The cause may also be due to direct thermal excitation of metal ions in the material [1,2]. The emission of a bright white light from CaO when heated by an oxygen/hydrogen flame was known as the limelight and its most frequent use was in theatres and during shows to illuminate the stage and performers.

The most frequently seen case where a substance is emitting bright white light even under the heating of a flame that does not exceed 2500 degrees Celsius, is the case of silicon dioxide (fused silica). It is possible perhaps to fuse together this luminescence phenomenon with the incandescent heating of a simple electric lamp and in a sense fuse together the archaic with the modern way of lighting. The fusion of silica and tungsten could provide means of generating bright white light even if the laws that govern incandescence and black body radiation would allow only for a warm/yellowish emission of light with low efficiency.

The authors have recreated the effect of the bright white luminescence of silica using electric heating instead of a flame assuming that a glowing filament is at a temperature similar to that of a hydrogen or methane mixed with oxygen (Methane (natural gas) in air 1950 °C, Hydrogen in air 2055 °C, Propane with air 1995 °C). In the case of hydrogen/oxygen mixture, a maximum temperature of about 2800 °C is achieved with a pure stoichiometric mixture, about 700 degrees hotter than a hydrogen flame in air. In the case of an incandescent lamp the filament is heated to about 2800 degrees (giving black body light emission of the same colour temperature) and in some cases such as the halogen lamp, the temperature can be raised to 3100 degrees. Therefore, both the electrically heated filament and a blow torch reach similar temperatures.

The author prepared lamps with silicon attached to the filament and operated them under DC mode at 20 watts. For the first time it was observed and recorded (see figures 1 and 2) the emission of bright light from the silica which was inserted inside the filament coil spiral in order to ensure good contact between them and reduction of heat losses. The filament glows with a warm white light after 15 watts of power input and at 20 watts which was the maximum power input, the source fluctuates with short bursts of bright white light of higher colour temperature (colder white). This was the effect of the silica reaching the right temperature for the luminescence effect to appear. The lamps gave us for the first time a demonstration of this luminescence effect, previously seen only under flame heating.

Since silicon, at a temperature of less than 3000 degrees, emits light of significantly higher colour temperature it could be possible to design an incandescent lamp that could offer, using this effect, light emission of much higher colour temperature and higher efficacy. Some problems to be addressed are the survival of the filament and the maintenance of silica near the hot spots.

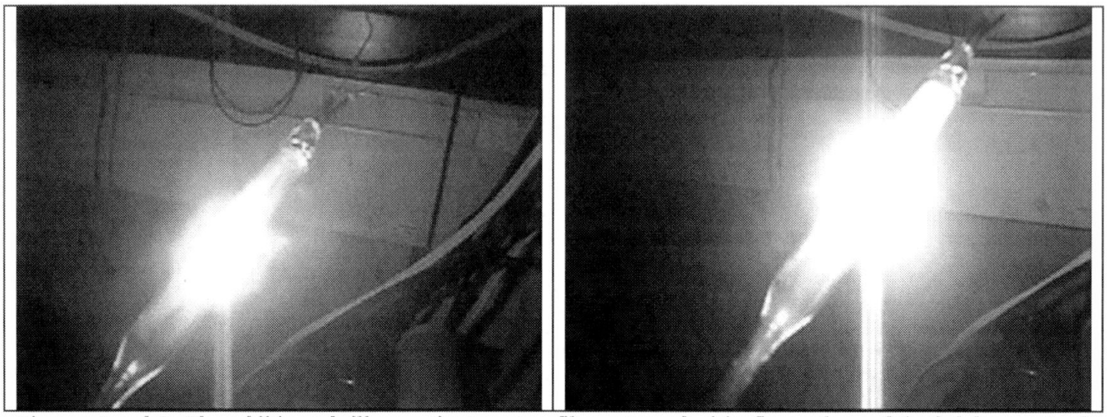

Figures 1 and 2. The addition of silica on the tungsten filament resulted in fluctuations of the brightness. In the left column one sees the brightness of the tungsten filament under 20 Watts. In the right column one sees the bursts of brightness increases when silica is added.

The many advantages that incandescent lamps offer such as the simplicity of their operation and the low cost will probably keep them in the market for a while longer. However if they want to compete with other existing and future technologies, especially in the efficiency game, then the incandescent lamp must undergo some major developments. We propose that **Electrocandoluminescence** or in general the use of materials capable of converting infrared energy to visible light is perhaps one way to go. Potentially more research on filament exotic materials that emit more visible light under heating could be another way.

REFERENCES

[1] High efficiency light source utilizing co-generating sources, US patent number 6,268,685 - D. Stark and A. Chen

[2] Candoluminescent electric light source, US patent number 4,539,505 – L. Riseberg

Luminance and efficacy improvement of mercury-free fluorescent lamps by synchronized high voltage application to auxiliary external electrode

Kojiro Oka, Kyosuke Miyamoto, Hideki Motomura, Masafumi Jinno
Department of Electrical and Electronic Engineering, Ehime University
3 Bunkyo-cho, Matsuyama 790-8577, Japan
Contact: hmoto@mayu.ee.ehime-u.ac.jp

ABSTRACT

The authors have been developing mercury-free fluorescent lamps and reported a luminance and efficacy improvement method by using an auxiliary external electrode (AEE) [1,2]. In this study, instead of simply grounding the AEE, another HV pulse was applied to the AEE, which was synchronized to the driving HV pulse and further luminance and efficacy improvement was achieved.

INTRODUCTION

The authors have been developing mercury-free fluorescent lamps and reported a luminance and efficacy improvement method by using an auxiliary external electrode (AEE) [1,2]. In this method, an aluminum strip was wound as an AEE on the outside of the bulb near the high voltage (HV) electrode. By grounding the AEE, the positive column was expanded. As a result, positive column contraction and self absorption of vacuum ultraviolet (VUV) emission were suppressed and the luminance and the efficacy were improved. In this study, instead of simply grounding the AEE, another HV pulse was applied to the AEE, which was synchronized to the driving HV pulse for further improvement.

EXPERIMENTAL SETUP

Figure 1 (a) shows a schematic of the mercury-free fluorescent lamp used in this study. The bulb was made of Pyrex and its inside diameter was 26 mm. The distance between main electrodes was 50 mm. An aluminum tape of 3 mm wide was wound 7.5 mm from the HV electrode (cathode) on the outside of the bulb as the AEE. The lamp was filled with xenon with 50 Torr at room temperature. This lamp was driven by pulsed power supply shown in Fig. 1 (b). The negative polarity driving HV pulse to the main electrodes was generated by switching of the FET 1. Another HV pulse with opposite positive polarity was generated by FET 2 (push-pull module, Behlke, HTS-81-06-GSM) and applied to the AEE. The frequency and the pulse width of both HV pulses were set at 30 kHz and 3 μs. Waveforms of the applied voltages and the injected charge were measured with an oscilloscope (Tektronix, TDS3054) and HV (Tektronix, P6015A) and low voltage (Tektronix, P6139) probes. Two 2.3 nF capacitors were used to measure the injected charge. The input power was calculated using *V-Q* Lissajous figure. Luminance distribution was measured with luminance meter (Topcon, SR-3).

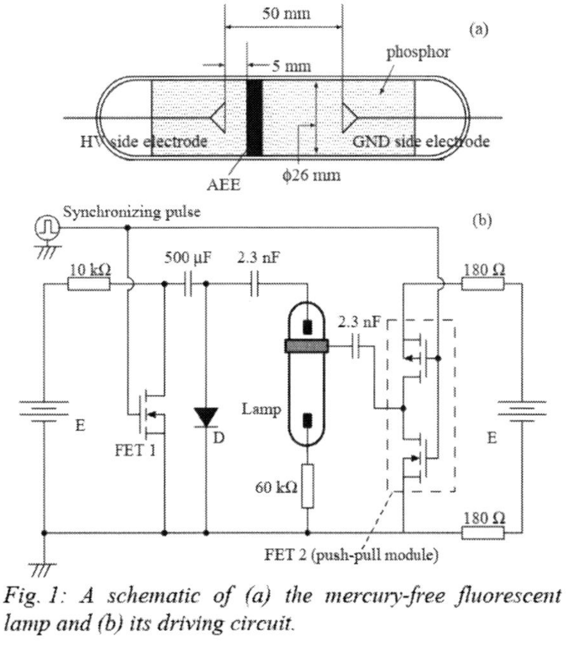

Fig. 1: A schematic of (a) the mercury-free fluorescent lamp and (b) its driving circuit.

RESULTS AND DISCUSSION

Figure 2 shows the luminance at the lamp center versus input power. The voltage of the HV pulse to the AEE was set at 3 kV. For reference, same measurements were done for the lamp without the AEE and for

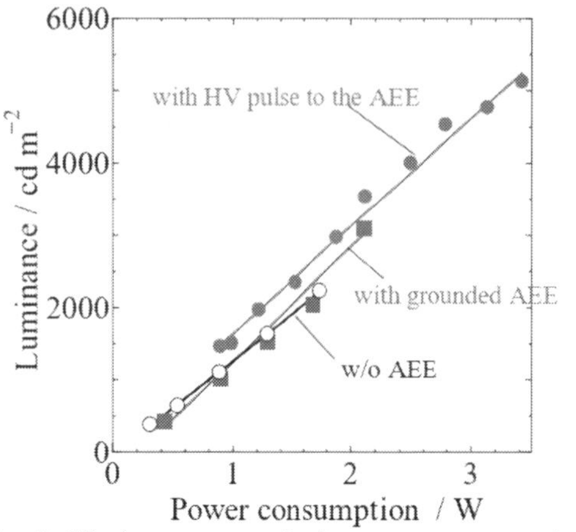

Fig. 2: The luminance at the lamp center versus the input power.

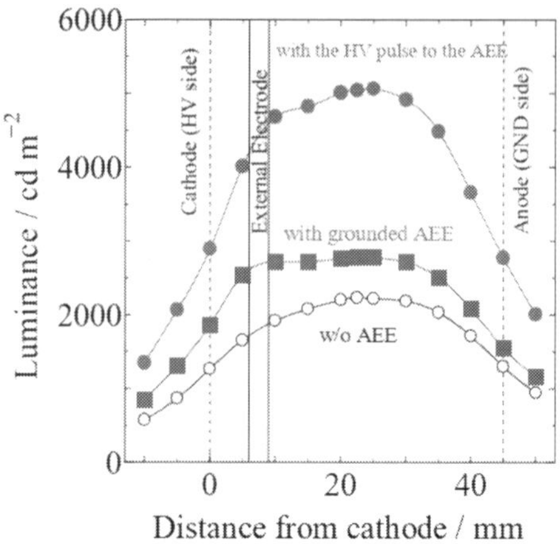

Fig. 3: The luminance distribution at the maximum input power for each lamp.

the lamp with the AEE that was grounded through a 100 Ω resistor. The input power was controlled by changing the applied voltage of the driving HV pulse to the main electrodes within the range where a diffuse positive column was obtained; the discharge disappeared below minimum input power setting and the positive column contracted over the maximum input power setting. By just using the grounded AEE, the maximum luminance was improved by almost 50% but the efficacy was almost same. On the other hand, by applying the HV pulse to the AEE, the luminance was enhanced by more than two times and the efficacy was also improved. Note that the luminance was enhanced for all position between the main electrodes as shown in Fig. 3. The discharge appearances of these three lamps are shown in Fig. 4. It is found that the positive column most expands to the radial direction in the lamp with the HV pulse application to the AEE. Moreover, the contracted discharge between the HV side electrode and the AEE is seen. It is found from the time resolved imaging using an ICCD camera that this "sub" discharge between the HV side electrode and the AEE triggers the expanded main discharge between main electrodes, which will be shown precisely at the symposium. The authors tried improving the luminance and efficacy by applying "sub" HV pulse to the main electrodes; lower voltage sub pulse was applied during the off-period of the main pulse (in between the main pulse train), which will be also shown at the symposium.

CONCLUSION

The luminance and the efficacy of the mercury-free fluorescent lamp was improved by using the AEE and the HV pulse was applied to the AEE, which was synchronized to the driving HV pulse for main electrodes. It is found that the "sub" discharge between the HV side main electrode and the AEE triggers the expanded main discharge between main electrodes.

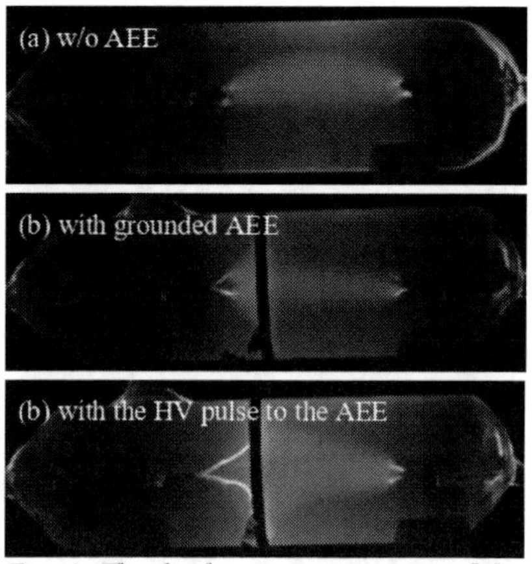

Fig. 4: The discharge appearances of the lamp (a) without AEE, (b) with grounded AEE and (c) with the HV pulse application to the AEE.

ACKNOWLEGMENT

This work was partly supported by NEDO through Industrial Technology Research Development Project (06A39007d).

REFERENCES

[1] M. Jinno *et al.*, Proc. 11th International Symp. Sci. Technol. Light Sources, CP-196, p. 605 (2007).
[2] M. Jinno *et al.*, J. Phys. D: Appl. Phys., **40** (2007) 3889.

Line-core and line-wing features in the temperature-dependent MgHe photoabsorption spectra

L. Reggami[1], M. Bouledroua[1,2]
[1]*Laboratoire de Physique des Rayonnements, Badji Mokhtar University, B.P. 12, Annaba, 23000 Algeria*
[2]*Faculté de Médecine, Badji Mokhtar University, B.P. 205, Annaba, 23000 Algeria*
Contact: boulmoncef@netscape.net

The present study deals with the collisional broadening of the monoatomic magnesium, evolving in a ground helium gas, in the wavelength and temperature ranges 260-310 nm and 100-3000 K, respectively. The emission and absorption spectral profiles, which might be generated from our theoretical computations, are based on the most recent potential-energy curves and transition dipole moments. The required interatomic Mg($3s^2$)-He and Mg(3s3p)-He potentials are constructed from two different sets [1]. The purpose of this treatment is twofold. First, using the quantum-mechanical Baranger impact approximation [2,3] the width and shift of the line core of the spectra are determined and their variation law with temperature is examined. Then, the satellite structures in the blue and red wings are analyzed quantum mechanically [4-6]. The calculations show especially that the free-free transitions are the most contributive in the MgHe photoabsorption spectra and a satellite structure is observable beyond the temperature 2000 K around the wavelengths 272 or 276 nm, depending on the used potential set. For both cases, the obtained results agree quite well with those already published.

References

[1] L. Reggami and M. Bouledroua, to be submitted (2010).
[2] M. Baranger, Phys. Rev. **111,** 481 (1958).
[3] M. Bouledroua, A. Dalgarno, and R. Côté, Phys. Rev. A **65,** 012701 (2001).
[4] K. Sando and A. Dalgarno, Mol. Phys. **20,** 103 (1971).
[5] H.-K. Chung, K. Kirby, and J.F. Babb, Phys. Rev. A **63,** 032516 (2001).
[6] K. Alioua and M. Bouledroua, Phys. Rev. A **74,** 032711 (2006).

Session Five

Advances in InAlGaN-based deep UV light emitting diode technologies

M. Kneissl[1,2], T. Kolbe[1], N. Lobo[1], J. Stellmach[1], A. Knauer[2], V. Küller[2], H. Rodriguez[2], S. Einfeldt[2], M. Weyers[2]

[1] Institute of Solid State Physics, Technische Universität Berlin, Hardenbergstr. 36, D-10623 Berlin, Germany
[2] Ferdinand-Braun-Institut, Leibniz-Institut für Höchstfrequenztechnik, Gustav-Kirchhoff-Str. 4, D-12489 Berlin, Germany
Contact: kneissl@physik.tu-berlin.de

ABSTRACT

Applications for UV light emitting diodes (LEDs) are discussed and the state-of-the-art in the area of InAlGaN UV LED technology is being reviewed. Performance limiting factors for high-efficiency UV LEDs are discussed and advances in the development of deep UV emitters are presented.

INTRODUCTION

Semiconductor-based ultraviolet (UV) light emitting diodes (LEDs) have gained considerable interest due to applications, e.g. in water purification, UV curing, medical diagnostics, counterfeit detection, and phototherapy. InAlGaN materials are hereby ideally suited since they cover the entire UV-A, UV-B and even large parts of the UV-C spectral range. A number of groups have reported AlGaN and InAlGaN multiple-quantum-well (MQW) light emitting diodes with emission wavelength from 300 nm to 350 nm and maximum external quantum efficiencies (EQE) ranging from 2% to 6% [1, 2, 3, 4]. Even shorter wavelength LEDs have been reported in the spectral range below 300 nm with maximum EQE around 1% [5, 6, 7, 8, 9, 10, 11]. However, the performance of these UV LEDs is still inferior to their visible wavelength counterparts. Fig. 2 summarizes the current state-of-the-art in UV nitride LED efficiency. As can be seen from the data, the EQE of AlGaN-based UV LEDs with emission wavelength <360 nm is at least one order of magnitude below the best devices in the near UV and violet wavelength range. There are multiple causes that are currently limiting the EQE of nitride based UV LEDs. A significant portion can be attributed to the relatively high defect densities in AlGaN materials as well as the poor injection efficiencies leading to reduced internal quantum efficiencies. Furthermore, most of the deep UV LEDs chip designs do not incorporate any advanced methods for light extraction, like surface texturing techniques or high reflective contacts, resulting in poor light extraction efficiencies.

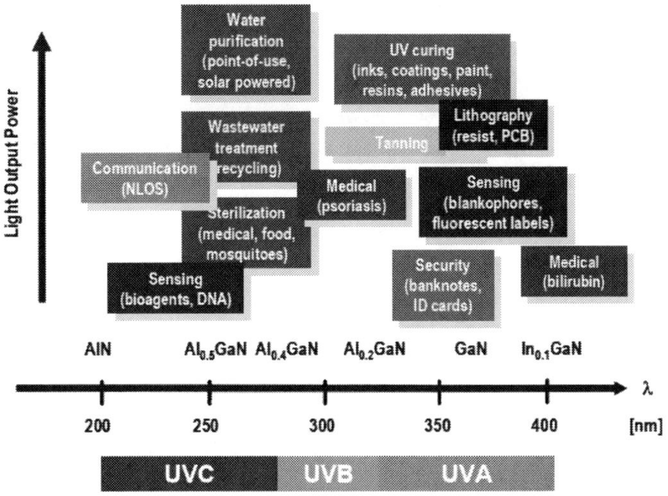

Fig.. 1:Applications for LEDs in the UVA, UVB, and UVC spectral range.

EXPERIMENTAL RESULTS

The UV LEDs in our studies were grown on (0001) sapphire substrate by metalorganic vapor phase epitaxy (MOVPE). Since the light is extracted through the sapphire substrate, the LED heterostructure is typically grown on UV transparent AlGaN and AlN buffer layers. However, due to the low surface mobility of the aluminum species and the large lattice mismatch between AlN and sapphire the growth of Al(Ga)N layers is very challenging. AlGaN and AlN layers grown on sapphire substrates typically yield dislocation densities in the mid 10^9 cm^{-2} to 10^{10} cm^{-2} range. This is more than one order of magnitude larger than in comparable GaN layers on sapphire. The dislocations in the material are typically threading dislocations that penetrate through the active region of the device. These can form strong non-radiative recombination pathways for the injected carriers leading to significantly reduced internal quantum efficiencies (IQE). Different approaches to reduce the dislocation densities in Al(Ga)N films on sapphire have been examined. These include, for example, migration enhanced MOVPE [9], epitaxial lateral overgrowth, and growth on low dislocation density bulk AlN substrate [12]. Using a modified reactor setup suitable for operation at very high temperatures (Tg >1500° C) we were able to demonstrate smooth AlN layers on sapphire with thicknesses of 700 nm and an rms roughness of less than 1 nm [13]. It was found that the structural quality of the AlN layers strongly depends on the sapphire treatment before the initiation of epitaxial layer growth. High resolution X-ray diffraction (HRXRD) shows peak widths of <500 arcsec for the (0002) reflection and a FWHM <1000 arcsec for the (10-12) reflection. From the X-ray data a screw dislocation density of $8x10^8$ cm^{-2} and an edge dislocation density of 10^{10} cm^{-2} can be derived. These values are consistent with threading dislocation densities obtained by cross-sectional transmission electron microscopy. Further details on the optimization of the growth process and layer characteristics can be found in Refs. 13 and 14.

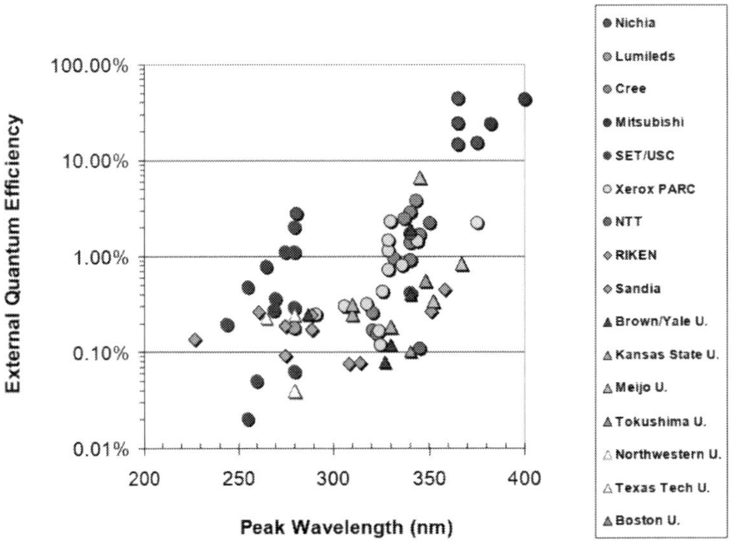

***Fig.2** : State-of-the-art in external quantum efficiencies for near and deep UV LEDs.*

A schematic of an UV LED device heterostructure on sapphire substrate is plotted in Fig. 3. After the growth of the AlN buffer layer an AlN/AlGaN short period superlattice is deposited in order to mitigate the large lattice mismatch between the Si-doped AlGaN current spreading layers and the AlN base. Next the active region is grown comprised of InAlGaN or AlGaN quantum wells (QWs). Finally an Mg-doped AlGaN electron blocking layer is deposited followed by a p-GaN cap layer. One of the motivations to use quaternary InAlGaN MQW active regions is the possibility to control the polarization fields in the MQWs by the appropriate choice of quaternary materials. In previous studies, aimed at the development of near UV LEDs, we were able to demonstrate that strain-free InGaN/InAlGaN MQW active region can be realized with almost zero polarization field in the quantum wells [15].

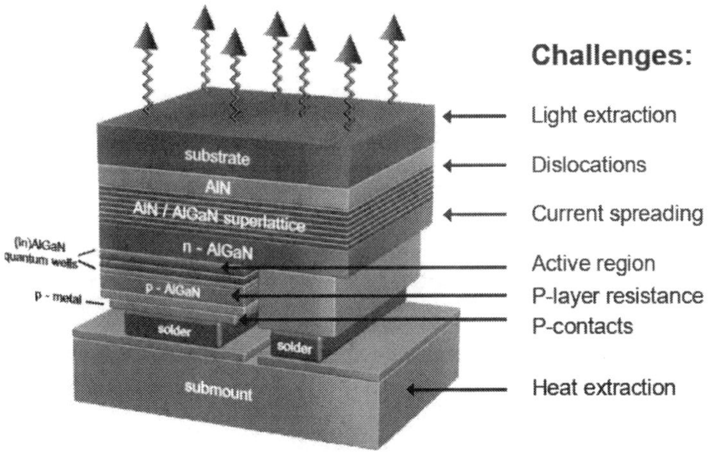

Fig.3 *:Schematic of a deep UV LED heterostructure flip-chip mounted onto a heatsink..*

Another critical issue for the realization of high efficiency UV LEDs is improving the internal quantum efficiency (IQE). The internal quantum efficiency is a product of the carrier injection efficiency η_{inj} and the radiative recombination efficiency η_{rad}. Simulations [16] of device heterostructures indicate that the injection efficiency η_{inj} is ~70% for UV-B LEDs. In addition, non-radiative recombination processes can be identified as one of the major causes for the low IQE in UV LEDs. Threading dislocations are currently the most prevalent defects in AlGaN-based heterostructures and largely responsible for non-radiative recombination processes. Based on a model from Ref. 17 a correlation between the defect density in the materials and the internal quantum efficiency of the device can be derived. For typical dislocation densities in the order of $\sim 5*10^9$ cm^{-2} non-radiative recombination lifetimes of less than 600 ps can be expected, resulting in radiative recombination efficiencies below 5%. For IQE > 20% non-radiative recombination lifetimes in the order of 3 ns are required, corresponding to threading dislocation densities below 10^9 cm^{-2}. Finally, improving light extraction efficiency ($\eta_{extract}$) is another critical issue. Since InAlGaN compounds are a high refractive index material, a large portion of the UV light cannot leave the chip. Estimates of the light extraction efficiency from ray-tracing simulations indicate that for bare UV LED chips on sapphire $\eta_{extract}$ is less than 8%. Therefore it will be essential to develop new designs for advanced light extraction in UV LEDs, e.g. employing surface texturing techniques [18] and nanopixel LED structures for enhanced light out-coupling [19].

ACKNOWLEDGEMENT

This work was partially supported by the German Federal Ministry of Education and Research within the "Deep UV LED" project under contract numbers 13N9933 and 13N9934 .

REFERENCES

[1] T. Nishida, N. Kobayashi, T. Ban, "GaN-free transparent ultraviolet light-emitting diodes", Appl. Phys. Lett. 82, 1 (2003).

[2] J. Edmond, Amber Abare, Mike Bergman, Jayesh Bharathan, Kristin Lee Bunker, Dave Emerson, Kevin Haberern, James Ibbetson, Michael Leung, Phil Russel, David Slater, "High efficiency GaN-based LEDs and lasers on SiC", Journal of Crystal Growth 272, 242(2004).

[3] Michael Kneissl, Zhihong Yang, Mark Teepe, Cliff Knollenberg, Noble M. Johnson, Alexander Usikov and Vladimir Dmitriev, "Ultraviolet InAlGaN Light Emitting Diodes Grown on Hydride Vapor Phase Epitaxy AlGaN/Sapphire Template", Jpn. J. of Appl. Phys. 45, 3905 (2006).

[4] Hirotoshi Tsuzuki, Fumiaki Mori, Kenichiro Ta-keda, Tomoki Ichikawa, Motoaki Iwaya, Satoshi Kamiyama, Hiroshi Amano, Isamu Akasaki, Haru-masa Yoshida, Masakazu Kuwabara, Yoji Yama-shita, Hirofumi Kan, "High-performance UV emitter grown on high-crystalline-quality AlGaN underlying layer", phys. stat. sol. (a), 1 (2009), DOI 10.1002/pssa.200880784

[5] J.P. Zhang, A. Chitnis, V. Adivarahan, s. Wu, V. Mandavilli, R. Pachipulusu, M. Shatalov, G. Simin, J.W. Yang, M.A. Kahn, "Milliwatt power deep ultra-violet light-emitting diodes over sapphire with emission at 278 nm", Appl. Phys. Lett. 81, 4910 (2002).

[6] V. Adivarahan, S. Wu, J. P. Zhang, R.A. Chitnis, M. Shatalov, V. Mandavilli, R. Gaska, M.A. Khan, "High-efficiency 269 nm emission deep ultraviolet light-emitting diodes", Appl. Phys. Lett. 84, 4762 (2004).

[7] J. Zhang, X. Hu, A. Lunev, J. Deng, Y. Bilenko, T.M. Katona, M.S. Shur, R. Gaska and M.A. Khan, "AlGaN Deep-Ultraviolet Light-Emitting Diodes", Jpn. J. Appl. Phys. 44, 7250 (2005).

[8] Y. Taniyasu, M. Kasu, T. Makimoto, "An alu-minium nitride light-emitting diode with a wave-length of 210 nanometres", Nature 441, 325 (2006).

[9] Asif Khan, Krishnan Balakrishnan, Tom Katona, "Ultraviolet light-emitting diodes based on group three nitrides", Nature Photonics 2, 77 (2008).

[10] Takayoshi Takano, Hideki Hirayama, Sachie Fujikawa, Kenji Tsubaki, "Realization of 270 nm band AlGaN based UV-LEDs on large area AlN templates with high crystalline quality", phys. Stat. sol. (c), 1 (2009) / DOI 10.1002/pssc.200880929

[11] Akira Fujioka, Takao Misaki, Takashi Murayama, Yukio Narukawa, and Takashi Mukai, "Improvement in Output Power of 280-nm Deep Ultraviolet Light-Emitting Diode by Using AlGaN Multi Quantum Wells", Applied Physics Express 3, 041001 (2010).

[12] Michael Kneissl, Zhihong Yang, Mark Teepe, Cliff Knollenberg, Oliver Schmidt, Peter Kiesel, Noble M. Johnson, Sandra Schujman, Leo J. Schowalter, "Ultraviolet semiconductor laser di-odes on bulk AlN", J. Appl. Phys. 101, 123103 (2007).

[13] F. Brunner, H. Protzmann, M. Heuken, A. Knauer, M. Weyers, and M. Kneissl, "High-temperature growth of AlN in a Production Scale 11x2" MOVPE reactor", phys. stat. sol. (c), 1 (2008).

[14] O. Reentilä, F. Brunner, A. Knauer, A. Mogi-latenko, W. Neumann, H. Protzmann, M. Heuken, M. Kneissl, M. Weyers, G. Tränkle, "Effect of the AlN nucleation layer growth on AlN material qual-ity", J. Cryst. Growth 310, (23), 4932 (2008).

[15] A. Knauer, H. Wenzel, T. Kolbe, S. Einfeldt, M. Weyers, M. Kneissl, G. Tränkle, "Effect of the bar-rier composition on the polarization fields in near UV InGaN light emitting diodes", Appl. Phys. Lett. 92, 191912 (2008).

[16] Simulations are based on the SiLENSe modeling software package by Semiconductor Tech-nology Research, Inc.

[17] S. Karpov, Y.N. Makarov, "Dislocation effect on light emission in gallium nitride", Appl. Phys. Lett. 81, 4721 (2002).

[18] L. Zhou, J. E. Epler, M. R. Krames, W. Goetz, M. Gherasimova, Z. Ren, J. Han, M. Kneissl, N. M. Johnson, "Vertical injection thin-film AlGaN/AlGaN multiple-quantum-well deep ultraviolet light-emitting diodes", Appl. Phys. Lett. 89, 241113 (2006).

[19] N. Lobo, H. Rodriguez, A. Knauer, M. Hoppe, S. Einfeldt, P. Vogt , M. Weyers and M. Kneissl, "Enhancement of light extraction in UV LEDs using nanopixel contact design with Al reflector", Appl. Phys. Lett. 96, 081109 (2010) DOI: 10.1063/1.3334721.

Modern automotive headlamps with improved light sources

M. Sasaki
R&D Dept, Koito Manufacturing Co.,Ltd.,500 Kitawaki, Shimizu-ku, Shizuoka-shi, 424-8764 Japan
Contact: msasaki@koito.co.jp

Abstract: *Automotive headlamps has been evolved with its light sources. From Tungsten-Halogen (TH) bulbs to High Intensity Discharge (HID) bulbs and then Light Emitting Diode (LED) is rising. The first LED headlamp was put in the market in 2007 and the second generation was released in 2009 to enhance the energy saving feature of LEDs. But LED headlamps are facing to tough competition with HID where they made the evolution for removing mercury and then going to reduce its system cost and power consumption.*

INTRODUCTION

As one of important social responsibility, reducing CO_2 emission has recently been crucial to prevent global warming. Developing and introducing low fuel consumption vehicles have been important policy not only for Japanese government but also for Europe and US. The electricity on the vehicle is produced by a generator which driven by an engine and it increases fuel consumption. So, we have strong demand to reduce power consumption of headlamps for reduction of fuel consumption.
On the other hand, the primary mission of headlamps is to provide sufficient visibility for the driver to improve safety during night-time driving. Therefore, it is important to cope with improved visibility and energy saving. The great milestone for this region is an introduction of HID headlamps in early 90s which doubles the lighting performance with reduced power consumption. Today, white LED is going to overtake the great predecessor but we still need time for deep penetration of this new face because of its higher cost.

HID BULBS AND HID HEADLAMPS

The first generation of automotive HID bulbs are classified as D1 and D2 where D1 has an integrated igniter and D2 doesn't. Technical challenge for automotive HID was instantaneous start up against its relatively small wattage while maintaining the life. They have 3,200lm flux, about three times greater than conventional TH bulbs, at 35W operation but 42 to 45W for the system including electronic ballast. They have experienced a first evolution for removing mercury as D3 and D4 correspondingly.
HID is a great solution for improving lighting performance with less consumption. It has been reported that the excellent lighting performance of HID headlamps could contribute for reducing traffic accident[1], but the fruit is limited because of their low taking late such as less than 10% world wide whereas 30% in Japan for new cars. The second evolution is planned for further reduction of power consumption. New D5 and D6 bulbs are expected to be introduced in 2011 with 2,000lm flux at 25W operation. Fully integrated system with integrated ballast is also planned as D7.

BENEFITS OF LED HEADLAMPS

Using white LEDs for headlamps results in several improvements not only for energy saving but also for safety driving. The life span of automotive white LEDs is much longer than average vehicle usage against extreme condition such as high temperature and vibration, so, users no longer experience sudden burnout of light sources. Instantaneous rise up of luminance has advantage to HID especially for entering tunnels in daytime. Another important aspect of LED headlamps is styling flexibility since headlamps are the key feature to characterize front look of the vehicle. LED is a not-so-hot light source comparing to conventional bulbs, so it enables to use plastic optics which could be molded for complex shape and they could be looks like jewels. This is a good challenge for optical engineers and molding specialists, and we just have started to create a unique front mask of the vehicle.

WHITE LED FOR HEADLAMPS

The important feature for headlamp LED is high luminance while keeping sufficient flux and efficacy. We have developed dedicated LEDs for headlamps[Fig.1] with Nichia Corporation. These LEDs are based on blue InGaN chip covered by yellow phosphor. They have 2 or 4 of $1mm^2$ large chip and 200 or 400 lm at 700mA operation for the first product. The performance is currently improved and then we have reduced the number of LEDs for the second generation headlamps.

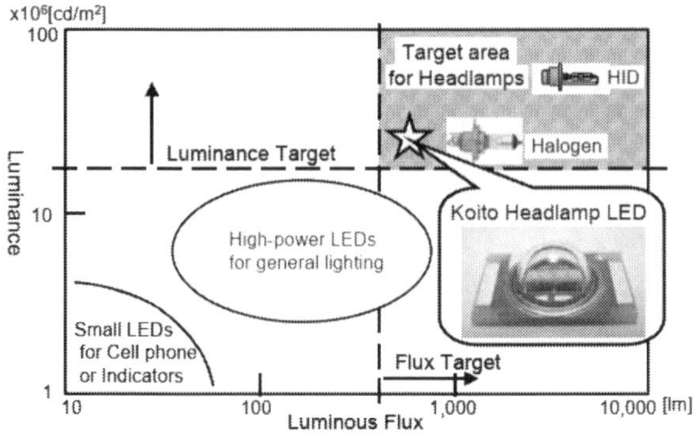

Fig. 1: Performance target for headlamp LEDs

THE FIRST PRODUCT AND EVOLUTION FOR PENETRATION

Koito has completed the development for LED headlamps and introduced the world's first mass produced LED headlamp with Toyota for Lexus LS-600h hybrid vehicle in May 2007. The lamp uses white LEDs for the low-beam, the main function of headlamps[Fig.2].

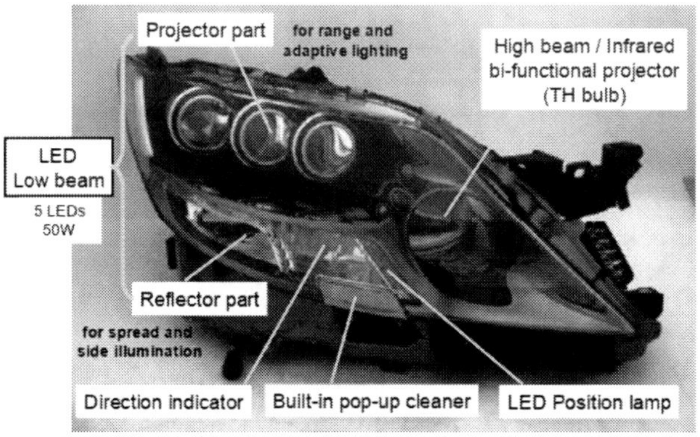

Fig. 2: The first LED headlamp for mass production. (Lexus LS-600h)

The low-beam consists of two different parts of optical configuration, such as a projector part and a reflector part. These three projectors have one LED each, and they work together for producing range of illumination. The reflector part contains two LEDs and produce spread and side illumination. The lamp achieved same lighting performance to the brightest HID headlamp and equipped with Adaptive Front-lighting System (AFS) for curved or bend road. Three projectors rotate together along with load curvature for this function.

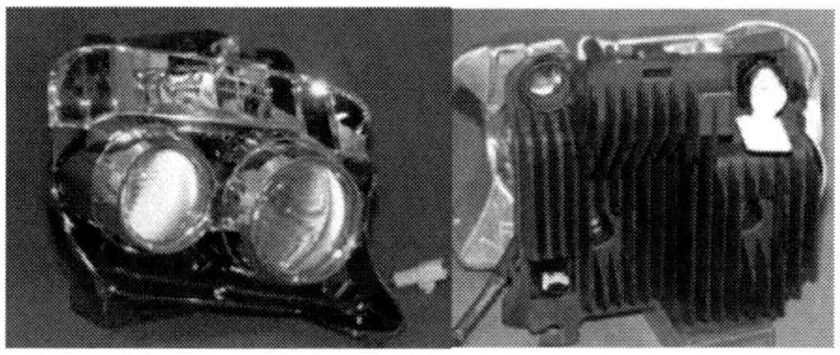

Fig. 3: The low beam unit for Toyota Prius and its heatsink on back.

Power consumption of the lamp is 50W for each low-beam including driver electronics. It is lower than 60W of typical halogen bulb but slightly greater than 45W of HID system since we emphasize performance and functionality rather than saving energy for this product. Then we have developed the second version for reducing power consumption to 35W while maintaining same lighting performance. This lamp has been installed onto four models such as Toyota Prius in 2009.

One of the key development for the LED headlamp is to keep LEDs below the maximum rated temperature against its internal heat and the heat from an engine just behind the lamp. We have succeeded to provide sufficient cooling with maximized air flow of natural convection instead of using electric fans[Fig.3].

CURRENT ISSUES

It is discussed that the head lamps with higher correlated colour temperature seems to have higher effective lighting performance compared with their photometric performance[2,3]. They also seem to have an benefit for finding obstacles in peripheral vision[4]. We will continue research for these aspects to improve safety with modern headlamp technology.

REFERENCES

[1] H. Schabe, F. Schierge, Proc. Int. Sympo. on Automotive Lighting, Sept. 25-26, 2007, p1
[2] P. Bodrogi, C. Schiller, T.Q. Khanh, Proc. Int. Sympo. on Automotive Lighting, Sept. 29-30, 2009, p65
[3] M. Okawa, S. Kobayashi, K. Okajima, Proc. Int. Sympo. on Automotive Lighting, Sept. 29-30, 2009, p53
[4] J.V. Derlofske, D. Dyer, J.D. Bullough, SAE 2003-01-0930

MICROCAVITY PLASMA LIGHTING: THIN SHEET LAMPS FOR GENERAL ILLUMINATION

J.G. Eden and S.-J. Park
Department of Electrical and Computer Engineering, University of Illinois, Urbana, IL 61801 USA

Eden Park Illumination, Inc.
903 N. Country Fair Dr., Champaign, IL 61821 USA
Contact: jgeden@illinois.edu

ABSTRACT

Thin, lightweight lamps comprising arrays of low temperature, microcavity plasmas driving a phosphor are being developed for general illumination applications. Having a packaged thickness of ~ 4 mm and currently available in sizes of 6 × 6 sq. in. (~225 cm^2) and 8 × 8 sq. in. (~400 cm^2), these "light tiles" generate luminance values up to 17,000 cd/m^2 (white) at a CRI of 80. Panel efficacies beyond 30 lumens/W are obtained with Xe_2 as the emitter and other ultraviolet-emitting sources are under investigation. The performance of these lamps, currently in limited production by a semiautomated process, relative to a conventional dielectric barrier discharge structure will be described.

INTRODUCTION AND BACKGROUND

For more than a decade, the properties of low temperature, nonequilibrium plasma confined within a cavity have been examined. Such microcavity plasmas [1] exhibit characteristics quite distinct from those of glow discharges of larger volume, including the ability to operate stably at atmospheric pressure and beyond. Furthermore, arrays as large as tens to hundreds of thousands of microcavity devices have been fabricated repeatedly in a wide variety of materials systems, including Si, ceramics, glass, and even aluminum foil. Flat, thin, and lightweight lamps based upon technology originally demonstrated at the University of Illinois are being developed and manufactured by Eden Park Illumination. Currently available as "lighting tiles" in sizes of 6 × 6 sq. in. (~225 cm^2) and 8 × 8 sq. in. (~400 cm^2), these lamps provide homogeneous emission over the entire lamp surface with a CRI of 80 and CCT readily adjustable from 3000 K to 4000 K.

BRIEF DESCRIPTION AND TECHNICAL DETAILS

Eden Park lamps incorporate a micro-dielectric barrier discharge (μ-DBD) structure in which low temperature, nonequilibrium plasma is confined, or partially confined, within cavities having a characteristic dimension < 1 mm. The electrodes are typically fabricated from aluminum onto which nanoporous alumina has been grown [2,3]. A thin layer of phosphor is printed onto the inside faces of two 400 μm thick sheets of glass between which the lamp electrodes are hard-sealed. The sealed "light engine" accounts for approximately one-half of the ~4 mm overall thickness of the packaged lamp. For the lamps described here, the excimer molecule Xe_2 provides vacuum ultraviolet fluorescence to drive the phosphor but other ultraviolet (UV) emitters are also currently under investigation.

Extensive testing and characterization of lamps ranging in size from 2 × 2 sq. in. (~25 cm^2) to 8 × 8 sq. in. (400 cm^2) have focused on electrical and optical properties. For a fixed value of E/N, the lamp luminance is found to increase linearly with the frequency of a sinusoidal voltage driving waveform. Furthermore, increasing the radiative output of the lamp has little impact on efficiency. With the current lamp design, for example, panel efficacy declines < 10% when the luminance is increased from 1500 cd/m^2 to 6000 cd/m^2. Luminance shows no sign of saturation as the driving voltage

Fig.1. Photograph of a ~400 cm^2 (6 × 6 sq. in.) lamp capable of delivering a luminance of 6000-12000 cd/m^2 with a CRI of 80.

frequency is increased from 2 kHz to the highest frequency examined to date, 21 kHz. Although the current microcavity electrode structure and the associated driving electrodes, and the voltage waveform are far from optimized, the incorporation of a microcavity array into the lamp improves panel efficacy by more than 50% relative to that for a conventional DBD lamp structure. Panel efficacies above 30 lumens/W are achieved with Ne/Xe gas mixtures and considerably higher values are expected with improved microcavity designs and other emitters.

Figure 1 is a photograph of a 6 × 6 sq. in. lamp producing a luminance of ~ 6000 cd/m^2. Experiments conducted with CCD cameras and a radiometer show the luminous intensity to be uniform over the entire lamp to within ± 10 %. Detailed studies of the response of red, green and blue phosphors (acquired from several sources) have resulted in R-G-B mixture films that consistently yield a CRI of 80. Also, the CCT can be reproducibly specified to within ±100 K in the 3000-4000 K interval. Accelerated aging tests of 6 × 6 sq. in. lamps have been in progress for months, and preliminary indications are that lamp lifetimes in excess of 20,000 h are quite feasible.

SUMMARY AND CONCLUSIONS

Thin, flat lamps incorporating arrays of microcavity plasmas have been demonstrated at the University of Illinois and are now in limited production by a semiautomated process at Eden Park Illumination. The near-atmospheric pressure, glow discharge operation of the lamp allows for efficient UV emitters to be synthesized *in situ* and continuously within the lamp. With the Xe dimer as the initial emitter, panel efficacies above 30 lumens/W have already been achieved with luminance values above 10,000 cd/m^2. The overall thickness of the lamp is nominally ~ 4 mm, the cost of materials is low, and all constituents of the lamp are non-toxic and fully recyclable. Consequently, these light "tiles" or "sheets" are comparatively inexpensive, environmentally sound, and versatile from the perspective of lighting design. These considerations suggest that microcavity plasma lamps afford opportunities in general illumination and niche lighting applications that have not been available previously. Further details regarding the structure, performance, and anticipated applications of these novel lamps will be described.

REFERENCES

[1] K. H. Becker, K. H. Schoenbach and J. G. Eden, *J. Phys. D.*, 39 (2006) R55

[2] S.-J. Park, J. D. Readle, A. J. Price, J. K. Yoon, and J. G. Eden, *J. Phys. D.*, 40 (2007) 3907

[3] K. S. Kim, J. K. Yoon, S.-J. Park, J. G. Eden, *IEEE Trans. Plasma Sci.,* 36 (2008) 1258

MOCVD for Solid State Lighting:
- Recent Advances Towards Larger Wafer Sizes -

A. Boyd, R. Puesche, F.Schulte, O. Schoen, B. Schineller, M. Heuken
AIXTRON AG, Kaiserstrasse 98, D-52134 Herzogenrath, Germany
Contact: m.heuken@aixtron.com

ABSTRACT

Price reduction for optoelectronic devices depends strongly on the availability of large wafer area substrates, processes and production tools. In AIXTRON Planetary Reactor- and CRIUS Close-Coupled-Showerhead (CCS) MOCVD tools the growth-performance of InGaN-based optoelectronic structures was assessed in the 6x6 inch and 1x200 mm configurations, respectively. In-situ bow measurements, emissivity corrected pyrometry and the Argus® multi-channel temperature scanner were used to optimize the growth and speed up the process development.

GENERAL

According to "Haitz' Law" the cost per lumen is bound to reduce by a factor of 10 per decade to keep up with the demand for low-cost LED replacements of light bulbs [1]. This decrease must be met by a substantial reduction in the Cost of Ownership (CoO) of the production tools. MOCVD as one of the most important steps in mass production of today's devices faces the challenges of high-reproducibility and yield, low-operating cost, high-precursor efficiency and high-uptime.

We achieved the goals of CoO-recduction by introducing the IC Platform design for different types of MOCVD reactor types, offering a common standard and lower maintenance and spare-part stock costs. In addition, the Planetary Reactor® design for nitride and arsenide-phosphide based applications as well as the well-proven Crius® design for nitrides were further optimized with respect to throughput and precursor utilization. In-situ metrology such as a dual-beam deflectometer for wafer bow analysis or the Argus dual wavelength pyrometry scanner offers valuable information on the growing layer, thus, speeding up the development process and allowing easy monitoring during the production process.

RESULTS ON PLANETARY REACTORS

Expanding the reactor geometry from the industry proven 42x2 inch configuration to 6x6 inch leads to a 28.6% increase total wafer area, thus in productivity [2]. This AIX 2800G4 HT Planetary Reactor® in the 6x6 inch configuration features a 3-fold central gas inlet with the group-III metal-organic precursors being sandwiched between an upper and lower NH_3 group-V hydride inlet (cf. Fig. 1). This allows extending the depletion profile of the gas phase across the entire 6 inch wafer area whilst offering an additional degree of freedom for growth profile tuning [3].

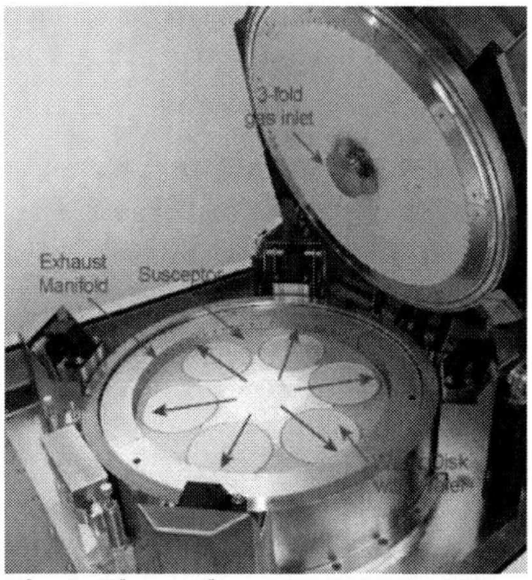

Fig. 1: Photo of an AIX 2800G4 HT in the 6x6 inch configuration with the main features marked.

To assess the performance of the tool in the new

configuration InGaN multi-quantum-well (MQW) test structures were grown and characterized. At a targeted emission wavelength of λ_{mean} = 450 nm a standard deviation of σ_λ = 2.4 nm was achieved on a 6 inch wafer in the Planetary Reactor. The thickness of the wafer was 5.3 µm with a standard deviation of σ_d = 1.9%. Fully loaded runs yielded wafer to wafer deviations of $\Delta\lambda_{max-min} < \pm 1$ nm at a mean wavelength of λ_{mean} = 475 nm.

RESULTS ON CRIUS® CLOSE-COUPLED SHOWERHEAD (CCS) REACTORS

The Crius® CCS reactor features a dual-plenum showerhead for separate supply of the group-III and group-V precursor species through interdigitated inlets (cf. Fig. 2). Due to its close proximity this gas supply is closely coupled to the growth surface, yielding a very uniform boundary layer with a well controlled growth front. Additionally, a three zone tungsten heater with central, middle and outer zones allows for temperature fine-tuning in concentric wafer configurations like 1x200 mm.

Fig. 2: Crius® CCS reactor with single 200 mm wafer positioned centrally on the wafer platter.

This temperature tuning using three radial heater zones could be used to optimize the wavelength uniformity, taking the readings from the Argus® pyrometry scanner as input. Similar MQW structures as for the Planetary Reactor were used to assess the performance of the Crius® tool. Using the optimized temperature profile a wavelength distribution of σ_λ = 3.3 nm at λ_{mean} = 512 nm was achieved on a 200 mm sapphire wafer. The thickness of the layers (dominated by the GaN buffer) was 4.8 µm with a standard deviation of σ_d = 2.6%.

CONCLUSIONS

Two MOCVD reactor configurations for high-volume mass production of SSL devices were discussed. Both are based on a common integration concept, thus reducing the amount of spare-parts and reactor maintenance.

Both reactors were tested with InGaN MQW structures and the results yielded that both reactors are highly capable of delivering the industry's production needs for solid-state lighting. Further results on the application of in-situ tools and the potential of cost reduction using Si substrates will be presented.

REFERENCES

[1] R. Haitz, Plenary Talk held at the Meeting of the "Deutsche Physikalische Gesellschaft", March 2003, Dresden, Germany

[2] R. Schreiner, A. Alam, H. Protzmann, B. Schineller, M. Heuken, to be presented at CS-Mantech 2010, May 17-20, Portland, Oregon, USA

[3] B. Schineller, C. Geng, M. Heuken, presented at 6th International Symposium on Solid-State-Lighting, Oct. 14-16, 2009, Shenzhen, China

Novel Molecular Discharges

R. Hilbig , A. Koerber , S. Schwan , D. Hayashi
Philips Research Europe, Weisshausstr. 2, 52066 Aachen, Germany
Contact: achim.koerber@philips.com

ABSTRACT

A novel type of highly efficient gas discharge light source is described which is employing diatomic *transition metal oxides* as visible radiators. Zirconium monoxide (ZrO) has been identified as very promising candidate combining highly attractive green and red emission band systems with very high dissociation energy (8.2eV) which assures that the molecule is stable even in the hot plasma center. Thus far, however, it has been *impossible* to keep ZrO in the gas phase of a closed discharge vessel, because at wall temperature usually compounds are formed which have negligible vapor pressures. We succeeded in establishing a regenerative chemical cycle by filling ZrX_4 (X=Cl, Br, I) together with a stable, but volatile oxygen compound (like $MoOX_4$) and realized thus highly attractive, novel gas discharge light sources.

INTRODUCTION

The aim of any lamp engineer is to develop a white light source with good color rendering properties ($Ra_8 > 80$) and highest luminous efficacy η. Current discharge lamps – emitting mainly *atomic* radiation – are reaching only about *half* of the theoretical efficacy limit of 200-230 lm/W. The possible efficacy rise by increasing atomic radiation is limited by self-absorption of the atomic lines (= radiation trapping). This limitation does not apply for *molecular* radiation, since the molecular emission is distributed over a huge number of ro-vibronic lines which is several orders of magnitude higher than the corresponding number of atomic lines. Therefore, we strive for a breakthrough efficacy improvement by introducing novel, *molecular* radiators for discharge lamps.

TRANSITION METAL ATOMS AND MOLECULES

Transition metals (TM) are known to have strong p-s transitions leading to radiation in the *near UV* or *blue* spectral range. Therefore, we need additional, highly efficient *molecular* radiators in the *green* and *red* spectral range to realize a discharge lamp emitting white light with good color rendering properties.

It is well-known that certain *transition metal oxides* (e.g. NbO) emit interesting red or near IR band systems [1]. For a systematic overview we simulated visible emission spectra of all Group IVb-VIb transition metal oxides using spectroscopic constants from literature [2][3]. This overview taught us that the spectrum of *zirconium oxide* (ZrO) is most promising, since it has broad, strong band systems in the green ($e^3\Pi$–$a^3\Delta$) and red ($d^3\Phi$–$a^3\Delta$) spectral region, but almost no emission in the IR [4].

TM oxides also show very *high dissociation energies* (8.2eV for ZrO), because they have a *double bond* between metal and oxygen. This gives a huge efficacy advantage compared to TM halides, because TM oxides will remain *undissociated* in the hot center ($\approx$ 4500K) of a high intensity discharge (HID) lamp.

REGENERATIVE CHEMICAL CYCLE

The volatility of TM oxides is generally extremely low. For TM with valences $\geq$ 5 the corresponding *TM oxide halides* (like e.g. $NbOBr_3$ or WO_2I_2) have reasonably high vapor pressures and can be used to bring *both* the TM *and* the oxygen into the vapor phase. Group IVb oxide halides ($ZrOX_2$), however, *cannot be evaporated* at reasonable coldest spot temperatures ($\leq$ 1500K), because their vapor pressures are below 1Pa. These limitations have thus far *excluded* the use of Group IVb oxides (e.g. ZrO) as gas phase radiators in closed systems like a discharge lamp.

We use *two different* "precursors" to bring the Group IVb metal and the oxygen into the vapor phase. By dosing the very volatile Group IVb tetra-halides (like ZrX_4) together with a volatile oxygen compound (like MoO_2X_2 [5] or P_2O_5 [6]) we succeeded in obtaining a regenerative chemical cycle. In this cycle the "precursor" molecules are dominating at low temperatures near the vessel wall (thereby excluding the formation of nonvolatile TM oxides or TM oxide halides), but the Group IVb monoxide dominates at the high temperatures of the plasma center (thereby generating the desired radiation). Amazingly, this "two precursor" trick works so well that the necessary cold spot temperatures become extremely low (500-700K).

SAMPLE RESULT

A tubular quartz envelope with 24mm inner diameter and about 30mm inner length was filled with $ZrCl_4$, $MoOCl_4$ and Xe. An excitation coil was directly wound onto the quartz vessel and used to couple 269W of 14.0MHz RF power into an inductive discharge torus within the lamp. Under these conditions the lamp emitted the spectrum shown in black in Figure 1 below. This figure also shows spectral simulations of the contributions of the various radiation emitters present at the plasma center.

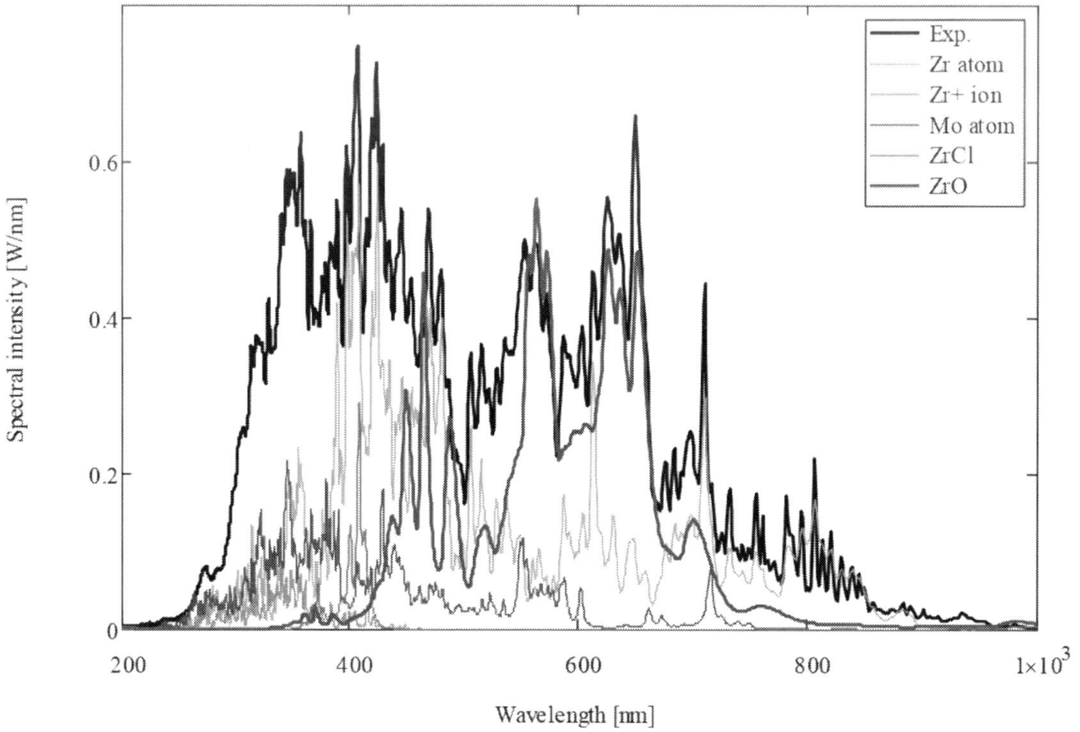

Figure 1: Spectrum of a lamp dosed with $ZrCl_4$ and $MoOCl_4$. Measurement (black) plus simulations for ZrO (red), ZrCl (magenta), Zr atom (light blue), Mo atom (dark blue) and Zr^+ ion (green) radiation.

The total radiation output of this lamp is 182W – thus corresponding to a plasma radiation efficiency of 68%. The visible radiation output (400-780nm range) is 128W **(~48%)** corresponding to a plasma efficacy of 102lm/W, a correlated color temperature of 5436K (21SDCM below the black-body locus) and a color rendering index Ra_8 = 87.8. If the UV radiation below 400nm would be converted by a suited phosphor into white light a plasma + phosphor efficacy of about **160lm/W** might be expected.

REFERENCES

[1] W.P. Lapatovich, W.M. Keefe, N. Brates, R.W. Liebermann, J. Maya, "Metal Oxide Containing High Intensity Discharge Lamp", paper Nr. 81, LS 4, Karlsruhe (1986).

[2] K.P. Huber, G. Herzberg, "Molecular Spectra and Molecular Structure. IV. Constants of Diatomic Molecules", Van Nostrand Reinhold Company, New York (1979).

[3] P.F. Bernath and S. McLeod, "DiRef, A Database of References Associated with the Spectra of Diatomic Molecules", J. Mol. Spectrosc., **207** (2001) 287. http://diref.uwaterloo.ca

[4] S.P. Davis, P.D. Hammer, "Energy levels of zirconium oxide", Astrophys. J. 332 (**1988**) 1090-1091.

[5] A. Koerber, R. Hilbig, D. Hayashi, S. Schwan, Int. App. Pub. WO 2008/126014 A2.

[6] R. Hilbig, A. Koerber, S. Schwan, D. Hayashi, Int. App. Pub. WO 2008/126020 A2.

Magnetron-powered, Electrodeless, HID Light Source

D.B.Eeles,[1] A.Neate,[1] F.Agnello,[1] E.C.Odell,[1] B.Preston,[1]
K. Lidström[2] and W. Rönnblom[2]
[1] *Ceravision Limited, The Mansion, Bletchley Park, Milton Keynes MK3 6EB, UK*
[2] *Dipolar AB, Gymnasievägen 16, 931 57 Skellefteå, Sweden*
Contact: david.eeles@ceravision.com

ABSTRACT

A new, microwave-powered, electrodeless HID light source has been developed. The light emitting plasma is established inside a small, cylindrically-shaped, void located centrally inside a cylindrical, transparent, quartz resonator. The microwave power source is a 2.45GHz magnetron driven by a power supply designed optimally for long magnetron operating lifetime and high power conversion efficiency. The source can be dimmed to 20% and has excellent rapid-start and re-strike performance.

INTRODUCTION

The development of electrodeless light sources (ELLS) continues to be a growth area in plasma-based light source technology. ELLS have long featured in the LS symposia [1] and their present status is reviewed in an invited paper at this meeting [2].

The scientific bases for the various important advantages of ELLS in comparison with mercury-based discharge lamps, for example, their more environmentally acceptable dose chemistries, much longer lifetimes, and greater lumen maintenance and colour stability, are well founded. However, concerns persist relating particularly to the perceived high relative cost and complexity of the associated power circuitry and their impact on price-performance comparisons with conventional discharge lamps.

The lamp development featured in this paper addresses the latter issue by coupling a novel configuration microwave ELLS HID source with a high performance, low cost, power supply which is designed specifically to exploit fully the long operating life and high energy conversion efficiency potential of low-cost, rugged, compact, magnetron microwave power sources.

A familiar source of microwave power in the range from several 100W upwards, the range of interest in many HID lamp application areas, is the 2.45 GHz magnetron found in domestic microwave ovens [3][4]. These devices provide microwave power at a cost of as low as 1US ¢ / Watt, currently some hundred times less than the cost of power originating from solid-state-based microwave power sources. The challenge is to operate a magnetron in a manner which conserves the life of the filament (cathode), the primary magnetron life-determining component, to make the lifetime of the device compatible with the anticipated long operating lifetime of the bulb-resonator component, namely, several 10,000s h. Despite being part of a vacuum tube, life-management issues for magnetron filaments have similarities with those for cathodes in low-pressure compact and linear fluorescent lamps where cathode emitter-loss is life-determining. An acceptable operating methodology must ensure also that the magnetron is operating at high efficiency which equates to operation as far as possible in the design π-mode and not in other, different frequency, modes which dissipate power inside the tube causing serious, life-shortening, damage.

RESULTS AND DISCUSSION

Light emission from the new ELLS HID system originates from a small, cylindrically-shaped, metal halide dosed, void located coaxially and centrally in a transparent, cylindrical, quartz resonator [5]. The resonator may usefully be regarded as a dielectric-filled resonant cavity of type intermediate between the air-filled cavity of a sulfur lamp [6] and the ceramic-filled cavity of a Ceravision Continuum™ 2.4 lamp [1]. The electrical, optical and thermal properties of the quartz resonator, and the high efficiencies both of power transfer to the plasma and of light generation, result in a compact source whose power is scalable but which operates typically near 0.5 kW. To power the quartz resonator we have designed a 2.45 GHz magnetron-based power supply whose primary characteristics are long operating life and high energy conversion efficiency. A schematic of the resonator, transition, magnetron and power supply sub-assembly of a 400W system is shown in Figure 1 below [7].

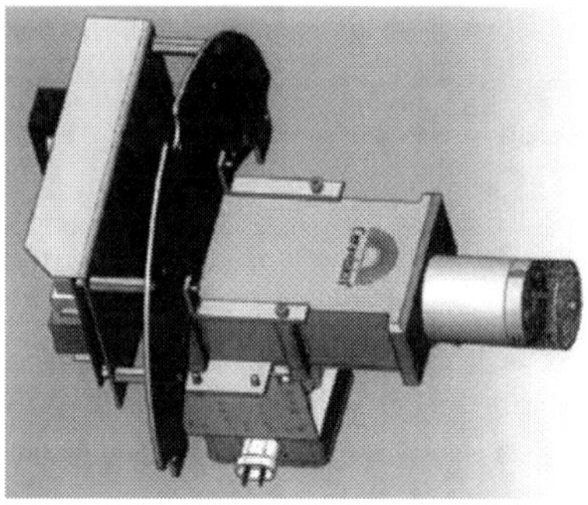

Figure 1. Magnetron-powered, quartz-resonator HID ELLS

The general principles of magnetron operation are well described in the literature [3][4]. The power supply design goal is to identify and maintain benign operating points for the magnetron, in particular for the filament, for the various different load conditions presented by the lamp which include: lamp-off; striking; warm-up; and operation at a range of different powers. It is essential that good matching between lamp and power supply is maintained for each of these different load conditions. Features of the power supply's design include: power factor control; separate filament and anode control circuits with feedback; high quality DC power; lamp ignition with feedback; temperature sensing; and WiFi communication, all controlled/monitored by an on-board processor. The many converters present in the design are each optimised for maximum efficiency at the different magnetron operating points. The design accommodates both "soft"-starting, and dimming which is linear in the range 100% - 20%.

The desired control of the magnetron's mode and filament emission are both evidenced by the frequency spectrum. Fig. 2a is the power spectrum near 2.45GHz of an *in situ* domestic microwave oven magnetron; the power wastefully dissipated in low efficiency parasitic modes is evident. Fig. 2b is the power spectrum (*3x expanded scale of Fig. 2a*) of a magnetron powered by the new supply; the narrow-bandwidth, single-mode operation is indicative of stable, long-life operation.

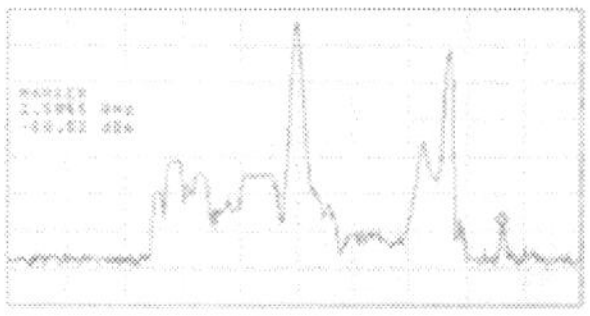

Figure 2a: Power spectrum: domestic microwave

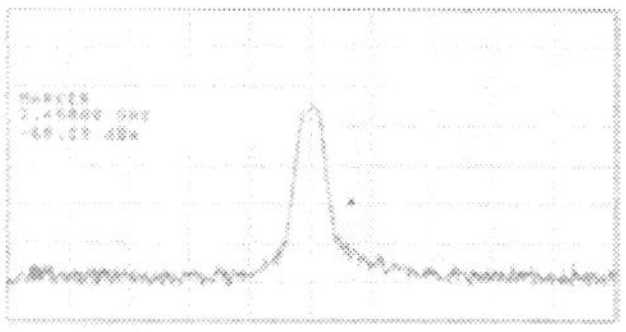

Figure 2b: Power spectrum: new supply

The new ELLS HID system [7] provides flicker-free light with CRI>80, excellent rapid start (<10s) and hot re-strike (<60s) performance, and dimming with only modest colour shift. Independent testing and examination of magnetrons taken from systems operated for long periods in installations shows only small shifts in operating points, and low deterioration levels of internal surfaces, supporting our belief that system lifetimes will be several 10,000s h with little performance deterioration.

REFERENCES

[1] V. Babykumar, A. Neate, E. Odell, B. Preston, A. Sadiq and R. Devonshire 2007 *Light Sources 2007*, eds. M.Q. Liu and R. Devonshire (Sheffield, UK: FAST-LS) pp. 109-110. "*A Compact Microwave Resonant Cavity Electrodeless Light Source System*"

[2] W.P. Lapatovich 2010 *Light Sources 2010*, eds. G.M.W. Kroesen and M. Haverlag (Sheffield, UK: FAST-LS) (this Proceedings volume). "*Electrodeless Lamp Technology Overview*"

[3] L. Sivan 1994 *Microwave Tube Transmitters* (London: Chapman & Hall) pp 155-193

[4] R. Meredith 2007 *Engineers' Handbook of Industrial Microwave Heating* (London: The Institution of Engineering and Technology) pp 247-303

[5] A.S. Neate, WO patent application 2009063205A2, November 2008

[6] J.T. Dolan, M.G. Ury and C.H. Wood, US Patent 5,404.076, April 1995

[7] Ceravision Alvara® 400 system

Molecular constants and standard enthalpies of formation for the lanthanide monochlorides, LnCl<g>

S.A.Mucklejohn[1]
[1] *Ceravision Limited, The Mansion, Bletchley Park, Milton Keynes MK3 6EB, UK*
Contact: stuart.mucklejohn@ceravision.com

ABSTRACT

Molecular constants and dissociation energies reported in the literature have been used to derive the heat capacities, enthalpies and standard enthalpies of formation for the lanthanide mononhalides, LnCl<g>, Ln = Ba, La to Lu.

INTRODUCTION

The lanthanide monohalides, LnX<g>, are important radiating species in high pressure plasmas derived from doses containing lanthanide(III) halides. To understand the characteristics of these plasmas and to determine the role of the lanthanide monohalides it is necessary to calculate the chemical composition of the plasma from the coolspot temperature (typically 1000 to 1500 K) to the arc centre temperature (typically in the range 4000 to 6000 K). To calculate the plasma composition, thermodynamic functions for all of the species under consideration must be available and held in the appropriate database format. None of the readily available thermochemical databases contain a complete set of self consistent values for the lanthanide monohalides. This paper outlines the determination of the standard molar enthalpies of formation for the series LnCl<g> from dissociation energies. In lamps with electrodes the use of lanthanide halides is restricted to the bromides and iodides. Electrodeless plasma lamps, however, allow chloride-containing systems to be used. Promethium, Pm, is an unstable element and does not occur naturally but it is included in all theoretical considerations of lanthanide compounds. Although Ba is not a lanthanide element it is often included in studies of the lanthanide series because of its position in the Periodic Table.

Scheme 1. Derivation of standard enthalpy of formation from dissociation energy

Ln<r> + 1/2Cl_2<r> = LnCl<g>	Standard enthalpy of formation of LnCl<g>	{1}
LnCl<g> = Ln<g> + Cl<g>	Dissociation of LnCl<g>	{2}
Ln<r> = Ln<g>	Standard enthalpy of formation of Ln<g>	{3}
1/2Cl_2<r> = Cl<g>	Standard enthalpy of formation of Cl<g>	{4}

{1} = {3} + {4} - {2}
where <r> is the standard state of the element at the reference temperature

The dissociation energies at 0 K for LnCl<g> were taken from reference [1], the standard enthalpies of formation for Ln<g> and Cl<g> were taken from the SGTE Substance Database [2].

The heat capacities of the LnCl<g> series were calculated from the molecular constants published by A.D.Chervonnyi & N.A.Chervonnaya [3] and listed in table 1. The energy levels used for these calculations were those for the corresponding Ln(II) species [4]. These calculations also provide the molar entropies at 298.15 K.

The standard molar enthalpies of formation for LnCl<g> were derived from scheme 1. Molar heat capacities for Ln<g> and Cl<g> were calculated using energy levels for Ln(I) and Cl(I) as listed in [4].

$$\Delta Cp(T) = C_p(Ln,g,T) + C_p(Cl,g,T) - C_p(LnCl,g,T) \qquad \{5\}$$

$$\Delta H(T) = \Delta H(298.15\ K) + \int \Delta C_p(T)dT \qquad \{6\}$$

Enthalpies of dissociation at 298.15 K were determined from equations {2}, {5} and {6}.

From the standard enthalpies of formation, the entropies at 298.15 K and the temperature variation of the molar heat capacities a dataset for each of the LnCl<g> series can be derived in the SGTE format for use with integrated thermochemical databank packages.

*Table 1. Molecular constants for LnCl<g>, * This work*

Ln	Molar mass	ω_e	$\omega_e x_e$	B_e	α_e	R_e	Reference
	g.mol^{-1}	cm^{-1}	cm^{-1}	cm^{-1}	cm^{-1}	cm	
Ba	172.793	279.3				2.567E-08 *	5
La	174.363	341.603	0.9797	0.1 *	3.6000E-04	2.498E-08	6
Ce	175.573	344.3	1.00	0.1 *	3.6600E-04	2.486E-08	3
Pr	176.360	346.9	1.02	0.1 *	3.7200E-04	2.475E-08	3
Nd	179.693	349.1	1.05	0.1 *	3.7800E-04	2.464E-08	3
Pm	180.453	351.0	1.06	0.1 *	3.8400E-04	2.453E-08	3
Sm	185.803	352.6	1.09	0.1 *	3.5600E-04	2.442E-08	3
Eu	187.413	288.7	0.89	0.1 *	3.9400E-04	2.571E-08	3
Gd	192.703	354.9	1.12	0.1 *	4.0000E-04	2.423E-08	3
Tb	194.377	355.6	1.14	0.1 *	4.0500E-04	2.413E-08	3
Dy	197.953	356.0	1.15	0.1 *	4.1000E-04	2.404E-08	3
Ho	200.383	356.1	1.17	0.1 *	4.1500E-04	2.396E-08	3
Er	202.713	355.9	1.18	0.1 *	4.2000E-04	2.388E-08	3
Tm	204.387	355.3	1.18	0.1 *	3.8000E-04	2.379E-08	3
Yb	208.493	291.1	0.92	0.1 *	2.3660E-04	2.488E-08	3
Lu	210.423	353.4	1.19	0.1 *	3.6399E-04	2.366E-08	3

Table 2. Enthalpies of dissociation ($\Delta_{dis}H_m$) for LnCl<g> at 298.15 K and standard molar enthalpies of formation ($\Delta_f H_m$) for LnCl<g> at 298.15 K

Ln	$\Delta_{dis}H_m$(LnCl,g,298.15 K)/J.mol^{-1}	$\Delta_f H_m$(LnCl,g,298.15 K)/J.mol^{-1}
Ba	439272	-132970
La	524380	22772
Ce	459956	78420
Pr	425654	51461
Nd	421142	27289
Pm	416994	-36529
Sm	422051	-95152
Eu	408445	-110956
Gd	453874	65939
Tb	474193	33165
Dy	395078	11476
Ho	411603	9890
Er	450971	-15842
Tm	380303	-26242
Yb	377080	-103292
Lu	383265	165899

REFERENCES

[1] L.A. Kaledin, M.C. Heaven & R.W. Field, *J. Mol. Spec.*, 193, 285 (1999).

[2] I. Hurtado, D. Neuschutz, 'Thermodynamic Properties of Inorganic Materials', compiled by SGTE, Landolt Börnstein, 19, Springer-Verlag, Berlin (1999–2005).

[3] A.D. Chervonnyi & N.A. Chervonnaya, *Russ. J. Inorg. Chem.*, 52, 1230 (2007).

[4] http://physics.nist.gov/PhysRefData/Handbook/index.html

[5] http://webbook.nist.gov/chemistry/

[6] X. Cao & M. Dolg, *J. Theoretical & Computational Chem.*, 4, 583 (2005).

Continuous self-absorption correction in sphere photometry measurements

M. Kotrebai[1], P. Mathews[2], and R. Malnar[2]
[1] *GE Lighting, Technology Product Testing, 1975 Noble Road, East Cleveland, OH 44112, United States*
[2] *Mathews, Malnar, & Bailey, Inc., 217 Third Street, Fairport Harbor, OH 44077*
Contact: Mihaly.Kotrebai@ge.com

ABSTRACT

Knowing about self-absorption differences during sphere photometry measurements is important, but the need to correct for it depends on factors like the lamps and hardware, the size of the sphere and the reflectivity of the paint used. Traditional correction methods are simple substitution and the short-term use of an auxiliary lamp. The ideal solution is to know the self-absorption factor for each lamp measured at any interval. Continuous self-absorption correction solves all the issues associated with the correction. It will provide a factor for every lamp tested at any time during its life cycle. The size of the change in hardware is irrelevant. The correction is spectral and will yield correct color measurement even with large colored fixtures. The correction will also take care of the changes in sphere responsivity as the system calibration ages. Extra requirement are the need for a very stable auxiliary lamp and additional quality control measures.

INTRODUCTION

Self-absorption correction is important in sphere photometry, especially when small spheres are used for the measurement or high reflectance paint is used for a sphere of any size. The system responsivity (including the detector) will change when any part of it is even slightly altered, such as when there are size, geometry or color differences between calibration and test lamps or when there are differences in the lamp mounting hardware between calibration and test. Ignoring such differences can result in errors of up to 40%.

There are various well-known methods for dealing with the self-absorption correction problem [1]. The simplest one is the substitution method when all hardware stays in the sphere for the calibration and the test lamp measurement, either in their measurement configuration or on the bottom of the sphere. In order to get a quantitative correction factor, an auxiliary lamp must be operated in the sphere. The change of the detector signal with the auxiliary lamp on between the calibration setup to the measurement setup provides the self-absorption correction factor that can be applied to measurements. These factors are often determined once for a lamp type and then that correction is applied whenever lamps of that type are measured. This approach ignores the variability from lamp to lamp and over time as lamps age. A better approach is to determine and apply a unique correction factor for each test at each measurement interval from a subset of the lamps under test, but there can be considerable self-absorption differences between lamps within the same test.

Measuring individual correction factors for each lamp in an off-line setting is possible but very time consuming since each lamp must be placed and measured in the sphere twice and then the correct factor must be used when the actual measurement takes place. This time consuming solution is especially inappropriate in laboratories where volume throughput is very important. This situation can be further complicated when using multi-head photometry systems with high reflectance paint, when each head might need its own correction factor in addition to the test lamp specific correction factor.

RESULTS AND DISCUSSION

A solution to all of the shortcomings of the different self-absorption correction methods is to make all measurements during calibration and test with an auxiliary lamp burning continuously in the sphere. Two measurements must be made for each test lamp, one with the test lamp on and another with the test lamp off. These measurements can be made in either order and are sufficient to provide a unique self-absorption correction for each test lamp measurement. Equation 1 gives the formula for determining the self-absorption-corrected intensity of a test lamp I_T given the intensity of the calibration standard lamp I_S:

$$I_T = I_S\left(\frac{y_T^1 - y_T^0}{y_S^1 - y_S^0}\right)\left(\frac{y_S^0}{y_T^0}\right) \quad (1)$$

Where y is the signal from the detector, subscripts T and S indicate test and standard lamps, respectively, and superscripts 0 and 1 indicate measurements taken with the test or standard lamp off and on, respectively. In the software implementation, all of the terms involving the standard lamp are combined into a single correction factor that is multiplied by the terms involving the test lamp.

It is possible to use this self-absorption correction algorithm with photo-detectors when the spectral shapes of the lamps involved (calibration, test, and auxiliary) are the same, such as for incandescent lamp measurements; however, when the sample lamp's spectrum is different from the auxiliary lamp's spectrum, correct lumen and color results can only be obtained using spectral measurements. While the simplicity of using a photo-detector for self-absorption measurements in both on-line and off-line environments is very tempting, it can be the source of very serious color measurement errors even though the lumen is corrected well. This error is especially apparent when lamps must be operated in large fixtures, such as LED street lighting, because these fixtures cause significant changes in the spectral reflectance and geometric illumination properties of the sphere. The continuous spectral self-absorption correction method automatically solves these problems.

Another problem that is resolved by the continuous self-absorption measurement method is changes in the sphere paint over the calibration cycle. This issue can be especially critical with overhead lamp pre-heaters and automated lamp feeders when dirt and oil from the mechanisms overhead fall in the sphere changing its reflection properties. Even without pre-heater, a sphere with high reflectance paint can change due to dust or particulates that fall off of test lamps.

The stability of the auxiliary lamp is very important in the continuous self-absorption correction method because any change in light output from the auxiliary lamp between calibrations propagates errors in test lamp photometric responses, so special auxiliary lamps must be used. As a general guideline in our laboratory, self-absorption correction was ignored when the correction factor was less than 0.5% for all lamps types measured on a photometry system. Thus, we have set the stability requirement for the auxiliary lamp to be less than 0.5% change in lumen output over a calibration interval, which is typically one workweek or about 40 hours of auxiliary lamp burn time. We have found that on this schedule our auxiliary lamps typically survive at least one year.

In the continuous self-absorption correction method the light output from the auxiliary lamp is assumed to be constant and any change in detector signal is interpreted as a change in sphere setup or other sphere conditions. Because the two types of changes cannot be distinguished from each other, special methods must be used to guarantee that the auxiliary lamp's output remains constant. To this end, we monitor the stability of the auxiliary lamp between calibrations using a photo-detector (the same photo-detector that was used for photo-detector based photometry prior to making the shift to spectral photometry) under a standardized sphere configuration. When this signal deviates from the value established at the beginning of the calibration cycle by more than a tolerance that is set for the system (typically 1-1.5%), a lamp with known lumen value, such as a calibration or control lamp, is measured. If the lumen reading of this lamp is within the measurement uncertainty (limit set to 0.5%), the auxiliary signal change is considered to be the result of system and not auxiliary lamp changes. Limits can be reset and measurements continued. If this limit is exceeded, the current calibration is declared to be invalid and the photometry system is recalibrated.

SUMMARY

While there are certainly more checks in place when using the continuous self-absorption correction method than with other correction methods, the payback in productivity and quality of measurements justifies the extra effort and enables seamless frequent changes from one type of product to another without the need to recalibrate the photometry system or maintaining an ever changing database of lamp and system specific correction factors.

REFERENCES

[1] Subcommittee of Light Sources of the IES Testing Procedures Committee, IESNA Approved Method for Total Luminous Flux Measurement of Lamps Using an Integrating Sphere Photometer, LM-78-07. New York, NY: Illuminating Engineering Society of North America, 2007

Visual performance of enhanced chroma lighting solutions

K. Tóth[1], I. Deme[1], L. Balázs[1], W. W. Beers[2], G. R. Allen[2], B.V. Nagy[3], Gy. Ábrahám[3]
[1]*GE Lighting , 77 Vaci ut,Budapest H-1044, Hungary*
[2]*GE Lighting , 1975 Noble Road, Cleveland OH 44112-6300, USA*
[3] *Department of Mechatronics, Optics and Instrumentation Technology of Budapest University of Technology and Economics , 4-6 Bertalan L, Budapest, H-1521, Hungary*
Contact: katalin1.toth@ge.com

ABSTRACT

Recently more and more emphasis has been put on the correct measurement of light quality. Current measures for that generally used in the lighting industry are the CIE color rendering index (CRI) and the correlated color temperature (CCT). However it's easy to find light sources (especially in the case of narrow band sources) with the same CCT and the same or similar CRI but with different spectral power distribution (SPD), which look very different to the eye[1]. In our project we have compared the physiological effects of several fluorescent light sources, having the same CCT, on human subjects. By applying similar photometric conditions visual acuity, color vision, perceived brightness and other subjective psychophysical test methods were used to evaluate the differences between the illuminants. Our results indicate the necessity for additional information beyond CRI and CCT in order to reflect back the results of the psychophysical tests and help customers to choose the appropriate light source to fulfil their needs.

INTRODUCTION

There has been a continuous effort for improving, modifying and providing additional information beyond CRI. One possible way is to expand the test colors set for which the color differences between the light source and a reference source are calculated in a certain color space[2]. The other way is to consider not just color fidelity but color preference as well. The color quality scale (CQS) NIST proposed considers the fact that observers usually prefer enhanced chroma light souerces and therefore increase of chroma is not penalized by CQS[3, 4]. They also suggest additional indexes e.g. color preference scale (Qp where increased chroma is rewarded) and gamut area scale (Qd). The third possible way is to apply the most recently accepted uniform color space and color appearance model to calculate color rendering: CRI-CAM02UCS[5]. Others consider the fact that color-opponency plays a role in color vision and brightness sensation, for example Hashimoto et al. introduced the feeling of contrast index (FCI) based on relative gamut area of four opponent colors[6].

RESULTS AND DISCUSSION

Since both Ohno et al.[4] and our earlier study[8] showed that enhanced chroma light sources are generally preferred, we further investigate its effect in other psychophysical experiments. In the tests we compared enhanced chroma lamps with their conventional counterparts having the same CCT with same illuminance level.

We tested visual acuity by means of regular plate tests viewed from five-meter distance with 0.1 visus resolution. Half of subjects in age group of 40-55 years showed above average visus values in the case of enhanced chroma lamps compared with the conventional lamps.

We evaluated human task performance (i.e. productivity) with a typing task of given text. We counted the amount and the errors in the typing. We have found higher performance under enhanced chroma fluorescent lamps compared to the reference triband source in the above age group.

To test perceived brightness we have designed and built a novel light booth where a reference source is used to compare the different illuminations. We showed nine different white and colored samples to 20 subjects of normal color vision within the age group of 20-35 to compare under the test and the reference illuminants. The individuals were asked to adjust the luminance level of the reference light source until it matched that of the test lamp. The results indicate significantly higher perceived brightness under enhanced chroma illumination than with its counterpart (Fig.1).

We also checked subjects with defective color vision under the different illuminants with general color vision screening processes such as the Ishihara plates and the Farnsworth D-15 color discrimination method. Both methods have shown better performance with the enhanced chroma type fluorescent lamps for anomalous trichromats when compared to standard D65 conditions (Fig.2). However anope types have remained seemingly unaffected when tested with any type of enhanced chroma light solution.

CONCLUSION

As was shown above the enhanced chroma light sources have better psychophysical performance, which is not reflected by the usual metric of color quality (CRI) as opposed to the recently proposed scales such as Qp, Qd and FCI (see Table 1.). Our results indicate the necessity for additional information beyond color rendering index (CRI) in order to help customers to choose the appropriate light source to fulfil their needs. We also found that enhanced chroma light sources are especially good for environment where visual acuity and brightness are important regarding human productivity.

REFERENCES

[1] A.M. Dugar, LS11 proceedings, 603 (2007).
[2] P.J.M. van der Burgt *et al.,* LS11 proceedings 557 (2007).
[3] W. Davis *et al.,* "Color Quality Scale" submitted to CIE TC1-69, Sep. 2009.
[4] Y. Ohno *et al.,* Visual Evaluation Experiment on Chroma Enhancement Effects in Color Rendering of Light Sources submitted to CIE TC1-69, Sep. 2009.
[5] C. Li *et al.*, Assessing Colour Rendering Properties of Daylight Sources Part II: A New Colour Rendering Index: CRI-CAM02UCS, submitted to CIE TC1-69
[6] K. Hashimoto *et al.*, A new method for specifying color rendering properties of light sources based on feeling of contrast. Submitted to CIE TC1-69, Sep. 2009
[7] W.W. Beers *et al.*, US Patent No. 2009/0102348A1, (2006).
[8] K. Toth *et al.*, CIE conference 2009

Fig. 1

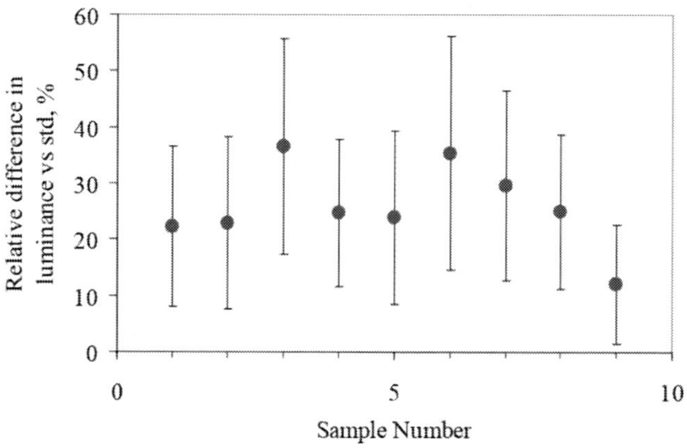

Fig. 2

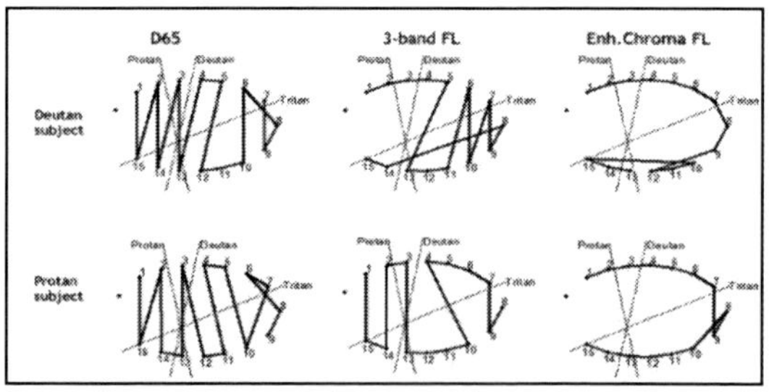

Table 1

	3-band fluorescent (FL)	Enhanced chroma FL[7]	Neodymium doped incandescent
CRI	83	63	77
CQS	77	68	88
Qp	80	86	101
Qd	92	116	112
CIECAM02	82	70	83
FCI	111	135	143

Comparison between current controlled and voltage controlled pulsed drivers for a Dielectric Barrier Discharge (DBD) lamp

A. El-Deib[1], F. Dawson[1], G. van Eerden[2], S. Bhosle[3], G. Zissis[3]

[1]*ECE Department, University of Toronto, 10 King's College Rd., Toronto ON, M5S 3G4, Canada*
[2]*Light Controls, Nedap, Groenlo, 7141 DC, The Netherlands*
[3]*LAPLACE, Universite de Toulouse, Toulouse, UMR 5213, France*
Contact: amgad.eldeib@utoronto.ca

ABSTRACT

The performance of a DBD lamp driven by a current controlled and a voltage controlled driver is examined to determine the effect of each driver approach on the lamp efficiency and the overall system efficiency. The current controlled driver produced higher lamp efficiency over the whole operation range but the overall system efficiency was only higher at high UV output densities.

1. INTRODUCTION

DBD lamps have been usually excited using sinusoidal voltage waveforms with electrical to UV conversion efficiencies less than or equal to 20% for Xe, 15% for $KrCl$ and 12% for $XeCl$. Recently it has been shown that the DBD lamp operating under pulsed excitation conditions will result in higher optical efficiencies. Using high frequency excitation pulses which are interrupted by idle times, it has been possible to increase the efficiencies to about 60% for Xe.The enhanced performance under conditions of pulsed excitation results largely from the ability for pulsed excitation to generate a diffuse discharge at near atmospheric pressures [1]. Due to the capacitive nature of the DBD lamp, it is more natural to inject current into the lamp as compared to imposing a certain voltage across its terminal. Imposing a voltage across a capacitance will result in a current spike which is not controlled. It has been also mentioned in [2] that there might be a direct relation between the injected current waveform and the optical output temporal behavior. Yet, most of the pulsed power supplies suggested in the literature like in [3] are based on a voltage controlled driver rather than a current controlled one.

In this paper, the results of an experimental comparison between a current controlled and a voltage controlled driver for the pulsed operation of a $XeCl$ DBD lamp are reported. The lamp is cylindrical with a length of $70cm$ and a gap distance of $8.5mm$. The efficiencies of the lamp and the overall system are shown for the two different drivers.

2. EXPERIMENTAL RESULTS

A newly developed power electronics driver was designed to drive the DBD lamp with a controlled pulsed current. The produced current waveform has two opposite pulses followed by a dead time for the remainder of the cycle. The details of this driver are given in [4] including the design procedure. A voltage controlled DBD driver was also implemented. This voltage controlled driver is based on the circuit topology described in [3] which consists of a full bridge converter connected to the DBD lamp through a step-up transformer.

Figure 1(a) shows a comparison between the lamp input power obtained for the two drivers for the same UV output power density. All measurements were performed at a $435V$ DC bus voltage and a $25kHz$ operating frequency. The lamp produces higher UV power density for the same input power when excited by the current controlled driver. The driver efficiency of the voltage controlled driver is however higher than the current controlled driver due to the lower number of components and the soft switching of the MOSFETs in the voltage controlled driver. Figure 1(b) shows the system efficiency for both drivers based on the assumption that the UV power density is uniform over the whole lamp surface area for both drivers. The DBD voltage under the voltage controlled driver oscillates until the next pulse because there is always a direct coupling between the lamp and the transformer magnetizing inductance as shown in figure 2(a). The DBD current is shown in figure 2(b). The peak current in the voltage controlled driver is not predetermined as it depends on the breakdown taking place in the lamp while it is already chosen during the design procedure of the current controlled driver.

3. CONCLUSION

The current controlled driver produced a higher lamp efficiency over the whole operation range and a higher system efficiency at higher UV power densities ($> 32mW/cm^2$ at $25kHz$). However, it should be mentioned that the gain obtained in the system efficiency may not be enough due to the increased cooling costs for the current controlled driver, which is due to the lower efficiency of the employed driver. The UV output of the DBD lamp showed some saturation behaviuor under the voltage controlled driver as shown in figure 1(a), which did not show up under the current controlled driver.

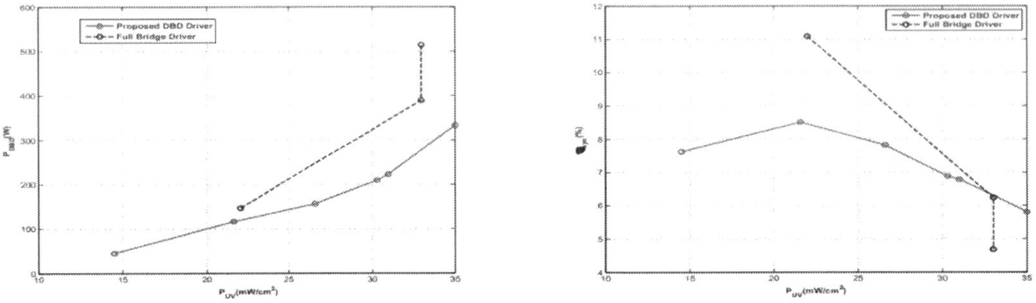

(a) (b)

Figure 1. Comparison between the current controlled driver (red solid line) and the voltage controlled driver (black dashed line) for (a) Lamp input electrical power and (b) System efficiency

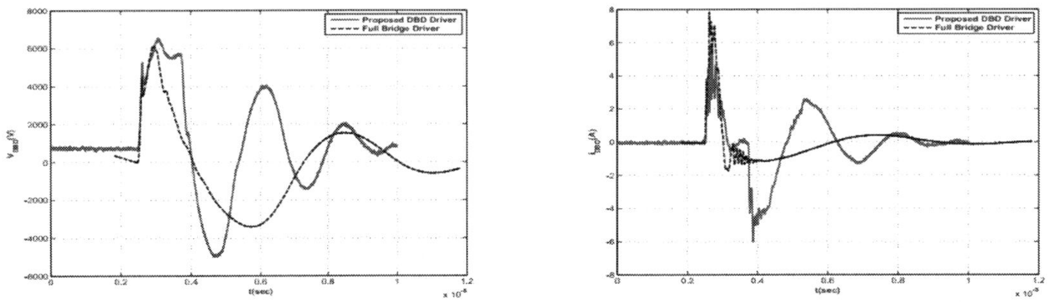

(a) (b)

Figure 2. Comparison between the current controlled driver (red solid line) and the voltage controlled driver (black dashed line) for (a) DBD lamp voltage and (b) DBD lamp current

REFERENCES

[1] R. P. Mildren and R. J. Carman, "Enhanced performance of a dielectric barrier discharge lamp using short-pulsed excitation," *Journal of Physics D: Applied Physics*, vol. 34, no. 1, pp. L1–L6, 2001. [Online]. Available: http://stacks.iop.org/0022-3727/34/L1

[2] R. Diez, H. Piquet, S. Bhosle, and J.-M. Blaquiere, "Current mode converter for dielectric barrier discharge lamp," in *Power Electronics Specialists Conference, 2008. PESC 2008. IEEE*, June 2008, pp. 2485–2491.

[3] W. Shin, J. Choi, and T. Kim, "Bidirectional pulse plasma power supply for treatment of air pollution," in *Power Electronics Specialists Conference, 2006. PESC '06. 37th IEEE*, June 2006, pp. 1–6.

[4] A. El-Deib, "Modeling of and driver design for a dielectric barrier discharge lamp," Ph.D. dissertation, University of Toronto, 2010.

Photometric characteristics of metal halide lamp

M. Hamady[1] , M. Aubès[1] , G. Zissis[1]

[1] *Universite de Toulouse, UPS, INPT, LAPLACE (Laboratoire Plasma et Conversion d'Energie) 118route de Narbonne, F-31062Toulousecedex9, France*

Contact: mohamad.hamady@laplace.univ-tlse.fr

ABSTRACT

The radiation transport that takes place in light sources represents the base of all photometric characteristics of the source. Radiation transport of mercury lamp doped by thallium iodide is calculated by solving the radiative transfer equation using the ray tracing technique. Pressure and temperature profile are deduced from the experimental results. Plasma composition of the lamp is calculated after solving eleven non linear equations representing the eleven species that exist in the discharge. Neutral Van der Waals and charged particles Stark broadening are taken into account while considering the line broadening.

GENERAL

The plasma in metal halide lamp is considered at local thermodynamic equilibrium (LTE). In this work, pressure and local temperature are determined from spectroscopic measurements. Temperature profile across the discharge is fitted by the following relation:

$$T(r) = T_c - (T_c - T_p)\left(\frac{r}{R}\right)^{\beta} \tag{1}$$

Where T_c is the axis temperature, T_w is the wall temperature, β is a parameter that describes the radial distribution of the discharge and R is the discharge radius. On the other hand, for line profile, only pressure broadening is taken into account (Doppler broadening is considered negligible). The resulting total line profile is obtained by the convolution of the Lorentzian and Quasi-static profiles [1]. In case of pure HID lamp (Hg, Na), by applying the equations of state for ideal gas (Boltzmann and Saha law) we are going to be able to calculate the plasma composition of the discharge. The interest of using metal halide lamps is the ability, following the chosen doping halide, to enrich the line spectrum and to improve the photometric characteristics of the lamp. In this work, we study the case of mercury lamp doped by thallium iodide. The presence of Thallium adds to the spectrum the green line 535 nm and also the red line 655 nm. However, the plasma composition depends of the quantity of each matter introduced at the moment of fabrication. Its calculation is done by solving eleven non linear equations. Among these equations we have two atomic ratios concerning the mercury/thallium (R1) and thallium/iodine (R2). These ratios are given by the constructor. Once the density of each species at every temperature in the discharge is known, the radiation transport is going to be solved by using a technique of ray tracing. First of all, the discharge is going to be meshed into small elements. N isotropic rays are going to be launched by each element. Each ray contributes to the out put power flux. The power flux that goes out from the emitting element volume of each ray is calculated and saved. The total output flux Φ_{out} that goes out from the discharge represents the summation of all power flux coming from all rays. A complete description of the model work is already described by previous work [2]. To obtain the photometric results of the lamp, we calculate the flux coming out of the lamp from each line in the visible domain. Then the luminous flux is calculated by adding the contribution of each line. The luminous flux is calculated by the following equation:

$$\Phi_{vis} = K_m \int_{\lambda} V(\lambda)\Phi_{\lambda}(\lambda)d\lambda \tag{2}$$

Where V(λ) is the curve of photopique vision and K_m is a constant equals to 683 *lm* / W. Then, the luminous efficacy is calculated by:

$$\eta = \frac{\Phi_{vis}}{P_{electric}} \tag{3}$$

Where $P_{electric}$ represents the electric power applied to the lamp. From the radiated flux of each line and using the color-matching functions we were able to calculate the Correlated Color Temperature (CCT) and the chromaticity coordinates.

RESULTS

The photometric results of a metal halide HID lamp are going to be calculated. Table 1 shows the characteristics of the discharge that are taken from the lamp constructor and also the pressure and temperature profile that is fitted by experimental results.

Radius (cm)	Distance between electrodes (cm)	Mercury Mass (mg)	Atomic ratio R_1=Hg/Tl	Atomic ratio R_2=Tl/I	Electric Power (W)	Average total pressure (bar)	T_c (K)	T_w (K)	β
1	10	120	148	1	1200	4.65	5800	1000	2.6

Table 1: characteristics of the studied discharge

The luminous flux of each line in the discharge is calculated. We noticed the great contribution of the thallium line 535 nm as it is very near of the maximum sensibility of eye in photopique vision. Applying equation (3), we have found that the luminous efficacy of this discharge equals to 142lm/W. This relative high luminous efficacy come from the fact that the discharge characteristics in table 1 correspond to the moment when the discharge current is maximum. This also verified by the proximal temperature that we found it to be equal to 6350K.

The photometric curve of this discharge is also calculated. Figure 1 represents the photometric curve of the lamp at the midpoint of the lamp's height

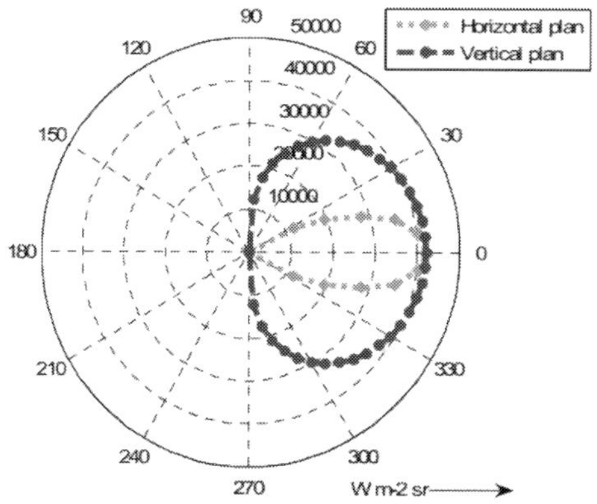

Figure 1: Photometric curve of the studied discharge

The luminous efficacy of the lamp is inversely proportional to the mercury/thallium ratio (R1). The quantity of Thallium in the discharge is decreased till R_1=200 in table 1. We have found a luminous efficacy equals to 131 lm/W.

REFERENCES

[1] G. Traving, "Interpretation of line broadening and line shift" , *Plasma Diagnostics,* ed. W. Lochte-Holtgreven, North-Holland Publishing Company, Amsterdam, 1968

[2] : M. Hamady, "Ray tracing calculation of radiation transport in HID lamps". Conference of XVII International Conference on Gas Discharges and their Applications, Cardiff, United Kingdom, Septembre 2008.

Angle of incidence and viewing angle for biologically reasonable lighting

K. Barkowsky[1], R. Kozakov[1], D. Kunz[2], C. Stoll[2], D. Uhrlandt[1], H. Schöpp[1]
[1] *INP-Greifswald, F-Hausdorff-Str. 2, Greifswald, D-17489, Germany*
[2]*Charité – University medicine, Sleep Research & Clinical Chronobiology, Gr. Hamburger-Str. 5-11, Berlin, D-10115, Germany*
Contact: barkowsky@inp-greifswald.de

ABSTRACT

Circadian action of lighting is of rising importance, but there is still much more information needed to develop new reasonable lighting concepts. To determine the importance of incidence and viewing angle in naturalistic situations, different lighting situations were investigated with participants lying in supine position similar to situations in hospitals and old-age homes. Commercially available lamps were used in fixtures overhead, below feet and as uplight. The colour temperature was chosen above 5000 K, with a good efficacy and colour rendering index. Melatonin was controlled by saliva probes. The direct comparison of different viewing angles showed, that larger viewing angles produce stronger melatonin suppression. The incidence angle has also importance but must be compared with the intensity.

INTRODUCTION AND EXPERIMENTAL DESIGN

Sleep problems and disorders of patients are a common problem in hospitals and old-age homes. Changing lighting to ensure appropriate intensity and timing of signals for the circadian rhythm is a promising approach to solve this problem. Up to now no relevant data are available for lying persons under naturalistic lighting situations. To investigate this problem we choose 3 lighting situations with high colour temperatures to ensure sufficient melatonin reaction. Melatonin is an appropriate marker for biological action of light because the increase of melatonin levels in evening hours may influence the following sleep phase.

The first lighting situation consists of wall lights fixed at the top end of the bed. For the second the same lamps and fixtures were used but positioned at the foot end of the bed. These two situations differ strongly in the incidence angle. For these two situations fluorescent lamps with a colour temperature of 6500 K were used. The luminance for overhead lighting was 387-456 lx, for foot end illumination 108-112 lx.

The third situation consists of an uplight which was partly concealed to ensure comparable luminance and to avoid glare. The used lamp had a colour temperature of 5600 K, the luminance 337-346 lx. Here we have the effect that the whole ceiling acts as luminaire. This result in a very large viewing angle especially compared to the other situations. The colour rendering index for all lighting situations was above 90. Luminance, illuminance and spectral distribution for the different lighting situations were carried out with a calibrated luminance camera (LMK), illuminance meter and compact spectrometer in the sleep laboratory. The measurements were executed at the later positions of the test subjects' eyes.

The test persons were aged between 25 and 35 years, healthy and of both sexes. Sleep quality was assessed with Pittsburg Sleep Quality index and no sleep disorders were allowed. Additionally test persons had to keep constant bedding times and to wear a so called actimeter for one week, which recorded their activity and resting times to assess good sleep hygiene. Their normal daily life was not restricted. Only extreme behaviour like excessive sport or alcohol usage were forbidden.

After this test persons entered the test period, which consists of a standardized timetable, which was only applied during the winter season. Test persons arrived in the lab at 7 pm. They stayed in a dim light situation (<1 lx) for some hours to ensure melatonin rise. Saliva probes were collected every half an hour. After this they placed themselves in a supine position at the intended points of the lighting situation which was switched on for half an hour one hour prior to habitual bedtime. Within this period as well as ten minutes before and afterwards saliva probes were taken every ten minutes. Thereafter they stayed in dim light for another hour with probe collection every half an hour. After this procedure they went home. Saliva probes were analyzed by an external lab to quantify melatonin concentration. In the first and one other random evening, no light treatment was applied, although test persons had to lie down and probes were collected as usual. The first so called adaptation night was needed to accustom test persons to the procedure, the other night was used as baseline for the experiments. In both cases applied light intensity was less then 1 lx.

RESULTS

The obtained results are given in Fig. 1. They show that dim light situation and adaptation night have only a minor effect on the melatonin levels as was to be expected, while all other situations show a strong suppression of melatonin. Here lighting from below causes the smallest effect, followed by lighting from above and finally diffuse lighting.

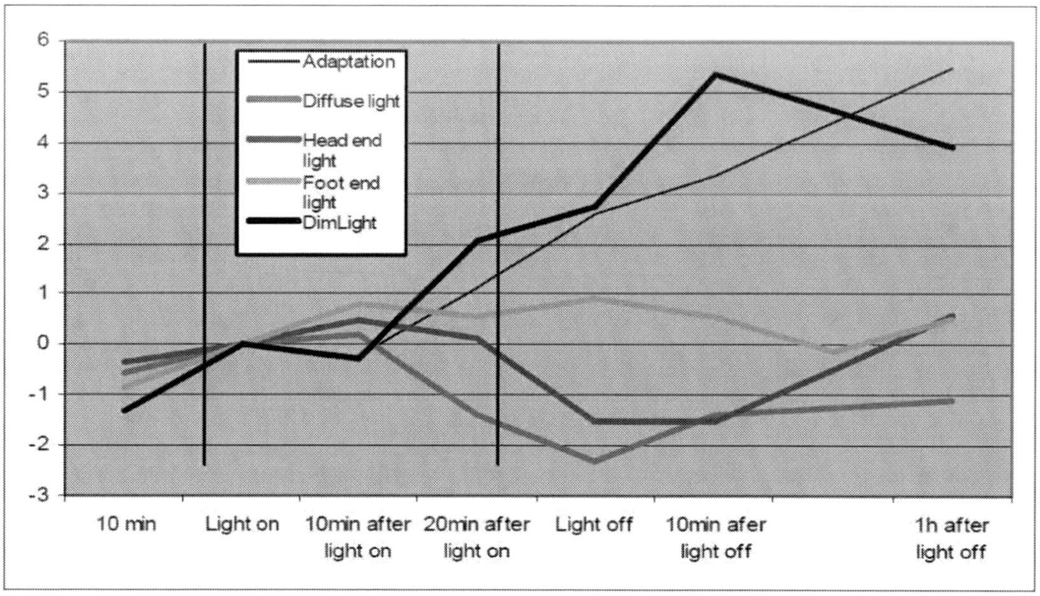

Fig. 1: Changes of melatonin concentration during and after light treatment

For further conclusions a comparison of the different incidence and lighting angles is needed. It is known from strict laboratory experiments, that light from the upper visual field causes more melatonin suppression than light from the lower visual field. In our experiment the same seemed true, as overhead lighting caused the higher melatonin suppression. But the effect cannot be distinguished from the influence of intensity without further mathematical processing.

This problem did not arise for the comparison of overhead lighting and diffuse lighting via uplight. Although the intensity and colour temperature of uplight were lower than those of direct overhead lighting, the diffuse lighting situation showed a stronger melatonin suppression, which could only be caused by the difference in viewing angle. Therefore it is shown, that viewing angle, i.e. the observed size of light source, has a perceivable influence on melatonin suppression. With growing viewing angle the influence on melatonin suppression rises too.

SUMMARY

Investigation of test persons lying in supine position under naturalistic lighting situations with different commercially available lamps and fixtures show that viewing angle influences melatonin secretion in the evening and hence the biological rhythm. The same could not yet be shown for incidence angle. The results can be used to develop new concepts of lighting for clinical and nursery facilities.

This work is supported by BMBF under grant FKZ 13N9779

185 nm radiation in the positive column of T2 low-pressure Ar-Hg discharge

Qiuyi Han[1], Shaolong Zhu[1], Graeme G. Lister[2], Shanduan Zhang[1]
[1] *Institute for Electric Light Sources, Fudan University, 220 Handan Road, Shanghai 200433, People's R China*
[2] *CRSL, OSRAM SYLVANIA, 71 Cherry Hill Drive, Beverly, MA, 01915, USA*
Contact: sdzhang@fudan.edu.cn

ABSTRACT

185 nm and 254 nm radiation were measured and compared for cold spot temperatures 20-60 °C and discharge currents 80-200 mA in the positive column of T2 low pressure mercury vapour discharges with inner/outer diameter 5/7 mm and argon fill pressure 5 Torr. It is found that 185 nm radiation power increases with discharge current and cold spot temperature, and the fraction of electrical power converted to 185 nm radiation also increases slightly with these parameters.

The main contributor to the luminous efficacy of fluorescent lamps, and detailed measurements of 254 nm radiation, to study the effect of mercury vapour pressure (controlled by cold spot temperature) and discharge current on low pressure Ar-Hg T2 discharges, have previously been reported [1]. Since 185 nm radiation in these discharges is also significant, its quantitative contribution to the energy balance in the positive column also needs to be evaluated.

EXPERIMENTAL SETUP

The geometry of the T2 sample lamp is shown in Fig. 1. The lamp, made of Synthetic fused quartz tubing with wall thickness of 1 mm, was filled with 10 mg mercury and 5 Torr (667 Pa) argon. Two electrostatic probes were sealed in the narrow tube at a distance of 100 mm, to measure the floating potential and calculate the electric field intensity of positive column. The cold spot was maintained by a brass water jacket, whose temperature was controlled by a recycling bath thermostat from 20 to 60 °C.

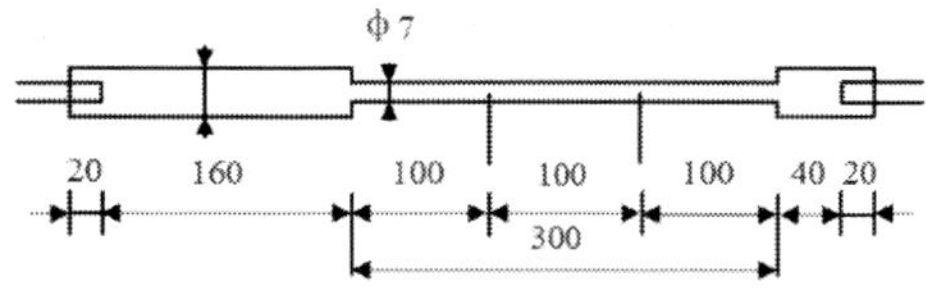

Fig.1 Geometry of the T2 discharge tube (unit: mm).

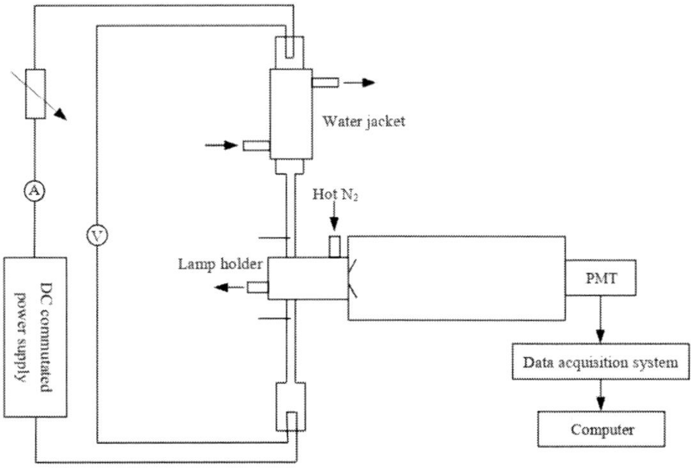

Fig. 2 Experimental set-up for radiation measurement.

The experimental set-up is shown in Fig. 2. The discharge was driven by a 50 Hz dc-commutated power supply (ranging ± 600 V) so that axial mercury cataphoresis could be avoided. The filament heating current was 110 mA and a 4.6 kΩ adjustable resistance provided the ballast. The discharge current was varied from 80 to 200 mA for each cold spot temperature. The radiance for each set of parameters was measured by a vacuum monochromator (Acton Research VM-505) and calibrated through a deuterium lamp radiance standard. The discharge tube was fixed in the lamp holder sealed to the entrance slit of the monochromator, and a flow of heated N_2 was used to expel the oxygen from the air. The tube outside the lamp holder was wrapped with black paper, to ensure that the wall temperature was higher than the cold spot temperature. During the measurement, the wall temperature was monitored continuously by a set of thermocouples. The radiant powers were then derived from the measured radiances using the relevant Koedam factors [1].

VARIATION OF RADIATION WITH CURRENT AND COLD SPOT TEMPERATURE

Fig. 3 shows the current dependence of 185 and 254 nm radiant powers for a cold spot temperature of 50 °C. It indicates that both 185 nm and 254 nm radiant power saturate with increasing current. The increase of current causes the electron temperature to fall, so that the atoms are more likely to de-excited through exoergic inelastic collision with slow electrons, which leads to the saturation of 254 nm radiant power. But for 185 nm, it probably results from mercury depletion near the axis of the discharge tube, due to the very high current density. Ambipolar diffusion is exteremely important in such a narrow tube [2]; the ground state mercury density decreases at very high current, and since the energy of Hg 6^1P_1 excited-state is 1.8 eV higher than that of Hg 6^3P_1 state, it is more difficult for mercury atoms to be excited to the Hg 6^1P_1 state, causing 185 nm radiation to saturate. The ratio of radiant power at 185 nm to that at 254 nm increases slightly with discharge current.

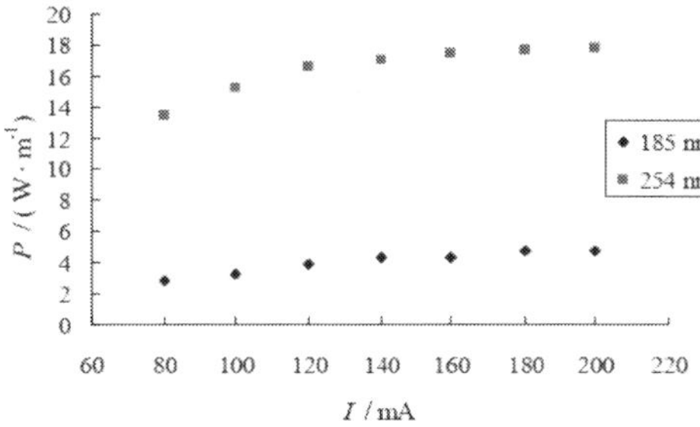

Fig.3 Current dependence of radiant power in T2 lamp filled with 5 Torr Ar.

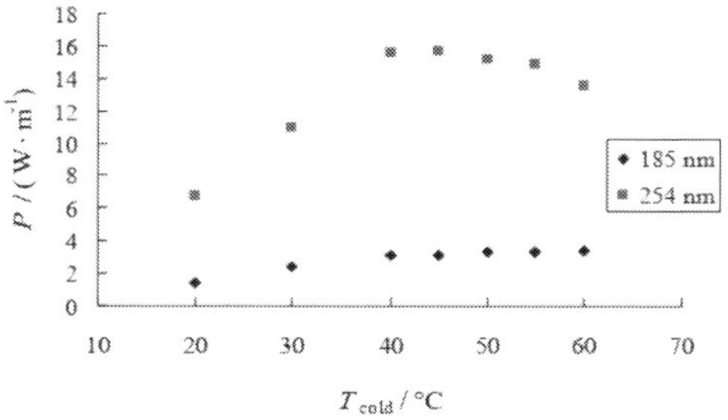

Fig.4 Cold spot temperature dependence of radiant power in T2 lamp filled with 5 Torr Ar.

The effect of cold spot temperature on 185 and 254 nm radiant power at 100 mA is shown in Fig. 4. As cold spot temperature increases, the 185 nm radiant power keeps rising, which means that the optimum cold spot temperature is higher than 60 °C. It is consistent with the tendency that 30 °C for T12 [3], 42 °C for T8 [4], and 50 °C for T4 [5]. The ratio of radiant power at 185 nm to that at 254 nm also increases slightly with rising cold spot temperature.

ACKNOWLEDGEMENT

The experimental programme was financially supported by OSRAM GmbH.

REFERENCES

[1] Han Q Y, Jiang H, Zhu S L, Lister G and Petrova T 2008 *J. Phys. D: Appl. Phys.* **41** 144008
[2] Fang D Y and Huang C H 1988 *J Phys D: Appl Phys.* **21** 1490
[3] Koedam M, Kruithof AA and Riemens J 1963 *Physica.* **29** 565
[4] Wani K 1994 *J Appl Phys.* **75** 4917
[5] Anderer P 1991 *J Appl Phys.* **69** 1224

Thermodynamics processes in DBD driven excilamps estimated by jump of pressure method: investigation of acoustic waves and heat generation

A.A. Pikulev[1], V.M. Tsvetkov[1], E.A. Sosnin[2], V.F. Tarasenko[2]

[1] *Federal State Unitary Enterprise – Russian Federal Nuclear Centre-All-Russian Research Institute of Experimental Physics, 37, Mira ave., 607190, Russian Federation*
[2] *Optical Radiation laboratory, High Current Electronics Institute SB RAS, 2/3, Akademicheskii ave., 634055, Russian Federation*
Contact: badik@loi.hcei.tsc.ru

ABSTRACT

In this report it is proposed to use the fast part of pressure jump (characteristic time 20-200 ms) for studying of thermodynamical and acoustic processes in excilamps of barrier discharges. The investigation results of KrCl excilamp of barrier discharges in pressure diapason 30-380 Tor ($Kr:Cl_2 = 300:1$) are presented.

INTRODUCTION

One of instruments suitable for investigations of thermodynamical processes occurring in excilamps of barrier discharges is a pressure jump measurement. Investigations show that pressure jump appearing in excilamp bulb after its turning on consists of two components: 1) fast component (characteristic time 20-200 ms) and slow component (~ 100 sec) [1]. The fast component of pressure jump is resulted from acoustic waves generation and direct gas heating during process of discharge. The slow component of pressure jump is caused by heating of excilamp bulb.

Earlier authors of this report proposed to use the amplitude of fast pressure jump component for UV-radiation power control during prolonged excilamp exploitation and slow component – for investigation of excilamps heat working regimes depending on cooling conditions [1].

In this report it is proposed to use the fast part of pressure jump for studying of thermodynamical and acoustic processes in excilamps of barrier discharges, in particular for determination of discharge power portions, which cause the acoustic waves generation and direct gas heating.

RESULTS

The investigations were conducted at KrCl excilamp of barrier discharges (mixture $Kr:Cl_2 = 300:1$) in pressure diapason 30-380 Tor. The excilamp working length was 330 mm, diameter - 43 mm and discharge gap was equal to 8.5 mm. The excilamp was powered by bipolar power supply with voltage pulses amplitude 4.5 kV and pulse repetition rate 100 kHz. To measure pressure jump in excilamp bulb there was used inductive differential pressure sensor DMI 0.1-2.

The analysis of experimental results shows that the following function gives a very good approximation of pressure jump fronts after discharge turning off

$$\Delta p = \Delta p_0 \left(\beta \exp\left\{ -\frac{t}{\tau_{ac}} \right\} + (1-\beta) \exp\left\{ -\frac{t}{\tau_T} \right\} \right), \qquad (1)$$

where Δp_0 – pressure jump amplitude; β, τ_{ac}, τ_T – approximation parameters. Dependences of parameters β, τ_{ac}, τ_T from pressure are presented in table. The relative error of approximation (2) doesn't exceed 2-3 %.

Table. Dependences of parameters β, τ_{ac}, τ_T from pressure

p, Tor	β	τ_{ac}, ms	τ_T, ms
60	0.18	27	30
90	0.29	25	54
180	0.34	15	104
210	0.35	13	115
240	0.36	13	140
300	0.3	12	150
400	0.2	11	160

Parameter τ_T, determined from experimental results is approximately equals to characteristic time of gas cooling in excilamp [2]

$$\tau_T = \frac{Vd}{6\gamma\chi S}, \tag{2}$$

where γ - adiabatic exponent; χ - thermal diffusivity; V – gas volume of excilamp; S – surface area of inner walls of excilamp bulb; d – discharge gap. Parameter τ_{ac} equals to characteristic time of acoustic vibrations damping with frequencies 40-65 kHz.

It may be shown that parameter β is connected to part of discharge power W, resulting in acoustic waves generation α_{ac}, by the following equation

$$\beta = \frac{\alpha_{ac}\tau_{ac}W}{V\Delta p_0}\left\{\delta + \frac{\gamma-1}{\tau_0}\frac{\tau_T\tau_{ac}}{\tau_T-\tau_{ac}}\right\}, \tag{3}$$

where τ_0 – the characteristic time of energy transfer from acoustic waves to heat; $0 \le \delta \le 1$ – acoustic vibration pressure sensor sensitivity coefficient.

The calculations show that for pressures under 120 Tor the greater part of discharge power goes on direct gas heating and for greater pressures – on generation of acoustic waves. At that parameter α_{ac} is maximal for pressure 300 Tor. This fact is explained by the following observation: for pressures 100-300 Tor discharge occurs in form of single filaments, which are the effective source of acoustic vibrations.

CONCLUSION

The results of pressure jump investigation in KrCl excilamp of barrier discharges have shown that fast pressure jump component is connected to acoustic waves generation (characteristic time 10-30 ms) and direct gas heating (30-160 ms) during discharge process; at that filaments are the main source of acoustic vibrations.

ACKNOWLEDGMENTS

This work was supported by the ISTC (Project No. 3583p).

REFERENCES

[1] V.M. Tsvetkov, A.A. Pikulev, The IX Int. Conf. "Atomic and Molecular Pulsed Lasers", Abstracts, Tomsk, 2009, p. 97.

[2] L.D. Landau, E.M. Lifshitz, Theoretical Physics. Vol. VI. Hydrodynamics, Moscow (1988) 736 p.

Efficiency improvement of ultra high pressure projector lamps by reflecting the unused UV radiation back into the lamp

K. C. Paul[1] , T. Takemura[2] , M. Nakayama[3] , and T. Igarashi[2]
[1] R&D Center, Ushio Inc., 1-90 Komakado, Gotenba, Shizuoka 412-0038, Japan
[2] R&D Center, Uhsio Inc., 1194 Sazuchi, Bessho-cho, Himeji, Hyogo 671-0224, Japan
[3] Optics Design Section, Uhsio Inc., 1194 Sazuchi, Bessho-cho, Himeji, Hyogo 671-0224, Japan

Contact: k.paul@ushio.co.jp

ABSTRACT

Visible region of the electromagnetic spectrum, generated by an extremely high pressure projector lamp, is used in the projection purpose. But a considerable amount of energy of the ultraviolet (UV) region is also generated and filtered out from the system. This work is devoted in an attempt to reuse this collectable UV light in enhancing visible lights and thus to improve lamp performance. A model is developed to account for the UV light of the 350–430 nm spectral band that is reflected back to the plasma of the light source to see its effects on the lamp properties. Simulations are carried out for a typical mercury lamp of 10 MPa operating pressure and of 1.4 A discharge current. Such a lamp usually operates horizontally by an alternating current but simplifications in the model are exercised neglecting natural convection. Ultimately, a 2D and axisymmetric assumptions for a direct current discharge are adapted. Radiation complexity and huge computational cost are avoided for the trade-off in which fundamental UV effects are essentially retained. 10% of a 100 W lamp or 10 W is assumed to return to the lamp, of which 65% or 6.5 W is absorbed in the discharge. This reflected UV radiation is predicted to raise the plasma temperature and to increase the visible radiation by about 6%.

INTRODUCTION

Ultra high pressure mercury lamps are in significant applications for projection purposes. Visible lights from these lamps are useful although the lamps naturally produce lights from other regions of the electromagnetic spectrum too. Our estimation suggests that only 31% of the total radiation from such a lamp is from the visible spectral band and remaining 69% is unused and lost in some form of energy. Optical setup of these lamps would make it viable of reusing the ultraviolet (UV) light if it appears effective in improving lamp performance through the enhancement of visible lights. In this perspective, this work is devoted to study the lamp performance by reflecting UV light back into the lamp. A powerful numerical model is developed to account for the reflected UV feedback into the lamp through the glass and thus to predict the plasma properties. Electrodes and gas-filled region of the lamp are modelled and the primary attention of this article is to determine reflected UV effects on plasma properties, especially on the visible light and electrodes temperatures. To avoid huge cost and tremendous complexity, calculations are done for the 2D and axisymmetric assumptions. Operating gas of the lamp is mercury of 10 MPa. A direct current (DC) of 1.4 A is used to supply a 100 W lamp.

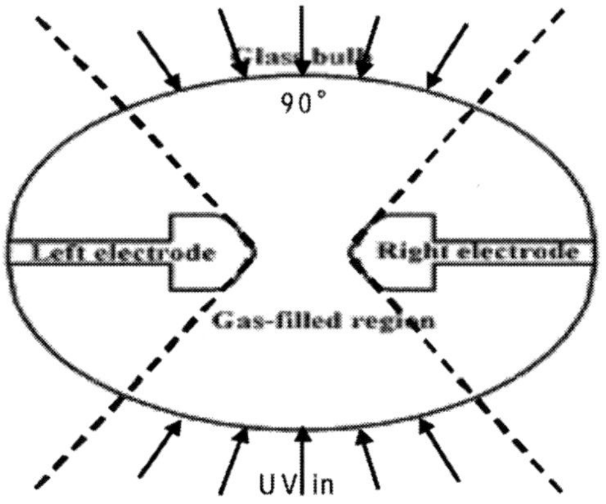

Fig.1 Schematics of the lamp and reflected UV irradiation

MODEL FEATURES

The complete set of magnetohydrodynamics (MHD) – conservation equations of mass, momentum and energy supplemented by the Maxwell's electromagnetic interactions– has been solved simultaneously in the fluid region of the lamp and the energy conservation equation has been used in predicting thermal situations of the electrodes. Although a 3D model is necessary to describe situations of a horizontally-burned projector lamp, we have adapted the model on 2D and axisymmetric assumptions to avoid complexity, in the preliminary stage of basic understandings. These assumptions prefer neglecting the buoyancy in order to retain plasma-plume in the central regions of the lamp as is seen in a practical projector lamp that burns horizontally. Exclusion of natural convection benefits the simulation work in terms of using a direct current (DC) supply although these lamps are usually operated by alternating current (AC).

Fig. 1 shows the schematics of the calculated domain. Reflected UV radiation, generated by the lamp, has been returned into the lamp through the glass. Reflected UV radiation has been assumed to enter the lamp from an angular glass window of 90° (+/- 45° with respect to the center between electrodes). Radiation has been modelled using the discrete ordinates (DO) method and the reflected UV band, 350–430 nm, has been treated to account for the irradiation. Other considered radiation bands are: 100–119, 119–350, 430–700, 700–4500 and 4500–20000 nm. An overall view of the model for a discharge lamp simulation can be seen elsewhere [1].

PREDICTED RESULTS

An analysis of the radiation spectrum, obtained from a projector lamp, suggests that about 10% of the supply power converts into the UV radiation of the 350–430 nm band. Thus for the 100 W lamp, about 10 W would be the UV radiation and has been assumed to return into the lamp. If we assume that 50–70% of the UV radiation is absorbed into the plasma, then a power of 5–7 W would be available for altering plasma situations. Directional behaviour, preciseness of angular discritization and glass-surface conditions affect the quantity of UV radiation entering the lamp. Since these gauges are difficult to control accurately, we have rather given attention to the UV absorbed power inside the lamp due the irradiation of the band. Adjustments have made a situation when 6.5 W of the irradiated UV power has been found to be absorbed in the lamp. In this absorbed power level, 30 W has been supplied at the glass surface. All results and discussions of this article are based on these situations.

Fig. 2 and Table 1 present the UV irradiation effects on the lamp properties. The average gas temperature of the lamp has been increased by about 35 K and the radiation from the visible band has been found to increase by about 2 W; it is a 6% rise compared to the normal operation.

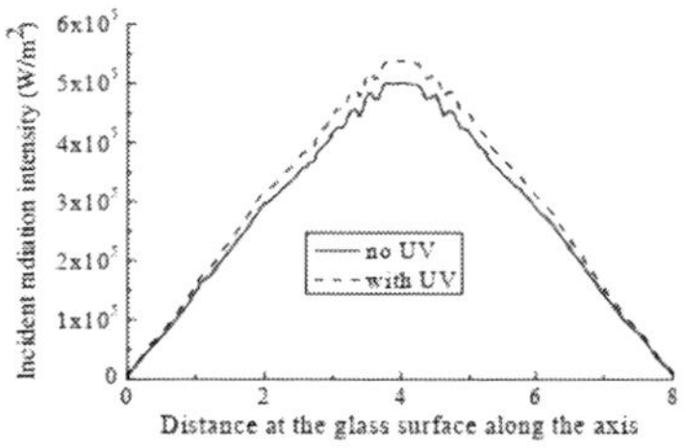

Table 1 *Effects of UV irradiation on the plasma properties*

Evaluating parameter	No UV	30 – W UV at glass
UV (350-430) band power	13.12 W	19.65 W [6.53 W input]
430-700 nm (visible) radiation: gas	29.18 W	31.3 W [2.12 W rise]
430-700 nm (visible) radiation: glass	30.87 W	33.11 W [2.24 W rise]
Average gas temperature	2616.16 K	2651.6 K [35.44 K rise]

Fig.2 Effect of UV irradiation on the incident radiation of the visible band at glass surface

CONCLUSIONS

A mathematical model has been developed to predict the UV irradiation effects on the ultra high pressure projector lamp. A projector lamp may produce an UV power, which is about 10% of the supply power. The attempt here in this article is to use this unnecessary UV for increasing the visible lights from the lamp. For fundamental understandings, simplifications have been exercised: a 2D and axisymmetric model has been used for the purpose. Simulations on a 100 W mercury lamp have shown that average gas temperature rises by about 35 K and thus the radiation from the visible band also increases. About 6% increase of the visible band radiation has been found in this work and we consider that this rise of output is highly significant in improving lamp performance.

REFERENCES

[1] K.C. Paul, T. Takemura, T. Hiramoto, M. Yoshioka, T. Igarashi, Y. Horikawa, T. Tsukamoto, and A. Sugitani, IEEE Trans. Plasma Science, 34 (2007) 188.

The Effect of Hydrogen Iodide on the Ignition of Hg-free Metal-halide Lamps

E. Estupiñán[1], R. Pereyra[1], Y-M. Li[1], W. Lapatovich[2]
[1]CRSL, OSRAM SYLVANIA, 71 Cherry Hill Drive, Beverly, MA 01915 USA.
[2]OSRAM SYLVANIA (retired)
Contact: edgar.estupinan@sylvania.com

ABSTRACT

Trace amounts of moisture adsorbed on metal-halide salt pellets during lamp fabrication may react with the salts during initial lamp light-up to form volatile hydrogen iodide (HI). Hg-free lamps can develop high ignition voltages due to the presence of HI, which adversely affects lamp operation especially ignition [1,2]. This work discusses the thermochemistry and spectroscopic measurements which predict and demonstrate that HI is indeed formed in the reaction of salts and moisture. Experimental confirmation of increased breakdown voltage with increasing concentration of HI (at the ppm level) was obtained in discharge cell experiments. Computer simulations of xenon and HI gas mixtures show qualitative agreement with experimental results. Our hypothesis is that the average electron energy gain and avalanche multiplication is retarded in the presence of HI, with increasing voltage needed for successful breakdown in moisture contaminated lamps.

INTRODUCTION

During the manufacturing of metal halide lamps, impurities and contaminants are to be avoided due to the effects these may have on lamp performance. Moisture, in particular, has a detrimental effect on the ignition characteristics of mercury-free lamps due to the formation of hydrogen iodide (HI) as a byproduct of reactions with hydrophilic mercury-free chemistry. Detailed thermochemistry calculations show that nanograms of water are enough to create fractions of a torr of HI in mercury-free lamps. HI is known to be a fast scavenger of electrons [3], with the forward reaction $HI + e^- = H + I^-$ occurring at every collision. This process depletes electrons from developing avalanche during ignition and consequently leads to a rise in the breakdown voltage needed to ignite the lamp.

EXPERIMENTAL PROCEDURE AND RESULTS

To test our hypothesis, it was necessary to do the following four things: (a) generate HI in a mercury-free chemistry burner, (b) confirm its formation, (c) measure changes in breakdown voltage due to the presence of HI, and (d) verify the regenerative nature of HI. Modified 250 W quartz arc tubes were dosed with mercury-free chemistry, xenon and boehmite (hydrated alumina salt), which permits dosing with controlled amounts of water. The samples were then placed in an oven and heated at 700 C to release the water and force chemical reactions. After removing the samples from the oven, the HI concentration was measured by UV absorption spectroscopy followed by a DC breakdown voltage and a second HI concentration measurement. The samples were then re-heated and a final HI concentration measurement was taken.

Figure 1 summarizes some key experimental results. The measured HI concentration shows a quasi-linear increase in pressure with increasing water vapor pressure and displays good agreement with the thermochemical predictions. Similarly, the DC breakdown measurements show increasing voltage required for breakdown with increasing water contamination and HI pressure. For example, a small increase of 2 torr in the HI pressure is enough to nearly triple the breakdown voltage of the lamp. Additional experimental work in cells containing only Xe and HI also shows significant increases in ignition voltage with trace amounts of HI concentration. Upon reheating the samples, the HI concentration that was depleted by the breakdown measurements was fully regenerated.

MODELING RESULTS

The simulation model is based on the solution of the drift-dominated transport equations for electrons and Xe ions with highly nonlinear Townsend ionization source terms controlled by the local electric field over gas pressure (E/p). The local electric field is determined by solving the Poisson equation. The secondary emission coefficient for the electron due to Xe ion bombardment at the cathode is taken to be 0.02. The DC breakdown voltage is taken to be the discharge voltage at very low discharge current. The basic arc tube geometry and electrode dimensions for 250 W quartz arc tubes is shown in Figure 2, where the electrode gap is 35 mm, and the radius of curvature of the electrode tip is assumed to be 250 microns. The simulation results for 300 torr Xe are also shown in Figure 2. The experimental DC breakdown voltage in the 250 W arc tube is in the range of 4.7 to 5.5 kV, very close to the simulation result shown in Figure 2. To further test the model, the DC breakdown voltage for the high pressure Xe model arc tube described in [4] is performed. The basic lamp parameters in [4] differ substantially with our 250 W arc tube geometry. Good agreement between the simulated breakdown voltage and the experimental value [4] is also found.

The simplest possible model is adopted to describe the effects of adding HI. Electrons are strongly attached to HI and the resulting HI^- is dissociated to H and I^-. The transport equation for iodide negative ion is added to the basic model. HI is assumed to have no effect on the excitation and ionization of Xe. The dissociative attachment cross-section data for HI is taken from [5]. The complete model for Xe with HI is solved numerically. The calculated DC breakdown voltage increases with partial pressures of HI, agreeing only qualitatively with the experimental data.

CONCLUSIONS

Contaminants in mercury-free metal halide lamps have an adverse effect on lamp performance. Trace amounts of water will react with metal halide salts to form hydrogen iodide (HI), as shown in thermochemistry calculations and confirmed by UV absorption spectroscopy. The presence of HI in small quantities can lead to an increase in breakdown voltage of over 10 kV. HI partially dissociates during ignition but may be become a recurrent problem due to regeneration after reheating (or restarting) of the lamp.

Simulation models are developed to calculate DC breakdown voltages in the actual HID lamp setting with Xe and HI. Good agreement between the modeled and measured DC breakdown voltages are found for pure Xe fill. For Xe with HI, the comparison is less favorable and points to the need for a better model.

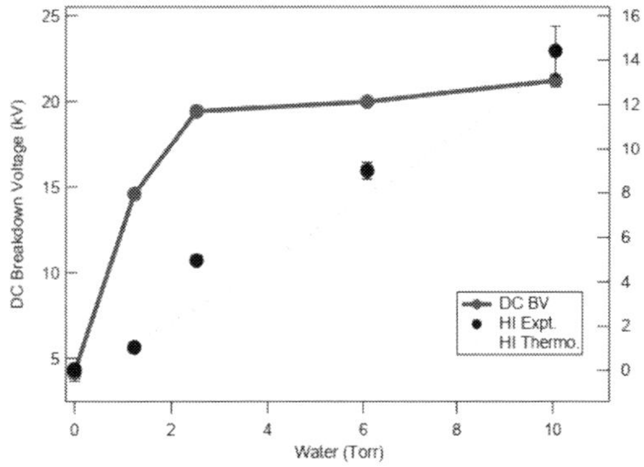

Figure 1. DC breakdown voltage and HI concentration as a function of water contamination. Error bars on the breakdown data are not large enough to be observable on the plot.

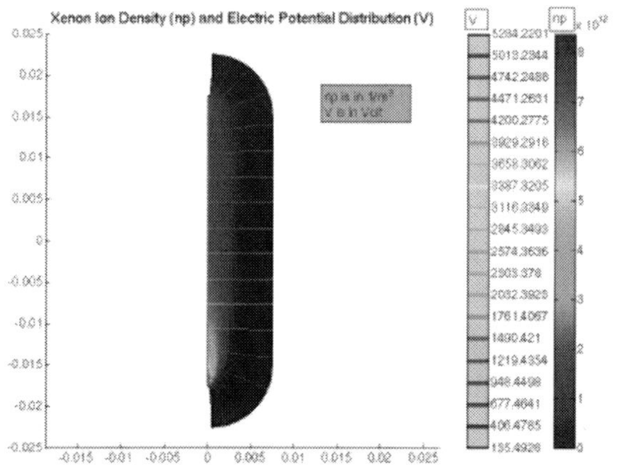

Figure 2. Xenon ion density and voltage distribution in the low current discharge (0.1 nA) with 300 torr Xe is shown. The peak ion density is close to the anode (bottom electrode). The top electrode is the cathode. The electrode tip radius is 250 microns and the breakdown voltage is 5.42 kV. Units are given in MKS.

REFERENCES

[1] A. Inouye, J. Light & Vis. Env., **2** (1978) 8-11.
[2] J. F. Waymouth, Electric Discharge Lamps, MIT Press, Cambridge, MA (1971) 257.
[3] K. Houfek, M. Cizek, and J. Horacek, Phys. Rev. A, **66** (2002) 062702/1-062702/9.
[4] M. Wendt, S. Peters, D. Loffhagen, A. Kloss and M. Kettlitz, J. Phys. D, Appl. Phys., **42** (2009) 185208.
[5] D. Klar B. Mirbach, H. J. Korsch, M.-W. Ruf, and H. Hotop, Z. Phys. D, **31** (1994) 235.

Use of refractory abhesives in protection of molybdenum foils and leads

N. Voggenauer, T. Emilsson and T. R. Brumleve
APL Engineered Materials, Inc., 2401 N Willow Road, Urbana, IL 61802-7332 USA
Contact: timothy_brumleve@adlt.com

ABSTRACT

Coatings containing a refractory "abhesive" such as boron nitride on the surface of Mo foil and lead-foil weld junctions are effective in preventing oxidation and shaling failures in quartz lamp pinch seals. By mixing silica or silicate glass with the refractory materials, the high temperature of the pinch process itself can be used to fuse the mixture to provide additional oxidation projection.

INTRODUCTION

Thin molybdenum foil with tapered edges and lenticular cross-section are used for making the gas tight metal-to-glass pinch or shrink seals needed for the construction of quartz metal halide and tungsten halogen lamps. When operated in air, the exposed foil and lead, and particularly the welded lead-foil junction, are vulnerable to oxidation. The resulting expansion by MoO_x reaction products forces the seal apart, causing cracks and failure of the seal. As the Mo metal oxidizes, its electrical resistance increases, leading to increased ohmic heating, higher temperatures and higher oxidation rates. Another common problem in pinch and shrink seals is the phenomenon known as "shaling" in which uneven stresses in the pinch area are caused by the quartz sticking to the molybdenum metal surfaces. This results in minute cracks which can severely weaken the glass and can lead to failure from very modest strains.

Previous approaches for protecting Mo foils and lead wires from oxidation have been described in [1] and [2]. In methods which involve only protective silica coats, the coatings stick to the quartz. Upon cooling and thermal contraction, the protective coat peels off the metal and shaling of the glass is often observed. By applying a refractory *abhesive* (RA) (*i.e.,* a high-temperature material such as boron nitride, BN, that *prevents adhesion* of the bulk quartz to the foil) to the silica coated structures, the integrity of the protective coat can be maintained, and shaling can be prevented [3]. Even without an underlying silica coat, the RA can improve the life of a seal by preventing weakening of the glass, and by slowing the access of air to the vulnerable metal. The permeability to air can be further decreased by fusible additives such as colloidal silica, alkali metal silicates or borosilicates in the RA formulation. These remain bonded to the metal. We have also observed that certain combinations of silica-RA (*e.g.*, silica-BN) may react chemically with Mo metal, producing surface reaction products that may further enhance oxidation protection.

EXPERIMENTAL

Mo sealing foils, 6 × 0.040 mm, (Shanghai Y&L Lighting), welded to 0.7 mm Mo wires were coated by dipping or brushing the "green" coating suspensions to cover the weld junction and the exposed end of the foil. These assemblies were then pinched in 14 × 11 mm quartz tubing in a fashion similar to a 2000 W quartz metal halide lamp, thereby fusing the silica or silicate components. These were placed in an oven at 400 C for up to 5500 hours, and removed periodically to visually record the extent of oxidation and damage. The three coating formulations described here are listed in Table I. The coating thicknesses for all 3 formulations were typically in the range of 0.03 to 0.30 mm. In some cases the assemblies were carefully freed from the glass using a diamond saw to allow analysis by optical and scanning electron microscopy (SEM) and by Secondary Ion Mass Spectrometry (SIMS) (Cameca ims 5f with depth profiling capability).

Table I Formulations used to prevent molybdenum oxidation

Test	Coating type	Formulation
A	BN only	BN Aerosol Brushable (Zyp Coatings), as received
B	$BN+K_2SiO_3$	BN mixed with aqueous K_2SiO_3 and organic binder
C	$BN+SiO_2+NaBO_2$	BN, colloidal silica (Nissan ST-OUP), aqueous $NaBO_2$, trace Al_2O_3 and organic binder

RESULTS AND CONCLUSIONS

Table II Damage levels

Damage Level	Appearance
0	No damage
1	Interference colors in small areas appear
2	Large colored stress rings observed
3	Small fractures inside of stress rings
4	Large cracks through the pinch

Visual and microscopic observations of the test pinches were made at regular intervals, and the level of oxidation damage was assigned a Level of 0 through 4 as noted in Table II. In test A (BN only), the RA suspension was applied directly to the foil and weld junction without an underlying silica layer and without any additional silica or silicates in the RA suspension. As can be seen in Fig. 1, the performance was superior to the uncoated case, with pinch seals approaching 4000 hours before large cracks (Level 4 damage) were observed. In test B (BN + K_2SiO_3) only Level 1 damage was observed after 5500 hours at 400 C, thus providing superior performance, as shown in Fig. 1. In test C (BN+SiO_2+$NaBO_2$) the foils were removed from the test pinches immediately after pinching, excess coating material was removed, and the foils were examined by optical microscopy, SEM and SIMS. Large areas of the foil surface which had been subjected to the high temperature of the pinch process exhibited a reflective, mirror-like finish, unlike the bare foil which exhibited 1 to 2 μm surface roughness. The SEM micrographs of the uncoated and the coated reflective surface are shown in Fig 2. SIMS analysis, seen in Fig. 3, of the reflective region shows significant boron levels penetrating the surface, deeper than the Si, Al and Na levels which are part of the exterior silicate film. This is consistent with formation of MoB_x compounds, and with the known low melting MoB_x phases possible at the pinch temperature [4]. Reaction products such as MoB_x may be partially responsible for the protection improvements of the BN refractory abhesives of formulations A and B shown in Fig. 1. Formulation B is shown to be particularly advantageous in preventing oxidation and shaling failures in quartz lamp pinch seals.

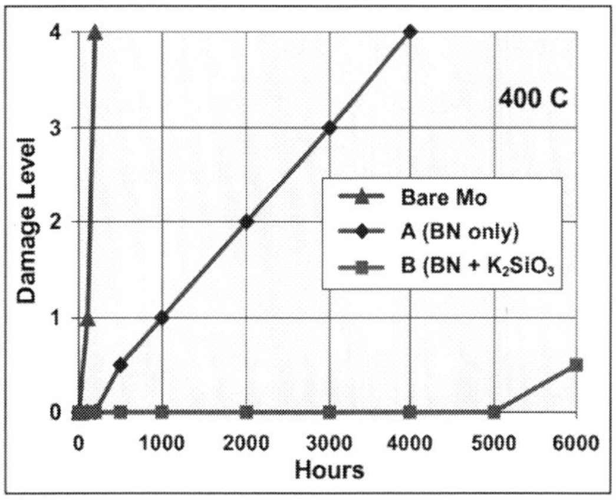

Figure 1. Damage level vs. time for coatings A, B and bare Mo.

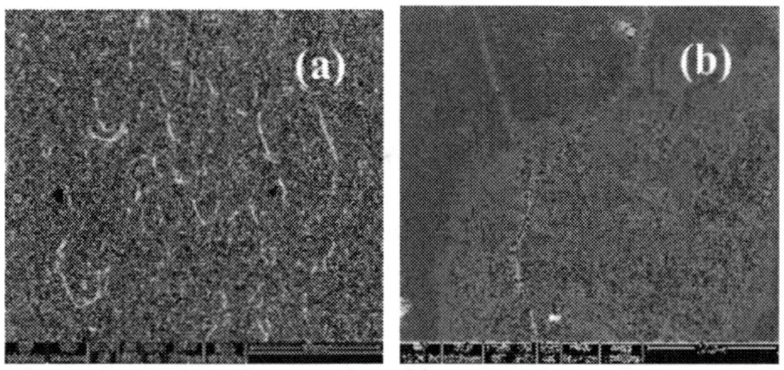

Figure 2. SEM micrographs of bare Mo (a) and cleaned Mo foil after coating with formulation C (b).

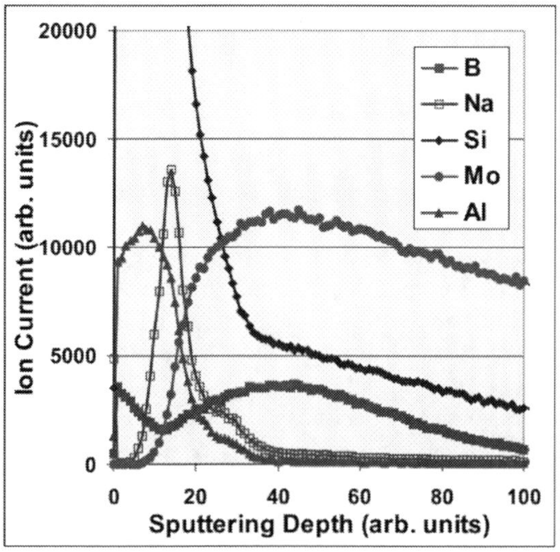

Figure 3. SIMS depth profiles of reflective reaction product of formulation C.

REFERENCES

[1] T. Emilsson, C.V. Varanasi and T.R. Brumleve, Light Sources 2004, G. Zissis, ed., Institute of Physics Conference Series No 182, 2004, p. 503.

[2] T. Emilsson, US Patent 7,153,179, and US Patent Applications 20040124759 and 20070082576.

[3] T. Emilsson and N. Voggenauer, US Patent Application 20090295291.

[4] Phase Equilibria Diagrams, Vol. X, A.E. McHale, ed., American Ceramic Soc., Westerville, OH, 1994, Fig. 8812, p. 136.

Diagnostics of planar XeI* UV exciplex lamp excited by diffuse coplanar surface discharge

M. M. Guivan[1], A. Brablec[2], P. St'ahel[2], P. Slavíček[2], J. Janča[2]
[1] *Department of Quantum Electronics, Uzhgorod National University, Pidgirna 46, Uzhgorod 88000, Ukraine*
[2] *Department of Physical Electronics, Masaryk University, Kotlářska 2, Brno 61137, Czech Republic*
Contact: m_guivan@rambler.ru

ABSTRACT

In this contribution to the research, electrical, optical, as well as energetic properties of the diffuse coplanar surface discharge (DCSD) operated with the Xe/I_2 and $Kr/Xe/I_2$ mixtures at pressures of $p(Xe) \leq 50$ kPa and $p(Kr/Xe) = 10 - 100$ kPa are presented. Varying the gas composition, pressure and pump pulse repetition rate makes it possible to control the discharge pattern and achieve a diffuse discharge or completely filamentary DCSD. The optimal UV output was obtained using Xe/I_2 = 20/0.04 kPa and $Kr/Xe/I_2$ = 46/4/0.04 kPa mixtures. The krypton admixture enhances the $XeI(B - X)$ emission for about 15% in comparison with a binary Xe/I_2 mixture. The average radiation power and efficiency of 8 mW/cm^2 and 3% were measured at a frequency of 100 kHz.

INTRODUCTION

Gas-discharge plasma on mixtures containing xenon and iodine vapours is considered to be an efficient source in the UVC (200–290 nm) range [1, 2]. The strong $XeI(B^2\Sigma_{1/2} - X^2\Sigma_{1/2})$ emission at λ_{max} = 253 nm is an attractive candidate for substituting the low-pressure mercury lamp that emits at 253.7 nm. As presented in our previous studies concerning surface barrier discharge [3, 4], krypton addition to the binary 'xenon/halogen' mixture resulted in UV output and discharge homogeneity enhancement. In this contribution we sought to determine whether that effect takes place for the DCSD. The aim of this study is to investigate the generation of the XeI* emission by means of diffuse coplanar surface barrier discharge operated with the Xe/I_2 or $Kr/Xe/I_2$ mixtures.

RESULTS AND DISCUSSION

The DCSD electrode module was manufactured on a glass plate. Strip–like electrodes had a width of 2 mm and the distance between strips was 1 mm, the thickness of the glass layer above electrodes was 0.14 mm. The electrode system was mounted inside a chamber of 1.5 l in volume. To excite a working mixture a generator of high voltage AC sinusoidal pulses was used; peak-to-peak voltage up to 10 kV, f = 1 – 100 kHz. The voltage U applied to the DCSD was measured using a high voltage probe (Tektronix P6015A). The waveforms of current I and charge Q were registered by means of a 50 Ω resistor or low-inductance 8 nF capacitor placed in series with the DCSD module and monitored by digital oscilloscope HP Infinium (500 MHz, 2 GSa/s). The radiation emitted by the discharge was registered and analyzed in the spectral range of 200–900 nm. The spectra were recorded by the Jobin Yvon TRIAX 550 monochromator (grating 1200 grooves/mm, quartz optical fiber, the spectral resolution of about 50 pm) with the high-speed CCD Spectrum ONE detector. The radiation power of the DCSD in absolute units was measured by a UV power meter Lab Master Ultima equipped with two calibrated exchangeable sensor heads (for 250 – 400 nm and 400 – 1100 nm).

In each half-cycle of the applied voltage U_0, the current waveform consisted of a displacement current and conductivity current peaks, with characteristics depending on the operating frequency and applied voltage. The peak current did not exceed 55 mA and their duration was in the range 0.7–1 µs.

The plasma consisted of numerous H–shaped elementary discharges with brighter streamer–like filamentary plasma that bridges two clouds of diffuse plasma formed over the embedded electrodes. In the case of DCSD at $p(Xe)$ = 20 kPa this research has revealed that the discharge view is dependent on the operating frequency. At a frequency of 6 kHz the diffuse part of DCSD has a greater brightness than the filaments. However for the f = 100 kHz situation this is reversed: the filaments are much more intense than the diffuse plasma. From an engineering viewpoint, considering DCSD as a radiation source, it is possible to regulate the discharge pattern with the Xe/I_2 mixture by means of the operating frequency. Figure 1(*b*) illustrates the dramatic changes in the discharge appearance when krypton was added to the Xe/I_2 mixture. The discharge visually appears to be a homogeneous plasma layer with the more intensive parts over the electrodes. Brightness was also greater in comparison with the binary mixture.

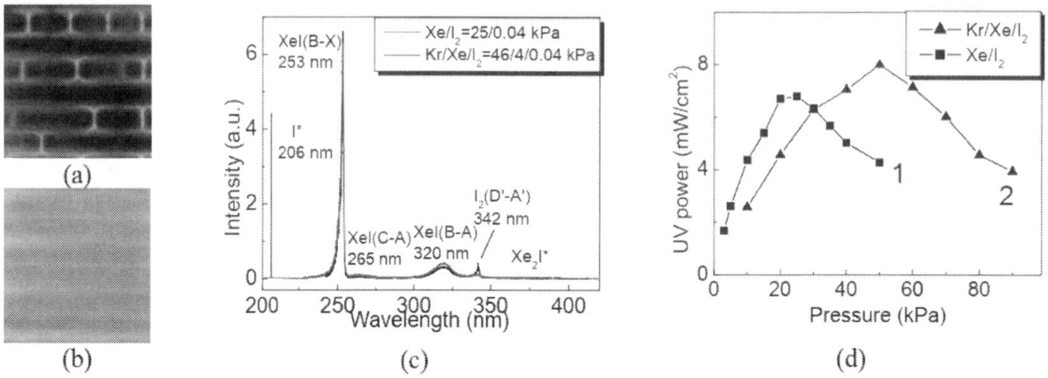

Figure 1: *(a) photo of DCSD in the Xe/I_2 = 20 kPa/40 Pa, f = 100 kHz, electrical input power 15.4 W; (b) photo of DCSD in the $Kr/Xe/I_2$ = 55 kPa/5 kPa/40 Pa, f = 6 kHz, electrical input power 8.1 W; (c) emission spectra of DCSD with the Xe/I_2 = 25/0.04 kPa and $Kr/Xe/I_2$ = 46/4/0.04 kPa mixtures; (d) optimization of working pressure for the Xe/I_2 (1) and $Kr/Xe/I_2$ (2) mixtures at f = 100 kHz.*

Measurements of the UV–visible emission spectra of the DCSD revealed the strong XeI*(*B–X*) exciplex radiation at λ_{max} = 253 nm. Weaker emissions of XeI*(*B–A*) (λ_{max} = 320 nm), XeI*(*C–A*) (λ_{max} = 265 nm), and I_2*(*D′–A′*) (λ_{max} = 342 nm) were also observed. Atomic iodine is represented in the spectrum by the emission at λ = 206.2 nm ($^2P_{3/2}$–$^2P^{\circ}_{1/2}$). The full width at half maximum (FWHM) of XeI($B \rightarrow X$) band decreased from 2.6 to 1.5 nm within the pressure range 5 – 100 kPa. Unlike the Xe/I_2 mixture, in the emission spectra of $Kr/Xe/I_2$ mixture it was observed to exhibit greater intensity of $I_2(D' \rightarrow A')$ band in comparison with the XeI($B \rightarrow A$) radiation (figure 1(c)).

The output DCSD power was measured as a function of the xenon or 92%Kr + 8% Xe pressure for the Xe/I_2 or $Kr/Xe/I_2$ mixtures with a fixed I_2 pressure $p(I_2)$=40 Pa. For a constant applied voltage, the UV output increased to a maximum value with gas pressure, and then falls when the gas pressure exceeds p(Xe)>25 kPa in the Xe/I_2 mixture and p(Kr/Xe)>60 kPa in the $Kr/Xe/I_2$ mixture (figure 1(d)). The average radiation power of 8 mW/cm^2 has been achieved with the $Kr/Xe/I_2$ = 46/4/0.04 kPa mixture at f = 100 kHz. The conversion efficiency of the DCSD varied with the electrical power, achieving maximum values of around 3%.

SUMMARY

The optimal composition of Xe/I_2 = 20–30/0.04 kPa and $Kr/Xe/I_2$ = 40–50/4/0.04 kPa working mixtures from the point of view of maximum yield of XeI*(*B–X*) exciplex emission at λ = 253 nm and the effect of the excitation conditions on the DCSD have been established. As to the output radiation and discharge uniformity the addition of krypton enhances the XeI* intensity at 253 nm to 15% in comparison with the Xe/I_2 mixture. It was possible to regulate the discharge pattern with the Xe/I_2 mixture by means of the operating frequency. UV radiation power density of 8 mW/cm^2 and the conversion efficiency from electrical input to maximum UV output of 3% were obtained with the $Kr/Xe/I_2$ = 46/4/0.04 kPa mixture at f= 100 kHz..

REFERENCES

[1] J.Y. Zhang, I.W. Boyd, J. Appl. Phys., 84 (1998) 1174.
[2] G.A. Volkova, G.N. Zvereva, Optics and Spectroscopy, 96 (2004) 373.
[3] M.M. Guivan, P. St'ahel, A. Brablec, J. Janca, H. Motomura, M. Jinno, Proc. 11th Int. Symp. on the Science and Technology of Light Sources LS-11, Shanghai (China) (2007) 547.
[4] N.N. Guivan, J. Janca, A. Brablec, P. St'ahel, P. Slavicek, L. Shimon, J. Phys. D: Appl. Phys. 38 (2005) 3188.

Efficient Driver for Dimmable LED Lighting

C. S. Moo, S. N. Lin, W. C. Yang, W. Hong
Department of Electrical Engineering, National Sun Yat-Sen University, Kaohsiung, 80424, Taiwan
Contact: mooxx@mail.ee.nsysu.edu.tw

ABSTRACT

A high efficiency driver is proposed for LED lamps with dimming feature. The current regulation is accomplished by processing partial power of the power conversion circuit so that a high overall efficiency can be realized. The dimming feature can be accomplished by means of current regulation or pulse-width modulation (PWM). Test results from an exemplar design for a 45 W LED lamp demonstrate that the circuit efficiency can be as high as 95.5% at the rated output and deteriorates slightly to 91% as the lamp is dimmed to 10% of the rated power.

INTRODUCTION

With better and better efficiency, the light-emitting diode (LED) has been becoming a kind of prominent light source in many high brightness applications. High light output is usually obtained from a large amount of individual LEDs stacked in series, which are driven by an identical current [1]. Unfortunately, driving a string of high power LEDs with an accurate current is at odds with simplicity and efficiency. To promise a high overall lighting efficiency, an efficient driver is required not only for the reduction of circuit losses, but also for the lumen maintenance of the LED itself.

Conventionally, a better solution is the use of a switching power conversion circuit, typically a fly-back converter for such an application, to convert a voltage regulator into a current regulator [2, 3]. This approach, however, is inefficient especially when the dimming function is performed [4]. To further improve the efficiency of LED lamps, this paper proposes an efficient driver circuit by processing a fraction of the converted power.

CIRCUIT CONFIGURATION

The main concept of the proposed configuration is illustrated by Fig. 1, in which an LED lamp is powered virtually from a dc voltage source in series with a current regulator. The voltage source is designed to afford the forward bias of the LEDs, while the current regulator provides the desired lamp current. As the forward voltage on the LED lamp is higher than the cut-in voltage, the lamp current turns out to be very sensitive to the applied voltage, meaning that the voltage on the current regulator varies in a small range. As a result, the current regulator handles only a small part of operating power while most power is supplied from the dc voltage source.

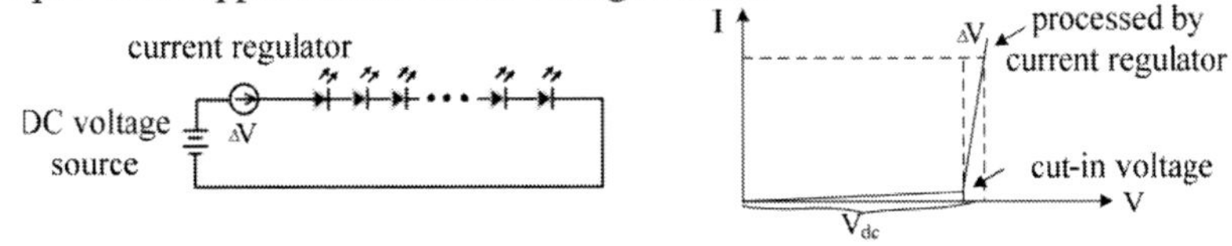

Fig. 1 Design concept of the proposed driver circuit

Fig. 2 shows an implementation example of the LED driver circuit for a 45 W LED lamp supplied from an AC line source of 110 V and 60 Hz. The LED lamp is formed by 40 LEDs in series. All LEDs are rated at a same current of 350 mA. The voltage source can be simply obtained by a diode bridge rectifier with a filter capacitor C_1. A fly-back converter functioning as a current regulator is interposed between the dc voltage source and the LED lamp. In this case, the operating voltage of the LED string is slightly less than the capacitor voltage, V_{C1}. The lamp current is regulated by adjusting the duty-ratio of the fly-back converter. With such a design, only a small amount of power has to be processed for current regulation, leading to high circuit efficiency.

The brightness the LED lamp can be controlled simply by changing the current flowing through LEDs. This approach, however, is not appropriate for some applications since the colour spectrum of an LED is related to its current. To avoid colour shift, a dimming switch is added as shown in Fig. 3. The flyback converter regulates the LED current at a constant value and the lamp brightness is controlled by varying the effective LED current with pulse width modulation (PWM).

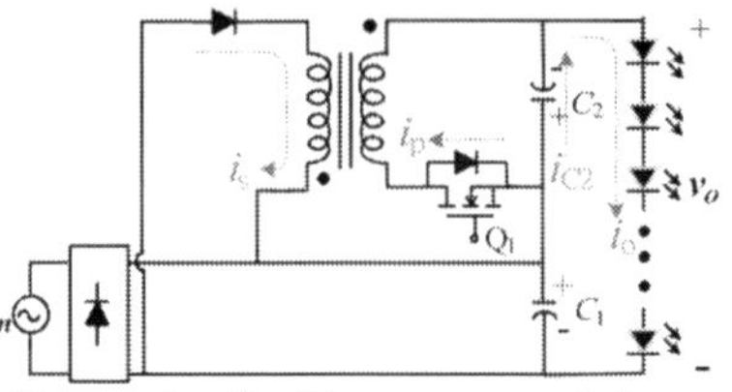

Fig. 2 Driver circuit with current regulation

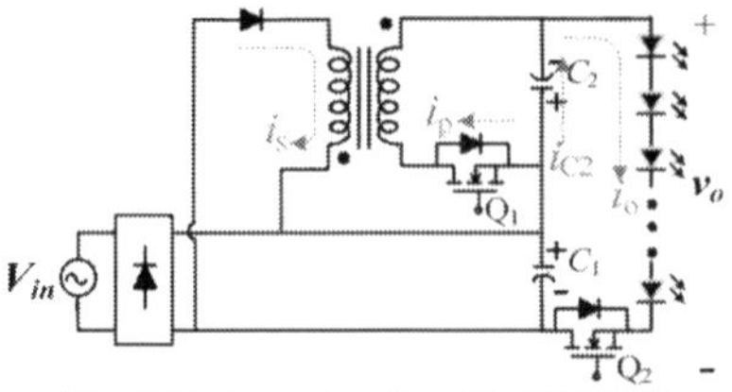

Fig. 3 Driver circuit with PWM dimming

EXPERIMENTAL RESULTS

Figs. 4 and 5 show the voltage and current waveforms on the key components under the rated operation and at 10% output power with PWM dimming, respectively. The fly-back converter draws a current from C_2 when Q_1 is turned on, and returns the energy back to C_1 as it is turned off. The voltage on C_2 varies from 19.30 V up to 42.56 V as the LED lamp is dimmed down to 10% of the output power. On the other hand, this voltage remains almost at the same level to retain a constant current magnitude when the lamp is dimmed by PWM control. For both dimming methods, the losses on the fly-back converter are relatively small since only a small portion of the lamp power is processed. As shown in Figs. 6 and 7, the circuit efficiencies are always than higher 91%, and can be as high as 95.5% at the rated output.

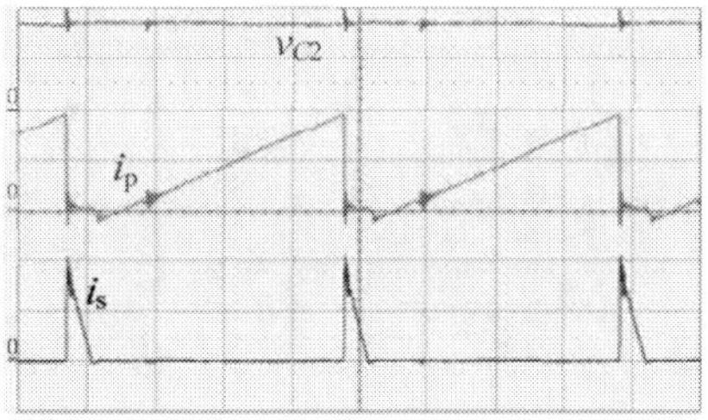

i_p, i_s:500 mA/div ; v_{C2}:10 V/div ; Time:5 μs/div

Fig. 4 Voltage and current waveforms with current regulation

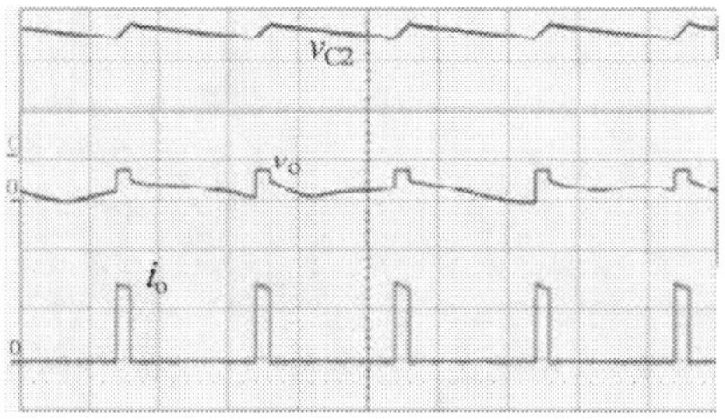

i_o:200 mA/div ; v_{C2}:5 V/div ; v_o:200 V/div ; Time:5 μs/div

Fig. 5 Voltage and current waveforms with PWM dimming

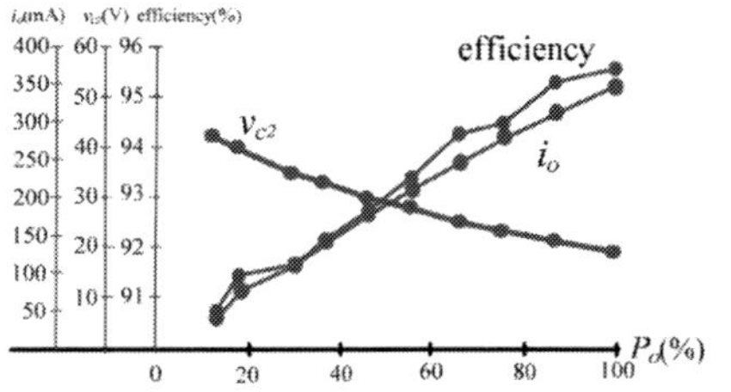

Fig. 6 Circuit performances with current regulation

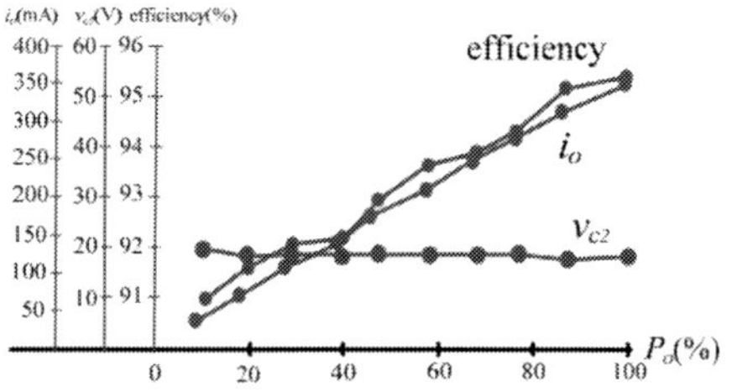

Fig. 7 Circuit performances with PWM dimming

REFERENCES

[1] T. F. Pan, H. J. Chiu, S. J. Cheng, and S. Y. Chyng, "An Improved Single-Stage Flyback PFC Converter for High-Luminance Lighting LED Lamps," in Proc. ICEMI, no. 4, pp. 212–225, (2007)

[2] H. van der Broeck, G. Sauerlander,and M. Wendt, "Power driver topologies and control schemes for LEDs," in Proc. APEC, pp. 1319–1325, (2007)

[3] L. Yu and J. M. Yang, "The Topologies of White LED Lamps' Power Drivers," in Proc. PESA, pp. 1–6, (2009)

[4] H. J. Chiu, Y. K. Lo, J. T. Chen, S. J. Cheng, and C. Y. Lin, "A High-Efficiency Dimmable LED Driver for Low-Power Lighting Applications," IEEE Trans. on IE, vol. 57, no. 2, pp. 735–743, (2009)

Enhanced light extraction in light-emitting diodes with photonic crystal structure selectively grown on *p*-GaN

Chu-Young Cho[1,2,3], Se-Eun Kang[1,2,3], Ki Seok Kim[1], Sang-Jun Lee[1,2,3], Yong-Seok Choi[1,2,3], Sang-Heon Han[1,4], Gun-Young Jung[1], and Seong-Ju Park[1,2,3] 1

[1] *Department of Materials Science and Engineering, Gwangju Institute of Science and Technology, Gwangju 500-712 Republic of Korea*

[2] *Department of Nanobio Materials and Electronics, Gwangju Institute of Science and Technology, Gwangju 500-712, Republic of Korea*

[3] *Center for Extreme Light Applications(CELA-NCRC), Gwangju Institute of Science and Technology, Gwangju 500-712, Republic of Korea*

[4] *Samsung LED Co. Ltd., Suwon 443-743, Republic of Korea*

Contact: ccy06@gist.ac.kr

ABSTRACT

We report on the properties of green light-emitting diodes (LEDs) with a photonic crystal (PC) structure on *p*-GaN. A PC structure was fabricated by the selective area epitaxy (SAE) of *p*-GaN using SiO_2 nano-pillars. The electrical characteristics of LEDs with PC were not degraded and the optical output power of green LEDs with PC was increased by 70% at 20 mA of injection current compared with that of conventional LEDs without PC. This enhancement of optical output power was attributed to the improvement in light extraction efficiency by the SiO_2/*p*-GaN PC layer on *p*-GaN.

MAIN TEXT

Incorporating photonic crystals (PCs) into light-emitting diodes (LEDs) is an effective method for improving the light extraction efficiency ($\eta_{extraction}$). Usually, the PC structure for enhancement of $\eta_{extraction}$ in GaN-based LEDs has generally been formed directly in the *p*-GaN layer. However, the incorporation of a PC structure in the *p*-GaN layer requires the use of a plasma etching process, resulting in plasma damage to the *p*-GaN layer. Many groups have reported that the electrical characteristics of LEDs with PC structures were degraded due to the etching damage to *p*-GaN [1-5]. To overcome the problem of plasma damage in *p*-GaN and electrical degradation in LEDs with a PC structure, we realized a PC structure on a *p*-GaN layer using a selective area epitaxy (SAE) method. In this study, we report on the electrical and optical properties of green LEDs with a PC structure formed on *p*-GaN. The $\eta_{extraction}$ of LEDs with a PC structure on a *p*-GaN layer was significantly enhanced by the PC effect and showed no degradation of electrical characteristics by avoiding the use of the plasma etching process on *p*-GaN.

The green InGaN/GaN LEDs were grown on a *c*-plane (0001) sapphire by metalorganic chemical vapor deposition (MOCVD). The five periods of InGaN/GaN multiple quantum wells (MQWs) were grown on *n*-GaN at 730 °C, followed by the growth of a 150 nm-thick *p*-GaN layer at 970 °C. To fabricate the PC structure, a 100 nm-thick SiO_2 layer was deposited as an SAE mask on *p*-GaN layer using plasma-enhanced chemical vapor deposition (PECVD). SiO_2 nano-pillars with a period of 320±7 nm and a radius of 100±5 nm were formed on *p*-GaN using a two-beam laser holography setup equipped with a 325-nm He-Cd laser and reactive ion etching (RIE). After selective overgrowth of *p*-GaN at 970 °C for 4 min, the *p*-GaN filled the surrounding region of SiO_2 nano-pillars. As shown in Fig. 1, a periodic PC structure of SiO_2 and *p*-GaN with different refractive indices was formed on the *p*-GaN layer. Figure 1(a) shows the schematic of a green LED with a selectively grown PC structure. Figure 1(b) is a cross-sectional SEM image of a *p*-GaN layer, showing that the SiO_2 nano-pillars were entirely surrounded by *p*-GaN.

Figure 2(a) shows the current-voltage (*I-V*) characteristics of green LEDs with and without a PC structure. As shown in Fig. 2(a), the forward voltage of the LED with a PC structure is 3.8 V, which is same as that of LED without a PC structure. The series resistances of both LEDs are also same. These results indicate that a PC structure selectively grown on a *p*-GaN layer does not cause the degradation of electrical properties. Figure 2(b) shows the optical output power of green LEDs with and without a

PC structure as a function of injection current. The optical output power of each LED was measured from the top side of LEDs using a 2 cm-diameter Si photodiode connected to an optical power meter. As shown in Fig. 2(b), the optical output power of green LEDs with a PC structure is increased by 70% at 20 mA of injection current compared with that of conventional LEDs without a PC structure. This improvement of optical output power is a result of the increase in the $\eta_{extraction}$ due to the PC effect, which inhibits the emission of guided modes and redirects trapped light into the radiated modes [1-4].

In conclusion, we demonstrate the green LEDs with a PC structure formed on a *p*-GaN layer. The electrical characteristics of green LEDs with a PC structure are not degraded and the optical output power of LEDs with a PC is increased by 70% at 20 mA compared with that of conventional LEDs without a PC. The enhancement of optical output power is attributed to an increase in $\eta_{extraction}$ by a SiO_2/*p*-GaN PC layer on *p*-GaN.

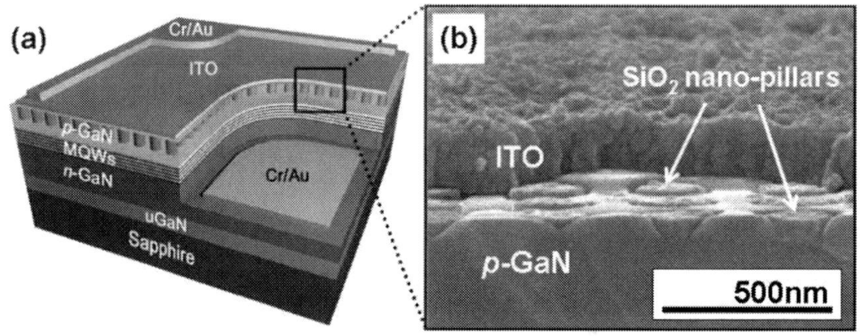

Figure 1. (a) Schematic of green LEDs with a selectively grown PC structure. (b) Cross-sectional SEM image of a SiO_2/*p*-GaN PC structure on *p*-GaN.

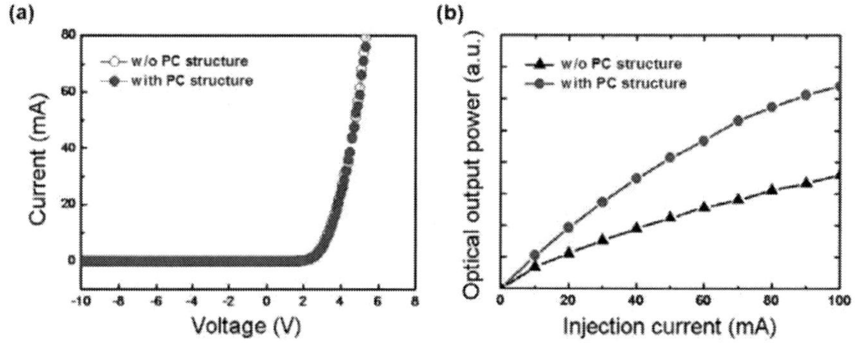

Figure 2. (a) *I-V* characteristics and (b) optical output power of green LEDs with and without a PC structure.

REFERENCES

[1] T. N. Oder, H. S. Kim, J. Y. Lin, and H. X. Jiang, Appl. Phys. Lett. 84 (2004) 466.
[2] J. Shakya, K. H. Kim, J. Y. Lin, and H. X. Jiang, Appl. Phys. Lett. 85 (2004) 142.
[3] D. H Kim, C. O. Cho, Y. G. Roh, H. Jeon, Y. S. Park, J. Cho, J. S. Im, C. Sone, Y. Park, W. J. Choi, and Q. H. Park, Appl. Phys. Lett. 87 (2005) 203508.
[4] J. Y. Kim, M. K. Kwon, K. S. Lee, S. J. Park, S. H. Kim and K. D. Lee, Appl. Phys. Lett. 91, (2007) 181109.
[5] X. A. Cao, S. J. Pearton, A. P. Zhang, G. T. Dang, F. Ren, R. J. Shul, L. Zhang, R. Hickman, and J. M. Van Hove, Appl. Phys. Lett. 75 (1999) 2569.

UV-C emitting phosphors

J.M.A. Caiut[1,3] , J. Dexpert-Ghys[1] , B. Caillier[2], Ph. Guillot[2]

[1]*CEMES-CNRS, BP94347, 31055 Toulouse, France*
[2] *DPHE, Université Jean François Champollion, place de Verdun, 81012 Albi cedex 9, France*
[3] *Photonic Materials Lab., Institute of Chemistry, UNESP, CP 355, Araraquara-SP, 14801-970, Brazil*
Contact: jdexpert@cemes.fr

ABSTRACT

The broad band interconfigurational ($4f^15d^1 \rightarrow 4f^2$) emission of Pr^{3+} doped in lanthanum orthophosphate ($LaPO_4$) and in calcium pyrophosphate ($Ca_2P_2O_7$) has been investigated under plasma excitation. The synthesis by spray pyrolysis at moderate temperature followed by a controlled annealing proves to be a very efficient way to produce good quality UVC phosphors.

INTRODUCTION

The emission of praseodymium Pr^{3+} depends greatly upon the nature of the inorganic matrix in which it has been introduced. As recently reviewed by A.M. Srivastava [1], the interest in efficient intraconfigurational ($4f^2 \rightarrow 4f^2$) and interconfigurational ($4f^15d^1 \rightarrow 4f^2$) emission transitions of Pr^{3+} in solids arises from a variety of practical applications. Part of the $4f^2$ and of the $4f^15d^1$ electronic configurations are schematized in Figure 1. In Quantum Splitting Phosphors, each ultra-violet incident photon produces more than one visible photon: this may be achieved by a controlled cascade process between the $4f^2$ sub-levels. Quantum splitting may be realized only if the lower level of $4f^15d^1$ is above $4f^2$ ($^1S_0$). In the other case: i.e. $4f^15d^1$ is below $4f^2$ ($^1S_0$) then, interconfigurational emission transitions occur. These emissions are parity-allowed: they give rise to broad bands, with fast emission decays. Based on these qualities, $Lu_3Al_5O_{12}:Pr^{3+}$ has been developed as a scintillator emitting near 310 nm [2]. The broad-band ultra-violet emission occurs in the so-called "UV-C" range and could find other applications as intermediate phosphors between the VUV excitation and the visible emission, or in UVC light emission for germicidal purposes. In this paper are reported the observations performed on lanthanum ortho-phosphate ($LaPO_4$) and on calcium pyrro-phosphate ($Ca_2P_2O_7$) doped with Pr^{3+}. The broad-band emission between 200 and 350 nm has been particularly investigated, under plasma excitation by the discharge produced in Ne/Xe 50% mixture.

EXPERIMENTAL

The experimental set-up developed to record the emission spectra and compare phosphors efficiencies has been described previously [3, 4]. The experiment chamber has been filled with a mixture of Ne/Xe 50% at a pressure of 250 mbar. In these conditions, the VUV excitation is mainly issue from xenon excimer states (173 nm) in comparison with resonant states of xenon (147nm). The phosphors emissions have been collected through a MgF_2 window by an optical fiber and analysed with an Ocenoptics Spectomer USB4000 (177-890nm).

Samples with formulations $La_{0.98}Pr_{0.02}PO_4$ ($LaPO_4$:Pr2%), $Ca_{1.96}Pr_{0.04}P_2O_{7.04}$ ($Ca_2P_2O_7$:Pr2%), $Sr_{1.96}Pr_{0.04}P_2O_{7.04}$ ($Sr_2P_2O_7$:Pr2%), or $Ca_{1.92}Pr_{0.04}Na_{0.04}P_2O_7$ ($Ca_2P_2O_7$:Pr2%, Na2%) have been synthesized by conventional methods of co-precipitation and drying. The compositions $La_{0.98}Pr_{0.02}PO_4$ and $Ca_{1.92}Pr_{0.04}Na_{0.04}P_2O_7$ have therefore been synthesized by spray-pyrolysis at 700°C, following the lines described for instance in [4]. The precursor solutions have been chosen in function of the wanted final compositions. All samples have been annealed under air after synthesis: the orthophosphates for 4 hours at 1000°C, and the pyrophosphates for 12 hours at 1200°C. The powder samples have been characterized by X-Ray diffraction, imaged by SEM.

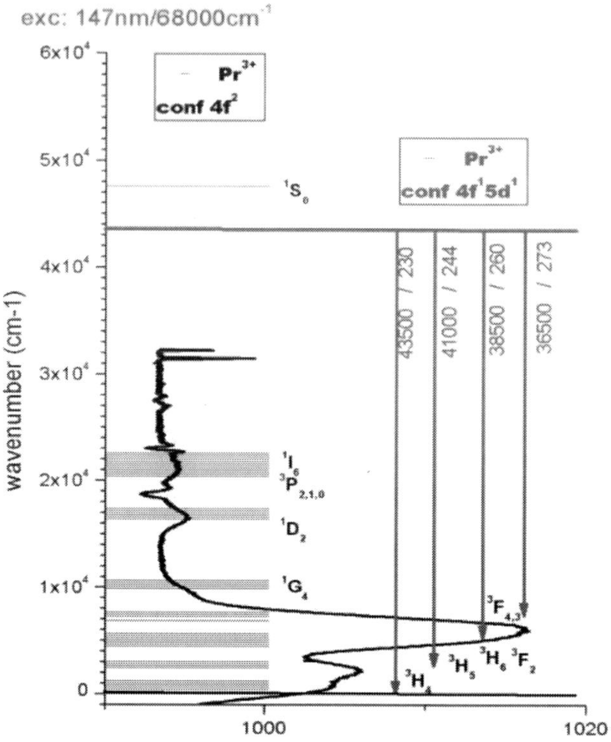

Figure 1: UVC-emission spectrum of sample $Ca_2P_2O_7$:Pr2%, Na2%. The Pr^{3+} electronic levels are schematized by horizontal lines. In blue: interconfigurational transitions with wavenumbers (cm^{-1}) /wavelengths (nm).

RESULTS

For all the samples investigated, the UVC emission consists of four components grouped in two broad bands centred respectively at 240 and 265 nm (an example is shown on Figure 1). Emission bands are shifted by 5-10 nm towards longer wavelengths in the pyrophosphates compared to the orthophosphates. Annealing of the pyrophosphates has a great effect on the UVC emission intensity, since the Pr^{3+} ($4f^15d^1 \rightarrow 4f^2$) emission is more efficient in the high temperature α- $Ca_2P_2O_7$ than in the low temperature β -$Ca_2P_2O_7$, as already pointed out in [5]. The synthesis of praseodymium- doped lanthanum orthophosphate and of calcium pyrophosphate by spray-pyrolysis at moderate temperature followed by a controlled annealing proves to be very efficient in the production of good quality UVC phosphors, appearing as elongated chips of about 20 μm length.

REFERENCES

[1] A.M. Srivastava, Journal of Luminescence (2009).
[2] http://www.furukawakk.co.jp.
[3] T. Beaudette, B. Caillier, Ph. Guillot, N. Joffin, J. Dexpert-Ghys and J. Galy 'Experimental characterization of phosphors under plasma excitation, Euro display 2005, pp 209-211, Scotland, Edinburgh, 2005/09/19-22
[4] J. Dexpert-Ghys, R. Mauricot, B. Caillier, P. Guillot, T. Beaudette, G. Jia, P.A. Tanner, B.-M. Cheng, J. Phys. Chem. (2010), 10.1021/jp909197t.
[5] European Patent EP1842892A1.

To delimit the far-field region of LEDs

Che-Chu Lin[1], Ching-Cherng Sun[1], Wei-Ting Chien[1], Ivan Moreno[2] and Yi-Chien Lo[1]

1 Department of Optics and Photonics, National Central University, Chung-Li 320, Taiwan

2 Unidad Academica de Fisica, Universidad Autonoma de Zacatecas, 98060, Zacatecas, Mexico

Contact: ccsun@dop.ncu.edu.tw

ABSTRACT

In the design of optical systems based on light-emitting diodes (LEDs), precision and accuracy are two factors among the key issues. In the analysis of systems using LEDs, it is useful to know if the distance is far enough from the LED to allow the radiation pattern to be simulated by the point source approach. Different types of LEDs have different far-field ranges. We analyze these differences by modulating key parameters like geometrical structure of encapsulating lens, chip position, and package errors. We find that far-field region considerably depends more on the shape of both lens and chip than all other parameters.

INTRODUCTION

LEDs have been regarded as the most promising light source for next-generation lighting [1]. However, there are special challenges in the optical design of a system containing LEDs, depending on the working distance, the optical modeling of an LED must be performed in a different way. There are three working conditions in LEDs: near-field, mid-field, and far-field [2-3]. The near-field is a very short region near the chip, where the electro-magnetic wave effects strongly change. In the mid-field region, the light pattern significantly varies from one distance to another due to the finite size of chip and LED package structure. Thus mid-field is a region for verifying optical models of extended light sources. Farther than the mid-field, in far-field the angular intensity distribution remains practically constant with distance. The LED can be defined as a directional point source, and it is important to make sure where does the far field begin.

TO DELIMIT THE FAR FIELD REGION

To determine the similarity between the simulated light pattern and the corresponding measured pattern, we apply the normalized cross correlation (NCC) [4]

$$NCC = \frac{\sum_n \left[I(\theta_n)_e - \bar{I}_e\right]\left[I(\theta_n)_s - \bar{I}_s\right]}{\sqrt{\sum_n \left[I(\theta_n)_e - \bar{I}_e\right]^2 \sum_n \left[I(\theta_n)_s - \bar{I}_s\right]^2}}, \quad (1)$$

where I_s and I_e are the simulated and experimental values of relative intensity, respectively. θ_n is the *n*-th angular displacement, and $\bar{I}_s$ and $\bar{I}_e$ are the mean values of simulation and experiment across the angular range. As the detected distance (D) increases the NCC approaches 100%. An NCC equal to 100% means that I_s and I_e are exactly the same. Here we choose three NCC values to delimit a working distance range. As the lower bound that we suggest, the minimum far-field (MFF) distance as that distance at which NCC = 99%; and we define the upper bound as the far-field (FF) distance for NCC = 99.90%. The third condition is the quasi far-field (QFF) distance for NCC = 99.50%. We recommend for precise designs to use an extended source model if the distance is shorter than the QFF distance. In general, if the extended source model is used at longer distances than QFF, a little more precise optical design is obtained, but after the FF distance the extended source model will give the same results than the point source approach.

In our research, we find that the far field distance changes in function of the shape and size of emitting area and encapsulating lens. In the following discussion, we will consider a hemispherical encapsulating lens and chip position. Figure 1(a) shows the NCC of the LED radiation pattern in function of distance *D* for several lens diameters. Figure 1(b) shows FF, QFF, and MFF. The FF distance is about 10~20 times larger than the MFF distance. The analysis of the lens size dependence leads to a basic condition: once increasing the lens diameter, the MFF distance also will be increased by 3 times of the lens diameter, and QFF will be about 5 times the lens diameter. This fact is very useful for precise optical modeling of LED systems. And Figure 1(c) shows the NCC of the LED radiation pattern in function of distance *D* for several chip shifts. The QFF and MFF distance show a nearly linear increment with chip displacement (Fig. 1(d)). Any LED suffers from tolerances like all optical devices. A common manufacturing error is the chip displacement from a nominal position. We consider a chip moved in backward direction. From observation of the far-field radiation patterns one can note that beam directionality increases as the chip is displaced in back direction. This is the only case in which the QFF and MFF increases as the far-field radiation pattern is more directional. The angular intensity distribution is narrowed when the chip is displaced in this way. The analysis showed some interesting aspects about the effect of these parameters. For example, in all cases the FF distance was between 1.6 and 23 times larger than MFF distance. And the FF distance was between 1.3 and 11 times larger than QFF distance. Also, the far-field region considerably depends more on the shape of both lens and chip than all other parameters.

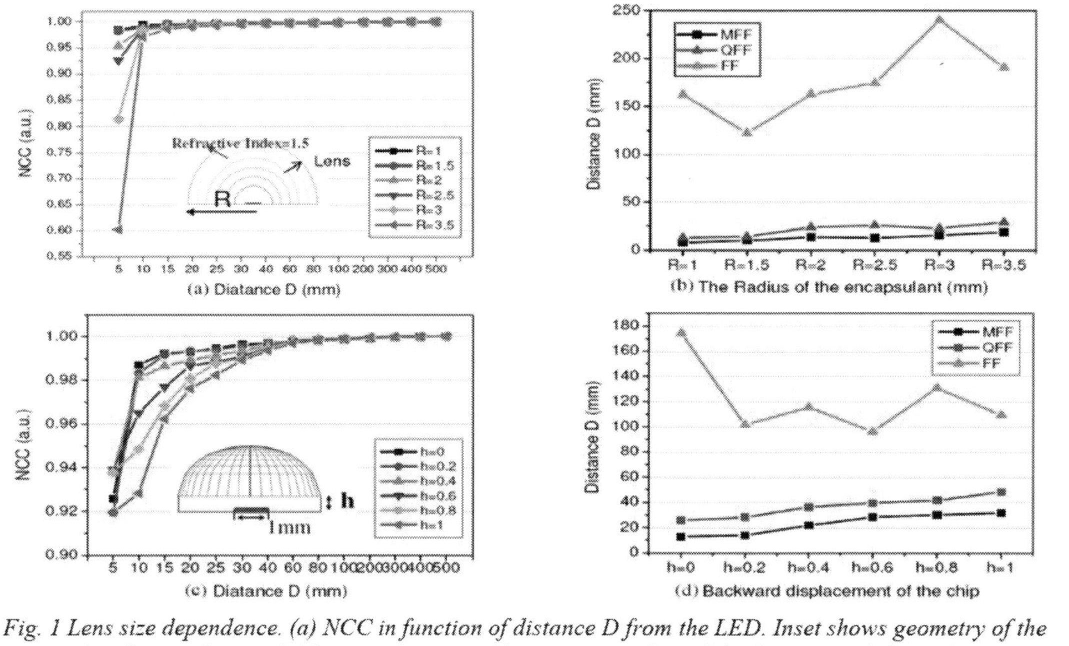

Fig. 1 Lens size dependence. (a) NCC in function of distance D from the LED. Inset shows geometry of the encapsulated LED. (b) Far field region for several lens sizes; Backward displacement of chip. (c) NCC in function of distance D from the LED. (d) Far field region for several chip displacements h.

REFERENCE

1. D. A. Steigerwald, J. C. Bhat, D. Collins, R. M. Fletcher, M. O. Holcomb, M. J. Ludowise, P. S. Martin,and S. L. Rudaz, "Illumination with solid state lighting technology," IEEE J. Sel. Top. Quantum Electron, **8**, 310 (2002).
2. C. C. Sun, T. X. Lee, S. H. Ma, Y. L. Lee, and S. M. Huang, "Precise optical modeling for LED lighting verified by cross correlation in the midfield region," Opt. Lett. **31**(14), 2193–2195 (2006).
3. I. Moreno, C. C. Sun, and R. Ivanov, "Far-field condition for light-emitting diode arrays," Appl. Opt. **48**(6), 1190–1197 (2009).
4. W. T. Chien, C. C. Sun, and I. Moreno, "Precise optical model of multi-chip white LEDs," Opt. Express **15**(12), 7572–7577 (2007).

Studies on the Effect of Impurities in Mercury-free Discharge for Lighting Application

A. N. Dagang[1], S. Bhosle[2], G. Zissis[1], A. Corazza[3], V. Massaro[3]

[1] *Université de Toulouse, UPS, INPT, LAPLACE (Laboratoire Plasma et Conversion d'Energie), 118 route de Narbonne, F-31062 Toulouse Cedex 9, France*
[2] *LED Engineering Development, 11 rue du Mont Dore, F-31100 Toulouse, France*
[3] *SAES Getters S.p.A, Viale Italia, 20020 Lainate, Milan, Italy*
Contact: ahmad.dagang@laplace.univ-tlse.fr

ABSTRACT

The common and typical impurities of H_2, N_2, CO_2 and CH_4 were added in varying concentration percentages into a xenon DBD lamp. Their discharge condition and spectroscopic behavior were studied. The plasma generation differs from one type of impurity to another. With two different concentrations of hydrogen, at 5% concentration it would stabilize the discharge and at 0.5% concentrations it would give an additional collision process where a strong peak of atomic lines were observed. Nitrogen with two types of concentration of 1% and 0.1% gives two kinds of characteristics in emission spectroscopy.

INTRODUCTION

The purpose of the study is to improve the understanding on the working mechanism of mercury-free discharge lamp, in particular to understand how the presence of impurities could effect the lamp's performance. In any discharges, the existence of impurities could give a severe change or bring a negative or positive effect to the discharge and plasma characteristics. The main impurities are typically from hydrogen (H_2), nitrogen (N_2), carbon monoxide (CO), carbon dioxide (CO_2), methane (CH_4), etc. Particularly in lamps, the presence of hydrogen even in a few parts per million (PPM) in the tube can cause an increase in starting voltage and ignition problems especially in high intensity discharge (HID) lamps[1]. Xenon based discharge has been identified as an alternative candidate to mercury, and due to this the authors used xenon based discharge as an effort to develop mercury-free lamps. In order to avoid any contamination from the internal electrodes, an electrodeless dielectric barrier discharge (DBD) was applied. The studies are focused on the effect of impurities on xenon DBD lamp, and the possibility of the mix gases (impurities with xenon) to be applied as a light source. The effects of typical type of impurities i.e. H_2, N_2, CO_2 and CH_4 are evaluated. Their discharge condition and emission spectra are discussed.

EXPERIMENTAL SETUP

The lamps used are a cube type quartz tube (12×12 mm) with a glass thickness of 1 mm, and was ignited using a DBD source. The DBD was supplied with pulse voltage with the amplitude ranging from 5 to 6 kV, at a pulse width of 1 μs, and the frequency is changed from 70 to 150 kHz. The lamps composition as below.

- Xenon reference lamp (lamp 1: 150 Torr)
- Xe-H_2 lamps (lamp 2: 95%-5%, lamp 3: 99.5%-0.5%, 150 Torr)
- Xe-N_2 lamps (lamp 4: 99%-1%, lamp 5: 99.9%-0.1%, 150 Torr)
- Xe-CO_2 lamp (lamp 6: 99.5%-0.5%, 150 Torr)
- Xe-CH_4 lamp (lamp 7: 95%-5%, 150 Torr)

Emission spectra were measured using Ocean Optics HR4000, which can provide an optical resolution at 0.025 nm and is responsive from 200 to 1100 nm. Fibre used is 400 μm in diameter, a UV-VIS High OC content type with its most efficient wavelength at 300 to 800 nm. The integration time of the spectrometer was changed accordingly from 200 to 2000 ms, the longer time for the less intense plasma.

RESULTS AND DISCUSSION

For each discharge there are differences in plasma intensity and stability. As can be seen in figure 1, lamps 1 and 5 produced a thick filament-like discharge compared to the rest of the lamps. For lamps 3, 4 and 6, the discharge was getting unstable when the frequency is increased. The most stable plasma produced was from lamp 2, where the position of a filament remained unchanged when the frequency was adjusted. Instead, its filament becomes diffuse when frequency is increased. Lamp 2 is a lamp with a high concentration of H_2, and this condition can be considered one of the reasons for obtaining stable plasma compared with lamp 3 with low concentration of H_2 which was unstable. Lamp 7 with high concentration of CH_4 is the most difficult to ignite the plasma, the intensity is the lowest among the lamps but the discharge was stable. Emission spectra for the lamp with pure xenon and the lamps with each impurity (H_2, N_2, CO_2, CH_4) at the frequencies of 100 kHz is shown in figure 2. For these four types of impurities, the molecular spectra only can be observed in the lamp with nitrogen at 1% concentration (lamp 4) where typical 2nd positive band of N_2 molecular spectra can be seen at 300~400 nm. However for the lamp with nitrogen at 0.1% concentration (lamp 5), there are no particular atomic or molecular lines that can be observed except xenon atomic 764.20 nm and in this case the waveform is similar to lamp 1, where a weak continuum emission can be seen at 350~450 nm and 500~550 nm. For the other lamps, most of the lines that can be seen are from xenon atomic spectra. The lines are mainly assigned to xenon atomic lines of 450.10, 467.12, 473.41, 491.65 nm that belong to transitions originating from $5p^5 6p$ ($2p_1$ to $2p_{10}$) configuration at energy levels between 9.58 eV and 11.98 eV. In the case of 0.5% hydrogen contained lamp (lamp 3), the typical hydrogen atomic lines of 656.28 nm, 486.13 nm and 434.05 nm (for hydrogen α, β, γ) cannot be observed. Instead, xenon atomic lines of 467.12 nm, 688.21 nm, 711.96 nm and 764.20 nm are strongly observed. The same observation goes to the 5% hydrogen contained lamp (lamp 5) but the atomic lines are weaker. In case of lamps with CO_2 and CH_4 (lamps 6 and 7), these have relatively low intensity and it is difficult to identify the spectra lines due to the noise that was simultaneously observed.

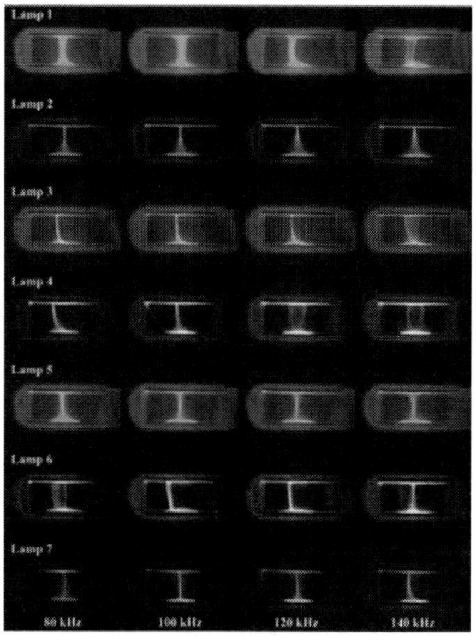

Figure 1. Discharge images for lamps 1~7, at 80~140 kHz

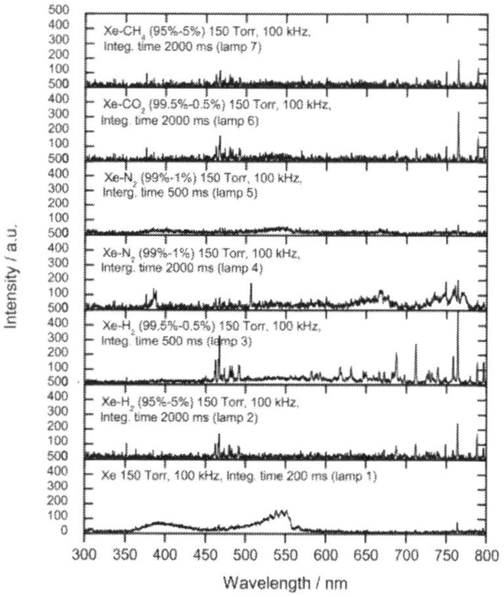

Figure 2. Emission spectra for lamps 1~7 at 100 kHz

CONCLUSION

The basic studies on the effects of typical impurities have been done. The use of H_2 at 5% of concentration could help to diffuse and stabilize the plasma and at another concentration of 0.5% it could generate an additional collision process. Two types of N_2 concentration lamps give a different characteristics of emission spectra. In the case of CO_2 and CH_4 concentration lamps, typical molecular spectra cannot be observed.

REFERENCES

[1] C. Boffito, A. Corazza *Proc. 10th Int. Symp. on Science and Technology of Light Sources* 95-104 (2004)

Towards a predictive model for organic light-emitting diodes: current density, recombination and light emission

R. Coehoorn[1,2], P.A. Bobbert[3], R.A.J. Janssen[2], H.-P. Loebl[4], J.J.M. van der Holst[3], F.W.A. van Oost[3], J. Cottaar[3,5], S.L.M. van Mensfoort[1,2], M. Carvelli[1,2,5] and R.J. de Vries[1,2,5]

[1] *Philips Research Laboratories, High Tech Campus 4, 5656 AE Eindhoven, The Netherlands*
[2] *Molecular Materials and Nanosystems, Department of Applied Physics, Eindhoven University of Technology, P.O. Box 513, 5600 MB Eindhoven, The Netherlands*
[3] *Theory of Polymers and Soft Matter, Department of Applied Physics, Eindhoven University of Technology, P.O. Box 513, 5600 MB Eindhoven, The Netherlands*
[4] *Philips Research Laboratories, Weisshausstrasse 2, D-52066 Aachen, Germany*
[5] *Dutch Polymer Institute (DPI), P.O. Box 902, 5600 AX Eindhoven, The Netherlands*
Contact: reinder.coehoorn@philips.com

ABSTRACT

In this paper, we present the theoretical, numerical modeling and experimental validation steps that we have recently made towards the realization of a predictive device model for white small-molecule based OLEDs for lighting applications.

INTRODUCTION

The application of organic light-emitting diodes (OLEDs) as large-area white light sources is expected to further extend the solid state lighting portfolio, with properties which complement those of inorganic LEDs. In the past five years, the luminous efficacy of white OLEDs has increased to 50-60 lm/W, and to more than 100 lm/W if special methods such as macroextractors for light-outcoupling are used. This is achieved by making use of complex small-molecule layered stacks, such as shown in figure 1(a). We envisage that the availability of a numerical OLED device model that predictively describes its functioning would strongly support a further enhancement of the efficiency, lumen maintenance and color stability. A key element of such models is a proper treatment of the disordered nature of the organic semiconductors used. This leads to a laterally non-uniform current density and recombination rate, which can only be fully understood from three-dimensional (3D) simulations [1]. We argue that, nevertheless, the results from 3D-simulations can be "translated" to effective one-dimensional (1D) models. For the case of blue-emitting polymer-based single-layer OLEDs, we have recently demonstrated that an experimentally validated predictive OLED model can be developed [2]. For the case of white small-molecule multilayer OLEDs with a structure such as shown in figure 1, the construction of such a model is in progress. In this contribution, we present the theoretical, numerical modeling and experimental validation steps that we have recently made towards the realization of such a model.

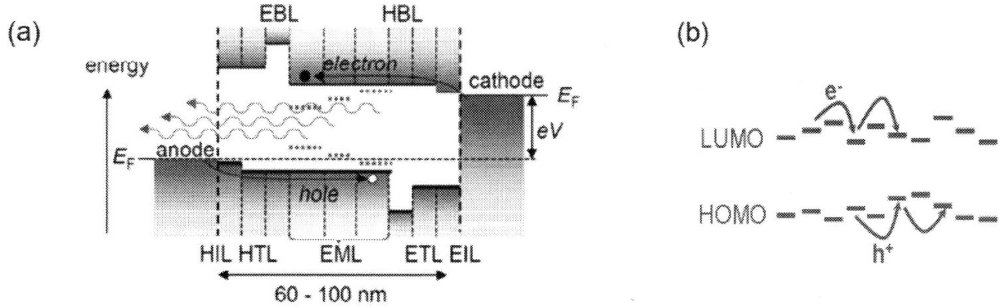

Fig. 1. (a) Schematic layer structure of a multilayer white OLED, containing hole and electron injection layers (HIL, EIL), transport layers (HTL, ETL), blocking layers (EBL, HBL), and an emissive layer (EML) containing red, green and blue emitting dye molecules (with Lowest Unoccupied Molecular Orbital (LUMO) and Highest Occupied Molecular Orbital (HOMO) energies indicated by dashed lines). E_F is the Fermi energy, and V is the applied voltage. (b) The electron (e^-) and hole (h^+) hopping process (schematic) in between the disordered LUMO and HOMO states on each molecule.

NOVEL MOBILITY AND RECOMBINATION MODELS

Master Equation (ME) and Monte Carlo (MC) modeling was used to calculate the temperature, electric field and carrier density dependence of the charge carrier mobility in the structurally disordered organic semiconductors used in OLEDs. Structural disorder gives rise to a non-uniformity of the localized molecular electron and hole energy levels in between which the electrical transport takes place via a hopping process (see figure 1(b)). We have developed novel mobility models for materials within which the energy levels show a Gaussian distribution [3], but also for materials with a more complex density of states as a result of the presence of a small concentration of trap states. This is highly relevant, e.g. to the electron and hole transport in the red, green and blue emissive layers where generally a small concentration of phosphorescent dye molecules is present of which the energy levels act as trap levels. Another novel theoretical finding is the discovery that the recombination rate is only well described by the "standard" Langevin formula if so-called "bipolar" mobilities are used, i.e. if the effect of the mixed presence of electrons and holes on the mobilities is taken into account [4].

NOVEL OLED DEVICE MODELS

It is shown that a full OLED device model can be realized in two ways. On the one hand, the mobility and recombination models described above can be used to develop a one-dimensional OLED model, by solving the coupled transport-recombination-Poisson equations. Ultimately, such models are a suitable basis for fast and user-friendly OLED PC-modeling software [5]. On the other hand, MC modeling using supercomputers can be used to develop a three-dimensional OLED model. This shows in detail the filamentarity of the current density and the occurrence of "recombination hot spots", thereby providing important additional information which is expected to be relevant to the understanding of lifetime-limiting processes and color stability. A comparison between the results from both approaches is given.

EXPERIMENTAL VALIDATION

From systematic studies of the current-voltage characteristics of various hole and electron transporting small-molecule materials, we have found that for various OLED-relevant materials good descriptions can be obtained of the mobility functions in terms of the Gaussian disorder models discussed above. We have obtained evidence of so-called "correlated disorder", and have found that the electron transport is for all cases studied affected by the presence of trap states. As a next step, extensive experimental and device modeling studies of α-NPD/BAlq bilayer OLEDs have been carried out, using a recently developed method for the determination of the light-emission profile across the OLED from the full angle, wavelength and polarization dependence of the light-emission [6]. The analysis reveals that the device performance is the result of a delicate balance of electron and hole transport, recombination, exciton diffusion, exciplex formation and the position-dependent light-outcoupling. Using this case as an example, the remaining challenges towards the realization of a predictive full multilayer OLED device model will be discussed.

REFERENCES

[1]. J.J.M. van der Holst *et al.*, Phys. Rev. B 79 (2009) 085203.
[2]. S.L.M. van Mensfoort, PhD Thesis Eindhoven University of Technology (2009).
[3]. W.F. Pasveer *et al.*. Phys. Rev. Lett. 94 (2005) 206601; M. Bouhassoune, S.L.M. van Mensfoort, P.A. Bobbert and R. Coehoorn, Organ. Electr. 10 (2009) 437.
[4]. J.J.M. van der Holst, Phys. Rev. B 80 (2009) 235202.
[5]. E.g. the SEFTOS (www.fluxim.com) and simoled (www.sim4tec.com) OLED modeling tools.
[6]. S.L.M. van Mensfoort *et al.*, Nature Photonics (published online, March 2010), DOI 10.1038/NPHOTON.2010.32.

Preparation and luminescence properties of Eu^{2+}-doped calcium silicon nitride by the carbothermal reduction and simultaneous nitridation of spray-pyrolyzed oxide powder

K. Yamaguchi[1], H. T. Hintzen[2], A. C. A. Delsing[2], S. Koda[1] and K. Itatani[1]
[1] *Department of Materials and Life Sciences, Faculty of Science and Engineering, Sophia University 7-1 Kioi-cho, Chiyoda-ku, Tokyo 102-8554, Japan*
[2] *Materials and Devices for Sustainable Energy Technologies, Department of Chemical Engineering and Chemistry, Eindhoven University of Technology, P.O. Box 513, 5600 MB Eindhoven, The Netherlands*
Contact: itatani@sophia.ac.jp

ABSTRACT

In this work, Eu^{2+}-doped calcium silicon nitride ($Ca_2Si_5N_8$: Eu^{2+}) was prepared by the carbothermal reduction and simultaneous nitridation of oxide powder prepared by an ultrasonic spray pyrolysis. The aqueous solution containing 0.71 mol·dm^{-3} colloidal SiO_2, 0.28 mol·dm^{-3} $Ca(NO_3)_2$ and 2.86×10^{-3} mol·dm^{-3} $Eu(NO_3)_3$ was sprayed using an ultrasonic vibrator into the furnace heated at 600°C in order to obtain the oxide powder. After the spray-pyrolyzed powder was heated at 800°C for 10 min in air, the resulting powder (amorphous materials) was mixed with carbon. Single phase $(Ca_{0.99}Eu_{0.01})_2Si_5N_8$ could be obtained when the mixed powder of oxide and carbon (the molar ratio of C/2CaO·5SiO_2: 12.0) was nitrided at a temperature as low as 1300°C (heating time: 2 h). The irradiation of light with the wavelength of 374 nm to $(Ca_{0.99}Eu_{0.01})_2Si_5N_8$ resulted in the emission of orange light with the wavelength of 588 nm.

INTRODUCTION

A light emitting diode (LED) is a promising illumination source of the next generation. Compared to the most popular fluorescent lamp currently used in our life, it has noted advantages of energy saving, long operation life and mercury free. The commercial white-LED generally consists of a blue emitting LED and a yellow-emitting phosphor (*e.g.*, Ce-doped yttrium aluminate garnet ($Y_3Al_5O_{12}$); YAG: Ce) and, therefore, has problem of poor red components. Recently, many researchers pay attention to the nitride phosphors from the viewpoints of (i) absorption of visible light and emission of longer wavelength light, and (ii) excellent thermal and chemical stabilities. Eu^{2+}-doped calcium silicon nitride ($Ca_2Si_5N_8$) phosphor, which has three dimensional networks consisting of a SiN_4-based tetrahedral structure, is a promising material as an orange emitting component in white-LED [1]. The phosphor particles are mainly prepared by the solid-state reaction of nitride compounds, *e.g.*, Ca_3N_2 and Si_3N_4. The technique of solid state reaction may have the advantage to control the chemical composition but have some problems regarding the chemical homogeneity at the atomic level, regular particle shape and narrow particle size distribution.

On the other hand, ultrasonic spray pyrolysis (USP) is a technique to instantaneously form the particles through (i) the introduction of droplets containing the desired kinds and amounts of components in the heating zone, (ii) the co-precipitation of solid materials due to the evaporation of water, and (iii) the solid-state reaction to form desired compounds. The resulting powder has the characteristics of homogeneous chemical composition and nano-sized particles, both due to the flash preparation route [2]. Unfortunately, the research on the preparation of phosphors in this way is still limited, chiefly due to the difficulty in the direct preparation of nitride compounds by the carbothermal reduction and simultaneous nitridation (CRSN) technique. On the basis of such background, this paper describes the research on the low temperature preparation of $Ca_2Si_5N_8$: Eu^{2+} phosphor through the combination of USP and CRSN techniques.

EXPERIMENTAL PROCEDURE

The spraying solution (250 cm^3) was prepared by adding 0.714 mol·dm^{-3} colloidal silica (SiO_2), 0.287 mol·dm^{-3} calcium nitrate tetrahydrate ($Ca(NO_3)_2 \cdot 4H_2O$) and 2.86×10^{-3} mol·dm^{-3} europium nitrate hexahydrate ($Eu(NO_3)_3 \cdot 6H_2O$) into water. The mixed solution was atomized with an ultrasonic spray vibrator with the frequency of 2.4 MHz and the resulting droplets were introduced into a furnace heated at 600°C. After the heat treatment of spray-pyrolyzed powder in air at 800°C for 2 h, the resulting powder was mixed with various amounts of carbon (the molar ratio of C to $2CaO \cdot 5SiO_2$: 12.0). After mixing, the compact of mixed powder was heated at a temperature between 1250 to 1550°C for 2 h in N_2 atmosphere.

The crystalline phases of the prepared powder were checked using an X-ray diffractometer (XRD, RINT2100V/P, Rigaku) with CuKα radiation (40 kV, 40 mA). The morphology of the particles was studied using a field-emission scanning electron microscope (SEM, S-4500, Hitachi). The photoluminescence properties were measured using a fluorescent sprectrophotometer (F-4500, Hitachi).

RESULTS AND DISCUSSION

The spray-pyrolyzed and then heat-treated powder (amorphous materials) was composed of spherical agglomerates with mean diameter of 0.96 μm. We then examined the conditions for CRSN in order to prepare $(Ca_{0.99}Eu_{0.01})_2Si_5N_8$ from the oxide powder. Phase changes during the CRSN process of oxide with carbon were examined on the basis of the XRD examination; note that the heating time was fixed at 2 h. The carbon started to reduce at 1100°C and became almost zero at 1200°C. In addition, $CaSiO_3$ and crystalline SiO_2 were present in the ranges of 900 to 1300°C and 1100 to 1300°C, respectively. The intended $Ca_2Si_5N_8$ started to form above 1200°C and the single phase material was obtained at 1300°C. The reaction process of oxide and carbon may be expressed as: $2CaO \cdot 5SiO_2 + 12C + 4N_2 \rightarrow Ca_2Si_5N_8 + 12CO$.

SEM micrograph of the powder heated at 1100°C for 2 h showed the presence of spherical agglomerates with diameters of around 1 - 2 μm, whereas that of the powder heated at 1200°C showed the presence of irregular-shaped agglomerates. The formation temperature of $Ca_2Si_5N_8$ prepared by the combination of USP and CRSN (*i.e.*, 1300°C for 2 h) may be 250°C lower than that prepared by the CRSN technique, using mechanically-mixed powders of $CaCO_3$, Si_3N_4 and Eu_2O_3 (*i.e.*, 1550°C for 2 h). This phenomenon indicates that the component atoms in the powder prepared by USP are homogeneously dispersed at the atomic level, and the CRSN process to form $Ca_2Si_5N_8$ is effectively achieved at temperature as low as 1300°C.

The photoluminescence spectra of the $(Ca_{0.99}Eu_{0.01})_2Si_5N_8$ (CRSN conditions; 1300°C, 2 h) indicated the broad excitation band ranged from 200 to 500 nm. When the light with the wavelength of 374 nm was irradiated to this compound, the emission band appeared with the center at 588 nm. The excitation band of $(Ca_{0.99}Eu_{0.01})_2Si_5N_8$ is attributed to the Eu^{2+} ion ($4f^7 \rightarrow 4f^6 5d$ transition), whereas the emission peak is ascribed to the $4f^6 5d^1 \rightarrow 4f^7$ transition of the Eu^{2+} ion [1].

CONCLUSIONS

Eu^{2+}-doped calcium silicon nitride ($Ca_2Si_5N_8$:Eu^{2+}) was prepared by the CRSN of spray-pyrolyzed oxide powder. The results of the CRSN process of oxide powder, as well as the luminescence properties, were summarized as follows:

(i) The spray-pyrolyzed and heat-treated powder (amorphous materials) was mixed with carbon (the molar ratio of $C/2CaO \cdot 5SiO_2$ value: 12.0). The single phase of $(Ca_{0.99}Eu_{0.01})_2Si_5N_8$ could be obtained when the mixed powder of oxide and carbon was nitrided at a temperature as low as 1300°C (heating time: 2 h) in N_2 atmosphere. This formation temperature was 250°C lower than that required for the CRSN, using mechanically-mixed powders of $CaCO_3$, Si_3N_4 and Eu_2O_3, reflecting the homogeneity of component atoms.

(ii) The irradiation of light with the wavelength of 374 nm to $(Ca_{0.99}Eu_{0.01})_2Si_5N_8$ (CRSN conditions: 1300°C, 2 h) resulted in the emission of orange light with the wavelength peaking at 588 nm.

REFERENCES

[1] Y. Q. Li, J. E. J. van Steen, J. W. H. van Krevel, G. Botty, A. C. A. Delsing, F. J. DiSalvo, G. de With, H. T. Hintzen, J. Alloys Comp., 417(2006)273.

[2] K. Itatani, M. Aizawa, J. Soc. Inorg. Mater. Japan, 10(2003)285.

Investigation of water photolysis by means of VUV excimer lamps

G.Zvereva
S.I.Vavilov State Optical Institute, St-Petersburg, 199034, Russia
Contact: zvereva@soi.spb.ru

ABSTRACT

Calculations of liquid water decomposition products under the influence of vacuum ultraviolet (VUV) emission (Xe excimer lamp, 172 nm) have been carried out. High reactive radicals OH^{*} are shown to have been appeared effectively and destroyed pollutants added to water. Movement of water relatively light source allows to increase thickness of VUV emission absorbing layer.

INTRODUCTION

It is known, that under influence of emission with wavelength less than 190 nm water is effectively decomposed with high reactive $^{*}OH$ radical formation. Quantum yield of this reaction at wavelength 172 nm is equal to 0.42 and increase up to 1. at 120 nm [1]. Absorption coefficient of liquid water in wavelength region λ<172 nm has values 10^3-10^4 cm^{-1} [2]. These high values provide 90% emission absorption in 10^{-3}-10^{-4} cm water layer. Penetration to such thin layer is one of the main disadvantage of VUV water decomposition process.

Radical $^{*}OH$ is a powerful oxidant which oxidation potential (2.8 eV) exceeds oxidation potentials of such widely used oxidants as chlorine (1.4 eV), hydrogen peroxide (1.8 eV) and ozone (2.1 eV). Modern oxidation technologies based on advanced oxidation processes (AOPs) use for the toxic organic compounds decomposition mostly hydroxyl radical [3]. In AOPs, the organic compounds can be completely decomposed to CO_2 and H_2O. Depending on AOPs, $^{*}OH$ radicals are generated by different methods: by chemical oxidation using H_2O_2 , O_3, in Fenton's reactions; by radiation methods including UV radiation, x-rays, e-beam; by photocatalysis using TiO_2 etc [4].

At present time technologies based on VUV water decomposition are in the stage of research [5], this in particular is connected with absence of acceptable VUV light sources . However, at last two decades intensive research and development of excimer VUV light sources based on barrier discharges in high pressure rare gases were carried out [6,7]. Emission spectra of such excimer lamps are occupied wavelengths region of 120-180 nm , efficiency of conversion of electric power to emission one can reaches 60%. In this way excilamps give a new methods of radical $^{*}OH$ formation.

INVESTIGATIONS OF LIQUID WATER PHOTOLYSIS

Following products is known to be generated during liquid water decomposition: H^{*}, $^{*}OH$, e_{aq}^{-}, H^{+}, H_2, OH^{-}, O^{-}, H_2O_2, O_2^{*-}, HO_2^{*} , O_2^{3-}, HO_2^{*-} [8,9]. In our work calculations of photolysis products concentrations were carried out. Unknown concentrations were found by solving of system of differential equations:

$$\frac{\partial \rho_i(x,t)}{\partial t} + \frac{\partial(\rho_i(x,t)v_i)}{\partial x} = S_i(x,t) \qquad i=1,..., L$$

where $\rho_i(x,t)$ - concentration of i-th component , v_i-water velocity relatively light source , $S_i(x,t) = \sum_l F_{il}^{+} - \sum_m F_{im}^{-}$ – sources of formation and decay of i-th component, L=12 number of photolysis products. Change of concentrations due to photo processes was taken into account by means of term: $F = -\Phi(1/h\nu)\, dI(x)/dx$, where Φ – quantum yield of the reaction , $I(x)=I(0)10^{-kx}$- intensity of the light at point x, k- absorption coefficient.

As a media distilled water with addition of model pollutants with concentration 10^{13} cm^{-3} was considered. Pollutants were oxidized by $^{*}OH$ radical with velocity $k=10^{-11}$ cm^3s^{-1}. To increase thickness of the absorbing VUV emission layer, water flows out the VUV light source (Xe excimer lamp, wavelength 172 nm, intensity I=1 W/cm^{-2}) with velocity v_i =100 cm/s was considered .

With time concentrations reach fixed values at every point *x*. In the figure.1 concentrations of radical $^{\bullet}OH$ and pollutants are shown. Effective production of $^{\bullet}OH$, pollution destruction and increasing of absorbing water layer are observed.

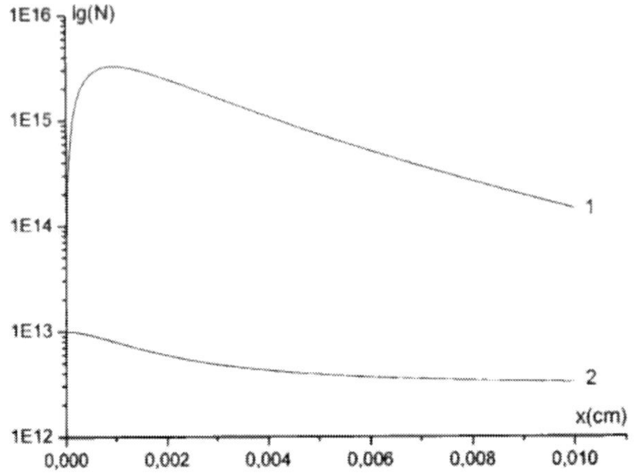

Fig.1 Concentrations of radical $^{\bullet}OH$ (1) and pollutants (2) as a function of distance from light source.

REFERENCES

[1] G. Heit, A. Neuner, P.-Y. Saugy, and Andre M. Braun. , J. Chem. Phys. A., v.102.,pp.5551-5561,(1998).

[3] J.L.Weeks, G.M.A.C. Meaburn and S.Gordon, Rad.Research, v.19, 1963, pp.559-567.

[2] Th.Oppenlander, "Photochemical Purification of Water and Air", Wiley-VCH, 2003, Weinheim, 368 p.

[3] Handbook of Environmental Engineering , v.4, Advanced Physicochemical Treatment Processes, ed. By L.K.Wang, Y.-T Hung, and N.K.Shammas, NJ, 2006.

[4] Handbook on advanced photochemical oxidation processes, EPA/625/R-98/004, 1998, p.97

[5] G.A.Volkova,N.N.Kirillova, E.N.Pavlovskaya,and I.V.Podmoshenskii, Bill. Tov. Zn. (rus.), N41, p.179, (1982).

[6] U.Kogelschatz, Plasma Chem. and Plasma Process., v.23,N1, pp.1-46,(2003).

[7] George V. Buxton, Clive L. Greenstock, W. Phillip Helman, and Alberta B. Ross., J. Phys. Chem. Ref. Data., v.12, N2. , p.513-531, (1988).

[8] B.H.J.Bielski, D.E.Cabelli, R.L.Arudi, J.Phys.Chem. Ref. Data, v.14, N4, p.1041, (1985).

LED-VILLE project: Traffic roundabout LED lighting system. Technical and social approaches

M. Aubès[1], J. Bance[2], S. Bhosle[3], L. Massol[3], P. Sajous[4], G. Zissis[1]
[1] *LAPLACE, Université de Toulouse, 118 rte de Narbonne, 31062 Toulouse Cedex 9, France*
[2] *Syndicat Départemental d'Energies du Tarn, 2 rue Gustave Eiffel - Zone AlbiTech, 81000 Albi, France*
[3] *Led Engineering Development, 10 avenue de l'Europe, 31520 Ramonville Saint-Agne, France*
[4] *Consultante en Aménagement urbain, 24 rue de la Grosse Borne, 27200 Vernon, France*
Contact : michel.aubes@laplace.univ-tlse.fr

ABSTRACT

We report a global study of a LED lighting system for an urban roundabout (LED-VILLE). A study of the photometric performances of the LED used has been performed in our laboratory. Furthermore, average illuminance, uniformity and luminance have been measured and compared to recommended data. Environmental footprints of this lighting system (carbon operation, life cycle analysis and light pollution) has been evaluated and compared to equivalent solution widely used to date. Finally, a social approach of the led street lighting has been achieved : The visual perceptions felt by driving or walking users through the roundabout have been collected and analyzed.

THE LED-VILLE PROJECT

The lighting research and development has accelerated to implement light sources more energy efficient and less polluting. Among these techniques, the LED technology seems a promising solution. However, if the LEDs begin to be widely used in interior lighting, many difficulties have still to be overcome before an extended use of LEDs can be extensively used in outdoor public lighting.

After the installation of a functional equipment using LED technology to illuminate a roundabout located on Le Séquestre (Tarn in France), the "Syndicat Departemental de l'Energie du Tarn" (SDET) has proposed to pilot an applied research project on the theme of LEDs street lighting.

This project, called LED-Ville, aims to produce:

- Quantitative studies conducted in laboratory and real conditions of operation
- Qualitative studies to provide data on the perception of road users (service rendered lighting LED)
- A "product" analysis of the technology and the possibilities of its evolution.

In 2007, a demonstrator has been installed. It consists of 4 luminaries, each equipped with 9 opto-units. Each opto-unit is a set up of three 4.7 W LEDs (Stanley).

RESULTS

The characterization of Leds

Laboratory measurements on LEDs [1] have been performed with a type C goniophotometer [2] and a spectroradiometer SPECBOS 1201 from JETI. From spectral illuminance measurements we have deduced energetic and luminous flux, photometric curves, correlated color temperature and colorimetric coordinates in CIE chromaticity diagrams. Photometric data are put on IES format [3] for use with lighting software.

The LEDs consisting of four chips are used with three different beam shaping lens in order to get uniform lighting. Bare LEDs have a quasi-lambertian light distribution.

The LEDs are supplied at constant current (350 mA) with a voltage of 13,5V. The correlated color temperature is about 4900 K with a dominant wavelength of 570 nm. The value of colour rendering index is 63 as evaluated by the usual CIE method [4].

For an opto-unit, the following mean values have been measured:

- Energetic flux: 1,75 W
- Luminous flux: 620 lumens

- Luminous efficacy: 45 lumens/W

The three LEDs of an opto-unit are fixed on a cooling flange. The operating chip temperature is about 60°C. Without cooling, temperature is higher and the luminous flux of a single LED lowers to 190 lumens [5].
Furthermore, measurements have been performed on LEDs after about two years working. The reduction of flux is about 14%.

Evaluation of the lighting system

Illuminance measurements have been made one year and two years after the installation. The mean illuminance E_m is respectively 10 lx and 9 lx. The value of the illuminance uniformity is U_0=0,3 and the mean luminance of the roadway 0,53 cd/m^2.
According to the AFE recommendations [6] or the European norm EN 13-201 [7] the values of E_m and U_0 should be 20 lx and 0,4.

Environmental impact

For the carbon-simplified operation, uptime of the lighting installation has been evaluated as 4424 hours a year. For each luminary the power consumption is 127W and the power supply dissipation 8W. Following French recommendations, we use a carbon footprint of 119 g of CO_2 for the electric kWh. Then, the operation of the lighting system leads to a CO_2 emission of 290 kg a year. A full Life cycle assessment performed shown that for the moment that the environmental impact of LEDs far beyond an equivalent ceramic metal halide based lighting system with the same nominal flux.
Furthermore we have estimated the light pollution [8]. ULOR is about 0.037 and UFR between 3.5 and 4.9.

Social study

The social study, carried out in 2009, aims at collect visual perceptions and opinions of users. In a first time, we have sent at each household leaving in Le Séquestre a poll. Then, 23 persons accepted participate at a survey on the roundabout. They were with an interviewer and they had to make a pedestrian trip and a car trip by night.
The results of the social study split up into 2 aspects. Some of them are interested in treatment of visual phenomena as glare, uniformity, and bulb-colour. Others carry reflexions about the mean to integrate at best LED street lighting into urban areas.

CONCLUSION

Data characterizing the performances of LEDs systems for public lighting are lacking. The purpose of this work is to evaluate all the aspects of a specific product in operation. If the LED technology does not perform quite adequately the lighting recommendations, there are promising trends with the fast growing of LEDs products.

REFERENCES

[1] C.I.E., Measurements of LEDs, Technical Report CIE 127-2007
[2] I.E.S., Goniophotometer types and photometric coordinates, IES LM-75-01, 2001
[3] I.E.S., Standard Filers Format for the Electronic Transfer of Photometric Data and Related Information, LM-63-02, 2002
[4] C.I.E., Colour Rendering Index of White LED Light Sources, CIE 177-2007
[5] Mottier P., Les diodes électroluminescentes pour l'éclairage, Lavoisier, Paris, 2008
[6] A.F.E., Recommandations relatives à l'éclairage des voies publiques, Paris, 2002.
[7] A.F.N.O.R., Norme NF EN 13-201-2, Eclairage public : exigencies de performances, Paris, 2005.
[8] A.F.E., Les nuisances dues à la lumière, Lux, Paris, 2006.

Attempt to reconstruction of cross-sectional distribution of emission spectra for automotive mercury-free metal halide lamp

H. Motomura, S. Inagaki, M. Jinno
Department of Electrical and Electronic Engineering, Ehime University
3 Bunkyo-cho, Matsuyama 790-8577, Japan
Contact: hmoto@mayu.ee.ehime-u.ac.jp

ABSTRACT

In order to realize mercury-free metal-halide lamp, it is important to know the spatial distribution of emission spectra in the lamp. However, since the arc shape is not symmetrical, it is impossible to know the spatial distribution using the Abel inversion. Therefore, the authors employed computed tomography technique to create the tomogram image from measured spectra. From the comparison between the tomogram images for Sc (508 nm) and Na (589 nm) emissions, it is found that the Sc emission shrinks into the arc center, whereas the Na emission diffuses. It is thought that the reason for this phenomenon is less diffusion for Sc because of higher excited level and larger mass of molecule compared to Na.

INTRODUCTION

Since demands for safe products are expanding, many efforts for research and development of mercury-free automotive metal halide lamps (MHLs) have been done [1,2]. Although these mercury-free lamps are already in the market, it is important to know the spatial distribution of emission spectra; that information is required to design projection optics, driving circuit, etc, which are suitable for mercury-free MHLs. Since automotive MHLs are operated by horizontal burning, the arc bends and becomes axially asymmetric; therefore simple method such as the Abel inversion is not applicable for derivation of spatial distribution of emission spectra. Thus, the authors employed computed tomography technique and reconstructed cross-sectional distribution of emission spectra.

EXPERIMENTAL SETUP

A commercial D4 metal halide lamp for automotive purpose was used as a test sample. The inside diameter of the arc tube is 2.6 mm and the discharge gap between two electrodes is 3.8 mm. Xenon gas is filled at 9.8 atm as a buffer gas and salts of ScI_3, NaI, ZnI_2 and InBr are filled as radiators. The lamp was operated by rectangular voltage with frequency of 300 Hz. Figure 1 shows the experimental setup to measure emission spectra for computed tomography (CT). The lamp is placed on a three-dimensional (*xyz*) translation stage. Two apertures, a plano-convex lens (f=10 cm) and an optical fiber head are fixed on an arm in line. This arm is placed on a rotation (θ) stage and rotates around the pivot that can be set at an arbitrary point in the lamp. The emission image is projected to the optical fiber head by ten times magnitude through two apertures and the lens. The apertures are used to set depth of field larger than the lamp diameter. The projected image is guided to a CCD array spectrometer (Ocean Optics, USB2000) through the optical fiber. Thus, projected spectra across the lamp axis from any azimuths (θ) can be measured.

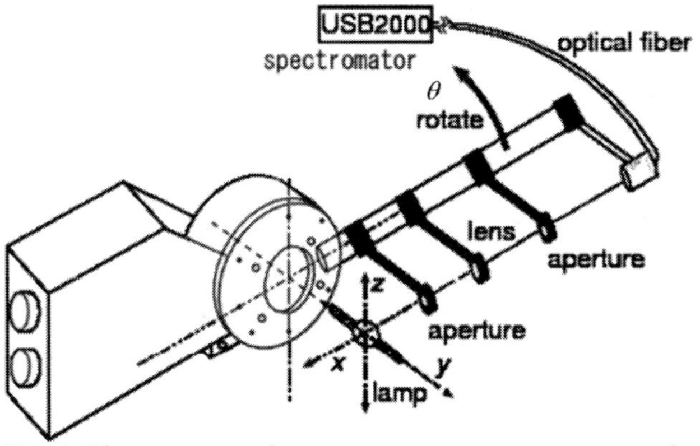

Fig. 1: The experimental setup to measure emission spectra for computed tomography.

A spectral emission coefficient $\varepsilon(\mathbf{r},\lambda)$ at the position of $\mathbf{r}=(x,y,z)$ was derived from the data set of the projected spectra for several azimuths θ when the arm rotated around r by using the radon transfer facility in Matlab. Typically, the step angle for θ was set at 4°, which means a spectral emission coefficient was reconstructed from 45 projected spectra.

RESULTS AND DISCUSSION

Figures 2 and 3 show emission intensity distribution in the cross-sections near the electrode and at the lamp center respectively. Intensities at three wavelengths are shown and they are assigned for Sc (508 nm) and Na (589 and 819 nm). It is obviously seen from these figures that the emissions move above, shrink vertically and diffuse horizontally at the lamp center, resulting from the arc bending. Especially the Sc emission at 508 nm most shrinks among three. The excitation energy of the upper state for the Sc emission at 508 nm (3.89 eV) and that for Na emission at 819 nm (3.62 eV) is similar, whereas the each atomic mass is 44.96 and 22.99. This mass difference can be one of the reasons for the Sc emission shrink resulting from the less diffusion of Sc atoms. The excitation energy of the upper state of the Na emission at 589 nm (2.10 eV) is the lowest among three, which means easy excitation at low temperature. Moreover, the Na emission at 589 nm is resonance one and self absorption occurs. These facts mean Na atoms are easier to excite to the upper level of 589 nm emission outside the arc center. Reconstructed image shown in Figs. 2 and 3 show most diffuse emission for 589 nm. Analysis other emission lines and derivation of temperature distribution by line pair method is under progress.

CONCLUSION

The authors employed computed tomography technique and reconstructed cross-sectional distribution of emission spectra of the arc in a mercury-free MHL for automotive purpose. It is found that the Sc emission at 508 nm shrinks into the arc center, whereas the Na emissions at 589 nm and 819 nm diffuse. Analysis other emission lines and derivation of temperature distribution by line pair method is under progress.

REFERENCES

[1] M. Born, Plasma Sources Sci. Technol. **11** (2002) A55.

[2] M. Born and M. Strösser, J. Phys. D: Appl. Phys., **40** (2007) 3823

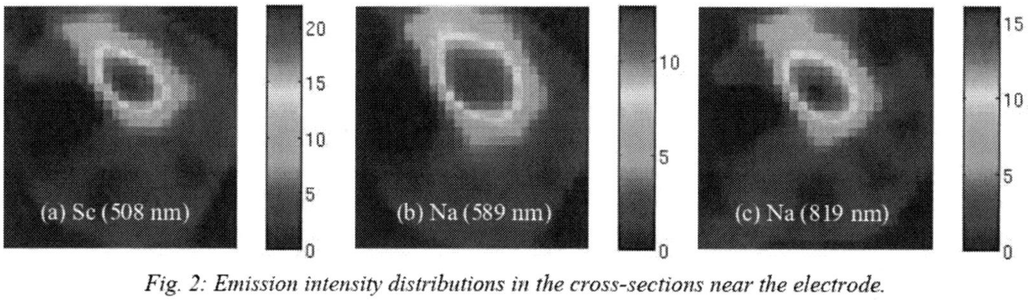

Fig. 2: Emission intensity distributions in the cross-sections near the electrode.

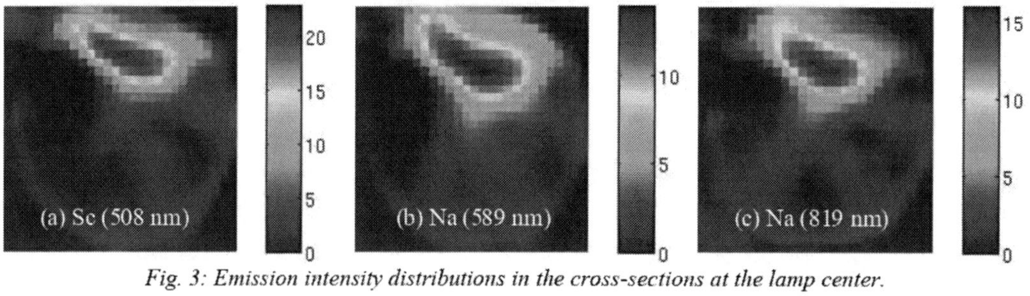

Fig. 3: Emission intensity distributions in the cross-sections at the lamp center.

Life Cycle Assessment of LED lamps and CFLs as the ideal light source for "sustainable development"

N. Yabumoto[1], A. Hatta[2], M. Jinno[3], H. Hattori[4]

[1] Analysis Atelier Co., 4-36-4 Yoyogi, Shibuya-ku, Tokyo 151-0053, Japan
[2]Kochi University of Technology, Tosayamada-cho, Kami-city, Kochi 782-8502, Japan
[3]Ehime University, 3 Bunkyo-cho, Matsuyama-city, Ehime 790-8577, Japan
[4]The Illuminating Engineering Institute of Japan, 2-8-4 Tsukasa-cho, Kanda, Tokyo 101-0048, Japan
Contact:hattori@multi-task.co.jp

ABSTRACT

The Illuminating Engineering Institute of Japan has been devoting its attention to the life cycle inventory analysis of energy consumption of light sources including LED lamps and CFLs from a viewpoint of "sustainable development". It has been proposed to arrange and to assess the figures of total luminous flux, floor illuminance, color temperature for aiming at decision of a concrete criterion in life cycle inventory of energy consumption or carbon-dioxide emission. The small diverging angle of LED light source remarkably improves the illuminating efficacy lux/W. As for the total luminous flux efficacy lm/W, however, the present LED lamps commercially available in the markets have not always overcome the CFL light sources.

INTRODUCTION

LED attracts attention as an environment-friendly lighting, and the life cycle assessment (LCA) results are reported one after another. However, it might hesitate how to compare variegated illuminants when we try concretely to compare the energy consumption. In life cycle inventory (LCI) calculation of the energy consumption, we proposed assemble and assess the figures of total luminous flux, floor illuminance, color temperature, and lifetime for aiming at decision of a concrete criterion. The functional unit was set to be the total luminous flux from the light source and also the floor illuminance right under the light source. The first functional unit assumed to be the total luminous flux 800 lm and the lifetime 40,000 hours. The second one assumed to be the floor illuminance 100 lx directly under the source for 1 m and the lifetime 40,000 hours. The color temperatures were selected 3000 K of warm white and 5000 K of cool white. The life cycle stage was assumed to be four stages of manufacturing, transport, use and end of life. The evaluated illuminants were two types of light bulb shape LEDs (LED A and LED B) in the present market, the future expected LED (F-LED), the compact fluorescent lamp (CFL), and the high frequency fluorescent lamp (Hf-FL). F-LED is the expected one that would be realized in about 2015. The illumination photometry was executed to all evaluation objects excepting for F-LED.

RESULST AND DISCUSSION

Figures 1 and 2 show the LCI results of the total luminous flux for the warm white and cool white of LED A, LED B, F-LED, CFL, and Hf-FL. Figures 3 and 4 show the LCI results of the illuminance at 1 m from the source for the warm white and cool white of them. All LEDs and fluorescent lamps consumed over 97% energy in use stage. This result was same as many previous publications on LCI calculations about illuminance. As for the total luminous flux, the present marketing LEDs are not so energy saving to be said in general, but Hf-FL is excellent. However, the energy consumption of F-LED that practical use is expected within five years is almost equal to that of Hf-FL at warm white and reverses at cool white. The effect of the directivity of LED becomes remarkable for the energy consumption at the floor illuminance. Even one of the present LEDs is equal to the Hf-FL at warm white, in addition, all LEDs show the effect of energy saving at cool white.

As LED consumes less energy for whole lifetime, LED has been expected to spread rapidly. In the other hands, manufactures of fluorescent lamps have continued to improve luminance and lifetime of them and reduce mercury usage. Together with OLED which will enter the market in near years, both LED and fluorescent lamps are expected to reduce effects on the environment.

SUMMARY

Life cycle inventory analysis on energy consumption of light sources was carried out in the Illuminating Engineering Institute of Japan. The results were compared for the cases that the functional unit of the light source was assumed to be total luminous flux or the floor illuminance right under the source, and that the color of the light source was warm white and cool white.

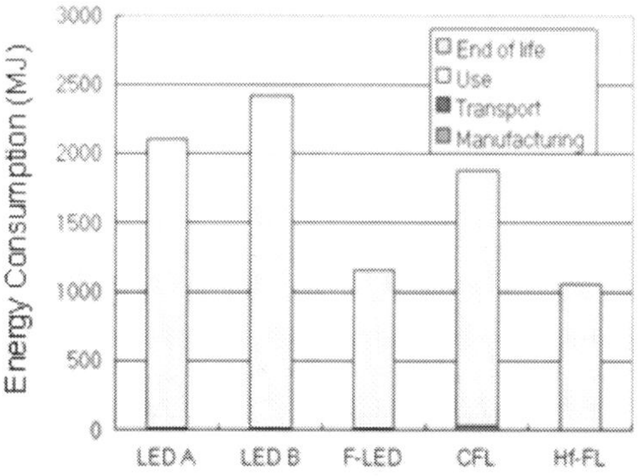

Fig. 1 Life cycle energy consumption of light sources for warm white under the functional unit of total luminous flux 800 lm for duration of 40,000hr.

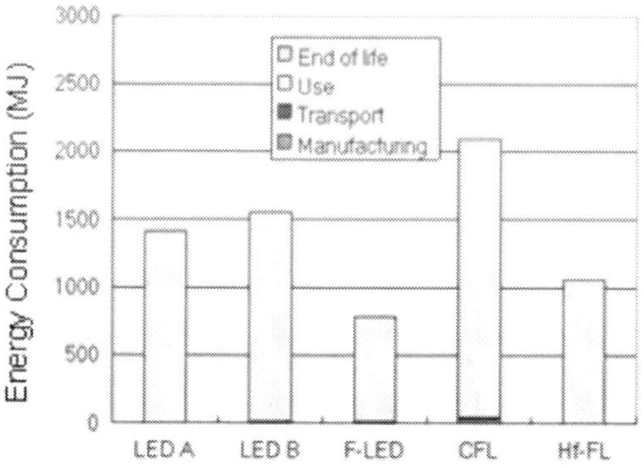

Fig. 2 Life cycle energy consumption of light sources for cool white under the functional unit of total luminous flux 800 lm for duration of 40,000hr.

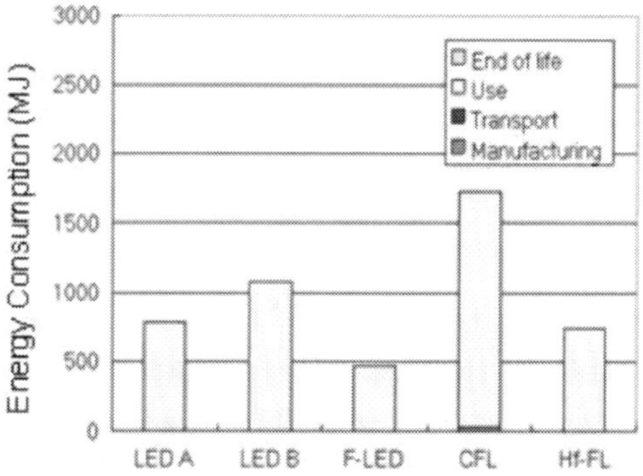

Fig. 3 Life cycle energy consumption of light sources for warm white under the functional unit of floor illuminance 100 lx at 1 m right under the source for duration of 40,000 hr.

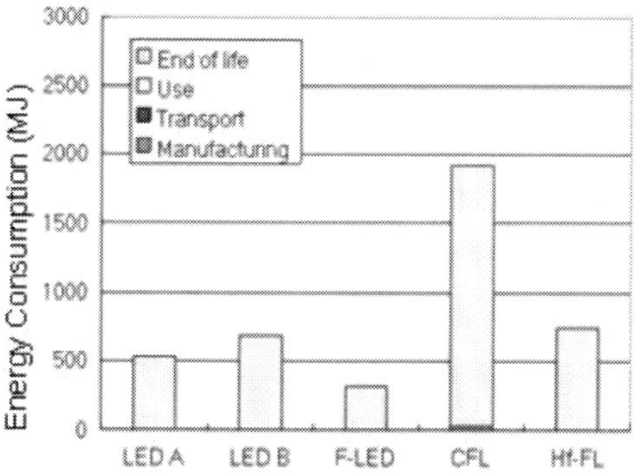

Fig. 4 Life cycle energy consumption of light sources for cool white under the functional unit of floor illuminance 100 lx at 1 m right under the source for duration of 40,000 hr.

ACKNOWLEDGEMENTS

This research was carried out in the Illuminating Engineering Institute of Japan by the special research committee for "preliminary assessment of standard to evaluate suitability of lighting system for sustainable development"and was financially supported by New Energy and Industrial Technology Development Organization (NEDO) as an Eco-innovation Promotion Project "Formulation of Sustainability Assessment Criteria of Lighting Systems". The authors gratefully acknowledge the invaluable contributions of all the committee members.

Method to Stabilize the Chromaticity Coordinate of White LED Composed of RGB-LEDs

Katsuhide Misono
Department of Electrical and Electronic Engineering, Tsuruoka National College of Technology, 104 Aza Sawada, Inooka, Tsuruoka-shi, Yamagata-ken, 997-8511 Japan
Contact: katsuhide_misono@yahoo.co.jp

ABSTRACT

We have developed a practical method to stabilize the chromaticity coordinate of a white LED composed of RGB-LEDs. Each of the peak wavelengths, FWHMs, and relative luminous outputs of RGB-LED populations were assumed to be normal distributions. RGB-LEDs were properly binned and mixed to produce an RGB-White LED. The results by Monte-Carlo simulation revealed that the variation of duv of white LEDs thus produced was less than 0.003, and all the RGB-LEDs in each population can be used. Even when RGB-LED populations are not normal distributions, our method, i.e. proper binning and mixing of RGB-LEDs taken from RGB-LED populations, is expected to be useful.

INTRODUCTION

White LEDs composed of Blue-LED+YAG phosphor or UV-LED+RGB phosphors are in practical use these days. Although these white LEDs have high efficiency or high color rendition, the intrinsic disadvantage is the energy loss caused by Stokes shift. Since one of the advantages of LED is the ease of dimming, we proposed the potential capability of a white LED composed of RGB-LEDs with high color rendering index as well as high luminous efficacy [1]. On the other hand, it was difficult to make an RGB-White LED with stability of chromaticity coordinate because the variation of chromaticity coordinate and luminous output of individual RGB-LEDs are generally large. We have developed a practical method to stabilize the chromaticity coordinate by proper binning and mixing of RGB-LEDs in populations. This method uses all RGB-LEDs in each population, and thus, leaves no useless LEDs.

SIMULATION PROCEDURES

We took into account the peak wavelengths, FWHMs, and relative luminous intensity as the optical characteristics of RGB-LEDs, and assumed that all these optical characteristics are described by normal distributions. The mixing ratio of RGB-LEDs was determined so that the color temperature and uv-deviation of white LED were Tc=6600(K) and duv=0.000, respectively. Once the mixing ratio of RGB-LEDs was found, this ratio was fixed. Then, the wavelengths, FWHMs, and luminous output were sampled by Monte-Carlo method, and xy-chromaticity coordinate of RGB-White LED was calculated. This process was repeated 1000 times, and the average and variation of chromaticity coordinate of RGB-White LED were calculated. We considered the following three cases.

Simulation 1 (baseline): RGB-White LED was produced by mixing RGB-LEDs selected randomly from each population.

Simulation 2: RGB-White LED was produced by mixing RGB-LEDs selected from limited range of each population, i.e. wavelength: median+/-2(nm), FWHM: median+/-0.5(nm), relative luminous output: median+/-5(%).

Simulation 3: All the optical characteristics of each RGB-LED were divided into 6 bins with the width of the standard deviation (σ), and then, RGB-LEDs were sampled from the individual bins one by one, and mixed to produce a white LED.

Table 1 shows the parameters of RGB-LEDs used for Simulation 1-3.

RESULTS

The calculated variations of the chromaticity coordinates for Simulation 1-3 are shown in Figure 1-3, respectively. In these figures, the ellipses colored red, green, and blue show duv=0.003, 0.005, and 0.010, respectively.

DISCUSSION

From Figure 1, it is seen that the variation of duv is over 0.010 in the case of Simulation 1, which implies that the color difference of RGB-White LED is easily distinguishable. On the other hand, in the case of Simulation 2, it is seen that duv<0.003 from Figure 2, which implies that the color difference is difficult to distinguish. The disadvantage in this case is that only a small portion (about 1/3) of the populations of RGB-LEDs is useful, and the rest of the populations are useless. When we look into the case of Simulation 3, it should be noted that duv<0.003 from Figure 3, and all RGB-LEDs in each population are useful. Therefore, our method of producing RGB-White LED by proper binning and mixing of RGB-LEDs can reduce the variation of the chromaticity coordinate to the level where the color differences of RGB-White LEDs are not distinguishable. Based on the principle of this method, we suggest that it is better not to mix a small number of high luminous output RGB-LEDs but to mix a large number of low/medium luminous output RGB-LEDs to effectively reduce the variation of the chromaticity coordinate of RGB-White LED with specified luminous output.

CONCLUSION

We performed Monte-Carlo simulations and discussed the stability of the chromaticity coordinate of white LED composed of RGB-LEDs properly binned and mixed. The variation of duv of RGB-White LED thus produced is less than 0.003. Another advantage of this method is that all RGB-LEDs in populations are useful. Even when the populations are not described by normal distributions, the method we proposed here is expected to be useful.

REFERENCES

[1] K. Misono: Optical Characteristics of White LED Composed of Ideal RGB-LEDs, paper presented in this symposium

Table 1 Parameters of Populations of RGB-LEDs used for Simulation 1-3

Simulation 1(100% of populations is used)

	Blue	Green	Red
Peak+/-σ (nm)	470+/-5	550+/-5	620+/-5
FWHM+/-σ (nm)	40+/-1	40+/-1	40+/-1
%Luminance+/-σ	1+/-0.3	1+/-0.3	1+/-0.3

Simulation 2(only 33% of populations is used)

	Blue	Green	Red
Peak(nm)	468-472	548-552	618-622
FWHM(nm)	39.5-40.5	39.5-40.5	39.5-40.5
%Luminance	0.95-1.05	0.95-1.05	0.95-1.05

Simulation 3(100% of populations is used)

		Blue	Green	Red
Peak(nm)	Bin1	-460	-540	-610
	Bin2	460-465	540-545	610-615
	Bin3	465-470	545-550	615-620
	Bin4	470-475	550-555	620-625
	Bin5	475-480	555-560	625-630
	Bin6	480-	560-	630-
FWHM(nm)	Bin1	-38	-38	-38
	Bin2	38-39	38-39	38-39
	Bin3	39-40	39-40	39-40
	Bin4	40-41	40-41	40-41
	Bin5	41-42	41-42	41-42
	Bin6	42-	42-	42-
%Luminance	Bin1	-0.80	-0.80	-0.80
	Bin2	0.80-0.90	0.80-0.90	0.80-0.90
	Bin3	0.90-1.00	0.90-1.00	0.90-1.00
	Bin4	1.00-1.10	1.00-1.10	1.00-1.10
	Bin5	1.10-1.20	1.10-1.20	1.10-1.20
	Bin6	1.20-	1.20-	1.20-

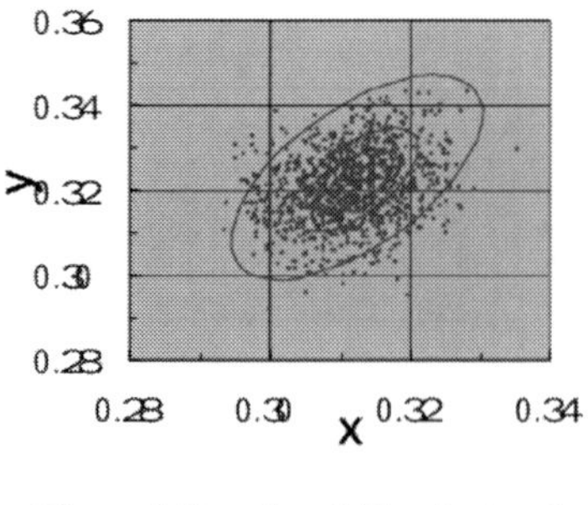

Figure1 Result of Simulation 1

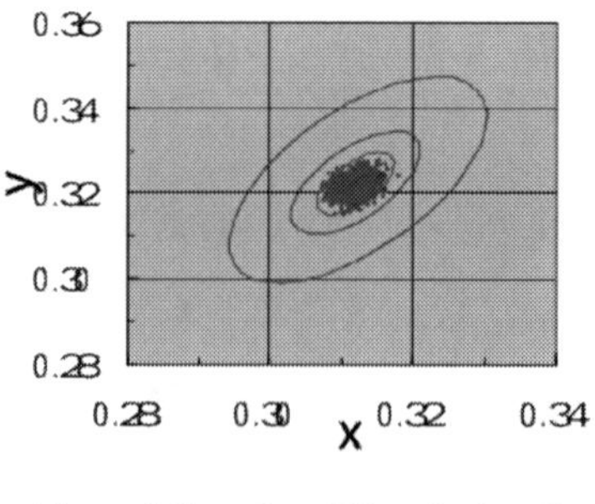

Figure2 Result of Simulation 2

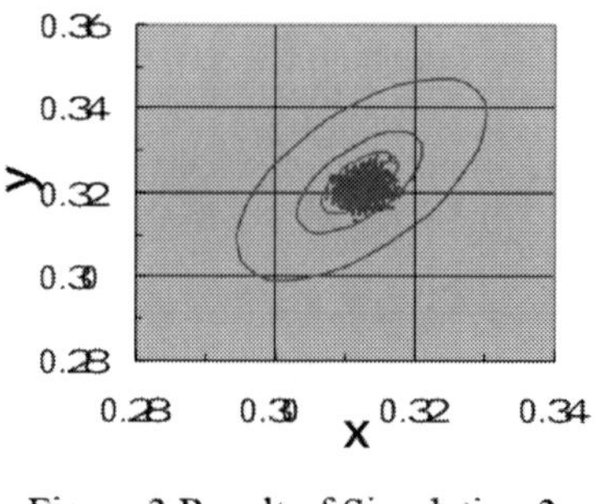

Figure 3 Result of Simulation 3

Emission characteristics of a capacitive discharge based on a mixture of water vapor with helium

A.K. Shuaibov, A.N. Malinin

Uzhgorod National University, 88000 Uzhgorod, Ukraine
E-mail: shuaibov@univ.uzhgorod.ua, mal@univ.uzhgorod.ua

ABSTRACT

The results of studies of plasma radiation of a repetitively pulsed capacitive discharge in water vapor with helium are presented. Optimization of the radical OH radiance bands, depending on the partial pressures of mixture components is made. The efficiency of excitation of radical OH* ($A^2\Pi$) in the discharge plasma for a mixture H_2O - He- = (5.3 - 94.7) % at a total pressure 2.85 kPa. The rate constant of this process varies in the range $1\cdot 10^{-16}$-$8\cdot 10^{-14}$ m^3/s as the parameter E/N (E - field strength, N- the mixture components concentration) in the range (1-300) Td. The energy losses of electrons on the excitation of the radical OH spectral bands ($A\ ^2\Pi \rightarrow X\ ^2\Sigma$) are optimal for the parameter E/N = 60-70 Td and their value is 25%.

GENERAL

Gas discharge plasma on the basis of water vapor is ecologically clean and effective working medium for the UV-VUV sources of spontaneous radiation on the hydroxyl OH vibronic bands systems (X-A; X-C) [1-3]. The most simple and sufficiently powerful ultraviolet lamps based on the mixtures of inert gases with water vapor are designed with the excitation of a longitudinal glow discharge. The partial pressure of helium or argon in such emitters is in the range 0.1 – 1.5 kPa, water vapor 50-150 Pa. They are mainly radiate in the spectral range 304-320 nm OH (X-A). UV - emission in this spectral range (especially at λ = 308 nm) is used in photo medicine. The hydroxyl OH (X-A) radiators can supplant the lamp with a more aggressive and costly gas medium on molecules XeCl (X-B) ($\Delta\lambda$ = 307-310 nm) and xenon chloride lasers.

To increase the resource of the low pressure UV – radiator on the water molecules dissociation products is a promising transition to the working media excitation by the capacitive discharge with electrodes separated from the working medium.

Sub microsecond repetitively pulsed capacitive discharge was ignited in a quartz tube with the length 40 cm and an inner diameter 0.5 cm. The distance between the outer cylindrical metal electrodes in a capacitive discharge was 16 cm. Amplitude of different polarity voltage pulses on the electrodes was 25-40 kV. The repetition rate of the excitation pulse was varied in the range 50- 500 Hz.

Registration of the emission spectra was carried out from the central part of the discharge tube. The spectral resolution of vacuum spectrometer was 0.7 nm. Figure 1 shows the plasma emission spectra of a capacitive discharge in the mixture H_2O - He - recorded in the spectral range 280-310 nm. Band with maximum at λ = 283.6 nm is well identified with the hydroxyl OH band (X-A, 0-1), the band with λ = 297.6 nm is identified with the OH band (X-A, 2-3). The optimization of the bands λ = 283.6 nm OH (X-A, 0-1) and λ = 310 nm OH (for the band 297.6 nm they are similar) radiance, depending on the partial pressure of water vapor and helium is carried out. The dependences of radiance of these bands in relative units on partial pressure of helium at a partial pressure of water vapor 133 Pa, as well as the partial pressure of water vapor at a partial pressure of helium 2.7 kPa are presented in Table 1.

To increase the efficiency of the hydroxyl OH* bands radiation the rate constants for excitation of the electron state (A^2P) were calculated, depending on the value of the parameter E/N. In the discharge in the mixture of water vapor with helium the excited OH* (A^2P) is formed mainly in the process of dissociative excitation in the collision of electrons with water molecules [1]:

$$e + H_20 \rightarrow OH^*(A^2P) + H(n=1) + e\ .$$

The rate constant (k) of this process was calculated by averaging the cross section of this reaction on the electron energy distribution function (EEDF). For the EEDF calculation in a mixture of water vapor with helium the program "Bolsing +" was used [2]. Elementary processes of electron collisions with water molecules and helium atoms, used in the calculations were taken from papers [1 - 4]. Fig. 2 shows the dependence of the rate constant of dissociative excitation for hydroxyl 0H* state (A^2P), depending on the size of the reduced field E/N.

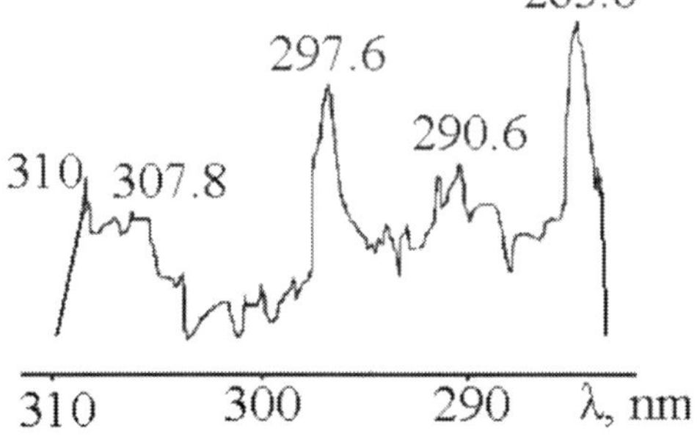

Fig. 1. The emission spectra of a capacitive discharge in the mixture p(He) - p(H_2O) = 7.9 kPa-133,3 Pa at f = 100 Hz.

Table 1

p(He), kPa	$J_{o.u.}$, λ 283,6 nm	$J_{o.u.}$, λ 310 nm
1,35	0.20	0.15
5.30	0.20	0.20
8.00	0.30	0.25
p(H_2O), Pa	$J_{o.u.}$, λ 283,6 nm	$J_{o.u.}$, λ 310 nm
15	0.25	0.15
70	0.30	0.10
150	0.40	0.20
200	0.25	0.15

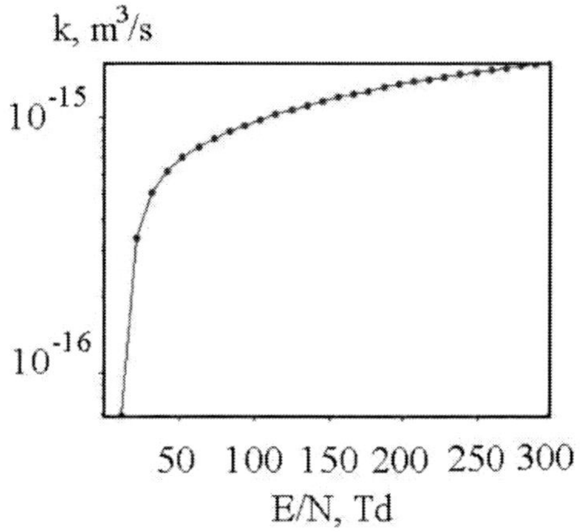

Fig.2. The dependence of the rate constant of dissociative excitation of radical OH in the collision of electrons with water molecules on the parameter E/N in the discharge based on a mixture of H_2O - He = (5.3 - 94.7) %. The total pressure of the mixture is 2.85 kPa.*

The rate constant is increasing rapidly with the parameter E/N increase to 40 Td, with further increase of E/N the rate of her growth is markedly reduced. It is expected that when the parameter E/N > 40Td, increase in the amplitude of voltage pulses will be less effective. According to the calculation of the energy loss of electrons on the excitation of the spectral bands of hydroxyl OH ($A\ ^2\Pi \rightarrow X\ ^2\Sigma$), they are optimal for the parameter E/N = 60-70 Td and they are 25%. In our experimental conditions, the parameter E/N varied in the range (227-363) Td, which is much higher than the optimum E/N for the excitation of the hydroxyl spectral bands. They are optimal for excitation of energy levels of hydrogen and oxygen atoms, which emit in the VUV - spectral range (H* 121.6 nm and O* 130.4 nm). The studies of plasma radiation in the mixture of water vapor and helium in this spectral range are necessary to confirm this result, which will be the subject of our subsequent studies.

[1]. Y. Itikava, N. Mason , J. Phys. Chem.Ref.Data, 34 (2005)1.
[2]. http://www.siglo-kinema.com/database/index.htm
[3]. J.D.Gorfinkiel, L.A.Morgan, J.J. Tennyson, J.Phys.B.At.Mol.Opt.Phys., 35(2002)543.
[4]. M.A. Khakoo, C.Winstead , V. McKoy, Physical Review , A 79(2009)052711-1.

Session Six

Performance of high-luminous efficacy white LEDs

K.Bando[1]

[1] *Nichia Corporation, Engineering Development Div., 491 Oka Anan-city Tokushima 774-8601, Japan*
Contact: kanji.bando@nichia.co.jp

ABSTRACT

The performance of InGaN/YAG white LEDs for lighting applications is reviewed. Advanced small-chip-technology has achieved high-luminous efficacy white LED of about 200lm/W by improving internal efficiency, extraction efficiency and electrical resistance. In addition, high-luminous flux device of 10,000lm by flip-chip array has been also developed by large-chip-technology.

INTRODUCTION

By the rapid improvement of LED technology, the luminous efficacy of white LEDs has been much higher compared to conventional incandescent and halogen lamps. Amid growing public interest in environmental issues, white LEDs are developing new applications in many fields such as illuminations, displays (LCD backlights), automobiles. More recently, markets of solid light sources such as LED-bulbs or LED-down lights are expanding sharply in Japan. According to calculations of METI (Japanese Ministry of Economy, Trade and Industry), the energy consumption could be reduced in half by changing all of existing incandescent bulbs and fluorescent lamps to high efficient lighting devices of 150lm/W. Energy-saving of illuminations will play an important role in carbon reduction and global warming prevention. With the spread of new lighting devices and fixtures by white LEDs, illumination industry is facing the historical turning point from conventional lighting to solid-state lighting.

HIGH EFFICACY LEDS BY SMALL-CHIP-TECHNOLOGY

The principle of white LED by InGaN/YAG is wavelength down-conversion by the YAG phosphor (Yttrium Aluminum Garnet, $(Y_a,Gd_{1-a})_3(Al_b,Ga_{1-b})_5O_{12}:Ce^{3+}$). The blue emission of peak wavelength 440nm-470nm by InGaN chip and the yellow emission of 540nm-575nm by YAG phosphor are mixed in the phosphor layer. The spectrum has two peaks of sharp blue and broad yellow. It is possible to make various white lights in the 2,850K-10,000K with color rendering of 60-85. In the first report of this structure, the performance was just 5lm/W, but by innovative technical

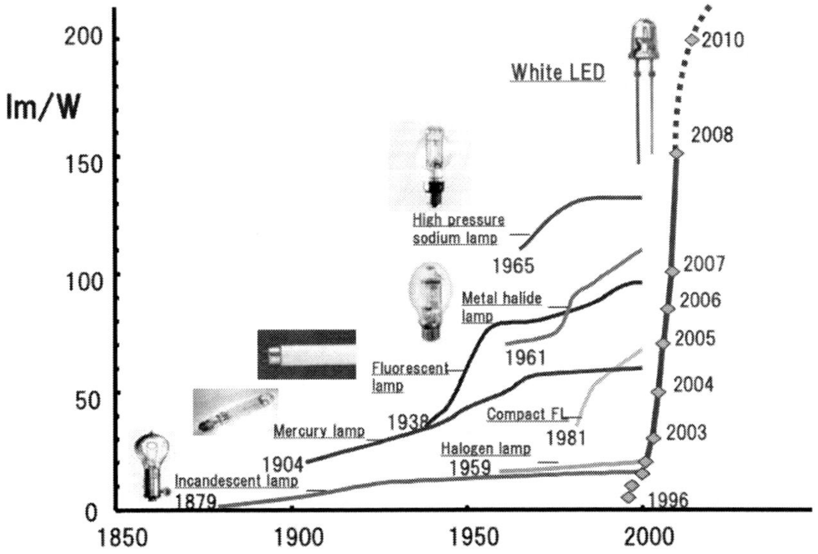

Figure 1. Historical progress of white LEDs and conventional white sources.

improvement of chips, phosphor and package, the high efficient LEDs of over 100lm/W are made for any applications in these days [1], [2]. The historical progress of InGaN/YAG white LEDs used small chips is summarized in Figure 1.

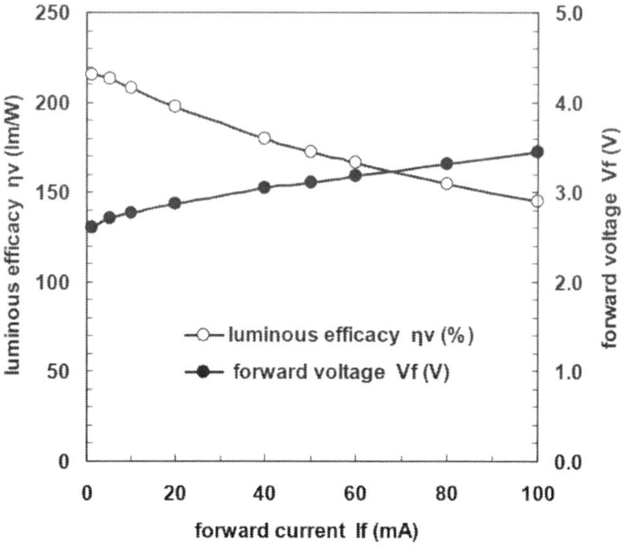

Figure 2. If- ηv and If-Vf characteristics of high efficacy white LED by single small chip (197lm/W at 20mA).

As shown in Figure 2, the performance of latest InGaN/YAG white LEDs has reached about 200lm/W (197lm/W, CCT 5,500K at 20mA). This efficacy surpasses any conventional white light sources which are commercially available. The performance of lighting devices is usually discussed in terms of ηv [lmW] (luminous efficacy of a source) and ηe [%] (radiant efficiency). These parameters are determined by the efficiency of blue LED chip, phosphor layer and package. The relationship of theses efficiencies can be shown via the following equations (1) (2) (3).

$$\eta v\ [\mathrm{lm/W}] = K \cdot \eta wpe = K \cdot \eta pkg \cdot (x + (1-x) \cdot \eta phos) \cdot \eta chip \quad (1)$$

K	[lm/W]	: luminous efficacy of radiation
η v	[lm/W]	: luminous efficacy of white LED
η wpe	[%]	: Poutput / Pinput of white LED
η chip	[%]	: radiant efficiency of blue chip
η phos	[%]	: efficiency of phosphor layer
η pkg	[%]	: external efficiency of package
Poutput	[W]	: light output (radiant flux)
Pinput	[W]	: electrical power input
x		: blue component of white LED spectrum

$$K\ [\mathrm{lm/W}] = Km \int Pwhite(\lambda) \cdot V(\lambda) d\lambda \,/ \int Pwhite(\lambda) d\lambda \quad (2)$$

K	[lm/W]	: luminous efficacy of radiation
Km	[lm/W]	: maximum luminous efficacy = 683lm/W
Pwhite(λ)		: spectral distribution of white LED
V(λ)		: spectral luminous efficiency

$$\eta wpe\ [\%] = \eta e \cdot Eg / Vf = \eta i \cdot \eta ext \cdot Eg / Vf \quad (3)$$

η e	[%]	: external quantum efficiency (EQE)
η i	[%]	: internal quantum efficiency (IQE)
η ext	[%]	: light extraction efficiency of chip
Eg	[eV]	: band-gap energy
Vf	[V]	: forward voltage

Technical challenges for theoretical limit of white LED are still continued aggressively worldwide and key improvement parameters summarize as follows.

1. Chip efficiency (ηwpe)
   High internal quantum efficiency, high extraction efficiency, epi crystal quality, lower Vf, chip size & design, current spreading, heat dissipation, low optical absorption.
2. Phosphor layer (light-converter) efficiency (ηphos)
   High light conversion efficiency, low temperature quenching, material quality & composition, particle size, optimized color & spectrum, coating technology.
3. Package efficiency (ηpkg)
   High light extraction efficiency, better light utilization for purpose, low thermal resistance, low optical losses, high reflective package body.

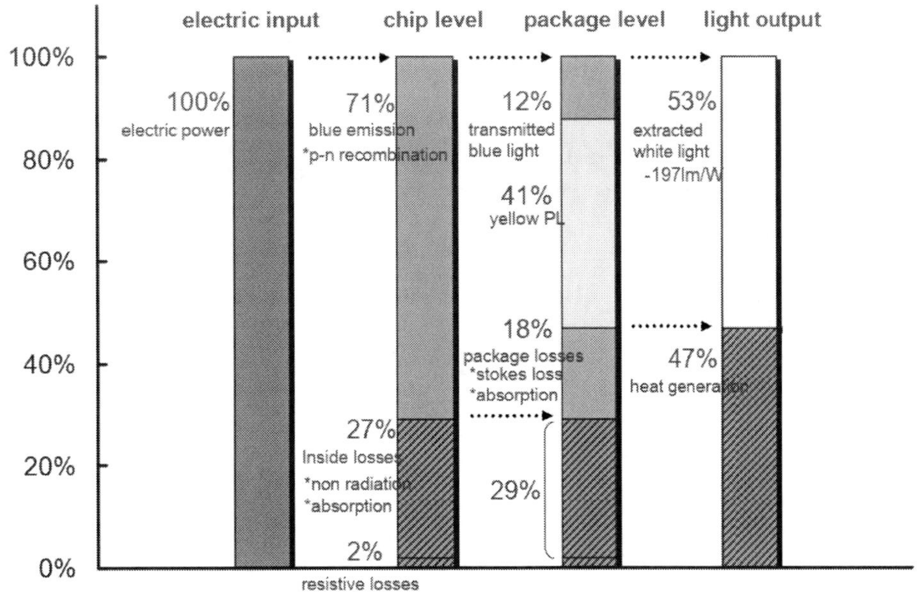

Figure 3. Energy conversion of 197lm/W high efficacy InGaN/YAG white LEDs at 20mA driving.

Figure 3 shows the detail of energy conversion of white LEDs from electrical input to light output, when taking 197lm/W-white LED as an example. The radiant efficiency (ηchip) of InGaN blue chips reaches 71% high, but final light output reduces 53% suffering from the inevitable Stokes loss and optical loss as scattering and absorbance in the package. The rest of 47% is lost to heat, which is dissipated through the package to the ambient. The performance of the latest blue chip used in this case is EQE (ηe) 73%, radiant flux 41.1mW, Vf 2.87V, λ p 442nm at 20mA. IQE (ηi) is assumed to reach to approximately 80% at least. In general, the efficiency of small chips is higher compared to large chips, because the light path from the active layer becomes shorter with fewer reflections. And the cost is decreasing by better semiconductor yield. But regarding the operating current, it needs to require careful attention, because the quantum efficiency of InGaN LED drops with increasing current density. It is important to decide a suitable operating current range in accordance with chip design, although the range is much expanding than before.

When making high flux power LEDs with small chips, multichip system is adopted from the restriction of the allowable current for one chip. By the spreading of the current to each chip, it can drive by high current simply in proportion to the number of the arrayed chips. In terms of thermal design, multichip array is easy to spread the heat to the large area. In addition, the latest multichip power LEDs reduce the heat generation by specially designed low-Vf chips. High efficient power LEDs such as 0.5W type (2 chips, 44lm at 150mA), 1W type (6 chips, 120lm at 300mA) or 3.6W type (9 chips, 360lm at 350mA) are commercially available. Figure 4. and Figure 5. show 3.6W type by 9 chips (3x3 circuit) and its characteristics. These devices are used as light engines of LED bulbs or down lights.

Figure 4. 3.6W type white LED by multichip system (9 chips, 360lm, 101lm/W at 350mA, 10.2V).

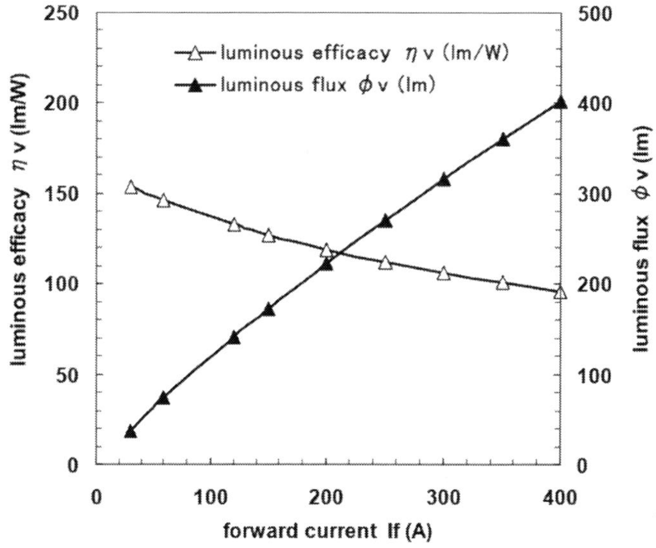

Figure 5. If- ηv and If- ϕv characteristics of 3.6W type white LED (9 chips)

HIGH FLUX LEDS BY LARGE-CHIP-TECHNOLOGY

When making small-size light source with high flux and high luminance, the use of large chips are more effective than small ones. Downsizing lighting source makes luminaries lightweight, stylish or compact designs.

Figure 6 shows a prototype of flip-chip array by 56 blue chips of 1mm x 1mm and YAG phosphor. The performance of it is 10,000lm, 105lm/W, 4,000K at 3.6A, as shown in Figure 7. The blue large chips are designed for high drive current considering current spreading and thermal dissipation. At 350mA, the radiant flux of the large chips is 600mW (Vf 3.25V) and EQE (ηe) is 62% (λ p 450nm), as shown in Figure 8. The source size of this array is 11.9mm x 10.6mm area which is relatively small to super-high flux of 10,000lm. The results of this experiment prove that LED can replace not only bulbs but also high flux HID lamps in future.

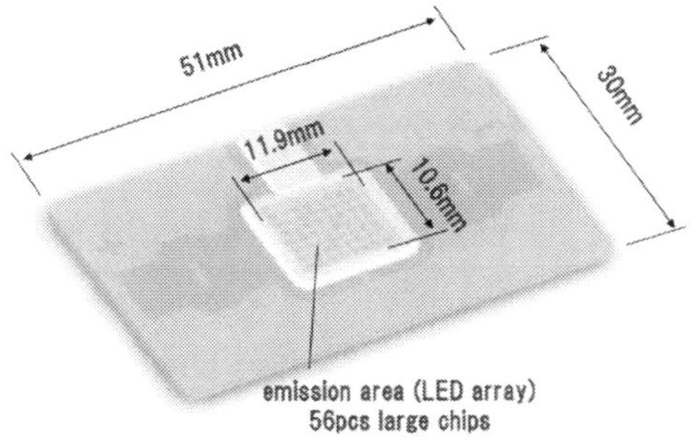

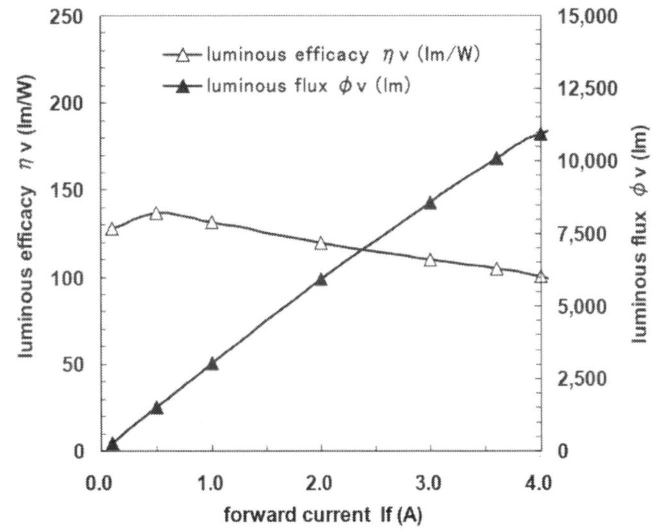

Figure 6. A prototype of super-high flux white LED by large chip array (56 chips).

Figure 7. If- ηv and If- ϕv characteristics of white LED by large chip array (10,000lm at 3.6A).

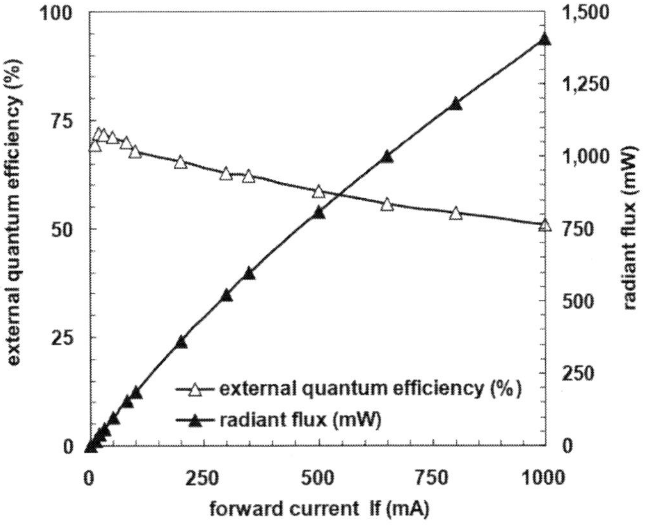

Figure 8. Characteristics of blue large chip for LED array (EQE 62%, radiant flux 600mW at 350mA).

CONCLUSION

Latest small-chip-technology and large-chip-technology are presented in this paper. The transition from conventional lightings to solid state lightings will be accelerated by the evolution of efficiency and the development of new LEDs suitable for each application.

REFERENCES

[1] K.Bando et al., Phosphor Res. Soc. 264th Meeting Tech. Digest, Nov.29(1996)5-14.
[2] A.Michiue et al., Proc. SPIE, 7216(2009)72161Z1-72161Z6.

Surface plasmon coupled InGaN/GaN quantum-well light-emitting diodes

Kun-Ching Shen, Chih-Feng Lu, Che-Hao Liao, Chih-Yen Chen, Zan-Yao Yu, Chieh Hsieh, Jyh-Yang Wang, Cheng-Hung Lin, Yean-Woei Kiang, and C. C. Yang
Institute of Photonics and Optoelectronics, National Taiwan University, 1, Roosevelt Road, Section 4, Taipei, Taiwan 10617 ROC
Contact: ccy@cc.ee.ntu.edu.tw

ABSTRACT
The output polarization ratio enhancement of an InGaN/GaN quantum-well light-emitting diode (LED) with surface plasmon (SP) coupling through the insertion of a SiO_2 layer between the p-GaN layer and a surface Ag nano-grating structure is demonstrated. The Ag nano-grating period for optimizing SP coupling is increased when a SiO_2 layer is added to the device that is consistent with the simulation results of surface plasmon polariton and localized surface plasmon resonances.

In this paper, the effects of an inserted SiO_2 layer (1.46 in refractive index) on the output polarization of an LED of SP coupling are demonstrated. InGaN/GaN QW LEDs with surface Ag nano-grating structures, including those with and without an intermediate SiO_2 layer, are fabricated to show the enlarged polarization ratio in an LED with the intermediate SiO_2 layer. To confirm the SP coupling effect, LEDs of the similar epitaxial structure but with a thicker p-GaN layer are also fabricated to show the results of weak SP coupling for comparison. The LED epitaxial structures include a 2-μm n-GaN layer, five InGaN/GaN QWs (emission at ~500 nm), a 15-nm AlGaN current-blocking layer, and a p-GaN layer on the top. Two epitaxial structures are prepared with the p-GaN layers at 60 and 120 nm by metalorganic chemical vapor deposition. In fabricating an LED of SP coupling, on the mesa of 300 μm x 300 μm, a central square of 200 μm x 200 μm is used for depositing a SiO_2 layer of 30 nm in thickness, with the outside area covered by Ni/Au of 5nm/5nm for current spreading. Within the SiO_2 area, a central square of 180 μm x 180 μm is used for fabricating Ag nano-gratings of 30 nm in metal height and of various periods, ranging from 250 through 600 nm with the step size of 50 nm, on the SiO_2 layer. The duty cycle of the fabricated Ag nano-gratings is fixed at 50 %. The schematic drawing of such a device is shown in the insert of Fig. 1. The Ag grating structures are fabricated with electron-beam lithography and liftoff process. The LEDs covered by the SiO_2 layer and Ag nano-gratings are assigned as samples A(x) (A'(x)) with x representing the grating period number (in nm) when the epitaxial structure of thin (thick) p-GaN layer is used. Those covered by Ag nano-gratings only (no SiO_2 layer) are assigned as samples B(x) and B'(x). Then, the LEDs covered only by SiO_2 layers (no Ag nano-gratings) are assigned as samples C and C'. Finally, those LEDs without SiO_2 or Ag nano-grating are assigned as samples D and D'.

Figure 1 shows the angle-integrated emission power variations with injection current of various samples, including the bottom emission ('-b' in label) and the summation ('-s' in label) of bottom and top emissions. Samples A(550) and B(500) are chosen for comparison in Fig. 1 because the grating period of 550 (500) nm leads to the maximum bottom or summation emission power among sample A (B) series, as to be demonstrated in Fig. 2. In Fig. 1, by adding the SiO_2 layer only (sample C), both bottom and top emissions (and hence the summation) are slightly enhanced due to light extraction enhancement (refractive index matching and slightly rougher SiO_2 surface), when compared with the reference sample (D). Without the SiO_2 layer, the fabrication of Ag nano-gratings on the p-GaN layer (sample B series) results in a significant increase of LED emission, either from top or bottom, due to the SP coupling effect. At the injection current of 60 mA, the emission of sample B(500) increases by 38 and 18 % from the bottom and top, respectively, when compared with sample D. Then, by fabricating Ag nano-gratings on the deposited SiO_2 layer (sample A series), the bottom emission of sample A(550) is further enhanced when compared with sample B(500) (increase by 46 % when compared with sample D). However, its top emission is reduced leading to a lower emission summation, as shown in the comparison between curves B(500)-s and A(550)-s. Hence, the emission summation of sample A(550) is increased only by 16 % when compared with sample D. Such a difference between top and bottom emission behaviors can be attributed to the change of SP radiation pattern when the Ag nano-grating period is increased.

Figure 2 shows bottom emission ratios as functions of Ag nano-grating period in samples A, B, and C series. The ratios are obtained by dividing the bottom-emitted powers of samples A, B, and C over that of sample D. It is noted that the emitted power of sample C is independent of grating period. The curve of sample C means to show the possible device non-uniformity over the epitaxial wafer. For a reasonable comparison among the devices at the same abscissa coordinate in Fig. 2, these LED samples are fabricated nearby at the same wafer radius for almost the same epitaxial structure. It is noted that the variation trends of curves A and B do not have a strong correlation with that of C, implying that the enhancement ratios of samples A and B series are not significantly affected by wafer non-uniformity. Curves A and B have the peaks around 550 and 500 nm in grating period, respectively. Figures 3(a) and 3(b) show the spectra of the x and y polarization components, including the top and bottom emissions, at 50 mA in injection current of samples A(550) and B(500), respectively. Their relative intensities are used for evaluating the polarization ratios (the stronger y-component integrated intensity over the weaker x-component) to give 1.82 and 1.56 for the top and bottom emissions, respectively, of sample A(550), and 1.31 and 1.33 for the top and bottom emissions, respectively, of sample B(500). Those results indicate that with the SiO_2 layer, the SP coupling leads to a stronger output polarization effect. Figure 4 shows the comparisons of integrated emission powers, including the bottom (-b) and top (-t) emissions, between samples A'(550), B'(500), and D', which were fabricated by using the epitaxial structure of thick p-GaN. In this epitaxial structure, the distance between the top QW and the p-GaN surface is about 140 nm, which is too large for generating significant SP coupling. In this situation, the effects of the Ag nano-gratings and the SiO_2 layer become unclear. For both top and bottom emissions, sample A'(550) shows slightly stronger outputs than sample B'(500). The weak SP coupling effect can be confirmed by the polarization results shown in the insert of Fig. 4, in which the spectra of the x and y polarization components, including the top and bottom emissions, at 50 mA in injection current of samples A'(550) and B'(500) are shown. Here, the polarization ratios of the top and bottom emissions of sample A'(550) (B'(500)) are only 1.07 and 1.04 (1.06 and 1.05), respectively. Such small polarization ratios indicate that the observed polarization ratios in Figs. 3(a) and 3(b) are due to the near-field SP coupling, instead of the far-field diffraction effect.

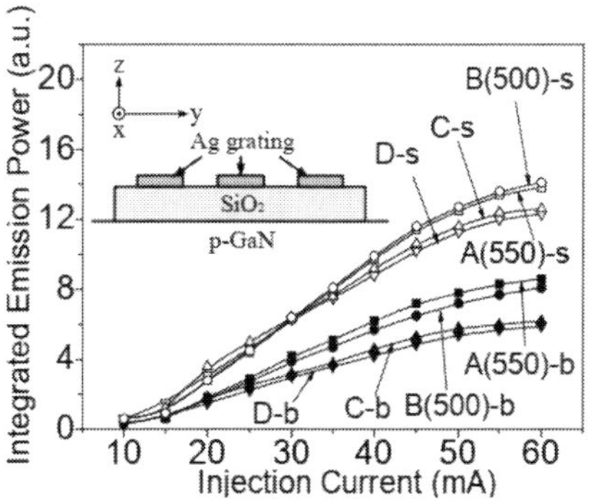

Figure 1

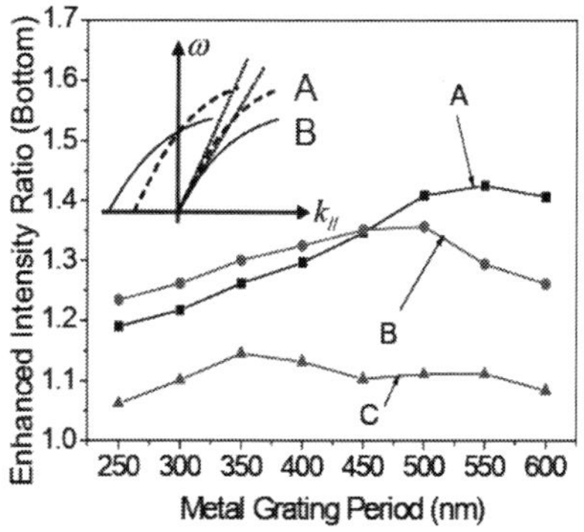

Figure 2

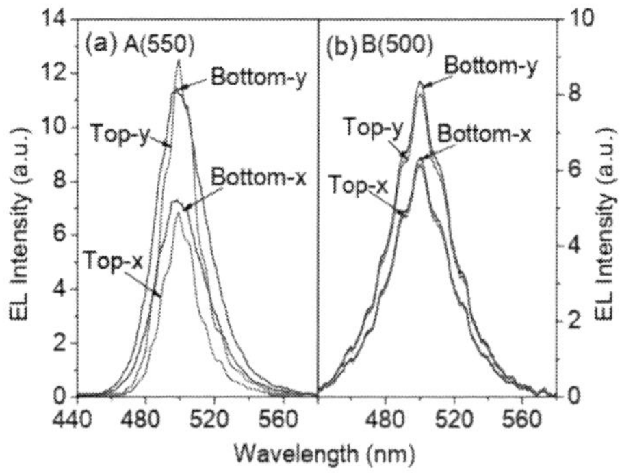

Figure 3

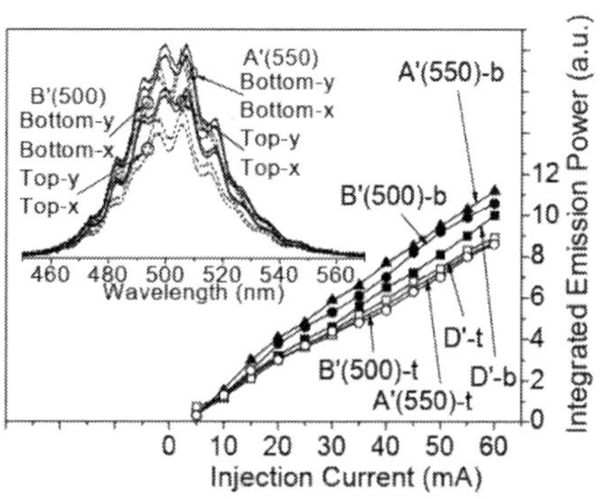

Figure 4

New Phosphor Developments

R. Withnall, G.R. Fern, T.G. Ireland, A.L. Lipman, R. Stone and J. Silver
Centre for Phosphors and Display Materials, Wolfson Centre for Materials Processing, Brunel University, Kingston Lane, Uxbridge, Middlesex UB8 3PH, UK

Contact: robert.withnall@brunel.ac.uk

ABSTRACT

An overview is presented herein on the development work of phosphors for phosphor conversion LEDs (pcLEDs) that has recently been carried out in laboratories worldwide. As blue emitting LEDs are generally preferred to UV LEDs for white light applications at the present time due to efficiency considerations, our focus is on LED phosphors that can be excited in the blue and emit at longer wavelengths in the visible. Broad band yellow and yellow-orange emitting phosphors are covered for single phosphor converted LEDs (1pcLEDs), as well as recent developments on red-emitting phosphors which can be used to enhance the quality of the white light in two phosphor converted LEDs (2pcLEDs). The trade-off between luminous efficacy and colour rendering index is investigated for 1pcLEDs, and the strengths and weaknesses of the various LED phosphors are surveyed based on criteria such as luminous efficiency, excitation wavelength range, emission wavelength range, temperature dependence of the emission, and chemical stability.

INTRODUCTION

Whereas lamp phosphors for mercury discharge light sources have been studied for well over three quarters of a century [1], the research and development of phosphors for blue LEDs is a relatively new area. This is because the blue-emitting LED itself was only developed as a high brightness light source towards the end of the 20th century [2,3], and it was first combined with a yellow emitting phosphor (YAG:Ce) to give white light by Nakamura and Fasol of the Nichia Company in 1996 [4]. Moreover the advent of the blue-emitting LED required the development of new types of phosphors, having strong blue absorptions and emissions at longer visible wavelengths, as there had been no compelling reason to develop such phosphors prior to that time. Thus, intense research and development activity was triggered worldwide in the area of phosphors for LEDs when the high brightness blue LED appeared on the scene. A comprehensive review of colour converted phosphors for LEDs has been given recently [5]. Herein, we will focus on some of the most promising photoluminescent phosphors being developed, by us and others, as colour conversion materials for LEDs, and assess their strengths and weaknesses.

GENERAL CONSIDERATIONS

As will be seen below, the majority of LED phosphors are activated by either Ce^{3+} or Eu^{2+} cations due to the high oscillator strength of their orbitally allowed *d-f* electronic transitions which can give rise to efficient light emission, unlike orbitally forbidden *f-f* electronic transitions which give rise to many visible emissions observed from rare earth cations. It should be borne in mind that such phosphors containing Ce^{3+} or Eu^{2+} ions usually need to be prepared under reducing conditions so that the activator cations do not occur in their higher oxidation states (Ce^{4+} and Eu^{3+}). A less widely used activator cation that is noteworthy is Mn^{4+} which is used for some recently developed red-emitting phosphors [6]. Another important consideration is that the degree of thermal quenching of a phosphor emission should be minimized in order to avoid losing emission efficiency. Some phosphors are more prone to the thermal quenching of their luminescence than others, so this will be taken into account below. It should be mentioned that it is best not to apply phosphors directly onto the LED chip, if they exhibit significant thermal quenching of their luminescence above ca. 100°C. Instead, it is possible to use them on a screen, remote from the LED chip, so that their operational temperature is lower.

PHOSPHORS FOR WHITE PHOSPHOR CONVERSION LEDS FOR GENERAL ILLUMINATION

LED/Single Phosphor Combinations

By far the most widely used single LED phosphor is cerium activated yttrium aluminium garnet, YAG:Ce^{3+} ($Y_3Al_5O_{12}$:Ce^{3+}). It was reported in 1967 as a cathodoluminescent phosphor, though the same paper also reports its photoluminescent properties [7]. The use of YAG:Ce^{3+} for fluorescent lighting was reviewed in 1987 [8], and methods of making it have been well documented in the literature [9,10].

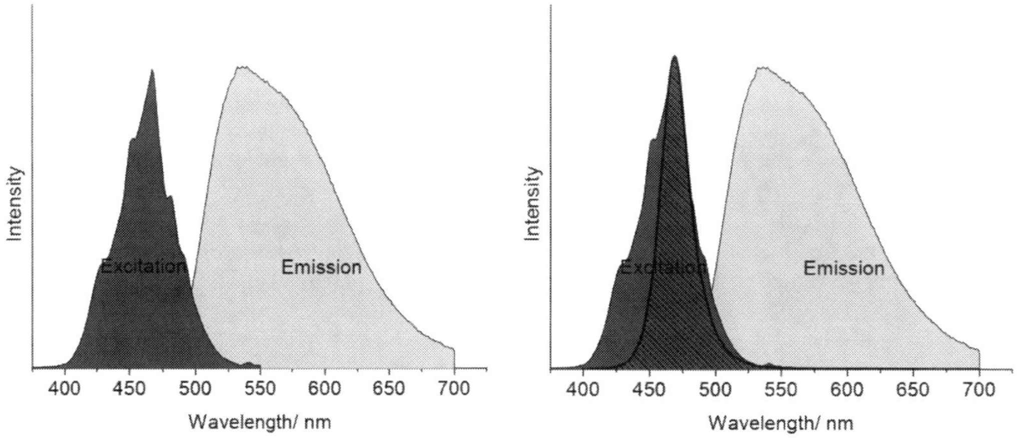

Figure 1. The excitation and emission spectra of YAG:Ce^{3+}(a) in the absence (left hand side), and (b) in the presence of the emission (shaded curve) from a blue LED (right hand side).

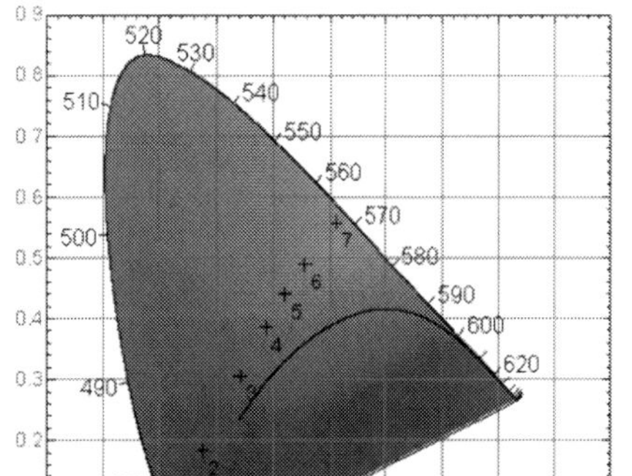

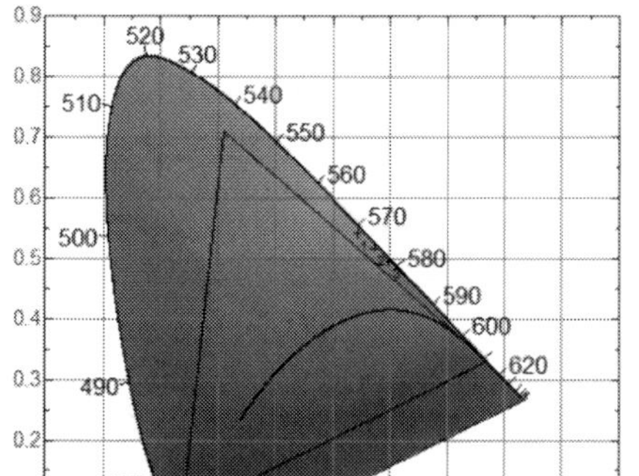

Figure 2 (left hand side). CIE chromaticity diagram showing the colour points of a blue LED (point 1) and the YAG:Ce phosphor (point 7) which are joined by a 'mixing line', the locus of colour points of combined LED/YAG:Ce phosphor emission with increasing amounts of phosphor from points 1to 7.

Figure 3 (right hand side). CIE chromaticity diagram showing the colour points of (Y_xGd_{1-x})AG:Ce phosphors with x equal to (a) 1.00 (point 1), (b) 0.75 (point 2), and (c) 0.50 (point 3). The Ce concentration was 2.45 mol% in each case.

An advantage of YAG:Ce is that it has a very broad emission band in the yellow (see Figures 1a and 1b). Thus, when the broad yellow emission mixes with the complementary blue exciting light (shown by the shaded emission curve in Figure 1b), a composite 'white' light is produced. This is a cold white light (having a high colour correlated temperature) and the colour rendering index does not exceed an Ra value of 80 due to the deficiency at the red end of the spectrum.

Figure 2 shows that a 'mixing line' can be traced out that connects the colour points of the blue LED (point 1) and the YAG:Ce phosphor (point 7); the more YAG:Ce phosphor that is applied to the blue LED chip, the closer the colour point to that of the phosphor itself. It can clearly be seen in Figure 2 that the CIE coordinates of points 1-7 on the 'mixing line' need to be shifted towards the red region of the chromaticity diagram in order to coincide with the Planckian which is the locus of points of a blackbody emitter shown by the solid black curve (in Figure 2). This is because the broadband emission from the YAG:Ce phosphor does not extend far enough into the red region of the spectrum in order for the composite emission (combined LED and phosphor emissions) to occur in the white region of the CIE chromaticity diagram. In order to achieve a better white colour point, some of the yttrium can be substituted for gadolinium in the garnet host lattice to give $(Y_xGd_{1-x})AG:Ce$ $(0 \leq x \leq 1)$ as shown in the chromaticity diagram presented in Figure 3. This has the effect of moving the $(Y_xGd_{1-x})AG:Ce$ colour point towards the red corner of the chromaticity diagram (from point 1 (x=1) to point 2 (x = 0.75) and to point 3 (x = 0.50)), and a mixing line can then be drawn between the blue LED colour point and colour points 2 or 3 that intersects the Planckian. Thus the colour point of the composite emission of the blue LED/$(Y_xGd_{1-x})AG:Ce$ $(0 \leq x \leq 1)$ phosphor combination can now come in the white region of the CIE chromaticity diagram. However, although YAG:Ce has an excellent luminous efficacy of ca. 275 lm/W when excited at 470 nm (c.f. a theoretical limit of luminous efficacy of ca. 346 lm/W), there is a loss in luminous efficacy of the $(Y_xGd_{1-x})AG:Ce$ $(0 \leq x \leq 1)$ phosphor as the percentage of gadolinium is increased in the garnet lattice, as can be seen in Figure 4.

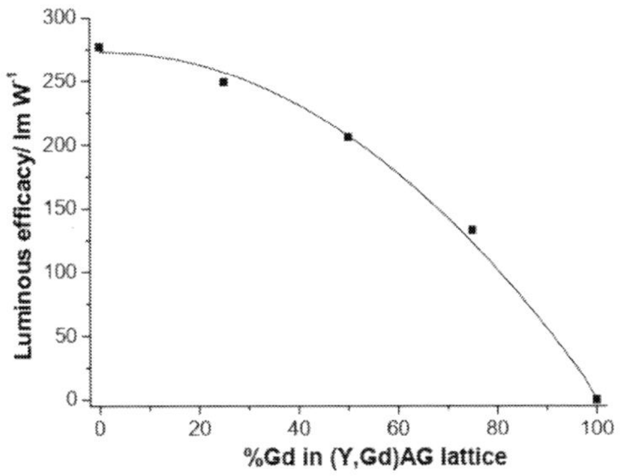

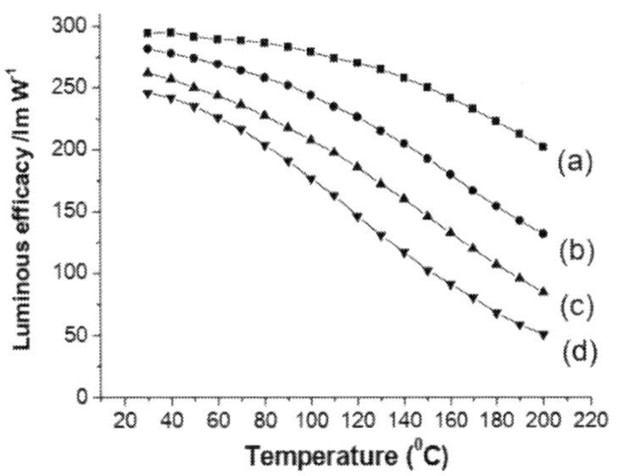

Figure 4 (left hand side). Luminous efficacies (λ_{exc} = 470 nm) for different lattice compositions of phosphors $(Y_xGd_{1-x})AG:Ce$ $(0 \leq x \leq 1)$.
Figure 5 (right hand side). Plots of luminous efficacy vs temperature for phosphors, $(Y_xGd_{1-x})AG:Ce$ $(0 \leq x \leq 1)$, having x values of (a) 1.0, (b) 0.90, (c) 0.80,and (d) 0.60.

In addition, it can be seen from Figure 5 that there is more thermal quenching of the luminescence as the concentration of Gd increases in the lattice, and this leads to a significant drop in luminous efficacy. This change in temperature dependence is related to the band structure of the phosphors, and this in turn relates to their lattice structure.

In conclusion, it has been shown that it is possible to shift the colour point of the (Y,Gd)AG:Ce phosphor towards the red by substituting Ce with Gd in the YAG lattice, and this also enables the colour point of the blue LED/phosphor composite emission to be shifted towards the white. However, this approach also results in a reduction in luminous efficacy of the phosphor. The red shift is due to the increased crystal field splitting of the 5*d* orbitals of Ce^{3+} when the larger Gd^{3+} ions replace the Y^{3+} ions on the dodecahedral sites of the garnet lattice. Conversely, a blue shift of the emission from 560 nm to 505 nm occurs for cerium activated yttrium gallium garnet, YGaG:Ce, when the larger Ga^{3+}

ions replace the Al^{3+} ions on the octahedral sites of the garnet lattice [11]. Another broad band yellow emitting phosphor that can be used in conjunction with a blue LED to give a white light is TAG:Ce, $Tb_3(Al,B)_5O_{12}$:Ce, which has a dominant emission wavelength of 570-576 nm [12-14]. This phosphor can be used in conjunction with a blue InGaN LED to generate white light having a colour correlated temperature in the range 4000-6000 K. The YAG:Ce and TAG:Ce phosphors have very good stability, but their principal disadvantages are their limited excitation ranges, their reduced efficiencies at elevated temperature in the range of 125-200°C, and the cost in the case of TAG:Ce as its host lattice contains terbium.

Other yellow emitting phosphors that can be used with blue InGaN LEDs for generating white light are the orthosilicates, $(Sr_xBa_{1-x})_2SiO_4:Eu^{2+}$. These are narrower band emitting phosphors than YAG:Ce or TAG:Ce which is why a pcLED consisting of a blue InGaN LED and a yellow emitting $Ba_2SiO_4:Eu^{2+}$ phosphor can only give white light having a colour rendering index of ca. 65 [15]. Other disadvantages of the orthosilicate phosphors is that they dissolve in deionized water, they are not chemically stable, and they have much reduced luminous efficiencies at elevated temperature in the 125-200°C range.

As mentioned earlier, pcLEDs comprising of blue LED/yellow phosphor combinations have a high colour correlated temperature (CCT), thus other efficient, broad-band emitting phosphors with lower CCTs are required for white light pcLEDs that have a 'warmer' white in order to satisfy consumer preferences. Yellow-orange emitting phosphors satisfy this requirement and one such phosphor is α-SiAlON, (Sr,Ca)SiAlON:Eu^{2+}, and another is the novel nitrido-YAG phosphor, $Y_3(Al,Si)_5(O,N)_{12}$:Ce which has a stoichiometry similar to that of YAG:Ce ($Y_3Al_5O_{12}$:Ce) with Si+N partly replacing Al+O [16-18]. These blue InGaN LED/single phosphor combinations are capable of generating warm white light having a colour correlated temperature of 3200 K.

It should be appreciated that there is a trade-off between CRI and luminous efficacy. This is illustrated by considering a hypothetical yellow phosphor emission centred on 565 nm and having a Gaussian band shape. As the full width at half maximum (FWHM) is increased, the luminous efficacy decreases and the CRI increases (see Figure 6).

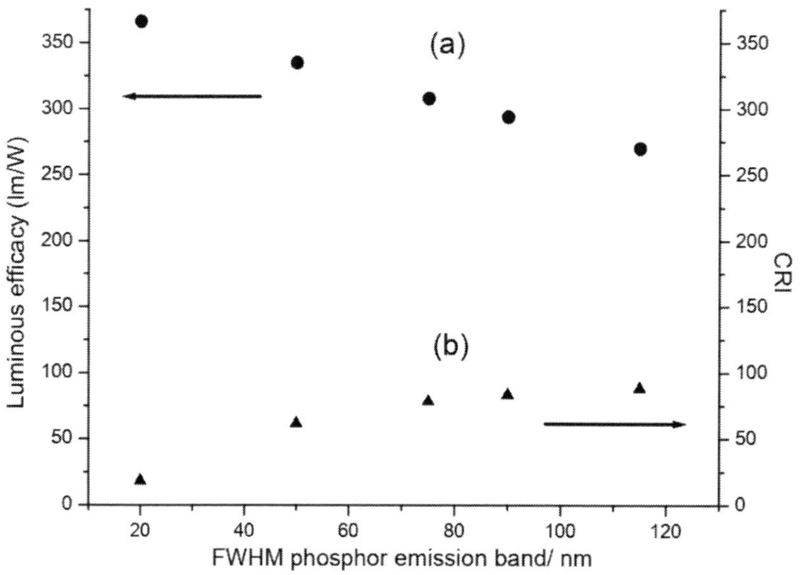

Figure 6. (a) Luminous efficacy versus FWHM phosphor emission band, and (b) CRI versus FWHM phosphor emission band. A Gaussian shape has been assumed for a phosphor emission band centred on a wavelength of 565 nm.

LED/Multiple Phosphor Combinations

The 1pcLED approach using a broad band yellow or yellow-orange emitting phosphor pumped by a blue LED can only give a modest quality white light. However, a better quality white light (higher CRI) can be generated with a 2pcLED approach using red and green emitting phosphors pumped by a blue LED. It should be appreciated that the luminous efficacy (L.E.) will be reduced on account of the aforementioned trade-off of CRI with L.E. Thus, the customer has the choice of paying for better colour rendering at the expense of luminous efficiency or, on the contrary, having better luminous efficiency at the expense of colour rendition. The market will decide how much loss in efficiency it will tolerate in order to have good illumination.

Up until relatively recently, there were no good red emitting phosphors available; the red emitting phosphors that were on the market, CaS:Eu^{2+} and SrS:Eu^{2+} (having maximum emission wavelengths of ca. 650 and 615 nm, respectively, under blue excitation), are sensitive to moisture and exhibit thermal quenching of their luminescence at elevated temperature (see Figures 7 and 8).

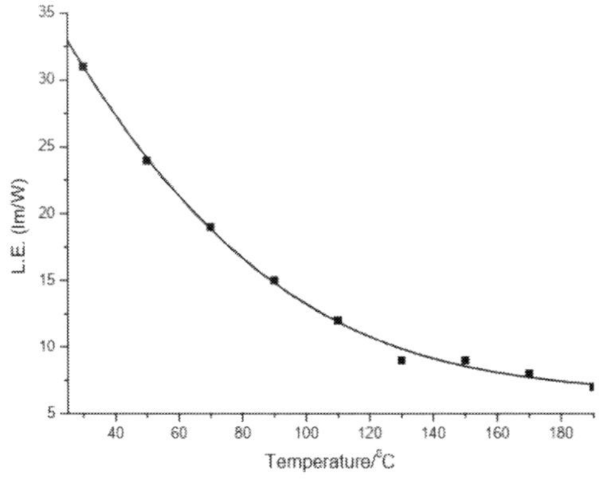

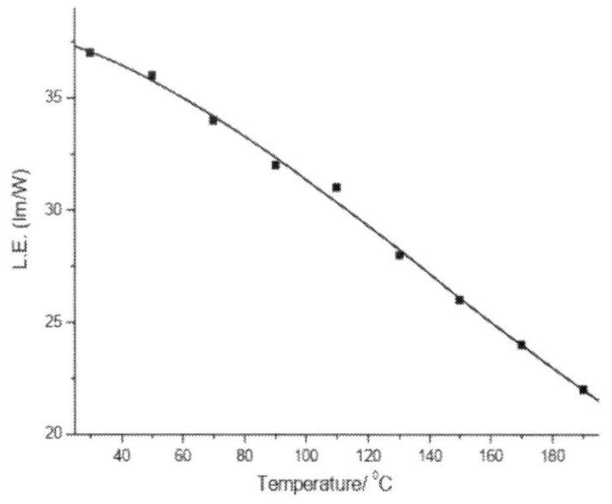

Figure 7 (left hand side). Luminous efficacy versus temperature for SrS:Eu red emitting phosphor under 470 nm excitation.

Figure 8 (right hand side). Luminous efficacy versus temperature for CaS:Eu red emitting phosphor under 470 nm excitation.

However, intense research activity on red emitting phosphors has led to the discovery of new nitride phosphors which are chemically stable and exhibit good emission at elevated temperature. One such phosphor is $(Sr,Ba,Ca)_2Si_5N_8$:Eu^{2+} [19,20], and another such phosphor is $CaAlSiN_3$:Eu^{2+} [21]. There are many possible ternary nitride phosphors that can be tried for solid state lighting, including M-Si-N phases (see Table 1) [22], M-Al-N, M-Ga-N and M-Ge-N phases, but a disadvantage is that their binary nitride precursors are usually sensitive to moisture and/or oxygen and consequently need to be handled in a glove box. Other disadvantages are that they need to be fired at high temperature and their particle sizes are difficult to control.

Table 1. Various M-Si-N phases that have been studied as host lattices for colour conversion phosphors [22].

M-Si-N	**[M = (Mg, Ca, Sr, Ba, Ln, Y)]**
$\mathbf{MSiN_2}$	$\mathbf{M_4SiN_4}$
$\mathbf{M_5Si_2N_6}$	$\mathbf{MSi_3N_5}$ $\mathbf{Y_2Si_3N_6}$ $\mathbf{Y_6Si_3N_{10}}$
$\mathbf{M_2Si_4N_7}$	$\mathbf{M_2Si_5N_8}$
$\mathbf{M_3Si_6N_{11}}$	$\mathbf{M_4Si_6N_{11}}$ $\mathbf{MSi_7N_{10}}$ $\mathbf{SrSi_7N_{10}}$ $\mathbf{M_9Si_{11}N_{23}}$

Osram and Nichia have used the red emitting $(Sr,Ba,Ca)_2Si_5N_8:Eu^{2+}$ phosphor in conjunction with YAG:Ce and a blue LED to achieve a warm white light in a 2pcLED [19,23].
A green emitting phosphor that can be used in a 2pcLED comprising of a blue LED pumping red and green emitting phosphors is the silicon-based oxynitride, $SrSi_2O_2N_2:Eu^{2+}$ [24]. The novel classes of silicon-based nitride and oxynitride phosphors for white LEDs have been reviewed by R.-J. Xie and N. Hirosaki [25] who reported recent developments in their preparation, crystal structures, luminescence and applications. Two alternative green emitting phosphors that can be used with a red nitride phosphor to generate white light from blue InGaN LEDs are the silicate garnet, $Ca_3Sc_2Si_3O_{12}$:Ce [26,27] and the orthorhombic $CaSc_2O_4$:Ce phosphor [28], although both phosphors are quite expensive. Yet another possible green emitting phosphor that can be used in a 2pcLED approach with a red phosphor and blue LED is the cerium activated lutetium aluminium garnet, LuAG:Ce [29]. Lutetium ions substitute the larger yttrium cations, thereby causing the emission to shift from ca. 560 nm (for YAG:Ce) to 545 nm (for LuAG:Ce) [30]. It should be noted that mixing or layering the phosphors on the LED chip can make a difference to the properties of the light that is generated because they the red and green phosphors can interact if their absorption and emission bands overlap.
Red line emitting phosphors, such as those activated by Eu^{3+}, Sm^{3+} and Pr^{3+}, are not usually chosen because they usually have insufficient absorbance for exciting with blue LEDs leading to lower device efficiency. For example, the advantages and disadvantages of using Eu^{3+} activated red-emitting phosphors as colour conversion phosphors for blue-emitting LEDs have been summarised for $Li(Eu_xY_{1-x})(MoO_4)_y(WO_4)_{2-y}$ phosphors [31]. However, a class of red line emitting phosphors consisting of complex fluoride lattices and activated by Mn^{4+}, $A_2[TiF_6]:Mn^{4+}$ (A = Na, NH_4) and $K_2[MF_6]:Mn^{4+}$ (M = Sn, Zr), show promise, as they have broad absorption bands in the near UV and blue regions [6]. Such red line emitting phosphors can apparently be used in a blue 2pcLED device with a YAG:Ce or TAG:Ce phosphor to give white light with a lumen equivalent of 330 lm/W_{opt} and a CRI of 90 at a colour correlated temperature of 2700 K [32].

Conclusions

Some newly developed phosphors for two versions of phosphor converted LEDs for white lighting have been covered. The 1pcLED consisting of a blue LED and a broad band green-yellow emitting phosphor gives a cool white light (CCT down to 4000 K) with a limited CRI of less than 80. In contrast, the 2pcLED approach consisting of a blue LED, and green and red phosphors can give a warmer white light (CCT 2800-3200 K) with higher CRI. However, it should be appreciated that the luminous efficacy (L.E.) of the 2pcLED is likely to be less than that of the 1pcLED on account of a trade-off of CRI with L.E. Thus, the customer has the choice of paying for better colour rendering at the expense of luminous efficiency or, on the contrary, having better luminous efficiency at the expense of colour rendition.
Red emitting $CaS:Eu^{2+}$ and $SrS:Eu^{2+}$ phosphors exhibit poor thermal quenching characteristics at elevated temperature, but red emitting silicon-based nitride phosphors have been developed recently which show much improved thermal properties. In the future it is hoped that the YAG:Ce phosphor will also be superseded by a nitride- or oxynitride-based phosphor that shows less thermal quenching.

Acknowledgement

We are grateful to the Technology Strategy Board-led Technology Programme for funding the NOVELELS programme.

References

1. *Phosphor Handbook*, 2nd Edition (Eds: W.M. Yen, S. Shionoya, H. Yamamoto), Taylor & Francis Group LLC, Boca Raton, FL, **2006**, Ch. 4.
2. For leading references, see N. Zheludev, "The Life and Times of the LED – A 100 Year History", *Nature Photonics*, **1**, 189-192 (2007).
3. E. F. Schubert, *Light-emitting Diodes*, Cambridge University Press (2003).
4. S. Nakamura and G. Fasol, *The Blue Laser Diode*, Springer (1996) pp 216-221.
5. J. Silver and R. Withnall, *Colour Conversion Phosphors for LEDs*, Chapter 3 in *Luminescent Materials and Applications* (Ed., A. Kitai), Wiley, Chichester, UK (2008).

6. E. Radkov and G. Ljudimil, US Patent 2006/0169998 A1.
7. (a) G. Blasse and A. Brill, Applied Physics Letters, 11, No 2, 53-55, 1967; (b) G. Blasse and A. Bril, J. Chem. Phys., 47(12) (1967) 5139-5145.
8. B.M.J. Smets, Materials Chemistry and Physics, 16, 283-299, 1987.
9. R. C. Ropp, *The Chemistry of Artificial Lighting Devices*, Elsevier, Sci. Publ. Amsterdam, London, New York **1993**, pp. 1-664.
10. R.C. Ropp, *Luminescence and the Solid State*, Elsevier Sci. Publ., Amsterdam, New York, **1991**, pp. 1-453.
11. W. Rossner, Phosphor Global Summit, Scottsdale, Arizona, USA. In the conference book, Paper 14, 1-18, 2003.
12. F. Kummer, F. Zwaschka, A. Ellens, A. Debray and G. Waitl, International Patent Application No. WO 01/08452 (2001).
13. M. Batentschuk, A. Osvet, G. Schierning, A. Klier, J. Schneider and A. Winnacker, Radiation Measurements, 38 (2004) 539-543.
14. C.-C. Chiang, M.-S. Tsai and M.-H. Hon, J. Electrochem. Soc. 155 (2008) B517-B520.
15. Chulsoo Yoon, Phosphor Global Summit, San Diego, California, USA. Contributed paper, 2009.
16. S.I. Sakata and T. Yamada, SiAlON-based phosphor and its production method, EP1498466, July, 15, 2004.
17. R.-J. Xie, N. Hirosaki, M. Mitomo, Y. Yamamoto, T. Suehiro and K. Sakuma,, J. Phys. Chem. B, 108 (2004) 12027.
18. R.-J. Xie, N.Hirosaki, K.Sakuma, Y.Yamamoto, and M.Mitomo, Appl. Phys. Lett., 84 (2004) 5404.
19. M. Yamada, T. Naitou, K. Izuno, H. Tamaki, Y. Murazaki, M. Kameshima and T. Mukai, Jpn. J. Appl. Phys. 42, L20- L23, 2003.
20. R. Withnall, J. Silver, G.R. Fern, T.G. Ireland, A.L. Lipman and B. Patel, J. SID, 16(2) (2008) 359-366.
21. K. Uheda, N. Hirosaki, Y. Yamamoto, A. Naito, T. Nakajima, and H. Yamamoto, Electrochem. and Solid-State Lett., 9, H22-H25, 2006.
22. A.K.Cheetham, Novel Phosphors for Solid State Lighting, Phosphor Global Summit 2006, San Diego, Ca, USA. In the conference book, Paper 7, 1- 13, 2006.
23. G. Bogner, G. Botty, B. Braune, H. T. Hintzen, J. W. H. van Krevel and G. Waitl, US Patent No. 6,649,946 November 18, 2003.
24. C.-C. Yang, C.-M. Lin, Y.-J. Chen, Y.-T. Wu, S.-R. Chuang, R.-S. Liu and S.-F. Hu, Appl. Phys. Lett. 90 (2007) 123503.
25. R.-J. Xie and N. Hirosaki, Science and Technology of Advanced Materials, 8(7-8) (2007) 588-600.
26. Y. Shimomura and N. Kijima, Proceedings of the 65th Meeting of Japan Society of Applied Physics, p. 1284 (2004).
27. Y. Shimomura and N. Kijima, US Patent Application 2004/0251809 A1.
28. N. Kijima, Phosphor Global Summit, Seoul, South Korea. Contributed paper, 2007.
29. X.-J. Liu, H.-L. Li, R.-J. Xie, Y. Zeng and L.-P. Huang, J. Lumin, 124 (2007) 75-80.
30. H. Winkler, Phosphor Global Summit, San Diego, California, USA. Contributed paper, 2010.
31. R. Stone, J. Silver, R. Withnall and G.R. Fern, Proceedings of IDW'09, Miyazaki, Japan, pp 359-361.
32. E. Radkov, Phosphor Global Summit, Seoul, South Korea. Contributed paper, 2007.

Automotive headlamps in the 21st century

S.M. Booij[1], M. Sikkens[2], J. Schug[3]
[1] Optics and Sensors group, Philips Applied Technologies, High Tech Campus 7, 5656 AE Eindhoven, the Netherlands
[2]LightLabs, Philips Lighting, Bldng EEC-210, 5600 JM Eindhoven, the Netherlands
[2]Business Center Automotive Aachen, Philips Technologie GmbH, Philipsstrasse 8, D-52068, Aachen, Germany
Contact: silvia.booij@philips.com

ABSTRACT

LEDs have proven themselves already in the automotive world in interior lighting and in signal functions (tail, brake, turn, daytime running lights). Recently LEDs are also beginning to appear in headlamps. This paper describes the advantages of LED headlamp systems and their functioning principle. Increasingly more advanced options become available with the introduction of LEDs which were not possible with conventional light-sources. The first implementations focus on styling differentiation, i.e. creating a unique and attractive appearance of the headlamp. Furthermore LEDs allow dynamic beam adaptation in adaptive front lighting systems (AFS). AFS systems dynamically change the illumination pattern on the road, depending on the driving conditions, which makes driving safer and more comfortable. Last but not least, we see a growing interest in basic energy efficient LED headlamp solutions resulting in lower fuel consumption and less CO_2 emission.

LED HEADLAMPS

LEDs are beginning to appear in automotive headlamps, replacing the systems that are currently dominant on the market: halogen and Xenon.
What are the advantages of LED headlamp systems?

- new styling possibilities
- energy efficiency, which is beneficial for the environment, and which can be very important in electric cars
- flexibility, since they are often built up out of several modules. This enables new styling features, but also adaptive front-lighting systems
- Lifetime is much longer than that of existing lamps

Although LED headlamp solutions can currently only be offered at a price premium, their benefits in the application are increasingly recognized and will stimulate further adoption in the various market segments over the next years. The harsh environment and high ambient temperature in automotive applications under the hood still present serious challenges for LEDs operating at the high power levels required for front lighting. We will later point out how they can be addressed with dedicated optimized solutions.

REGULATIONS

Strict regulations exist for the radiation pattern on the road that is produced by an automotive headlamp, both for the high beam and the low beam pattern. For the low beam or passing beam hardly any light may be present above the horizon on the side of the road where on-coming traffic drives. On the driving side light may illuminate above the horizon along the so-called kink-line, see figure 1.

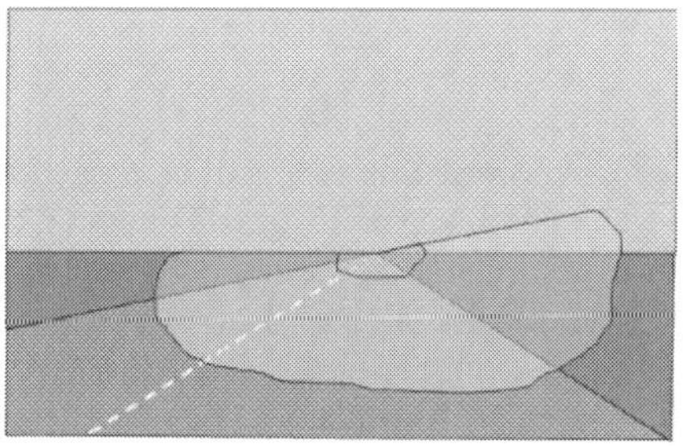

Fig.1 Schematic picture of area of the road that is illuminated by a low beam

LUXEON ALTILON SOURCE TECHNOLOGY

The LUXEON® Altilon incorporates Philips Lumileds' most advanced LED technology to achieve:

- highest luminance
- robust performance in a challenging thermal environment
- excellent color control

Thin Film Flip Chip (TFFC) technology allowed a breakthrough in light extraction from the package. White light is generated through a partial conversion of blue light in an accurate Lumiramic ceramic phosphor plate, which allows for an excellent targeting of the desired color point.

The carefully designed thermal stack consisting of an AlN submount and a copper heat slug allows for a low thermal resistivity towards the heat sink.

Fig. 2 Philips Luxeon Altilon source

LED HEAD LAMPS

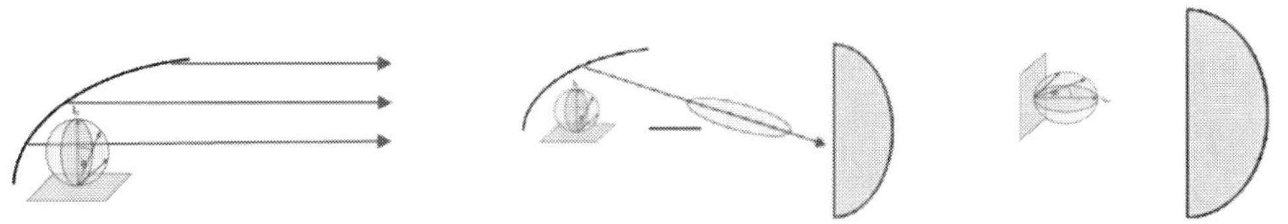

Figure 3 a) Example of a reflection system, b) Example of a indirect projection system, c) Example of compact direct projection system

Headlamps can be divided in reflector systems (today mainly used with halogen lamps) and projection systems (today mainly used with Xenon lamps). A similar classification holds for LED systems.

The basic principle of these systems is illustrated in figure 3.

The reflector system of fig. 3a consists of an LED array placed in a more-or-less parabolic reflector which "images" the edge of the array onto the horizontal cut-off line. True imaging is used in a conventional projection system as shown in fig. 3b. An elliptical reflector creates a light patch in the focal plane of a projection lens which is then imaged onto the road by the lens; a shield in the focal plane creates the cut-off and kink-line. An interesting variation on this projection system is shown in fig. 3c, where the LED is directly imaged by the projection lens aperture. If a small collimator is added, the projection lens can also image the collimator exit aperture onto the road. This exit aperture can be shaped to include both the horizontal cut-off line and the kink-line.

The first LED headlamps on the market date back to 2007, the Lexus LS600h uses an indirect projection system for the LED low beam. The Audi R8 with full-LED headlamps was launched in 2008, which makes use of both reflector optics and of a variant of a projector system.

Basic energy-efficient LED headlamp solutions

Due to legal requirements (see ECE regulations 112 and 123, see www.unece.org/trans/main/wp29/wp29regs.html), the installed luminous flux for the creation of a basic low beam pattern needs to exceed 1000 lm. In the near future, single automotive front lighting LED sources providing this light output will become available. Already today the LUXEON Altilon is offered with a luminous flux in excess of 850 lumen.

Due to their unique directional emission characteristic LEDs allow to build headlamp optics with over 60 % overall efficiency, which is almost a factor of two higher than for conventional light sources. This results in more than 600 lumen on the road for the LED system while standard halogen systems generate about 500-600 lumen. At the same time, the total power consumption does not need to exceed 20 W per headlamp while typical halogen sources consume 55 W to 60 W.

Thus, basic LED headlamp solutions start to compete with headlamps using halogen light sources. The planned penalties for car makers who exceed the limit of 130 g CO_2 / km set forward by the European Commission will be a further incentive to invest in this environmentally friendly technology.

Visible Light Communications Based on Modulation of White LEDs

J. Vučić, C. Kottke, L. Fernández, K. Habel, A. Paraskevopoulos and K.-D. Langer
Fraunhofer Institute for Telecommunications, Heinrich-Hertz Institut, Einsteinufer 37, 10587 Berlin, Germany
Contact: jelena.vucic@hhi.fraunhofer.de

ABSTRACT

Colour light-emitting diodes (LEDs) are already prevalent in signalling applications, while the white LEDs present a major candidate for future general illumination. Visible light communications (VLC) consider simultaneous modulation of these LEDs (while "on") and thus offer an additional service at comparably low effort. We demonstrated transmission rates of hundreds of Mb/s at illumination levels recommended by the standards for working environment.

INTRODUCTION

Using illumination LED sources simultaneously for data transmission presents a relatively new area of research, referred to as visible-light communications, and it is widely believed that many applications for this type of optical communications are still to be discovered [1].

The presence of visible light is obvious, allowing an easy establishment of a roughly directed link from transmitter (Tx) to receiver (Rx) and its monitoring. The major issue of using such LEDs for broadband applications is their restricted modulation bandwidth, being typically a few MHz for phosphorescent white LEDs. We have, however, demonstrated a tenfold enhancement of the bandwidth, when the phosphorescent portion of the optical spectrum is suppressed at the detector [2]. Our experimental efforts involving a blue-filtered detection and on-off keying (OOK) featured data rates in excess of 100 Mb/s [3], and significantly higher data rates when using discrete multitone (DMT) modulation [4].

DEMONSTRATION OF VLC

The experimental setup is shown in **Fig. 1a**. In all measurements, a commercially available phosphorescent white LED module was used (Osram, OSTAR LE CW E3). The LED driver was a custom-designed amplifier, whose ac output was superimposed onto the dc bias current and supplied to the LED module by aid of a bias T. Our studies have shown that power consumption of the transmitter is ~30 % larger for a VLC system with respect to the pure illumination function, due to the addition of the LED driver.

The ac-coupled analogue receiver combined several key components on a single board. A blue optical filter was mounted in front of the photodiode for suppressing the phosphorescent component of the white light. A commercially large-area silicon PIN diode assembled with a polymer lens (Hamamatsu S6968) was followed by a low-noise trans-impedance stage.

Two sets of experiments were performed: the first one using NRZ-OOK transmission and subsequently the second using DMT transmission via offline processing. Since the primary function of the applied optical source is general illumination, the brightness level at the receiver was considered as the parameter of relevance. Due to the use of a single LED module (with relatively wide radiation angle) in our experiments, collimation optics was applied in the free-space channel, to achieve the targeted brightness level in front of the photodiode. This is also the reason why the Tx-Rx distance in the experiment is relatively short (< 1 m). In a practical scenario, both required brightness and typical indoor distances would be achieved by combining several luminaries.

On-off Keying

We used NRZ pseudo-random bit sequences (PRBS 2^{23}-1) and a modulation index of ~ 0.35. For an illuminance level of ~1100 lux at the Rx (in the recommendation range of the lighting standard [5] for well-lit office environment), a transmission rate of 125 Mb/s was achieved with a BER of $1.8 \cdot 10^{-3}$.

Discrete Multi-Tone Modulation

In this case, time traces of a DMT signal were software-generated and then fed into an arbitrary-waveform generator, whose output drove the Tx configuration. At the Rx, time traces of the received signal were recorded by a storage oscilloscope and demodulated by software. We considered a DMT signal consisting of 32 subcarriers within a bandwidth of 50 MHz. Subcarriers were modulated with quadrature amplitude modulation (QAM) of different orders, i.e. bit-loading was applied.

The measurements were performed at brightness levels of ~ 550 lx and 1100 lx at the Rx. The resultant subcarrier QAM orders (**Fig. 1b**) were manually adjusted to achieve the maximum transmission rate, while keeping the overall BER within the forward error correction (FEC) coding limits ($< 2\cdot10^{-3}$) [6]. The transmission rates achieved were 231 Mb/s for 1100 lx, and 202 Mb/s for 550 lx. Furthermore, we showed that 100 Mb/s with a BER of $\sim 7\cdot10^{-4}$ is possible at only ~180 lx.

Our group is currently working on the realization of a real-time VLC system, with digital signal processing implemented on FPGA [7].

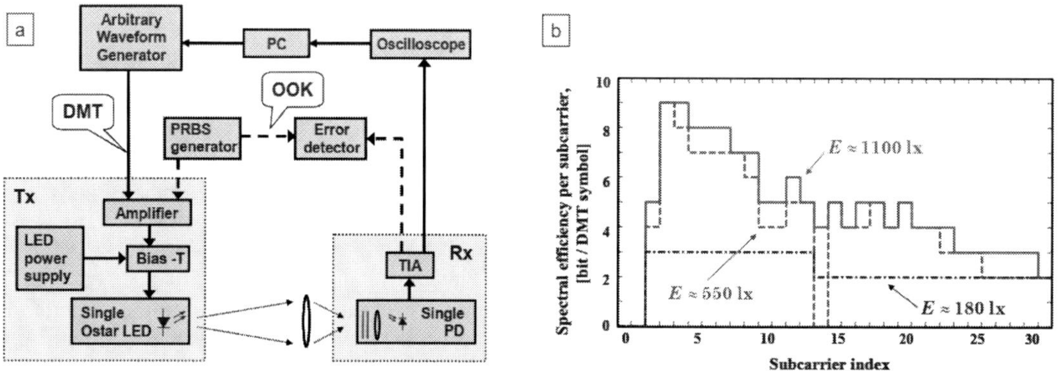

Figure 1. a) Experimental setup, b) Bit-loading masks for DMT measurements.

CONCLUSION

Recent advances in white LED technology have opened new application fields for the simultaneous implementation of lighting and visible light communications. We demonstrated an indoor visible-light link based on phosphorescent white LEDs operating at 125 Mb/s with OOK and 230 Mb/s when using DMT with off-line processing at an illuminance level of about 1100 lx (recommended for a well-lit office environment) at the receiver. All demonstrated overall BERs lie below $2\cdot10^{-3}$, which can be completely compensated with FEC. Overhead which would be needed in the system (e.g., for FEC or OFDM guard interval) is ~10%. With deployment of a high sensitivity receiver (avalanche photodiode) doubling in rate performance can be expected.

ACKNOWLEDGEMENT

Part of this work received funding from the EU FP7 Programme, referred to as OMEGA. The authors would like to thank S. Nerreter and J. W. Walewski from Siemens Corporate Technologies, Germany for providing the analogue front-ends for the experiments.

REFERENCES

[1] Visible Light Communications Consortium, http://www.vlcc.net
[2] J. Grubor, S. Randel, K.-D. Langer, and J.W. Walewski, "Broadband information broadcasting using LED-based interior lighting," *IEEE JLT*, vol. 26, no. 24, pp. 3883-3892, 2008.
[3] J. Vučić, *et al.*, "125 Mbit/s over 5 m wireless distance by use of OOK-modulated phosphore-scent white LEDs," *in Proc. of ECOC 2009*, Paper 9.6.4.
[4] J. Vučić, *et al.*, "White-Light wireless transmission at 200+ Mbit/s net data rate by use of discrete-multitone modulation," *IEEE PTL*, vol. 21, no 20, pp. 1511-1513, Oct. 15, 2009
[5] European standard EN 12464-1: Lighting of indoor work places, 2003.
[6] ITU-T Recommendation G.975.1 (02/2004).
[7] Website www.ict-omega.eu

Improvement of the quantum efficiency in green light-emitting diodes with pre-TMIn flow treatment

Ya-Ju Lee[1], Yi-Ching Chen[1], Chia-Jung Lee[1], Chun-Mao Cheng[1], Shih-Wei Chen[2] and Tien-Chang Lu[2]
Institute of Electro-Optical Science and Technology, National Taiwan Normal University, 88, Sec.4, Ting-Chou Road, Taipei 116, Taiwan
Department of Photonics, National Chiao Tung University, 1001 Ta Hsueh Road, Hsinchu, Taiwan
Contact:yajulee@ntnu.edu.tw

ABSTRACT

The detail characteristics of green Light-Emitting Diode (LED) grown by metal-organic chemical vapor deposition with the pre-TMIn flow treatment has been studied. The investigation reveals that for the sample with the assistance of pre-TMIn flow treatment, the internal quantum efficiency (the radiative efficiency *B*) and the temperature dependence on the external quantum efficiency have be significantly improved. The improvement is thought to be attributed to the preferential formation of In-rich dots with pre-TMIn flow treatment that effectively suppresses the excitons from being trapping by threading dislocations and overflowing outside the active regions at elevated temperatures.

INTRODUCTION

The developments of InGaN alloys are most important for the achievement of high efficient light-emitting diodes (LEDs) across the visible spectrum to match the color quality of traditional light sources. However, the peak EQE of green LEDs is significantly lower than that of InGaN-based blue and AlGaInP-based red LEDs, a feature referring to the so called "green gap".[1] Moreover, the efficiency of green LEDs droops with the increasing of injected currents more severely than in blue LEDs.[2] Thus extensive approaches were developed over the past decade to overcome above issues in InGaN-based green LEDs. In this letter, we report the detail properties of InGaN-based green LEDs with pre-TMIn flow treatment and discuss the possible mechanism for its superior performances over the conventional LED. As a result, we conclude the pre-TMIn flow treatment is a promising approach for bridging the "green gap".

RESULT & DISCUSSION

The green LED structures were grown on *c*-plane sapphire substrates by low-pressure metal-organic chemical vapor deposition. Following growth of a GaN nucleation layer at 520°C, a 2-µm-thick undoped GaN layer and a 2.5-µm-thick Si-doped n-type GaN cladding layer were grown at 1050°C. The active region consists of nominally undoped six periods of 3-nm-thick $In_{0.30}GaN_{0.70}N$/14-nm-thick GaN MQWs. An interruption between the well and barrier layers was given for the stabilization of reactor temperature and gas flows. During the growth interruption, a TMIn flow of 60sccm was introduced into reactor prior to the growth of InGaN well for 10s duration. No introduction of TMIn flow prior to well layers was also grown for comparison (the conventional sample). Finally, a 300-nm-thick p-type GaN cladding layer were grown on top. Figure 1 shows cross-sectional transmission electron-microscopy (TEM) images of the MQWs prepared (a) without and (b) with pre-TMIn flow treatment, respectively. The boundary of InGaN/GaN QWs for both samples are clear and well defined, suggesting the pre-TMIn flow treatment causes a minor influence in terms of interface abruptness. Segregations of indium atoms along growth direction of (0001) were observed on both samples. It suggests the introduction of TMIn flow onto the GaN surface prior to the growth of InGaN well did not alleviate its desorption of indium composition and subsequent penetration into GaN barrier layer. The IQE as a function of carrier concentration for both samples are shown in Fig.2 (a). About 9.2% enhancement of IQE was achieved for the sample with pre-TMIn flow treatment at a carrier concentration of typical value of $2x10^{18}cm^{-3}$. The relation between measured IQE and carrier concentration for both samples is well described by $IQE=Bn^2/(An+Bn^2)$, where *A* and *B* are nonradiative and radiative coefficients, respectively. If one assumes a reasonable value of A= $3x10^7 s^{-1}$, the value of B and be obtained as well.[3] The calculated values are B=$1.9x10^{-11}$ cm^3s^{-1} and B=$3.34x10^{-11}$ cm^3s^{-1} for the samples without and with pre-TMIn flow treatment, respectively. Fig.2 (b) shows the radiative coefficient B as a function of etch pit density (EPD).The pits might be originated by local lattice expansion due to existence of In-rich dots within MQWs, rather than connected to the threading dislocations.[4] Therefore, the strong spatial localization of In-rich dots effectively prevents the excitons from being captured by threading dislocations, leading to a larger B value, as well as higher IQE.

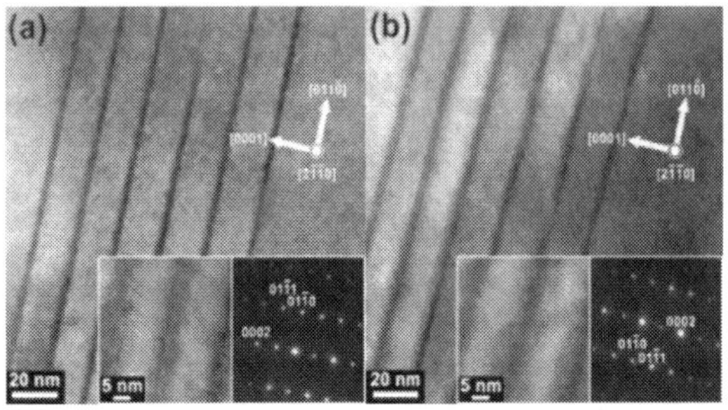

Fig.1 Bright-field TEM images of InGaN/GaN MQWs grown (a) without and (b) with pre-TMIn flow treatment.

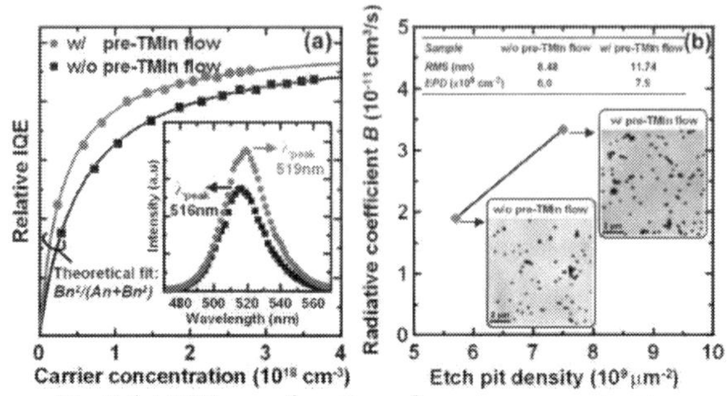

Fig.2 (a) IQE as a function of carrier concentration. (b) Radiative coefficient B as a function of etch pit density.

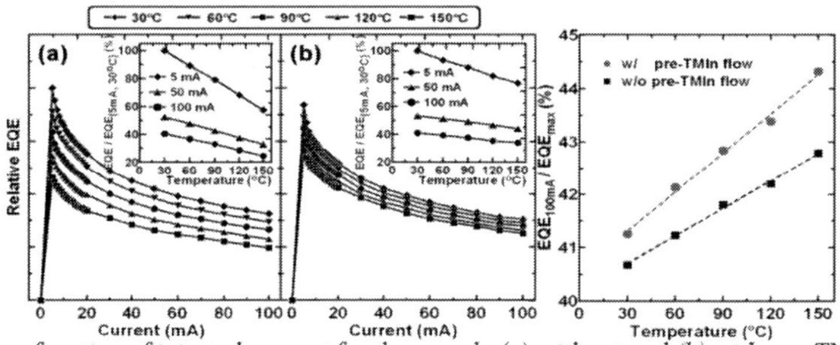

Figure 3 EQE as a function of injected current for the sample (a) without and (b) with pre-TMIn flow treatment from T=30°C to T=150°C. (c) EQE droop as a function of temperature for both samples.

Figure 3 plots the EQE as a junction of injected current for the samples (a) without and (b) with the pre-TMIn flow treatment from T=30 °C to T=150°C. On average, the EQE value is enhanced by 7.5% at T=30 °C, and further increases to 44.5% at T=150°C. In addition, the EQE values for the sample with pre-TMIn flow treatment fall slightly with increasing temperatures, demonstrating a more stable temperature characteristic. Studies have suggested that the localization of injected carriers within In-rich dots operates as excellent radiative recombination centers, which effectively suppresses the injected carriers from overflowing outside MQWs at elevated temperatures. The EQE droop as a function of temperature for our samples were plotted in Fig.3 (c). For both samples, the peak efficiency falls more rapidly with increasing temperature than the efficiency at high current, thus the magnitude of the EQE droop actually decreases with increasing temperature. Nevertheless, less efficiency droop was observed for the sample with pre-TMIn flow treatment at all temperatures.

In summary, we have demonstrated the pre-TMIn flow treatment being a promising scheme for the improvement of output performances of InGaN-based green LEDs and bridging the "green gap" in solid-state lighting.

REFERENCE

[1]M. R. Krames, O. B. Shchekin, M. M. Regina, G. O. Mueller, L. Zhou, , G. Harbers, M. G. Craford, IEEE Journal of Display Technology, **3**, 160-175 (2007).

[2]T. Mukai, M. Yamada, and S. Nakamura, Jpn. J. Appl. Phys., Part 1 **38**, 3976 (1999).

[3]Y. C. Shen, G. O. Mueller, S. Watanabe, N. F. Gardner, A. Munkholm, and M. R. Krames, Appl. Phys. Lett. **91**, 141191 (2007).

[10]K. Watanabe, J. R. Yang, S. Y. Huang, K. Inoke, J. T. Hsu, R. C. Tu, T. Yamazaki, N. Nakanishi, and M. Shiojiri, Appl. Phys. Lett. **82**, 718 (2003).

Development of free-standing InGaN LED devices on ALD-Al_2O_3/Si substrate by wet etching

Muhammad Jamil[1,*], Tianming Xu[1,2], Andrew Melton[1], Balakrishnam Jampana[1], Tahir Zaidi[1], Sheng Liu[2], Ian Ferguson[1,3]

[1] *School of Electrical and Computer Engineering, Georgia Institute of Technology, Atlanta, GA, 30332, USA*
[2] *School of Mechanical Science and Engineering, Huazhong University of Science and Technology, Wuhan, Hubei Province, 430074, China*
[3] *Dept. of Electrical and Computer Engineering, University of North Carolina at Charlotte, 9201 University City Boulevard, Charlotte, NC 28223-0001*
[*]*Corresponding author: E-mail: jamil@gatech.edu, Phone: 1-(404)-385-2888*

Abstract:

Free-standing InGaN-based LEDs grown on Al_2O_3/Si (111) have been achieved using selective area wet etching. The silicon substrate was selectively removed by wet etching while the chip was mounted in a dual in-line package (DIP), which prevented the thin film from cracking or warping. No significant change in electrical characteristics, peak emission wavelength, or EL intensity versus drive current was observed. This substrate-removal process and the challenges involved are discussed here. Such packaging techniques could be beneficial for commercial-scale production of InGaN-based LEDs grown on silicon substrates.

Introduction:

The last decade has seen significant advances in efficiency and total lumen output of nitride-based light emitting diodes (LEDs). However, several challenges remain, if LEDs are to make a significant penetration into the general lighting market. One of the main challenges is cost, as LED production and fabrication costs remain high compared to other sources [1]. Metal organic chemical vapor deposition (MOCVD) of GaN on Si has been investigated to overcome this issue, as large-area, high quality Si substrates are both readily available and inexpensive [2,3]. This work was aimed at developing a low cost device fabrication and Si substrate removal process realized by wet etching. Recently reported results of the transfer of GaN-based LEDs on Si (111) to copper substrates are also encouraging [4]; however, more work is needed in this area to reduce the process complexity while achieving the benefits of using Si substrate technology.

Experimental Results and Discussion:

The GaN-based device structures in this study were grown using a modified commercial MOCVD system with a vertical injection, rotating disk configuration. Al_2O_3 interlayers were deposited using a custom-built ALD system with trimethyl aluminum (TMAl) and H_2O as Al and O precursors, respectively. LED structures were grown on 1.5 μm-thick GaN/Al_2O_3/Si (111) templates and consisted of 1.0 μm of n-type GaN, a three-period $In_xGa_{1-x}N$/GaN MQW structure as the active region, and ~160nm of p-type GaN on top. These structures were fabricated into 350 × 350 μm^2 devices using standard techniques. Finally, the electrical and optical characteristics of the devices were measured before and after the Si substrate removal process by wet etching. Figure 1 (a) shows a 20x optical microscope image of a LED structure on ALD/Si. The majority of the film was crack-free. The use of Al_2O_3 layers can fully eliminate the cracks while maintaining the higher crystal quality in the GaN layer, as compared to films grown on bare Si. The top view of the fabricated LED devices is shown in Figure 1 (b,c). A nickel oxide current spreading layer was used to ensure uniform current spreading distribution through the device. The uniform light emission from the device surface, as shown in Figure 1(c), indicates good current spreading over the device structure area. Free-standing LED devices were achieved using selective area wet etching. Conventional device design was used in our LEDs fabrication, in which p-type and n-type contacts are located at the same side of wafer. LED devices were bonded to the dual in-line package (DIP). The front side of the devices was coated with epoxy to protect the bonding wires during etching. Vias were drilled in the DIP to expose the Si substrate to the etchant solution. The Si substrate was then selectively removed by wet chemical isotropic etching in HNA solution (HNO_3:HF:CH_3COOH = 3:1:1) at room temperature. Free-standing LED epilayers were successfully revealed after the etching processes. The mounted LED on Si substrate lit up with 10 mA forward drive current, as shown in Figure 2(a). Front side image of free-standing GaN LED thin films can be seen in Figure 2(b).

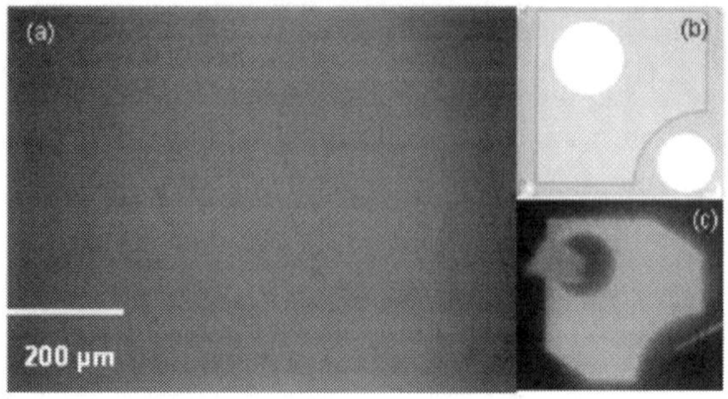

Figure 1: *(a) Optical Microscope image of LED film, (b) Top view of fabricated LED device, and (c) LED on ALD/Si lighted under normal operating condition.*

Figure 2 (c) shows the I-V curves before and after the Si (111) substrate was selectively removed. There is no noticeable difference between the two curves, indicating the removal of silicon substrate does not affect the electrical properties of the LEDs. Electroluminescence (EL) characteristics of the GaN LED devices grown on ALD/Si were also measured before and after the selective removal of Si substrate. EL intensity was also measured as a function of drive current for the same LED devices. The devices before and after wet etching both reached peak luminescence intensity with a drive current of 50~60mA. The efficiency at higher drive current of LEDs after etching was slightly higher after etching, which may be due to better thermal dissipation after the removal of silicon substrate. However, total radiation flux was reduced, which is due to the absorption by the epoxy adhesive glue on the front side of the device. These properties show promise for further development of substrate transfer and packaging techniques necessary for the commercial-scale production of InGaN-based LEDs on silicon substrates.

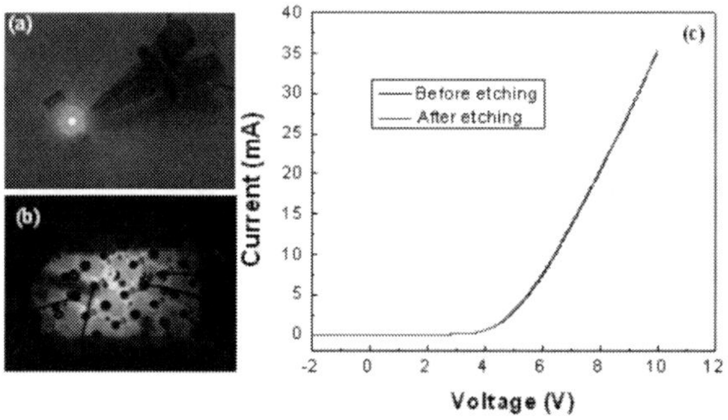

Figure 2: *(a) GaN LEDs lit up under forward driving current. (b) free-standing transparent LED devices from front side, and (C) I-V curves before and after wet etching.*

REFERENCES:

[1] T. Whitaker, ed., in LEDs Magazine (Cabot Media, Bristol, UK, 2007), Vol. Feb. 2007, p. 38.

[2] A. Dadgar, J. Blasing, A. Diez, A. Alam, M. Heuken, and A. Krost, Japanese Journal of Applied Physics 39, (2000) L1183.

[3] S.D. Hersee, D. Zubia, X. Sun, R. Bommena, M. Fairchild, S. Zhang, D. Burckel, A. Frauenglass, and S.R.J. Brueck, IEEE Journal of Quantum Electronics 38 (8), (2002) 1017.

[4] K. M. Wong, X. Xou, P. Chen, and K. M. Lau, IEEE Electron Device letters, Vol. 31, (2010)132.

Phosphor converted white LED with improved efficiency

B.Kolodin, G.Kuenzler, J.Li, P.Brown, A. Deshpande, J. Reginelli
GE Lighting, 1975 Noble Road, East Cleveland, OH 44112, USA
Contact: boris.kolodin@ge.com

ABSTRACT
Proposed short pass filter (SPF) employment to improve efficiency of phosphor converted white LED based on near UV LED dies and remote RGB phosphor architecture. SPF is transparent for near UV radiation in the 390-430 nm spectral range and possesses high reflectance for visual spectral range. SPF is placed between near UV LED and phosphor layer and redirects part of down warding white light to the outside of the package. Demonstrated 26 % average improvements for the conversion efficiency from 10.9 to 8.0 mW/Lm and 24 % average increase for luminous efficacy to hit 50 Lm/W level for 3000 °K CCT and 85 CRI.

INTRODUCTION
Phosphor converted White LEDs based on near UV LED dies and RGB phosphors demonstrate outstanding long term color stability but possess lower efficacy than White LEDs based on so called Blue platform mainly due to Stokes shift losses. As an example, Vio® White LED power package shows less than 75 K color change over 50,000 hour rated life and 70 % lumen maintenance [1]. Vio® chip on board (COB) package and remote phosphor architecture provide unique opportunity for the efficiency improvement by short pass filter (SPF) employment [2], [3].

EXPERIMENTAL RESULTS

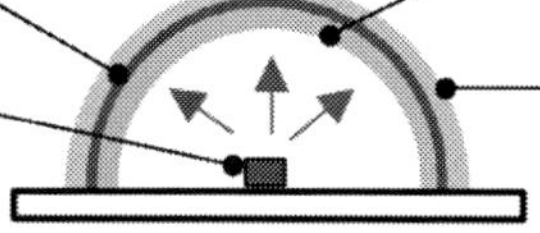

Figure 1. Phosphor converted White LED with SPF

White light emitted by the phosphor layer propagates upwards and downwards and about 50 % of the light emitted by the phosphor layer could be potentially absorbed by LED dies and PCB coating [Fig.1]. By design SPF is transparent to near UV radiation emitted by the LED but partially reflects the White radiation emitted by phosphor layer, bound to get absorbed by LED dies and PCB layer, to the outside of the package. Reflected White light passes through the phosphor layer directly or via scattering by the phosphor particles and thus increases the total lumens output.
The generation of White light by the UV LED/ RGB phosphor technology has very attractive feature: LED excitation spectra and White light spectra are not overlapped, therefore we do not expect significant White light spectra disturbance for the phosphor converted White LED with incorporated SPF [Fig. 2].
Excitation LED light engine consists of one 400 nm LED die with close to Lambertian light intensity distribution. Multilayer SiO_2/Ta_2O_5 SPF was fabricated on the glass cap by CVD process. Fig. 1 demonstrates ideal SPF spectra, measured SPF spectra had ~ 10 % optical losses. SPF cut off wavelength was chosen near 435 nm for LED with 400 nm peak wavelength and HWFM 17 nm to have whole LED spectra within high transparency spectral area of SPF. SPF transparency depends on incident angle and cut off wavelength experienced 0.58 nm/ angle degree shift according our measurements.

Since the cut off wavelength is strongly dependent on the incident angle of the UV LED radiation, it is desired to hold the range of incident angles within 0- 20 degrees to have engineered high SPF transparency. Non-imaging collimating reflective or refractive optics required if excitation light engine represents LED array consists of several LED dies.
The phosphor down converting element is based on commercial silicate LED grade phosphor engineered for Warm White light with 3000 CCT and 85 CRI [4]. A molded phosphor/silicone cap with embedded specified phosphor blend was used for direct comparison between White LED packages with and without SPF.

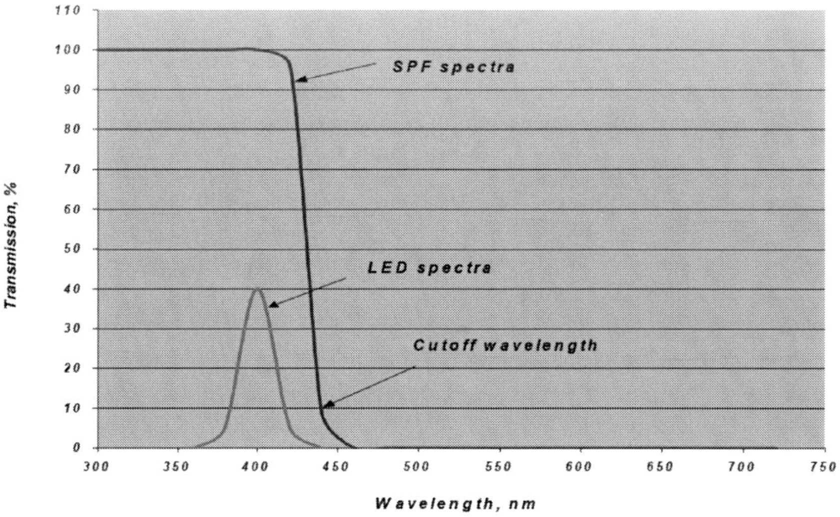

Figure 2. SPF spectra

Optical measurements included simultaneous monitoring of the conversion efficiency and luminous efficacy for the optical setup according to Fig.1, for reference White LED package without SPF contains clear glass cap. SPF incorporation resulted in 26 % average improvement for the conversion efficiency from 10.9 to 8.0 mW/Lm and 24 % increase for luminous efficacy to hit 50 Lm/W level for 3000 °K CCT and 85 CRI.

ACNOWLEDGMENT
The authors thank Emil Radkov, Srinath Aanegola and Jerry Duffy for a number of stimulating discussions and support on this work.

REFERENCES

[1] www.gelighting.com/led, Vio® High-Brightness LED, data sheet.
[2] T. F. Soules, S. E. Weaver, Jr., C. -L.Hsing, B. Kolodin, M. Sommers, A.A. Setlur, and T.E. Stecher, US Patent # 7,479,662 (January 20, 2009).
[3] S. K. Aanegola, J. T. Petroski, E. Radkov and S.E. Weaver, Jr., US Patent # 7,224,000 (May 29, 2007).
[4] D. Hancu, M. Adyam, E. Radkov, P. Kumar, A. A. Setlur, A. M. Srivastava, H. A. Comanzo, G. Chandran, M. V. Shankar, US Patent # 7,501,753 (March 20, 2009).

Determination of Thermochemical Parameters for the Tin(II) Halides, SnXX' (X, X' = Cl, Br, I), in Solid, Liquid and Gaseous Phases

G. Siddons[1], H. Donald Brooke Jenkins[2], S. A. Mucklejohn[3], R. Devonshire[1]
[1] *Department of Chemistry, University of Sheffield, Sheffield S3 7HF, UK*
[2] *Department of Chemistry, University of Warwick, Coventry CV4 7AL, UK*
[3] *School of Architecture, Design & Built Environment, Nottingham Trent University, Nottingham NG1 4BU, UK*
Contact: h.d.b.jenkins@warwick.ac.uk

ABSTRACT

The tin(II) halides, SnXX' (X, X' = Cl, Br, I), are important additives in high intensity discharge (HID) lamps. Unfortunately, previously published thermochemical data for the binary tin halides, SnX_2 (X = Cl, Br, I) are incompatible with the experimentally measured vapour pressures - key properties in simulations of operating HID lamps. Consequently, the derivation of a self-consistent set of thermodynamic data ($\Delta_f H^o$, $\Delta_f G^o$, S^o and C_p^o) for the tin(II) halides, SnXX' (X, X' = Cl, Br, I) that is also consistent with vapour pressure measurements is, accordingly, a challenging, timely and worthwhile goal which would fill a gap in the current literature. The study has provided data for both binary and mixed halogen tin(II) halides which, in the case of the latter species, is the first time such data has been reported.

INTRODUCTION

The modelling of complex chemical processes, such as those in operating discharge light sources, requires databases of reliable, self-consistent thermochemical parameters. There are several well established software packages that can calculate the temperature variation of the chemical compositions of multi-component, multi-phase systems, an example being MTDATA [1]. However, their use with respect to particular processes is often restricted by the availability and accuracy of the necessary thermochemical data. Our particular interest in the tin(II) halide system stems from their use in HID lamps [2].

METHOD

In this study we have carried out critical assessments of the physicochemical properties and standard thermochemical data for the tin(II) halides, with the exception of the fluoride-containing species. We have employed a range of methods in this work including: quantum chemistry calculations to determine the structure and vibrational frequencies of all gas phase species; the Jenkins-Glasser entropy-volume correlation [3]; the Benson and Hisham Approach; Latimer's method; the Thermodynamic Difference Rule [4]; and Metathetical Approximations [5], to derive a self-consistent set of parameters for the respective solid, liquid and gaseous phases. The selection of thermochemical parameters for the tin(II) halides SnX_2 (X = Cl, Br, I) has been based on those combinations of the standard molar enthalpies of formation at 298.15 K, standard molar entropies at 298.15 K, standard molar enthalpies of fusion, and temperature variations of molar heat capacities, that best predict the experimentally measured vapour pressures above the respective liquid phases.

RESULTS AND CONCLUSIONS

A set of self-consistent thermodynamic parameters has been derived for the tin(II) halides, SnXX' (X, X' = Cl, Br, I), according to the above expressed benchmark. The molecular parameters of the gaseous species were calculated in modest time (< 2 hours per molecule) using quantum chemistry methods.

The thermodynamic data determined for the gas phase species at 298.15 K are given in Table 1 with, in the case of the binary tin halides, the data being significantly different from that which is presently available. The important practical consequences of the new data generated in this study is demonstrated in Figure 1, which compares the experimentally measured partial pressure of $SnBr_2(g)$

above $SnBr_2$(l) with predictions based upon the old and new data sets. The data for the mixed tin(II) halides is to our knowledge the first such data to be published. A more detailed account of this work has been published recently [6].

Species	$\Delta_f H^o$ / kJ mol^{-1}	S^o / J K^{-1} mol^{-1}	$C_p{}^o$ / J K^{-1} mol^{-1}
$SnCl_2$(g)	-205.443	305.716	54.579
$SnBr_2$(g)	-128.974	328.522	56.309
SnI_2(g)	-14.600	344.429	56.986
SnClBr(g)	-167.002	322.900	55.451
SnClI(g)	-110.000	330.945	55.806
SnBrI(g)	-71.787	342.287	56.652

Table 1. Thermochemical parameters for gaseous binary and mixed tin(II) halide species at 298.15 K determined in this study. N.B. Parameters are given to three decimal places by convention.

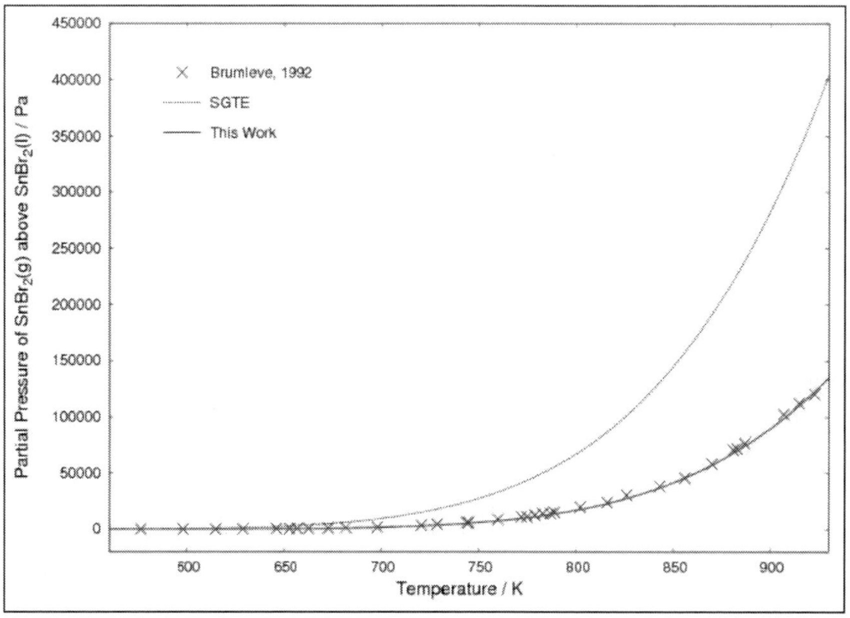

Figure 1. Comparison of the partial pressure of $SnBr_2$(g) above $SnBr_2$(l) measured experimentally by Brumleve [7] with modelling predictions based upon both the thermochemical data determined in this study and that found in the SGTE database [8].

REFERENCES

[1] R. H. Davies et al., CALPHAD, 26(2) (2002) 229.
[2] D.S. Payne, G.J. Wilson, S.A. Mucklejohn, G. Siddons, R. Devonshire, in Light Sources 2007, eds. M.Q. Liu and R. Devonshire, FAST-LS, (2007) 421.
[3] H. D. B. Jenkins and L. Glasser, Inorg. Chem., 42 (2003) 8702.
[4] H. D. B. Jenkins and L. Glasser, Inorg. Chem., 41 (2002) 4378; H. D. B. Jenkins and L. Glasser, L. J. Am. Chem. Soc., 126 (2004) 15809; L. Glasser and H. D. B. Jenkins, Inorg. Chem., 46 (2007) 9768; H. D. B. Jenkins and J. F. Liebman, J. Chem. Eng. Data., 54 (2009) 351.
[5] H. D. B. Jenkins, (2010), manuscript in preparation.
[6] G. Siddons, H. D. B. Jenkins, S. A. Mucklejohn, R. Devonshire, J. Chem. Eng. Data., 54(8) (2009) 2153; G. Siddons, PhD Thesis, University of Sheffield, (2010).
[7] T. R. Brumleve, S.R Walston, S.A. Mucklejohn and N.W. O'Brien., J. Chem. Therm., 24(7) (1992) 773.
[8] SGTE Substance Database v11.0. Scientific Group Thermodata Europe. (2006).

A reduced order model for the Dielectric Barrier Discharge (DBD) lamp using the Proper Orthogonal Decomposition (POD) technique

A. El-Deib[1], F. Dawson[1], S. Bhosle[2], G. Zissis[2]

[1]*ECE Department, University of Toronto, 10 King's College Rd., Toronto ON, M5S 3G4, Canada*
[2]*LAPLACE, Universite de Toulouse, Toulouse, UMR 5213, France*
Contact: amgad.eldeib@utoronto.ca

ABSTRACT

The power supply design process for a nonlinear load like the DBD lamp should be based on an accurate mathematical model for the lamp which requires a resonable simulation time. Therefore, a reduced order model for the DBD lamp is implemented by applying the Proper Orthogonal Decomposition technique on the PDE model of the lamp which is descritized by the Finite Volume Method. The reduced order model results are compared to the full order model results and good agreement between them is obtained.

1. INTRODUCTION

The circuit models of the DBD lamp suggested in the literature give a macroscopic description of the discharge process as they show only the temporal evolution of the lamp $i - v$ characteristics based on empirical formulations supported by experimental observations for limited operating conditions. On the other hand, a Partial Differential Equation (PDE) based physics model of the lamp describes the discharge in more detail as it offers both temporal and spatial evolutions of the different variables like electron, ion and neutral species densities in addition to the electric field within the discharge gap. Therefore a model based on the PDE system would describe the behaviour of the lamp more accurately without any experimental curve fitting procedures. However, the PDE model requires long simulation times, therefore in this paper a reduced order model for the DBD lamp based on the PDE system is suggested. The PDE system is converted into an Ordinary Differential Equation (ODE) system using the Finite Volume Method (FVM) and then the order of this ODE system is reduced using the Proper Orthogonal Decomposition (POD) technique.

2. DBD REDUCED ORDER MODEL USING POD

The PDE model of the DBD lamp used in this work is the same as the model described in [1] with the boundary conditions described in [2]. This model consists of the continuity equations for each charged particle in the plasma (electrons and positive ions in the Xe gas case which is used in this paper) coupled to Poisson's equation to solve for the electric field simultaneously. The PDEs describing the continuity equations are transformed into a number of ODEs using the FVM. An analytical solution for Poisson's equation is also derived as shown in [3]. The POD technique, also known as the Karhunen-Loeve method [4], is used in this paper to reduce the order of the nonlinear ODE model of the DBD lamp. To describe the main idea behind this technique, consider the following nonlinear ODE system resulting from the discretization of the original nonlinear PDE:

$$\dot{m}(t) = f(m(t), u(t)), \; y(t) = h(m(t), u(t)) \tag{1}$$

where $u(t)$ is the input to the system, $y(t)$ the output and $m(t)$ is the system states vector. The algorithm of the POD technique is as follows:

1. For a certain input $u(t)$, the state trajectory at certain instances of time $m(t_i)$ (called snapshot) is obtained by solving the full order system of equation 1. These snapshots are collected in one matrix M given by

$$M = [m(t_1), m(t_2), \ldots\ldots, m(t_m)] \tag{2}$$

2. The Singular Value Decomposition (SVD) of the matrix M is performed. If the singular values of this matrix M fall off rapidly, a low order approximation of this system can be computed. The SVD will result in the following decomposition

$$M = U\Sigma V^T \approx U_k \Sigma_k V_k^T \tag{3}$$

k corresponds to the order of the smallest singular value that will be included in the approximation process.

3. To obtain the reduced order model, the direct projection method can be used [4] where the matrix U_k is used as the transformation matrix, therefore the reduced system is given by

$$\dot{\eta}(t) = U_k^T f(U_k \eta(t), u(t)), \; y(t) = h(U_k \eta(t), u(t)) \tag{4}$$

3. RESULTS

The POD technique has been implemented using the direct projection approach in MATLAB. The program takes the results obtained from the FVM model and performs the reduction algorithm. The program output is the transformation matrix used to convert the actual states of the original system into the reduced states. As proof of concept, the results of two different cases are shown in figures 1(a) and 1(b). The figures show the waveform obtained from the FVM model using the specified number of volumes and the waveform obtained from solving the reduced order model. In figure 1(a), the solution obtained by the FVM model with 100 volumes corresponding to 400 variables (there are four equations being solved) is nearly matched with the solution using a reduced order model of only 200 variables.

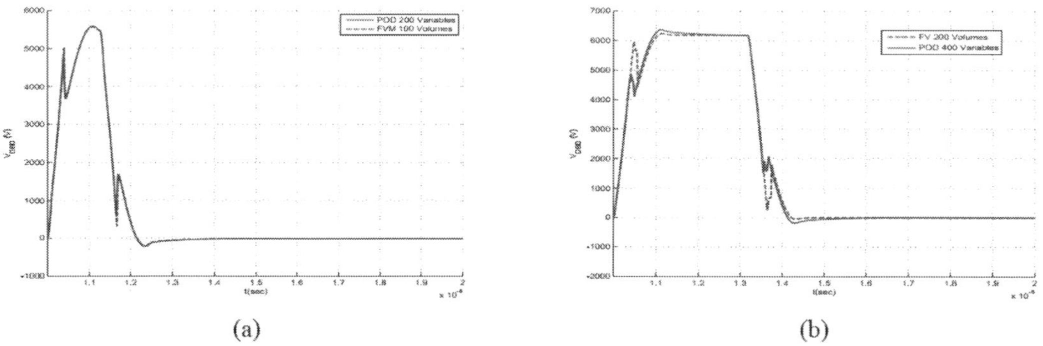

(a) (b)

Figure 1. Comparison between FVM and POD (a) case 1: 400 variables based FVM solution is reduced to a 200 variable solution and (b)case 2: 800 variables based FVM solution is reduced to a 400 variable solution

4. CONCLUSION

A reduced order model for the DBD lamp has been implemented by performing the Proper Orthogonal Decomposition (POD) technique on the FVM based ODE model. The results obtained show good agreement and it was possible to reduce the system by nearly one half in terms of the number of variables solved for. The advantage of the POD technique is that it can be applied to high-complexity linear as well as nonlinear systems. The disadvantage of the POD technique is that the resulting reduced system depends on the input signal used to find the snapshot matrix M and also it depends on the time instances at which the snapshots are taken.

REFERENCES

[1] U. Kogelschatz, "Dielectric-barrier discharges: Their history, discharge physics, and industrial applications," *Plasma Chemistry and Plasma Processing*, vol. 23, no. 1, pp. 1–46, March 2003.

[2] S. Bhosle, G. Zissis, J. Damelincourt, and A. Capdevila, "A new approach for boundary conditions in dielectric barrier discharge modeling," in *13th International Conference On Gas Discharge and their Applications GD2006*, Sept. 2006.

[3] A. El-Deib, "Modeling of and driver design for a dielectric barrier discharge lamp," Ph.D. dissertation, University of Toronto, 2010.

[4] S. Tan and L. He, *Advanced Model Order Reduction Techniques in VLSI Design*, 1st ed. New York: Cambridge University Press, 2007.

Solid-state lighting with tailored colour quality

A. Žukauskas[1] , R. Vaicekauskas[2] , A. Tuzikas[1] , M. S. Shur[3]
[1]*Institute of Applied Research, Vilnius University, Saulėtekio al. 9, bldg. III, Vilnius, LT-10222 Lithuania*
[2]*Department of Computer Science, Vilnius University Naugarduko g. 24, Vilnius, LT-03225 Lithuania*
[3]*Department of Electrical, Computer, and System Engineering, Rensselaer Polytechnic Institute, 110 8th Street, Troy, NY 12180 USA*
Contact: arturas.zukauskas@ff.vu.lt

ABSTRACT

We report on a prototype red-amber-green-blue (RAGB) tetrachromatic white lighting source that operates within a dynamical trade-off between several colour rendition characteristics, such as the ability to render colours with high fidelity and the ability to render colours with increased or decreased chromatic saturation. The SPD of the source is varied as a weighted sum of two trichromatic end-point SPDs with the maximized ability to render object colours with increased chromatic saturation (RGB source) and decreased chromatic saturation (AGB source), respectively. Such RAGB sources with tailored colour quality can provide "smart" lighting to fit individual needs and preferences of colour vision.

INTRODUCTION

Solid-state lighting technology, which is based on direct-emission and phosphor-converted light-emitting diodes (LEDs), allows to dynamically control spectral power distribution (SPD) of light. Colour rendition properties of a white multiwavelength lighting source can be tailored by varying the average driving current of individual LEDs within a cluster containing four and more coloured components.

Here we report on a tetrachromatic solid-state lighting source, which can be dynamically adjusted to enable a trade-off between different (and even opposing) colour rendition characteristics. The traded off characteristics are colour fidelity, colour saturating, and colour dulling.

TRADE OFF BETWEEN COLOUR RENDITION CHARACTERISTICS

Basically, two most important colour rendition properties of illumination are the ability to make object colours appear "natural" (colour fidelity) and the ability to make colours appear "vivid" and easily distinguishable (colour saturating) [1]. The two colour rendition properties are opposite, since when a multiwavelength lighting source is adjusted for rendering object colours with high fidelity, most colours do not have an increase in chroma and when it is adjusted for increasing chromatic saturation, few colours are rendered as natural. Therefore, a trade-off between colour fidelity and colour saturating can be performed. An additional colour rendition property considered here is the ability to make object colours appear desaturated (colour dulling). The later property opposes both colour fidelity and colour saturating and can be introduced into the trade-off as well.

In order to quantify the three colour rendition properties, we use the statistical metrics, which measures colour quality of illumination in the percentage of test colour samples with different behaviour of colour rendition vectors when a reference illuminant is replaced by a source under assessment [2]. In this approach, 1269 Munsell test colour samples are computationally sorted with regard to each of three colour rendition properties. The test colour samples with the colour shifts remaining within 3-step MacAdam ellipses are scored to colour fidelity index (CFI); and the samples with the colour rendition vector escaping from the 3-step MacAdam ellipses are scored to either colour saturation index (CSI) or colour dulling index (CDI) depending on whether the projection of the vector has an excess magnitude directed toward increased or decreased chroma, respectively.

A general approach to designing a lighting source with traded off colour rendition properties is as follows. Consider a multiwavelength LED cluster with the radiant fluxes of the constituent coloured LEDs adjustable in such a way that any of three colour rendition abilities (colour fidelity, colour saturating, or colour dulling) can be maximized. The corresponding SPDs are $S_{\mathrm{maxCFI}}(\lambda)$, $S_{\mathrm{maxCSI}}(\lambda)$, and $S_{\mathrm{maxCDI}}(\lambda)$, respectively. A dynamical trade-off between the three colour rendition properties is achieved by varying the SPD of the cluster as a weighted sum of the three SPDs with the maximized CFI, CSI, and CDI, respectively,

$$S(\lambda) = \sigma_1 S_{\max \mathrm{CFI}}(\lambda) + (1 - \sigma_1)[\sigma_2 S_{\max \mathrm{CSI}}(\lambda) + (1 - \sigma_2) S_{\max \mathrm{CDI}}(\lambda)], \quad (1)$$

where σ_1 and σ_2 are the weight parameters that control the trade-off ($0 \le \sigma_{1,2} \le 1$).

TETRACHROMATIC LED CLUSTER WITH TAILORED COLOUR QUALITY

In order to perform a dynamical trade-off between colour rendition properties, the cluster must contain independently driven coloured LEDs that generate white light of controlled chromaticity with different SPDs. The principle of additive colour mixing implies that the number of the coloured LEDs in such a cluster must be at least four.

A set of four LEDs that allow adjusting the composite SPD with any of three statistical colour rendition indices (CFI, CSI, and CDI) in excess of 50% using a single cluster can be found among commercial coloured LEDs. For instance, an appropriate set can be provided by Philips Lumileds Lighting Luxeon® Rebel LED family: direct-emission royal-blue (B, 452 nm peak wavelength) and green (G, 523 nm) InGaN LEDs; InGaN-based phosphor converted amber (A, 589 nm) LED, and direct-emission red (R, 637 nm) AlGaInP LED, respectively. For a prototype RAGB lamp composed of these LEDs, the highest colour fidelity ability is attained when all four LEDs are used (full RAGB source with optimal partial fluxes of the coloured components). Extinguishing the amber LED (RGB source) results in the highest colour-saturating ability, whereas the highest colour-dulling ability is obtained by excluding the red LED (AGB source).

For a tetrachromatic source, any SPD is a linear combination of two different SPDs with the same chromaticity. If the trichromatic spectral distributions $S_{\max\mathrm{CSI}}(\lambda)$ and $S_{\max\mathrm{CDI}}(\lambda)$ are defined, a single non-negative weight parameter σ ($0 \le \sigma \le 1$) is sufficient to run over all possible tetrachromatic SPDs, including the blend with the highest colour fidelity,

$$S(\lambda) = \sigma S_{\max \mathrm{CSI}}(\lambda) + (1 - \sigma) S_{\max \mathrm{CDI}}(\lambda). \quad (2)$$

Figure 1 shows the statistical colour rendition indices of the prototype RAGB LED cluster as functions of weight parameter σ. At $\sigma = 0$, the cluster has the highest colour-dulling ability (CDI = 55%), whereas CFI and CSI have low values of 26% and 1%, respectively. With increasing weight parameter to 0.35, CDI decreases to 4%, CSI remains low (2%), and CFI arrives to the highest value of 80%. With further increase of weight parameter to 1, CDI drops to 1%, CFI decreases to 10% and CSI attains the maximum value of 73%. Since the relative partial radiant fluxes of the RAGB LED cluster are varied as the weighted sums of the end-point fluxes with the same weight parameter, tailoring of the colour quality of the cluster can be performed using a single control.

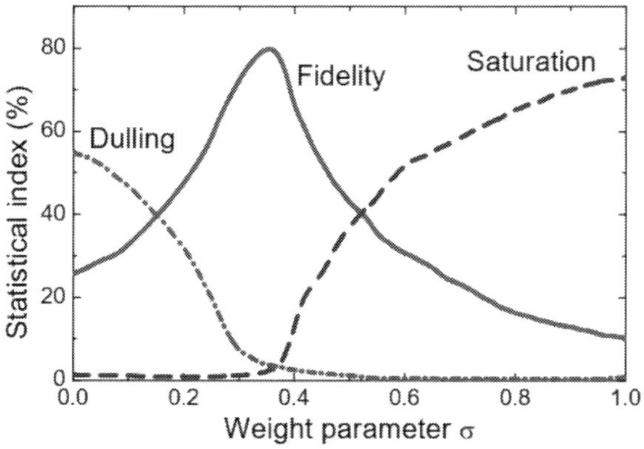

Fig. 1. Statistical colour rendition indices of colour fidelity (red line), colour saturation (blue line), and colour dulling (green line) as functions of trade-off weight parameter for the prototype RAGB LED cluster at a correlated colour temperature of 4000 K.

The proposed approach to colour rendition engineering can be used for meeting individual needs and preferences of colour vision. "Smart" lighting with tailored colour quality can find numerous applications in merchandise, architectural, entertainment, medical, and landscape lighting, for highlighting specific object colours, as well as in other colour-sensitive applications, such as phototherapy, filming, photography, and design.

REFERENCES

[1] M.S. Rea, J.P. Freyssinier-Nova, Color Res. Appl., 33 (2008) 192.
[2] A. Žukauskas, R. Vaicekauskas, F. Ivanauskas, H. Vaitkevičius, P. Vitta, M.S. Shur, IEEE J. Sel. Top. Quantum Electron. 15 (2009) 1753.

Electric Characteristics of Dielectric Barrier Discharge Lamp with Coaxial Construction

Yoshio Watanabe, Tomohiro Yamaguchi and Masatoshi Kobayashi
Department of Electrical, Electronics and Information Engineering,
Kanagawa University, Yokohama, 221-8686, Japan
Contact: watalab@kanagawa-u.ac.jp.

ABSTRACT

Electric characteristics of DBD lamp with a coaxial construction are investigated. Uniform irradiation is obtained. Then its equivalent circuit is derived. The analysis using the equivalent circuit indicates that the most of the discharge power is consumed in the plasma sheath formed on the inner surface of the tube. That might be the reason why uniform irradiation is produced in the DBD lamp.

Introduction

A cold cathode fluorescent lamp is commonly used as a backlight for a liquid crystal display in TV. A large-sized display panel needs a long-sized lamp of which discharge voltage is high. A dielectric barrier discharge (DBD) lamp has a long life due to no electrode inside the lamp tube and its discharge voltage is independent of the tube length. These features are convenient to fabricate the long-sized backlight. However, an ordinary DBD lamp in which a pair of electrodes is mounted on the outer surface of the lamp tube has the disadvantage of high discharge voltage, since the electric field for discharge must be formed through the glass wall of the lamp tube. Moreover, ignition voltage of DBD lamp will be high due to the absence of electron emission sources inside the tube [1].

In this paper, DBD lamp with a coaxial construction of low discharge voltage and low ignition voltage is developed and its characteristics are investigated.

Experimental Lamp

The tested lamp is shown in Fig.1. One of the electrodes made of a nickel wire is located on the axis of the tube. The other electrode made of a copper tape is mounted on the outer wall of the tube shown in Fig.2. A small quantity of mercury and 3 Torrs of argon gas are filled in the tube. The operation frequency is 100-250 kHz of sinusoidal wave form.

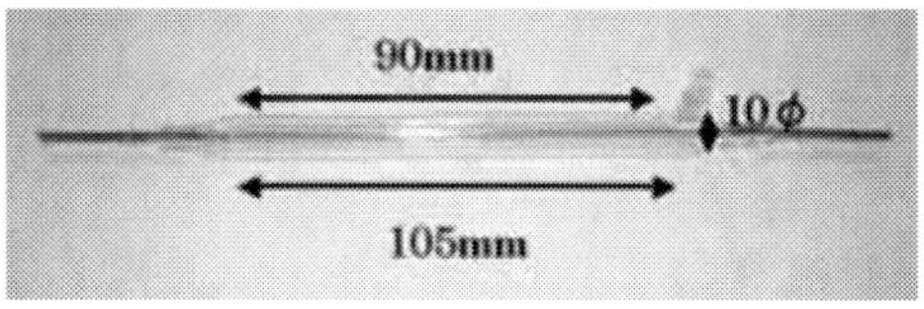

Fig.1 Overview of the tested lamp

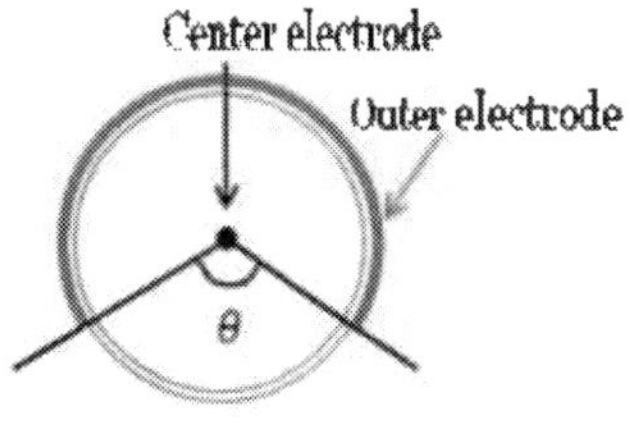

Fig.2 Electrode configuration

Experimental Result and Analysis by Equivalent Circuit

The overview of the lamp under operation is shown in Fig.3. Uniform intensity is observed by eyes.

Fig.3 Overview of the lamp under operation

To analyze the voltage and wattage characteristics, an equivalent circuit shown in Fig.4 is assumed where C_g is a capacitance due to the glass wall of the tube, C_s is a capacitance due to the plasma sheath formed on the inner surface of the tube, R_s is a plasma resistance in the plasma sheath and R_p is a plasma resistance in the discharge region. The values of C_g, C_s, R_s and R_p are determined by the following method. At first, four values are assumed at each discharge current and calculated the voltage and wattage characteristics as a function of operation frequency. Hereafter, the calculated results are compared with experimental results. Then the values are modified until two characteristic curves are fitted each other. The obtained result curves for the voltage and wattage as a function of operating frequency are shown in Fig.5 and Fig.6, respectively. The dots indicate experimental data and lines show calculated data. The estimated

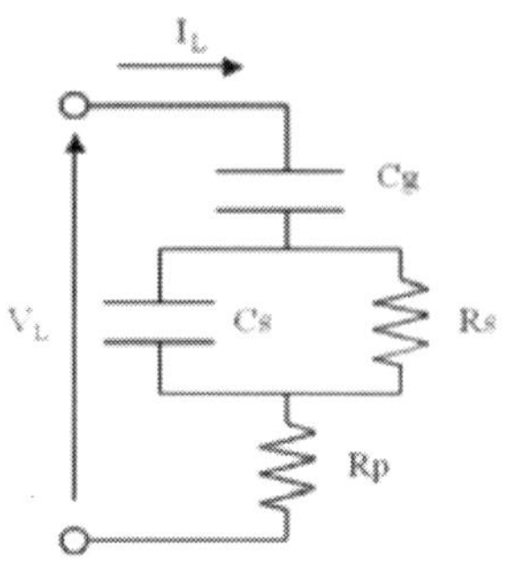

Fig.4 Assumed equivalent circuit

values, *C*g, *C*s, *R*s and *R*p, for calculation are shown in Fig.7 as a function of discharge current. The dependence of each parameter on the operation current seems to be reasonable. Using this equivalent circuit, discharge power at *R*s and *R*p are calculated. The results indicate that the discharge power in the tube is mainly consumed in the plasma sheath formed on the inner surface of the tube. For instance, the consumed power at *R*s is 2.64W while the power at *R*p is 0.001W, in the case of 100kHz and 10mA of discharge current. This might be the reason why uniform irradiation is established in the DBD lamp, since most of the discharge power is consumed in the plasma sheath.

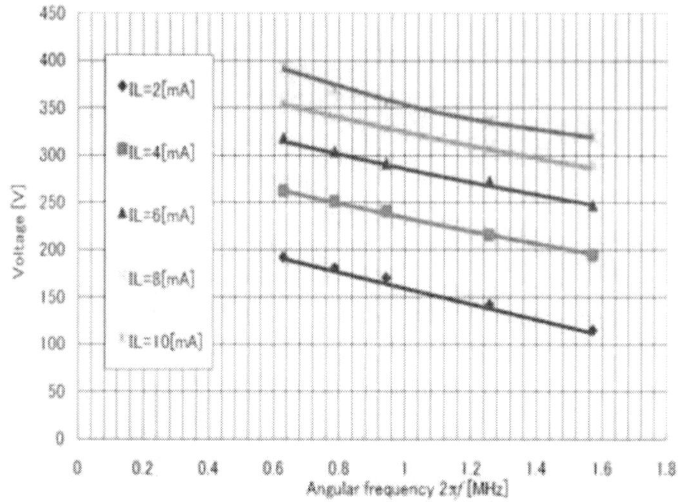

Fig.5 Calculated voltage characteristics vs. operation frequency

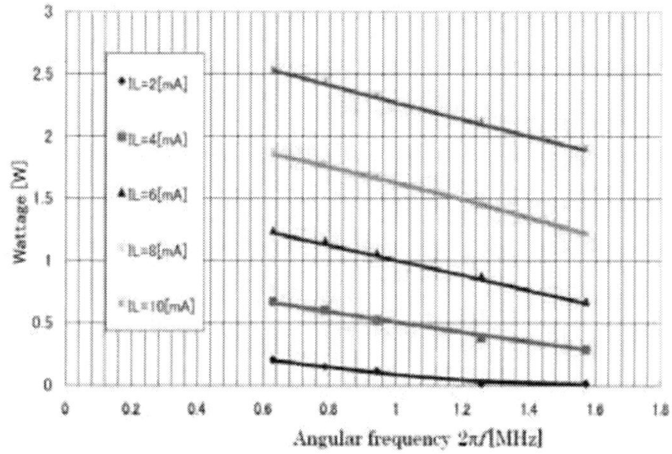

Fig.6 Calculated wattage characteristics vs. operation frequency

Conclusion

Electric characteristics of DBD lamp with a coaxial construction are investigated and its equivalent circuit is derived. The results indicate that the most of the discharge power is consumed in the plasma sheath that produces the uniform irradiation in the DBD lamp.

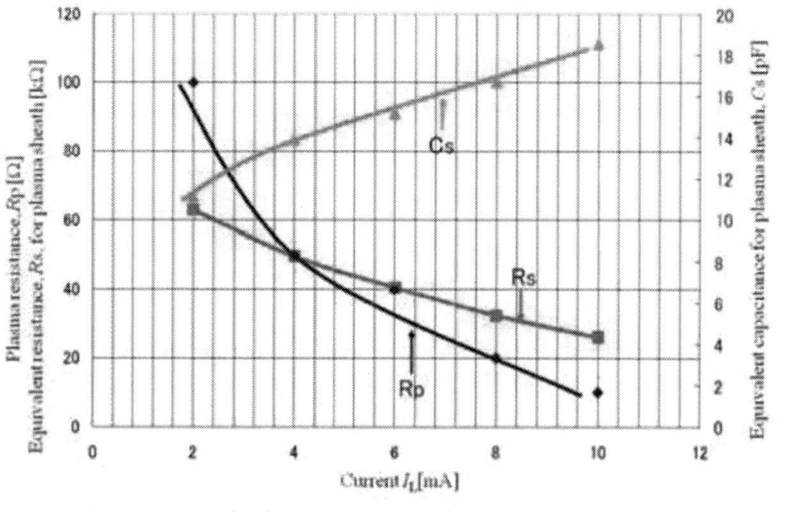

Fig.7 Estimated values, Cg, Cs, Rs and Rp, in calculation

References

[1] R.Lecheler, O.Schllmoser, W.Sowa, *Proc.* LS10, pp.465-466. (2004)

High Temperature Reliability of Driver for Solid State Lighting

S. Tarashioon[1,2], S.W. Koh[1,3], C. W.D. van Driel[3], G.Q Zhang[3]
[1]*Delft Institute of Microelectronics and Submicron Technology, Delft University of Technology, Mekelweg, 2628CD Delft, The Netherlands*
[2]*Material Innovation Institute, Mekelweg, 2628CD Delft, The Netherlands*
[3]*Department of Precision and Microsystem Engineering, Delft University of Technology, Mekelweg, 2628CD Delft, The Netherlands*
[4]*Philips Lighting, LightLabs, NL-5611BD, Eindhoven, The Netherlands*
Contact: S.Tarashioon@m2i.nl

ABSTRACT

Due to their long lifetime and high efficacy, solid state lighting (SSL) has the potential to revolutionize the illumination industry. It is generally known that the LED driver is one of the weakest components in the LED luminaries therefore the useful lifetime of the LED luminaries may be shorten by the failure of the LED driver [1]. In addition, different applications may require different topologies of LED drivers. Hence, there is a need to investigate and understand the reliability of LED drivers in solid state lighting application. Therefore, the goal of this study is to investigate the high temperature performance and reliability of LED drivers.

INTRODUCTION

Recently, solid state lighting has been the subject of considerable interest due to their potential to revolutionize the illumination industry. One of the key advantages of LED emitter is their high efficacy as compared to their incandescent counterpart. For example, they are known to have an efficacy of up to 150 lm/W, as compare to only about 15lm/W for a conventional 60–100 W incandescent light bulb [2,3]. Furthermore, LED luminaires also claimed to have a life of more than 50,000 hours and this exceeds the life of nearly all other light sources [4].

These long lifetime claimed by the manufacturer is often based solely on the estimated depreciation of lumen in the LED used in the product. However, a LED lighting system is composed of many components working together and their total system reliability is strongly dependent on the reliability of the weakest components. Since it has been known that the LED driver is one of the weakest components in the LED luminaries[1] and they degrade rapidly as temperature increases[5], there is a need to investigate the high temperature reliability of these drivers. However, a driver is again made of different components. Adding complexity to these failure modes, different applications, such as dimming function and ENERGY STAR requirements [6-7], may require different topologies of LED driver as shown in Figure 1. In addition, failed LED driver may have more than 1 failure modes. Typical driver's failure modes are as follows :-

- Short circuit
- Open circuits
- Current surges
- Over voltage
- Over current
- High noise level

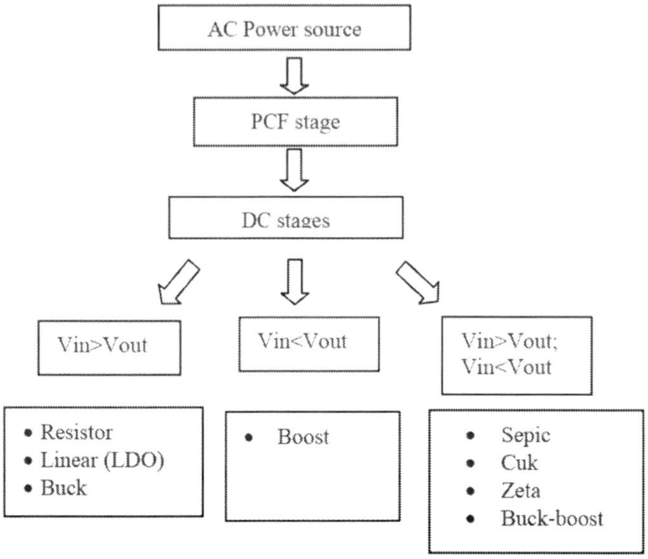

Figure 1: LED driver topologies for different input and output voltages of the system[1]

FUTURE WORK

The failure of the system may be induced by the degradation of the sub-components/interfaces or the interaction of the different sub-components. However, limited studies have tried to experimentally investigate the degradation process of these sub-components at harsh environments such as elevated temperature or overstress voltages. To the best of author's knowledge, there is only 1 high temperature study on the reliability of LED driver [5]. Therefore, the objective of this study is to investigate the reliability of the LED drivers and their subcomponents at harsh environments and to correlate the performances of the drivers with external ambient variables. In order to achieve this, a test board is designed to investigate the driver's reliability and the results will be presented in the conference.

REFERENCES

[1] G. Archenhold, "Driving Responsibly," in Mondo arc United Kingdom: Mondiale Publishing Ltd, 2009, pp. 93-94.

[2] "http://en.wikipedia.org/wiki/Light-emitting_diode."

[3] Patrick Mottier, "LEDs for Lighting applications", John wiley & sons, Inc., 2009

[4] U. S. DOE, "LED Measurement Series: LED Luminaire Reliability." vol. 2009.

[5] L. Han and N. Narendran, "Developing an accelerated life test method for LED drivers," 2009, p. 742209-742209-8

[6] Steve Winder, "Power supplies for LED driving", Elsvier Inc., 2008

[7] IEC 61347-2-13, "Lamp controlgear, Part 2-13: Particular requirements for d.c. or a.c. supplied electronic controlgear for LED modules", 2006

Thermodynamics processes in DBD driven excilamps estimated by jump of pressure method: thermal power controls

V.M. Tsvetkov[1], A.A. Pikulev[1], E.A. Sosnin[2], V.F. Tarasenko[2]

[1] Federal State Unitary Enterprise – Russian Federal Nuclear Centre-All-Russian Research Institute of Experimental Physics, 37, Mira ave., 607190, Russian Federation
[2] Optical Radiation laboratory, High Current Electronics Institute SB RAS, 2/3, Akademicheskii ave., 634055, Russian Federation
Contact: badik@loi.hcei.tsc.ru

ABSTRACT

In this report for determination of heat power of dielectric barrier discharges (DBD) driven KrCl-excilamp it is proposed to use slow part of pressure jump (characteristic time ~ 100 sec). The investigation results of excilamp heat working regimes versus cooling conditions are presented.

INTRODUCTION

Nowadays DBD driven excilamps have obtained a wide area of applications in technology, ecology and medicine. That results from following advantages of these UV-radiation sources: 1) high efficiency (10-15 % for exciplexes rare gas-halogen and 40-60 % for rare gases dimers); 2) narrow radiation line spectral bandwidth (5-10 nm); 3) long working time (more than 5 thousands hours for KrCl-excilamp) [1], [2].

During prolonged exploitation of excilamps the task of excilamp temperature control is appears for purpose of working gas overheating excluding. The excilamp bulb temperature is determined by two competitive processes: 1) heat power produced during discharge process and 2) thermo conduction processes.

To determine excilamp heat power one can use discharge power minus UV-radiation power of excilamp [3]. In this report for determination of heat power and average temperature of excilamps' bulbs it is proposed to use slow part of pressure jump (characteristic time ~ 100 sec).

RESULTS

The investigations have been conducted on KrCl-excilamp of barrier discharges (mixture $Kr:Cl_2 = 300:1$, total pressure is 240 Torr). The excilamp working length was 330 mm, diameter - 43 mm and discharge gap was equal to 8.5 mm. The excilamp was powered by bipolar power supply with voltage pulses amplitude 4.5 kV and pulses repetition rate 100 kHz.

To measure pressure jump in excilamp bulb there was used inductive differential pressure sensor DMI 0.1-2; the oscilloscope Tektronix TDS-3014B was used for measurements conduction.

The temporal change of excilamp heat energy E when working is described by equation

$$\frac{dE}{dt} = W - kS\Delta T, \tag{1}$$

where W – heat power; k – heat-transfer coefficient; S – square of excilamp cooled surface; ΔT – differential temperature between excilamp surface and ambient medium.

From equation (1) one can obtain that after excilamp switching on the slow part of pressure jump is changing by the following way

$$\Delta p = \Delta p_0 \left\{ 1 - \exp\left\{ -\frac{t}{\tau} \right\} \right\}, \quad \tau = \frac{mc_V}{kS}, \quad \Delta p_0 = \frac{\tau W p_0}{mc_V T_0}, \tag{2}$$

where Δp - the slow part of pressure jump; τ - characteristic time of heat exchange; m, c_V – mass and heat capacity of excilamp bulb; p_0, T_0 – initial gas pressure and temperature.

Equations (2) allow to find heat power W, heat-transfer coefficient k and characteristic time of heat exchange τ from pressure jump measurements. To define excilamp bulb temperature one can use the following relation

$$\Delta T = \frac{\Delta p T_0}{p_0}. \quad (3)$$

To define the heat power there were conducted pressure jump measurements during 100 sec after excilamp turning on without cooling. The increasing of pressure appeared to be almost linear and for 100 sec made up 25 Tor (temperature increasing is equals to 31 K). The calculated heat power equals to 48 W. The calculated Δp_0 value make up 160 Tor that corresponds to temperature jump about 200 K.

The characteristic time of heat exchange without cooling was found from pressure jump oscillogram after excilamp turning off. Before turning off lamp had been working during 100 sec. In 100 sec after turning off the pressure almost linearly dropped by 5 Tor (temperature decreased by 6 K). The characteristic time of heat exchange without cooling equaled to 500 sec.

The using of cooling by means of pressured air blowing over excilamp outward and inward surfaces hastens the heat-transfer process by an order of magnitude and decreases τ up to 30-50 sec. The cooling limits the amplitude of pressure jump (slow part) by value of 16Tor, i.e. maximal excilamp bulb temperature increasing equals only to 20 K.

CONCLUSION

In this report it is shown that slow part of pressure jump gives us possibility to determine such important parameters as heat power and excilamp temperature, heat-transfer coefficient and characteristic time of heat exchange between excilamp surface and ambient medium. Experiments, conducted on KrCl excilamp of barrier discharges, have shown that the using of cooling by means of pressured air hastens the heat-transfer process and decreases maximal temperature of excilamp bulb in ~ 10 times.

ACKNOWLEDGMENTS

This work was supported by the ISTC (Project No. 3583p).

REFERENCES

[1] Boyd I.W., Zhang J.Y. // Nuclear Instr. and Meth. in Phys. Research B, 121 (1997) 349.
[2] E.A. Sosnin, T. Oppenländer, V.F. Tarasenko, J. Photochem. Photobiology. C, 7 (2006) 145.
[3] The encyclopaedia of low-temperature plasmas. Vol. XI-4. Gas and plasma lasers. Moscow, FIZMATLIT, 2005, p.530, 531. [in Russian]

A "Class-A2" Ultra-Low-Loss Magnetic Ballast with improved power factor for T5 Fluorescent Lamps

W.M. Ng , D.Y. Lin, S.Y.R. Hui, *Fellow IEEE* and W. Yan, *Member IEEE*
Centre for Power Electronics
City University of Hong Kong
Hong Kong, China
Email: eeronhui@cityu.edu.hk

***Abstract*—This paper shows that magnetic ballasts with power factor correction are feasible for high-voltage T5 28W lamps. Compared with the Class A2 of energy-efficient electronic ballast, the magnetic ballast system has lower ballast loss, lower product and maintenance costs, longer lifetime and higher system efficacy. Therefore, the patent-pending magnetic ballast proposal presented in this paper may provide a truly sustainable lighting solution to the lighting industry.**

I. INTRODUCTION

T5 fluorescent lamps have been increasingly used in both office and industrial applications. Compared with T8 lamps, T5 lamps can, therefore, have less impact on the global environment than T8 lamps [1].

The sustainable lighting technology should satisfy the criteria of (1) energy saving, (2) long lifetime, (3) recyclability. This patent-pending proposal [4] presents a novel T5 electromagnetic ballast, LLC, that meets the three criteria for sustainable lighting technology.

Since T5 lamps have on-state voltage close to the mains voltage, series inductive-capacitive (LC) ballasts (Fig.1) for high-voltage lamps can be used [2,3]. To precisely determine the LC ballast parameters for a T5 28W lamp, the physical discharge lamp model reported in [5,6] is adopted for this study. To increase the system power factor, an inductor is connected across the AC mains in (Fig.2).

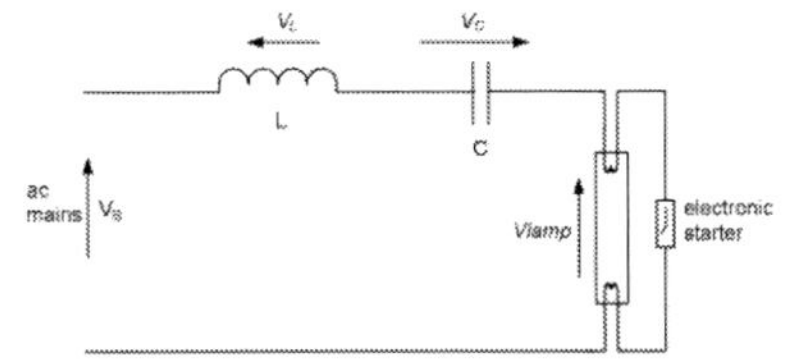

Fig.1 Circuit diagram of the magnetic ballast (LC ballast) for T5 28W lamps

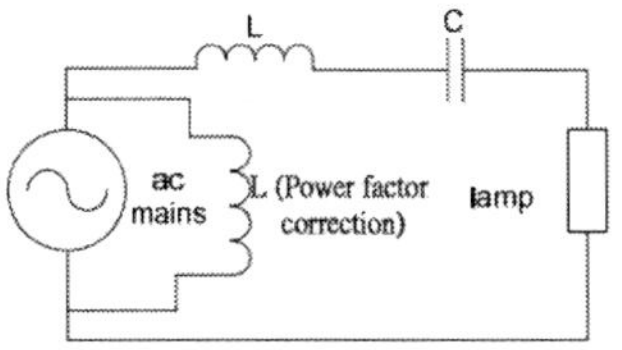

Fig.2 An LLC ballast

II. EXPERIMENTAL VERIFICATION

The inductor-capacitor magnetic ballast has been applied to a 28W T5 lamp. The experimental results for full power operation at input voltage of 230V are shown in Fig.3 and Fig.4. Experiments are then repeated to operate the LLC ballast to evaluate its performance with power factor correction. The corresponding measured waveforms are shown in Fig.5 and Fig.6.

Center for Power Electronics, City University of Hong Kong.

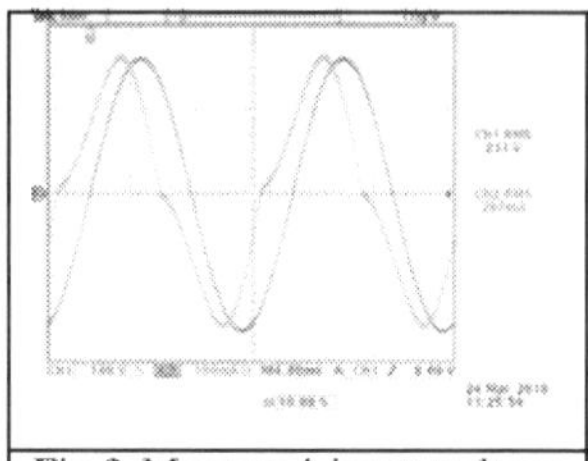

Fig.3 Measured input voltage Vs & input current I_s without power factor correction inductor

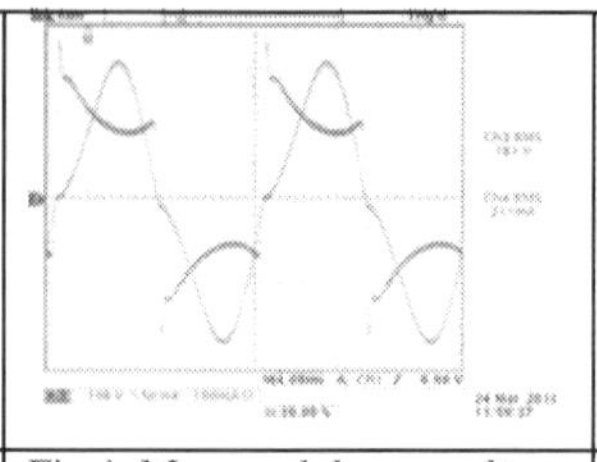

Fig.4 Measured lamp voltage V_{lamp} & lamp current I_{lamp} without power factor correction inductor

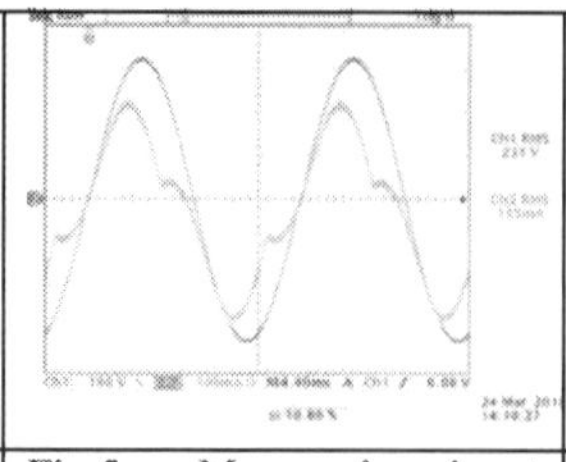

Fig.5 Measured input voltage Vs & input current I_s with power factor correction inductor

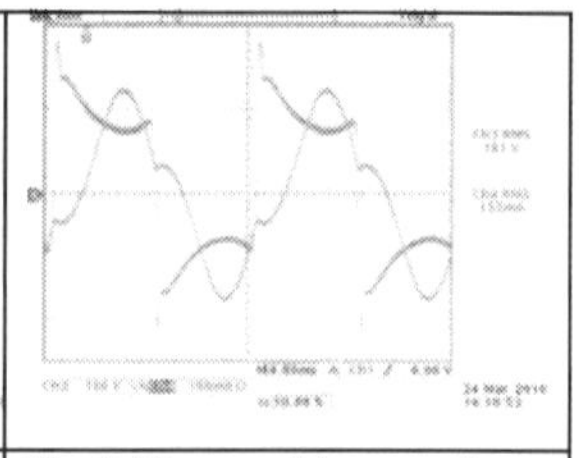

Fig.6 Measured lamp voltage V_{lamp} & lamp current I_{lamp} with power factor correction inductor

The electrical and luminous performance of the ultra-low-loss LLC ballast has also been compared with electronic ballasts commercially available in the market. Using the same Philips TL5 28W/865 lamp, the LLC ballast and two electronic ballasts are tested and compared. All tests are carried out in a photocolorimeter with an integrating sphere. Table 1 includes the measurements obtainied when the three systems are tested at V_s=230V using the same lamp. While the input power values of the three systems are almost the same, it is noted that the system loss of the LLC ballast (3.23W) is much smaller than those of the two electronic ballasts (4.65W and 3.28W). This practical result confirms the inherent low loss feature of using magnetic ballast for high-voltage and low-current lamps because the "low-current" feature of the lamp enables low conduction and core losses in the magnetic choke. Consequently, the energy efficiency (90%) and luminous efficacy (75.38 Lumen/Watt) of the LLC ballast are higher than those of the other two electronic counterparts, despite that fact that high-frequency operation of the lamp would yield a higher luminous output.

Table 1:Comparison of electrical and luminous performance based on the use of the same Philips T5 28W lamp

Model	Input Power (W)	Lamp Power (W)	Ballast Loss (W)	System power factor	Luminous Flux (Lumen)	Energy Efficiency (%)	System Luminous Efficacy (Lumen/Watt)
Ultra-low-loss LLC ballast	32	28.766	3.23	0.8908	2415.5	90	75.38
Philips EB-S128 TL5 230	30.95	26.30	4.65	0.9422	2188.1	84.98	70.70
Osram QT-FH 1X14-35 230240 CW	30.90	27.62	3.28	0.973	2263.8	89.39	73.26

III. Conclusion

Explanations and practical verification are provided in this paper to confirm that LLC ballast can match or even outperform electronic ballast in terms of energy efficiency and luminous efficacy for high-voltage and low-current lamps such as T5 28W fluorescent lamps.

References

[1] "T5 Fluorescent lamp system", National Lighting Information Program, Vol. 6, Issue 1, Rensselaer Polytechnic Institute, Troy, NY 12180 USA, website http://www.lrc.rpi.edu/programs/nlpip/publications.asp

[2] Dan E Rothenbuhler;. Samuel A. Johnson; Glenn A. Noble; Jon P. Seubert; "Preheating and starting circuit and method for a fluorescent lamp" US patent 5,736,817, April 7, 1998.

[3] Dan E Rothenbuhler and. Samuel A. Johnson, "Resonant voltage multiplication, current-regulating and ignition circuit for a fluorescent lamps", US patent 5,708,330, Jan. 13, 1998

[4] S.Y.R. Hui, D.Y. Lin and W.M. Ng, "A Passive LC Ballast and Method of Manufacturing a Passive LC Ballast", Patent Application No. PCT/IB2009/007289, 03 Nov 2009

[5] W. Yan and S.Y.R. Hui;, ""A semi-theoretical fluorescent lamp model for time-domain transient and steady-state simulations", IEEE Transactions on Power Electronics, Volume 22, Issue 6, Nov. 2007 Page(s):2106 – 2115

[6] D. Lin, W. Yan, G. Zissis, and S. Y. R. Hui, "A simple physical low pressure discharge lamp model," in Energy Conversion Congress and Exposition, 2009. ECCE. IEEE, 2009, pp. 2051-2058

Temperature measurements at thoriated cathodes in a model lamp and its interpretation by numerical simulation

A. Bergner[1,2], M. Westermeier[2], C. Ruhrmann[2], J. Mentel[2], P. Awakowicz[2],

[1]*Institute for Theoretical Electrical Engineering, Universitätsstr. 150, 44801 Bochum, Germany*
[2]*Institute for Electrical Engineering and Plasma Technology, Universitätsstr. 150, 44801 Bochum, Germany*
Contact: Andre.Bergner@ruhr-uni-bochum.de

ABSTRACT

This work deals with temperature measurements at thoriated cathodes in a model lamp and the emitter effect of thorium. The cathode is operated in the diffuse mode with direct current (DC) and its temperature is recorded by the two-dimensional 1λ-method. Numerical simulations of the cathode temperature in dependence on the work function are presented and compared with the experimental results. The reduction of the work function and cathode temperature by thorium is shown and a physical explanation is given, too.

GENERAL

Since many years high intensity discharge (HID) lamps are used in applications where high lumen output levels are requested such as sports, industrial or automotive lighting. To enlarge the efficacy of the lamps and thereby their life time scientific reserach is spent on the emitter effect which can be induced by elektropositive metals such as thorium. The emitter effect reduces the work function of tungsten effecting that the electrodes operate at lower temperatures and power losses. In a paper by Kühn and Kock [1] the diffuse mode of a thoriated cathode is investigated. The arc is operated in argon under atmospheric pressure. Many important plasma parameters such as the electron and gas temperature are measured in front of the cathode by emission spectroscopy. Unfortunately the lateral electrode temperature profile was not determined within this work. So we measured the lateral temperature profile of a thoraited cathode under the same conditions in the Bochum model lamp described by Dabringhausen *et al.* [2]. The cathode was a thoriated (1.8% ThO_2) tungsten rod (diameter 0.6 mm), the anode a pure tungsten rod (diameter 2 mm). Both had a length of 10 mm and the discharge gap was 10 mm, too.

TEMPERATURE DETERMINATION

To determine the lateral electrode temperature profile we used a special version of the so called 1λ-method given by Luijks *et al.* [3]. A two-dimensional picture of the electrode is taken with an interference filter for 890 nm and a FWHM transmission of 10 nm. The advantage is that the arc attachment becomes visible in contrast to spectroscopic methods. For the temperature measurements the electrode is focused on a CCD camera with the filter in front. By calibration with a tungsten ribbon lamp it is possible to determine absolute temperatures from these pictures. Furthermore the lateral temperature profile of the electrode is fitted with the least squares method to reduce measurement inaccuracies. In this way we determined the electrode tip temperature $T_{c,tip}$ for DC currents of 6 A and 8 A. The extrapolated value of $T_{c,tip}$ is 2904 K for 6 A and 2957 K for 8 A, respectively. These results shall be interpreted by numerical simulation.

NUMERICAL SIMULATION

We performed numerical simulations of the cathode for different work functions to find out the influence of thorium on the temperature of the cathode. The simulation method used in this work was given by Lichtenberg *et al.* [4] and Dabringhausen *et al.* [5]. A review of the theory of cathodes of high pressure arc discharges is given by Benilov [6]. For the simulation we used a commercial finite element tool to solve the stationary heat conduction equation ($-\nabla \cdot (\kappa \nabla T_c) = 0$) on a two-dimensional grid. It has the dimensions of the experimentally investigated electrode. For the solution of the differential equation a constant foot point temperature was chosen as one boundary condition (Dirichlet). Another boundary condition (Neumann) is the heat flux density on the electrode surface. It can be devided into two parts namely the heat flux density from the plasma (q_p) to

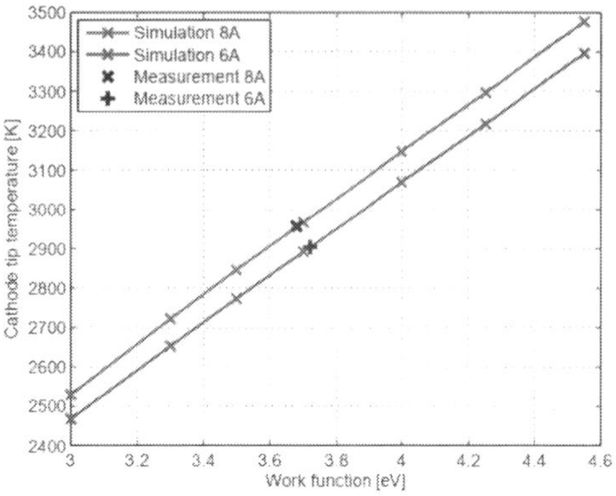

Figure 1. Simulated cathode tip temperature in dependence on tungsten work function for 6 A and 8 A, additionally measured tip temperature for these two currents

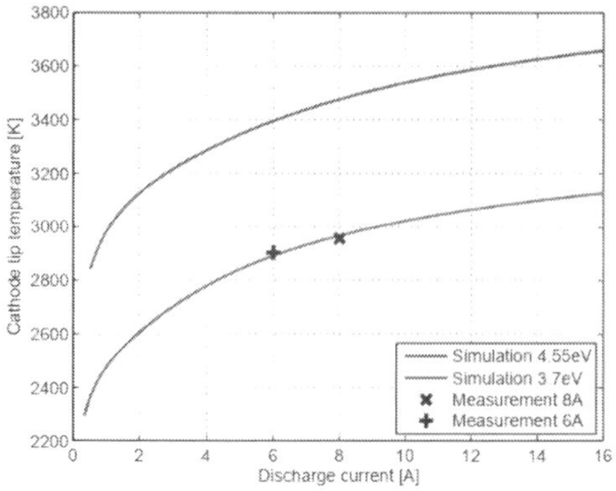

Figure 2. Simulated cathode tip temperature in dependence on discharge current, additionally measured tip temperature for 6 A and 8 A

the electrode and the thermal radiation from the electrode to the ambient. The heat conduction equation was solved for a few work functions between 3 eV and 4.55 eV. Results for $T_{c,tip}$ are given in figure 1 and 2 where the the measured data are also marked. As expected the cathode tip temperature increases with increasing work function and increasing current, respectively.

INTERPRETATION

As shown in figure 1 and 2 the work function of a thoriated electrode is reduced by the emitter effect. The numerical value is approximately 3.7 eV but the work function of pure tungsten is 4.55 eV. The corresponding values of $T_{c,tip}$ are 2892 K (3.7 eV) and 3396 K (4.55 eV) for 6 A, 2968 K (3.7 eV) and 3477 K (4.55 eV) for 8 A. Investigations of thoriated tungsten by Brüche and Mahl [7] have shown that a monolayer of thorium atoms can not be maintained on tungsten surfaces for temperatures above 2500 K by an equilibrium between evaporation of thorium and its replacement from the interior of the electrode body. But the monolayer will be sustained if the evaporated thorium atoms are brought back as thorium ions to the cathode surface. It can be shown that the thorium ion current is much smaller than the argon ion current. So the thorium ion current can be disregarded within the simulation. By this thorium ion current a monolayer of thorium atoms is deposited on the cathode end face. It reduces the work function of tungsten. Thereby the cathode operates at reduced temperatures and power losses, respectively.

ACKNOWLEDGEMENT

This work was supported by DFG Research Training Group 1051.

REFERENCES

[1] G. Kühn, M. Kock , J. Phys. D: Appl. Phys. **39** (2006) 2401-2414.

[2] L. Dabringhausen *et al.*, J. Phys. D: Appl. Phys. **35** (2002) 1621-1630.

[3] G. M. J. F. Luijks, S. Nijdam, H. v. Esveld, J. Phys. D: Appl. Phys. **38** (2005) 3163-3169.

[4] S. Lichtenberg *et al.*, J. Phys. D: Appl. Phys. **38** (2005) 3112-3127.

[5] L. Dabringhausen *et al.*, J. Phys. D: Appl. Phys. **38** (2005) 3128-3142.

[6] M. S. Benilov, J. Phys. D: Appl. Phys. **41** (2008) 144001.

[7] E. Brüche, H. Mahl, Z. tech. Phys. **12** (1935) 623-627

One dimensional modelling of XeCl barrier discharge exciplexe lamp

L.T Doanh[1], S. Bhosle[1], G. Zissis[1]

[1]*Université de Toulouse; UPS, INPT; LAPLACE (Laboratoire Plasma et Conversion d'Energie)*
118 route de Narbonne, F-31062 Toulouse cedex 9, France
CNRS; LAPLACE; F-31062 Toulouse, France
Contact: lethanh@laplace.univ-tlse.fr , doanhniceeve@yahoo.com,

ABSTRACT

Basing on the previous simulation models for Dielectric Barrier Discharge (DBD) lamps [1][2] and the knowledge on their boundary conditions [2] we present a one –dimensional modelling of a DBD excilamp for the mixture of Xenon and Chlorine at high pressure. The model presented here permits us to describe the space and time evolution of all species considered in the volume of the discharge when the DBD lamp is supplied with various waveforms.

Keywords: Dielectric Barrier Discharge (DBD), Excilamp, Excimers, Exciplexe, Ultraviolet, Power Supply.

1. INTRODUCTION

Excilamps are artificial Ultraviolet sources based on the emission of excimers or exciplexes. These ultraviolet are widely used in several industrial applications such as ozone synthesis, surface treatment or modification, thin film deposition as well as lighting and medical technology.

In this paper we have developed a drift diffusion model of a plane excilamp, in the mixture of Xenon and chlorine, which is implemented in COMSOL. Depending on the boundary condition and the power supply mode applied for the XeCl excilamp, the densities of all gas species during the discharge are computed according to the space and time.

2. DESCRIPTION OF THE MODEL

2.1 Geometry of a DBD excilamp

The figure 1.1 presents the discharge geometry of the XeCl DBD lamp and the associated resolution domains considered in our model.

The figure 1.2 presents the list of chemical reactions considered in our model. Here, all reaction rates are set according to the literature [3,4]. The transport coefficients and the direct ionisation coefficient are calculated owing to Bolsig software [5]

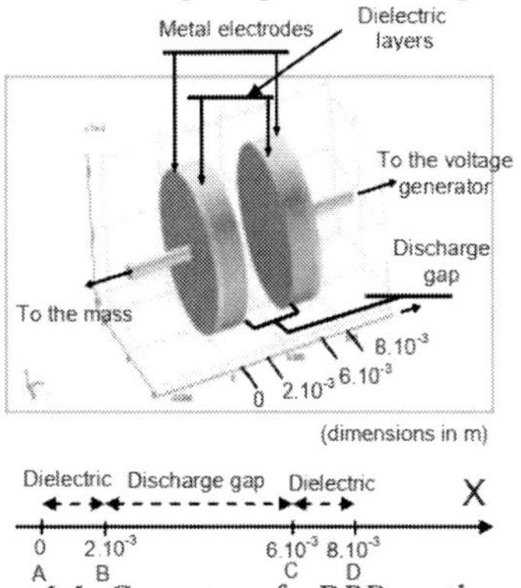

Figure 1.1: Geometry of a DBD excilamp

Reactions	Reaction rates
$Xe + e^- \rightarrow Xe^*_{exc} + e^-$	K1
$Xe + e^- \rightarrow Xe^*_{res} + e^-$	K2
$Xe + e^- \rightarrow Xe^*_{met} + e^-$	K3
$Xe^*_{met} + e^- \rightarrow Xe^+ + 2e^-$	K4
$Xe + e^- \rightarrow Xe^+ + 2e^-$	K5
$Xe^*_{exc} + Xe \rightarrow Xe^*_{res} + Xe$	K6
$Xe^*_{exc} + Xe \rightarrow Xe^*_{met} + Xe$	K7
$Xe^*_{res} + Xe \rightarrow Xe^*_{met} + Xe$	K8
$Xe^*_{met} + Xe \rightarrow Xe^*_{res} + Xe$	K9
$Xe^*_{res} + 2Xe \rightarrow Xe^*_2(O^+_u) + Xe$	K10
$Xe^*_{met} + 2Xe \rightarrow Xe^*_2(^3\Sigma^+_u) + Xe$	K11
$Xe^*_2(O^+_u) + Xe \rightarrow Xe^*_2(^1\Sigma^+_u) + Xe$	K12
$Xe^*_{res} + Xe^*_{res} \rightarrow Xe^+ + Xe + e^-$	K13
$Xe^*_{met} + Xe^*_{met} \rightarrow Xe^+ + Xe + e^-$	K14
$Xe^*_{res} + Xe^*_{met} \rightarrow Xe^+ + Xe + e^-$	K15

Reactions	Reaction rates
$Xe^+_2 + e^- \rightarrow Xe^*_{exc} + Xe$	K16
$Xe^*_2(^1\Sigma^+_u) \rightarrow 2Xe + 172nm$	K17
$Xe^*_2(^3\Sigma^+_u) \rightarrow 2Xe + 172nm$	K18
$Xe^*_2(^3\Sigma^+_u) \rightarrow 2Xe + 172nm$	K19
$Xe^*_2(O^+_u) \rightarrow 2Xe + 152nm$	K20
$Cl_2 + e^- \rightarrow Cl^*_2 + e^-$	K21
$Cl_2 + e^- \rightarrow Cl + Cl^-$	K22
$Cl_2 + e^- \rightarrow Cl^+_2 + 2e^-$	K23
$Xe^+ + Cl^- \rightarrow XeCl^*$	K24
$Xe^+_2 + Cl^- \rightarrow XeCl^* + Xe$	K25
$Cl^+_2 + Cl^- \rightarrow Cl_2 + Cl$	K26
$XeCl^* \rightarrow Xe + Cl + 308nm$	K27
$Cl^*_2 \rightarrow 2Cl + 259nm$	K28
$XeCl^* + Xe \rightarrow Xe + Xe + Cl$	K29
$XeCl^* + 2Xe \rightarrow Xe_2Cl^* + Xe$	K30
$Xe_2Cl^* \rightarrow 2Xe + Cl + hv$	K31

Figure 1.2: List of the reactions in Xenon/Chlorine discharge

2.2 Governing equations

The equation system in our model is described under the coupling of the drift –diffusion equations and Poison's equation. These systems are considered in local field approximation, consequently, the transport coefficients are considered as exclusively depending on the electric field.

$$\begin{cases} \vec{\nabla}.(-\varepsilon\vec{\nabla}V) = e(n_{i+} - n_{i-} - n_e) \\ \dfrac{\partial n_e}{\partial t} + \vec{\nabla}.(-D_e\vec{\nabla}n_e + \mu_e n_e \vec{\nabla}V) = S_e \\ \dfrac{\partial n_{i+}}{\partial t} + \vec{\nabla}.(-D_{i+}\vec{\nabla}n_{i+} - \mu_{i+} n_{i+}\vec{\nabla}V) = S_{i+} \\ \dfrac{\partial n_{i-}}{\partial t} + \vec{\nabla}.(-D_{i-}\vec{\nabla}n_{i-} - \mu_{i-} n_{i-}\vec{\nabla}V) = S_{i-} \end{cases}$$

Here :

n_{i+}, n_{i-}, n_e : Density of electron, positive and negative ions

D_e, D_{i+}, D_{i-}: Diffusion of electron, positive and negative ions

μ_e, μ_{i+}, μ_{i-}: Mobility of electron, positive and negative ion

S_e, S_{i+}, S_{i-}: Source term for electron, positive and negative ions

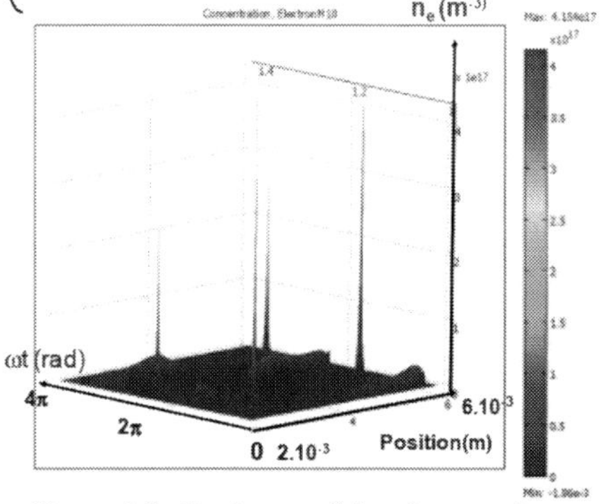

Figure1.3: Evolution of the electron density

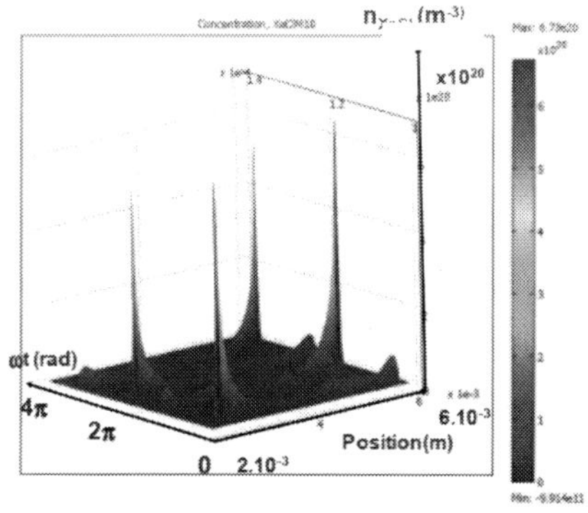

*Figure 1.4: Evolution of the exciplexe XeCl**

2.3 Boundary conditions

The boundary conditions for the electrostatic potential and all species on the dielectric walls in our model are defined according to [2].

3. RESULTS AND DISCUSSIONS

The figure 1.3 and 1.4 present the evolution, in the discharge volume , respectively of the electron density and XeCl* exciplexes density on two periods of sine wave 50kHz, 8kV peak, voltage source.

The obtained results show that the density peaks of species occur at each half period, on both sides of the gas gap. They are slightly distant from the surface of the dielectric. The maximum charge density achieved on the anode (left side on figure 1.3 and 1.4) is much lower than the cathode; this is affected by the polarity of the supply voltage characteristics.

4. CONCLUSION

The model presented in this paper describes the evolution of a DBD excilamp for various power supplies. It brings an important tool for the investigation of the influence of different power supplies on the production of the UV sources. Moreover, it will help in the design of innovative power supply topologies in order to achieve a high level of UV power and efficiency of the excilamp

REFERENCES

[1] S. Bhosle, R.V. Barrientos, G. Zissis, M.P. Pacheco, J.P. Sotelo, July 15-20, 28th ICPIG, Prague, Czech Republic, 2007

[2] S. Bhosle, R. Diez, H. Piquet, L.T. Doanh, B.Rahmani, D. Buso, 2008, COMSOL Conference, Hanover, 2008

[3] M.M. Turner, P.W. Smith, April, IEEE Transactions on Plasma Science, 1991

[4] A. Oda, Y. Sakai, H. Akashi, H. Sugawara, June, J.Phys.D: Appl.Phys, 32, 1999

[5] "The Siglo Database", CPAT and Kinema Software, 1995

Cathode sheath formation in Xe_2^*-DBDs – simulation and experimental investigation

M. Paravia[1], Sz. Beleznai[2], J. Ensslin[1], R. Kling[1], W. Heering[1]

[1] *Light Technology Institute, Karlsruhe Institute of Technology, Engesserstr. 13, 76131 Karlsruhe, Germany*
[2] *Dept. of Atomic Physics, Budapest University of Technology and Economics, Budafoki ut 8, 1111 Budapest, Hungary*

Contact: mark.paravia@kit.edu

ABSTRACT

Xenon excimer dielectric barrier discharges (DBDs) can be used as mercury-free flat light sources. But for DBDs the sheath regions are important due to small gas gap distances within several millimeters. In this work we present measured and simulated spatial and temporal resolved NIR emission. From this cathode sheaths between 200 µm to 400 µm in front of the barrier are derived.

INTRODUCTION

The breakdown of pulsed Xenon excimer DBDs is an avalanche breakdown within less than 100 ns. During breakdown a positive space-charge region moves form anode to cathode and Xe excitation shows high efficiencies of up to 71 % in the positive column [1]. However after breakdown and until current commutation or self-extinguishing a lossy cathode sheath is sustained. By this the discharge becomes an inefficient glow discharge. The small gas gap distances influences the total efficiency of DBDs by the ratio of cathode sheath to gas gap.

It is known that NIR emission (Xe^{**}-Xe^*) can be used for spatially resolved characterization [2]. In this work we show the development of cathode sheath by side on measurements of NIR emission combined with numerical simulation.

EXPERIMENTAL SETUP AND SIMULATION

We use a plane DBD with gas gap $d = 10$ mm at a Xenon pressure $p_{Xe} = 100$ mbar. The DBD is driven with a resonant pulse gear at a repetition rate of $f = 20$ kHz [3]. The temporal and spatial breakdown is detected by ICCD camera PCO Dicam Pro combined with SCHOTT RG780 filter. The electrode area is 3.5 cm in width and 0.7 cm in depth to keep depth of field.

In addition to experimental results a self-consistent numerical model, as presented in [4], is used. Among others the one-dimensional model describes temporal and spatial evolution of 14 Xenon plasma species, particle densities and mean electron energy. In simulation the spatial resolution is 40 µm. In order to combine experiment and simulation, the experimental waveform is oscilloscopically measured and is linearly interpolated for smaller time steps. The data are used as input for the simulation.

RESULTS

The time-resolved lamp voltage and NIR emission for experiment (a) and simulation (b) are depicted in figure 1. In experiment a small pre-ignition is observed at $t = 1.0$ µs. The reason is that ignition voltage is barely reached but ignition is not supported by high current density. The DBD ignites homogeneously at 1.3 µs in the negative half swing of pulsed excitation. A reignition starts at 2.2 µs with opposite electrode polarity. In simulation the DBD ignites at 1.5 µs and 2.1 µs. Besides missing pre-ignition simulation matches experiment in temporal behavior and emission intensity.

The NIR emission is depicted as a temporally and spatially resolved streak image in figure 2. The cathode sheath region is formed by development of ionisation wave front coming from anode (top at 1.3µs) to cathode (bottom). The main emission is located adjacent to the sheath region, while the positive column emits weakly. During glow phase after breakdown the sheath region decreases to a minimum width of 250 µm in experiment.

For the reignition (at ~ 2μs) the previous cathode sheath is neutralized by fast electrons, but is still visible. The breakdown remains as avalanche type forming a new cathode sheath due to opposite electrode polarity. The cathode sheath region reaches a minimum width of 200 μm. Again the positive column emits weakly.

The simulation shows the development of ionization wave front more clearly. We observe a good spatial agreement with experiment, but the simulated minimum width is approximately 400 μm. Consistent with experimental data the sheath width after reignition with approximately 320 μm is smaller than the first sheath width. The positive column during first ignition is stronger than that during reignition, which is in consistence with the experiment.

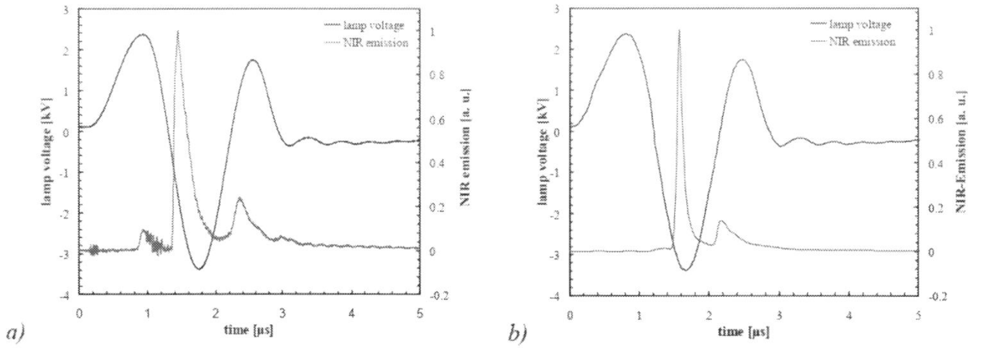

Figure 1Lamp voltage and NIR emission in experiment (a) and simulation (b)

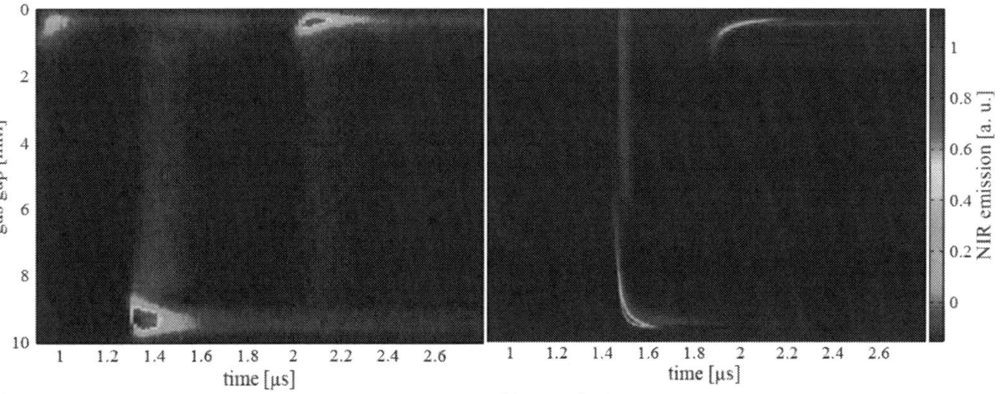

a) experiment *b) simulation*

Figure 2 Temporal and spatial resolved NIR streak image of a homogeneous Xe_2^*-DBD experiment a) experiment with time resolution of 10 ns and b) simulation

DISCUSSION

Using the experimental excitation waveform as input for the simulation, we could show good agreement of breakdown and formation of cathode sheath in experiment and simulation. A temporally and spatially resolved measurement of the lamp NIR emission is a simple way to characterize the breakdown. However it should be noted that this radiation origins from excited Xenon atoms and not from the excimer itself. At constant pressure excitation is strongly dependent on mean electron energy, a lower NIR emission implies lower electron energy.

REFERENCES

[1] Carman R J and Mildren R P, 2003, *J. Phys. D.,* **36** 19-33
[2] Mildren R P, Carman R J, and Falconer I S, 2001, *J. Phys. D.,* **34** 3378
[3] Paravia M, Trampert K E, and Heering W, ICOPS, Karlsruhe, Germany, 2008
[4] Beleznai S, Mihajlik G, et al., 2006, *J. Phys. D.,* **39** 3777-87

Melt-Quench synthesis using arc imaging furnace for phosphor materials

T. Ishigaki[1], K. Toda[1,2], M. Yoshimura[3]
[1] *Center for Transdiciplinary Research, Niigata University, 8050 Ikarashi 2-no-cho, Nishi-ku, Niigata 950-2181, JAPAN*
[2] *Graduate School of Science and Technology, Niigata University, 8050 Ikarashi 2-no-cho, Niigata 950-2181, JAPAN*
[3] *Materials and Structures Laboratory, Tokyo Institute of Technology,4259 Nagatsuta, Midori, Yokohama,Kanagawa, 226-8503, JAPAN*
Contact:tishigaki@eng.niigata-u.ac.jp

ABSTRACT

In order to synthesize phosphor materials rapidly, we have applied a novel "melt synthesis technique" rather than conventional solid state reactions. In the melt synthesis, the mixture of oxides or their precursors is melted in a short period of time (10-60 sec) by a strong light radiation in an arc imaging furnace on a water-cooled copper hearth to form a globule with 1-7mm in size, then cooled rapidly by shutting off the light radiation. We have succeeded to synthesize several compounds of phosphor materials, $Y_3Al_5O_{12}$:Ce(YAG) and $SrAl_2O_4$:Eu,Dy.

GENERAL

The complex oxides, phosphors, are not easy to synthesize by conventional solid state reaction techniques, because the reaction rates among oxides are so slow by solid state diffusions that they form homogeneous compounds or solid solutions. On the other hand, melt reactions are very fast (10-60 sec) and homogeneous due to liquid mixing and fast diffusion in the liquid phase. Therefore, these melt synthesis techniques are suitable for synthesizing multi-component compounds and solid solutions in which homogeneous cation mixing is essentially required.

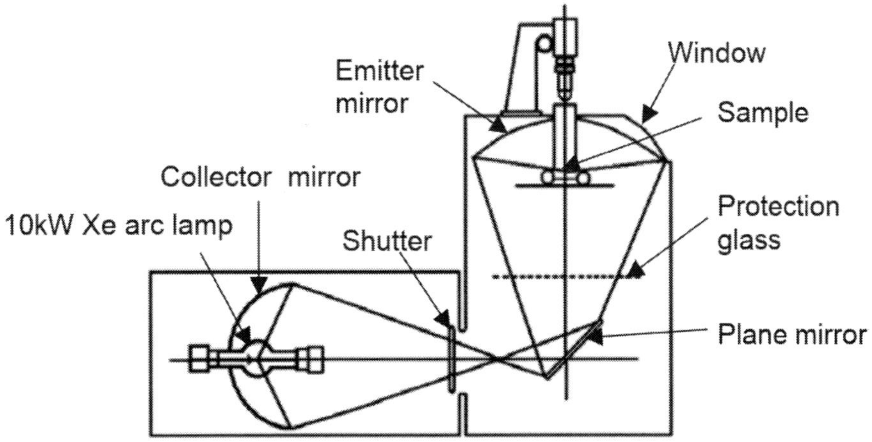

Fig. 1. Optical setup of the arc-imaging furnace. The light source is Xenon arc lamp. The light is reflected a collection mirror, plane mirror and emission mirror to heat a sample on a sample stage.

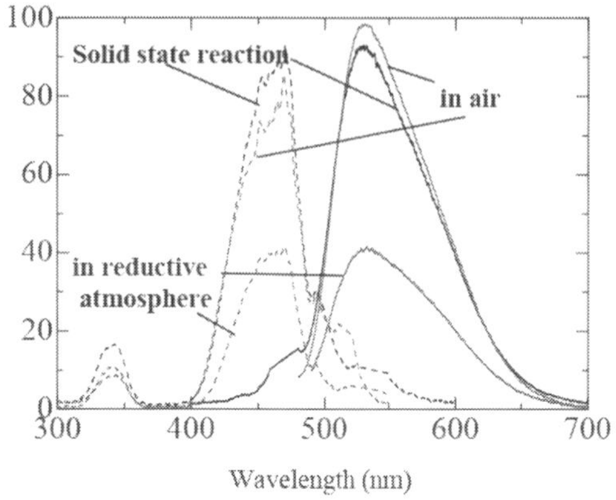

Fig. 3 Excitation (dotted line) and emission (solid line) spectra of melted YAG:Ce samples

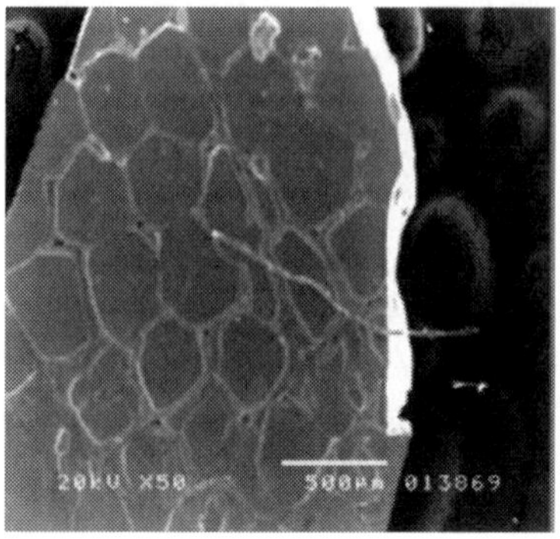

Fig. 4 SEM image of melted YAG:Ce sample

The YAG:Ce, $SrAl_2O_4:Eu^{2+}$, Dy^{3+} melted sample could be synthesized by the melt technique within a short period of time, i.e., 5–60 s, in a single step from the mixed powders. The sample was of globular shape with about 5 mm diameter. Figure 3 shows excitation and emission spectra of melted YAG:Ce samples. Melted in air and reductive atmosphere, both of the samples show almost as same spectra behaviours as solid state reacted sample. According to SEM image of the sample (Fig.4), one of the advantages of this method, it is easy to obtain big size drain samples. It is because sample was reacted in molten salt.

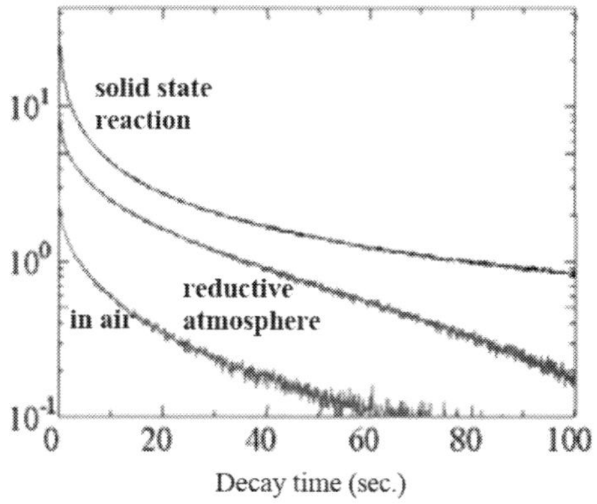

Fig. 4 Afterglow decay curves of melted $SrAl_2O_4:Eu^{2+}$, Dy^{3+} samples

Figure 5 shows after grow decay curves of melted $SrAl_2O_4:Eu^{2+}$, Dy^{3+} samples. It is also same trend properties between solid state reaction sample and melted samples.

As a result, the melt synthesis using arc imaging furnace is powerful technique for phosphor material research.

REFERENCES

1. T. Ishigaki, K. Seki, E. Nishimura, T. Watanabe, and M. Yoshimura, J. Alloys and Comp., 408-412, (2006), 1177
2. T. Ishigaki, E. Nishimura, K. Seki, T. Watanabe and M. Yoshimura, J. Electroceramics, 17, (2006), 885
3. T Ishigaki, K. Seki, S. Araki, N. Sakamoto, T. Watanabe and M. Yoshimura, Ceramic Transactions, 190, (2006), 21
4. T. Ishigaki, K. Toda, T. Watanabe, N. Sakamoto, N. Matsushita, M. Yoshimura, J. Mater. Sci., 43, (2008), 4749

A study on Minimization of Optical Interference in LED Visible Light Communication based on Power Line Communication

J.H. Yun[1] , G.B. Hong[1] , Y.K. Kim[1]
[1]Department of Information and Communication, Wonkwang University
344-2 Shinyong-dong, Iksan Chonbuk 570-749, South Korea
Contact: asdfx3105@naver.com

ABSTRACT

In this paper, we made the optical transmitters and receiver with white LED and optical sensor. In addition, depending on the distance of 10cm ~ 50cm, the voltage waveform and the frequency were measured using the oscilloscope. As results, the light intensity decreases and the voltage and loss of data occurred in the farther distance. Therefore, continued research and improvements are needed in data loss.

INTRODUCTION

Visible Light Communication is a communication technology for transmit the information based on the Power Line Communication to meet current wireless communication by using "Visible light(380nm ~ 780nm)", which is not the electromagnetic effects. The visible light communication has characteristics to be ubiquitous, transmitted at ultra high speed and harmless for human body and electronic devices, compared to those by radio and infrared communications.

The data can be transmitted by the visible light communication even through a high voltage of home/office lighting. Also, The lightings are set everywhere. Then, the data wireless transmission system can be easily established through the visible light communication device attached to the lightings. In this paper, we have confirm the potentialities to applies in next-generation wireless networks.

EXPERIMENTS AND RESULTS

1) Configuration of system

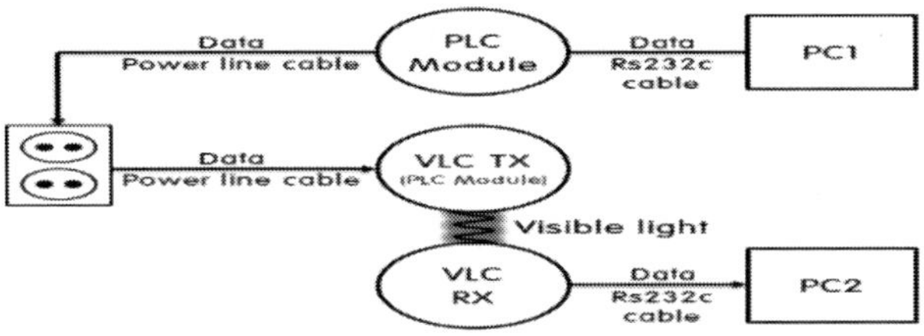

Fig. 1 : Configuration of Visible Light Communication System

Fig. 1 shows the configuration of the short range visible light communication system with white LED, which are needed several technical details for visible light communication system. the system consist of the power line communication modules and White LED transmitter module, receiver module with optical sensor.

Fig. 2 : Transmitter and Receiver of Manufactured Visible Light Communication

Fig. 2 show the Transmitter and Receiver of Manufactured Visible Light Communication. 3 piece of the high brightness white LED was used for the transmitter use and 3 piece of extra LED was used for confirm the data communication.

2) Experiments

First of all, we confirmed the waveform and signal from receiver data by using the oscilloscope in visible light communication system. and we checked the relation between distance of receiver from transmitter and signal intensity.

3) Results

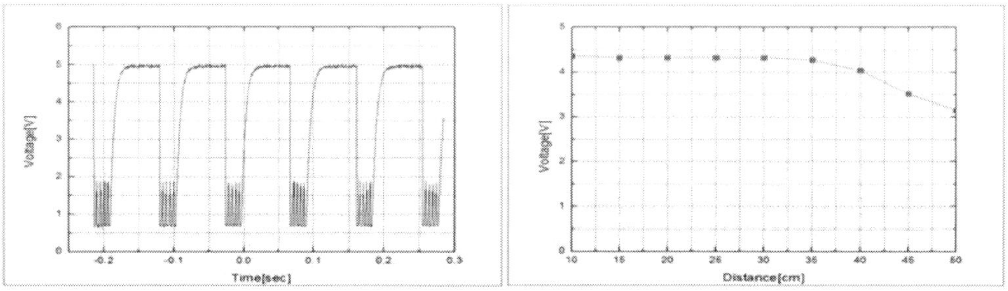

Fig. 3 : Waveform and voltage in the distance from 10 to 50cm

The signal depending on the distance shows in fig. 3. The intensity in the distance from 10 to 30 cm is almost constant but the signal decreased over 35 cm. the range of voltage decrease from 4.36 [V] in 10 cm to 3.16 [V] in 50cm, and the signal period of transmitter shows about 0.1 [Sec]. Also, the signal noise increased in 50cm.

3. CONCLUSION

In this paper, we manufactured the system of Visible Light Communication with white LED and checked the relation between distance of receiver from transmitter and signal intensity. the signal decreased in the distance over 35 cm and extra noise increased according to increase the distance. Although, by using the visible light for the data transmission, many problems related to radio and infrared communications are solved, the interference of other visible light should be considered in the future.

4. ACKNOWLEDGEMENT

This work was financially supported by Small Business Administration through Industry-Academic Joint Technology Development Project in 2009.

5. REPERENCE

[1] I.E. Lee M.L. Sim F.W.L. Kung, "Performance enhancement of outdoor visible-light communication system using selective combining receiver", IET Optoelectronics, Vol. 3, No. 1, Feb (2009)

[2] Miki. M, Asayama. E, Hiroyasu. T, "Intelligent Lighting System using Visible-Light Communication Technology" cybernetics and Intelligent Systems, 2006 IEEE Conference on, pp. 1~6, June (2006)

[3] T. Komine, S. Haruyama, M. Nakagawa, "A study of Shadowing on Indoor Visible-Light Wireless Communication Utilizing Plural White LED Lightings", Wireless Personal Communications 2005, Vol. 34, No. 1-2, pp. 211~225, July (2005)

[4] T.Komin, M. Nakagawa, "Performance Evaluation of Visible-Light Wireless Communication System Using White LED Lightings", in Proc. of the Ninth IEEE Symposium on Computers and Communication, pp. 258~263, June (2004)

[5] T. Komine, M. Nakagawa, "Intergrated system of white LED visible-light communication and power-line communication" , IEEE Trans. Consumer Electronics, Vol. 49, no. 1, pp. 71~79, Jan (2003)

Why does the lumen maintenance of sodium-scandium metal halide lamps improve by VHF operation?

W. van Erk[1], G.M.J.F. Luijks[2], W. Hitchcock[3]

[1] *Retired employee Philips Lighting, Sonderrick 47, 5505 NB Veldhoven, The Netherlands*

[2] *Advanced Development Lighting, Philips Lighting, P.O. Box 80020, 5600 JM Eindhoven, The Netherlands*

[3] *HID Lamp Development, Philips Lighting Company, 7265 Route 54, Bath, NY 14810, USA.*

Contact: gerard.luijks@philips.com

ABSTRACT

The question why sodium scandium metal halide lamps perform better on VHF (Very High Frequency) drivers than on LF (Low Frequency) CWA ballasts will be addressed with focus on burner aspects. It is argued that at high frequency operation sodium loss is less and the absence of thermal fluctuations in the electrode tip causes less damage to this part of the electrode. The low rate of tungsten evaporation shows that the electrode temperatures are relatively low, indicating that under these conditions thorium is functioning as emitter, both in the anode and the cathode phase.

INTRODUCTION AND EXPERIMENTAL RESULTS

Conventional sodium scandium metal halide lamps perform very well on Very High Frequency (VHF) systems [1,2]. Especially the lumen per watt (LPW) maintenance on VHF is better than on the conventional magnetic Constant Wattage system (CWA).

This paper will address the question why these lamps perform better on VHF compared to CWA and will focus on the lamp side of the system, considering physical, chemical and mechanical effects of VHF operation.

In all tests described, the lamps burned vertically with base up and the lamps contained an excess scandium metal. A first test with 10 MS400/BU/PS lamps per set resulted in a LPW maintenance on VHF (Advance Dynavision^{TM}) of 83% at 10,000 hrs with a standard deviation of 4%, much better than the controls on CWA: 51% and standard deviation 6 %. Striking is also the difference in lamp voltage rise at 10,000 hrs: 12 V on VHF and 22 V on CWA. At 10,000 hrs the salt composition was analysed of one VHF and 2 CWA lamps. The CWA lamps had lost about 10% of their sodium, whereas for the VHF lamp the loss was not measurable. This is corroborated by the Na/Sc ratio: sodium loss will lead to a lower Na/Sc ratio and indeed the Na/Sc mole-ratio is lower for the two CWA lamps: 19/1 and 22/1 versus 28/1 for the VHF lamp [3].

Lumen maintenance of these lamps is strongly correlated with wall blackening, i.e. deposition of tungsten on the burner wall, and that is linked to electrode performance. Out of a second test of 10 MS400/BU/PS M155 lamps that had reached 4000 hrs, for 3 lamps the amount of tungsten on the burner wall was measured (wet chemical analysis) and the electrodes were examined using a SEM (Scanning Electron Microscope). Striking is the very large difference in the amount of tungsten that has reached the burner wall: 1.4 (0.3) mg W for CWA and 0.09 (0.02) mg W for VHF. In fig. 1 examples of how the electrode tips look like after 4000 hrs are shown. The lower electrodes have a rather smooth surface with the typical "mushroom"shape. In the CWA lamps cracks are found in the tip surface, whereas the lower electrodes of the VHF lamps hardly have any cracks. On one of the VHF electrode tips a thorium deposit was still present [4]. For the upper electrodes holds that on the CWA electrode tips protrusions are seen whereas with VHF the surfaces are smoother.

In an earlier experiment in which the role of thorium was examined it was found that the presence of thorium in the burner is essential to obtain a good maintenance: At 1000 hrs the average maintenance of the set without any thorium was about 70% whereas with thorium, either in the salt or in the electrodes or in both, the maintenance was above 90%. In principle scandium can be a good gaseous emitter but apparently here its pressure is not high enough to yield sufficient coverage and thorium is necessary to do the job [4].

DISCUSSION

Sodium loss: The results indicate that the rate of sodium loss is much lower on VHF than on CWA. Calculations show that at the very high frequency the direction of the field switches so fast that electrons, emitted by the field wire do not have enough time to travel through the nitrogen filled outer bulb to the arc tube surface. So with VHF the flux of electrons to the arc tube surface will be less and this will reduce sodium loss.

Surface structure: The crack formation at CWA operation is probably due to "thermal ratcheting" by the periodic heating in the cathode phase (spot like attachment) and cooling in the anode phase (diffuse attachment) [5]. Eventually, on the longer term cracks can lead to breaking up of the tip structure and to more than normal burn back of the electrode. This periodic heating and cooling is absent at very high frequency operation.

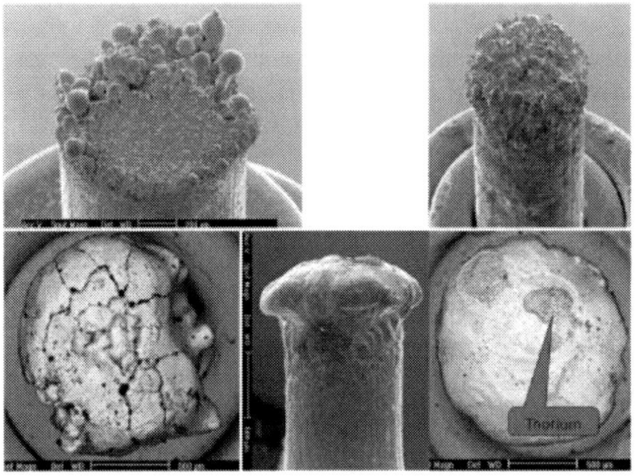

Fig. 1: Electrode surfaces after 4000 hrs burning vertically base up. Upper pictures are from the upper electrodes and the lower pictures from the lower electrodes. The electrodes on the left have burned on CWA and those on the right on VHF. On the lower electrode of one of the 3 VHF examined burners a macroscopic thorium deposit was still present. Below in the middle is a side view of the lower VHF electrode

Emitter: Our results show that on VHF the electrodes perform well: the tungsten removal rate is very low compared to CWA. It must be concluded that under VHF conditions thorium acts even better as an emitter than under CWA conditions. A distinction must be made between the lower and the upper electrode: At the lower electrode there is during the first period of burning a macroscopic thorium deposit, a liquid film or pool. So then the emitter is present in both anode and cathode phase. The deposit will disappear when the iodine pressure increases due to sodium loss and as argued above this will take longer with VHF operation. After that the supply of emitter to the electrode surface is via the gas phase: the gas-phase emitter mechanism [4]. This applies to the upper electrode, too. Here, part of the emitter flux to the cathode surface is due to cataphoresis of the emitter ions and part is normal adsorption. At high frequency operation the cataphoresis flux will be almost zero. However, thorium adsorbs very strongly on a tungsten surface. Apparently with thorium the normal flux of neutral emitter atoms is sufficient to produce the thorium monolayer necessary for the lowering of the electrode temperature. If so, then this layer will be present in the anode phase of VHF too, because of the absence of the cataphoresis flux away from the surface. This means that with VHF the emitter thorium will be present in both the cathode and anode phase either as film or as monolayer.

REFERENCES

[1] A. Lamouri, A. Naruka. J. Sules, LS-11 Shanghai 2007, Proceedings The 11th International Symposium on the Science & Technology of Light Sources, May 20th – 24th, Shanghai, China, p. 543

[2] A Lamouri, A Naruka, J Sulcs, C V Varanasi and T R Brumleve, J. Phys. D: Appl. Phys. 38 (2005) 3028-3032

[3] W. van Erk, J. Tu, J. Suijker, J. Phys. D: Appl. Phys. 38 (2005) 3040-3046

[4] G M J F Luijks, S Nijdam and H v Esveld, J. Phys. D: Appl. Phys. 38 (2005) 3163-3169

[5] E. W. Parkes, in Thermal Stress, edited by P. P. Benham and R. Hoyle (Pitman, London, 1964)

The effect of iris colours to the energy flux distribution on the retina

Chong-Jhih Jiang*, Yi-Chun Chen, Ching-Cherng Sun
Department of Optics and Photonics, National Central University, Jhongli 320, Taiwan
Contact: s982406018@dop.ncu.edu.tw

ABSTRACT

The optical reflectivity spectra at the iris are different among ethnic groups. The effect of iris reflectance to the energy flux distribution on the retina was simulated by using a biometry-based human eye model with a transparent or cataractous lens. The results indicated that the eye with a light-coloured iris and a cataractous lens would suffer more irradiation from the iris-cornea multi-reflection and hence be more susceptible to the short-wavelength spectral content in the light sources.

INTRODUCTION

White-light LEDs based on a blue chip plus single phosphor have an apparent peak in their spectra at the original emission line of the blue chip. However, possible biohazards of blue light had been reported, such as causing more serious glare, affecting the biological clock, and damaging the cornea and crystalline lens [1]. As light sources with more short-wavelength content caused more discomfort glare [2], it is important to investigate the glare effect of white-light LEDs. In addition, when light rays enter the human eyes with different iris colours, the optical reflectivity spectra at the iris are significant different among ethnic groups, especially at the short wavelengths in the visible range. Some of the light reflected from the iris gets back into the human eye due to Fresnel reflection at the corneal anterior and posterior surfaces. Therefore, the correlated colour temperature (CCT) of the same light source varies when observed by people with different iris colours and may lead to distinct glare effects. This study aims to investigate the effect of iris colours to the flux distribution on the retina, in particular when using white-light LEDs with blue-peak spectra.

METHOD

A biometry-based human eye model was used in this simulation study. This model reassembles the visual acuity, the chromatic dispersion behaviour, and the intraocular scattering of the human eye [3]. This study investigated the effects of two independent variables, one is the lens condition (*i.e.* transparent or cataractous), and the other is the iris reflectance to represent the effect of iris colour. The concentration of the scattering particles in the crystalline lens was adjusted to mimic the transparent and cataractous lenses. The iris had an outer diameter about 11.2 mm, and its inner diameter was adjusted to have an entrance pupil of 4-mm diameter. The curvature of the iris coincided with the anterior surface of the crystalline lens. The iris reflectance was varied from 0 to 0.6 and the iris surface was assumed rough to reflect light rays in random directions.

Figure 1 shows the optical configuration of the simulation study. A point light source with an emission line at 455 nm was put on the visual axis of the human-eye model and was 6 meters away from the cornea anterior surface. Light rays passing through the cornea either went through the pupil directly or hit on the iris. The simulation separated the light rays underwent multiple reflections between the iris and cornea from those directly going through the pupil, in order to observe the energy flux contribution on the retina due to different iris reflectance.

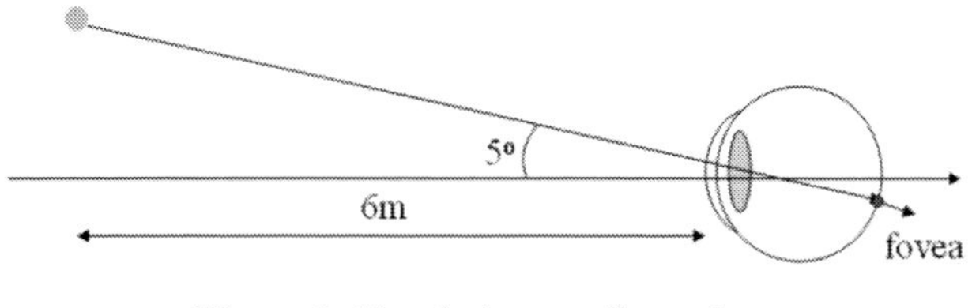

Figure 1. Simulation configuration.

RESULTS AND DISCUSSIONS

The total fluxes on retina and in the 1-mm-diameter fovea region are shown in Figures 2(a) and 2(b), where the blue lines correspond to a transparent lens and the gray lines correspond to a cataractous lens. It is evident that light reflected from the iris did not have significant contribution to the total flux in the fovea region, but had significant effects on the retina. Due to the multiple reflections between the iris and cornea, the total flux on the retina increased linearly with the iris reflectance. Because of the scattering effect in the crystalline lens, the total flux in the fovea region corresponding to a cataractous lens was about two-third of that corresponding to a transparent lens. Similarly, the total flux on retina with a cataractous lens was lower than that with a transparent lens, but this time a smaller drop of about 10%. Consequently, the image signal-to-noise ratio would be lower for the eye with a cataractous lens. Figure 2(c) shows the flux distribution on the retina for a cataractous lens when obstructing light rays passing through the pupil directly. It is observed that very little energy contributes to the visual field up to 40 degrees and most energy fell on the peripheral visual field. The joint effects in Figures 2(a)-2(c) imply that people with cataract will have nosier visual perception due to noise irradiation in the peripheral visual field and weaker flux distribution on the fovea.

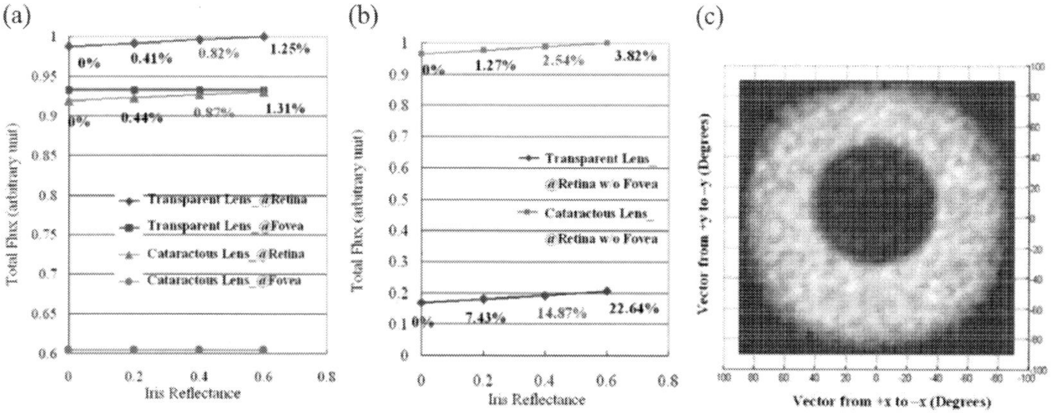

Figure 2. (a) Total flux on fovea or the whole retina, where the horizontal axis is the hypothetic iris reflectance; (b) Total flux on the retina without fovea; (c) Flux distribution on the retina.

SUMMARY

The effect of iris reflectance to the energy flux distribution on the retina was simulated for the human eye with a transparent or cataractous lens. The results showed that light reflected from the iris did not contribute to the total flux in the fovea region, but greatly increased the total flux on the retina. Hence eyes with blue-coloured irises would suffer noisier irradiation from the blue spectral content in the light sources. The pronounced scattering effect of a cataractous lens greatly reduced the total flux at fovea, and hence the image signal-to-noise ratio would be much lower than that of a transparent lens. Further investigations on the CCT calculations and the corresponding glare effects are in progress.

ACKNOWLEDGEMENT

This work was supported in part by the National Central University's Plan to Develop First-class Universities and Top-level Research Centers Grants 975928 and 985947 (Visual Comfort and Cognition in Artificial Lighting).

REFERENCES

[1] L. A. Newman, M. T. Walker, R. L. Brown, T. W. Cronin, and P. R. Robinson, "Melanopsin forms a functional short-wavelength photopigment," *Biochemistry* **42,** 12734-12738 (2003).

[2] J. D. Bullough, J. Van Derlofske, P. Dee, J. Chen, and Y. Akashi, *An Investigation of Headlamp Glare: Intensity, Spectrum and Size*, NHTSA Publication DOT HS 809 672 (2003).

[3] C. J. Jiang, Y. C. Chen, T. H. Yang, and C. C. Sun, "Development of a human eye model for visual performance assessment," *Proc. SPIE* 7427, 742706 (2009).

Properties of a ceramic metal halide lamp at different frequencies and reduced power

Wei Li, Shuijun Shi, Jie Liu, Ling Chen and Shanduan Zhang
Institute for Electric Light Sources, Fudan University, 220 Handan Road, Shanghai 200433, People's R China
Contact: sdzhang@fudan.edu.cn

ABSTRACT

Ceramic metal halide (CMH) lamp with power 70 W and colour temperature 4000 K is driven by a square waveform power supply with frequency range 50 Hz to 50 kHz. The parameters are measured for luminous flux, colour rendering index, colour temperature, voltage and current. The results show CMH lamps can be operated within a wide range of frequencies, but cautions should be taken to avoid voltage rise when higher frequencies (~10 kHz) are selected. At a moderately high frequency (5 kHz), the CMH lamp can remain its good performance when dimming depth is 60%.

INTRODUCTION

Ceramic metal halide (CMH) lamps powered by electronic ballasts have higher efficacy and colour rendering index (CRI) than those by inductance ballasts. Ballast designers usually focus on the optimal electric parameters of a ballast. Among those factors, two will be of practical significance, namely the frequency of output power and dimming properties of CMH lamp driven by a ballast. There have been reports on dimming properties of CDM lamps [1-3], but none on the frequency. In this experiment, a 70 W CMH lamp with CCT 4000 K is driven by a square waveform power, with a 150 ohm resistance as a ballast. First, the lamp is operated at the rated power, but the frequency is set at 50, 100, 1 k, 10 k, 30 k and 50 kHz. Then the frequency is fixed at 5 kHz, and the power is adjusted at 60%-100% of the rated power. Electric, photometric, colour and spectral parameters are measured under different conditions. The derived trend can serve as a reference for ballast designers.

EXPERIMENT

Figure 1 is a schematic of the experimental set-up. A DC power supply, of which the output voltage can be adjusted, is connected to an inverter with waveform generator in order that a commutated DC is created. A power resistance (150 Ω), in series with the lamp, serves as the ballast. The spectrometer and the integrating sphere are used to obtain photometric, colour and spectral parameters. Electric parameters can be gotten by a digital storage oscilloscope coupling with voltage and current probes.

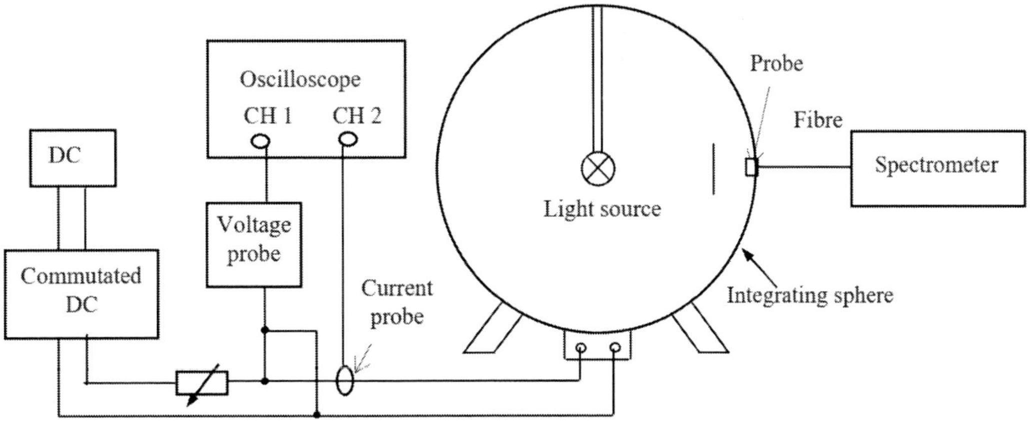

Fig.1 Schematic of the experimental set-up

Supported by Key Technologies R & D Program of Shanghai Science and Technology Committee (Grant No. 08DZ1140802)

RESULTS

(1) Frequency properties of the CMH lamp at rated power

Table 1 shows the ratio of the maximal value to the minimal value of the parameters concerned. The effect of frequency on luminous flux (<3.4%), CRI (<1.4%) and CCT (<1.0%) is negligible. But such effect on voltage and current deserves consideration.

Figure 2 and 3 illustrates the variation of voltage and current over frequency. On the whole, the voltage increases while current decreases with the rise of the frequency. The reason may be as follows: The higher the frequency is, the faster the electric field changes direction between the electrodes and less time is available for electrons to accelerate. In order to maintain discharge, higher voltage is needed to speed up electrons.

Table 1 maximal and minimal values of parameters concerned

	Φ (lm)	CRI	CCT	U_{rms} (V)	I_{rms} (A)
max	6150	94.0	3842	94.4	0.791
min	5949	92.7	3805	88.7	0.746
max/min	1.034	1.014	1.010	1.064	1.060

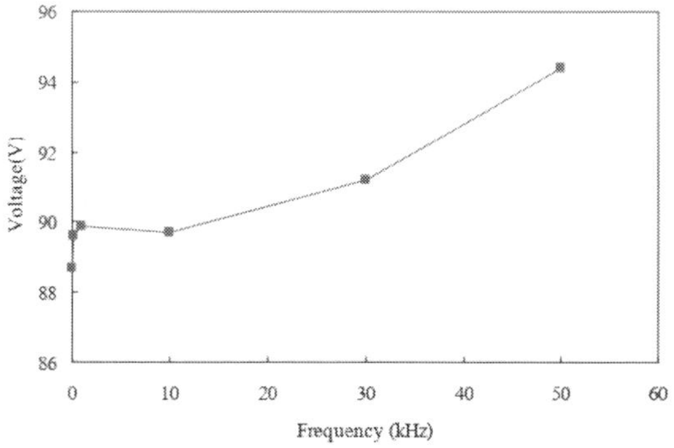

Fig.2 Frequency dependence of voltage.

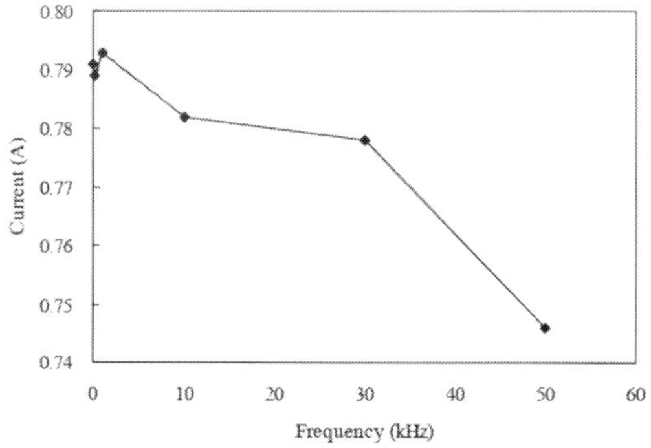

Fig.2 Frequency dependence of current.

(2) Dimming properties of the CMH lamp at 5 kHz

To explore the dimming properties of the CMH lamps at a relatively high frequency, which is not the case with traditional ballasts, 5 kHz is selected as the dimming frequency.

Figure 4 shows the spectral power distribution of the CMH lamp under dimming conditions at 5 kHz. The spectrum of such a CMH lamp is mainly composed of two peaks (at about 535 nm and 595 nm) and strong continuum. The peak at 535 nm remains almost the same at reduced power, so it affects the lamp little. The peak at 595 nm will weaken while shift to the shorter wavelength, which means approximately the same luminous flux and CRI, but higher CCT. In terms of the continuum, the central part (540~580 nm) contributes most to the luminous flux because the visual function reaches maximum at 555 nm. The continuum at long wavelength range weakens greatly when the lamp is dimmed, which mainly results in the decrease of CRI and increase of CCT because less "warm" light is available at reduced power. As a result, the efficacy, CRI and CCT of the CMH lamp changes within the range of tolerance because of the reasons mentioned above.

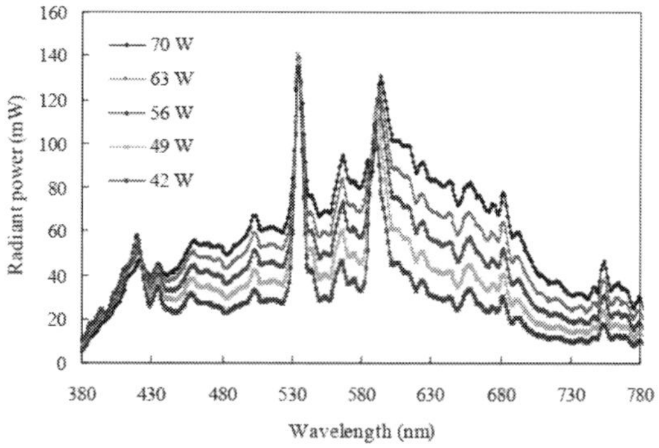

Fig. 4 Spectral Power Distribution of the lamp at reduced power and at 5 kHz

REFERENCES

[1] Guest E C, Girach M H, Mucklejohn S A, *et al* 2008 *Lighting Res. Technol.* **40** 333

[2] Kappen T G M M 2004 *Institute of Physics Conference Series* **182** 43

[3] Stockwald K 2004 *Institute of Physics Conference Series* **182** 331

LED retrofit light bulb technology for up to 60W eq. incandescent replacement

I. Bakk[1], F. Wagner[1], P. Pachler[1], H. Hoschopf[1], S. Tasch[1], F. P. Wenzl[2], P. Petz[3], A. Favarolo[4], I. Wilson[5], F. Stevens[5], F. Baldauf[5], P. Hartmann[1].

[1]LEDON Lighting Jennersdorf GmbH, Technologiepark 10, A-8380 Jennersdorf, Austria

[2]Institute of Nanostructured Materials and Photonics, Joanneum Research Forschungsgesm.b.H., Franz-Pichler Straße 30, A-8160 Weiz, Austria

[3]Ledon Lamp GmbH, Höchster Strasse 8, A-6850 Dornbirn, Austria

[4]Zumtobel Lighting, Schweizerstraße 30, A-6850 Dornbirn, Austria

[5]Tridonic UK, Merrington La Ind Est, Spennymoor DL16 7UR, United Kingdom

Contact: Franz-Peter.Wenzl@joanneum.at

ABSTRACT

We address recent advancements of LED retrofit light bulb development. A range of high-performance LED modules has been optimized to meet the requirements of general lighting and LED retrofit bulb applications. Based on a red-white approach using efficiency-optimized phosphor-converted and red monochrome-LEDs, these modules deliver warm-white emission of highest quality and efficacy. LED bulbs equivalent to 60W, 40W incandescent bulbs and a $25W_{eq.}$ candle type one are presented.

GENERAL

There is no doubt that a replacement of incandescent light bulbs by LED light bulbs will reduce energy consumption, save resources due to their long life-time and will also have a beneficial climate impact [1]. These advantages are also reflected in the decision of the EU and many other countries all over the world to ban energy inefficient incandescent light bulbs within the next years. Although compact fluorescent lamps (CFL) are the prevalent alternative at the moment, LED retrofit bulbs, which do not contain mercury and which already surpass CFLs with respect to energy efficiency, experience an increasing pull of the market. There are two major challenges which are crucial for a successful market penetration of LED retrofit bulbs within the next years. On the one hand the light flux has to be increased in order to be able to replace the whole portfolio of incandescent light bulbs, at least up to 100 $W_{eq.}$. On the other hand, as known from the market launch of compact fluorescent lamps (CFL) that are commonly associated with rather poor light quality, for a successful market penetration it is essential that LED bulbs deliver light of a quality people are used to and are convinced by.

In this contribution we discuss an approach for the realization of LED modules and LED retrofit bulbs fabricated therefrom which are applicable for the replacement of already up to 60 W equivalent incandescent bulbs along with excellent color perception and color rendering properties. The modules basically consist of a number of red-emitting monochrome-LEDs in combination with phosphor converted LEDs. A set of these so called "XED" modules has been developed and implemented successfully into retrofit LED-lamps of Lemnis Lighting (Netherlands) and Ledon (Austria)

Fig. 1: Examples of Ledon XED Modules XED3 (left) and XED5 (right, concept drawings)

In dependence of the number of phosphor converted and red LEDs a set of different XED modules (for concept drawings of 2 examples see Fig. 1), which deliver light fluxes in between 300 and 800 lm,

is available. Based on this, LED retrofit bulbs delivering up to typ. 650 lm at a color temperature of 2900K (at thermal equilibrium and at 25°C ambient) using only 10W for a 60 W incandescent bulb replacement are now offered to the market. Further types comprise a $40W_{eq.}$ with typical flux of 360lm and 6W power consumption (see Fig.3b), and a $25W_{eq.}$ candle with typ. 210 lm at 5W.
The phosphor type and concentration for the color converted LEDs is chosen in such a way that the color coordinates of the mixed blue and phosphor emission (for the emission spectrum see Fig. 2) are situated well above the Planckian locus in the color gamut [2], and optimized for high efficacy [3]. The final color mixing of the phosphor converted and the red LEDs is achieved by a milk glass forming the LED lamp housing. The technology ensures color rendering of R_a>90 throughout a broad CCT-range of warm white color temperatures (2000-3500K). This is accomplished by using appropriate combinations of 3 different phosphors for the specific CCTs targeted at (see Fig. 2). For the standard products a CCT of 2500K is achieved in the cold state. Dependent on the size and quality of the heat sink (see Fig. 3 for the temperature distribution for one possible arrangement to transfer the heat from the system to the ambient air) this value shifts to a CCT of approx. 3000 K in the temperature-equilibrated state due to the different temperature-characteristics of the red and blue/white LEDs, respectively.

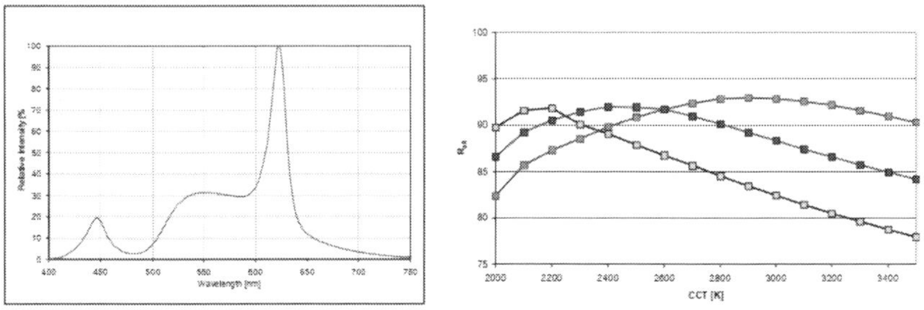

Fig2: Left side: Emission spectrum of the XED-5 module, Right side: Color rendering index R_a of XED modules achieved with three different phosphors at target color temperatures.

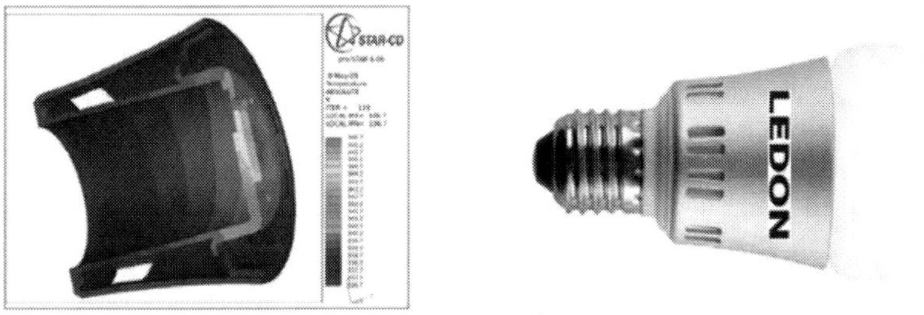

a b

Fig. 3a: Temperature distribution for one possible arrangement to transfer the heat from a 40 W replacement bulb to the ambient air. b) Retrofit LED Bulb with E27 socket for 40W incandescent replacement

To further reduce these variations, advanced module types of this series will utilize a simple on-board compensation circuit including an NTC (a ceramic resistor with negative temperature characteristics) to maintain the color coordinates within a McAdams 3 ellipse at the target value throughout the operation including warming up.
Based on these outstanding characteristics, the XED-family is ideally suited to support the upcoming revolution of the general lighting and especially the retrofit LED lamp markets and on the best way to replace the whole portfolio of incandescent bulbs within the next years.

[1] E. F. Schubert, J. K. Kim, H. Luo, J. Q. Xi, Rep. Prog. Phys. 69, 3069 – 3099 (2006)
[2] A. van de Ven, et al., Proc. of 1st Int'l. conf. on White LEDs and Solid-state Lighting, 153-156 (2007).
[3] DE 102007043355, DE 102008025864.4, patents pending

Development of high correlated color temperature, 12000K FL

Hiroshi Yagi[1] , Masanori Shimizu[1] , Takeshi. Morita[2]
[1] *Lighting Company, Panasonic Corporation*
[2] *Fukuoka women's University*
Contact: yagi.hiro@jp.panasonic.com

ABSTRACT

To utilize the non-visual effect of short wavelength light, we developed new fluorescent lamp with very high correlated color temperature(C.C.T.), 12,000K which corresponds to that of fine skylight. This lamp has high color rendering property, CRI>90, R9>80, Ga (relative gamut area)>100. We conducted field tests at a factory, and revealed that the participants showed 3 ~ 4% higher performance under 12,000K lighting than conventional 5,000K lighting. In addition, we found that the body temperature on awakening rose more rapidly under 12,000K than conventional 5000K.

INTRODUCTION

Traditionally, the lighting has been developed and designed to contribute the human vision, and in other words, they have been the devices to illuminate objects and the living space.

Recently, a large number of studies have clarified that light has non-visual effects (1)(2). The light is supposed to affect the 24h-cycle biological rhythm, which is called circadian rhythm. Many researchers have shown that the light can suppress the secretion of melatonin, which is the hormonal material related to sleep, and that the light helps to maintain the circadian rhythm properly. They also confirmed that short wavelength light is more effective to suppress the circadian rhythm and the existence of retinal pigment "melanopsin", which relates to the secretion, was recognized.

Based on the above physiological researches, the developments of new fluorescent lamp(FL) with very high correlated color temperature(C.C.T.) have been accelerated all over the world. 8000K and 17000K FLs have been already launched with the asserting that people will be more energetic and/or alerted under high C.C.T. lighting. Some researchers report, in some factory work, the performances improved under the high C.C.T. lighting(3). But, on the other hands, as C.C.T become higher and higher, the component of reddish spectra is involved less and less, and the light color tends to be blueish white. So, there seemed to be difficulties associated with the appearance of object colors especially reddish object and human skin under high C.C.T. lighting.

DESIGN CONCEPT AND THE DECISION OF C.C.T.

Many researchers have submitted the action spectrum of melatonin suppression by the light. The typical three examples of them are shown in Fig1. The peak wavelength of each spectrum is shorter than that of the relative luminous efficiency curve. The large amount of the irradiation power in this short wavelength region means high C.C.T and the melatonin suppression efficiency tends to saturate around 15,000K as shown in Fig2. It must be noted that extremely high C.C.T. causes the degradation of lighting environment. Therefore, we set up the target C.C.T. at around 12,000K which corresponds to the C.C.T. of fine skylight. As human beings encounter this C.C.T. on a daily basis, 12,000K lighting is supposed to realize high melatonin suppression efficiency without uncomfortable feeling for the lighting environment.

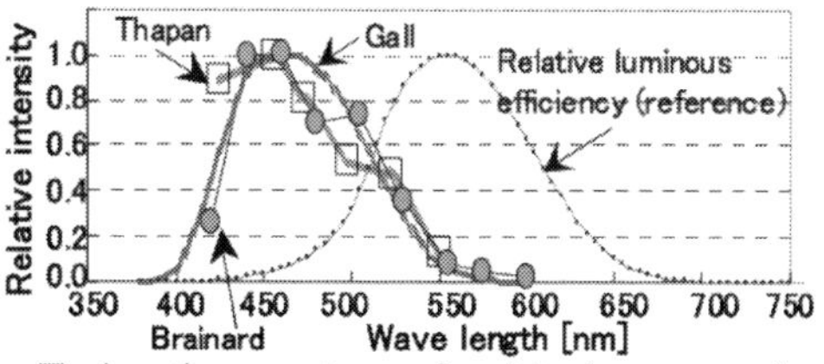

Fig.1 action spectrum of melatonin suppression

DESIGN OF SPECTRAL DISTRIBUTION AND THE COLOR RENDERING

We found that basic spectrum design of 12,000K based on three band type FLs leads to insufficient color rendering especially for reddish color. Although the simplest way to improve color rendering is to add long wavelength reddish light, we found another solution., adding 520nm irradiation to the conventional three band type FLs. The 520nm irradiation corresponds to dark green color which is opponent color of red. So, the increase of 520nm can improve the color rendering of red. In addition, the increase of 520nm leads higher melatonin suppression efficiency.

Fig4 shows gamut area (U*,V*) of medium chroma samples(R1~R8) for the new FL. Three cases of 12,000K light source are shown; the standard illuminant(S1), conventional three band FL(S2), and new lamp with increase of 520nm irradiation(S3). Fig4 means, as the point is farther from the origin, each color is perceived more vivid under the light source lighting. High C.C.T. standard light illuminant lighting, S1 has poor color rendering property for reddish color. Under S2 lighting, the color rendering for red is poorer than S1 lighting. On the other hand, it is clear that S3 lighting enables better color rendering for red than S1 and S2. In Fig5 the gamut area (U*,V*) of higher chroma samples(R9~R12) also shows similar properties to Fig.4.The color rendering indices are CRI>90, R9>80, Ga (relative gamut area)>100. And it can create refreshing atmosphere like fine skylight.

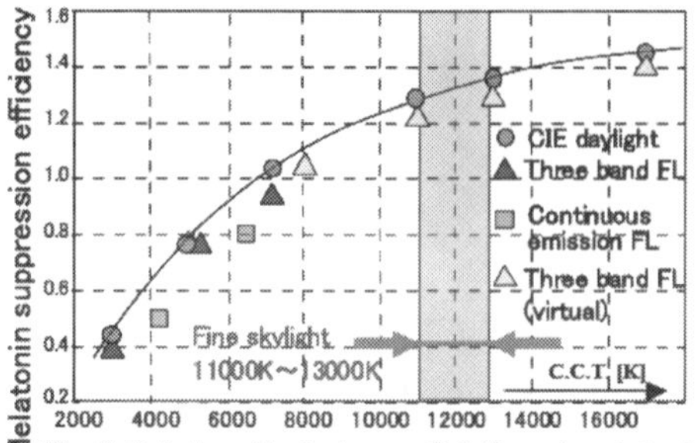

Fig.2 Relationship between C.C.T. and melatonin suppression efficiency per unit luminous flux

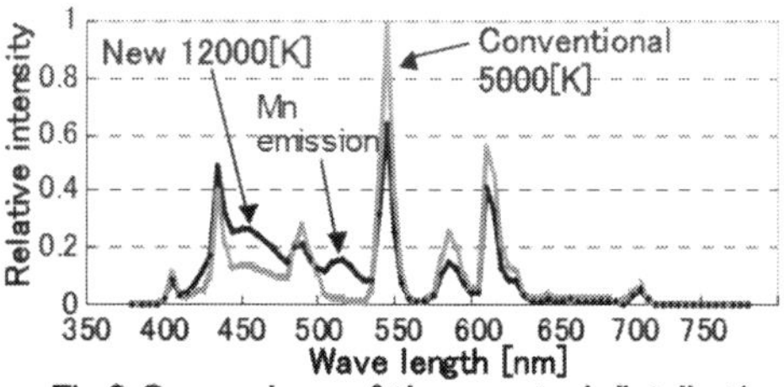

Fig.3 Comparison of the spectral distribution

LAMP APPLICATION TEST

To investigate the physiological effect of the new 12,000K lamp lighting, we tested the ascent rate in core temperature on awakening. As shown in Fig6, the core temperature rose more rapidly with 12,000K lamp lighting than conventional 5,000K lighting. It is expected that the temperature rise is related to the sympathetic nerve activation and that 12,000K lighting is more effective for awakening.

In addition, we conducted a field test for factory work to examine the alertness and/or activity. The working process is assembling of wiring, the test span is 4 months and total man-days are 240. We revealed that the performance of the work rose up for 3~4%. And also as the result of POMS (Profile of mind state), which is executed at the same time, the subjective value of vigour mood rose, too. (POMS is used in industrial, medical, and sport field as the global standard.)

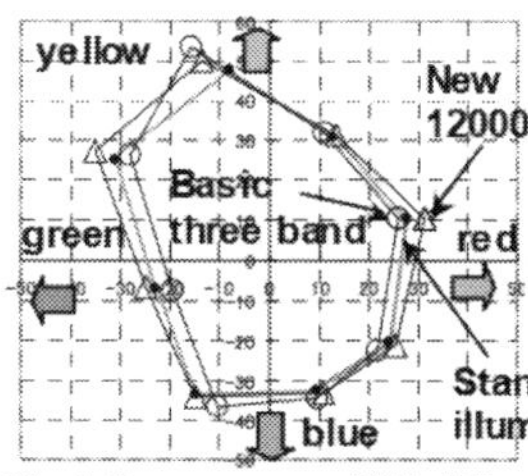

Fig.4 Gamut area of R1-R8

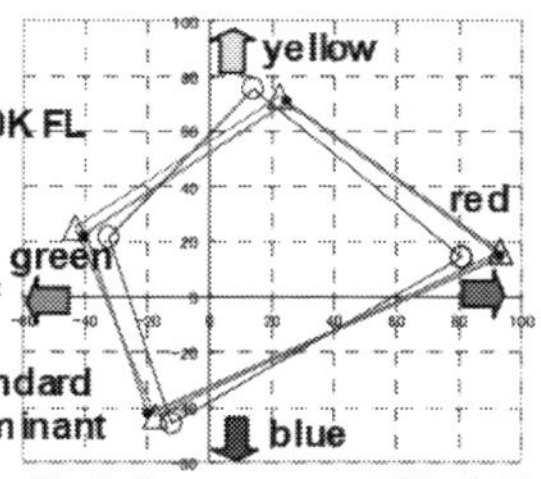

Fig.5 Gamut area of R9-R12

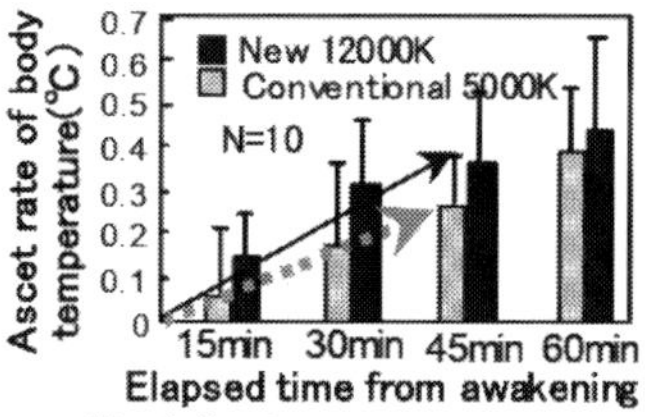

Fig.6 Body temperature rise for each lamp

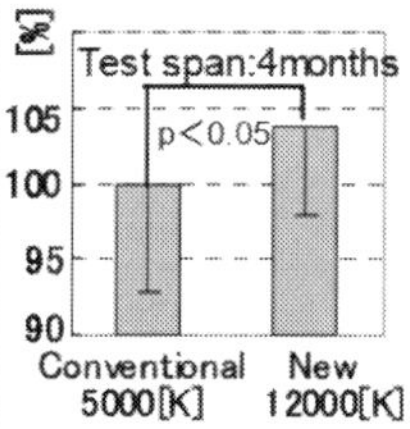

Performance (relative scale) [%]
Fig.7 Field test for a facory work

CONCLUSION

- We have developed high C.C.T. (12,000K) fluorescent lamp, which is suitable for high melatonin suppression and is able to create refreshing atmosphere like fine skylight.
- We have investigated lamp application tests and revealed the followings;
    1. It is supposed that 12,000K is more effective for awakening than conventional 5000K lamp.
    2. In a factory work (assembling of wire), the performance of the work rose up for 3~4%.
- Taking advantage of the active illuminated space and refreshing atmosphere like fine skylight, the expected applications are such as gymnasium, factory, office, school, underground city, and so on.

REFERENCES

1)CIE Symposium: Light and Health: non-visual effects, 2004
2)The 2nd CIE Expert Symposium on Lighting and Health, 2006
3)Peter R. Mills et al.: The effect of high correlated colour temperature office lighting on employee well being and work performance, J. Circadian Rhythms,5.2, 2007

Simulation and Optical Characteristics of Secondary Optics for High Efficiency LED lighting

A. Seung-Min Lee[1] . B.Gi-Seung Choi[1] . C. Jong-Chan Lee[1]
[1] *Kumho Electric, Inc. 211-1, Jangji-ri, Dongtan-Myeon, Hwaseong-City, Gyeonggi-Do, 445-812, Rep of Korea*
Contact: chaos@kumho.khe.co.kr

ABSTRACT

This paper aims to distribution about 60° that influence the performance of 2nd optic reflector was designed for LED lighting. The maximum size of reflector height of 6 [mm], width of 6 [mm] by setting were designed using OptisWorks. When Reflector height decreases, the distribution angles were narrow. While changing of aspheric element in reflector, the intensity of the side portion at reflector was changed. Those were satisfied the each goal and uniform illuminance distribution. In addition, the designed reflector had optical characteristic which were 60° distribution and 96.3% luminaire efficiency.

1. INTRODUCTION

Optical characteristics of LED lighting are an important factor which is influencing the performance of a lighting distribution. The distributions are controlled by optic lens, reflector, diffusion and optical engine.
In the paper, 2nd optic were designed for LED lighting with high efficiency. Therefore, the computer simulating software, OpticWorks was used for structure design, optic analysis and optical performance including lighting distribution. A high efficiency reflector, which had 60° distribution, was designed with 1W High Power LED.

2. SUBJECT

2-1 Simulation condition

Table 1 is showing the simulation condition. LED pack-age is 1 [W] power, 100 [lm] luminous flux and 110° lambertian distribution. Al of 90% reflectance and 10% absorption was used for reflector material at simulation condition. The LED package distribution was shown in Figure 1.

Table 1. Simulation condition

LED Package	Luminous: 100lm Distribution: 110°
Reflector	Material : Aluminum Refractive index :1.440 Reflectance: 90% Absorptance:10%

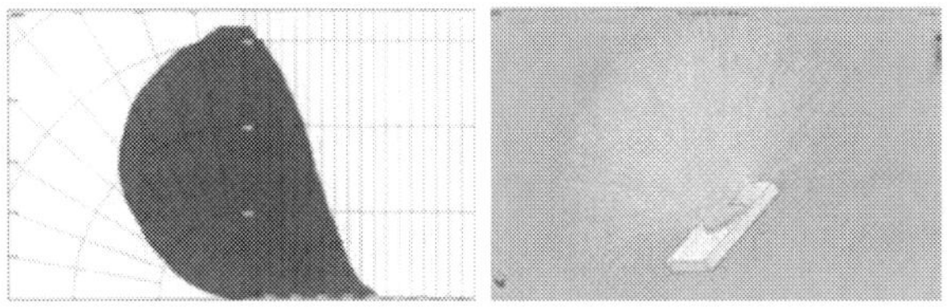

Figure 1. LED package distribution

2-2 Secondary optics design

The parabola reflector is commonly used that is determined by a formula which is $y^2=4ax$. Most important characteristics of the parabola reflector is that assuming the point source, parallel beam of light can be obtained to the focus area.

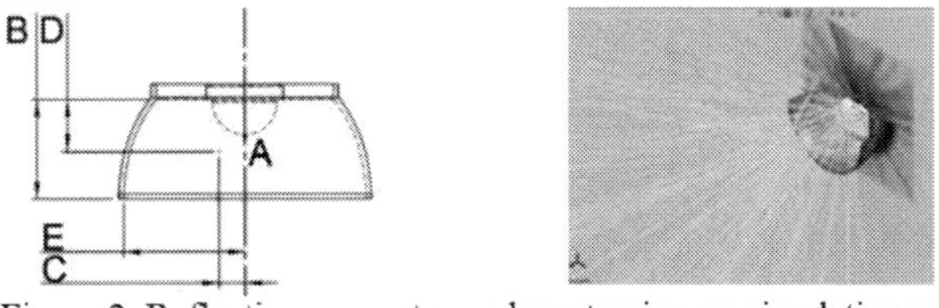

Figure 2. Reflection geometry and ray tracing on simulation program

The shape of the curve and the intensity curve depend on the focal distance and distance E of reflector. In addition, which are depended LED source of the size and shape. Height of reflector can change on the angle of reflection. [2][3]

2-3 Simulation and Test results

The simulation results were confirmed similar results to the target distribution. Figure 3. is optimal distribution and illumination which are 60° distribution and 96.0% luminaire efficiency.

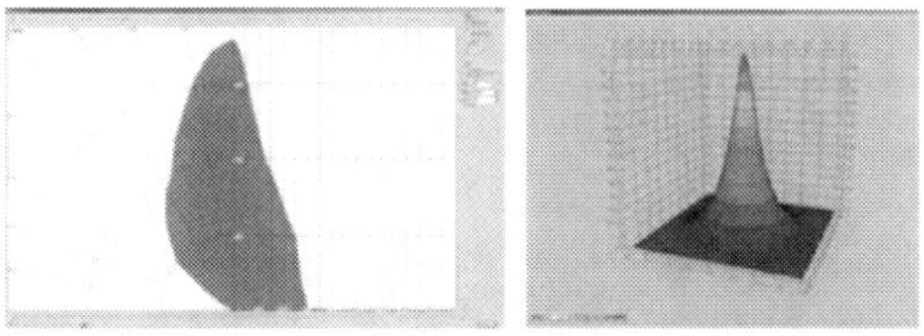

Figure 3. Reflector of characteristics distribution and illumination

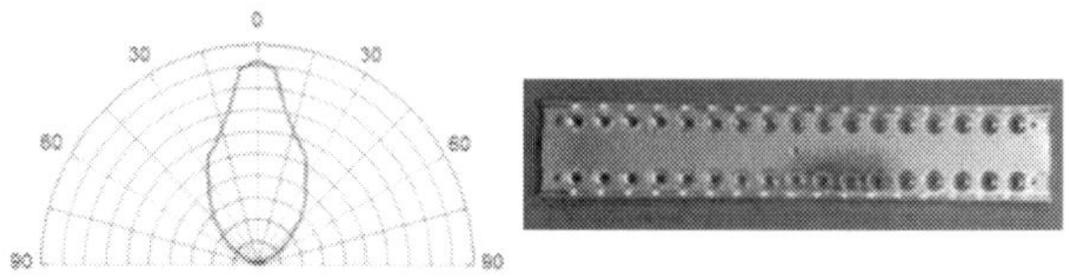

Figure 4. High efficiency reflector of LED light made by applying optimization factor

Figure 4. is reflector of LED light which is designed by the simulation results and measured results by goniophotometer [LG-1.3], as results 60° distribution and 96.3% luminaire efficiency. In this result, matched ratio 99.6% was gained from between the simulation result and the measuring result.

3. CONCLUTION

This paper, reflector which is the distribution 60° with the attributes was designed for the distribution control. OptisWorks optical design program was used to predict performance. According to change the element of aspherics, target curve similar with distribution curve could be obtained. At the results of optical characteristic were 60° distribution and 96.0% luminaire efficiency. Therefore, high efficiency LED light was manufactured by the simulation result which was matched ratio 99.6%.

REFERENCES

[1] Minuk Lee, hun Kim, Fall Institute Conference on KIIEE, vol.9, No.3, p.67, 2009, 10.
[2] R. H. Simons and A.R. Bean, MPG Books Ltd, Bodmin, Cornwall, pp. 206-224.
[3] J.R. Coaton and A. M. Marsden, John Wiley & Sons, Inc., New York, Toronto, pp. 354-362.
[4] Jong-Chan LEE, Gi-Seung CHOI, Proceedings of The 4th International Conference of LED and SSL., Seoul, Korea, pp. 86-89, 2010.

ACKNOWLEDGMENTS

This research paper had been performed by Ministry of Knowledge Economy project, "Development of High Luminous Intensity 200W Class LED Lamp System for High Ceiling" and "Development of LED lighting technology for 110 lm/W of luminaire efficacy employing HWLP LEDs"

Power supplies for excilamps – a review of structures for UV emission control

D.V. Schitz[1], H. Piquet[2], R. Diez[3], M. Djibrillah[2], S. Bhosle[2], V. Tarasenko[1], E. Sosnin[1]
[1]*Institute of High Current Electronics SB RAS, Akademichesky Ave. 2/3, Tomsk, 634055, Russia,*
[2]*LAPLACE Laboratory, UMR CNRS-INPT-UPS, Toulouse University, 2 rue Camichel, 31071 Toulouse, France*
[3]*Electronics Department, Pontificia Universidad Javeriana, Bogotá, Colombia*
Contacts: Hubert.Piquet@laplace.univ-tlse.fr – schutz@loi.hcei.tsc.ru

ABSTRACT

Dielectric barrier discharge excimer lamps (excilamps) are well known powerful and efficient sources of UV emission. The UV emission is tightly related to the electric power transferred to the lamp and very different performances are obtained, depending upon the characteristics of the power supply which is used. This paper proposes a review of power topologies, which have been used by the authors for various purposes (laboratory physic domain investigations and UV emission process oriented studies). For each of them, the topology and its operating principle, as well as a discussion of the merits concerning the performances (UV emission, design pitfalls, efficiency, reliability, waveforms …) are proposed.

POWER SUPPLY REQUIREMENTS FOR DBD EXCILAMPS

The design of power supplies for DBD excilamp [1] relies on their electrical characteristics, described by the equivalent circuit [2] and the static electrical characteristic of the gas which fills the bulb – see Fig. 1. To establish a power transfer to the lamp, a high voltage ($V_{threshold}$ # 5kV) has to be applied across its terminals.

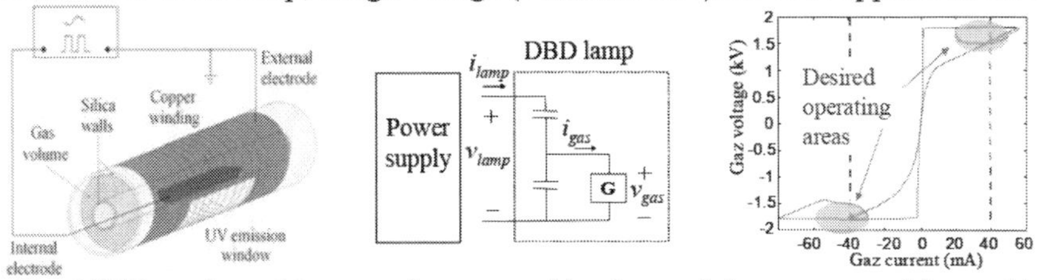

Fig. 1 DBD excilamp (a) - equivalent circuit (b) - electrical characteristic of the gas (c)

Once the breakdown is reached, according to the capacitive nature of the lamp, a bidirectional lamp current is needed (frequency range 20 kHz – 150 kHz). UV emission being tightly correlated with current, power or current control of the electric generator is a desirable feature; the studied power range is from 50W up to several 100W. The high voltage needed across the lamp requires the installation of a step-up transformer between the power stage and the bulb. The transformer, according to the frequency range, uses ferrite cores. The high voltage on the secondary implies a large turn number and careful insulation (Teflon interlayer).

SINE VOLTAGE POWER SUPPLY

Physicists frequently use sine high voltage for excilamp excitation [2][3] in investigations. Usually such supplies have 2 main parts: a power amplifier and a step-up transformer, which may be assembled as a self-excited oscillator: the inductance of the transformer and the added capacitor define the oscillating frequency.

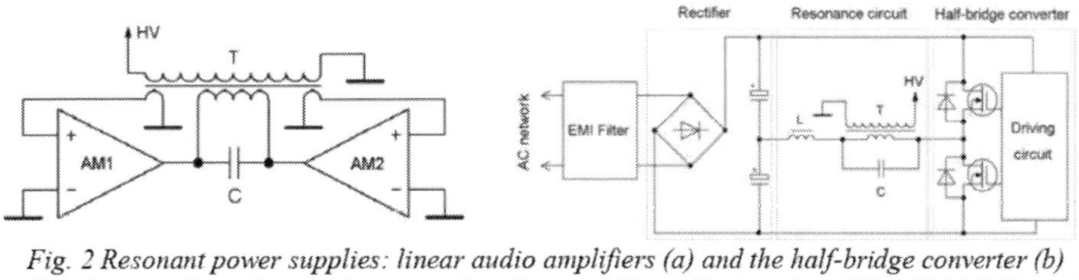

Fig. 2 Resonant power supplies: linear audio amplifiers (a) and the half-bridge converter (b)

For basic realizations, radio valves or integrated audio amplifiers (for instance TDA7294 [5]) can be used, as on Fig. 2(a). TDA7294 have short circuit protection and thermal shutdown, making this resonant supply very failsafe. The main drawback is its poor efficiency, due to linear operation of the amplifiers, but regarding the power level (typ. 150W) and the application (lab. investigations on plasmas), this issue is not critical. A 6.5 kV magnitude sine waveform is obtained on the secondary of the ferrite core transformer.

For applications where efficiency is important, static converters using power transistors (MOSFET, IGBT) in switched mode (see Fig. 2(b)) are used: rectified AC network voltage supplies a half or full bridge topology; the switching legs associate two MOSFETs with anti-parallel diodes. The obtained square voltage excites the resonant circuit: a ferrite core choke L and a capacitor C associated with a transformer T. Therefore a high voltage of sine shape (amplitude 3-6 kV) is obtained on the secondary of T. The magnitude of the output voltage is adjusted by means of the driving frequency, relatively to the eigen frequency of the resonant loop L-C-T [4]. This simple supply is failsafe: output overload only leads to quenching of resonance oscillations, but primary current remains controlled. Over voltages and voltage spikes arising in the step-up transformer (due to sparks, bad contacts ...) that could damage the supply are filtered by the LC circuit; anti-parallel diodes cut-off over-voltage. Thanks to zero current switching of transistors (ZCS) [4], efficiency is improved.

PULSE POWER SUPPLIES

Efficiency increase of DBD excimer lamps [6][7][8] at high-voltage short pulse excitation and several tens of kHz has been established with a diffuse discharge (great number of cone-like micro-discharges). Bridge circuits Fig. 3(a) are widely used for high voltage pulsed power (application range 20-300W, up to 1kW): rectified AC network voltage (eventually with power factor correction - PFC) feeds a half or full-bridge converter. Each switching leg has two MOSFETs or IGBTs with anti-parallel diodes, controlled by a pulse-width modulator (PWM). High-voltage pulses with 20-150 kHz frequency are produced at the secondary of the transformer (its primary winding, connected to the bridge, uses RF cable because current slope belongs to MHz range). The voltage applied to the bulb causes spike currents, giving very efficient UV emission spikes. Current controlled topologies [9] use thyristor-like synthesized semiconductors. Operating in discontinuous mode (resonance between L and the lamp) - Fig. 3(b), they offer direct and instantaneous lamp current control; this has proven to efficiently handle UV emissions. Continuous mode topologies are also considered.

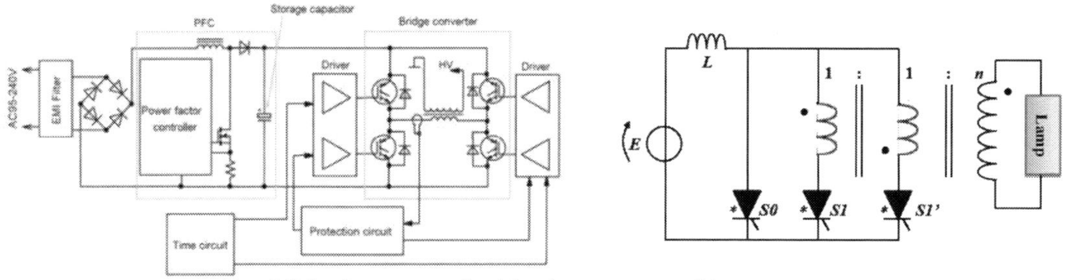

Fig. 3 Pulsed power supply: (a) voltage source – (b) current source

Acknowledgement: part of this work is French-Colombian cooperation, supported by ECOS-Nord / COLCENCIAS / ICETEX program.

REFERENCES

[1] M.I. Lomaev, E.A. Sosnin, V.F. Tarasenko, D.V. Schits, V.S. Skakun, M.V. Erofeev, A.A. Lisenko. *Capacitive and Barrier Discharge Excilamps and Their Applications (Review)*. Instruments and Experimental Techniques, 2006, Vol. 49, No. 5, pp. 595-616

[2] R. Diez; JP. Salanne; H. Piquet; S. Bhosle; G. Zissis: *Predictive model of a DBD lamp for power supply design and method for the automatic identification of its parameters* – E.P.J. App. physics, 2007.

[3] Kogelschatz U. *Silent discharges for the generation of ultraviolet and vacuum ultraviolet excimer radiation* // Pure &App/. Chem., Vol. 62, No. 9, pp. 1667-1674,1990.

[4] Y. Cheron, *Soft commutation*, Chapman and Hall, 1992

[5] STMicroelectronics, *TDA7294, Application note*. http://www.st.com

[6] Mildren R.P., Carman R.J. *Enhanced performance of a dielectric barrier discharge lamp using short-pulsed excitation* // J. App. Physics D. – 2001. – №34. – L1-L6.

[7] Vollkommer F., Hitzshke L. *Dielectric Barrier Discharge* // Proc. of the 8th International Symposium on Science and Technology of Light Sources (Greifswald, Germany, 1998), pp. 51–60.

[8] Lomaev M.I., Schitz D.V., Skakun V.S., Tarasenko V.F. *Influence of excitation pulse form on barrier discharge excilamps efficiency* // Proc. of SPIE. – 2002. – V.4460. – P.38-45.

[9] R. Diez; Hubert Piquet; Sounil Bhosle : *Control of UV Emission of an Excimer Lamp by Means of a Current-Mode Power Supply* – IEEE - IECON 2009 Conf. – Porto, 3-5 nov. 2009

Photon cascade emission of Pr3+ in strontium aluminate

Y. Ikeda , K. Masada , H. Kurokawa, H. Motomura , M. Jinno , K. Tachibana
Department of Electrical and Electronic Engineering, Ehime University
3 Bunkyo-cho, Matsuyama, Ehime 790-8577, Japan
Contact: mjin@mayu.ee.ehime-u.ac.jp

ABSTRACT

The photoluminescence and excitation spectra of Pr^{3+} activated strontium aluminate (SAO), which are Hot Isostatic Pressing (HIP) processed and normally sintered, has been investigated at the room temperature (300K). The luminescence excited by Xe 172nm consists of optical transitions emanating from the 4f5d and the $^1S_0$ states of Pr^{3+}. As for the HIP-processed samples, the emission from $^1S_0$ and $^3P_0$ were observed. On the other hand, as for the samples which were made in a tubular furnace, no emission from $^1S_0$was observed and UV emission of 250-350nm ($4f^15d^1 \rightarrow 4f^2$) was observed. The results of energy-dispersive X-ray spectroscopy (EDS) analysis show that $SrAl_{14}O_{25}$ is made by HIP-processed, and that $SrAl_2O_4$ was made with a tubular furnace. These results show that quantum-cutting phosphors can be obtained by Pr^{3+} activated $SrAl_{14}O_{25}$.

INTRODUCTION

As blue phosphor for vacuum ultraviolet excitation, BAMs ($BaMgAl_{14}O_{23}:Eu^{2+}$ or $BaMgAl_{12}O_{17}:Eu^{2+}$) is well known [1] [2] and used for plasma displays etc. In BAMs Eu^{2+} is a luminescent centre, and a wide luminescence is generated in the region of 400-550nm by $4f^65d^1 \rightarrow 4f^7$ transition of Eu^{2+}. However, the peak of excitation spectrum of BAM lies around 250nm and absorption efficiency is low VUV emission of Xe excimer at 172nm in comparison with resonance radiation of Hg at 254nm [2]. In addition, there is a problem that stokes-loss is large to obtain visible blue light. Therefore we pays attention to Pr as an activator, and in this research we tried to produce a blue phosphor that has high performance for Xe VUV. This phosphor , a quantum cutting phosphor, is expected to absorb one photon in VUV region and radiates 2 blue photons of transitions $^1S_0 \rightarrow {}^1I_6$ and $^3P_0 \rightarrow {}^3H_4$. [3] [4].

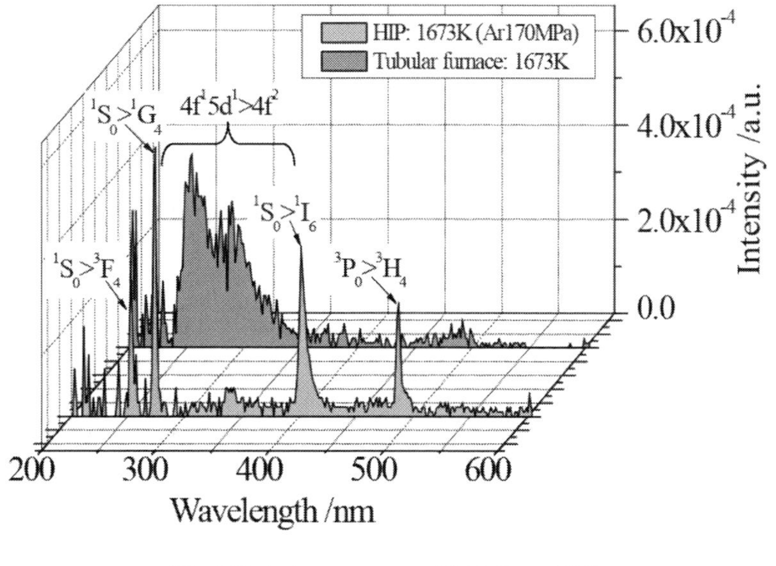

Fig. 1 Emission spectrum of SAO:Pr at 300K

EXPERIMENTAL

We used Pr for activator in strontium aluminate (SAO) which is the host crystal of the phosphor. Two kinds of sintering methods are examined. One uses HIP-process, and the other uses a conventional tubular furnace. In both sintering processes the temperature was kept at 1670K (temperature start-up 2Hrs - constant 2Hrs - cooling 2Hrs) and the ambient gas was Ar. In the case of HIP-process the pressure was 170MPa, and in the case of tubular furnace the flow rate of Ar was 1.0 l/min. Each phosphor was excited by a Xe excimer lamp with 172nm emission and the photoluminescence was measured by a spectroscope at room temperature (300K). The crystal condition of each phosphor is observed by FE-SEM and qualitative analysis was done by EDS.

Element	HIP Atom%	Tubular furnace Atom%
O K	63.1	49.4
Al K	34.4	34.9
Sr L	2.5	15.8
Total	100	100

Fig. 2 SEM and EDS Analysis data of SAO:Pr

RESULTS AND DISCUSSIONS

The emission spectra of produced SAO:Pr phosphors are shown in Fig. 1. As shown in this figure, the phosphor produced by HIP-process shows the emissions of Pr^{3+} in UV region at 253nm ($^1S_0 \rightarrow {}^3F_4$), 273nm ($^1S_0 \rightarrow {}^1G_4$), and in visible region at 402nm ($^1S_0 \rightarrow {}^1I_6$), 486nm ($^3P_0 \rightarrow {}^3H_4$). After the first photon emission by the $^1S_0 \rightarrow {}^1I_6$ transition with blue emission at 402nm, the $^1I_6$ state trances to the $^3P_0$ state. This transition is a rapid non-radiative multiphonon relaxation. Then the second photon at 486 nm is emitted by the $^3P_0 \rightarrow {}^3H_4$ transition. Thus, it is shown that cascade emission from $^1S_0$ state of Pr^{3+} can yield two visible photons in sequential two-step radiative process by HIP-processed.

On the other hand the phosphor produced by a tubular furnace shows not a line spectrum which is known as f-f transition. However a broad band emission in the region of 250-350nm is observed. The reason why this Pr^{3+} band emission is obtained is that the lowest energy level of $4f^15d^1$ lies lower than the upper level of $^1S_0$ (47000cm^{-1}), and this corresponds to the Pr^{3+} $4f^15d^1 \rightarrow {}^3H_4$, $^1G_4$ transition.

In order to characterize the morphology structure and chemical composition of the of SAO:Pr phosphors prepared by combustion synthesis, scanning electron microscopy and EDS measurement were carried out. The typical results are shown in Fig.2. The phosphor produced by HIP-process has an irregular morphology and the particles look like square boards. On the other hand, phosphor generated by a tubular furnace consists of small particles. The EDS analysis of two kind of phosphors shows that the composition ratio of the HIP-processed phosphor is about Sr:Al=1:14, and that of the tubular furnace produced phosphor is about Sr:Al=1:2. This difference between these host crystals (SAO) may bring the difference of luminescence property.

CONCLUSION

The photoluminescence and excitation spectra of Pr^{3+} activated SAO have been investigated. This experiment shows that the quantum cutting is obtained by HIP-processed phosphor, and the cascade emission from $^1S_0$ state of Pr^{3+} can yield two visible photons in sequential two-step radiative process. As the future subjects, we should optimize the composition ratio of Pr and the sintering condition.

REFERENCES

[1] R.Ohtsuka, ITE Technical Report 25(5), 103-108, 20010130

[2] *Phosphor Handbook*, Ohmsha, 330-335, 1987

[3] *Phosphor Handbook*, CRC Press, 509, 2007

[4] A.M.Srivastava, W.W.Beers, Journal of Luminescence 71, 285-290, 1997

10klm multichannel LED light source for hard edge color variable spot with excellent color mixing

R. Kurt[1], E. Lenderink[1], J. de Vries[1], T. Tukker[1], M. Sipkes[1], M. van der Lubbe[1], S. Verbrugh[2], A. van den Brandt[2]
[1] *LightLabs, Philips Lighting, Mathildelaan 1, P.O. Box 80020, 5600 JM Eindhoven, The Netherlands*
[2]*Special Lighting, Philips Lighting, Steenweg op Gierle 417, B-2300 Turnhout, Belgium*
Contact: ralph.kurt@philips.com

ABSTRACT

Based on 120 closely packed thin film flip chip LEDs on small individual sub-mounts combined with Lumiramic (ceramic phosphor conversion platelet) color conversion technology, we have realized a high brightness LED array with a diameter of 29 mm with 6 individual addressable color channels. Equipped with our advanced, high quality color mixing and efficient light collimation optics we demonstrated a hard edge LED spot with gobo projection for atmosphere creation and scene setting in theater and entertainment lighting. With a total input power of about 350W we have measured 10.000 lm white light out of array, which is considered to be an interesting entry point to meet the demanding requirements of entertainment lighting applications including color tunable theatre spot lighting.

INTRODUCTION

Entertainment lighting is about scene setting with (colored) light and sophisticated creation of atmospheres and moods with specific light effects. The promise of LEDs adjusting light settings very fast and with infinite variations comes with major technical challenges: efficient color mixing and sufficient source brightness. In order to realize a hard edge spot allowing projection of gobos (patterned frames) without any color fringing, the color mixing has to be realized in the light source in an efficient way.

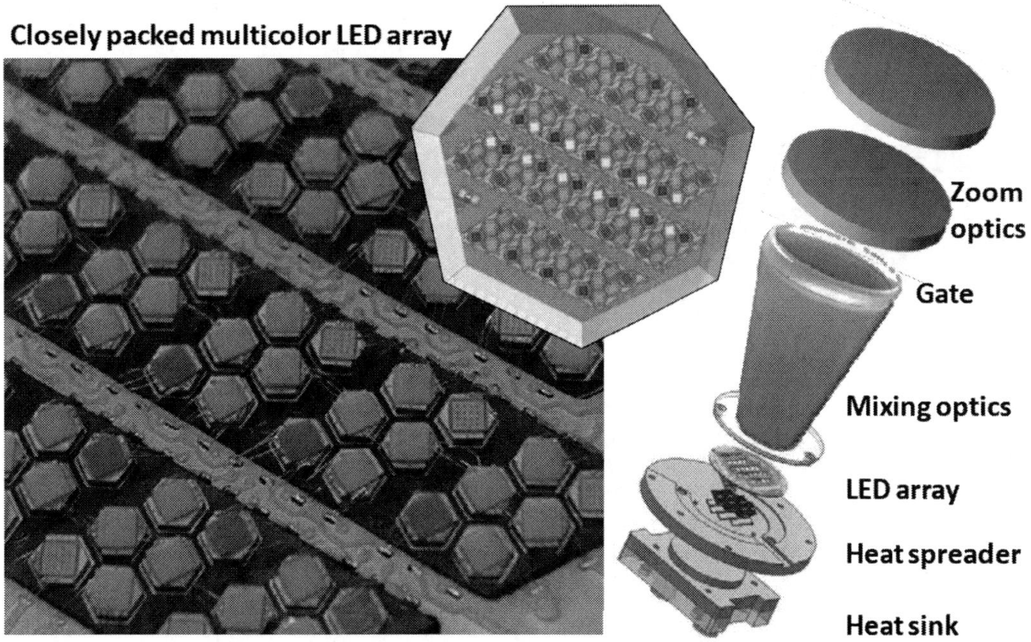

Figure 1: The closely packed multicolor LED array comprising 120 LEDs in a diameter of 29mm is arranged onto a heat spreader and emits its light (~10klm) into the entrance aperture of a reflective mixing and collimation optics, which guarantees good color mixing at the exit aperture i.e. in the gate of a projection apparatus, the later is projected by the zoom optics as a hard edge spot onto the stage.

SYSTEM DESIGN

The requirements on color homogeneity can only be met when the different colored light sources are packed densely together. We soldered thin film flip chip LEDs closed together onto AlN substrate tiles, which have been subsequently soldered onto a Cu heat spreader for thermal management. Subsequently red, green, amber, cyan and white Lumiramic phosphor conversion plates [1] were attached directly onto the blue emitting dice as seen in Figure 1. Electrical connection is realized by multiple wirebonds. The inset shows the layout of the LED array with a diameter of 29 mm comprising 120 LEDs at 6 different colors (RAGBCW) arranged in a heptagonal entrance aperture of a mixing reflector unit [2].
We have demonstrated the potential of our LED based light source for spot beams by modifying a commonly used fixture in theatre lighting i.e. the Source Four [3]. This is a projection system to enable gobo projection, zooming and Hard Edge spot. We replaced the lamp and ellipsoidal reflector by our custom made LED array and our etendue-limited collimation and mixing optics [2].

RESULTS AND DISCUSSION

The LED array was subjected to optical and thermal characterizations before it was built into the lighting fixture. The array was found to produce 9930 lm when operated at 750 mA per die, corresponding to a total electrical input power of 327 W. Due to the relatively wide phosphor emission bands, the spectrum is much smoother and less peaked than conventional RGB color variable systems with direct emitters. Using infrared thermography, the temperature of the phosphor plates was measured during operation. When the base plate was kept at a constant temperature of 25°C, the highest temperature measured anywhere in the array was found to be 96°C. Most phosphor plates in the array were 10-20°C cooler than this value.
The LED array was then attached to the collimation and color mixing optics, which is in turn attached to the projection optics of a Source Four fixture. Figure 2 shows a color photograph of the hard-edge beam onto a white wall. No color artifacts are visible in this beam, and no color artifacts appear around the shadows of objects in the beam (middle panel of Figure 2) and around the white-dark edges of a projected gobo (right panel of Figure 2). These tests show that the color mixing optics are very well capable of mixing all the colors at the projection gate inside the Source Four, both in position and in direction, for this application. The only color artifacts still visible are in the far edge of the beam, when the stop used to create the hard-edge beam is removed. But with the stop in place, color mixing in the beam is good enough for the application, which is the first time ever demonstrated for systems of this kind.

Figure 2: Demonstration of the obtained beam quality of the hard edge spot with excellent color mixing in the gate allowing gobo projection without color shadows or other disturbing artifacts.

REFERENCES

[1] H. Bechtel et al, Proc. SPIE 7058, 70580E (2008).
[2] T.M. Marshall et al, US patent 6200002B1 (2001).
[3] Source Four, registered trademark of ETC (www.etcconnect.com)

Optical Characteristics of White LED Composed of Ideal RGB-LEDs

Katsuhide Misono
Department of Electrical and Electronic Engineering, Tsuruoka National College of Technology, 104 Aza Sawada, Inooka, Tsuruoka-shi, Yamagata-ken, 997-8511 Japan
Contact: katsuhide_misono@yahoo.co.jp

ABSTRACT

We investigated the optical characteristics, i.e. color rendering index (Ra), color gamut, and luminous efficacy, of white LEDs composed of ideal RGB-LEDs. The proper peak wavelengths and FWHMs of RGB-LEDs were obtained that achieve white LED with high Ra in Group 1B (80 ≦Ra <90) as well as luminous efficacy for color temperatures ranging from 2800(K) to 6500(K). The obtained Ra and luminous efficacy exceed those of Blue-YAG white LEDs and tri-band fluorescent lamps. The peak wavelengths and FWHMs of RGB-LEDs that give high color gamut, which is the possible measure of color preference, were also obtained. The use of L18 orthogonal array reduced the number of computation to find solution efficiently. Our results suggest the potential capability of RGB-White LEDs.

1. INTRODUCTION

The white LEDs based on blue-LED+YAG-phosphor and UV-LED+RGB-phosphors are in practical use these days. Although the advantages of these white LEDs are their high efficacy and/or Ra, the intrinsic disadvantage is the energy loss caused by Stokes shift. Since the advantage of LEDs compared with other light sources are their ease of dimming, it is possible to produce a white LED that one can flexibly control the color temperature and color preference by additive color mixing of RGB-LEDs. In this paper, we discuss the optical characteristics such as color rendering index (Ra), color gamut, and luminous efficacy (=lumen/visible radiation power) of RGB-White LEDs produced by additive color mixing of RGB-LEDs

2. SIMULATION PROCEDURES

The color temperatures were set to 5 levels: 2800(K), 3500(K), 4200(K), 5000(K), 6500(K), which are widely used for general lighting. The spectral power distributions of RGB-LEDs were assumed to be Gaussian, and the peak wavelengths and FWHMs were selected by consulting Reference [1]. We followed the calculation procedure of Ra and efficacy described in Reference [2]. We also investigated the color preference of white LED by paying attention to the color gamut of the octagon composed of R1-R8 on uv chromaticity space by consulting References [2]-[4]. Here, we adopted %Gamut1 as the measure of color preference, which is defined by the following equation.

Table 1 Peak Wavelengths and FWHMs of RGB-LEDs Allocated in L18 Orthogonal Array

	-	-	B-Peak	B-FWHM	G-Peak	G-FWHM	R-Peak	R-FWHM
Level 1			450	10	530	10	600	10
Level 2			460	20	540	20	610	20
Level 3			470	40	550	40	620	40

No.	-	-	B-Peak	B-FWHM	G-Peak	G-FWHM	R-Peak	R-FWHM
1			450	10	530	10	600	10
2			460	20	540	20	610	20
3			470	40	550	40	620	40
4			450	10	540	20	620	40
5			460	20	550	40	600	10
6			470	40	530	10	610	20
7			450	20	530	40	610	40
8			460	40	540	10	620	10
9			470	10	550	20	600	20
10			450	40	550	20	610	10
11			460	10	530	40	620	20
12			470	20	540	10	600	40
13			450	20	550	10	620	20
14			460	40	530	20	600	40
15			470	10	540	40	610	10
16			450	40	540	40	600	20
17			460	10	550	10	610	40
18			470	20	530	20	620	10

$$\%\text{Gamut1} = \frac{\text{Gamut of Octagon by Test Light Source}}{\text{Gamut of Octagon by Reference Light Source}} \quad (1)$$

L18 orthogonal array was applied to reduce the number of calculation and find the main effect efficiently. Table 1 shows the peak wavelengths and FWHMs of RGB-LEDs allocated in L18 orthogonal array.

3. RESULTS

The relationship between Ra and luminous efficacy for 5 color temperatures is shown in Figure1, where test numbers from 1 to 18 are taken as parameters. Figure 2 shows the relationship between Ra and %Gamut1, where test numbers from 1 to 18 are taken as parameters.

4. DISCUSSION

The candidate of the peak wave-lengths and FWHMs of RGB-LEDs that achieve Ra in Group 1B (80≦Ra <90) and high efficacy for 5 color temperatures is determined from Figure 1, i.e. L18-03. The estimated Ra and luminous efficacy of this combination of RGB-LEDs exceed those of Blue-YAG white-LEDs and tri-band fluorescent lamps, and this suggests the potential capability of RGB-White LEDs.

According to Reference [3], %Gamut1 of 140 is expected to give the best preferred color. Figure 2 shows that L18-08 is expected to give almost best color preference.

Our results suggest that the RGB-White LED gives a degree of freedom to design a white LED that places a high value on Ra or a white LED that places a high value on the color preference.

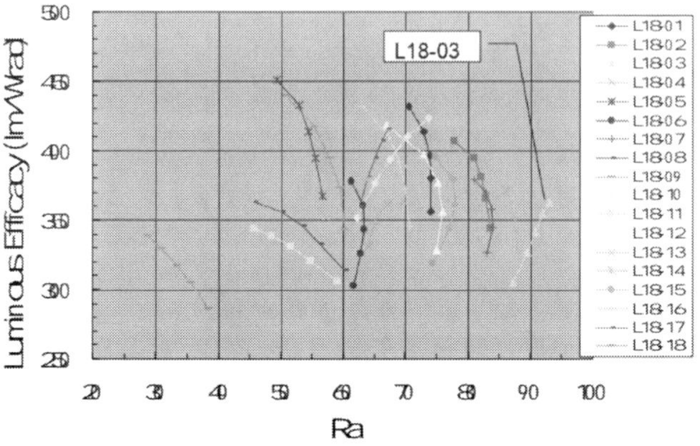

Figure 1 The Relationship between Ra and Luminous Efficacy

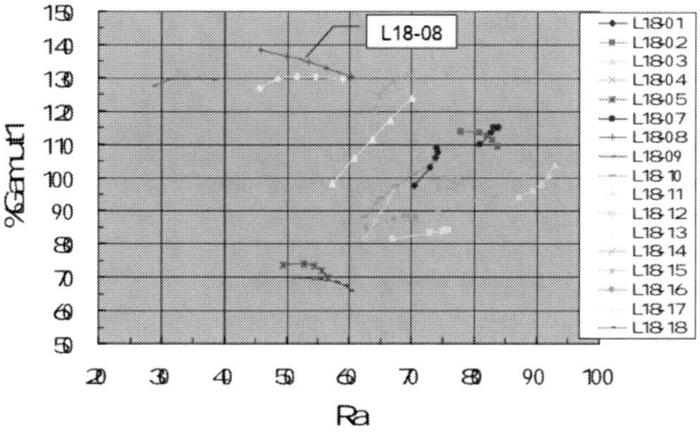

Figure 2 The Relationship between Ra and %Gamut 1

5. CONCLUSION

We carried out the simulations to investigate the effect of the spectral power distributions of RGB-LEDs on the photometric and colorimetric characteristics of RGB-White LEDs.

We determined the proper spectral power distributions of RGB-LEDs to achieve both high Ra and high luminous efficacy for practical color temperatures. We also determined the proper spectral power distributions of RGB-LEDs to realize almost best color preference. Thus, the researchers and engineers in LED industries can apply our results as a useful reference to develop RGB-White LEDs for their needs.

The application of the orthogonal array reduces the number of calculation of a model with many design parameters, and one can find the proper spectral power distributions of RGB-LEDs efficiently.

When an RGB-White LED is designed, it is important to ensure the stability of color coordinate, because the variations of peak wavelengths and luminous intensities of RGB-LEDs are generally large. This practically important issue is tackled in another paper by the author in this symposium [5].

REFERENCES

[1] M. Koedam and J.J.Opstelten, Lighting Research and Technology, **3**-3(1971)pp.205-210

[2] CSAJ: Handbbok of Color Science, University of Tokyo Press(1998) (in Japanese)

[3] D.B.Judd, J.IES, **62**(1967)pp.593-598

[4] W.G.Pracejus, J.IES, **62**(1967)pp.531-534

[5] K.Misono: Method to Stabilize the Chromaticity Coordinate of White LED Composed of RGB-LEDs, paper presented in this symposium

Session Seven

Unsaturated ceramic metal halide lamps
A new generation of HID lamps

J. Hendricx[1] , J. Vrugt[2] , C. Denissen[1] , J. Suijker[1]
[1] *GBU HID, Philips lighting B.V., Building EEC, P.O.Box 80020, Eindhoven, 5600 JM, The Netherlands*
[2] *Current address: Het Ronsel 63, Bergeijk, 5571SX, The Netherlands*
Contact: j.hendricx@philips.com

ABSTRACT

Ceramic metal halide lamps contain a liquid salt pool in the burner during operation. This limits the light technical characteristics (efficiency, colour quality) and life time and leads to colour point shift in case of variations in power, environmental temperature and burning position. These disadvantages can be avoided by making the burners unsaturated: with all salt in the gas phase. Therefore specific burners have been designed with iridium feedthroughs, a special filling plug and dimensions to give a sufficient high and homogeneous wall temperature. These lamps outperform the current lamps and offer new application possibilities: e.g. dimming. The excellent colour characteristics of the unsaturated lamp can be maintained over a wide dimming range. Other opportunities, made possible by unsaturated burner design, are fast run up, mercury free fillings, hot restrike and miniaturisation. They open the way to a unique generation of lamps which are appropriate for new application fields.

1. INTRODUCTION

The last five years have brought numerous innovations in ceramic metal halide (CMH) lamps: Osram integrated HID reflector lamp, Philips CDM Elite, National MT and GE CMH Ultra [1, 2, 3 and 4]. All these CMH lamps are small high efficient light sources with a good colour quality and a high luminance, suited for reflector applications. The present burner designs have a long extended plug with a small crevice between inner plug wall and electrode feedthrough. The long plug is necessary to keep the sealing frit at a low temperature. This prevents crystallisation of the frit, which can cause cracks in the alumina plug. A second reason is to slow down the exchange of frit components with the lamp fill. The crevice is filled with an overdose of salt to obtain sufficiently high salt temperatures and thus partial pressures of the salt components. The salt pool in these saturated lamps will be partly present in both the crevice and the discharge tube (Fig.1). Both the crevice and overdosing of salt lead to a number of disadvantages.

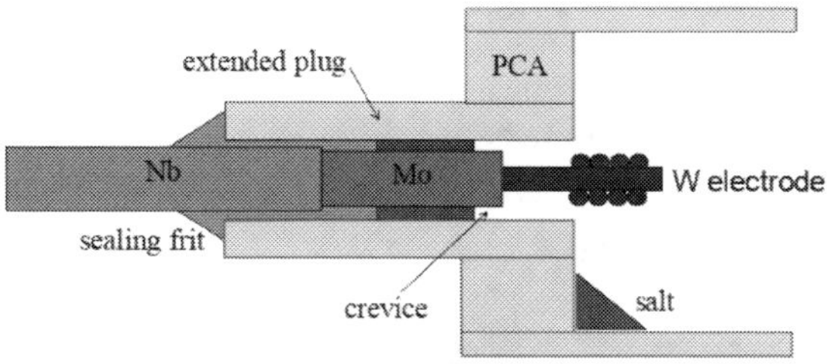

Fig.1 End construction of CMH lamp with crevice between extended plug and electrode feed through combination.

Consequences of the crevice are:

- Colour instabilities and colour spread, caused by liquid movement and evaporation, both in the discharge tube (dt) and the crevice, leading to changes in partial salt pressures in the plasma.
- A high dependence of the colour point on burning position, outer bulb dimensions and compact fixtures due to changes in the lowest wall temperature and thus in the partial salt pressures.
- Limitations in salt types to obtain optimal light technical characteristics, due to preferential deposition of salt components in the crevice.

Consequences of the liquid salt pool are:

- A high dependence of the colour point on lamp power variations due to corresponding changes in wall temperature and thus in partial salt pressures.
- Coloured areas in the beam spot due to discoloration of transmitted and reflected light by the salt pool.
- A limited life time due to alumina wall corrosion by the liquid salt.

- Limitations in salt type and composition to obtain optimal light technical characteristics, due to excessive wall corrosion when utilizing a relatively high concentration of rare earth (RE) halides.
- A long run-up time, caused by the fact that the whole discharge tube has to be heated up to the nominal burning temperature before a stable light and colour output is achieved.
- Limited possibilities to apply gas filled outer bulbs for burner miniaturisation due to a too high sensitivity of partial salt pressures for environmental temperature variations.

Numerous alternative end constructions have been investigated to avoid the negative consequences of the crevice. An obvious solution is to minimise the length of the crevice, which implies a few constraints for the sealing frit. A high temperature resistant sealing frit has to be applied or a frit containing the same elements as the lamp fill. In the last case the effect of elements exchange between lamp fill and frit at a high temperature can be limited. Examples of high temperature resistant frits are AlDy and AlY, both without SiO_2 [5]. Examples of sealing frits, containing the same RE elements as applied in the lamp fill, are AlDySi and AlCeSi [6]. However if the complete crevice is filled up with such frit this can lead to serious problems. The frit can be attacked (Fig.2) or can crystallise which might lead to a leaky burner. In addition such high temperature resistant frits often have a high melting point which poses quite a challenge for an industrial sealing process.

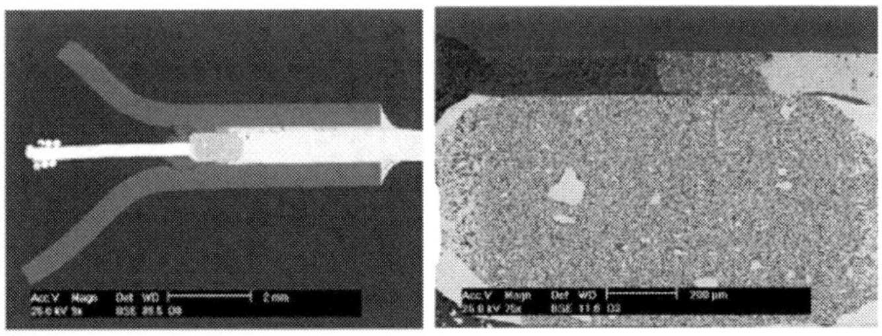

Fig.2 Example of a crevice, filled up with sealing frit containing the same elements as the lamp fill after a few hundreds of burning hours (left). The blow-up of the frit area (right) clearly shows severe leaching, leading to leakage over time.

Another option is to apply a braze to seal the burner, partly or completely filling up the crevice. Examples of brazes of metal alloys are MoAlB [7] and PtNb [8 and 9], respectively shown in Fig.3 and 4. In Fig.4 a schematic overview and a cross section of an end construction, sealed with a PtNb alloy, are given. Nb, with the same expansion coefficient as Al_2O_3, is used as feedthrough metal. After the sealing process the PtNb braze covers the Nb rod completely. This PtNb layer is expected to be sufficiently resistant against the halide filling above 1000K, while the Nb rod is not. However, Fig.4 shows that after 1400 burning hours the PtNb layer and Nb rod were attacked, which will lead to early lamp failure.

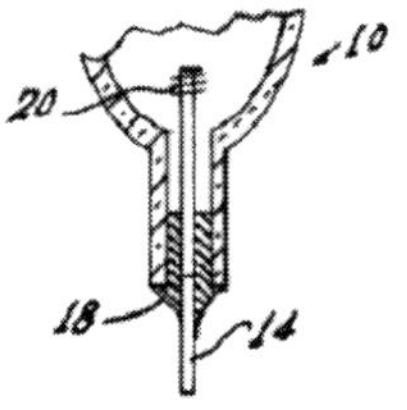

Fig.3 A MoAlB brazing alloy, resistant at higher temperatures than standard sealing frits [7].

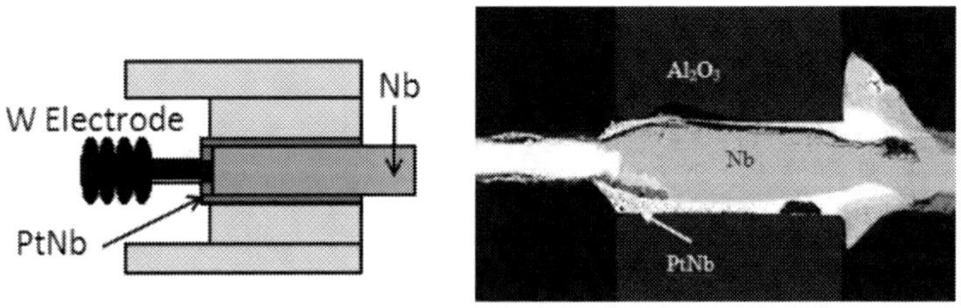

Fig.4 Schematic overview (right) and SEM photograph of a cross section (left) of an end construction with a completely filled crevice with PtNb braze [8 and 9].

Various other lighting companies have investigated crevice-less feedthroughs for CMH burners. A few examples are given in Fig.5 [10, 11, 12 and 13]. Fig.5a shows MoV end caps with at one side a W feedthrough rod. Alternative materials such as carbides, borides or nitrides, which could be halide resistant, are claimed in [14]. Fig.5b shows an end construction with end caps of e.g. Mo cermets with Mo or W feedthrough wires and an additional MoRe filling tube. Fig.5c shows one of the numerous end constructions with Mo end caps, co-sintered with Al_2O_3 or cermet plugs. Fig.5d shows a frit-less end construction with an extremely small crevice between feedthrough and co-sintered Mo tube to inhibit pooling of the discharge medium.

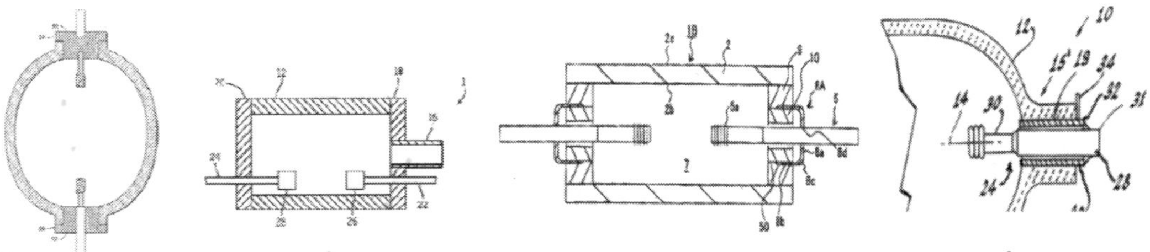

a b c d

Fig.5 Recent patents of creviceless end constructions. a: MoV end plug with a W rod as feedthrough. Sealing performed by laser welding of electrode to MoV plug [10]. b: Halide resistant end cap consisting of e.g. Mo cermet with MoRe filling tube and Mo or W based feedthroughs [11]. c: Mo end cap co-sintered between Al_2O_3 or cermet plugs[12]. d: Mo tube, shrink fit in the receptor, with a crevice width less than 0.01 mm [13].

These end constructions can also be used to avoid a salt pool and to make unsaturated burners, where all the salt is in the vapour phase during lamp operation. Of course, in addition to an end construction without a crevice, the burner design must be such to create a sufficiently high minimum wall temperature.

None of the described end construction options have been translated into products in the market as far as we know. In the next sections we will describe a new burner design which is not a rarefied academic wonder but which can be a mass-producible product.

2. BURNER DESIGN

An end construction for an unsaturated CMH burner, without crevice and use of frit, requires a vacuum tight connection between alumina and current conducting feedthrough. The vacuum-tightness has to be maintained over life and the feedthrough material has to resist the halide filling at a relatively high temperature.

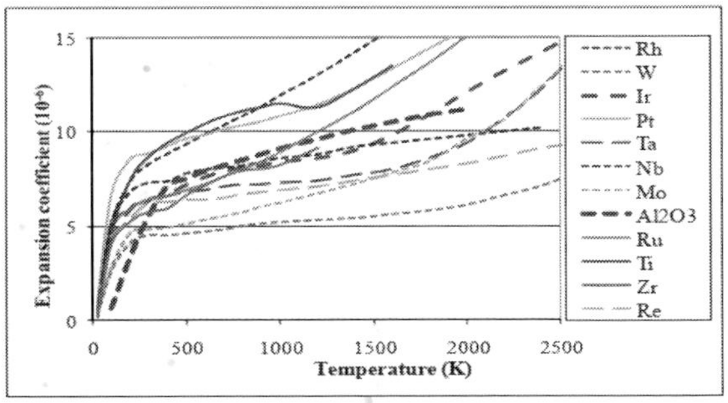

Fig.6 Expansion coefficient of Al_2O_3 and various metals as function of temperature [15].

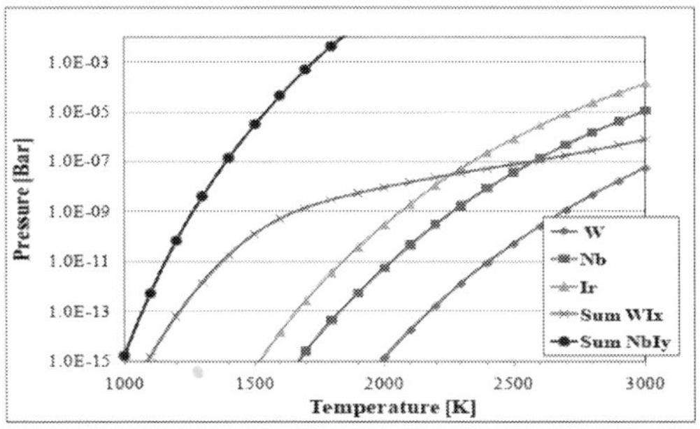

Fig.7 Vapor pressures above various metals as function of temperature (fill: metal+10 Bar Hg+0.1 Bar HgI_2).

A first condition for vacuum-tightness of the alumina feedthrough interface is that the thermal expansion coefficient of the feedthrough material matches Al_2O_3 over the whole temperature range of the burner, from ignition to full power. In Fig.6 an overview of the expansion coefficient of various metals is given as function of temperature in comparison with Al_2O_3 [15].

A second condition is that the vapour pressure of the feedthrough material is sufficiently low and its melting temperature sufficiently high. In Fig.7 the vapour pressures of some metals and metal iodides are given as function of temperature. The third condition is that the material is resistant against the halide filling of the burner. In table 1 an overview of relevant metals, fulfilling these three conditions

Metals with expansion coefficient matching Al_2O_3	Ir, Ru, Zr, Nb,
Metals with expansion coefficient nearly matching Al_2O_3	Pt, Re,Ta, Rh, Ti
Metals with T_{melt} > 2200 °C	Ir, Ru, Nb, Mo, Ta, Os, Re, W
Metals with lower vapour pressure at 1550K than W at 3000K	Ir, Ru, Nb, Mo, Ta, Os, Re, W
Halide resistant metals	Ir, Ru, Mo, Re, W, Pt, Pd, Rh, Os
Applicable metals for creviceless feedthrough	Ir, Ru, Re

Table 1 Overview of metals fulfilling several conditions for a creviceless feedthrough in a CMH burner. Possible alloys are not mentioned here.

separately and all together, is given. The best options for such a feedthrough are Ir, Ru and Re or alloys of elements with Mo, W, Pt and/or Rh. According to table 1, Ir is one of few metals which can be applied best as feedthrough material. In Fig.8 a picture and drawing of a real burner with such an Ir feedthrough are given [16]. The vacuum tight connection between Ir and alumina is realized by sintering alumina around the Ir. The interface between Ir and alumina appears to be very strong. The adhesion strength Ir-alumina has been tested (Fig.9) and was found to be stronger than 200 MPa.

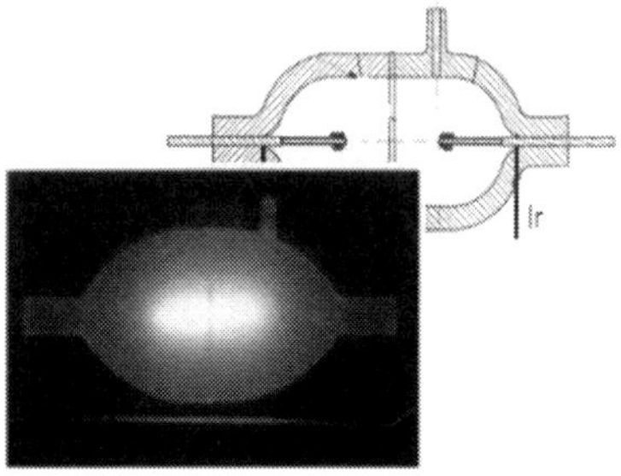

Fig.8 A CMH burner with directly co-sintered Ir feedthroughs

Fig. 9 Three point bending test (left upper) with sample consisting out of an alumina plug with Ir rod (left lower) to determine the adhesion strength of Ir-Al_2O_3 interface. If the sample breaks, the connection between alumina and Ir stays intact (right).

During lamp operation heat is conducted to the short extended plugs by the electrodes. The large temperature gradient along W electrode and Ir feedthrough (from 2800K to 1500K) creates stresses in the alumina plug which might lead to cracks. Fig.10 gives an indication of these stresses. The stresses are relatively high (max.: 250MPa) but are only present in a small part at the inner side of the cylindrical plug. The Ir feedthrough end construction has been tested in real burners over life. The concept proved to be vacuum-tight over more than 20000 switches or 17000 hours.

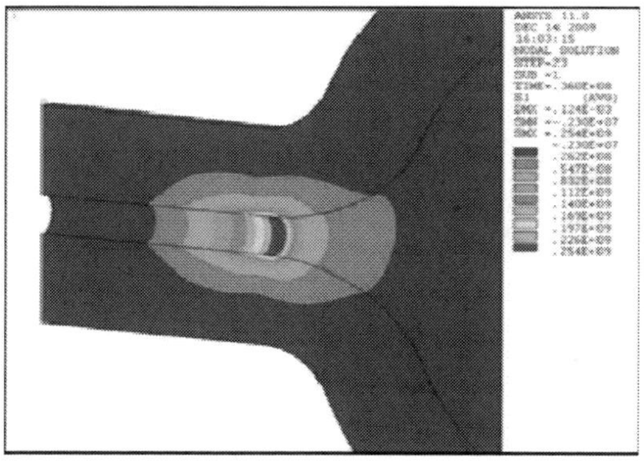

Fig.10 Detailed view of the calculated stresses in extended plugs of burners with Ir rod feedthroughs with a small feedthrough diameter.

Of course, such a creviceless burner concept with Ir feedthroughs has to be filled after end sintering. Therefore a small ceramic filling tube is present on the discharge tube (dt) as shown in Fig.11 [17]. Through this tube the burner is filled and closed.

To realize unsaturated lamps with as much attractive salts as possible, a high minimum inner wall temperature is required. However a too high wall temperature leads to alumina decomposition and deposition of Al on the inner side of the outer bulb. A homogeneous wall temperature distribution along the dt is thus preferred. Fig.12 shows that with short extended plugs, an optimized burner shape and an optimized wall thickness such a homogeneous temperature distribution can be realized. In 39W burners the wall temperature differences along the discharge tube wall could be limited to less than 50K. Fig.12 also shows that the temperature of the extended plugs can be even higher than the

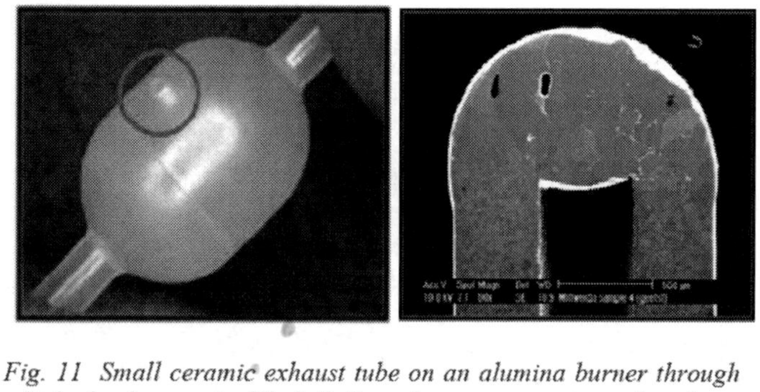

Fig. 11 Small ceramic exhaust tube on an alumina burner through which the burner is filled (left) and a cross section of such an exhaust tube after closing the tube with help of a laser (right).

maximum wall temperature of the discharge chamber itself. This is caused by the heat transport of the electrodes to the relatively short extended plugs in this design. In such burner designs for unsaturated lamps the minimum inner dt wall temperature (1450-1550K) is significantly higher than in normal saturated lamps (1200-1300K).

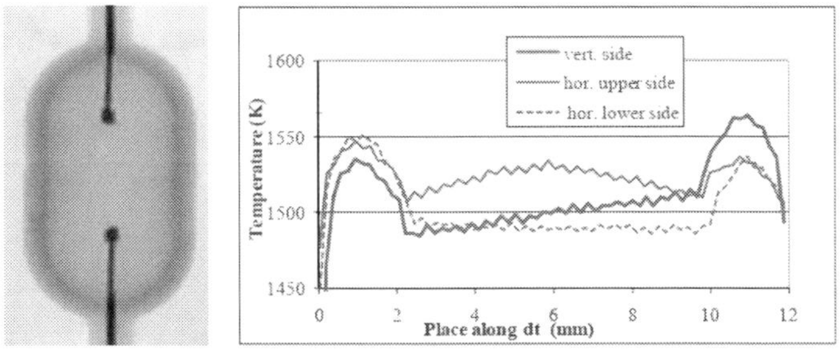

Fig. 12 Wall temperature distributions of an unsaturated 39W burner (left), measured along the side of a vertically burning and along the upper and lower side of a horizontally burning burner (right).

Therefore much higher metal iodide pressures can be realized which results in a better light technical performance. Fig.13 shows the pressures of some metal iodides at 1200 and 1450K. An additional advantage of these high metal iodide pressures is that they lead to better emitter coverage of the cathode, giving a significantly lower electrode temperature [18], which is advantageous to limit W evaporation and blackening of the inner discharge tube wall.

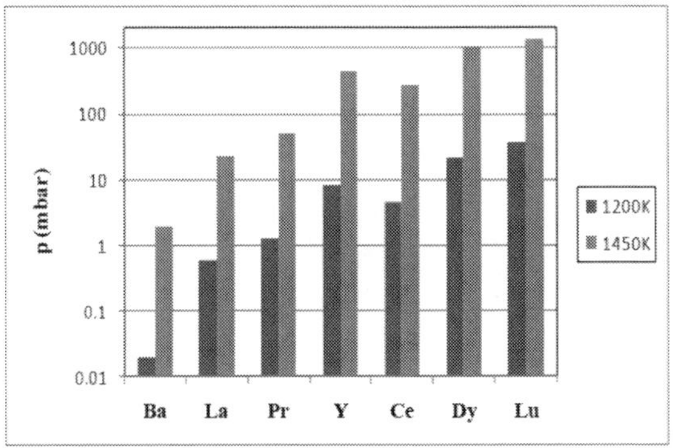

Fig.13 Vapour pressure of various metal iodides (mono salt) at 1200 and 1450K.

3. CHEMISTRY

Typically a liquid salt pool is present in the discharge volume of all metal halide lamps. Such a salt pool does not have only negative consequences but in fact this large pool of liquid salt helps to stabilize the discharge for small variations in composition of the salt mixture and towards variations in salt dose. Also impurities in the discharge tube will not influence the plasma composition mainly because they will dissolve in the large pool. For instance metal impurities will only have a small partial pressure in the plasma because of their dilution in the salt. Oxygen and water will react with the metal halides in the salt and cause only a small reduction of the metal halide pressures [19]. Since the liquid salt pool is relatively large, small movements of the salt will not completely change the mean temperature of the salt. This means that the plasma composition is relatively stable for these small salt movements. However, the saturated lamp still is quite sensitive for different burner orientations due to the change in minimum wall temperature in a different position.

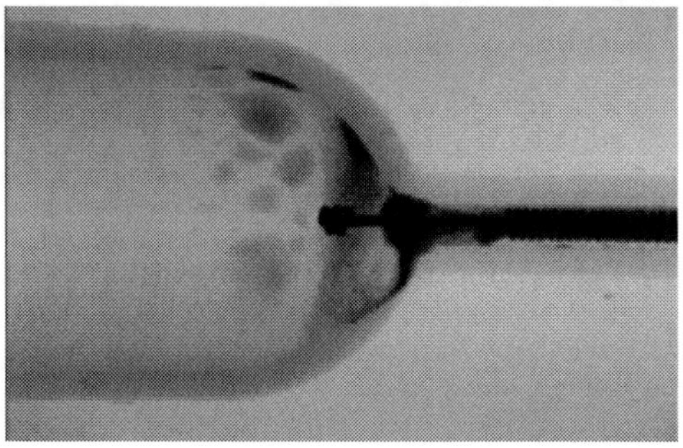

Fig.14 Transport of Al_2O_3 in a saturated burner with a salt pool containing RE salt after 500h (1.3 mole % Ce in the salt pool; T_{salt}= 1510K).

During the lifetime of a lamp the presence of the large salt pool can cause problems; especially the RE salts are aggressive for the alumina envelope. Due to a combination of chemical transport via the gas phase and the liquid phase, alumina is dissolved at some locations and deposited at other points (Fig. 14). The details of this process are still under investigation but the main reactions have been discussed earlier [20, 21, 22 and 23]. Transport via the gas phase involves AlI_3. The oxygen from Al_2O_3 is transported via the liquid phase. Both etching and deposition cause problems in the long run. The etching can create a hole in the alumina wall, resulting in a leaky discharge tube. The deposited Al_2O_3 can cause stress if it is deposited in the crevice between metal feedthough and alumina wall, which can also cause cracking and early lamp failure. When we dose the salt in such a small amount that all components will evaporate a new situation occurs [24]. The vapour pressure of the salts does not depend any longer on the temperature of the burner wall and high metal halide pressures can be reached. The life time of the lamp does not suffer anymore from the fact that the liquid salt damages the alumina wall. The new situation where all metal halides are evaporated rapidly shows that a metal halide lamp is a high temperature reactor with all kinds of chemical processes, both desired and undesired.

Of course the reactions mentioned for the saturated lamps will still occur: the RE halides will react with alumina to rare earth garnets or other aluminates:

$$3\ ReI_3 + 4\ Al_2O_3 \Leftrightarrow 3\ AlI_3 + Re_3Al_5O_{12} \qquad (1)$$

However these reactions are equilibrium reactions and the equilibrium will be established in the burner in a very short time, leaving enough RE iodide in the burner.

When these unsaturated burners are realised it becomes clear that the filling changes and needs further protection. Over time a clear potassium line becomes visible in the spectrum (Fig.15), originating from the electrode material. In saturated lamps this line is not very obvious since most of KI dissolves in the salt pool. The disappearance of rare earths by reaction with K can be controlled by limiting the amount of K in the electrode or by dosing a sacrificial metal halide like HgI_2. K then forms KI and will not react with the salts which are important for light generation in the discharge.

$$2K + HgI_2 \Leftrightarrow 2KI + Hg \qquad (2)$$

In addition to metallic contamination entering the discharge volume, oxide contamination can also be detrimental. Translucent alumina is manufactured by adding dopants such as MgO to the Al_2O_3. During lifetime the MgO will enter the discharge chamber in very small amounts. Again when it is dissolved in the liquid salt pool the metal is not very obvious in the spectrum but in unsaturated lamps the Mg presence is clear and careful firing of the Al_2O_3 is essential to limit the contribution of Mg in the spectrum.

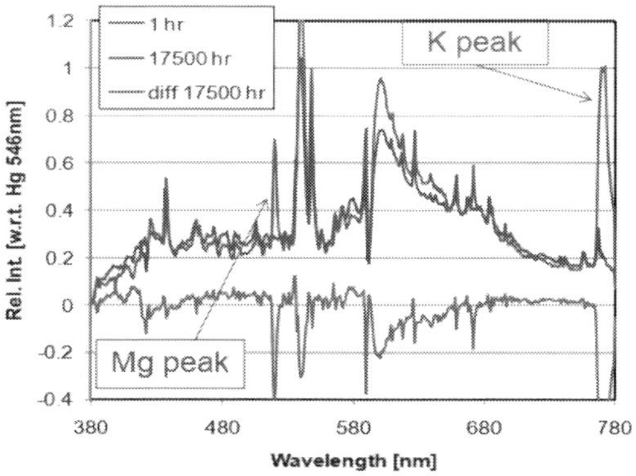

Fig.15 Relative spectral distributions (relative with respect to the Hg 546 nm line) of an unsaturated burner after 1 and 17500 burning hours and the difference of both spectra. Specific peaks appear over life.

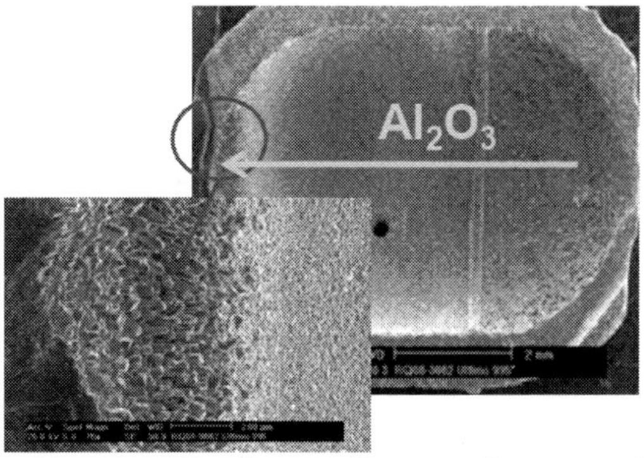

Fig.16: Inner side of a vertically operated unsaturated 39W CMH burner. Some Al_2O_3 is transported from top to bottom (detail in picture).

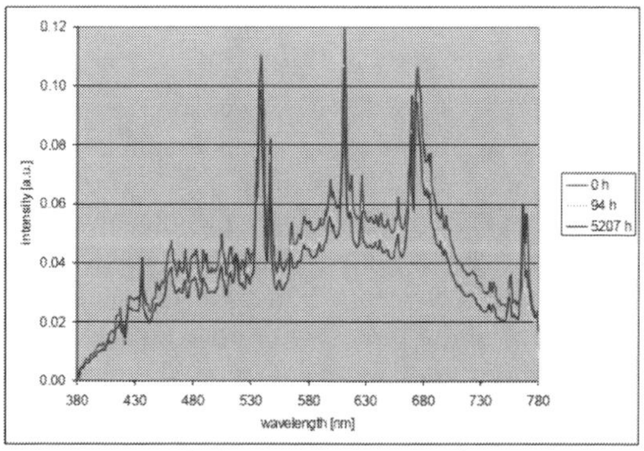

Fig.17: Spectral changes occur in a short time span. When the equilibrium is established, the spectrum is constant.

High wall temperatures can also stimulate low level transport of both Al and O atoms [25]. A thin layer of the alumina can be transported even without a liquid phase. Fig.16 and 21 show some alumina transport in a vertical burning unsaturated burner after 5000 burning hours. The most likely mechanism is transport by the reversible reaction:

$$AlI_3 + Al_2O_3 \Leftrightarrow 3AlOI \qquad (3)$$

However, chemical transport of the alumina is no longer the limiting factor for lamp lifetime.
Once all of the above mentioned chemical equilibriums are established the lamp fill will hardly change anymore and as a result no further spectral changes will be observed over life. An example of a spectrum of an unsaturated burner after 0, 94 and 5207 burning hours illustrates this (Fig.17).

4. COMPARISON OF SATURATED AND UNSATURATED BURNERS

A comparison is made between some characteristics of 4200K conventional (saturated) and new (unsaturated) CMH burners (39-70W) to illustrate the afore mentioned negative consequences of crevices and salt pool in saturated burners can be overcome with the new creviceless design. The data cited for the conventional burners are representative for burners, presently on the market.

Short term colour point spread and stability.

In Fig.18 the colour temperature (CCT) during lamp operation (over 10 burning hours) at constant lamp power is given for both lamp types. The figure shows that the unsaturated lamps are clearly more stable than the saturated lamps, which can be explained by the absence of liquid salt in crevice and dt. Also the initial spread is clearly lower. In unsaturated lamps this spread is determined by the amount of dosed salt; in saturated lamps by the largely undefined size and location of the salt pool in crevice and dt.

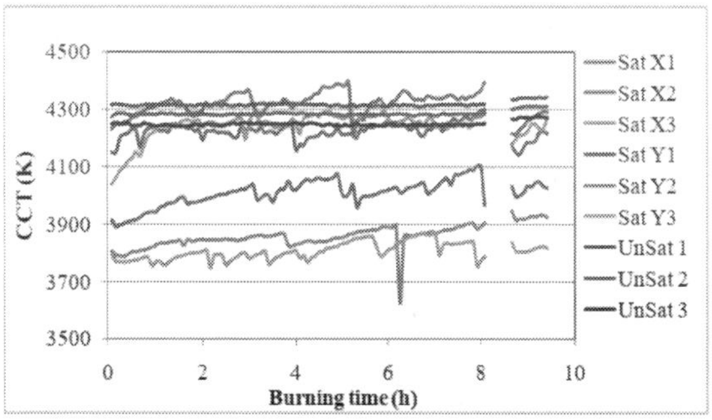

Fig.18 Colour temperature (CCT) as function of burning time of 6 saturated (Manufacturer X and Y) and 3 unsaturated 4200K burners. After 8h the lamp is switched off for 0.5h.

Long term colour point stability.

In saturated lamps there will be a continuous change in partial salt pressures due to wall temperature changes caused by blackening as well as loss of salt components caused by reactions with the sealing frit and the alumina envelope. Consequently the colour point will shift over life. In unsaturated lamps the equilibrium between salt fill and lamp components is already achieved in an early state and over life no further significant colour point shift will take place as long as all chemical reactions are controlled. The colour point will be very stable over life (< 3SDCM).

Colour point stability as function of burning position.

In saturated burners the transport of salt, especially into the long crevices, leads to a significant CCT shift (200-400K) for different burner orientations. In unsaturated lamps the colour temperature hardly changes (< 50K); none of the salt is transported due to changes in wall temperatures. Small differences in colour temperature between horizontally and vertically burning unsaturated lamps can be explained by gravity driven demixing of plasma components.

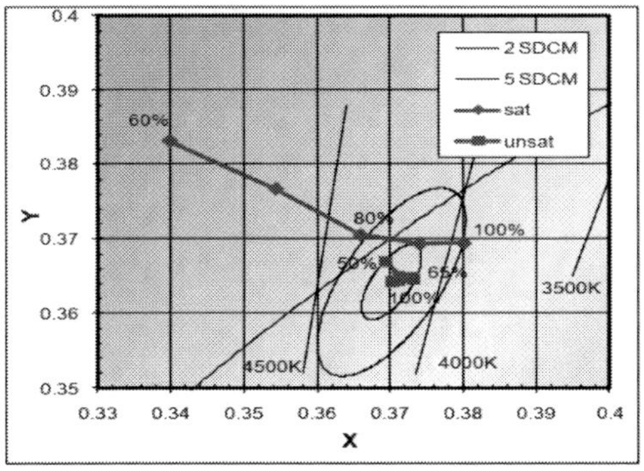

Fig.19 Colour point as function of relative lamp power for saturated and unsaturated 4200K burners

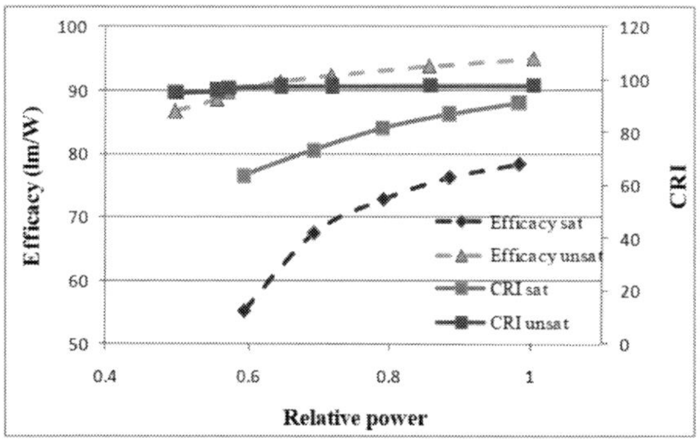

Fig.20 Efficacy and colour rendering (CRI) as function of relative lamp power for saturated and unsaturated 4200K burners.

Colour point stability, efficacy and colour rendering as function of lamp power (dimming).

In Fig.19 and 20 colour point shift, efficacy and colour rendering are given as function of lamp power. At nominal power (100%) the wall temperatures of these burners are sufficiently low (< 1550K) to guarantee long life times. The colour point shift, but also the change in efficacy and CRI of unsaturated lamps, up to 50% lamp power reduction, is almost inconsequential. As long as the burner stays unsaturated the spectrum hardly changes. This makes such a burner very appropriate for dimming applications. Minor colour point changes are caused by changes in current density (lamp voltage stays constant). The larger colour point shift in saturated lamps is caused by changes in wall temperature and thus in partial salt pressures.

Life time.

One of the main failure causes of saturated lamps is cracking of the alumina wall due to wall attack. It must be clear that in lamps without a salt pool this failure mode ceases to exist. Fig.21 shows the hottest part of the alumina wall of an unsaturated burner after 5000 burning hours. No serious wall attack is observed and less than 40 µm of the wall material has been transported. Long life times can be expected and more than 17000h have already been proven.

Run-up.

After switching the lamp on, it takes some time for a saturated burner to achieve its nominal light output but for an unsaturated burner this time is much shorter (Fig.22). The reason is the lower mass of the discharge tube, especially of the end construction.

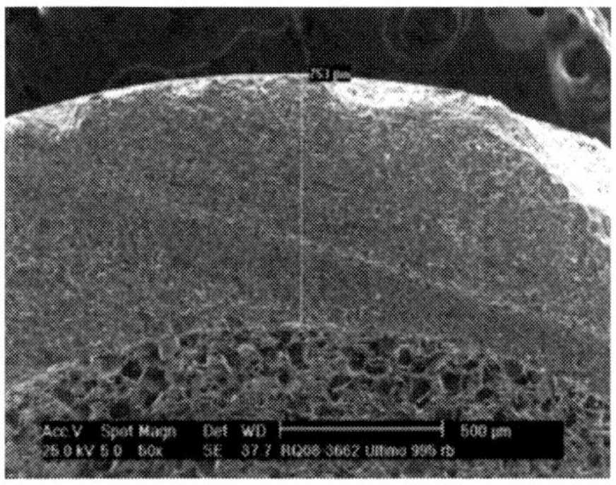

Fig.21 Cross section of the hottest part of the alumina wall of an unsaturated burner filled with a Dy containing salt after 5000 burning hours.

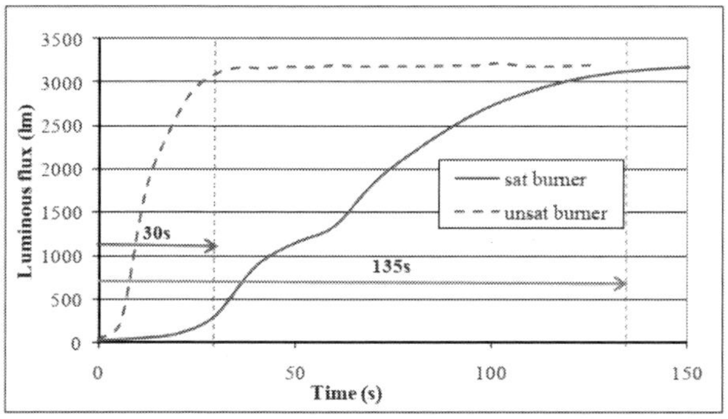

Fig.22 Luminous flux as function of time of 35W saturated and unsaturated burner operated at the same full electronic driver.

Colour homogeneity in beam spot.

Fig.23 and 24 compare the beam intensity and colour temperature distribution along a vertical cross section of the beam pattern of a PAR 30 reflector for saturated and unsaturated burners. With unsaturated lamps a more homogeneous colour temperature distribution over the beam pattern and a slightly higher burner flux can be achieved due to the absence of a salt pool. Yellowish stains, normally visible in the beam pattern of saturated lamps due to the salt pool, are absent in the patterns of unsaturated lamps.

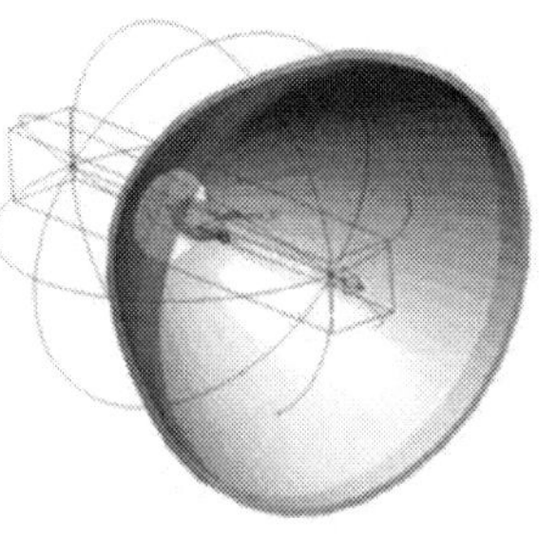

Fig.23 Burner optimally positioned in PAR 30 reflector appropriate for a small beam (10 degrees)

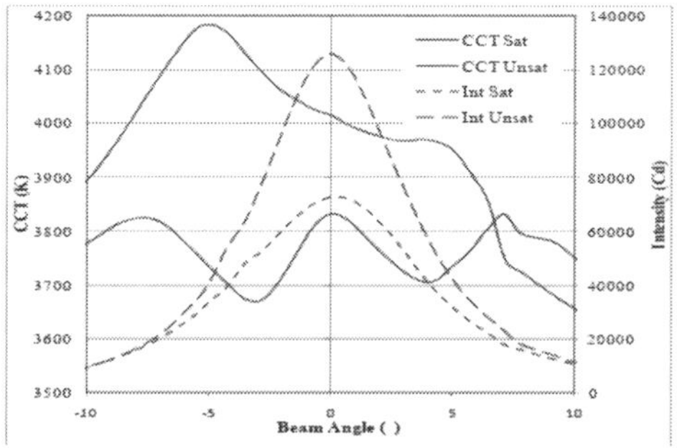

Fig.24 Measured intensity and colour temperature distribution along a vertical cross section of the beam pattern obtained with saturated and unsaturated burners, in a PAR 30 reflector.

Apart from overcoming the negative consequences of crevices and salt pool on lamp characteristics, unsaturated lamps offer some additional opportunities:

Applicability of an identical burner in different outer bulbs and reflectors.

Saturated burners are very sensitive to different outer bulbs or reflectors because of their high dependence on changes in wall temperature. Each outer bulb or reflector requires an adapted burner design, especially if deviations in the nominal colour point are not tolerated. An unsaturated burner can be positioned in an arbitrary outer bulb or reflector, without a change in colour point. This strongly facilitates factory logistics.

Application of high efficient salts.

Some salt compositions can lead to burners with a very high efficacy. However, saturated burners are very sensitive to selective deposition of salt components e.g. NaCe iodide [26]. During burning, more Na than Ce is transported into the crevice, leading to different compositions of the salt pool in crevice and burner. During run up salt that is deposited in the crevice, can be blown back again into the dt. Both effects lead to a significant colour point spread. Since in unsaturated lamps this is not possible, specific salts to obtain high efficacies can be applied in these lamp types.

Application of salts with a high RE composition.

Metal halide designers often prefer RE elements as additives to obtain burners with a good colour rendering and high efficacy. However in saturated burners high mole fractions of RE elements lead to severe alumina wall corrosion (see Fig.14). Since unsaturated lamps do not have a salt pool, there is little or no wall attack. High RE pressures can be implemented to improve efficacy and/or CRI. This is already illustrated in Fig.20 where the efficacy at nominal power is ~20% and the CRI ~7% better than for saturated lamps. The partial Dy pressures of the saturated and unsaturated burners shown in this figure are approximately 500 Pa and 5000 Pa. Fig.25 shows that with such high RE pressures combined with a blue absorbing salt element, even halogen-like spectra can be realised (fill: SnI_2, LiI, TlI, DyI_3: lamp power: 74W, CRI: 97, R9: 90, efficacy: 81 lm/W).

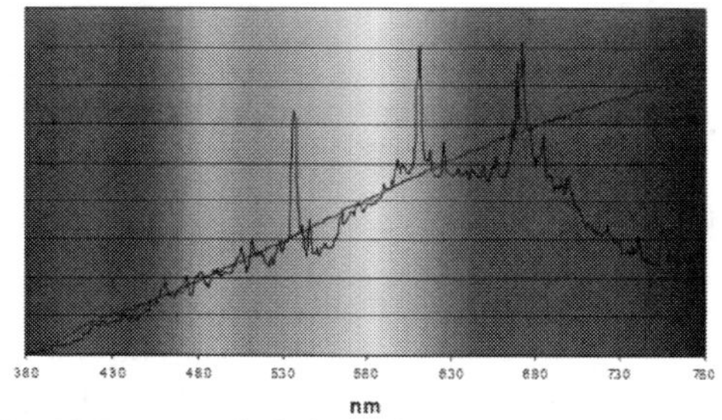

Fig.25 Spectra of a halogen lamp and an unsaturated CMH burner with a high RE pressure, leading to a high CRI (98) and a halogen like spectrum.

Of course not all improvements and opportunities of unsaturated lamps can be combined in one single lamp. Choices have to be made, depending on the required light technical characteristics and application.

5. FUTURE OPTIONS

In addition to the mentioned advantages in section 4 these new unsaturated burners offer new opportunities, which can enlarge their application fields. Some examples are:

Mercury free burners: One of the main consequences of leaving Hg out of the burner is the drastic drop in lamp voltage. This voltage drop can largely be compensated by high partial RE pressures or additives such as Zn or Zn iodide. The creviceless burners with high minimum wall temperatures make it possible to create these high pressures. To compensate for efficacy loss, in case of omitting Hg, higher pressures of efficiently emitting elements as Ce and Pr can be realized. In saturated lamps with crevices this is not possible due to wall attack by the salt pool and selective condensation of these salts in the crevices.

Miniaturization: The absence of long extended plugs makes it possible to apply the unsaturated burners in extremely small reflectors such as MR16, MR11 or MR8. Gas filled outer bulbs can be applied for further miniaturization of the discharge tube itself. Unsaturated burners are in contrast with saturated burners not sensitive to changes in fixture or environmental temperatures because salt pressures and thus light technical characteristics cannot be influenced by conduction of the gas filled outer bulb.

Instant light: Unsaturated burners show already a significantly faster run-up than saturated burners on standard electronic drivers (Fig.24). This run-up time can be shortened up to 2-10 seconds (depending on the lamp wattage) by increasing the run-up current and/or the start gas pressure. The run-up time can also be shortened by reducing the thermal mass further, by using burners with yet thinner walls. The low thermal gradients in these unsaturated lamps open up this design freedom.

6. CONCLUSIONS

To the best of our knowledge, we are the first to succeed in realizing mass producible unsaturated CMH lamps. These unsaturated designs overcome many of the disadvantages of the current generation of saturated lamps. The new design offers greatly improved light technical characteristics by implementing higher wall loading and more aggressive filling components which result in significantly higher salt pressures. Over life salt pressures can be controlled: colour point stability over more than 17000 burning hours has been proven for lamp fillings with relatively high (aggressive) RE pressures. Combined with other future options for unsaturated burners such as fast run up, mercury free fillings and miniaturisation, the way to a lamp generation for new and environmental friendly application fields is open.

Acknowledgement. This work is supported by the IWT-Vlaanderen, Belgium.

7. REFERENCES

[1] J.J. Li, A. Johnson, G. Freeman, H. Rothwell, et al., Osram Sylvania, US2009/0280713A1 (2009).
[2] J.J.F. Geijtenbeek, A.A. Duisters, T.G.M.M. Kappen, J.L.G. Suijker, V.M. Fischer. G.M.J.F. Luijks, H.A. van Esveld, V.L.E. Bruyndoncx, WO2005IB50746 A (2005).
[3] H. Nohara, A. Utsubo, Y. Kanazawa, N. Nishiura, et al., National-Matsushita, Proc. 11th Int. Symp. On the Science & Technology of Light Sources (LS11), Shanghai, China (2007), 107-108.
[4] T.D. Russell, M. Rahmane, P.J. Meschter, G.W. Utterback, General Electric Company, US2009/0146571A1 (2009).
[5] B.P. Bewlay, J.S. Vartuli, J.A. Brewer, General Electric Company, US2006/0008677 A1 (2006).
[6] S.M.R. Gelderland, T.G.M.M. Kappen, Koninklijke Philips Electronics N.V., WO2008020406 A2
[7] G.C. Wei, Y. Zheng, Osram Sylvania Inc., US 2009/0269523 (2009).
[8] G.F. Belder, M.J.M. Kessels, A.C. van Amstel, B.T. Verlinden, A.S.G. Geven, M. Bolech, H.P.M. Gubbels, US2005/0200281A1 (2005).
[9] C.J.M. Denissen, US2008/0297051A1 (2008).
[10] W. Lapatovich, J. Avallon, J. Serveran, Osram Sylvania Inc., US2009/0058300 A1 (2009).
[11] B.P. Bewlay, J.A. Brewer, S.S. Coulombe, S.M. Decarr, L.E. Iorio, A. Kebbede, T.J. Sommerer, and J.S Vartuli, General Electric Company, US2007/0159105 A1 (2007).
[12] K. Watanabe, T. Ohta, N. Masui, NGK Insulators Ltd, US 2006/022596 A1 (2006).
[13] V. Perez, J. Avallon, J.T. Neil, Osram Sylvania Inc., US2010/0026181 A1 (2010).
[14] R. Goihl, Osram Gesellschaft mit beschrankter Haftung, DE 10 2007 055 399 A1 (2007).
[15] Y.S. Touloekian, Thermal Expansion: Metallic Elements and Alloys, Plenum, New York (1975).
[16] J.C.M. Hendricx, P.J. Vrugt, F.J.G. Hakkens, C.J.M. Denissen, Koninklijke Philips Electronics N.V., US 2009/0267515A1 (2009).
[17] M.C. Raas, F.J.G. Hakkens, D.K. Dijken, A.G.M de Nijs, A.J.A.C. Dorrestein, J.C.M. Hendricx, P.J. Vrugt, Koninklijke Philips Electronics N.V., US2010/0026183 A1 (2010).
[18] O. Langenscheid, M. Westermeier, J. Reinelt, J. Mentel and P. Awakowicz, J. Phys. D: Appl.Phys. 41, 144005 (2008).
[19] L. Bruninx-Poesen and G. Solleveld, 2nd Int. Symposium on Incoherent Light Sources (LS-2), Enschede, Netherlands (1979), 105.
[20] W. van Erk , Pure Appl. Chem. ,Vol. 72, No. 11, 2159 (2000).
[21] Z. Toth and L. Lovas, Pure Appl. Chem. 79, 1771 (2007).
[22] S. Fischer, U. Niemann and T. Markus, J. Phys. D: Appl. Phys. 41, 144015 (2008).
[23] G.C. Wei, J. Phys. D: Appl. Phys. 38, 3057 (2005).
[24] P.J. Vrugt, C. Ning, Koninklijke Philips Electronics N.V .,WO 2008068666 A2.
[25] T.G.M.M. Kappen, Proc. 10th Int. Symp. on the Science & Technology of Light Sources (LS-10), Toulouse, France (2004), 43-52.
[26] J.J.F. Geijtenbeek and J.A.J.M. van Vliet, Koninklijke Philips Electronics N.V.,US5973453 A.

Emitter erosion in fluorescent lamps

K. Rackow[1], F. Sigeneger[1], J. Ehlbeck[1], D. Uhrlandt[1], K.-D. Weltmann[1], G. Lieder[2], M. Lieberer[2]

[1]INP Greifswald, Felix-Hausdorff-Str. 2, Germany, 17489 Germany

[2]R&D R-E, OSRAM GmbH, Augsburg, Germany

Contact: rackow@inp-greifswald.de

Abstract *- The erosion of emitter material at the electrodes and its transport in the plasma are important processes limiting the lifetime of fluorescent lamps. To understand these processes a detailed knowledge of the plasma processes in the electrode region is required.*

The emitter material, mainly barium, is released from the thermionic cathode due to sputtering and evaporation. Whereas the sputtering is mostly avoided during normal lamp operation the evaporation of barium is the dominant loss process. Various experimental methods combined with theoretical investigations were applied to study the erosion of barium in the electrode region in commercial lamps.

INTRODUCTION

Today, the lifetime of fluorescent lamps is mainly limited by the electrode durability. The electrode is the junction between plasma and solid and is therefore responsible for driving the lamp current. In general it consists of a tungsten coil coated with a work function lowering emitter mix of alkaline earth oxides with barium as the commonly most important component. During steady state operation, the electrode works normally in spot mode with temperatures in the range of 1200–1400 K.

Unfortunately, the emitter erodes from the electrode during lamp life due to evaporation and sputtering. At normal lamp operation, the erosion is in general evaporation of barium caused by the high temperature of the spot. Sputtering of barium due to ion bombardment requires cathode fall voltages larger than the threshold energy which is rarely reached during the normal lamp operation. In case of the commonly used preheating of coils, the loss of emitter material limits the lifetime of fluorescent lamps. Therefore, the determination of this loss, especially of barium, is of main interest for investigation of the lifetime of a lamp. Consequently, a detailed knowledge of the plasma in the cathode region is required.

For a detailed investigation experimental and theoretical investigations at a fluorescent lamp driven at 25 kHz have been performed. The densities of barium atoms and ions at different distances from the hot-spot centre have been measured time-resolved by laser induced fluorescence (LIF). The time-dependent cathode fall voltage was approximately determined using an improved band method. Furthermore, the electron density in the electrode region was measured by the use of a 150 GHz microwave interferometer. A model is applied to deduce the net flux of barium from the cathode surface during 25 kHz operation.

Furthermore, we developed a method for determination of barium evaporation and diffusion in the electrode region by reducing plasma effects. Therefore, a commercial fluorescent lamp is driven at a special operating mode where the lamp is switched off for a short time (approx. 5 ms) during barium density measurements. This allows determining the evaporation as well as the undisturbed diffusion of barium within some limitations.

THE MODEL

The model [1] is based on the assumption that the plasma is mainly spherically symmetric in the vicinity of the spot. It comprises the solution of the time-dependent particle balance equations of barium atoms and ions in ground and excited states. The main impact of the plasma on the barium transport is the ionization of barium atoms due to electron collisions which largely depends on the cathode fall voltage UCF. Therefore, the ionization frequency ν_{Ba}^{io} is determined by solving the space-dependent electron Boltzmann equation for a range of cathode fall voltages U_{CF} in spherically symmetric geometry [2].

The electron Boltzmann equation is transformed into a partial differential equation with shifted energy arguments which is solved as an initial boundary value problem from higher to lower energies. The spatial solution region spans from the approximated radius of the spot r_c up to the radius r_a of an assumed outer virtual spherical anode. From the solution of the Boltzmann equation the electron density n_e and the ionization rate coefficient ν_{Ba}^{io} are calculated.

The transport of the barium atoms and ions is investigated by solving the particle balance equations

$$\begin{aligned} \frac{\partial}{\partial t} n_a + \nabla \cdot \Gamma_a &= -\nu_{Ba}^{io}(r) \cdot n_a \\ \frac{\partial}{\partial t} n_i + \nabla \cdot \Gamma_i &= \nu_{Ba}^{io}(r) \cdot n_a \end{aligned}, \qquad (1)$$

which include the time- and space-dependent densities n_a and n_i of barium atoms and ions. The atom flux $\Gamma_a = -D_{Ba} \nabla n_a$ and the ion flux $\Gamma_i = \mu_i n_i E - D_{Ba} \nabla n_i$ are determined by the diffusion coefficient D_{Ba}, the ion mobility μ_i and the electric field strength E. Similar to the kinetic equation, the balance equations are solved in one dimension assuming spherical symmetry. In the anode phase the barium ionization frequency is assumed to be negligible.

Appropriate boundary conditions have to be chosen at the spot ($r = r_c$) and the virtual anode ($r = r_a$). At the spot a constant flux Γ_a^c is prescribed to the flux of barium atoms. This corresponds to the assumption of a time-independent temperature of the electrode surface in the spot. The determination of the barium atom flux from the spot surface requires in principle a detailed description of the barium diffusion through the emitter and of the evaporation process. To avoid possible errors due to the insufficiency of existing models and limited knowledge of corresponding data, the value of Γ_a^c is used as a scaling factor. Including the measured densities of barium atoms in absolute units, the whole solution is scaled to reproduce on average the measured atom densities. This procedure finally allows to determine the value of the evaporation flux independent of the description of the evaporation process itself.

EXPERIMENT

Fluorescent lamps similar to commercial T8 - 36 W lamps were used for the investigations. The lamps are filled with a mixture of 75% krypton and 25% argon at a pressure of 210 Pa. To enable light transmission for optical measurements, the glass tubes have been cleared of phosphor in the electrode region. A current controlled power supply working at 25 kHz was used for lamp operation. A set of experimental methods was used to study the emitter erosion.

Cathode fall measurement

The cathode fall, necessary as input for the modelling, was measured by the so-called band technique [3, 4]. A copper band was wounded around the outside of the glass tube in the electrode region and the voltage between band and electrode was measured. This voltage represents the capacitive coupled wall potential and includes the cathode fall, the voltage drop in the plasma from axis to wall and the wall sheath.

To deduce the cathode fall voltage from this measurement some considerations are required. Garner has investigated [3] the relation between the various voltage drops inside the discharge and the voltage measured at the band for conditions which are similar to those considered here. It was shown that the band voltage can be used as a measure for the cathode fall voltage if an appropriate correction factor is taken into account. At 25 kHz a correction factor of 1.2 was used. The high pass characteristic of the measurement circuit eliminates the DC part of the potential. Therefore we assume a cathode fall equal to zero at the beginning of the cathode phase by adding a corresponding DC voltage.

Absolute barium densities measurement

The density of barium atoms and ions in ground state was measured by laser induced fluorescence (LIF) in a short distance from the electrode. The LIF technique allows the detection of ground state or long-lived, excited-state atoms, molecules or radicals. The LIF process can be described as a two-step process [5]. First, a specific species in its ground state $|1\rangle$ (energy E_1) is excited by resonant absorption of laser photons ($h\nu_{laser}$) to a higher electronic energy level $|2\rangle$ of energy E_2 with $h\nu_{laser} = E_2 - E_1$. After laser excitation, this excited state relaxes by spontaneous emission to a lower state $|3\rangle$ of energy E_3 by emitting fluorescence photons with the energy $h\nu_{LIF} = E_2 - E_3$. This fluorescence radiation is emitted nearly isotropic [6]. For a spectral width of the laser $\Delta\nu_{laser}$ that is much broader than the line width of the exited transition $\Delta\nu_{12}$ a rate equations model could be used to describe the excitation and relaxation processes [7].

In the first approximation the time integrated fluorescent signal S_{LIF} is proportional to the ground state density n_1:

$$S_{LIF} = \int I_{LIF}(t)dt \propto S(\lambda)\cdot\frac{\Delta\Omega}{4\pi}\frac{A_{23}}{A_2}\cdot\sigma_{LIF}\cdot\frac{E_{laser}}{F_{laser}h\nu_{laser}}\cdot V\cdot n_1, \tag{2}$$

where $S(\lambda)$ is the sensitivity of the detection system at the fluorescent wavelength, $\Delta\Omega$ is the detection angle, the ratio A_{23}/A_2 of the Einstein transition probabilities is the relation between the detected and all possible transitions to lower energy levels, σ_{LIF} is the corresponding fluorescence cross section, F_{laser} is the cross sectional area of the laser beam and V is the observation volume.

To excite the barium atom from the ground state 6^1S_0 to the upper level 5^1P_0 a Nd:YAG pumped dye laser (350.1 nm) operating at 10 Hz was used. The excited level decays spontaneously via different transitions. The fluorescent radiation at 582.6 nm caused by the transition to the 5^1D_0 level was detected and used to determine the absolute population density of the ground state.

Electron densities measurement

The electron density was measured by a 150 GHz microwave interferometer (figure 1). A Gaussian microwave beam was focused via dielectric lenses and guided through the electrode region. The minimal diameter of the beam in the waist was about 4 mm.

The wave vector k of microwave in plasma depends on the electron density according to the dispersion relation of the plasma and could be formulated as:

$$k = \frac{\omega}{c}\sqrt{1-\frac{n_e}{n_c}}, \tag{3}$$

where ω denotes the microwave field frequency, c the speed of light , n_e the electron density and n_c the so called cut-off-density. The cut-off-density defines the maximum electron density where wave propagation is possible and is given by:

$$n_c = \omega^2\cdot\frac{\varepsilon_0 m_e}{e^2}, \tag{4}$$

where ε_0 is the vacuum permittivity, m_e the electron mass and e the elementary charge. For electron densities significantly smaller than the cut-off-density the change of the wave vector, according to a change of electron densities, is determined by detecting the phase of the transmitted microwave.

We use a heterodyne setup where the transmitted 150 GHz signal is mixed with a second 150.1 GHz signal from a local oscillator to an intermediate frequency of 100 MHz. The phase could be detected by comparing this signal with the signal of a reference path.

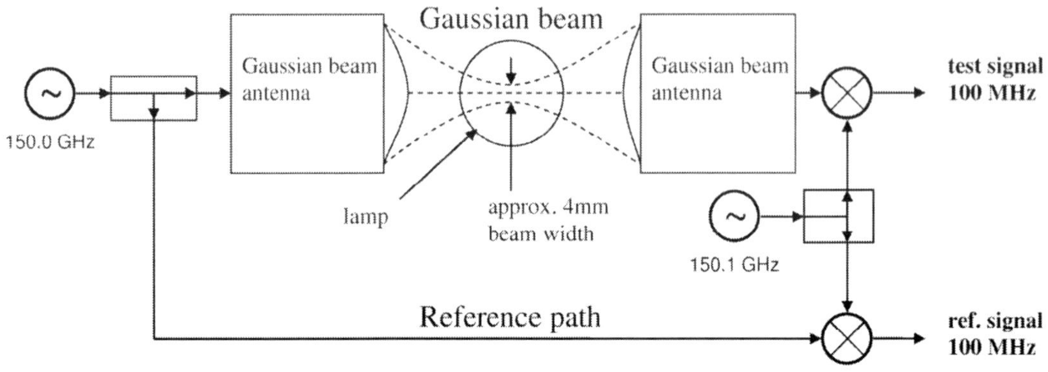

Figure 1: The used 150 GHz interferometer. The 150 GHz signals are generated by frequency doubling of phase is detected by comparing the phase of the down mixed test and reference signal.

Evaporation measurements (Blanking)

In addition to barium density measurements during steady-state operation combined with the model, we used a special electric operation mode to determine the barium evaporation and diffusion by suppressing plasma effects. During normal operation, the lamp is switched off for a short time (approx. 5 ms) and the barium densities as well as the electron density are measured. This so called blanking method was also used by Garner for measurement of electron densities in the electrode region [8]. The cathode fall and therefore the high energetic electrons decreases immediately after switching off the lamp. As a result, the evaporation of emitter material could be observed without ionisation and excitation caused by plasma electrons.

For switching off, we used a very fast IGBT-shorting circuit placed parallel to the power supply (see figure 2).

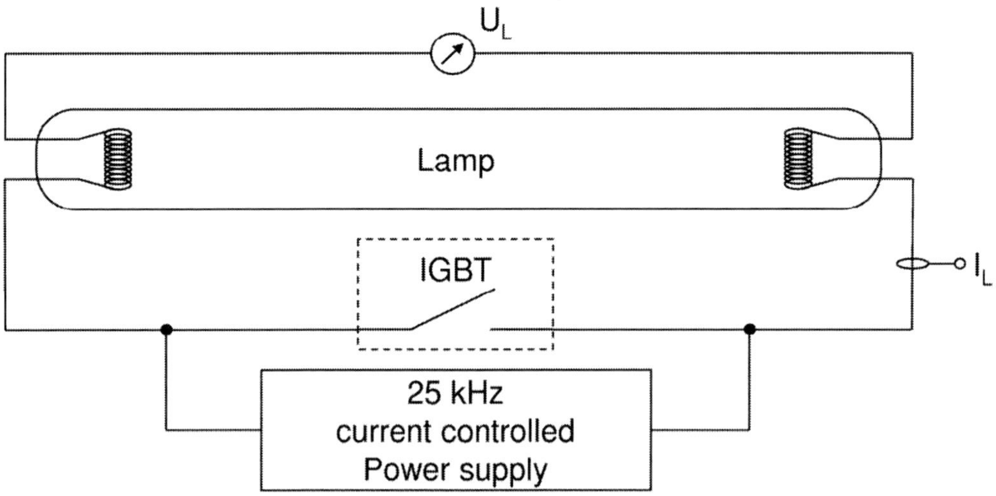

Figure 2: Electrical setup for lamp operation and cathode fall measurement. For shorting the lamp voltage a IGBT-shorting circuit is placed parallel to the power supply.

RESULTS

Typical results for a lamp driven at 25 kHz are presented in figure 3. During the cathode phase, the cathode fall voltage increases to sustain the lamp current through the electrode. Electrons gain energy in the cathode sheath and strongly ionize barium atoms resulting in a pronounced decrease of their density. Simultaneously, the electron density increases. During the anode phase, the ionization rate is strongly reduced and the barium atom density increases due to diffusion from the hot-spot.

The electron kinetic and transport equations are solved using the measured time-dependent electrode fall voltage. The solution is scaled to meet the atom densities measured by LIF on average.

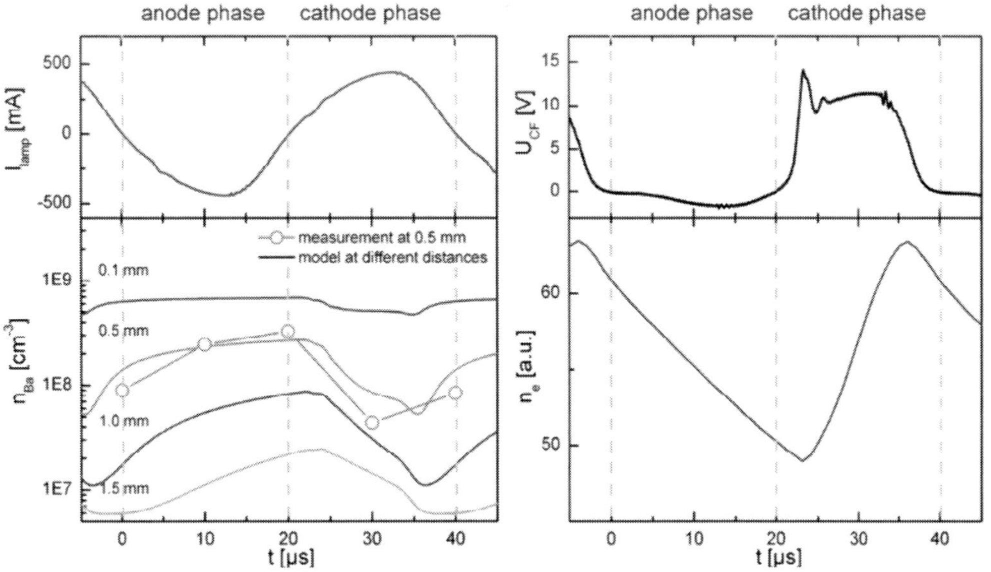

Figure 3: Time-dependent experimental and modelling results for a T8-lamp working at an rms-current of 320 mA and 25 kHz: lamp current (top left), measured cathode-fall voltage (top right), barium atom density calculated at various distances from the spot and measured at 0.5 mm (bottom left) and measured electron density (bottom right)

Figure 4 shows typical results of the blanking method. The lamp is driven with a discharge current of 320 mA at 25 kHz. At t=0 ms the lamp is switched off by shorting the voltage with the IGBT-shorting circuit. The cathode fall decreases immediately and there are no longer high energetic electrons (the so-called beam electrons) present in the electrode region. Accordingly, the strong ionization of barium atoms decreases immediately. The remaining plasma electrons relax in a µs timescale due to ambipolar diffusion. Hence, the undisturbed barium evaporation can be observed by measuring the barium density a few µs after switching off the lamp. The density increases due to evaporation from the hot-spot and diffusion into the electrode region. Correspondingly, the barium density increases in a ms time scale. After a certain time, which increases with the distance from the coil, the local diffusion reaches a stationary state and a saturated density could be observed.

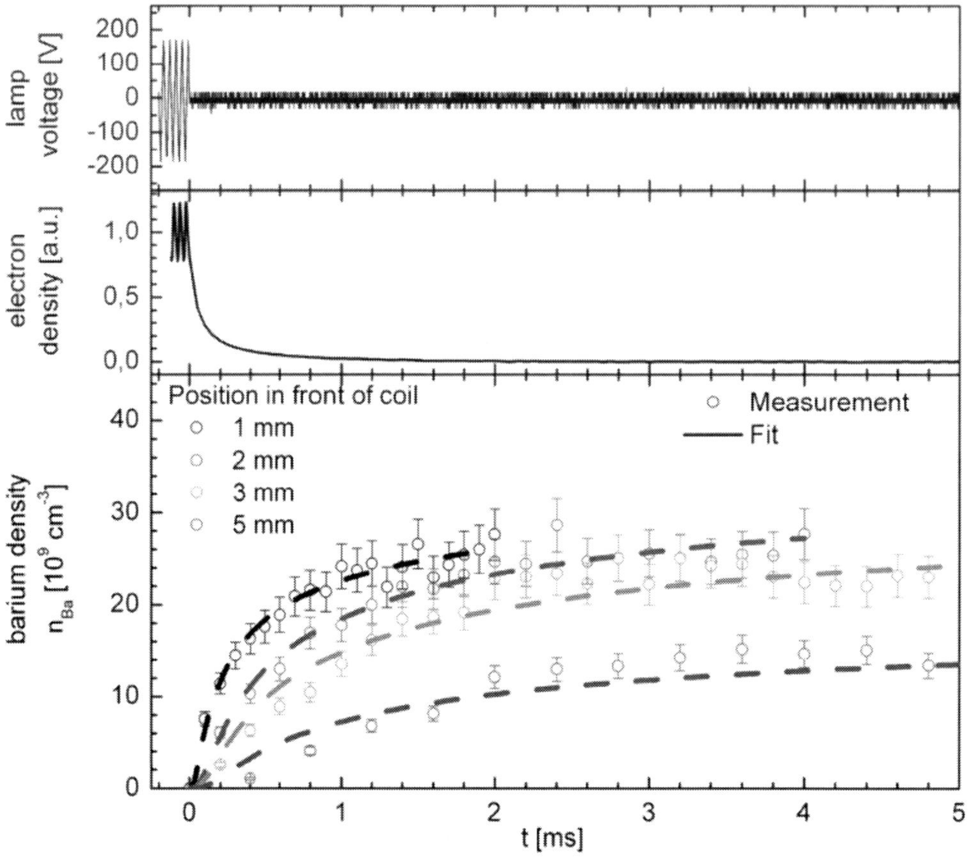

Figure 4: Time-dependent lamp voltage (top), electron density (middle) and measured barium densities (bottom) at various distances from the coil after switching off the lamp (blanking).

As an example for parameter studies, the barium density during the anode phase and after switching off the lamp is shown in figure 5. Results for different values of the lamp current I_{lamp} are shown as function of an additional heat current I_{heat}. In general, higher lamp current leads to higher hot-spot temperatures. Consequently, the barium evaporation rate and therefore the measured density, in the anode phase and during switching off, increases. Furthermore, the densities increase with the heat current due to the additional heating.

Only at low lamp current (I_{lamp}< 200 mA) a contrary tendency was found. In this case a larger cathode fall is required to heat the spot by ion impact. This leads to sputtering in the cathode phase. Without an additional heat current the density measured by blanking for a 200 mA lamp current are higher than for 100 mA. This indicates that the evaporation and therefore the hot-spot temperature are higher for 200 mA. In contrast, the densities measured in the anode phase are nearly the same for both lamp currents. This could be explained by an additional erosion of barium atoms in case of 100 mA by sputtering during the cathode phase.

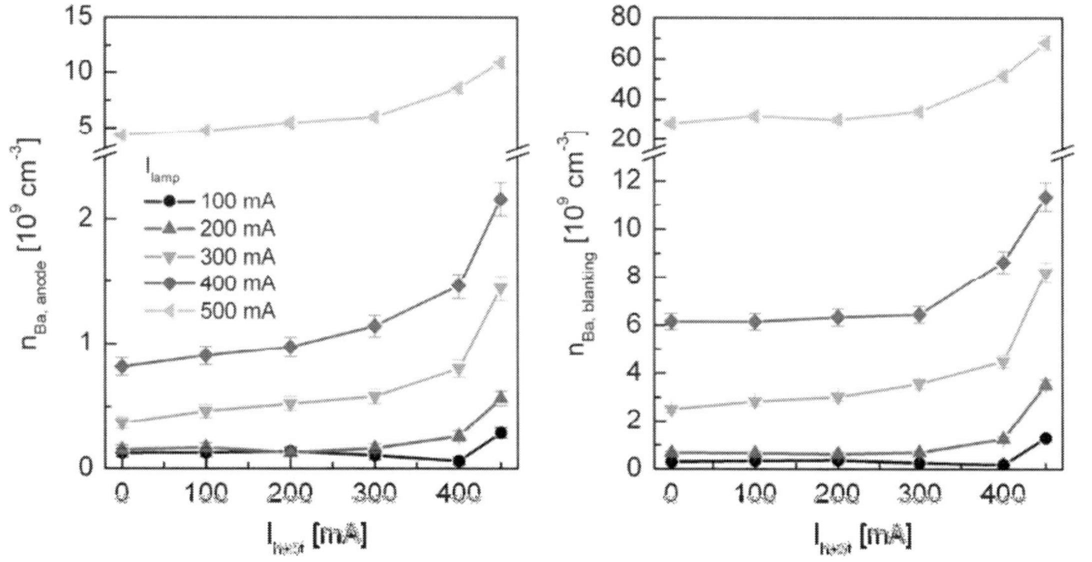

Figure 5: Barium density as function of the additional applied heat current I_{heat} for different values of the lamp current I_{lamp}. Left: Density during normal operation in the anode phase measured at a distance of 0.5 mm from the spot. Right: Density measured 4 ms after switching of the lamp at a distance of 3 mm from the spot.

An increase of the heat current decreases the cathode fall and correspondingly the amount of sputtering. This decrease is more pronounced than the additional evaporation caused by the heat current at heat currents I_{heat} of about 300-400 mA. Therefore, a corresponding minimum of the barium density is observed. This effect illustrates the complex response of the plasma-electrode interface to changing discharge parameters.

ACKNOWLEDGMENT

This research has been supported by R&D OSRAM, Augsburg.

REFERENCES

[1] F. Sigeneger, et al., Barium transport in fluorescent lamps, J. Phys. D: Appl. Phys. submitted
[2] F. Sigeneger, G. Sukhinin and R. Winkler, Plasma Chem. Plasma Process. 20 (2000) 87–110
[3] R. Garner, J. Phys. D: Appl. Phys. 41 (2008) 144009
[4] R. Nachtrieb, F. Khan and J.F. Waymouth, J. Phys. D: Appl. Phys. 38 (2005) 3226.
[5] J. Amorim, G. Baravian and J. Jolly, J. Phys. D: Appl.Phys. 33 (2000) R51–R65
[6] S. Hadrath, J. Ehlbeck, G. Lieder and F. Sigeneger, J. Phys. D: Appl. Phys. 38 (2005) 3285–3295
[7] K. Niemi, V. Schulz-von der Gathen and H.F. Dobele,
J. Phys. D: Appl. Phys. 34 (2001) 2330–2335
[8] R. Garner, J. Phys. D: Appl. Phys. 41 (2008) 144010

Control Methods for Enhancing Run-up of Metal Halide Lamps

Nancy H. Chen and Joseph A. Olsen
Central Research and Services Laboratory, OSRAM SYLVANIA, 71 Cherry Hill Drive, Beverly, MA 01915 USA
Contact: nancy.chen@sylvania.com

ABSTRACT

Control methods applicable to the run-up phase of metal halide lamp operation are described. Performance improvements include reduced wait time to reach full lumen output and control of light level during warm up. Added functionality includes lamp identification and detection of a compromised outer jacket.

INTRODUCTION

The advent of electronic ballasts with microcontrollers allows consideration of more sophisticated control methods which have potential for improving metal halide system performance and functionality. An interesting area of application is during the run-up phase of operation, due to the changes in lamp behavior that occur during warm up to steady state. A modern control and ballast allows response to these changes as well as analysis of the changes for lamp characterization.

LIGHT CONTROL DURING RUN-UP

The efficacy of a metal halide lamp can vary by an order of magnitude between a cold and fully warm lamp. During run-up, the decreased efficacy of a cold lamp can be compensated by temporary overpowering [1,2]. In this way the user is able to enjoy full lumen output sooner. As the lamp continues warming and the efficacy rises, the power can gradually be decreased towards the rated level in such a manner as to keep the lumen output constant, so that the approach to steady state is relatively transparent to the user. Key to the control is a way of estimating the instantaneous efficacy of the lamp without use of additional hardware such as photo sensors or thermocouples. One way of describing the relative efficacy η during run-up of a cold lamp is as a function of the cumulative energy E that has been deposited in the lamp arc since ignition [3]. The power dissipated in the arc heats the lamp up, which increases its efficacy. The efficacy vs. energy function $\eta(E)$ can be stored in the ballast in the form of a table or coefficients to a mathematical expression. An example $\eta(E)$ is shown in Fig. 1, along with an approximating exponential function as shown. Larger values of the coefficient E_1 correspond to a larger thermal mass of the arc tube.

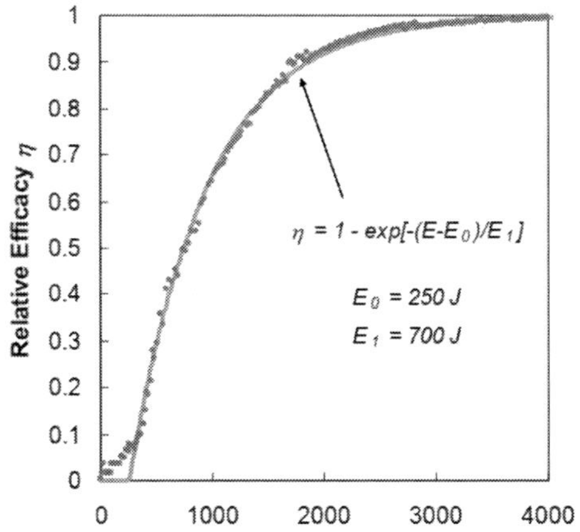

Fig. 1. Example relative efficacy η of a 35 W lamp vs. cumulative arc energy E since ignition.

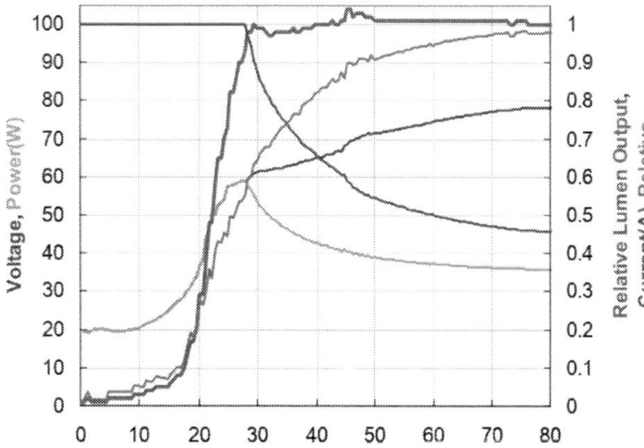

Fig. 2. Light output, efficacy, current, voltage, and power vs. time during run-up of a 35 W Lamp at 1 A max allowed run-up current.

During early run-up, when lamp efficacy is at its lowest, the required power to produce full light output may be very high and therefore unattainable due to an imposed current limit. The resulting run-up is then at constant current until full lumen output is reached, followed by a constant lumen output throughout the completion of warm up to steady state. An example of such a run-up path is shown in Fig. 2. as the green plot. Also shown are the corresponding power*(purple)*, current*(red)*, voltage*(blue)*, and efficacy*(orange)*. For a commercial OSRAM HCI-TC 35W/830 lamp with a modest allowed run-up current of 1 A, the run-up time to full lumen output is about 30 s. Note that an advantage of this energy-based control method over voltage-based control methods [4] is that the control is more robust against variations in voltage from lamp to lamp and over life. However, run-up must start from a cold (or other known) thermal state.

LAMP CHARACTERIZATION FROM RUN-UP BEHAVIOR

Analysis of lamp behavior during run-up can be used to characterize the lamp. The lamp voltage, which is dependent on the lamp temperature, can be analyzed as a function of cumulative energy deposited in the arc in order to estimate the thermal mass of the arc tube. As an example, for the run-up method described above, it has been observed for typical metal halide lamps that when the voltage derivative V' with respect to energy E reaches a peak, the relative efficacy is approximately 0.4, as shown in Fig. 3. By detecting the peak of $V'(E)$, the control can then calculate an approximate value of E_1, which is an indication of the thermal mass of the lamp. Since higher wattage lamps generally have higher thermal mass, a lamp of unknown wattage can be identified *during run-up*. Knowing the proper steady state power and an approximate $\eta(E)$, the run-up can then be completed according to the light control method above. This allows a single ballast to operate lamps of different wattages without requiring advance knowledge of which lamp is connected. An alternative method of lamp identification has been given by [5].

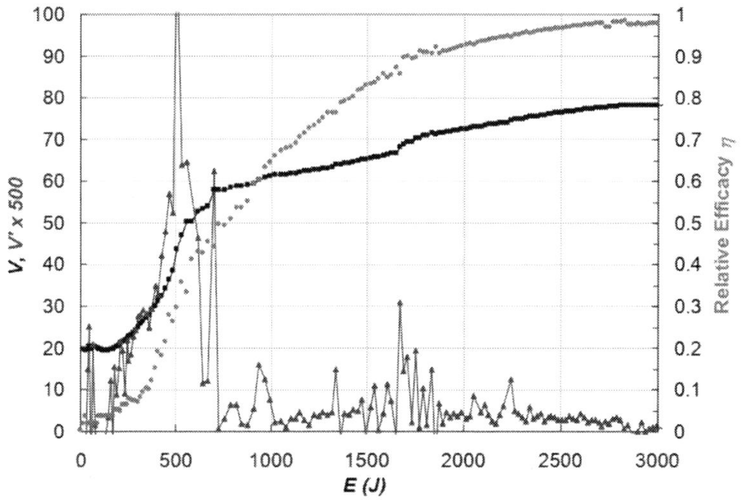

Fig. 3. Voltage and efficacy behavior as a function of cumulative arc energy during run-up of a 35 W lamp.

If the lamp selection is restricted to a single type, then a measured deviation in voltage behavior during run-up can be an indicator of an irregularity such as a mismatched or damaged lamp. For example, if the vacuum environment within the outer jacket is compromised, then increased thermal coupling manifests itself as an increase in apparent thermal mass of the arc tube with corresponding elevated E_1 values. Other examples of voltage-related monitors of compromised outer jackets are given in [6].

REFERENCES

[1] For example, automotive metal halide lamps have elevated run-up power to satisfy light requirements.
[2] C.-S. Moo, S.-Y. Tang, W.-T. Tsai, and H.-L. Cheng, *IEEE Transactions on Plasma Science*, 36 (4), pp. 1907-1912 (2008).
[3] N. H. Chen and J. A. Olsen, US Patent 7,589,477 (2009).
[4] For example, M. Sasaki, A. Murata, and G. Oda, US Patent 5,212,428 (1993); see also references in [3].
[5] S. Y. Tang, W. T. Tsai, C. S. Moo, C. R. Lee, and L. L. Lee, *Proceedings of the 11th International Symposium on the Science and Technology of Light Sources*, Shanghai, pp. 213-214 (2007).
[6] J. A. Olsen and N. H. Chen, US Patent 7,619,362 (2009).

Halide Penetration in Seal Glass for Ceramic Metal Halide Lamps

Raghu Ramaiah, Jianmin He
GE Lighting, Nela Park, 1975 Noble Road, Cleveland, OH, 44112, USA
Contact: raghu.ramaiah@ge.com

ABSTRACT

Halide penetration of seal glass in Ceramic Metal Halide lamps is one of the major life limiting mechanisms. This paper describes a method by which this penetration rate can be accurately measured, and a mathematical form for the penetration rate is derived.

INTRODUCTION

In order to properly function as a acceptable seal-glass material for Ceramic Metal Halide (CMH) lamps, the seal glass must have the following set of properties: a) Chemical compatibility of the seal glass material with lamp components such as polycrystalline alumina (PCA), metal halides, electrode materials (typically tungsten, molybdenum and niobium); b) Thermal compatibility, i.e. the matching coefficients of linear expansion of seal glass with alumina and electrode components; c) low vapour pressure such that the arc is not contaminated; d) low melting point to minimize manufacturing difficulties, yet high enough to provide adequate hermeticity during lamp operation. The most common seal glass material for Ceramic Metal halide lamps is a Dy_2O_3-Al_2O_3-SiO_2 (DAS) system[1]. Typical compositions by wt% are 65:15:20 respectively, but these can vary to provide incremental benefits in some of the above listed properties. The main shortcoming of the DAS system is its reduced chemical compatibility at elevated temperatures. During prolonged operation of CMH lamps with DAS system, the halides progressively react with the seal glass. This reaction front changes thermal compatibility locally, which, in turn, causes cracks to form and propagate through the seal material resulting in loss of hermeticity and eventual lamp failure.

EXPERIMENTAL METHOD

Key thermals and morphology of seal glass at the end of the seal for typical CMH lamp is shown in Fig.1.

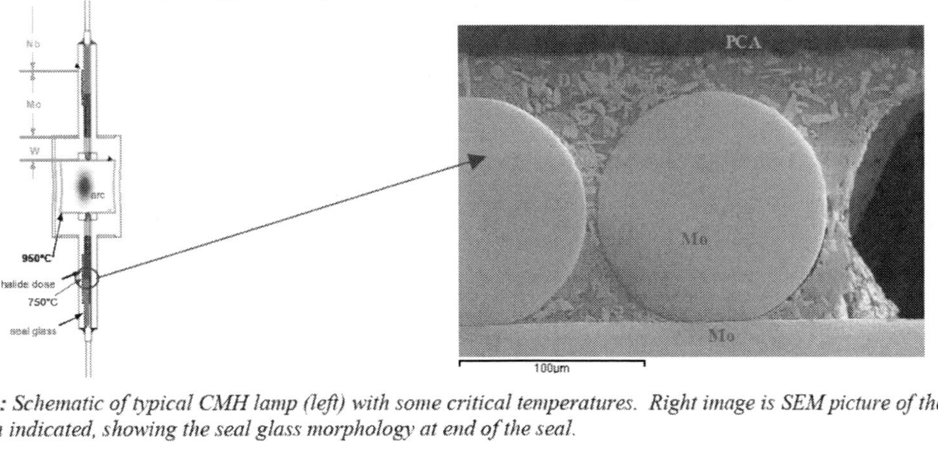

Fig. 1: Schematic of typical CMH lamp (left) with some critical temperatures. Right image is SEM picture of the region indicated, showing the seal glass morphology at end of the seal.

To measure extent of seal glass penetration by halides, we have attempted a halide species map using SEM/EDAX. A complete map typically requires several hours. We have found that, alternatively, partitioning the seal region as shown in Fig.2 provides equivalent information for halide penetration in the seal glass.

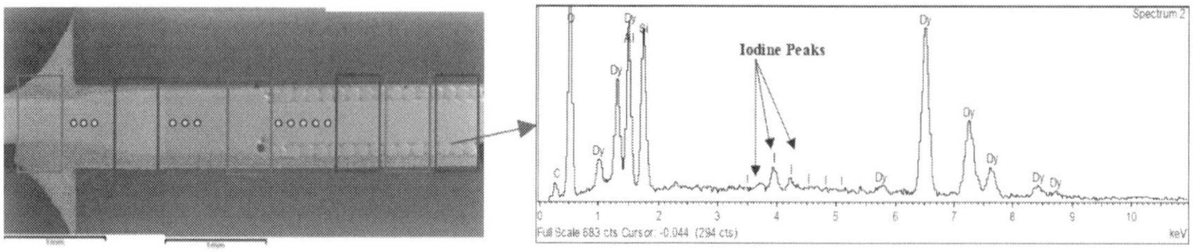

Fig.2: *Representation of partitioning the seal region into several rectangular sections (left) and quantifying the Iodine peak (right) by EDAX in each box. Box position gives the distance from a reference location (end-of-seal)*

An EDAX scan of one box typically takes less than a second giving a quantitative estimate of the amount of Iodine in the box. The box location with reference to the end of the seal gives the penetration depth of the reaction front. Typically, such maps will take ~ 1 minute per lamp.

RESULTS

Several samples of CMH lamps with DAS sealing glass and TmI_3-based dose and LaI_3-based [2] dose were operated in Open Air and in a simulated Hot Fixture environment, for several thousand hours. Samples from these 3 sets were removed at various intervals and SEM/EDAX analysis was performed as described above. Results are shown in Fig. 3.

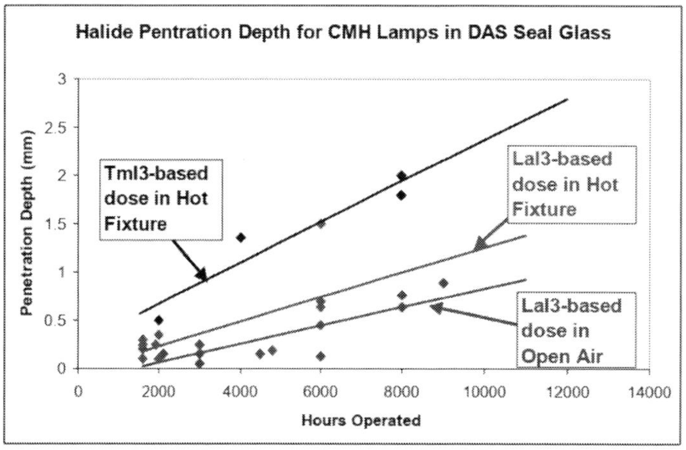

Fig.3: *Halide Penetration Depth measured at several different operating hours for three groups of CMH lamps, in cooler vs hotter environments, and two different types of CMH lamp chemistries.*

The seal temperatures in Open Air was ~50C lower than in Fixtures. The results above yield a penetration rate given by an Arrhenius type of equation, below, for the LaI_3-based chemistry:

$$\text{Rate (mm/hr)} = 21914\, e^{[-14400/T]}$$

In this equation, T is in deg C, and is the liquid halide pool temperature at the point of contact with the seal glass. The results further suggest that TmI_3-based halide are more aggressive than LaI_3-based halide dose chemistry for the CMH lamps, i.e. the rare-earth component of the CMH lamp chemistry should preferably come from Group I-III type vs group VI-VIII types to minimize the rate of progression of the reaction front in the DAS seal glass.

REFERENCES

[1]. S. DeCarr et.al.: *Characterizing Metal to Ceramic Seals for High Intensity Discharge Lamps*, Aug.12-16, 2001, LS-9 Proceedings, Ithaca, NY

[2]. R. Ramaiah et.al.: US patent , 20080129202

New Shadow-less getters for discharge lamps

A. Corazza, M. Riva, S. Giorgi
SAES Getters S.p.A., Viale Italia 77, Lainate (MI) 20020, Italy
Contact: alessio_corazza@saes-group.com

ABSTRACT

New miniaturized getters based on active alloys with excellent sorption performances have been developed for small High Intensity Discharge lamps. These getters have a tiny, filiform structure suitable for the mounting in the outer bulb of compact lamps in such a way that they do not create any shadow to the radiation generated by the burner.

INTRODUCTION

One important trend of modern light sources is the miniaturization of the lamps. Following this tendency current lighting technology is more and more pushing for the development of smaller and more efficient getter structures, especially for new compact High Intensity Discharge (HID) lamps. The reduction of lamps dimensions sets tight geometrical constraints and requires a miniaturization of the getters in order to allow an easy mounting in a correct position inside the outer bulb and to avoid any shadow effect on the light emitted by the central burner.
The availability of getter materials with enhanced properties, such as the St777 (Zr-Fe-Y) alloy [1] and the St787® (Zr-Co-Rare Earths) alloy, with large sorption capacity towards all the gaseous impurities [2, 3], has allowed to design innovative tiny getter structures suitable, in particular, for small and miniaturized lamps.

CHARACTERISTICS OF THE NEW SHADOW-LESS GETTERS

The St787® and St777 getter alloys combine very good hydrogen sorption capacity even at high temperatures, thanks to the ability to maintain low H_2 equilibrium pressures, with high efficiency in sorbing all the active gases (CO, CO_2, water vapour, hydrocarbons).
These excellent features make possible the use of a reduced amount of active material, compared to the more conventional getter solutions, enabling the design of tiny "shadow-less" getter structures, capable of meeting the needs of compact or even miniaturized lamps.
The structure of the "shadow-less getter" configuration is shown in fig. 1: it has a wire configuration with a trapezoidal cross section, formed by a metal sheath enclosing the getter powder; the filiform structure is provided with a longitudinal slit to allow a quick impurities sorption by the active alloy.

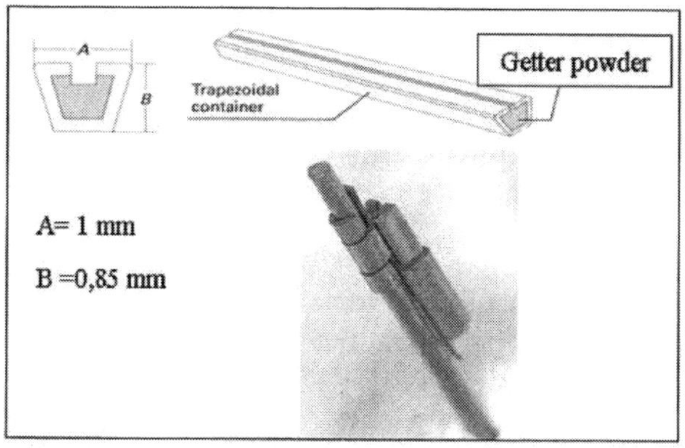

Fig.1: Shadow-less getter structure

Thanks to its configuration, the getter can be fixed onto the metallic frame present in the lamp outer bulb, as a support of the arc tube, so that it does not cause any shielding to the radiation and to the heat from the burner. The getter wire can be mounted by spot welding, or can be crimped exploiting the applications of small tabs, in a position aligned along the lamp metal frame and substantially hidden to the burner. This characteristic is particularly important for the new Low Wattage Metal Halides lamps that have a very small outer bulb, a few cm^3 or even less in volume.
The content of the active material in the getter wire is about 2 mg per mm of length: taking into account the performances of the materials, just some millimeters of wire are usually enough to cope

with the gas load evolved in the small lamps during lifetime.

SORPTION CHARACTERISTICS

It is well known that the Zr-Fe-Y and Zr-Co-Rare Earths alloys adopted in the small shadow-less getters have low temperature activation characteristics: getter activation can be obtained during the baking-exhaust process or can be achieved in the first minutes of lamp operation if the working temperature of the getter is above 350°C. As these materials start becoming active at about 300°C, it is advisable to avoid over-heating in air or in poor vacuum during lamp manufacturing in order to prevent damages to the getter alloys. Specific studies were carried out in order to evaluate the sorption performances of the filiform getter wires in comparison to those of the more conventional getter tablets that have bigger dimensions and usually create some shadow in the small lamps. Comparative sorption tests were performed on 5 mm long getter wires containing 10 milligrams of St777 alloy or 10 mg of St787® alloy and on 4 x 4 mm getter tablets containing 15 mg of St787® alloy. Static sorption tests with hydrogen were conducted heating the samples at a temperature of 400°C in a vessel volume of 5000 cm^3 filled with hydrogen at an initial pressure of 0,1 mbar. No specific getter activation was applied. As shown in fig.2, despite the lower mass of active material and the lower exposed getter surface, the pumping speed of the wires is better than that of the tablet. Especially in the case of the St777 wire the gettering activity and the gas sorption capacity are significantly higher than the performances of the St787® tablet. The Zr-Fe-Y alloy exhibits much larger H_2 sorption capacity than the Zr-Co-Rare Earths material.

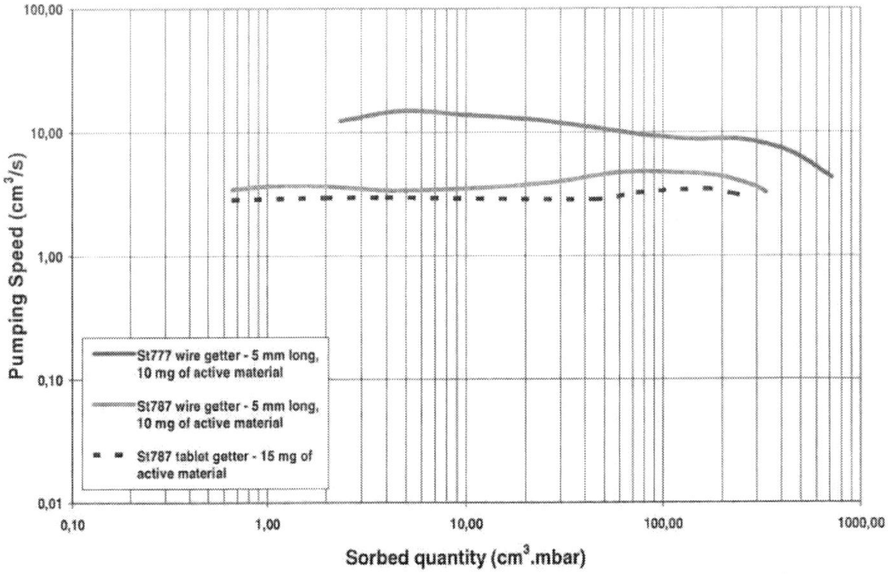

Fig.2: curves of hydrogen sorption tests carried out in static conditions at 400°C on different getters (Initial H_2 pressure on the getter: 0,1 mbar; chamber volume 5000 cm^3). No getter activation applied.

CONCLUSIONS

The newly developed shadow-less getters take advantage of the combination of the high performances of the St777 and St787® alloys and tiny structure of the getter wires: the large sorption capacity of the new alloys, especially of the St777, makes it possible to realize getter structures apt to satisfy the needs of miniaturized light sources. The shadow-less getters are particularly suitable for small HID lamps because they do not shield the light radiation emitted from the burner.

REFERENCES

[1] European patent EP 1817439.
[2] V.Massaro, A.Corazza, S.Giorgi, Proc. of the 11th Int. Symp. on the Science and Technology of Light Sources, Shanghai (2007) 203.
[3] A.Corazza, C.Boffito, Proc. of the 9th Int. Symp. Sc. and Tech. of Light Sources, Ithaca (2001) 387.

CeO_2 as an oxygen dispenser in ceramic metal halide lamps

Jianmin He[1], Timothy Russell[1], Zoltán Tóth[2]
[1]*GE Lighting, Nela Park, 1975 Noble Road, Cleveland, OH 44112, USA*
[2]*GE Lighting, GE Hungary, 1044 Budapest,Vaci ut 77, Hungary*
Contact: timothy.russell@ge.com

ABSTRACT

A tungsten-halide cycle in a ceramic metal halide lamp is essential for its superior lumen output over life. This paper describes the use of CeO_2 as an oxygen dispenser in a practical CMH lamp design.

INTRODUCTION

Ceramic Metal Halide (CMH) lamps have found widespread applications in recent years, with the superior color rendering capability and very high energy efficiency measured by lumen per watt. However, they often have a light output that diminishes over time due to blackening of the discharge vessel walls, caused by tungsten evaporated from the electrode to the wall. It has been proposed to incorporate a calcium oxide or tungsten oxide oxygen dispenser in the discharge vessel, as disclosed in US6362571B1 [1] and US6356016B1 [2]. Lamps produced according to the proposals allow for the development of a tungsten-halide cycle, which transport the deposited tungsten from the wall back to the electrode and strongly counteracts the blackening of the discharge vessel. The condition for the presence of a tungsten-halide cycle is the presence of a small amount of available oxygen. These proposals exclude the use of rare-earth halides because they form relatively stable oxides and would not allow for the presence of available oxygen. Lamps produced according to the proposals may not, however, simultaneously meet acceptable lamp efficiency, color point, color stability, lumen maintenance, and reliability values for a commercial lamp.

We have found that certain rare earth halides, such as cerium iodide and lanthanum iodide, do allow for the presence of a sufficient amount of available oxygen to enable a tungsten-halide cycle. It has also been found necessary to add oxygen to the vessel in a very controlled manner, so as to obtain sufficient oxygen under lamp burning condition for the development of the tungsten-halide cycle, but not so much as to cause corrosion and early failure of the tungsten electrode shank. This paper demonstrates CeO_2 as a practical oxygen dispenser [3].

METHOD

The amount of oxygen released under lamp operating condition and gas phase solubility of tungsten was evaluated with thermodynamics calculation. Lumen maintenance over lamp burning life was obtained to confirm the theoretical calculations.

RESULTS

In its most oxidized state, cerium oxide takes the stable form CeO_2. But when exposed to an oxygen depleted environment, CeO_2 readily releases oxygen, transforming to its most reduced, oxygen poor form Ce_2O_3.

In a ceramic discharge vessel with walls made of aluminum oxide, Al_2O_3, the addition of a controlled amount of CeO_2 leads to a reaction that releases a small amount of available oxygen, most probably by the reaction below:

$$2CeO_2 + 11Al_2O_3 = 2CeAl_{11}O_{18} + 1/2O_2 \quad (1)$$

This available oxygen will react with the rare-earth iodide, CeI_3, forming additional amounts of the cerium aluminate. If too much CeO_2 is added, all of the rare-earth iodide could be depleted from the dose. For this reason it is important to be able to add a controlled amount.

Some of the oxygen released from reaction (1) will also be available for the development of the tungsten-halide cycle that will prevent the darkening of the discharge vessel walls. In Fig.1, we show some thermodynamic calculations for the gas phase solubility of tungsten (SPW = summed pressures of all tungsten species) for various added amounts of CeO_2. Increasing amounts of added CeO_2 leads to increasing tungsten solubility at all points within the discharge vessel. When the added amount of CeO_2 exceeds about 0.72 μmoles, the tungsten gas phase solubility at the wall at 1200 K – 1400 K exceeds the tungsten solubility at the electrode tip at about 3000 K. This is the condition needed for an effective tungsten-halide cycle.

In FIG. 2 is shown the lumen maintenance curves for two sets of 70W CMH lamps that had been dosed with 1.0 and 1.4 μmoles of CeO_2. Also included is the LM% curve for another set of 70W CMH lamps to which no CeO_2 was added. The addition of 1.0 μmole of CeO_2, or more, achieves lumen maintenance of 100% after 1000 hrs of burning. With no added CeO_2, the lumen maintenance is only 93%.

The lamps with 1.4 μmoles of added CeO_2 illustrate an additional problem of controlling the amount of added available oxygen. Too much added oxygen will lead to further reaction of the excess oxygen with the REI_3 and a loss of lumens. The reduced LM% for the lamps with no added CeO_2 illustrates the problem with not adding sufficient amount of available oxygen.

An advantage of using CeO_2 as the source of available oxygen is the relative ease of dosing. 1.0 μmoles of CeO_2 has a mass of 0.18 mg and could be dosed into the ceramic discharge vessel using standard dosing equipment. The amount of WO_2 or WO_3 needed to add to yield an equivalent amount of available oxygen is 3x to 4.5x times less than the amount of CeO_2 required. It has not proven possible to dose directly the required amount of WO_2 or WO_3 with existing dosing equipment.

CONCLUSION

CeO_2 has been shown to be an effective oxygen dispenser that enables the design of CMH lamps with superior lumen maintenance. It has the additional practical benefit of being able to be dosed on standard manufacturing equipment.

REFERENCES

[1] J.L.G Suijker, M.C. Raas, US Patent 6,362,571

[2] J.L.G Suijker, M.C. Raas, US Patent 6,356,016 B1

[3] J. He, Z. Toth, T. Russell, US Patent Application 2009/0146570 A1

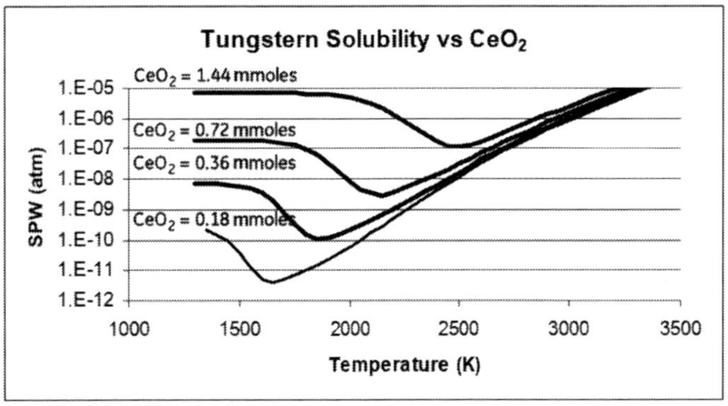

Figure 1-Tugnsten solubility vs. added CeO_2

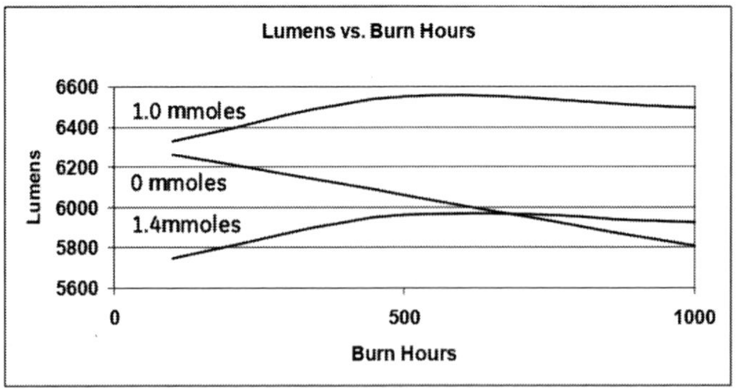

Figure 2-Lamp Lumens vs. Burn Hours for lamps with 3 levels of added CeO_2

Color Mixing within Optical Elements to Compensate for Non-Uniform Inherent within Phosphor Coated SSL Devices

Cao Yiyu
GE Lighting, GE China Technology Center, 1800 Cailun Road, Shanghai 201203, China
Contact: yiyu.cao@ge.com

ABSTRACT

Color separation after secondary optics is a common issue for LED comprised of blue dies and yellow phosphors, especially when the optics is placed in the near field of LED. The article introduces a method to solve the color separation issue through adding rational portion of diffusive material in a refractive optics.

COLOR SEPERATION PROBLEM

The mostly widely spread way of constructing white LED is to use blue LED with yellow phosphors. The spectral of this kind of white LED is shown in Fig 1. And a typical 6-die LED package example is shown in Fig 2. The source of blue rays and yellow rays were separated spatially.

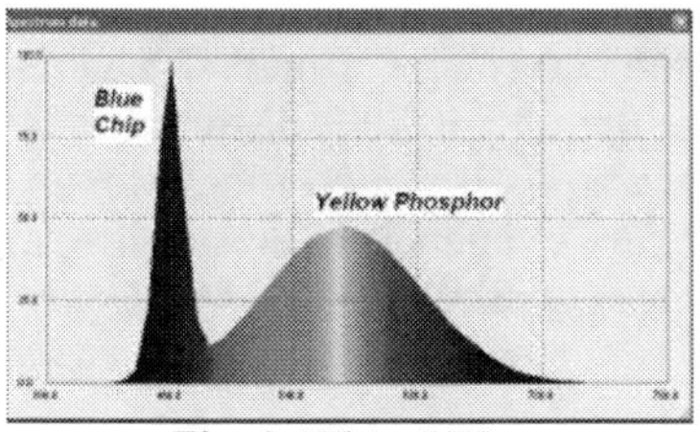

Fig. 1: *White LED spectra*

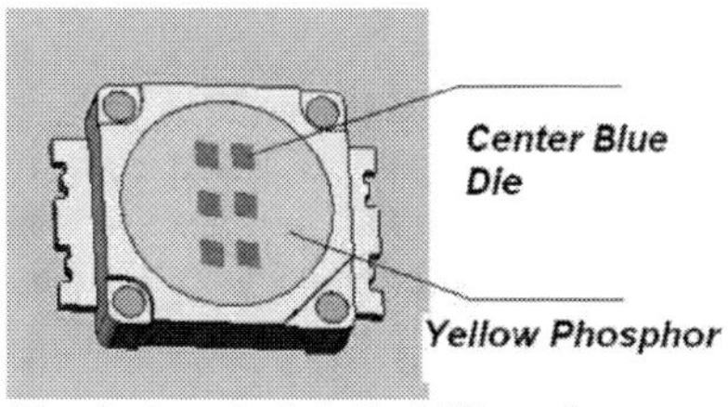

Fig. 2: *A typical 6-die LED package*

This tiny spatial separation mostly won't cause visible color separation. Unless the secondary optics component are adjusting the direction of the rays, and the optics are very near to the LED component (Fig. 3). Unfortunately, in a project when the dimension and height of optics was limited, the optics has to be placed near the LED, and the spatial source separation cannot be neglected.

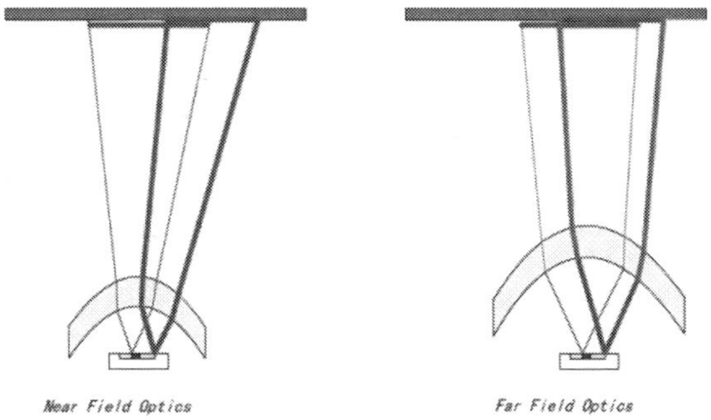

Fig. 3 *Near field optics are more likely to create visible color separation*

That was the problem we meet when a batwing lens was designed and mounted into a white LED at near field (Fig.4). The target of the batwing lens is to enlarge the dimension of the illuminated pattern. When the lens was built, besides the optical distribution as was designed for, we also got a side effect of a visible yellow ring on the illuminated surface (Fig 5).

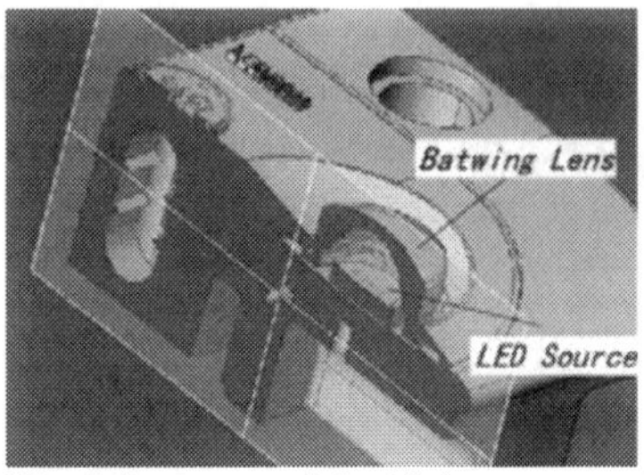

Fig 4 *Project model with lens at near field*

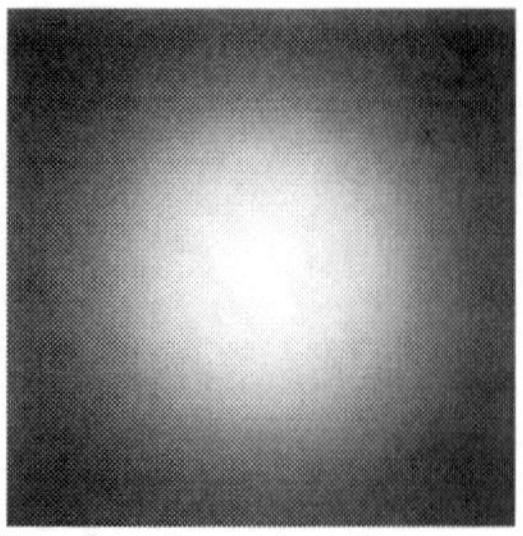

Fig.5 *Yellow ring illumination problem*

DIFFUSIVE MATERIAL SOLUTION

A common solution to that color separation problem is to apply diffusive texture or tiny scattering facet on the surface of the optics. But they do not apply to our project, because the diffusive level of texture is hard to control, and would affect our optical performance too much. Also after trying with modification of lens shape and tooling surface processes, it finally came to the solution of adding diffusive frosted material method, which is a new method of solving this problem.

The lens was originally constructed with clear polymethyl methacrylate (PMMA) material. For clear PMMA, all light goes straight through the material. There are also diffusive PMMA, which contain very small particle inside, so some light would be scattered when passing through. The diffusive PMMA was also called frosted PMMA, some of them can be 98% haze property. [2] By mixing up these 2 PMMA together and injected as the optics, a blended material optics can be constructed. The diffusive level of this blend material can be easily controlled by the percentage of these 2 materials. The general side effect of a diffusive material is that it changes the output light distribution, and decreases the illuminate pattern dimension that we desired to enlarge. Also, more light would be absorbed because of the diffusive PMMA added. Fig.6 shows the flux efficiency and resulted spatial color uniformity performance for optic lenses with various percentages of diffusive PMMA.

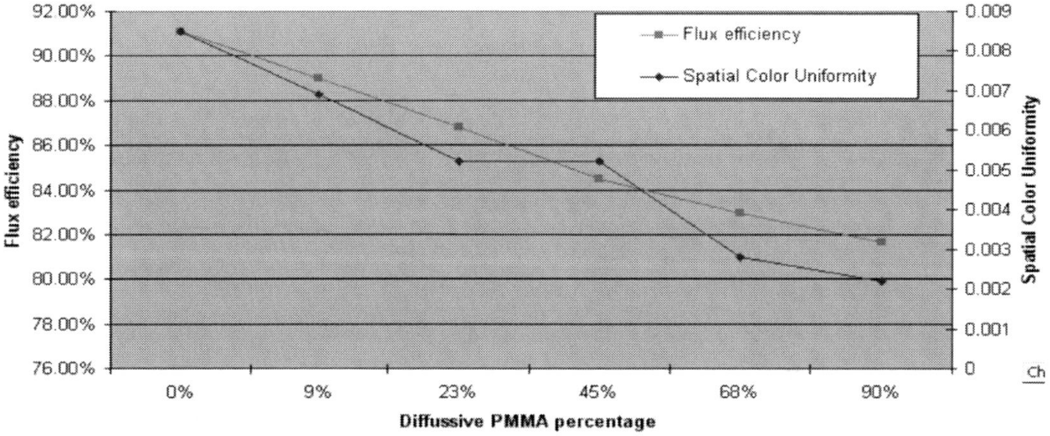

Fig.6 *Diffusive PMMA percentage & Flux efficiency & Spatial Color Uniformity*

As we don't have to decrease the spatial color uniformity to zero, we just need to make it a certain level that human eyes could not notice. So we injected a series of lens with different percentage of diffusive and clear PMMA, did some trade-off, and finally found one that met our design request.

REFERENCES

[1] Nichia Corporation, Specification for Nichia chip type White led, Model NS6W083AT
[2] Altuglas Group, Altuglas V045 frosted and DR frosted data sheet

Thermodynamic data and phase diagrams for the InX-InX', TlX-TlX' and InX-TlX systems, (X,X'=Cl,Br,I)

A.T.Dinsdale[1], R.H.Davies[1], and S.A.Mucklejohn[2]
[1] *National Physical Laboratory, Hampton Road, Teddington, Middlesex TW11 0LW, UK*
[2] *Ceravision Limited, The Mansion, Bletchley Park, Milton Keynes MK3 6EB, UK*
Contact: alan.dinsdale@npl.co.uk

ABSTRACT

Indium(I) and thallium(I) halides are widely used as dose components in metal halide discharge lamps. This paper reviews experimental reports for the phase diagrams of the (InX-InX'), (TlX-TlX') and (InX-TlX) systems where X, X' = Cl, Br, I and lists interaction parameters that can be used to model the chemical behaviour of these systems.

INTRODUCTION

Halides of indium(I) and thallium(I) are used in conjunction with other metal halides in numerous metal halide discharge lamps. Within the plasma these and their dissociation products give rise to atomic and molecular radiation. Atomic radiation from indium is predominantly from the transition at 451 nm while that from thallium arises from the transition at 535 nm. In lamps with electrodes the use of indium and thallium halides is restricted to the bromides and iodides. Electrodeless plasma lamps, however, allow these mixtures to be extended to include chloride-containing systems.

A survey of the published experimental reports for the phase diagrams of the (InX-InX'), (TlX-TlX') and (InX-TlX) systems with additional limited thermodynamic data allowed parameters to describe the interactions between ions in the liquid and other condensed phases to be derived. The phase diagram information was taken from the following: InCl-InBr [1]; InCl-InI [1]; InBr-InI [1]; TlCl-TlBr [2-6]; TlCl-TlI [7,8]; TlBr-TlI [9]; InCl-TlCl [10-12]; InI-TlI [13]. No experimental reports were found for the (InBr-TlBr) system. The phase diagram for the (InI-TlI) system is shown in figure 1.

The crystalline phases identified in the systems considered here are:

<CSCL>	Structure of TlCl, TlBr and the high temperature (red) phase of TlI
<INCL>	Distorted NaCl structure based on the low temperature form of InCl
<TLI>	Structure of the high temperature phases of InCl, InBr, InI and the low temperature (yellow) form of TlI

Thermodynamic data used in this work were in the format defined in equation {1}.

$$\Delta G(T)/J.mol^{-1} = A + B.(T/K) + C.(T/K).\ln(T/K) + D.(T/K)^2 + E.(T/K)^3 + F.(T/K)^{-1} \qquad \{1\}$$

Thermodynamic values from the SGTE Substance Database [14] for InCl, InBr, InI, TlCl, TlBr and TlI were used to derive data for the following species: InCl<CSCL>; InBr<CSCL>; InI<CSCL>; TlCl<CSCL>; TlBr<CSCL>; TlI<CSCL>; InCl<TLI>; InBr<TLI>; InI<TLI>; TlCl<TLI>; TlI<TLI>; InCl<INCL>; InCl<LIQUID>; InBr<LIQUID>; InI<LIQUID>; TlCl<LIQUID>; TlBr<LIQUID>; TlI<LIQUID>. Thermodynamic data for $InTlCl_2$<INTLCL2> were derived during the course of this work.

The format for the interaction parameters is shown in equation {2}.

$$G(T) = x_A.G_A + x_B.G_B + RT[x_A.\ln\{x_A\} + x_B.\ln\{x_B\}] + x_A.x_B.[L_0 + L_1 (x_A - x_B)] \qquad \{2\}$$

where: R is the molar gas constant, T is temperature/K, x is mole fraction and L_0 and L_1 are the parameters given in table 1.

Table 1. Values for interaction parameters as defined in equation {2}.

		Phase		
System	L	<TLI>	<CSCL>	<LIQUID>
InCl - InBr	L_0	1000		-1000
InCl - InI	L_0	6000		-6000
	L_1	5500		-3500
InBr-InI	L_0	2500		-2200
	L_1	3000		-4333
TlCl - TlBr	L_0		2670	0
TlCl - TlI	L_0		14500	2000
	L_1		8500	
TlBr - TlI	L_0		4000	0
InCl - TlCl	L_0		-3000	-5
InBr - TlBr	L_0		2000	4000
InI - TlI	L_0	8165	7000	8358

Figure 1. Phase diagram of the (InI-TlI) system.

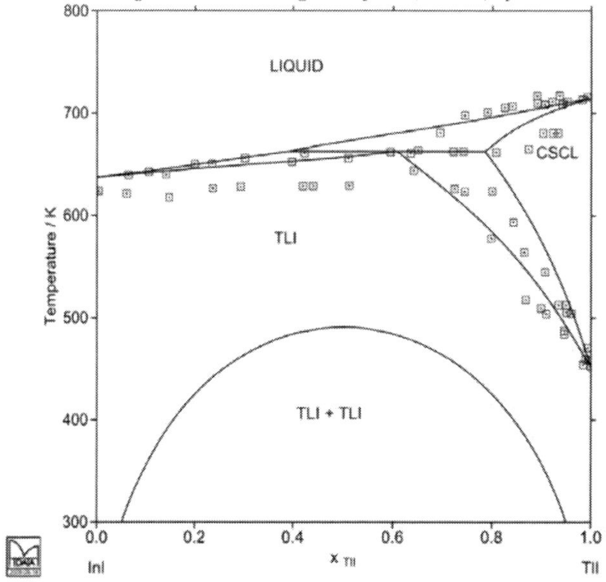

REFERENCES

[1] L.V. Povolotskaya and V.A. Smirnov, *Zh. Neorg. Khim.*, 24, 2291 (1979).

[2] K. Monkemever, *Neues Jahrb. Mineral., Geol.*, 22, 38 (1906).

[3] I.L. Favorski, *Isz. Sekt. Fiz. Khim. Anal., Inst. Obshch. Neor. Khim., Akad. Nauk SSSR*, 13, 281 (1940).

[4] I.I. Ilyasov, L.W. Rozhkovskaya and A.G. Bergman, *Zh. Neorg. Khim.*, 2, 1883 (1957).

[5] M.A. Ol'skaya, O.H. Postnikova and P.I. Fedorov, *Metallurgiya*, 3, 29 (1970).

[6] J. Kapala and K.Skudlarski, *Int. J. Mass Spec. Ion Phys.*, 40, 255-263 (1981).

[7] E.M. Chernotordik, *Zh. Obshch. Khim.*, 4, 459 (1934).

[8] G. Flor, C. Marghertis, G.C. Vigano and C. Sinistri, *Z. Naturforsch.*, A37, 1073-1076 (1982).

[9] K. Monkemever, *Neues Jahrb. Mineral., Geol.*, 22, 39 (1906).

[10] A.P. Palkin and T.N. Vigutova, *Zh. Neorg. Khim.*, 10, 194 (1965).

[11] T.N. Larionova and T.M. Pustyl'nik, *Zh. Neorg. Khim.*, 14, 3151 (1969).

[12] P.I. Fedorov, N.S. Malova and G.V. Vorob'eva, *Zh. Neorg. Khim.*, 15, 2853 (1970).

[13] O.N. Postnikova, Yu N. Denisova. P.I. Fedorov, N.S. Malova and L.A. Radushkevich, *Zh. Neorg. Khim.*, 18, 1436 (1973).

[14] I. Hurtado, D. Neuschutz, 'Thermodynamic Properties of Inorganic Materials', compiled by SGTE, Landolt Börnstein, 19, Springer-Verlag, Berlin (1999–2005).

High Luminance and Luminous Efficacy Mercury-free Flat Fluorescent Lamp (MFFL)

Ki-Woong Whang, In Woo Seo, Jae-Chul Jung, Byung Joo Oh
Department of Electrical Engineering, Display Technology Research Center, Seoul National University, Shilim-dong, Kwanak-gu, Seoul, 151-744, Korea
Contact: kwhang@snu.ac.kr

ABSTRACT

We realized a high luminance and luminous efficacy MFFL with Ne-Xe mixture as the discharge gas. The MFFL has a very simple structure which can be enlarged to any bigger size, and is inherently flat type. Recently we optimized the thickness and cross-sectional shape of phosphors which resulted in a big improvement in the luminous efficacy. In addition, we investigated various mixing ratios of red, green, and blue phosphors for the appropriate colour coordinates of general lighting and LCD backlighting. As a result, the MFFL showed the high luminous efficacy of 82.1 lm/W at the luminance of 3600 cd/m^2 with the colour coordinate around (0.36, 0.36).

INTRODUCTION

Mercury-free flat fluorescent lamps have been actively studied to get rid of the harmful mercury from the fluorescent lamps and to meet the growing demands for the energy efficient general lighting and LCD backlights [1-4]. The UV source of the lamps is Xe instead of mercury, which is environmentally friendly and has the merit of immediate response regardless of ambient temperature, but used to show the demerit of low excitation efficiency due to the shorter wavelength of UV than that of mercury. We devised a MFFL with an auxiliary electrode which shows a wide and stable operating voltage margin with a luminous efficacy of 60 lm/W at the luminance of 1550 cd/m^2 [5]. In this paper, the phosphor thickness was optimized and the surface area of phosphor was increased by the uneven cross-sectional structure for the high luminance and luminous efficacy. It has been shown that the colour coordinates of the lamps can be changed by the phosphor mixing ratio changes for various lighting purposes.

RESULTS

The MFFL has a simple structure composed of a pair of main electrodes covered with dielectrics. Its special feature is a rectangular-shaped auxiliary electrode, which does an important role to form the fully diffused glow discharge at reduced voltages. Alternating square voltage pulses were employed to make the glow discharge [5]. There are several factors affecting the luminous efficacy and luminance of the lamp: Xe partial pressure, phosphor thickness, phosphor surface area, and phosphor mixing ratio. High partial pressure of Xe in the Ne-Xe mixture gas showed high luminous efficacy due to the increase of the second continuum emission centred at 173 nm from Xe excimer [6]. In this study, a Ne-Xe mixture with the Xe content of 25% was used. The thicknesses of the rear and front phosphor were optimized, the rear phosphor of 70 μm and front phosphor of 5 μm showed the best luminance and luminous efficacy. The relatively larger thickness for the rear phosphor may be due to the fact that it also acts as the reflector for the light generated by the front phosphor. The front phosphor acts to convert the VUV into the visible light and transmit the visible light from the rear phosphor at the same time. Thus, an optimized thickness of front phosphor is critical to extract the maximum visible light from the lamp. Based on the optimized thickness of phosphors, we introduced the bumpy structure for the rear phosphor to increase the surface area of the phosphor as shown in Figure 1 which resulted in 50% increase [7]. Since the MFFL has the versatile application areas from the indoor illuminating to LCD backlighting, it requires various colour coordinates which can be controlled by the mixing ratio change of each phosphor. The white phosphor was made by mixing the red ($(Y,Gd)BO_3$:Eu), green ($LaPO_4$:Ce,Tb), and blue ($BaMgAl_{10}O_{17}$:Eu) phosphors with an appropriate mixing ratio to obtain the desired colour coordinates. The typical coordinates are achieved with the matched mixing ratios of phosphors as in Table 1.

Colour-saturating LED clusters

A. Žukauskas[1] , R. Vaicekauskas[2] , M. S. Shur[3]
[1]*Institute of Applied Research, Vilnius University, Saulėtekio al. 9, bldg. III, Vilnius, LT-10222 Lithuania*
[2]*Department of Computer Science, Vilnius University, Naugarduko g. 24, Vilnius, LT-03225 Lithuania*
[3]*Department of Electrical, Computer, and System Engineering, Rensselaer Polytechnic Institute, 110 8th Street, Troy, NY 12180 USA*
Contact: arturas.zukauskas@ff.vu.lt

ABSTRACT

The ability of trichromatic light-emitting diode (LED) clusters to increase chromatic saturation of object colours was optimized using statistical analysis of colour shifts for 1269 Munsell samples. The highest colour-saturating ability is attained with the wavelengths of the red and blue LEDs close to the edges of the visible spectrum. A trade-off between colour-saturating ability and luminous efficacy of radiation requires shifting the wavelengths of the red and blue components toward the centre of the spectrum. Another trade-off between colour-saturating ability and the ability to render object colours with high fidelity needs avoiding emission in the yellow part of the spectrum. Clusters composed of commercial red, green, and blue LEDs with the peak wavelengths outside the spectral range of 530 nm to 610 nm were shown to have a high colour-saturating ability that is stable in respect of the possible drift of peak wavelengths.

INTRODUCTION

Solid-state technology is capable of tailoring the spectrum of lighting sources using direct-emission and/or phosphor-converted light-emitting diodes (LEDs). This enables illumination with different colour-rendition properties. Until recently, a sole widely considered colour-rendition characteristic of white lighting sources was the ability to reproduce object colours with high fidelity. However, some composite lamps, which contain narrow-band LEDs, render object colours within an increased chromatic gamut rather than with high fidelity [1]. Similar colour-saturating effect has been known for years for neodymium-glass filtered incandescent lamps [2].

In this work, general design rules for colour-saturating solid-state sources of light [3] are considered using the statistical method of assessment of colour rendition properties of lighting [4].

COLOUR-SATURATING CAPACITY OF LIGHTING SOURCES

As a measure of the colour-saturating capacity of a lighting source, we use the percentage of test colour samples that gain a perceptually noticeable increase in chroma when a reference broad-band source (blackbody or a daylight phase illuminant) is replaced by the source under test [3,4]. For a reliable statistical output, a very large number (1269) of test colour samples from the Munsell palette was used. The colour shifts for each sample were examined in respect of interpolated 3-step MacAdam ellipses: the samples that have a colour shift escaping from the ellipse with an excess component directed towards an increased chroma were scored to the colour saturation index (CSI) and the samples that have a colour shift within the ellipse were scored to the colour fidelity index (CFI). CSI was used as a figure of merit that quantifies the colour-saturating ability of a lighting source.

Various spectral power distributions (SPDs) containing different number of narrow-band and wide-band components were assessed. High colour-saturating ability was found to be characteristic of trichromatic blends of narrow-band red, green, and blue coloured components. For instance, application of a blend of 30 nm wide coloured components peaking at 450 nm, 530 nm, and 630 nm results in an increased chromatic saturation for 70% to 80% of the test colour samples depending on correlated colour temperature (CCT). Increasing bandwidth to 70 nm decreased the number of object colours that gain in chroma to about 40% (making it equal to the number of object colours rendered with high fidelity). This evidences that direct-emission red AlGaInP and green and blue InGaN LEDs with the average bandwidth of 30 nm are good candidates for the development of colour-saturating lamps.

Trichromatic sources containing narrow-band components were found to attain the highest colour-saturating ability when the red and blue components are shifted to the edges of the visible spectrum. However, this results in a very small luminous efficacy of radiation (LER). Fortunately, LER can be increased to about

250-300 lm/W at a small expense of CSI when the red and blue components are somewhat shifted toward the centre of the spectrum (to about 625 nm and 440 nm, respectively).

The colour-saturating ability of trichromatic sources was found to be highly critical against the spectral content in the yellow region. For instance, when the red component was shifted to wavelengths shorter than 610 nm and the green component was shifted to wavelengths longer than 530 nm, the colour saturating effect tended to vanish; while the colour-fidelity effect increased (CFI has the highest values at the peak wavelengths of about 450 nm, 540 nm, and 610 nm).

COLOUR-SATURATING RGB LED CLUSTERS

The above basic considerations on maximizing the number of object colours with increased chroma imply that practical colour-saturating lamps are clusters containing red, green, and blue (RGB) LEDs. To demonstrate the feasibility of such an approach, SPDs of RGB clusters of LEDs provided by the Philips Lumileds Lighting Luxeon® Rebel family, which comprises direct-emission royal-blue (452 nm peak wavelength), blue (469 nm), cyan (512 nm), and green (523 nm) InGaN LEDs; and red-orange (625 nm) and red (637 nm) AlGaInP LEDs, were examined. High values of CSI (≥ 70%) can be obtained in eight types of clusters with the long-wavelength LED selected out of red or red-range ones, the medium-wavelength LED selected out of green or cyan ones, and the short-wavelength LED selected out of blue or royal blue ones. Table 1 presents data on colour quality of eight RGB clusters with a CCT of 4000 K. The highest values of CSI are reached for the clusters containing red 637-nm and green 512-nm LEDs, whereas the lowest CSI values are obtained for the clusters containing red-orange 625-nm and cyan 523-nm LEDs.

Table 1. Colour quality indices of trichromatic LED clusters (CCT = 4000 K).

LED peak wavelengths (nm)			Colour quality indices		
Red	Green	Blue	CSI	CFI	R_a
637	523	469	75	7	22
637	523	452	73	10	27
637	512	469	77	3	1
637	512	452	78	4	6
625	523	469	73	8	42
625	523	452	70	13	47
625	512	469	76	4	20
625	512	452	76	6	28

The stability of the colour-saturating ability of the practical RGB LED clusters with respect to possible scatter and thermal drift of the peak wavelengths was examined. Figure 1 shows the variation of CSI of the 637-523-452 nm LED cluster with shifting one of the three peaks when the remaining two being kept at the nominal position. Colour saturation ability is rather insensitive to the blue LED shift. However, CSI drops considerably with increasing the wavelength of the green LED and with decreasing the wavelength of the red LED by more than 10 nm, which is higher than typical technological scatter and thermal drift. Nevertheless, LEDs that enter the spectral range between 530 nm to 610 nm are to be avoided in colour-saturating clusters.

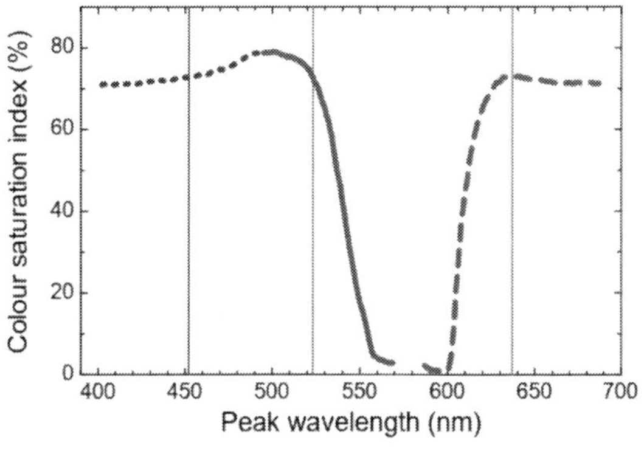

Fig. 1. Effect of shifting the peak wavelength on colour saturating ability of a trichromatic LED cluster composed of 637-nm, 523-nm, and 452-nm LEDs (CCT = 4000 K).

REFERENCES

[1] Y. Ohno, Opt. Eng. 44 (2005) 111302.
[2] P.J. Bouma, Philips' Technical Review 3 (1938) 47.
[3] A. Žukauskas, R. Vaicekauskas, M. Shur, Opt. Express, 18 (2010) 2287.
[4] A. Žukauskas, R. Vaicekauskas, F. Ivanauskas, H. Vaitkevičius, P. Vitta, M.S. Shur, IEEE J. Sel. Top. Quantum Electron. 15 (2009) 1753.

Efficiency evaluation of phosphor-white high power light-emitting diodes

A. Keppens[1,2], P. C. Acuña[3], H. T. Chen[4], G. Deconinck[2], P. Hanselaer[1,2]
[1] *Light & Lighting Laboratory, Catholic University College Gent, Gebroeders Desmetstraat 1, B-9000 Gent, Belgium*
[2] *ESAT/ELECTA, K.U.Leuven, Kasteelpark Arenberg 10, B-3001 Leuven, Belgium*
[3] *LABE, National University of Colombia, Carrera 30 - 45, 03 Bogotá, Colombia*
[4] *Fujian Engineering Research Centre for SSL, Xiamen University, Siming South Road 422, Xiamen, China*
Contact: arno.keppens@kahosl.be

ABSTRACT

High power light-emitting diode (LED) efficiency is strongly dependent on the device type and on operating conditions. Electrical and optical power measurements are therefore performed on a selection of phosphor-white LEDs at a number of currents and temperatures. The results allow for a comparative evaluation of their dominant loss mechanisms.

EXPERIMENTS

Five commercial phosphor-white high power LED packages, indicated as L1 to L5, have been selected for investigation through various current-voltage and spectral radiant flux measurements. Forward current-voltage characteristics at different junction temperatures have been determined by placing the LEDs in a Heraeus UT6 isothermal oven with active air circulation. The predefined oven temperatures (300, 320, 340, and 360 K) have been precisely measured with a PT100 thermistor. In order to avoid junction heating during measurement, a Keithley 2440 5A SourceMeter was pulse-operated by a LabVIEW 7.1 program.
Spectral radiant flux measurements at different currents and junction temperatures have been performed with a custom-made integrating sphere [1]. The LEDs were attached to the sphere surface using an aluminium mounting plate incorporating a Peltier element and PT100 thermistor. The Peltier element regulates the plate temperature until the LED junction temperature – determined by a forward voltage measurement [2] – shows the desired value.

RESULTS AND DISCUSSION

The measurement equipment described above allows determining the LED input power P_{in}, power loss into the internal series resistance ΔP_R, external quantum loss ΔP_{EQ}, Stokes shift power loss ΔP_S, eye sensitivity loss $\Delta P_{vis} = \Phi_{e,\lambda}(1 - V_\lambda)$ and luminous flux Φ [3]. All these quantities can be determined independently and are connected by the following equation, which was also used to verify the measurement results:

$$P_{in} = \Delta P_R + \Delta P_{EQ} + \Delta P_S + \Delta P_{vis} + \Phi / (683\ \mathrm{lm \cdot W^{-1}}) \qquad (1)$$

At first, the current and junction temperature dependence of each power loss has been examined at 300 K junction temperature and a 350 mA drive current, respectively (see Fig. 1).

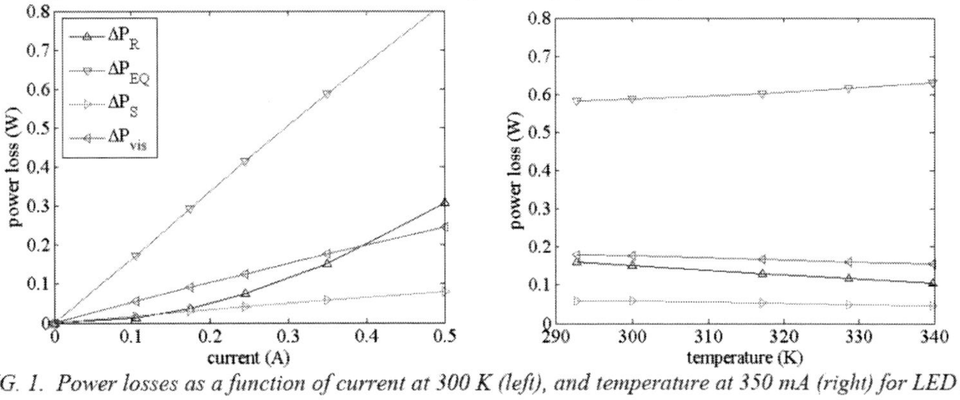

FIG. 1. Power losses as a function of current at 300 K (left), and temperature at 350 mA (right) for LED L1.

With increasing current, the quantum, Stokes and visual losses increase quasi linearly. This means that the quadratic increase of the power loss into the internal series resistance forms the main reason for the decreasing diode efficiency with increasing current. On the other hand, the decreasing diode efficiency with junction temperature is only due to the increasing external quantum loss with increasing temperature, as other losses slightly decrease.

Secondly, all five terms on the right-hand side of Eq. (1) have been compared for all LEDs at normal operating conditions (350 mA and 330 K junction temperature) and per watt input power (see Fig. 2). This way, the last term multiplied with 683 (lm/W) equals the luminous efficiency. From the data in Fig. 2, it is clear that the internal series resistance directly consumes about 10 to 20 % of the initial input power. As a result, the overall efficiency is remarkably higher for LEDs with a reduced resistance (e.g. L1). External quantum loss is clearly the dominant loss factor and can be related to the colour temperature of the LEDs. Indeed, L1 and L2 are cool white devices and exhibit the smallest quantum losses. L3 is neutral white, while L4 and L5 are warm white. Non-fluorescent absorption in the phosphor seems to be more important for these warm white LEDs. Also from Fig. 2, the luminous flux is found to be correlated to the external quantum loss. For this reason, the LED efficiency is always higher for cool white LEDs [4]. The Stokes shift losses equal about 5 % of the input power for all measured LEDs. Although the conversion loss per photon is larger for warm white LEDs, this effect is compensated by the smaller number of emitted photons. The remaining luminous flux part is always a few percent smaller than the visual loss for phosphor-white LED spectra.

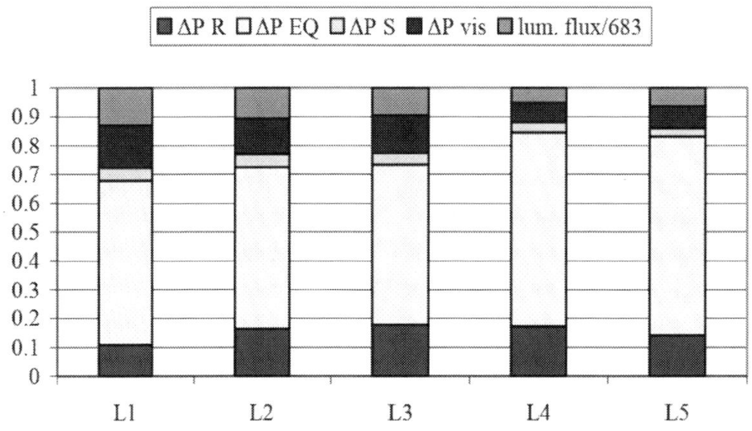

FIG. 2. Comparison of normalised power losses and flux for all LEDs at 350 mA and 330 K.

CONCLUSIONS

Different power loss mechanisms present in phosphor-white high power LEDs have been evaluated at different operating conditions and mutually compared. Power loss in the internal series resistance has been found to amount 10 to 20 % of the initial input power and is the main reason for the decreasing diode efficiency with increasing current. External quantum loss is clearly the dominant loss factor and can be related to the colour temperature of the LEDs. This quantum loss is also responsible for the efficiency decrease with junction temperature.

REFERENCES

[1] P. Hanselaer, A. Keppens, S. Forment, W. R. Ryckaert, and G. Deconinck, "A new integrating sphere design for spectral radiant flux determination of light-emitting diodes (LEDs)," Meas. Sc. & Tech. 20, 095111 (2009).

[2] A. Keppens, W. R. Ryckaert, G. Deconinck, and P. Hanselaer, "High power light-emitting diode junction temperature determination from current-voltage characteristics," J. Appl. Phys. 104, 093104 (2008).

[3] M. R. Krames, O. B. Shchekin, R. Mueller-Mach, G. O. Mueller, L. Zhou, G. Harbers, and M.G. Craford, "Status and future of high-power light-emitting diodes for solid-state lighting" IEEE J. Disp. Tech. 3, nr. 2, 160-175 (2007).

[4] E. F. Schubert, "Light-emitting diodes – second edition," Cambridge University Press (2006).

Thermodynamics processes in DBD driven excilamps estimated by jump of pressure method: radiation intensity control

E.A. Sosnin[1], V.M. Tsvetkov[2], A.A. Pikulev[2], S.M. Avdeev[1], V.F. Tarasenko[1]
[1] *Optical Radiation laboratory, High Current Electronics Institute SB RAS, 2/3, Akademicheskii ave., 634055, Russian Federation*
[2] *Federal State Unitary Enterprise – Russian Federal Nuclear Centre-All-Russian Research Institute of Experimental Physics, 37, Mira ave., 607190, Russian Federation*
Contact: badik@loi.hcei.tsc.ru

ABSTRACT

The thermodynamic processes in DBD KrCl- and Cl_2-excilamps in diffuse discharge form and in microdischarges form have been investigated by means of pressure-jump measurements. It have been shown, that the linear dependence between jump of pressure and radiation intensity can be useful for radiation intensity control.

EXPERIMENTAL

Excilamps are a class of spontaneous radiation sources based on transitions of rare gas excited dimers R_2^*, on halogen excited dimers X_2^* or on rare gas halide excited complexes RX* [1–3]. They emit incoherent radiation in the ultraviolet (UV, 200 nm < λ < 400 nm) and vacuum-UV (VUV, 100 nm < λ < 200 nm) region of the electromagnetic spectrum. In the literature, these lamps are usually designated as "excimer lamps", "exciplex lamps" or in general as "excilamps".

During longtime excilamp operation it is necessary to keep thermal mode of operation and control radiant power. For this aim the air or water cooling is applied. The choice of cooling system is connected to excilamp thermal power and with maximum permissible temperature of working gas. Usually for control of radiant power the physical and chemical photodetectors are used [4]. In the present study an alternative method is tested. This is the measurement of pressure jump fast part amplitude (characteristic time ~ 100 ms), arising in excilamp volume after switching of device.

Experiments were performed with two DBD driven coaxial lamps made of fused quartz (KU-2, OOO "Technoquartz"). The parameters of the lamps and power supply units presented in Table 1. The static pressure in the bulbs was controlled by vacuometer, and for registration of dynamic pressure jump the differential inductive sensor DMI-0.1-2 was used. The radiant power of lamps was measured with a H8025-222 photodetector (Hamamatsu Photonics K.K.) with the spectral sensitivity maximum at 222 nm and UV-818 detector (connected to Newport 1830-C Picowatt Digital Optical Power Meter). We also controlled the emission spectrum of the discharge and values of voltage and current pulses.

Table 1.

Operating molecule	KrCl	Cl_2
Radiation spectra	narrow band, maximum at λ = 222 nm	narrow band, maximum at λ = 257.8 nm
Diameters of coaxial bulb, mm	43/20	43/22
Radiating length, cm	30	12
Gas mixture	Kr-Cl_2 = (50÷400)-1	Ar-Cl_2 = (200÷400)-1
Pressure range, Torr	30÷380	90÷500
Power supply characteristics: polarity voltage amplitude maximum, kV PPR, kHz	 bipolar 4.5 100	 unidirectional 4 15÷70

We proposed the formula for heat power of discharge (HPD) in DBD-driven bulb with gas media by means of well-known relationship for temperature distribution between to parallel plates [5]

$$W = \frac{6\text{лД}pST_0}{p_0 d}.$$

where λ - heat-conduction coefficient of gas; S – surface area of the lamp discharge volume; W – HPD; d – thickness of discharge gap; T_0, p_0 – initial temperature and pressure of gas; Δp – pressure jump; V – full volume of gas in excilamp bulb, including buffer and discharge volumes.

With use of this formula for Cl_2-excilamp case it was shown, what on direct heating of gas goes about 40 % of discharge power, about 1.6 % is spent for $^3\Pi_{2g}\rightarrow{}^3\Pi_{2u}$ radiation band of Cl_2^* molecule, unknown part of energy goes on ArCl* radiation (we did not estimate it in our tests, because of bad optical transmittance bulb wall of vacuum ultraviolet radiation), and other part results in heating of bulb walls at electrons and ions bombardment acts.

We also have constructed dependences of amplitude via pressure jump fast component on radiant power $p_0(W_{opt})$ and the same one for HPD $p_0(W)$ for KrCl-excilamp. There are in a good agreement with linear functions such as $p_0 = A_{opt} + B_{opt}\ W_{opt}$ ($p_0 = A + BW$) (Table 2). Relative error does not exceed 20-30 %. The approximation were carried out by least squares method. For Cl_2-excilamp the experimental dependences $p_0(W_{opt})$ and $p_0(W)$ also in a good agreement with linear approximations.

Table 2.

$Kr:Cl_2$	Coefficient of linear approximation			
	A, Torr	B, Torr/W	A_{opt}, Torr	B_{opt}, Torr/W
50:1	-0.73	0.45	-0.9	6.3
100:1	-0.37	0.54	2.7	5.7
200:1	3.8	0.48	3.7	5.5
300:1	-1.9	0.57	-0.47	6.9
400:1	2.7	0.53	-2.1	8.5
average value	0.71	0.51	0.58	6.6

We also observed that it is possible to use the linear approximation in case of Cl_2-, Br_2-, I_2-excilamps in a wide range of pressure (approximately up to 600 Torr). In these conditions the discharge exists in diffuse form [6]. In KrCl-excilamp with growth of pressure up to 200-300 Torr takes place the microdischarge formation without a background luminescence. In this case heating of gas already is partially carried out in acoustic fluctuations energy dissipation, and therefore the linear approximation $p_0(W_{opt})$ is inapplicable.

CONCLUSION

The researches of thermodynamic processes in DBD-driven KrCl- and Cl_2-excilamps which has been carried out with the help of pressure jump measurement have shown, that at lamp switching the dependence of pressure jump via time consists from fast (100 ms) and slow parts (100 s). It is shown, that for the control of radiant power it is possible to use linear approximation between fast pressure jump and ultraviolet radiation intensity. The specific conditions where this method works were pointed.

ASKNOLEGMENTS

This work was supported by the ISTC (Project No. 3583p).

REFERENCES

[1] E.A. Sosnin, T. Oppenländer, V.F. Tarasenko, J. Photochem. Photobiol. C, 7 (2006) 145.
[2] B. Eliasson B., U. Kogelschatz, Appl. Phys. B, 46 (1988) 299.
[3] E.A. Sosnin, Light & Engineering, 15 (2006) 49.
[4] V.N. Batalova, E.A. Sosnin, E.A. Zakharova, V.F. Tarasenko, Instr. Exp. Techniques, 46 (2003) 73.
[5] L.D. Landau, E.M. Lifshitz, Theoretical Physics. Vol. VI. Hydrodynamics, Moscow (1988) 736 p.
[6] S.M. Avdeev, E.A. Sosnin, V.F. Tarasenko, Optics and Spectroscopy, 103 (2007) 526.

AC ignition of HID lamps - statistical and formative lag times

A Sobota[1], J H M Kanters[1], E M van Veldhuizen[1], F Manders[2], M Haverlag[1,2]

[1] *Department of Applied Physics, Eindhoven University of Technology, Postbus 513, 5600MB Eindhoven, The Netherlands*

[2] *Light Labs, Philips Lighting, Mathildelaan 1, 5611BD Eindhoven, The Netherlands*

Contact: a.sobota@tue.nl

ABSTRACT

Noble gas discharges are typically used in lighting applications, under a very large range of conditions. A lot of modelling and experimental work has already been done in order to describe the starting behaviour of lamps. Focusing on high-pressure systems, high voltages needed for the ignition of lamps pose a problem. The purpose of this work is to examine the mechanisms of AC lamp ignition, which was already proven to be efficient in lowering HID lamp ignition voltage. More specifically, we use electrical measurements and observe the discharge by means of iCCD imaging to determine the lag times associated with AC lamp ignition. We use sine voltage with frequencies ranging from 60 kHz to 1 MHz, with various rise rates of the voltage amplitudes. Lamps we use are standard 70 W Philips burners, filled with 0.3 or 0.7 bar Argon or Xenon. We find that the statistical lag times in the system can be as long as 0.15 s and that they are at least two orders of magnitude longer than formative lag times.

High intensity discharge (HID) lamps are used for many applications, all of which require high energy output from very small volumes. To meet these requirements, lamps are filled with noble gas at high pressure (0.1 to 15 bar) and salts. Noble gas gives light during the warm up phase, and salts are the main radiators during the stable operation of the lamp. High noble gas pressures make lamps difficult to ignite - it typically takes a 5 kV to 20 kV pulse to bridge the gap between electrodes, while the latter stages of operation require much lower potential difference.

Many of attempts have been made to lower the breakdown voltage in HID lamps. For example, it has been shown that small addition of Hg can ease the breakdown process. HID lamps, however, already contain mercury. Small addition of Xe in the background Ar atmosphere has also been shown to lower the breakdown voltage in HID lamps. Another method is using sine voltage in sub-MHz range for lamp ignition instead of pulses [1]. By using this method, the breakdown voltage can be reduced by 56% with respect to pulse ignition [1].

High breakdown voltage is not the only problem encountered in HID lamp ignition. Some initial electrons are needed to start the avalanche process which will in turn lead to breakdown and lamp ignition, and they are expected to be produced in the lamp by means of, for example, cosmic rays. Since the lamps are very small (our typical burner volume is 8×10^{-8} m^3), there is little probability that there will be sufficient electrons available in the burner at any given time. This will cause large statistical lag times. Here, we present measurements of statistical lag times for our lamp arrangement, and we compare them with estimates for formative lag times in the same lamps.

The basic setup we used is described in [1]. The setup consisted of a set of function generators that produced the sine voltage shape with rising amplitude, which we amplified using a high-frequency amplifier and a passive transforming element. We used frequencies in the range 60 kHz - 1 MHz. The rising rates of the voltage amplitude varied from 20 to 1280 V/ms. The voltage was brought to a standard 70W Philips HID burner, filled with Argon or Xenon at 0.3 or 0.7 bar. Electrodes were tungsten, 0.6 mm in diameter. Distance between the electrode tips in the burner was 4 or 7 mm. The lamp voltage and current were measured by a HV probe and a Rogowski coil connected to an oscilloscope. For certain experiments we used an influx of 254 nm UV light.

We measured breakdown voltage as a function of frequency and amplitude rise rate, with and without UV irradiation. By comparing them, we were able to obtain the lag times in the ignition process without UV irradiation.

To determine formative lag times, we observed the breakdown process with an iCCD camera. We observed the formation of the discharge at 60, 220, 400 and 800 kHz, with voltage amplitude rising with 320 V/ms.

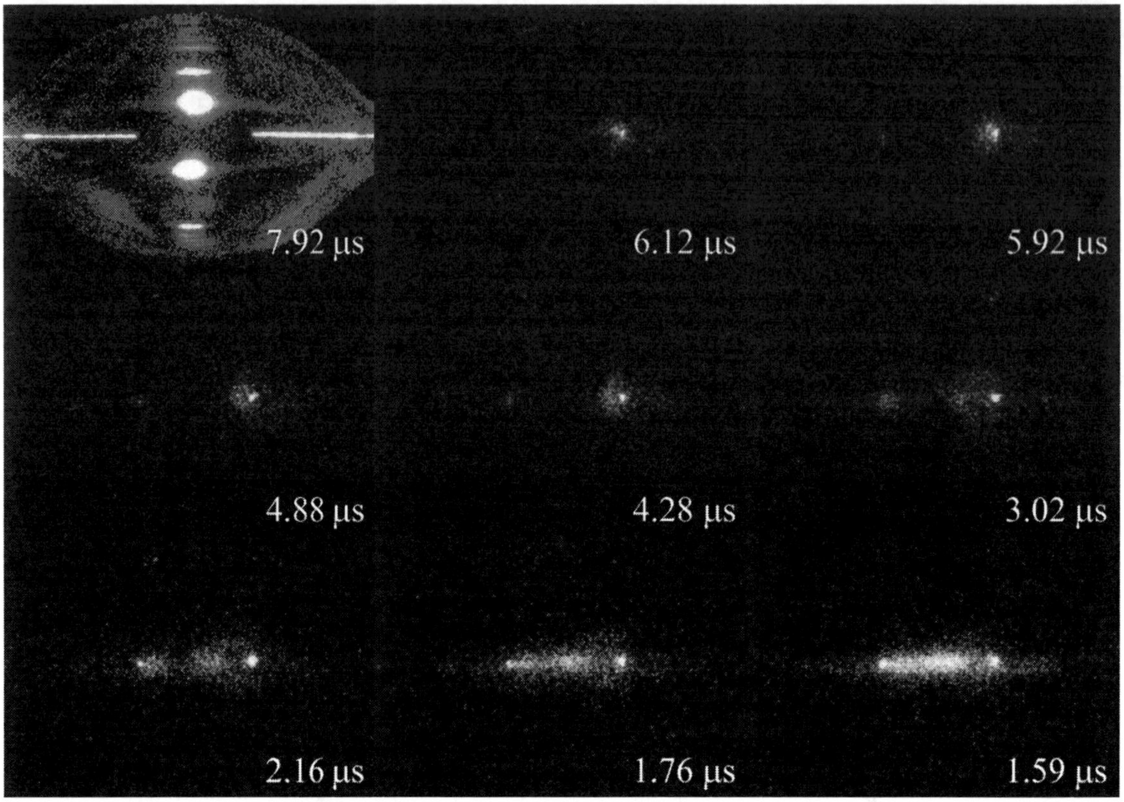

Figure 1. Development of an AC discharge in a typical 70 W burner. The lamp is filled with argon at 700 mbar. Electrode distance is 4 mm. The electrode on the right-hand side is charged with sine voltage with rising amplitude; voltage frequency is 800 kHz and the amplitude rises with 320 V/ms. Gate width for the camera was 1 voltage cycle (1.25 μs) long. Time shown in the bottom right corner of each picture denotes the amount of time between the beginning of the camera gate and breakdown. The top left picture shows the position of the electrodes and the burner wall in the lamp.

Mean statistical lag times are calculated in two ways for all lamp types, and various combinations of frequency and amplitude rise rates. We examine the relation introduced by Von Laue and Zuber, and later expanded by Cobine and Easton in 1943 and find that the relation still holds for AC voltage with rising amplitude. We also show that the statistical lag times in AC ignition of HID lamps are at least two orders of magnitude larger than the formative lag times.
In addition to lag times analysis, we would also like to comment on appearance of the discharge, which differs greatly over the range of frequencies we used. For 60 and 220 kHz, we observe streamer discharges that appear and cross the gap between electrodes during the last voltage cycle before breakdown. At 400 kHz, the discharge appears more diffuse, especially in Argon. When observing the lamps with 7 mm distance between electrodes, we see that the formation time takes more than one cycle, with complete lack of light emission around the voltage zero-crossings. At 800 kHz, the discharge is even more diffuse, and the formation of the charged channel takes more than 1 voltage cycle for every electrode distance.

In conclusion, we measured statistical and formative lag times during AC ignition of HID lamps. We found statistical lag times to be as high as 0.15 s and confirmed the validity of the relation for statistical spread of statistical lag times introduced by Cobine and Easton.

REFERENCES

[1] J Beckers et al, J. Phys. D: Appl. Phys., vol. 41, p. 144028, 2008.

Determination of the ground state density of Ce in front of the electrode of a YAG lamp by absorption spectroscopy

C. Ruhrmann, M. Westermeier, J. Reinelt, J. Mentel and P. Awakowicz
Ruhr-University Bochum, Electrical Engineering and Plasma Technology, D-44780 Bochum, Germany

Contact: ruhrmann@aept.ruhr-uni-bochum.de

ABSTRACT

The work function and with it the temperature of tungsten electrodes in high intensity discharge (HID) lamps can be lowered and the lifetime of lamps increased by the gas phase emitter effect. A determination of the emitter effect of cerium (Ce) is performed by electrode tip temperature measurements as well as Ce ground state density measurements by means of a broadband absorption spectroscopy setup [1, 2]. This spectroscopic method is independent of the plasma temperature which is determined up to now by using the emission coefficient of a mercury double line [3]. Because of this, these density measurements are important for future work with environmentally friendly mercury free HID lamps.

INTRODUCTION

HID lamps are used in large fields of application e.g. in automotive, street, stadium, store, plant or industrial lighting as well as in video projection systems. The improvement of lifetime is a particular interest of actual research into HID lamps. It can be achieved during lamp operation by a reduction of the temperature of the electrodes being accomplished by the "gas phase emitter effect" [4, 5, 6, 7, 8]. It is generated by a monolayer on the electrode surface of electropositive atoms of certain emitter elements (e.g. Ce or Dy) which are added to the lamp filling. This dipole layer reduces the effective work function and therefore the potential barrier for electrons leaving or entering the electrode. To quantify the gas phase emitter effect electrode tip temperature measurements and ground state density measurements of the emitter material in front of the electrode are necessary.

EXPERIMENTAL SETUP

The lamps being investigated are special research lamps with tungsten electrodes with a diameter of 360 µm and a discharge vessel made of yttrium aluminium garnet ($Y_3Al_5O_{12}$) shortened YAG [4]. This material is a transparent ceramic suitable for optical observations. The lamps are filled with a buffer gas which consists of an argon-krypton-mixture and with mercury to generate a pressure of 2 MPa during lamp operation and a sufficiently high power input into the arc. To utilise the gas phase emitter effect an emitter element is added to the lamp filling. The lamp is operated in a vertical position.

The lamps are operated with switched-dc current (400 – 800 mA) and different frequencies (1 – 1000 Hz). The electrical setup is described in detail in [5]. To determine the lateral electrode temperature profile the electrode is focused on a CCD camera with an interference filter in front. By calibration with a tungsten ribbon lamp absolute temperatures are deduced from the camera picture [5]. Spectroscopic measurements are realised by a spectrograph and a second CCD camera. Optical absorption spectroscopy with an ultra high pressure (UHP) lamp as backlight source [1, 2] is performed to measure ground state densities independently of the plasma temperature.

RESULTS AND DISCUSSION

Figure 1 shows the phase resolved cerium ground state atom density in front of the upper electrode of a YAG lamp filled with 1 mg Ce at a switched-dc current of 800 mA for different frequencies. The spectroscopic measurements are done at a resonance absorption line at 577.36 nm.

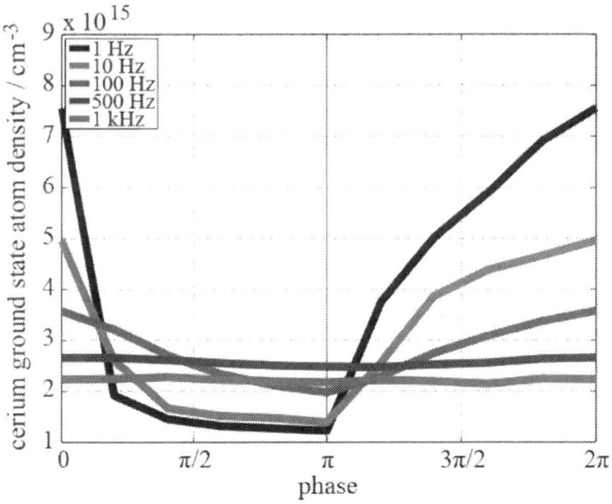

Figure 1: Phase resolved Ce ground state atom density in a pure Ce lamp (1 mg Ce) 125 μm in front of the electrode, i = 800 mA, switched-dc, f = 1, 10, 100, 500, 1000 Hz; first half period: anodic phase; second half period: cathodic phase

An increase of the atom density is found in the cathodic phase due to a rare earth ion current which is directed towards the cathode. At the cathode the ions are neutralized and an atom monolayer on the electrode surface occurs. This cathodic emitter effect reduces the work function and therefore the temperature of the electrode [4, 5, 6, 7]. A reduction of the atom density is found in the anodic phase due to a reversal of the ion current. The atom density is nearly constant over time for higher frequencies indicating that the ions are too immobile to follow the frequency above 500 Hz. They remain directly in front of the electrode, recombine and form a monolayer on the electrode surface even during the anodic phase. The emitter effect [8] remains active also within the anodic phase and because of this the anode temperature decreases with increasing frequency.

CONCLUSION

At low frequency an emitter effect occurs in the cathodic phase while at higher frequencies the emitter effect is extended to the anodic phase due to the inertia of ions of the respective emitter material. Therefore the work function and the temperature of the electrode are reduced and the lamp lifetime increases especially if operating frequencies in the order of some kHz are used.

REFERENCES

[1] M. Westermeier, *et al.*, J. Phys. D: Appl. Phys. **43** (2010), 124015.
[2] M. Westermeier, *et al.*, LS12 paper, "Phase resolved Dy density and plasma temperature measurements by absorption and emission spectroscopy of Dy spectral lines".
[3] J. Reinelt, *et al.*, J. Phys. D: Appl. Phys. **41** (2008), 144002.
[4] GMJF. Luijks, *et al.*, J. Phys. D: Appl. Phys. **38** (2005), 3163-3169.
[5] O. Langenscheidt, *et al.*, J. Phys. D: Appl. Phys. **41** (2008), 144005.
[6] M. Westermeier, *et al.*, IEEE Transactions on Plasma Science, Vol. **36**, No. 4 (2008), 1176-1177.
[7] GMJF. Luijks, *et al.*, J. Phys. D: Appl. Phys. **41** (2008), 144006.
[8] T. Hartmann, K. Günther, J. Mentel, J. Phys. D: Appl. Phys. **43** (2010), 025201.

This work was supported by Philips Lighting, the DFG (Graduiertenkolleg 1051) and the Ruhr-University Research School.

High efficiency T5 lamps with low mercury dose

[1]A.Corazza, [1]S.Giorgi, [2]J. Reichardt, [2]A.Hollstein, [2]C.Harzig
[1] *SAES Getters S.p.A., Viale Italia 77, Lainate (MI),20020 - Italy*
[2] *OSRAM GmbH, Berliner Allee 65, Augsburg , D-86136 - Germany*
Contact: alessio_corazza@saes-group.com

ABSTRACT

The paper presents the characteristics of LUMILUX T5 lamps produced with very low mercury doses thanks to the use of TiHg based dispensers able to introduce accurate mercury amounts in lamp.

INTRODUCTION

It is well known that the energy saving related to the use of Low Pressure Mercury Lamps instead of inefficient light sources provides environmental advantages that overcompensate the presence of a small amount of toxic element in these lamps. In principle, just few tens of micrograms of mercury should be sufficient to obtain the efficient radiation emission from fluorescent lamps; however several mechanisms of mercury consumption are present in the discharge tubes involving the need of higher quantities of that element to balance Hg depletion occurring during lamp operation. Main mechanisms of consumption are related to the interaction of mercury with glass, with phosphors, with emissive material from electrodes, with gas impurities (especially oxygen and oxygenated impurities inducing mercury oxidation) [1, 2].

Thanks to successful research efforts, specific counter-measures have been adopted in order to produce lamps with low mercury consumption:

- special coatings in order to minimize the interaction glass-mercury
- use of low consumption phosphors and mixtures of phosphors with protective materials
- reduction of sputtering phenomena from cathodes and use of metallic shields around the electrodes to intercept the sputtered emissive material (especially barium) that, otherwise, can form amalgam and trap mercury on the glass tube
- minimization of gaseous impurities in lamp as a consequence of improved processes and of the use of getters able to remove contaminants inside the lamps.

Nowadays the best in class Fluorescent Lamps need just a few milligrams of mercury and exhibit optimum performances both in terms of lumen output and operational lifetime. In the last years T8 tri-phosphor tubes resulted to be successful low mercury consumption and energy saving alternatives to the old T12 and T10 tubes.

Modern T5 lamps are particularly efficient and yield even greater energy saving than T8 lamps.

CHARACTERISTICS OF T5 LAMPS WITH LOW MERCURY DOSE

T5 lamps have been developed with a design suitable to enable slim luminaires and to assure a higher efficiency, also as a consequence of the optimization to maximize luminous flux at 34°-38°C, temperature usually achieved in compact luminaires. These 16mm lamps use less glass and phosphors than the 26 mm T8 lamps and, therefore, they need a lower mercury dose because Hg consumption is significantly lower.

OSRAM LUMILUX T5 lamps have very interesting performances with luminous efficacy up to 104 lm/W (at $T_{Optimum}$); the combination of special technologies and high quality three-band phosphors assures excellent lumen maintenance over lifetime, high color rendering and low mercury consumption.

In T5 Lumilux lamps main mechanisms of mercury depletion are minimized: in specific lifetime tests it has been measured that much less than 1,4 mg of Hg is consumed over 20000 hours of lamp operation.

Until some months ago OSRAM T5 lamps employed the low dose of 1,8 mg of mercury, precisely introduced in lamp through

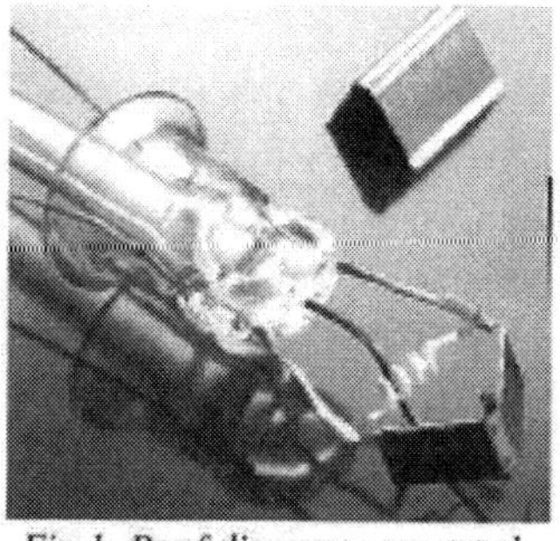

Fig.1: Roof dispenser mounted above the electrode

the SAES dispenser named "Roof". The "Roof" is an innovative Hg dispenser consisting of a metallic strip coated with a powder mixture able to release mercury and with a getter powder: it enables an accurate and reproducible dosing of small amounts of the metal in lamp. The dispenser has the shape of a "roof" and is mounted above the electrode (Fig.1). The TiHg compound present on the Roof is stable up to 500°C; it is mixed with a special copper based promoter so that, once the lamp is sealed, the dispenser can release all the contained mercury during a short activation at 850°-900°C, typically applied by means of RF induction heating.

A Zr-Al getter powder is usually combined with the Hg dispensing alloy: after the thermal activation, this getter alloy starts removing the gaseous impurities in the bulb and improves the lamp filling gas purity. It is effective in sorbing deleterious active gases such as hydrogen that it is known to increase lamp starting voltage and oxygenated gases which could lead to formation of mercury oxide.

As an example, some of the technical characteristics of a 54W LUMILUX T5 lamp are summarized in the following table:

	T5 HO 54W/840
Lamp Characteristics	With reference ballast 100 hours – 25°C
Actual Lamp wattage (W)	54
Luminous flux (lm)	4450
Luminous efficacy (lm/W)	82
Ignition voltage at RT (Vrms)	< 520
Hg content (mg)	1,8

Table 1: characteristics of 54W LUMILUX T5 lamp

FURTHER REDUCTION OF THE MERCURY CONTENT IN T5 LAMPS

Recently, as a result of specific development actions and production process improvements, a further step in the reduction of the mercury content has been done in LUMILUX T5 lamps. These lamps adopt the new roof dispenser with the very low dose of 1,4 mg of mercury. The low variability of the mercury amount released by the roof allows the use of a reduced dose of mercury without any problem for lamp operation and lifetime.

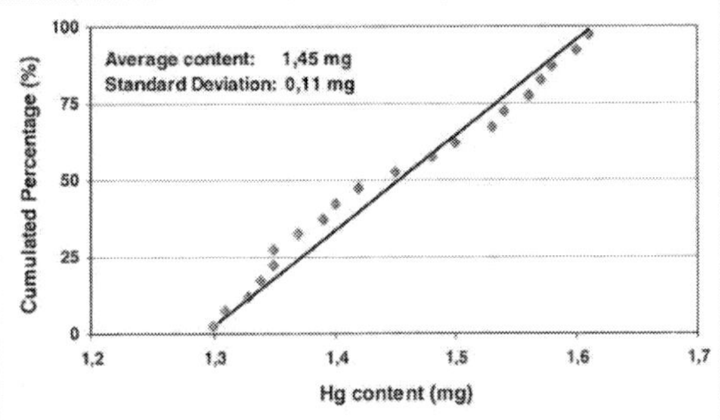

Fig.2: Probability plot for Hg content in1,4 mg roofs

	Hg content Measured in roof (mg)
Average content	*1,45*
St. Deviation	*0,11*

Table 2: Hg content in 1,4 mg roofs (determination by titration method)

	Hg amount measured in lamp (mg)
Average content	*1,44*
St. Deviation	*0,10*

Table 3: Hg amounts measured in T5 lamps after roof activation (cataphoresis technique)

CONCLUSIONS

LUMILUX T5 lamps are very efficient light sources with excellent performances and characteristics of very low mercury consumption. The use of "Roof" dispensers able to release an accurate and reliable dose of mercury assures the possibility to minimize the Hg amount in lamp without risks of lifetime problems. Last achievement was the development of T5 lamps with a content of 1,4 mg of Hg dosed by means of the new "Roofs".

REFERENCES

[1] M.Oomen, Proc. 8th Int. Symp. on the Science and Technology of Light Sources, LS8, (1998), 24.
[2] P.Bakk, N.Benko and L.Nyulaszi, J.Phys. D: Appl. Phys. **42** (2009) 1.

Resonance Radiation of Xenon Excimer Discharges

M. Paravia, J. Ensslin, R. Kling, W. Heering

Light Technology Institute, Karlsruhe Institute of Technology, Engesserstr. 13, 76131 Karlsruhe, Germany

Contact: mark.paravia@kit.edu

ABSTRACT

Xenon excimer discharges generate VUV radiation in the wavelength range from 147 nm to 200 nm. Usually the observed spectrum interferes with transmission edge of fused silica. To investigate Xenon resonance radiation a MgF_2 window is used as dielectric barrier. This work shows that even for a pressure of p_{Xe} = 400 mbar resonance radiation is 5 % of total radiation.

INTRODUCTION

Xenon excimer discharges based on dielectric barrier discharges (DBD) are promising candidates for efficient UV sources and mercury free light sources. The emitted radiation can be converted by means of phosphors into the desired wavelength range to use the excimer lamp as flat light source or e.g. for UV treatment of water. By relaxation from ($^3P_1$) into ground state ($^1S_0$) the xenon atom emits resonance radiation at λ = 146.961 nm. With increasing pressure, parts of vibrational energy can be dispensed and emission spectrum is dominated by first excimer continuum around λ = 151 nm and second excimer continuum with dominant wavelength of λ = 172 nm. Although resonance radiation can excite an internal phosphor coating, it can not be directly measured from outside if silica is used as barrier. For efficiency estimations and phosphor optimization the relative part of resonance radiation has to be known.

It is known, that the emission spectrum is independent from electrical excitation, frequency and power density but strongly pressure dependent [1, 2]. While below p_{Xe} = 40 mbar first of all resonance radiation is generated, above p_{Xe} > 100 mbar the second continuum dominates [3]. Resonance radiation for pressures above p_{Xe} > 70 mbar [3] or p_{Xe} > 100 mbar [4] has been not yet detected. This is only due to reabsorption in an unknown optical path of neutral xenon gas. This leads to underestimated resonance radiation. To overcome this we use a MgF_2 window as dielectric barrier and directly coupled into a VUV monochromator. By doing this, the measured emission spectrum is equal to the excitation spectrum for phosphor coated lamps.

EXPERIMENTAL SETUP

A McPherson VUV monochromator (234/302VM) and PMT are calibrated using a PTB traceable deuterium source. The spectral resolution is $\Delta\lambda$ = 0.8 nm. The entrance slit is connected with a grid electrode and nitrogen purged housing to the DBD. The DBD is permanently evacuated and flushed before measuring with Xenon 4.8. The pressure range was varied between 25 mbar < p_{Xe} < 500 mbar. The gas gap is d = 2.25 ± 0.25 mm. The DBD was operated using a resonant pulse ECG, as presented in [5]. Schematic drawing and electrical parameters are given in figure 1.

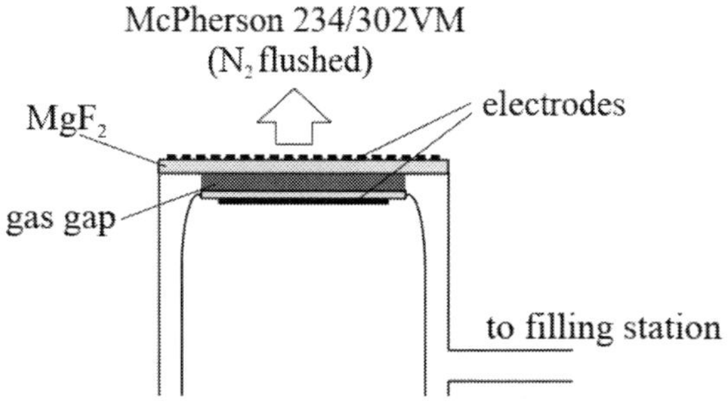

parameter	value
gas gap	d = 2.25 ± 0.25 mm
MgF_2 barrier	d_{MgF2} = 3 mm
lamp diameter	D = 40 mm
filling gas	Xenon 4.8
pressure	p_{Xe} = 25 – 500 mbar
operation	pulsed, t_{pulse} = 1 - 2 µs
frequency	f = 20 kHz - 70 kHz
lamp power	P = 0.4 W - 9.6 W

Figure 1 Schematic of experimental setup and electrical parameters

RESULTS

Figure 2 shows xenon excimer spectra for pressures between 25 mbar < p_{Xe} < 500 mbar. At a pressure of p_{Xe} = 25 mbar the spectrum is dominated by resonance radiation. Above p_{Xe} = 50 mbar the second continuum is in evidence. Starting from p_{Xe} > 100 mbar the second continuum dominates the spectrum. With increasing pressure the second continuum becomes narrower as the half-width declines slightly from 14.1 nm at p_{Xe} = 100 mbar to 13.0 nm at p_{Xe} = 500 mbar.

In contrast to [3, 4] a clear separation of the resonance radiation and the radiation of the first continuum for pressures up to p_{Xe} = 400 mbar is possible. This means that resonance radiation must be taken into account for phosphor development and plasma efficiency estimations.

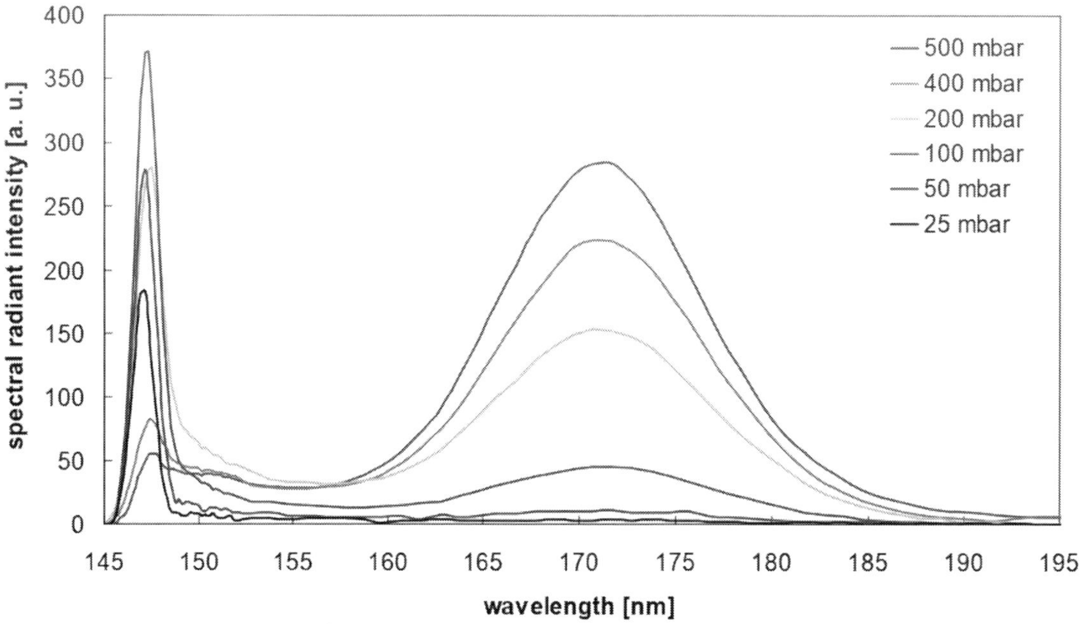

Figure 2 Spectral radiant intensity for Xenon pressure between 25 mbar < p_{Xe} < 500 mbar, d = *2.25± 0.25 mm*

The relative spectral distribution is split into distinct wavelength ranges and given in table 1. Here wavelength range from 145 nm to 149 nm is defined as resonance radiation, 149 nm < λ < 155 nm is defined as first continuum and 155 nm < λ < 197 as second continuum. Resonance radiation declines with increasing pressure from 70% to 3% of lamp total VUV output. Due to broadening of resonance line the first continuum only slightly decreases from ~ 8% to 4%. The second continuum dominates above p_{Xe} = 100 mbar with more than 50% of total VUV radiation and increases to 93% at highest pressure of p_{Xe} = 500 mbar.

Table 1 Relative spectral distribution vs. pressure

p_{Xe} (mbar)	25	50	100	150	200	300	400	500
resonance radiation (%)	70.1	61.8	41.3	27.7	16.6	9.0	4.9	2.7
first continuum (%)	7.7	8.1	8.6	8.9	7.8	4.8	5.0	3.9
second continuum (%)	22.2	30.1	50.1	63.5	75.6	86.2	90.2	93.4

REFERENCES

[1] Liu S and Neiger M, 2003, *J. Phys. D.,* **36** 1565-72

[2] Sewraj N, Merbahi N, et al., 2009, *J. Phys. D.,* **42** 045206

[3] Gellert B and Kogelschatz U, 1991, *Applied Physics B: Lasers and Optics,* **52** 14-21

[4] Jinno M, Motomura H, et al., 2005, *Journal of Light & Visual Environment,* **29** 91-98

[5] Paravia M, Trampert K E, and Heering W, ICOPS, Karlsruhe, Germany, 2008

Relationship between corrosion depth and coil dimension in the capillary part of mercury-free ceramic metal halide lamps

Yuichiro TAKAHARA, Takahito KASHIWAGI, Takuya HONMA, and Kozo UEMURA
Toshiba Lighting & Technology Corp., 1-201-1, Funakoshi-Cho, Yokosuka-shi, Kanagawa, 237-8510, Japan
Contact: yuuichiro.takahara@tlt.co.jp

ABSTRACT

We have studied the corrosion phenomena in the capillary part of mercury-free ceramic metal halide lamps, which have a higher fraction of thulium iodide. The corrosion depths at the inner wall of the capillary were found to depend on the structure of the electrode assembly in the capillary part, particularly the coil diameter. On the basis of the model calculation, it can be supposed that the corrosion depths depend on the temperature of the electrode assembly.

INTRODUCTION

We have developed mercury-free metal halide lamps, which improved the lamp voltage, efficacy, and color rendering index [1, 2]. These improvements have been accomplished by increasing the proportion of thulium halide or holmium halide in comparison with that of ordinary mercury-containing lamps. Meanwhile, the rare earth halide is one factor that promotes corrosion of the ceramic wall. The ceramic wall consisting of aluminium oxide is corroded by chemical transport, which is described in [3] by the following equilibrium:

$$Al_2O_3 + AlI_3 \leftrightarrow 3AlOI.$$

Aluminium iodide is produced by the reaction between aluminium oxide and the rare earth iodide. The presence of $Tm_3Al_5O_{12}$ as the molten corrosion product was confirmed by Raman spectroscopy, and the reaction is given as:

$$3TmI_3\ (l) + 4Al_2O_3\ (s) \leftrightarrow Tm_3Al_5O_{12}\ (s) + 3AlI_3\ (g).$$

The symbols g, l, and s denote gas, liquid, and solid phases, respectively.

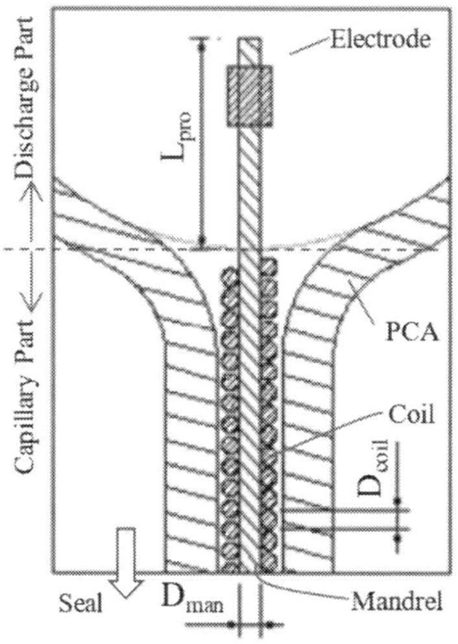

Figure 1. Structure of the capillary part of the test lamp.

RESULTS

Figure 1 shows the structure of the capillary part of the test lamp. D_{coil} and D_{man} are the coil and the mandrel diameter of the electrode assembly, respectively. L_{pro} is the length of the electrode assembly protruding from the end of the capillary. The discharge vessel of polycrystalline alumina (PCA) has an inner volume of 1.5 cm^3. The capillary has an inner diameter of 1.0 or 1.2 mm. The salt mixture consists mainly of TmI_3, NaI, and ZnI_2. The weight ratio of TmI_3 is 65% of the total weight of iodide except for the ZnI_2. The buffer gas is Xe. The lamps were vertically operated at 150 W.

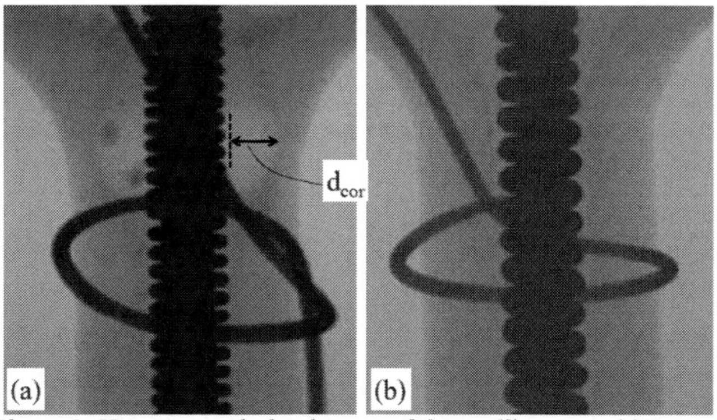

Figure 2. X-ray transmission images of the capillary part. (a)D_{coil} = 0.15 mm, L_{pro} = 4.2 mm, (b)D_{coil} = 0.30 mm, L_{pro} = 4.2 mm.

Figure 2 shows X-ray transmission images of the lamps observed after 1000 hours of operation. The inner wall of the capillary of the lamp (a) was corroded extensively. The maximum corrosion depth in each of the lamps (d_{cor}) was measured. The relationships between d_{cor} and D_{coil}, L_{pro} were investigated by using multiple regression analysis. Statistical significance was obtained when p-value < 0.05. Figure 3 shows the relationship between d_{cor} and each of D_{coil} and L_{pro}.

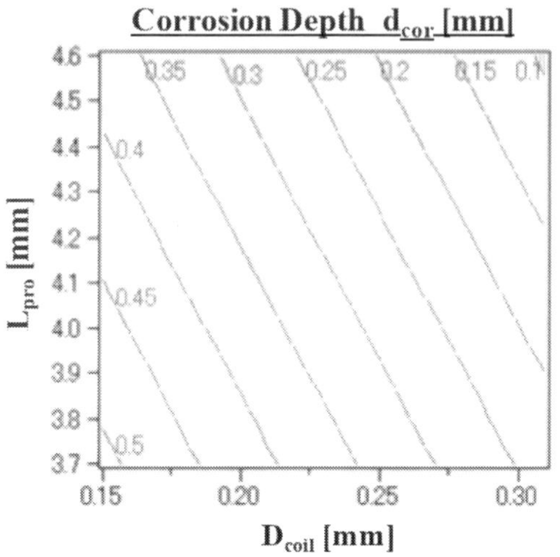

Figure 3. Contour plot of the corrosion depths d_{cor} as a function of D_{coil} and L_{pro}.

DISCUSSION

The temperature distribution of the capillary part was investigated using model calculations. In the heat transfer model, heat flows through the electrode assembly and the PCA capillary. Heat flows between them by conductive transfer through Xe gas and by radiative transfer. The heat sources are set at the electrode tip and the discharge side of the PCA capillary.

Figure 4 shows the calculated temperatures at the electrode assembly surface (i.e., the coil surface) and the PCA capillary surface, and the experimental temperatures at the PCA capillary surface. The ordinate means the temperature at a position 2 mm from the junction of the discharge part of the vessel with the capillary part. This corresponded to the position that was corroded in the experiment. In the calculated results, the temperature of the PCA capillary surface is nearly independent of D_{coil}. The temperature of the electrode assembly surface decreases with increasing coil diameter. In figure 3, the corrosion depths decrease with increasing coil diameter. It can be supposed that the corrosion depths depend on the temperature of the electrode assembly rather than that of the PCA capillary.

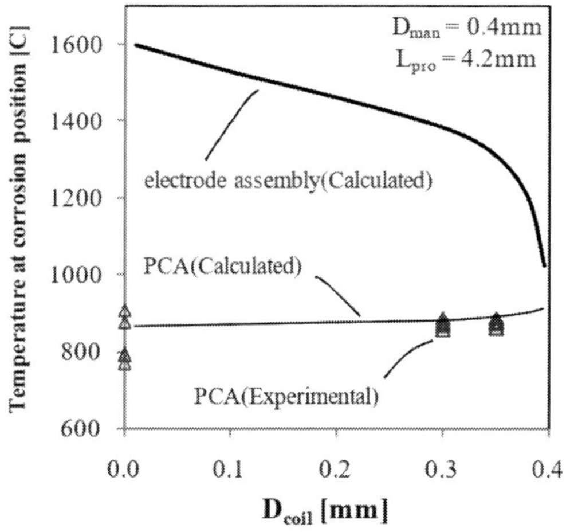

Figure 4. Temperature at the corrosion position versus the coil diameter D_{coil}. The continuous lines indicate the calculated data of the PCA capillary surface and that of the electrode assembly surface (i.e., the coil surface). The triangle marks indicate the experimental data of the PCA capillary surface. The inner diameter of the PCA capillary is 1.2 mm. The diameter of the electrode assembly becomes equal to the inner diameter of the PCA capillary at $D_{coil} = 0.4$ mm.

CONCLUSIONS

The corrosion depths were found to depend on the coil diameter D_{coil} and the protrusion length L_{pro}. On the basis of the model calculation, it can be supposed that the corrosion depths depend on the temperature of the electrode assembly in the capillary.

REFERENCES

[1] European Patent Application 2091304 A1

[2] PCT International Application WO 2009119612

[3] S. Fischer, U. Niemann and T. Markus, "Transport phenomena of aluminium oxide in metal halide lamps", J. Phys. D: Appl. Phys., 41(2008)144015

Reduction of defects in GaN by using InN islands

Sang-Jun Lee[1,2,3], Chu-Young Cho[1,2,3], Sang-Heon Han[1,4] and Seong-Ju Park[1,2,3]
[1] *Department of Materials Science and Engineering, Gwangju Institute of Science and Technology, Gwangju 500-712, Republic of Korea*
[2] *Department of Nanobio Materials and Electronics, Gwangju Institute of Science and Technology, Gwangju 500-712, Republic of Korea*
[3] *Center for Extreme Light Applications(CELA-NCRC), Gwangju Institute of Science and Technology, Gwangju 500-712, Republic of Korea*
[4] *Samsung LED Co. Ltd., Suwon 443-743, Republic of Korea*
Contact: ceag100@gist.ac.kr

ABSTRACT

We investigated the effect of InN islands on the crystal quality of GaN grown by metalorganic chemical vapor deposition (MOCVD). The two-step growth at low (630 °C) and high (1020 °C) temperature was applied to grow GaN on InN islands. The photoluminescence (PL) intensity of GaN grown on InN islands was increased by 25% compared to that of GaN without InN islands. This enhancement of optical properties was attributed to the improved crystal quality of GaN grown on InN islands which block threading dislocation.

MAIN TEXT

The GaN-based light-emitting diodes (LEDs) are epitaxially grown on sapphire substrates. However, the as grown GaN epitaxial layer has high threading dislocation (TD) density typically in the range of 10^{8-10} /cm^2 due to the large mismatches in lattice constants and thermal expansion coefficients between GaN and sapphire. These defects are nonradiative recombination centers and they diminish the LED performance. Therefore, the reduction of TD is of great importance for the development of highly efficient LEDs. Several epitaxial growth methods such as epitaxial lateral overgrowth (ELO), pendeo epitaxy and SiN_x masking have been proposed to improve the crystal quality [1-3]. Also, it was reported that InN islands tend to nucleate on the sites where TD intersect the GaN surface [4]. In this study, we demonstrate the improved crystal quality of GaN by using InN islands as defect blocking layer.

The experiments were performed in MOCVD using trimethylgallium (TMGa), trimethylindium (TMIn) and NH_3 as source materials. C-plane (0001) sapphire was employed as a substrate. Before epitaxial growth, a sapphire substrate was cleaned under H_2 at 1030 °C. After the growth of a 25 nm-thick low-temperature GaN buffer layer at 550 °C, a 2 μm-thick undoped GaN layer was grown at 1020 °C. Then, InN islands were deposited on it at the growth temperature varied between 550 and 600 °C. The growth parameters such as growth temperature, pressure and flow rate of MO source are changed to control the density and size of InN islands. In order to embed InN islands in GaN layer, two-step growth process was applied at low (630 °C) and high (1020 °C) temperature.

Figure 1 shows the schematic diagram of GaN growth with InN islands and SEM images of InN islands. The size of InN islands is the range of 100 ~ 200 nm. The density of InN islands is controlled from ~10^7 to ~10^8 with various growth parameters.

Figure 2 (a) is the photoluminescence (PL) intensity of overgrown GaN as a function of density of InN islands. As shown in Fig. 2 (a), the PL intensity of GaN on InN islands was higher than that of GaN without InN islands (reference). However, the PL intensity of GaN embedded InN islands with density of 3×10^7 was decreased compared to that of reference. This means the TD cannot be efficiently blocked due to the low density of InN islands. Figure 2 (b) shows the enhancement ratio of PL intensity of GaN grown on InN islands to that of GaN without InN islands. As shown in Fig. 2 (b), PL intensity of GaN that has InN islands with density of 2×10^8 and 5×10^8 was higher by 20% and 25% than that of reference, respectively. This is attributed to the improved crystal quality of GaN grown on InN islands which block TD.

In summary, we investigated the effect of InN islands on optical properties of GaN. The PL intensity of GaN that has InN islands with density of 2×10^8 and 5×10^8 was higher by 20% and 25%

than that of reference, respectively. This indicated that crystal quality of GaN was improved due to the high density of InN islands which block the propagation of TD in GaN.

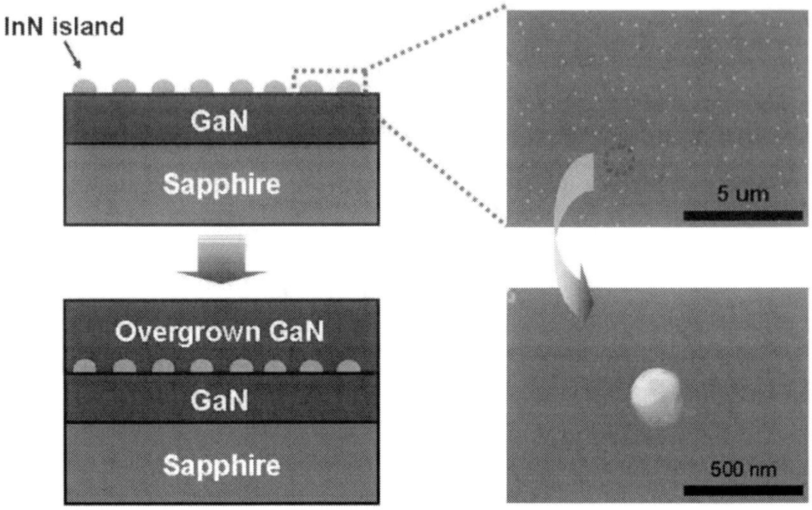

Figure 1. Schematic diagram of GaN growth with InN islands and SEM images of uncapped InN islands

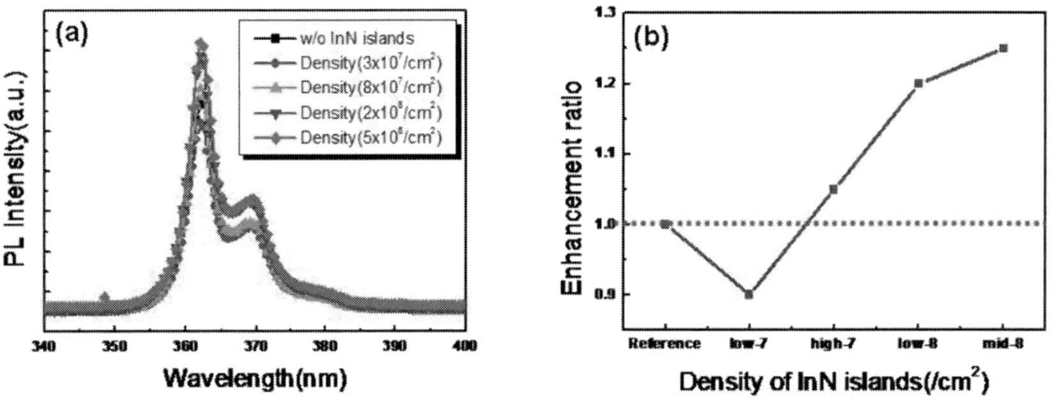

Figure 2. (a) PL intensity of overgrown GaN as a function of density of InN islands and (b) Enhancement ratio of PL intensity of GaN with InN islands based on that of GaN without InN islands

REFERENCES

[1] O. H. Nam, M. D. Bremser, T. S. Zheleva, and R. F. Davis, Appl. Phys. Lett. **71** (1997) 2638
[2] K. Linthicum, T. Gehrke, D. Thomson, E. Carlson, P. Rajagopal, T. Smith, D. Batchelor, and R. Davis, Appl. Phys. Lett. **75** (1999) 196
[3] A. Chakraborty, K. C. Kim, F. Wu, J. S. Speck, S. P. DenBaars, and U. K. Mishra, Appl. Phys. Lett. **89** (2006) 041903
[4] J. G. Lozano, A. M. Sanchez, R. Garcia, D. Gonzalez, D. Araujo, S. Ruffenach, and O. Briot, Appl. Phys. Lett. **87** (2005) 263104.

Novel Long-life Xenon Short-arc Lamp with AC-operation

M. Morkel , U. Hartwig , H. Rehn
OSRAM GmbH, Nonnendammallee 44, 13625 Berlin, Germany
Contact: m.morkel@osram.com

GENERAL

We introduce a novel powerful xenon short-arc light source, which shows excellent life-time behavior for more than 1000 h due to the implementation of a halogen cycle and a specially adapted AC driver operation mode. A CRI greater than 95, day light color temperature and a high lumen output for small apertures are achieved.

LAMP DESIGN

The current design is optimized for 75 W operation. The lamp consists of pure tungsten electrodes with an electrode gap of about 1 mm sealed in a vessel of quartz glass. The Xe filling pressure is around 20-25 bars. Due to a special filling and sealing technique a tip-less design could be established. Hence distortions in the optical characteristics are avoided, which is of particular importance for low wattage lamps with small bulb diameters.

The special AC driver operation mode leads to liquid electrode tips during operation. This would usually result in strong bulb blackening within minutes or a few hours. Therefore a halogen cycle was established by adding several hundred ppms of a halogen to the xenon. Tungsten deposition on the bulb wall is then effectively prevented as depicted in fig. 1. Moreover, the AC operation mode leads to the regeneration of the molten tips in the arc attachment area. The light spot does no longer move out of the ideal reflector position as observed with DC driven Xe short-arc lamps due to electrode burn-back, resulting in an excellent life-time behavior.

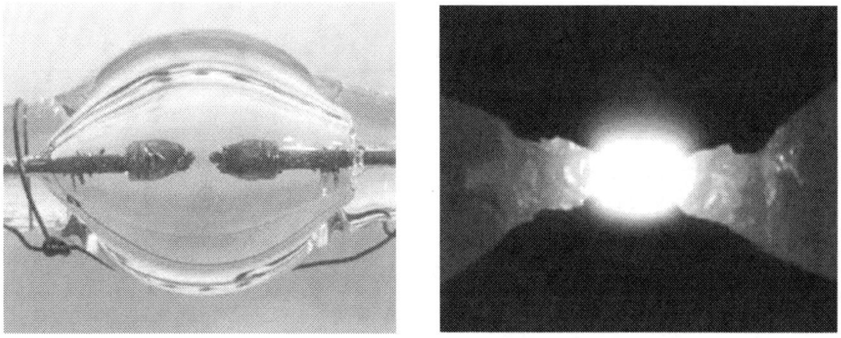

Fig. 1: left: test burner with a life time of 1960 h, right: tip development during operation

Light output fluctuations are not perceived because the operational frequencies are higher than 100 Hz and the stability is additionally increased by the special AC driving mode and the liquid tips during operation.

Finally, no elements of environmental relevance such as mercury or rare earth metals are used in the current design. Since the electrodes do not contain any emitter material, the lamp is also free of radioactivity.

INITIAL DATA

The spectrum shows the typical distribution for xenon light (fig 2.), therefore a very high CRI of more than 95 can be achieved. The infra-red radiation is suppressed by the dichroic reflector coating. As can be seen, the light can be focused on very small aperture spots. The reflector is optimized to take advantage of nowadays used fibre optics with a numerical aperture of greater than 0.6.

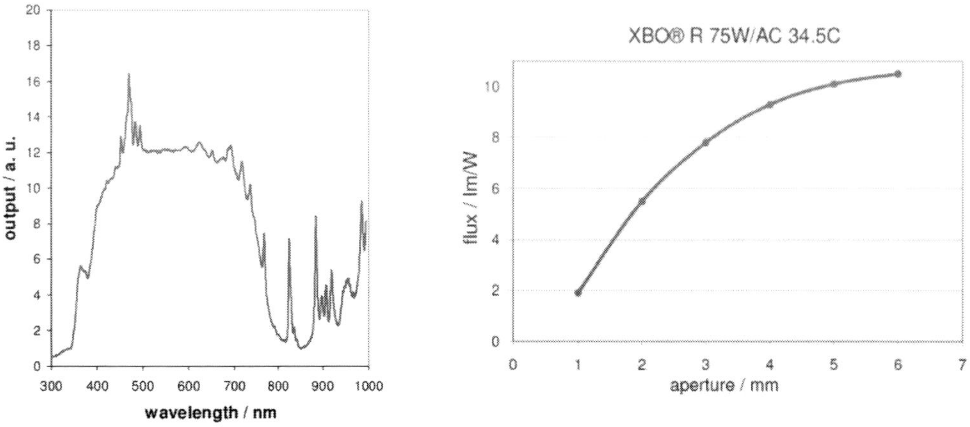

Fig. 2: left: typical spectrum for the reflector lamp, right: light output for different apertures

LIFE-TIME DATA

The life-time behavior of the lamps is excellent. In fig. 3 degradation curves are displayed for the reflector lamp. The maintenance corresponds to the initial lumen output in a 5 mm aperture. The lamps were operated in a typical operational cycle (1 h on / 15 min off). In addition, several test lamps were operated for more than 2000 h with only minor degradation.

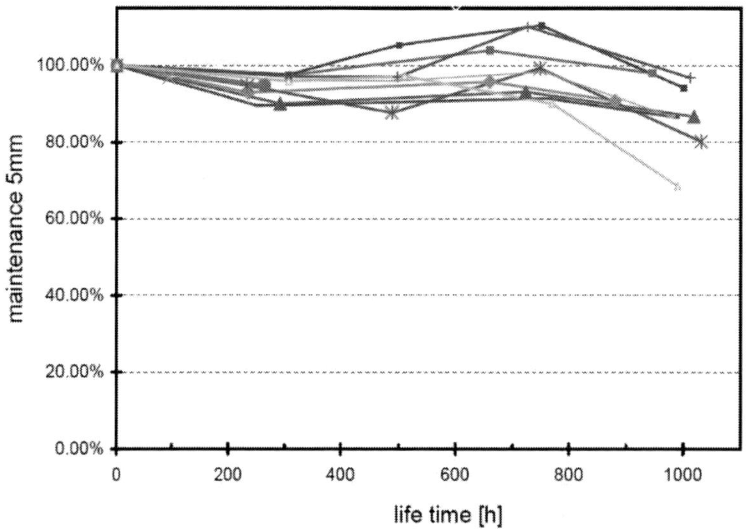

Fig. 3: XBO® R 75W/AC 34.5C: life-time behavior

Quantifying the environmental benefits of new energy efficient lighsources

S.A.Mucklejohn[1]

[1] *School of Architecture, Design & Built Environment, Nottingham Trent University, Nottingham NG1 4BU UK*
Contact: stuart.mucklejohn@ceravision.com

ABSTRACT

The contribution of energy efficient lightsource technology to energy saving, greenhouse gas emissions and sustainable development is illustrated by considering the impact of the introduction of metal halide discharge lamps with ceramic arctubes.

INTRODUCTION

Although there are many studies describing the energy savings of new lighting systems and installations there are few reports that quantify the contributions of new lightsources to energy saving and sustainable development by generic type. The 1987 Bruntland Commission (the World Commission on Environment and Development) defined sustainable development as 'development that meets the needs of the present without compromising the ability of future generations to meet their own needs'.

In this work the example of metal halide discharge lamps with ceramic arctubes is used to illustrate the contribution of recently introduced lightsources to energy savings, reduction of greenhouse gas emissions and sustainable development. The methods outlined here can also be applied to other newly introduced lightsources such as LEDs, OLEDs and high efficiency plasmas.

A preliminary account of this work has been presented previously as part of a review of the use of molten salts in high intensity discharge lamps [1].

Although the attractions of metal halide discharge lamps with ceramic envelopes (cmh lamps) have been widely recognised by the lightsources industry, lighting designers and lighting engineers, the contributions of these products to sustainable development have been largely overlooked. By comparing lightsources used in various lighting schemes, with and without cmh lamps, energy savings and reductions in greenhouse gas emissions due to the development of the cmh lamps can be quantified. The power range of cmh lamps now covers 20 to 400 W and several different scenarios have to be considered when comparing lighting schemes.

The model is based on the effects of substituting cmh lamps for less efficient sources that would have been used in the absence of cmh technology. For example, 20 W cmh lamps are considered to have replaced halogen lamps giving the same output while 70 W cmh lamps are considered to have replaced a combination of metal halide discharge lamps with silica arctubes, halogen and incandescent sources.

For lamp powers up to 150 W cmh lamp technology quickly became well established and widely used in the lighting market due to its efficient performance and high reliability. However, powers above 150 W have proved to be much more demanding for cmh lightsource developers and manufacturers with the consequence that to date a rather limited variety of products at 250 and 400 W are available. In this work, 250 and 400 W cmh lamps are considered to have replaced metal halide discharge lamps with silica arctubes of the same power.

Figure 1 shows the estimated annual reductions in CO_2 emissions due to the introduction of cmh lamp technology. These values are based on an estimate of the number of cmh lamps in use, the lightsources they have replaced and data for the emissions of greenhouse gases per kWh of electricity generated in a range of European countries.

It is important not to overlook the reduction in load requirements for the electrical supply network as a consequence of energy efficient lighsources. However, these reductions have to be considered on a local basis rather than an international basis because of the nature of power generation plants.

The use of highly efficient light sources with good colour rendering properties allows cities, for example, to improve the visual appearance of the night time environment by the introduction of additional lighting schemes such as highlighting architectural features while not placing undue burdens on power requirements and greenhouse gas emissions nor incurring excessive operating costs. An excellent summary of public lighting in cities is provided by Raynham [3]. This combination of improving the quality of life while managing the impact on the environment fits in perfectly with the vision and ideals of the European Commission.

The methods outlined here help to illustrate the impact that the lightsources and lighting community has in reducing energy consumption whilst maintaining or improving the quality and effectiveness of lighting installations.

Figure 1. Estimated annual reductions in CO_2 emissions due to the introduction of cmh lamp technology. CO_2 emissions are based on values contained in the Eco-Invent database for the generation of electricity in Europe.

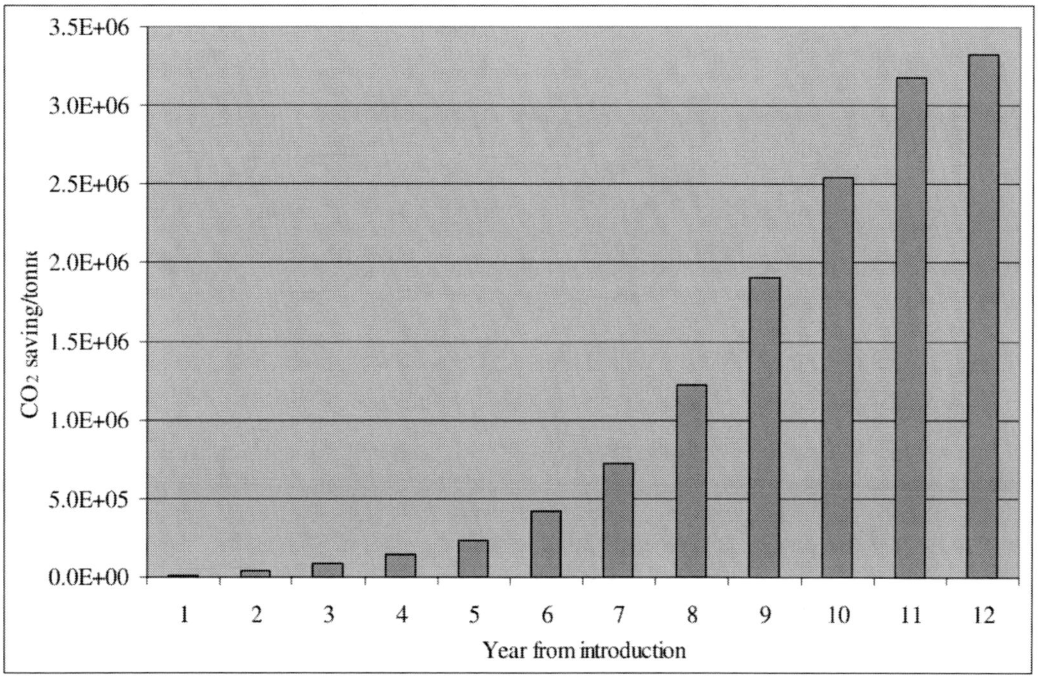

REFERENCES

[1] S.A.Mucklejohn, *Molten Salts Discussion Group Christmas Meeting*, London, Dec 2007.

[2] See: http://www.ecoinvent.org/

[3] P.Raynham, *4th Int. Conf. ILUMINAT*, Cluj-Napoca, Romania (2007).

Light Extraction Enhancement of GaN-based LEDs

Ching-Cherng Sun, Tsung-Xian Lee and Shang-Yu Tsai, Yi-Chien Lo, Che-Chu Lin and Wei-Ting Chien

Institute of Lighting and Display Science, Department of Optics and Photonics, National Central University, Chung-Li 320, TAIWAN
Contact: ccsun@dop.ncu.edu.tw

ABSTRACT

In this paper, we present a study of light extraction of GaN-based LEDs, including sapphire-based and Thin-GaN through Monte Carlo ray tracing. The mechanism in enhancing light extraction incorporated with implanting micro pyramid (lens) array and lens encapsulation through photon recycling is introduced and discussed.

INTRODUCTION

Light emitting diodes (LEDs) have been regarded as the most promising solid-state light source in next-generation lighting. One of the obstacles to spread LEDs into practical application is the cost issue. It arises from the limited lumen per watt which performed by a white LED operated in a high-power condition. To solve the problem, one of the approaches is to increase the emitting efficiency or the external quantum efficiency (EQE) as large as possible. The EQE is a measurable term which is defined as the ratio of the photon number to the number of injection electron-hole pair. EQE is a factor results from multiplication of internal quantum efficiency (IQE) and light extraction efficiency (LEE). Both IQE and LEE are difficult to measure. Fortunately, LEE can be effectively estimated by using Monte Carlo ray tracing to calculate the ratio of lights escaping from the LED die if the fine structure is much larger than wavelength of lights emitted by the LED.[1-7] There are various studies addressing how to increase the LEE through implanting specific micro structures, photonic crystal (diffraction structures) in an LED die or through specific packages. In this paper, we will present our study in LEE for GaN-based LEDs, where LEE enhanced by both cavity and quantum photon recycle will be introduced.

PHOTON RECYCLING THROUGH ABSORPTION AND REEMISSION

Light extraction could be enhanced through photon recycling with quantum effect. The mechanism of photon recycling for light extraction enhancement in LED dice can be described that the reflected photons could be absorbed by the active layer and partially be re-emitted so that the new emission photons may escape the LED die. Here we define a factor of photon recycling rate (PRR) to describe the re-emission rate of an absorbed photon. The final LEE through quantum photon recycling can be written

$$\eta_{LEE}=\eta_{LEE0}+\eta_{LEE1}+\eta_{LEE2}+\ldots=\eta_{LEE0}+\eta_{LEE0}(PRR\times R_{AP1})+\eta_{LEE0}(PRR\times R_{AP2})+\ldots, \quad (1)$$

where η_{LEEi} is the LEE for the i times recycling, and R_{APi} describes the ratio of absorbed photons to all the original photons for the i times recycling, and i is 1, 2, 3,…. From the equation, it is easy to understand that photon recycling enhances light extraction more dominant in an LED die with larger absorption coefficient in the active layer and with larger PRR. The simulations of LEE are based on four structures shown in Ref. 7, including sapphire-based LEDs with or without patterned substrate, and Thin-GaN with or without surface texture. The LEE with photon recycling at different absorption coefficients, eg., 200 cm^{-1} and 10^4 cm^{-1}, and in different package types are shown in Fig. 1, where the micro pyramid array with 30° slanted surfaces is implanted in each LED. The simulation shown in Fig. 1 also includes the lens encapsulation (LEC) effect. The simulation result can be described as follows. The first is that photon recycling is much helpful in the cases of higher absorption in the active layer. In the case of 200 cm^{-1}, the value of LEE reaches its maximum for 5 or less runs of quantum photon recycling. It is because that the active layer has less ability to capture the photons. In such a condition,

LEE enhancement structure such as micro pyramid (lens) array and the approach of LEC could be the most effective ways to increase LEE. The second is that the condition is different in the case of 10^4 cm^{-1}. The quantum photon recycling may play a major rule in LEE. In the condition of heavy absorbed active layer and the LED is without micro structure and LEC, the only way to increase LEE is photon recycling. The enhancement ratio of LEE could reach 200% for the PRR of 90%. The last is that in the cases of equipped with micro structure and/or LEC, quantum photon recycling makes less effect. It is obvious to find in Fig. 1 where the effect by photon recycling for the cases of lens encapsulation is not as dominant as those for bare chip.

SUMMARY

In this paper, through simulation of Monte Carlo ray tracing, we discuss and study different mechanisms for the enhancement of LEE in GaN-based LEDs, with photon recycling through absorption and re-emission in the active layer. In physics insight, an alternative way to extract photons outside is through quantum photon recycling. It could be possible, if there exits a condition of high PRR, e.g. 90 %, the LEE may be enhanced to 200% though there is without implanting micro structure array and lens encapsulation. Such condition needs further study since it is not easy to measure the PRR in a real LED die since photon recycling may couple other effects such as internal quantum efficiency and LED structure for light extraction.

Acknowledgment
This study is sponsored by the National Science Council of the Republic of China with the contract no. 97-2221-E-008-025-MY3. The authors would like to thank Breault Research Organization for the support of ASAP simulation program.

REFERENCE

[1] W. B. Joyce, R. Z. Bachrach, R. W. Dixon, and D. A. Sealer, J. Appl. Phys. 45, 2229 (1974).
[2] D. Z. Ting, and T. C. McGill, Opt. Eng. 34, 3545 (1995).
[3] S. J. Lee, Jpn. J. Appl. Phys. 37, 5990 (1998).
[4] S. J. Lee, Appl. Opt. 40, 1427 (2001).
[5] A. Badano, and J. Kanicki, J. Appl. Phys. 90, 1827 (2001).
[6] T. X. Lee, C. Y. Lin, S. H. Ma and C. C. Sun, Opt. Exp. 13, 4175 (2005).
[7] T. X. Lee, K. F. Gao, W. T. Chien, and C. C. Sun, Opt. Exp.15, 6670 (2007).

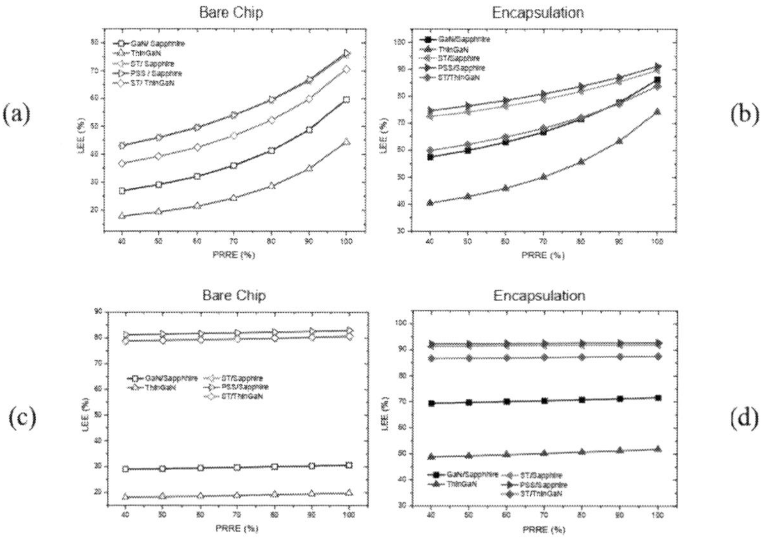

Fig. 1 LEE vs. PRR of quantum photon recycling, (a) without LEC (10^4 cm^{-1}), (b) with LEC (10^4 cm^{-1}), (c) without LEC (200 cm^{-1}), (b) with LEC (200 cm^{-1}).

Spatial resolution of the light-emission profile in single-layer organic light-emitting diodes

M. Carvelli[1,2,3], R.A.J. Janssen[2], R. Coehoorn[1,2]

[1] *Philips Research Eindhoven, High Tech Campus 4, 5656 AE, Eindhoven, The Netherlands*

[2] *Molecular Materials and Nanosystems, Department of Applied Physics, Eindhoven University of Technology, P.O. Box 513, 5600 MB Eindhoven, The Netherlands*

[3] *Dutch Polymer Institute (DPI), P.O. Box 902, 5600 AX Eindhoven, The Netherlands*

Contact: marco.carvelli@philips.com

ABSTRACT

In this paper, we investigate the precision to which the position of a sheet of excitons can be deduced from the analysis of the angular and polarization dependent emission spectra in organic light-emitting diodes (OLEDs).

INTRODUCTION

In the past five years, OLEDs have emerged as a promising option for energy-efficient solid-state lighting and for cheap light sources produced on flexible foils. The highest luminous efficacies, of more than 100 lm/W to date when using macroextractors for enhancing the light-outcoupling, are obtained using multilayer structures of evaporation-deposited small-molecule organic semiconductors. Flexible OLEDs are usually based on a single active layer deposited by spin-coating or ink-jet printing. Within both technologies, a key factor that needs to be controlled is the shape of the emission profile, i.e. the spatial distribution of the emitting excitons across the active layer thickness.

In single-layer OLEDs, the shape of the emission profile depends on the injection conditions and on a delicate balance between the electron and hole mobilities. A comprehensive novel method to determine the shape of the emission profile was recently published by our group [1]. Crucial elements of the method are the use of a combined classical and quantum-mechanical microcavity model for properly treating the radiative decay probability, the inclusion of optical absorption in the emitting layer ("selfabsorption") and of optical anisotropy, a self-consistent determination of the intrinsic electroluminescence (EL) spectrum and the use of the full angular dependence of the emission. This last feature is of essential importance for determining the emission profile with high resolution.

In this paper, we investigate the accuracy with which this method may determine the emission profile in OLEDs by determining the extent to which it would correctly provide the position of a infinitesimally thin plane of emitting molecules located at a varying distance (d) from the cathode.

SYSTEM STUDIED

The layer stack of the device for which the emission is modelled is: glass/ ITO (120 nm)/ PEDOT:PSS (100 nm)/ Light Emitting Polymer (160nm)/ barium (5 nm)/ aluminum (100 nm), where the light-emitting polymer is a fluorene-(7.5 mol-% triarylamine) co-polymer (PF-TAA). This material been used as the blue has emitter in the 13" full-color OLED-TV display demonstrated by Philips in 2005. The device is identical to one of the devices studied in ref. 1.

RESULTS

The emission spectra generated by a sheet of excitons (modeled as oscillating dipoles), located at a distance d from the Ba/Al interface, are simulated for a wide range of emission angles (from 0° to 70° in steps of 2°) and for both s- and p-polarized light. The analyzed wavelength range is 450-600 nm, resolved in 5 nm steps. In the analysis equal weight is given to each wavelength, in a similar way as described in ref. 1. An experimentally realistic 5%-Gaussian noise is added to the simulated spectra, for 100 different cases. The emission profiles determined are characterized by three parameters, as described in ref. 1: the peak position, the width and the asymmetry. A schematic example of a parameterized emission profile which represents the possible end result of the analysis for one specific "theoretical"set of noisy emission intensities is presented in Figure 1(a), together with the delta-function type starting profile (dashed line) which would be obtained if no noise would be added.

Actually, the figure greatly exaggerates the peak shift Δ obtained typically from the analyses, which does not exceed values of the order 1 nm. Furthermore, the peak widths and the asymmetry of the profiles obtained are even almost negligible. The emission profiles obtained are thus very close to delta-functions, shifted by a distance Δ. Figure 1(b) shows the dependence on d of the average value of Δ (closed spheres), obtained from an analysis of 100 simulated angle- and polarization-dependent spectra for each value of d. The figure also gives the standard deviation of Δ. Smaller values of Δ and of its standard deviation correspond to a higher spatial resolution.

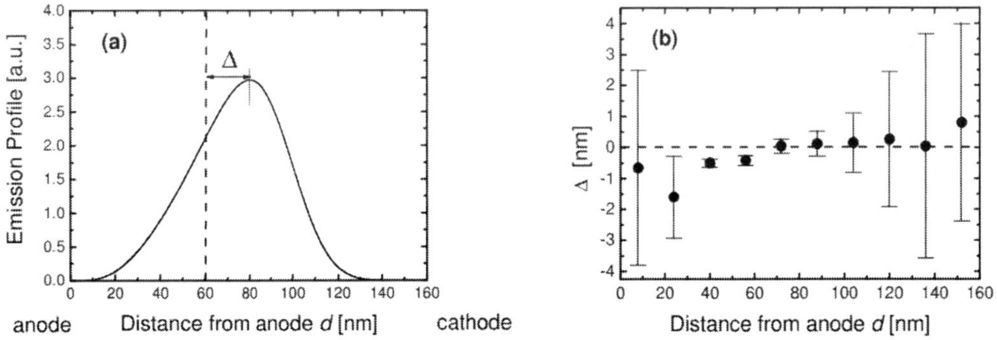

Fig.1: (a) General shape of the emission profile (parameterized by the peak position, width and asymmetry [1]) as obtained from theoretical emession spectra to which noise has been added, compared to the original delta-funtion type emission profile, located at a distance d *from the anode. (b) Distance Δ of the determined profile peak from the position* d, *as a function of* d. *The closed spheres indicate the mean value of Δ, while the vertical bars refer the standard deviation of Δ.*

The figure shows that the accuracy with which the position of an emitting sheet of excitons may be determined is highest in the central region of the emitting layer, while the uncertainty increases close to the electrodes. This may be explained by noting that with decreasing distance to one of the metallic electrodes the probability of non-radiative decay due to exciton quenching increases. As a result, the emission intensity decreases, leading to an increased uncertainty in the precise position of the plane of emitting dipoles.

CONCLUSIONS

We have investigated the spatial resolution of a recently developed method to determine the light emission profile in OLEDs, by analyzing the emission profiles as obtained for the case of a narrow (delta-like) emitting zone. The resolution is found to be on the order of 1 nm, provided that the electron-hole recombination is not too close to one of the electrodes. We expect that the method will provide the position of the emission zone with a similar (~1 nm) resolution for the case of state-of-the-art small-molecule based white emitting devices, such as discussed in ref. 2, where the electron-hole recombination region is confined to a central emissive layer far from the electrodes.

REFERENCES

[1] S.L.M. van Mensfoort *et al.*, Nature Photonics (published online, March 2010), DOI 10.1038/NPHOTON.2010.32.
[2] S. Reineke *et al.* Nature 459 (2009) 234-239.

Enhanced performance of a vacuum-ultraviolet light source based on Kr_2^* (λ=146nm) and Ar_2^* (λ=126nm) excimers using short-pulse excitation of a windowless dielectric barrier discharge

R.J. Carman and D.M. Kane
Department of Physics, Macquarie University, North Ryde, Sydney, NSW 2109, Australia
Contact: rcarman@science.mq.edu.au

ABSTRACT

The operating characteristics of a VUV light source based on Kr_2^* and Ar_2^* excimer emission from a windowless dielectric barrier discharge plasma have been investigated in detail for both short-pulse and sinusoidal voltage excitation methods.

EXPERIMENTAL

Incoherent light sources emitting in the vacuum ultraviolet (VUV) (λ<200nm) spectral regions are of topical interest for use in emerging applications including materials processing/deposition & nanofabrication [1] and surface modification & cleaning [2]. Dielectric barrier discharge (DBD) excitation of the rare gases provides a practical and efficient method for generating pulsed incoherent VUV radiation from excited diatomic molecules (excimers) at selected wavelengths λ=84nm (Ne_2^*), λ=126nm (Ar_2^*), λ=146nm (Kr_2^*), and λ=172nm (Xe_2^*). High-voltage sinusoidal, square-wave, or short-pulse waveforms are generally employed to drive DBDs at kHz repetition frequencies. Previously, a number of experimental/theoretical studies on Xenon DBDs [3][4] have established the critical importance of the shape of the applied voltage waveform for optimising the source performance, especially in relation to the electrical to VUV conversion efficiency. To date, analogous studies using the four other rare gases in DBD sources are relatively scarce [5]. In this paper, we investigate the electrical and optical characteristics of a DBD based VUV source with Ar, and Kr gas fills. Since part the emission wavelength of the Ar_2^* excimer is below the cut-off for transmission through standard VUV compatible optics such as MgF_2 (120nm), a windowless DBD geometry has been employed as shown schematically in fig.1(a). The barrier discharge electrodes were fabricated from two blind quartz tubes, each tube encased a water-cooled stainless-steel tube electrode. High-voltage (4-8$kV_{pk\text{-}pk}$) bi-polar pulse waveforms with ~200ns risetimes (300ns FWHM) were applied to

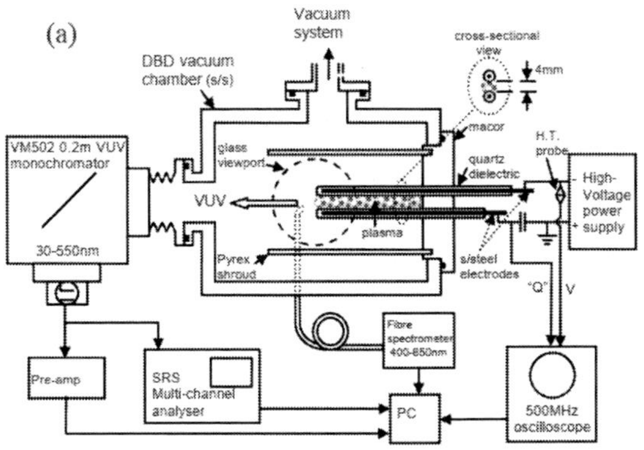

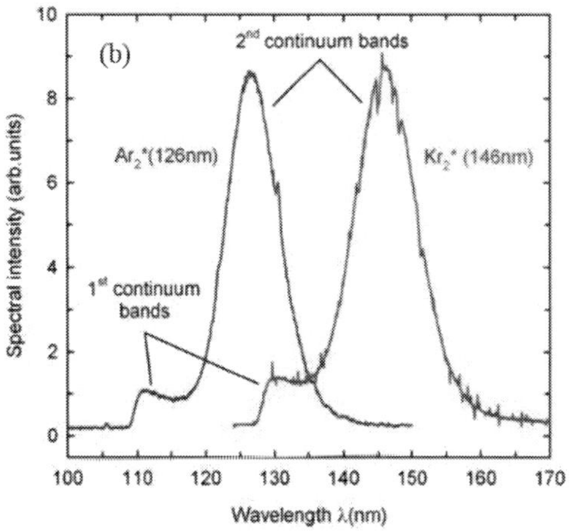

Fig. 1: (a) Schematic diagram of the DBD vacuum chamber containing the windowless VUV source, and associated equipment for electrical and optical diagnostics of the plasma; (b) VUV spectral output of the source using short-pulse voltage excitation (pulse rate = 32kHz, gas pressure = 800mbar).

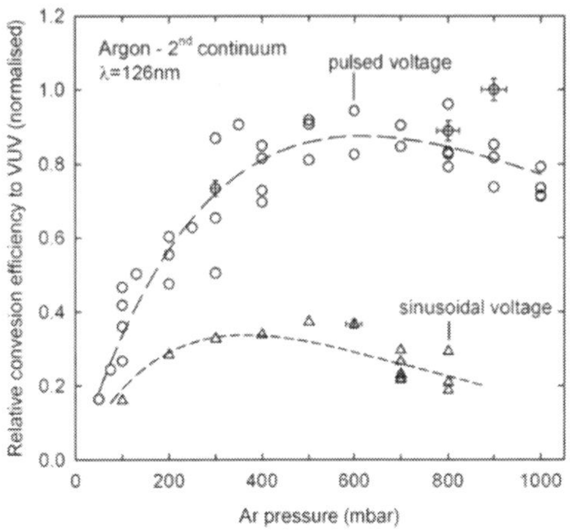

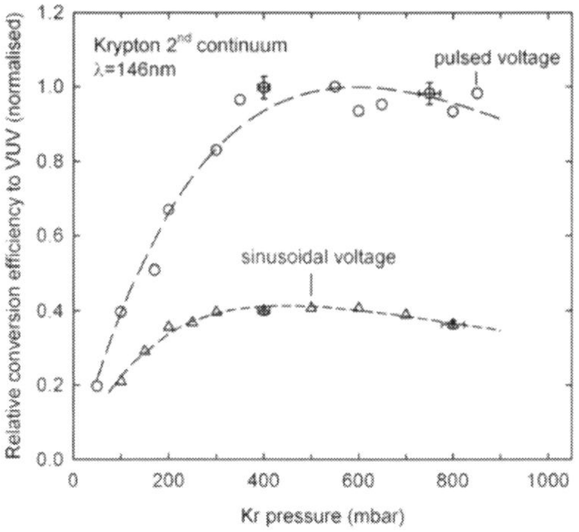

Fig. 2: Electrical to VUV conversion efficiency (normalised) as a function of gas pressure for short-pulse (circles) and sinusoidal (triangles) voltage waveforms. Waveform repetition frequency = 32kHz.

the upper electrode and were measured directly with a Tektronix P6015A HT probe. The displaced charge was determined from the voltage drop across a 3.0nF high-voltage ceramic capacitor connected in series with the grounded electrode (Tek P6139A 500MHz probe). Electrical waveforms were recorded by a Tektronix TDS3054B (500MHz) oscilloscope. VUV output from the ribbon-like plasma, viewed end-on, was detected by a 0.2m f/4.5 vacuum monochromator (ACTON VM502) fitted with an Iridium coated holographic grating (1200 l/mm). The temporal dependence of the intensity of key emission features (fixed λ) was measured by operating the PM tube in the photon counting mode. Due to the negligible absorption of the excimer emissions by its parent gas, the VM502 was operated in the same buffer gas environment as that used in the main vacuum chamber.

Figure 1(b) shows typical VUV output spectra from a short-pulse excited DBD plasma for both Ar and Kr gas fills at p=800mbar. In fact, for both gases at pressures p>100mbar, the 2nd continuum emission band of the excimer dominates the spectral emission over the entire ultraviolet wavelength range 30-400nm. The overall electrical to VUV conversion efficiency has been evaluated from the ratio of the VUV output energy (2nd continuum) per excitation cycle and the input energy per cycle as described in [5]. VUV efficiencies for both sinusoidal and short-pulse excitation methods are shown in figure 2 as a function of gas pressure up to ~1000mbar. For both gases, our results clearly show a significant increase in the VUV efficiency (~2.5x) when using short-pulse voltage waveforms compared to sinusoidal waveforms. With short-pulse excitation, the efficiency for both gases increases with rising pressure, reaches a maximum around ~600mbar, and decreases slightly at higher pressures. The initial rise and subsequent saturation of the conversion efficiencies with increasing pressure is in broad agreement with the recent experimental results of Morozov et-al [6] for an e-beam pumped rare-gas excimer source. Compared to sinusoidal voltage excitation, we have also found that short-pulse operation allows greater control of parameters associated with the temporal evolution of the VUV pulse shapes (risetime, instantaneous peak power, pulse width).

[1] I I Liaw and I W Boyd in *"Functionalized Nanoscale Materials, Devices and Systems,* NATO Science for Peace and Security Series B: Physics and Biophysics, ed. A Vaseashta and I N Mihailescu, Springer NL (2008)
[2] D M Kane, D Hirschausen, B K Ward, R P Mildren and R J Carman, in Laser Cleaning II, ed. D.M. Kane Chapter 13 243-256 World Scientific Publishing Co., Singapore, (2006)
[3] Vollkommer F and Hitzschke L, Proc. 8th Int. Symp. Science and Technology of Light Sources (LS-8), Greifswald, Ger, IL-07 (1998) 51-60.
[4] R.J.Carman and R.P. Mildren, J.Phys.D:Applied Physics, **36** (2003) 19-33.
[5] R J Carman, D M Kane and B K Ward, J. Phys.D: Applied Physics, **43** (2010) 025205
[6] Morozov A, Heindl T, Krucken R, Ulrich A and Wieser J, J. Appl. Phys. **103** (2008) 103301

Comparison of Microwave and ECG driven Ultra High Pressure Lamps

C. Kaiser[1], M. Paravia[1], R. Kling[1]

[1]*Karlsruhe Institute of Technology, Light Technology Institute, Engesserstr. 13, Karlsruhe, 76131, Germany*
Contact: christoph.kaiser2@kit.edu

ABSTRACT

The main failure criteria for ultra high pressure lamps in projection systems is the sputtering of the electrodes and the following processes. To overcome this failure, inactive electrodes could be used. One possible solution for this is the excitation with microwaves. This papers compares the luminous flux, the luminous efficiency and the color temperature of a microwave driven ultra high pressure lamp with a standard ECG operated one.

INTRODUCTION

In the last decades, major efforts have been made in the development of ultra high pressure lamps for projection applications. Beginning at huge VGA projection solutions with 400 screen lumens in the middle of the nineties systems were shrinked down in size but with enhanced light output up to 1000 screen lumens [1]. High lamp pressure up to more than 200 bar is gained during operation in such systems. By this, more light is radiated from the pressure broadened continuum than from line radiance itself [1] [2]. White light is achieved. The main failure criteria of ultra high pressure lamps is the sputtering of the electrodes and entailed processes, like blackening of the glass. To prevent this, a method to drive such lamps with inactive electrodes would be an enormous advance. This could potentially be done by exciting the lamps with microwaves. First experiments for exciting ultra high pressure lamps with microwaves are shown in [3]. The main focus on that work was the microwave excitation of such lamps. To ensure the outcome of the optical measurements the present work gives an improved method for a spectral distributed optical measurement of a microwave using driving mode. The results for the microwave driven lamp will be compared with the conventionally electronic control gear (ECG) driven one and those already covered by literature.

EXPERIMENTAL SETUP

Following conditions have to be met by the experimental setup. The radiation of microwave power hast to be suppressed to perform a measurement of consumed power. This is done by driving the lamp in a resonator cavity. A luminous flux measurement as well as the measurement of the spectral distribution of the light output has to be provided. Due to the resonator cavity, the microwave driven lamp can not be measured in an integrating sphere. For this a reference measurement is done with the an ECG driven lamp in an integrating sphere. After that, the ECG driven lamp can be placed inside the resonator cavity. Due to the same geometry of the lamps, the illumination of an optical detector can be taken as reference value for the luminous flux. In this work, the lamp Osram P-VIP 120W/132W was used, being an typical example for ultra high pressure lamps for projection systems. To anticipate thermal influences it was driven with an jacket tube made of fused silica.

Reference Measurement: For the reference measurement the ultra high pressure lamp is driven at 130 W with an ECG inside an integrating sphere with the diameter of $D = 500\,\mathrm{mm}$ and a photometer (LMT U1000 Lichtmesstechnik GmbH Berlin).

Microwave Setup: To obtain an antenna excitement of the lamp, it has to be shortened to roughly the size of $1/4\lambda \approx 3\,\mathrm{cm}$. For this, the current feedthroughs were cut in the molybdenum foil of the lamp. The lamp was placed in the middle of the resonator cavity at the same spot like in the reference measurement. The used resonator cavity of brass has an inner diameter of $95,4\,\mathrm{mm}$ and is featured with two viewing ports covered with copper mesh. The cavity containing the shortened lamp is numerically designed to be resonant for TM_{010} modes at the frequency of $\nu = 2,33\,\mathrm{GHz}$ with a low damping. The microwave was submitted to the resonator by a signal generator (Rhode & Schwartz SMT03) feeding an traveling wave tube amplifier (Amplifier Research

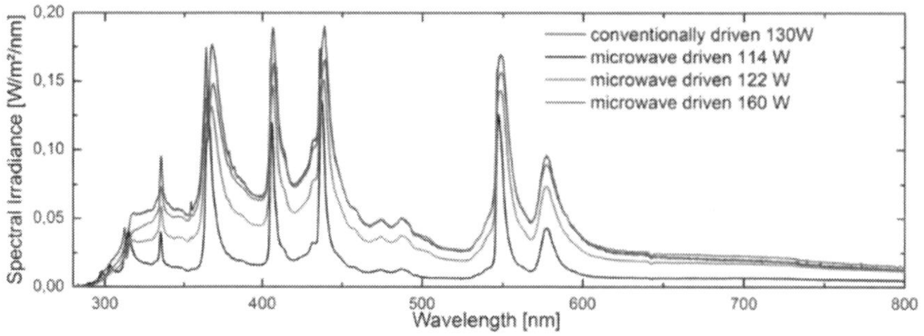

Figure 1. Spectral Distribution of Light Output

200T2G4). The measurement of the consumed microwave power is done with a 50 db bidirectional coupler in front of the resonator cavity connected two power sensors (Rhode & Schwartz NRV-Z51) in combination with a power meter (Rhode & Schwartz NRVD). Losses due to intrinsic power damping of the resonator, which are not contributing to plasma excitation, were estimated via a power measurement in idle mode. With the use of this, the resonator damping was compensated arithmetically.

Optical Measurements: In the following measurements, the lamp was driven in the middle of the resonator cavity. Every measurement was done with the help of an absolute calibrated detector-spectrometer system (Spectro 320a from Instrument Systems) with a distance of $d = 430\,\mathrm{mm}$ between the detector and the lamp center. To suppress stray light from inside the resonator three apertures were placed between the lamp and the detector. For this setup, it was possible to obtain the spectral distribution of emitted light and the illumination of the detector. In a first run, the conventionally driven lamp is measured. With the known luminous flux of the lamp, the geometry of it does not have any influence. Each luminous flux of the following measurements can be estimated using the proportionality of the illumination and the luminous flux. After this, the microwave driven lamps were investigated. Measurements where accomplished in different wattages, starting at 114 W up to 169 W with the increment of roughly 10 W.

RESULTS AND CONCLUSIONS

Some representative spectral irradiance measurements of the recorded ones are shown in figure 1. Referring to this, the lamp power, the spectral output as well as the luminous flux in the microwave driving mode compared to the ECG driving mode is lower at comparable wattages. Increasing of the lamp power of the microwave driven one, also increases the luminous flux. For this, at about 160 W microwave power the luminous flux of both driving modes was appointed to be nearly equal. Furthermore, a color shift to red could be measured, moving the color temperature from $T_C = 8942\,\mathrm{K}$ conventionally driven to $T_C = 8726\,\mathrm{K}$ microwave driven. After all, it has to be stated, that the luminous efficiency of the microwave driven lamp ($51,8\,\mathrm{lm/W}$) compared to the ECG driven one ($65,5\,\mathrm{lm/W}$) is only about $80\,\%$ at the same luminous flux. The results of this work are mostly in agreement with [3]. In this work also an lower luminous flux of microwave driven ultra high pressure lamps was recorded. But nevertheless, in that case only an luminous efficiency of $70\,\%$ compared to the conventionally driven one was stated. The higher efficiency in this work can be explained by the suppression of microwave radiation. Also a color shift was recognized in [3], but in that case to the blue region contrary to the present work. The investigation of the differences in luminous efficiencies of the both driving modes, as well as the color shifting behavior will be part of the future work in our RF laboratory.

REFERENCES

[1] G. Derra Journal of Physics D: Applied Physics, v38, p2995-3010, 2005
[2] J. Pollmann-Retsch, Journal of the Society for Information Display, v15, p789 2007
[3] M. Kando, Journal of Physics D: Applied Physics, v41, 144026 10pp, 2008

Effects of red-light-emitting diodes and ultraviolet-B radiation on growth and photosynthetic characteristics of mung bean

Fang-Min Li[1], **Ming Yue**[1], **G .Zissis**[2], **Zhi-Guo Lu**[3]

[1] *Key Laboratory of Resource Biology and Biotechnology in Western China (Northwest University), Ministry of Education; School of Life Science, Norhtwest University, Xi'an 710069, China*

[2] *LAPLACE UMR5213, Université Toulouse III, 31062 Toulouse Cedex 9, France*

[3] *Department of Physics, Northwest University, Xi'an 710069, China.*

E-mail: luzg@nwu.edu.cn

ABSTRACT

In order to determine the role of the different quality of light sources on growth and photosynthetic characteristics of mung bean, seedlings were exposed to red-light-emitting diodes (red LEDs), ultraviolet-B radiation (UV-B) and the ordinary fluorescent white light source (CK) which was the control. Changes in the growth parameters of seedling, the photosynthetic characteristics, the concentrations of chlorophyll a (chl a), chlorophyll b (chl b), and UV-B absorbing compounds were measured when the seedling were treated for 20d and 40d separately. The results showed that the values of the photosynthetic characteristics and the concentrations of chl a and chl b were enhanced by red LEDs for 20d, and the values of the photosynthetic characteristics, plant fresh weight, dry weight, root length and the concentrations of chl a and chl b were enhanced by red LEDs for 40d. It is suggested that red LEDs promotes the elongation of plant root growth and photosynthetic characteristics.

1. INTRODUCTION

Light quality have an important role in the regulation of plant growth, morphogenesis, photosynthesis, metabolism and gene expression [1]. Compared to the ordinary fluorescent light source, the LED light sources can provide a single wavelength of light quality, with high photoelectric conversion efficiency, wavelength fixed, low heat [2, 3]. All these advantages make it to be an new important light source in the field of plant physiology and plant cultivation. LEDs are a promising irradiation source for plant growth in space for long life, minimal mass, volume and being a solid state device. The red LEDs (wavelength 650nm) were used as a supplementary light source for the greenhouse tomato in 1982, which was reported earlier by the Japan's Mitsubishi Corporation [4]. During the process of laboratory cultivation, it were found that the ordinary fluorescent light lacked the ultraviolet part of the solar spectrum background, which was essential growth factors to play an important biological role [5].

It is necessary to find effective ways to replace or assist the ordinary fluorescent light source. In this study, we used the red LEDs and UV-B radiation as additional light sources for the process of plant cultivation in the laboratory. Our aims were to determine the role of the different quality of light source on growth and photosynthetic characteristics of mung bean.

2. EXPERIMENTATION AND RESULTS

The ordinary fluorescent light source (Philips Inc, power 40W) were made in Netherlands. The lights directly irradiated the seedling of mung bean for12h. The red LEDs (Xinli Inc, power 40W) were made in China. The UV-B radiation was provided by filter Qin brand (Baoji Lamp Inc, China) 30-W fluorescence sunlamps. The dose of UV-B irradiation was 0.861kJ·m^{-2} per day for5h the same as the red LEDs.

From the results of the growth parameters, it was observed that the values of mung bean seedlings irradiated for 20d by red LEDs and UV-B were not significantly different with that of the ordinary fluorescent light. However, the red LEDs treatment caused a significant increase ($p<0.05$) of the values of plant height, fresh weight, dry weight and root length compared with that of the ordinary fluorescent light for 40d. With the UV-B radiation for 40d, an obvious decrease ($p<0.05$) of plant height was observed, while it induced a marked increase ($p<0.05$) in fresh weight, dry weight and stem diameter.

The results of the photosynthetic characteristics showed that the red LEDs treatment induced a significant increase ($p<0.05$) in the values of mung bean seedlings for the two durations. However, a significant decrease ($p<0.05$) was detected in the values of the photosynthetic characteristics of the UV-B radiation for 40d.

The concentrations of chl a and chl b may affect the values of the photosynthetic characteristics to a certain extent. It were obviously that the red LEDs treatment induced statistically significant increases ($p<0.05$) not only in the values of the photosynthetic characteristics but also in the concentrations of chl a and chl b for the two durations. Compared with the ordinary fluorescent light the UV-B radiation didn't cause significantly differences.

3. DISCUSSION

In this study we used red LED and UV-B as supplementary light sources for the ordinary fluorescent light to study the role of these light sources on plant cultivation. The results showed that the effects of these light sources on growth and photosynthetic characteristics of mung bean were obvious. The red LED light source promoted the growth of the mung bean root, which was useful to absorb the nutrients and water of the soil.

Chlorophyll concentrated in the chloroplast grana is the main pigment to capture the energy for photosynthesis in green plants. The results showed that the red LED light source can significantly increase the concentrations of the chlorophyll, which effectively promoted the photosynthesis and water use efficiency. On the contrary, UV-B radiation induced a notably decrease in the photosynthesis and water use efficiency. It was suggested that the UV-B treatment will reduce the stomatal opening degree of mung bean and then affect the gas exchange in photosynthesis.

It will be a long time for the LEDs light sources to replace the ordinary fluorescent light completely during the process of plant cultivation, because different types and different developmental stages of plants require different light quality and intensity.

REFERENCES

[1] F Naoya, KY Mitsuko, U Masami, T Kenji, S Sadanori. Effects of light quality, intensity and duration from different artificial light sources on the growth of petunia (Petunia× hybrida Vilm.). J Jpn Soc Horticult Sci, 71 (2002): 509-516.

[2] D J Tennessen, R Bula, T D Sharkey. Efficiency of photosynthesis in continuous and pulsed light emitting diode irradiation. Photosynth Res, 44 (1995): 261-269.

[3] P R Poudel, I Kataoka , R Mochioka. Effect of red and blue-light-emitting diodes on growth and morphogenesis of grapes. Plant Cell Tiss Organ Cult, 92 (2008): 147-153.

[4] D J Barta, T W Tibbitts, J Bular. Evaluation of lighting-emitting diodes characteristics for a space-based plant irradiation source. Adv Space Res, 12 (1992): 141-149.

[5] F.M Li, J Wang, Y.P Chen, Z.R Zou, X.L Wang, M Yue. Combined effects of enhanced ultraviolet-B radiation and doubled CO2 concentration on growth, fruit quality and yield of tomato in winter plastic greenhouse. Fontiers of biology in China, 2 (2007): 414-418.

Microwave discharge performed at a low pressure of Ar/Hg mixture in the coaxial waveguide with a short-cut inner electrode. Experiments and numerical simulations.

E.M. Barkhudarov, N.V. Denisova[1], I.A. Kossyi, E.V. Kulumbaev[2], N.I. Malykh, M.A. Mysakyan

Prokhorov Institute of General Physics, Russian Academy of Sciences, Moscow, Russia
[1] Khristianovich Institute of Theoretical and Applied Mechanics, Siberian Branch, Russian Academy of Sciences, Novosibirsk, Russia
[2] Kyrgyz-Russian Slavic University, Bishkek, Kyrgyzstan
Contact: barkhudarov@yandex.ru

ABSTRACT

The results from experimental and theoretical research of the characteristics of microwave discharge performed at a low pressure of argon-mercury mixture in a coaxial waveguide are presented. The discharge is formed and maintained due to the transformation of microwave radiation into a surface wave propagating along the boundary between plasma and dielectric. The electron density is much higher than critical density $n_e >> n_{cr}$ (f=2.45GHz). The radial distribution of mercury atoms emitting radiation with wavelength λ=546.1 nm was obtained. Based on numerical simulations, the spatial distributions of the electric field intensity and electron density in discharge were obtained. The experimental results are in qualitative agreement with the results of numerical simulations.

GENERAL

The research of microwave discharge in a low pressure medium is of great interest in respect to the problems of the low temperature plasma as well as to the various applications, and to the creation of UV radiation sources in particular [1 and 2].

The microwave discharge in a low pressure medium ($2\text{-}2\cdot10^{-2}$ torr) performed in a cylindrical discharge tube which is sequential to the short-cut inner electrode of the coaxial wave guide [3] is investigated experimentally and theoretically in this paper. The plasma initiation is conditioned by transformation of microwave radiation into a surface wave propagating along the boundary between plasma and dielectric. The setup for microwave discharge ignition is presented in Fig. 1.

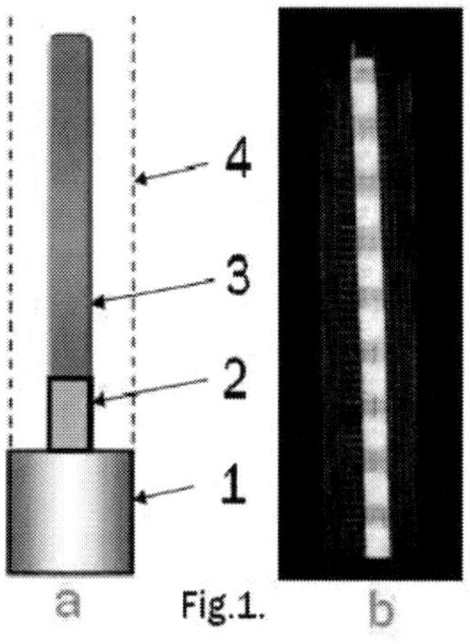

Fig. 1. The setup of microwave discharge ignition and its integral photo taken with neutral filter. (1) magnetron; (2) and (4) inner and outer electrodes of the coaxial waveguide, correspondingly; (3) discharge tube.

The standard generator from the microwave oven (f=2.45GHz; N≤800) serves as a source of microwave radiation in the experiment. The cylindrical quartz lamps 34 and 70 cm in length and d=2.5 cm in diameter were used. The tube wall has the temperature of about t≈40°C. To determine the electron density, the discharge area was explored in transversal direction by microwave radiation with wavelengths λ = 8 and 2.4 mm and by interferometry technique (λ=2.4 mm). The electron density exceeds the critical density corresponding to the critical frequency of microwave radiation igniting the discharge in all the range of pressure variation in discharge tubes. The longitudinal periodic structure of the discharge was detected having period of about ≈ 12 cm. Assuming the axial symmetry and applying Abel conversion, the radial distribution of mercury atoms radiating from 7S level (the upper level for the spectral line 546.1 nm) was obtained within argon pressure values $2.8\cdot10^{-2}$ and $2\cdot10^{-2}$ torr.
The characteristic feature of this distribution is a hollow at the discharge axis (see Fig. 2).

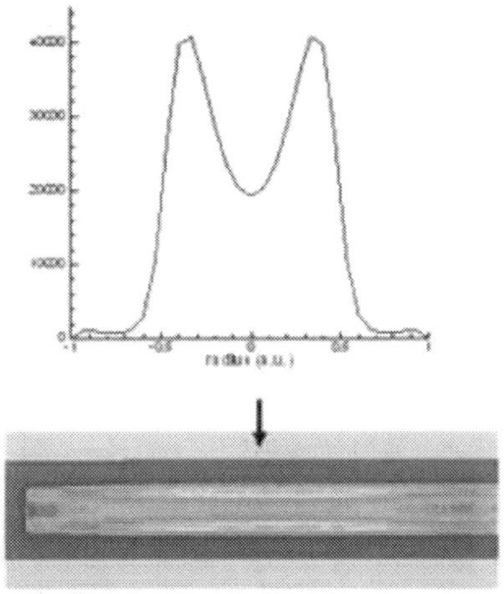

Fig.2.

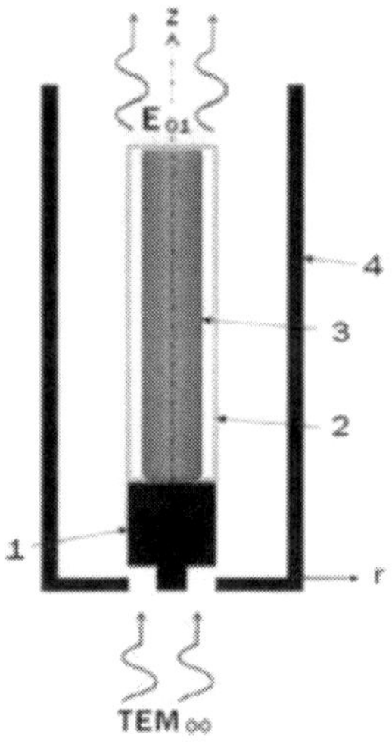

Fig.3.

Fig. 2. The radial distribution of mercury atoms emitting radiation with wavelength λ=546,1nm in the discharge cross section.

Fig. 3. The arrangement of microwave discharge performed in coaxial waveguide. (1) inner short-cut electrode; (2) quartz tube; (3) discharge plasma; (4) outer electrode.

Based on the self-consistent two-dimensional model, numerical simulations were carried out for the characteristics of microwave discharge in equilibrium (Ar/Hg) in the coaxial waveguide. The setup is presented in Fig. 3. The microwave power is supplied by TEM_{00} mode of coaxial waveguide. The system parameters provide existence and propagation conditions for the only fundamental wave E_{01} of the coaxial waveguide. The Maxwell equation system and the stationary electron balance equation are solved together. Some results from numerical simulations are presented in Fig. 4. The source data was the following: f=2.45 GHz; P=0.5 torr; the discharge tube length was 33 cm; and microwave power supply was 800 W.

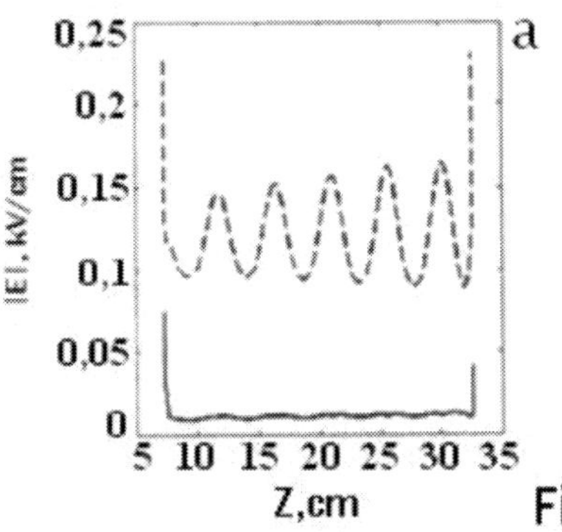

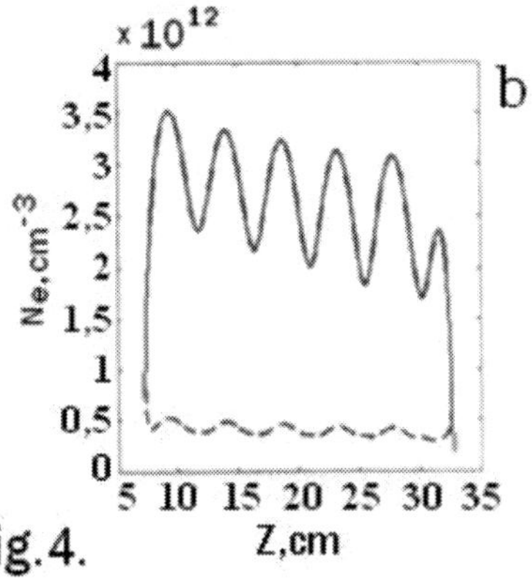

Fig.4.

Fig. 4. The axial distributions of electric field intensity module (a) and electron density (b) at the axis of the gas-discharge quartz tube (solid lines) and near its wall (dotted lines).

The obtained spatial distributions for the electric field intensity and electron density are periodic in longitudinal direction; the period occurred to be about ≈ 12 cm. The maxima of electric field intensity correspond to the minima of electron density and vice versa. The electron density is much higher than critical density for the microwave radiation (n_{cr}=7,7·10^{10}cm^{-3}). Let us point out that "resonant" kind of microwave discharge can be realized in this case. The obtained results are in qualitative agreement with the experimental results.

REFERENCES

1. AlAl – Samma'a, Pandithas and Lucas J.J. Phys D: Phys.,2001, 34, 2775-2781.
2. E.M.Barkhudarov, M.A. Misakyan, I.A. Kossy, and et. Microwave disharges. Proceeding, Ed By Yu. A. Lebedev, pp. 279-284, 2006, Yanus-K, Moscow.
3. E.M.Barkhudarov, S.I. Gritsinin G.V. Dreiden et all. Fiz. Plazmy 30, 575 (2004) I. Plasma Phys. Rep.30, 531, 2004.
4. The axial distributions of electric field intensity module (a) and electron density (b) at the axis of the gas-discharge quartz tube (solid lines) and near its wall (dotted lines).

Molecular radiation from overloaded Hg-free HID

R. Methling[1] , St. Franke[1] , L. Hitzschke[2] , M. Käning[2] , B. Schalk[2]
[1] *INP Greifswald, Felix-Hausdorff-Str. 2, D-17489 Greifswald, Germany*
[2]*OSRAM GmbH, Werner-von-Siemens-Str. 6, D-86159 Augsburg, Germany*
Contact: methling@inp-greifswald.de

ABSTRACT

Hg-free lamps are capable providing strong continuum emission. Therefore, lamps with Xe, AlI_3, TlI, and XI_3 were investigated, where X stands for the rare-earths Tm, Yb, Ce, Pr, Nd, and Gd. Typically, the spectra were dominated by a broad and intense continuum radiation originating from the rare-earth mono-iodide. The shape of the continuum changed with the ordering number of the rare-earth and with increasing power. At very high power input Planck-like spectra and self-absorption of the continuum radiation were observed.

EXPERIMENTAL RESULTS

Molecular radiation is known to play an important role in high intensity discharge (HID) lamps, which in modern HID lamps is mainly observed in the mantle region of the arc; see [1] and references therein. In [2, 3] we have shown that the contribution of molecular radiation to the total radiative output can be substantially increased for mercury-free metal-halide lamps. However, the presented molecular quasi-continuum was primarily based on the radiation of thulium mono-iodide, which could be characterized by two broad and intense maxima in the visible spectral range [4]. Hence the question arose, if other rare-earth mono-iodides own a continuum emission with comparable intensities and spectral structure.

In this contribution we present results on Hg-free discharges filled with different rare-earth iodides operated at very high power densities. In detail we investigated lamps filled with Xe, AlI_3, TlI, and XI_3, where X stands for the rare-earths Tm, Yb, Ce, Pr, Nd, and Gd. Cylindrical ceramic discharge tubes were used with an inner diameter of 9 mm, a tube length of 13 mm and an electrode distance of 10 mm. The nominal power was 150 W. However, the lamps were operated by a square wave current in vertical position, starting from 100 W up to an overload of 450 W. Due to rather low voltages the current rose up to 5 A, which is far beyond the limit for the long-time operation of the electrodes of about 2.5 A. Consequently, the measurements had to be carried out rather fast, within about one hour after one hour of burn-in at 150 W. Neither thermalisation of the discharge vessel nor stationary conditions of the evaporation was expected during power variation. The main goal was to observe the qualitative changes in the continuum emission when the input power density is strongly increased. Hence, all the electrical and photometric parameters (e.g., the luminous efficacy, colour rendering index, colour difference) can only be interpreted as a rough guideline for comparison between the lamps and not as long-term, steady-state parameters.

Figures 1a-c show spectra of lamps with CeI_3, PrI_3, and GdI_3 at 150, 200, and 400 W. The measurements were carried out in an integrating sphere of 1.5 m diameter with a compact fibre optic spectrometer working in the visible spectral range between 380 and 780 nm (CVI Melles Griot FOS200-SPL). Firstly, it should be mentioned that the spectrum is dominated by a continuum emission, which is not the result of a superposition of atomic spectral lines but that of molecular radiators. This can be validated by measurements with higher spectral resolution. Secondly, the continuum is shifted towards shorter wavelengths from CeI to PrI to GdI, as can be seen very clearly from Fig. 1a-c at 150 W. Thirdly, with increasing power, the spectral contour at shorter wavelengths resembles more and more that of a Planckian radiator of around 2000 K (cf. dashed black line). This indicates a very high optical depth of the plasma in this spectral range. Finally, a self-reversal of the continuum contour can be observed in the case of praseodymium at 400 W, which is revealed by the dip around 500 nm.

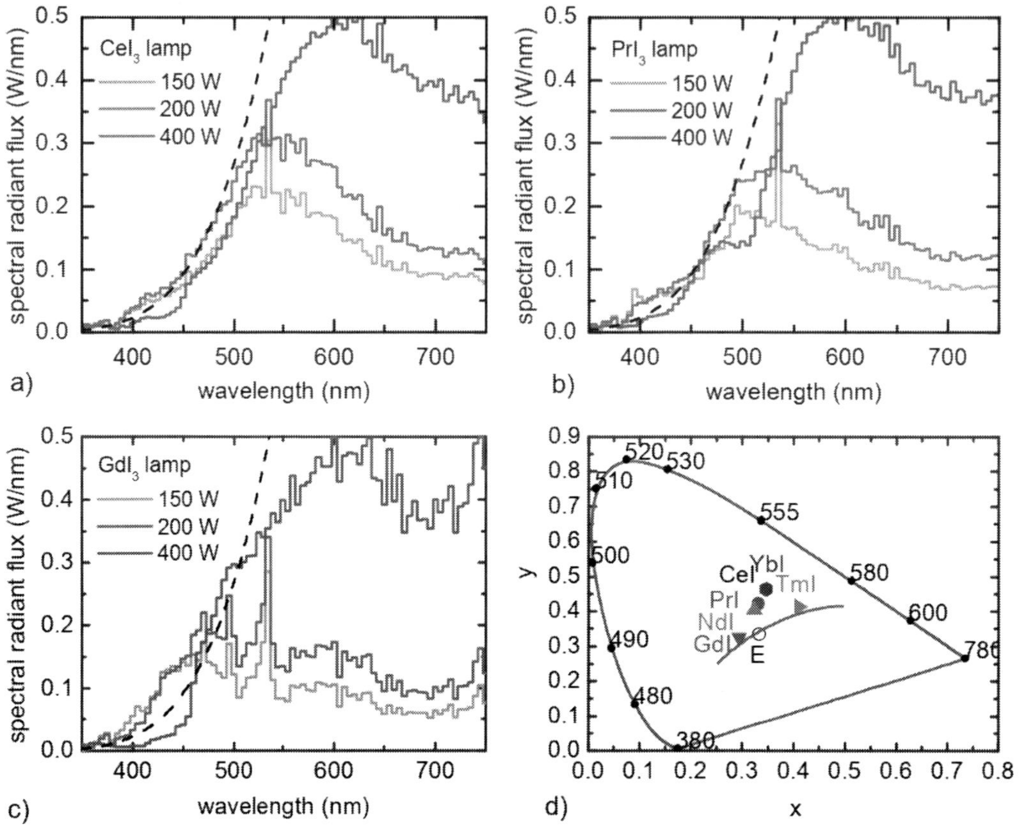

Figure 1a-c: Spectra of lamps with CeI_3, PrI_3, and GdI_3 at 100, 150, and 400 W. Power increase deforms the blue edge of the continuum towards a Planck like contour (cf. dashed black line, 2000 K blackbody radiator). Figure 1d: Chromaticity coordinates of all investigated rare-earth iodides at 150 W.

Fig. 1d shows the chromaticity coordinates of all investigated rare-earth iodides at 150 W. Each of them owns its specific continuum structure leading to different chromaticity coordinates. Some of the rare-earth iodides show a blue shift of the emission with increasing element's ordering number (Ce, Pr, Nd, Gd). Though, a common feature is the occurrence of Planck-like spectra and of self-absorbed continuum emission.

With this contribution it was demonstrated that several rare-earth iodides are capable providing strong continuum emission in Hg-free HID lamps which offers new perspectives for designing lamp spectra.

REFERENCES

[1] M. Käning, B. Schalk, H. Schneidenbach, J. Phys. D: Appl. Phys., 40 (2007) 3815-3822.

[2] St. Franke, H. Hess, R. Methling, H. Schneidenbach, H. Schöpp, L. Hitzschke, M. Käning, B. Schalk, Proc. LS-11 (2007) 341-342.

[3] St. Franke, R. Methling, H. Hess, H. Schneidenbach, H. Schöpp, L. Hitzschke, M. Käning, B. Schalk, J. Phys. D: Appl. Phys., 40 (2007) 3836-3841.

[4] M. Käning, L. Hitzschke, B. Schalk, M. Berger, St. Franke, R. Methling, Hg-free HID lamps, Proc. LS-12/WhiteLED-3 (2010).

Convection velocity in HID lamp including acoustic resonance

Arezki Toumi[1], John Hirsch[2], Bernd Baumann[3], Marcus Wolff[3], Sounil Bhosle[4] and Georges Zissis[1]
[1]LAPLACE Laboratory, 118, route de Narbonne, 31062 Toulouse, France
[2]Lightlabs, Philips Lighting, Mathildelaan 1, 5600 JM Eindhoven, the Netherlands
[3]Hamburg University of Applied Sciences, Berliner Tor 21, 20099 Hamburg, Germany
[4]Led Engineering Development, 10, avenue de l'Europe, 31520 Ramonville St Agne, France
Contact: arezki.toumi@laplace.univ-tlse.fr

ABSTRACT

This paper shows the fluid fluxes calculated in 2D axi-symmetric model using finite element method for a HID lamp operated at high frequency including the influence of acoustic resonances.
Keywords: HID lamp, acoustic resonance, acoustic velocity, modelling, convection velocity

INTRODUCTION

Electronic ballasts used to supply High Intensity Discharge (HID) lamps are usually operating at low frequency (some hundreds hertz) square waves in order to avoid the appearance of arc instabilities. This phenomenon is known as acoustic resonance. Indeed, at high frequency (from the kilohertz to some hundreds of kilohertz), the acoustic resonances appear when a standing pressure wave develops inside the lamp burner.
The advantages of a ballast working at higher frequencies than the usual ones is a gain on its weight, size and cost. It was shown that the best working frequency is around 300 kHz [1]. For this reason, a lot of studies on acoustic resonance have been performed and more solutions have been proposed to avoid it, however the interaction between acoustics resonances and the arc is still poorly understood. Consequently, a model including this interaction is required to improve the physical comprehension and to evaluate the impact of acoustic resonances on arc distortion in function of the frequency.

ACOUSTIC STREAMING

Afshar [2] attributes the coupling of the arc instabilities and the acoustic wave to acoustic streaming. He explains that a vortex acoustic flow, called acoustic streaming, develops and it is due to a non-linear second order effect of the acoustic wave that has a time-averaged amplitude different to zero. Also, Lighthill [3] mentioned that this acoustic streaming is caused by the spatial variation of Reynolds stress by the generation a net force per unit volume and influences the fluid flux of the plasma in the burner of HID lamp. This force F_j, for the direction of axis j, is written:

$$F_j = -\frac{\partial \overline{\rho v_i v_j}}{\partial x_i} \tag{1}$$

With ρ is the mass density, v_i and v_j are the acoustic velocities and $\overline{\rho v_i v_j}$ is the temporal mean.
In order to find the convection velocity in the burner of a HID lamp, we implemented the force generate by the acoustic streaming in Navier-Stokes equation [Eq.2].

$$\rho(\vec{U}.\vec{\nabla})\vec{U} = \vec{\nabla}.\left[-p\vec{I} + \eta\left(\vec{\nabla}\vec{U} + (\vec{\nabla}\vec{U})^T\right)\right] + \vec{F} \tag{2}$$

With η the viscosity in the plasma, $\vec{F}$ the force per unit volume ($\vec{F} = -\rho\vec{g} + \vec{F}_{acoustic}$ with $-\rho\vec{g}$ the gravitational force and ρ is mass density), $\vec{I}$ is identity operator and p the stationary pressure.

RESULTS

On the figure 1.a shows the fluid velocity field generated by the gravitational force (convection) and on the figure 1.b the fluid velocity including acoustic streaming force can be seen. As we observe on figure 1.b, two vortex fluxes are developed which is in good agreement with the results of Afshar [2].

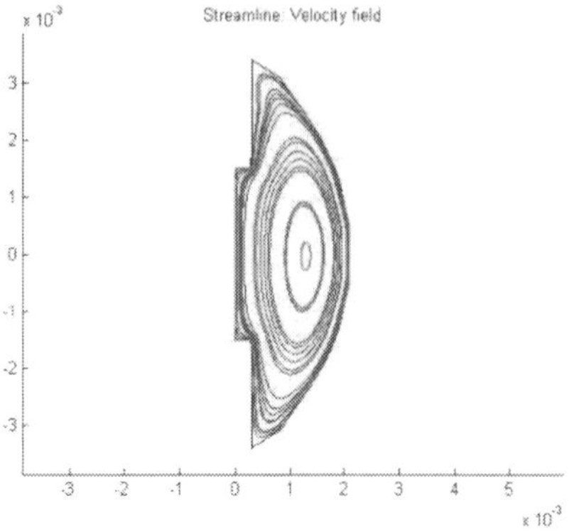

(a)

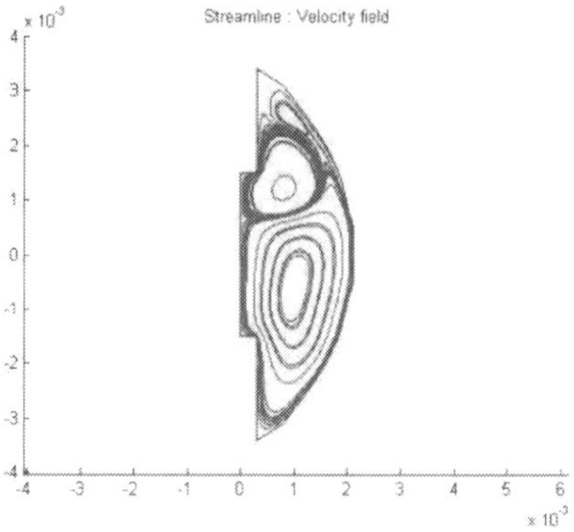

(b)

***Figure 1**. a)Fluid flux due to the natural convection velocity (generated by the gravitational force) and b) fluid flux including acoustic streaming computed for HID lamp for f=47 kHz corresponding to the first longitudinal resonance mode.*

CONCLUSION

In this paper, we presented the results of an acoustic streaming model, which provides some explanations on the mechanisms of acoustics resonances in HID lamps at high frequency. Indeed, the appearance of fluxes combining convection and acoustic streaming can be responsible for the instabilities in the burner of the HID lamp and explains the relation between the acoustic resonances and the arc.

REFERENCES

[1] G. Trestman, Minimizing Cost of HID Lamp Electronic Ballast. 28th Annual Conference of the Industrial Electronics Society, IEEE 2002 p 1214-1218.
[2] F. Afshar. The Theory of Acoustic Resonance and Acoustic Instability in HID lamps. LEUKOS, 20(1):27–38, 2008.
[3] J. Lighthill. Acoustic streaming. Journal of Sound and Vibration. 61(3), p 391–418, 1978.

Session Eight

Mercury-free HID lamps

M. Käning[1], L. Hitzschke[1], B. Schalk[2], M. Berger[1], St. Franke[3], R. Methling[3]

[1]Research Europe, OSRAM GmbH, Werner-von-Siemens Straße 6, 86159 Augsburg, Germany
[2]Vitec Group Videocom Division, Erfurter Strasse 16, 85386 Eching, Germany
[3]INP, Leibniz-Institut für Plasmaforschung und Technologie e. V., Felix-Hausdorff-Straße 2, 17489 Greifswald, Germany
Contact: m.kaening@osram.de

ABSTRACT

Lamp manufacturers are currently improving their efforts to remove hazardous mercury from their products. The present contribution is focussing on HIDs for general lighting applications. It starts with a summary concerning the attempts made to replace Hg by Zn. As an alternative to them, discharges dominated by molecular radiation are proposed. After introducing the main characteristics of molecular radiation, a detailed analysis will be done on molecular absorption for a thulium-iodide based fill system. The consequences of the interplay between emission and reabsorption on photometric data are discussed. It is shown that HIDs dominated by molecular emission are able to deliver efficacies of above 90 lm/W at a CRI of 90.

INTRODUCTION

During the past decade the general understanding for sensitive, environmental related issues has been sharpened substantially. One consequence of this is that lamp manufacturers have intensified their efforts not only to improve the energy efficiencies of light sources but also to eliminate hazardous substances in their products. For conventional HID lamps the latter relates first of all to mercury and thallium. In principle it would seem that the avoidance of thallium is not difficult, but the same can not be said of mercury. The reason for this is that due to the high vapor pressure of mercury, and its large momentum transfer cross section for colliding electrons, the operation voltage of HID lamps can be fixed at values of about 100 V. The latter helps a great deal in limiting the needs and inevitable efforts for the ECG, the electrodes and last but not least also for lamp construction.

In addition to all of this and not less important is the fact that mercury has also a decisive impact on the resulting temperature profile of HID lamps [1–3]. The reason for this is that an energy redistribution due to UV radiation emitted from the hot arc core regions occurs and subsequently reabsorption processes at outer radial positions at lower plasma temperatures follow. This way mercury influences not only the efficacy but also all other photometric data of the lamp. After consideration of these facts it becomes quite obvious that the elimination of mercury in HID lamps is anything but easy. The emphasis of the present contribution is on electroded systems for general lighting applications. Automotive lighting, see [4, 5], as well as electrode-less solutions, which have been discussed in detail in [6] and references therein, are not in the focus.

The straightforward strategy for mercury replacement is a direct substitution of mercury by a substance with similar high vapor pressure, corresponding momentum transfer cross-section as well as similar radiation properties. With this in mind the most successful attempt focused on general lighting applications turned out to be based on Zn [7–10]. The best data was obtained using a saturated Zn-NaI-CeI_3 filling with Xenon as starting gas. It resulted in an efficacy of 120 lm/W and a color rendering index of 79. Unfortunately it was noted that with increasing efficacies the color rendering index was decreasing. Moreover, investigations also revealed that the replacement of Hg by Zn had a direct impact on the PCA corrosion. The latter has to been seen in correlation with the findings that cold spot temperatures of more than 1350 K were necessary for gaining sufficient high Zn vapor pressures to guarantee the desired operation voltages.

A completely different concept to avoid mercury in HID lamps is introduced in the this approach [11]. It started from the fact that, for a given power, the resulting discharge voltage is higher the smaller the arc cross section is. Furthermore, from earlier investigations it was known [2, 3, 12] that molecular radiation emitted in the colder mantle region of the discharge is not only able to deliver the desired radiation in a quite efficient way, but also leads to the desired arc constriction. The explanation for the latter is as follows: due to the origin of the radiation in the mantle region this area is now cooling down. On the other hand, to allow a definite current to flow through the discharge, the core temperature has to be adequate. Therefore in case of rising power the core temperature is now increasing considerably. Based on this concept it is not only possible to remove mercury from HID discharges, but also to change their working principle substantially. The present findings were already given in part in [13]. This contribution deals with these findings in more detail.

MOLECULAR EMISSION OF THULIUM MONO-IODIDE IN HID

To understand the findings in the subsequent sections of this contribution the most important basics of molecular emission in HIDs shall be briefly summarized in this section.

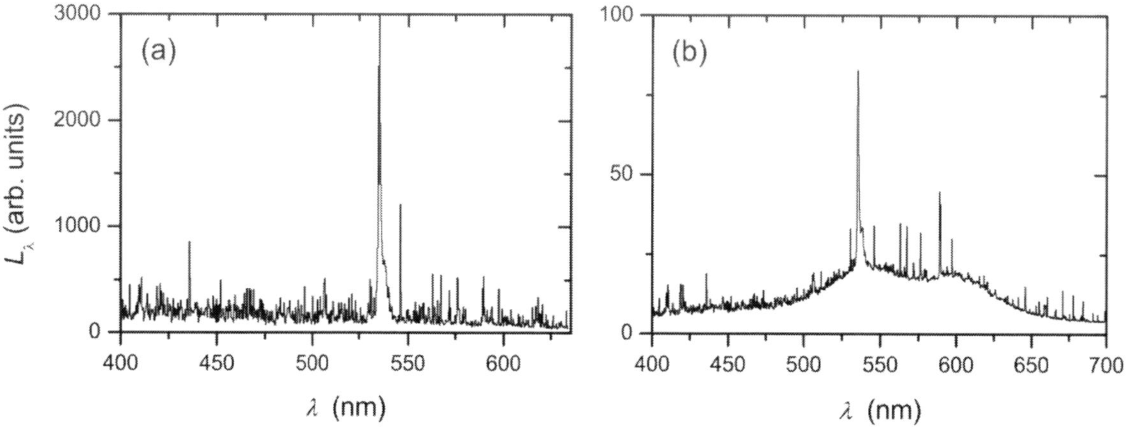

Figure 1. Side-on-spectra from core (a) and mantle (b) of a Hg-Ar-TlI-TmI$_3$-discharge at P = 240 W. Note, not only the factor 30 of intensity scales of these two plots, but also the continuum intensity at e.g. 550 nm is in core only about 5 times bigger than in mantle.

As described in [12] for Hg-Ar-TlI-discharges with only one rare-earth component (TmI$_3$, HoI$_3$, DyI$_3$) the origin of their quasi-continuous radiation was investigated in detail. For example, fig. 1 shows two spectra from core and mantle regions of an Hg-Ar-TlI-TmI$_3$-discharge burning in a typical 250 W quartz vessel at P = 240 W. The spectra in this figure were actually derived from fig. 1 of [12] by binning the intensities to 0.3 nm spectral resolution in order to be more comparable to all following figures of the present contribution.

The vast number of atomic lines present in these spectra contribute roughly half of the total intensity emitted by the discharge. The other half has to be attributed to quasi-continuous molecular emission of mono-iodides. It was supposed that the chemically more stable mercury iodide (HgI) is responsible for radiation from the core, whereas thulium iodide (TmI) was identified as emitting mainly in the mantle region. Since the integral intensities of these two contributions are comparable it follows that molecular TmI emission contributes actually about 25% to the total intensity. This amount of radiation contribution was considered to be unexpectedly high for this kind of discharge, which up to then was unknown.

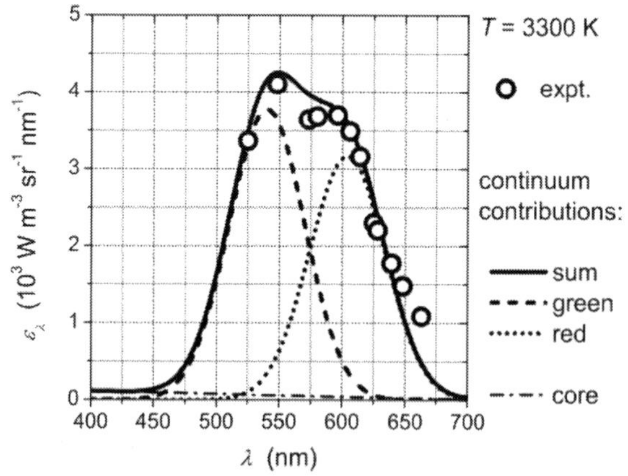

Figure 2. Emission coefficient of TmI radiation at 3300 K: circles are experimental data points, lines show the contributions according to the parametric approach approximating the TmI spectrum with two Gaussian profiles. Note, the contribution from the HgI core component is at this temperature still negligible.

In a next step effective excitation energies for the upper electronic states were determined. It turned out that the TmI emission could be assigned to two separate effective excited states. Furthermore, the data analysis revealed that the spectral distribution of the emission can be approximated by using two Gaussian line-shapes as a substitute for the very complex structure of the rotational vibrational electronic molecular transitions. Fig. 2 depicts, as an example, the spectral distribution of the emission coefficient at a temperature of 3300 K.

The proposed procedure not only allowed the determination of effective excited states, but also an estimation of corresponding transition probabilities for these components. The whole analysis depends on the quality of the temperature profile being used. In addition it is also necessary to know the densities of TmI at the given temperatures. The TmI density can only be gained using thermochemical modelling which starts from the evaporated amount of the TmI_3. A further complication arises from radial demixing processes which have been taken into account in the model. This whole procedure of determining parameters for molecular continuum emission has not only been done for TmI, but also for HoI and DyI.

MOLECULAR EMISSION FROM MERCURY-FREE DISCHARGES IN PCA BURNERS

The investigations described in the previous section have been performed in quartz burners because the possibility of measuring the arc temperature profile by means of optical emission spectroscopy should not be sacrificed. This, however, made experiments at high energy load - to attain high wall temperatures - practically impossible [14].

Therefore, in the attempt to further improve the photometric data, the transition towards PCA burners was inevitable. As a result standard cylindrical PCA burners for $P = 150$ W with a fill of AlI_3, TlI, TmI_3 and Xe as start gas were used to establish a discharge dominated by TmI molecular emission. For the lamps it was ensured that the additives AlI_3 and TlI were completely evaporated.

Fig. 3 shows irradiance spectra of such a discharge for three different power levels. Note that the spectrum for $P = 90$ W in fig. 3a resembles the one in fig. 1b. The typical thulium mono-iodide molecular quasi-continuum emission with some strong thulium atomic lines on top can clearly be seen. The main difference between both figures, however, is that fig. 1b was recorded only in the arc mantle region of the mercury containing discharge, whereas fig. 3a is more like a spatially integrated spectrum of the complete mercury-free discharge lamp. From fig. 3a it is obvious that the TmI molecular radiation in the PCA burner now dominates the major part of the total emission. Actually a similar spectrum could also be obtained in quartz burners, see e.g. fig. 3a in [14], but

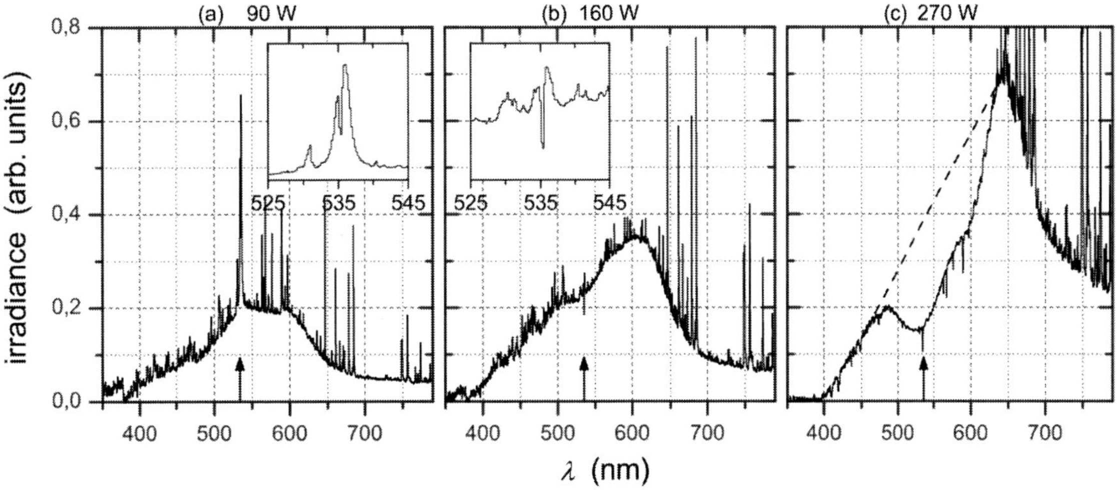

Figure 3. Irradiance spectra of a Xe-TlI-AlI_3-TmI_3 discharge in a PCA burner for three different power levels measured with optical fiber at a distance of about 15 cm from the arc.
Arrows mark the position of the prominent Tl line at 535 nm, which is zoomed in insets for (a) and (b).
(a) $P = 90$ W – continuum emission as known from discharges in quartz burners.
(b) $P = 160$ W – beginning of molecular absorption.
(c) $P = 270$ W – strong self-reversal for the whole molecular band between 460 and 640 nm. For meaning of dashed line see discussion in the text.

again, due to the high wall load, without further perspective to attain much higher power loads.
Note that in fig. 3a the most prominent self-reversed Tl line at 535 nm is located on top of the continuum. This line is often used to determine the radial temperature profile of Tl-containing discharges via the Bartels method (see [15] for more details). For the purpose of the following discussion it is relevant that the intensity of this line's self-reversal shoulders is proportional to the maximum temperature met along the line-of-sight. The inset of fig. 3a zooms the Tl line in order to show the absorption dip at the line center as well as the asymmetric line wings.
Starting from fig. 3a the power was increased. The assumption was that in doing so not only the evaporation of TmI_3 could be increased considerably, but also a corresponding increase of the emitted radiation would result. The latter part of this assumption turned out to be untrue as fig. 3b illustrates. First of all it is obvious that not only the general appearance of the original symmetric spectrum changed, but also that the Tl 535 nm line almost disappeared. The zoom (see inset in fig. 3b) proves that although the Tl line is still visible it now shows quite a different line-shape than before. The barely visible self-reversal shoulders and wings on top of the TmI continuum, indicate that the whole line must have been absorbed. The alternative interpretation of the Tl line-shape change would mean that the core temperature must have dropped sufficiently. However, since the power of the arc had been increased this interpretation is definitely unreasonable. Furthermore it is also obvious that all Tm atomic lines between 550 and 600 nm did not rise in intensity although all other lines above 600 nm did.
The physics behind this observations become evident if the power is increased even further. Fig. 3c eventually shows a very different spectrum for a power of $P = 270$ W. Now there are absolutely no lines visible on top of the continuum between 350 and 640 nm. The only indication of atomic transitions are absorption dips not only for Tl 535 nm, but also for strong thulium atomic lines between 400 and 600 nm. This means that all atomic radiation coming from the arc core gets absorbed and has no chance to escape the discharge.
The further comparison of fig. 3b and fig. 3c shows that the general shaping of the irradiance spectra is indeed very similar for wavelengths below 460 nm and above 640 nm. An exception, however, is the range between 460 and 640 nm where the spectral intensity actually further decreases. As before (from fig. 3a to fig. 3b), here it is very likely that absorption is also responsible for this unexpected behavior. To get some measure for this absorption, the dashed line – called "base-line" – is introduced in fig. 3c connecting the spectral irradiance at wavelengths 460 and 640 nm. As pointed out above these positions are the lower and upper margins towards similar spectral ranges.
The intensity along the "base-line", $I_{BL}(\lambda)$, is used to estimate the "absorption" according to

$$A(\lambda) = 1 - I(\lambda)/I_{BL}(\lambda).$$

The result for this "absorption" is plotted as the bottom trace in fig. 4. The shaping of the absorption trace resembles the well known TmI spectrum at a low power. For comparison the emission spectrum for P = 90 W from fig. 3a is included as the upper trace in fig. 4. Check also the similarity of the absorption trace to fig. 2.
As discussed in [12], quasi continuum emission of mono-halides consists of many merged electronic rotational vibrational bands which can barely be seen in radiation spectra in the case of TmI. In the absorption trace, however, molecular bands do show up as tiny humps (marked with little hooks in fig. 4). This is an additional clear evidence of molecular structures which further supports the assumption that molecules dominate the radiation properties of this kind of discharge.

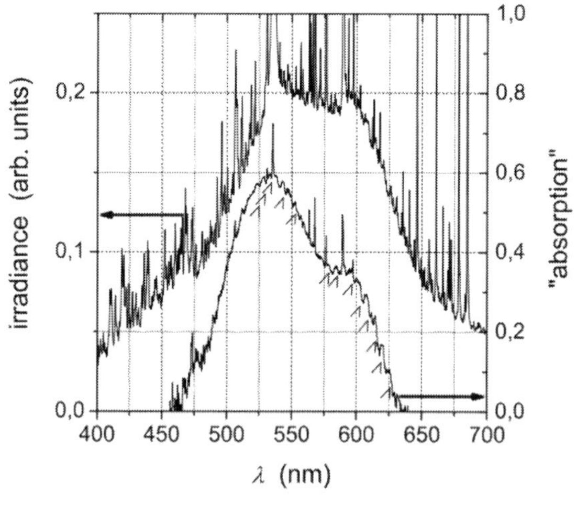

Figure 4. Left axis – top spectrum for P = 90 W (fig. 3a). Right axis – bottom absorption data for P = 270 W computed from fig. 3c (see text for discussion).

To summarize the above discussion it can be stated that the shaping of the spectra in fig. 3 are the result of the interplay of molecular emission and molecular reabsorption. In particular, the very distinct absorption feature

observed in fig. 3c is in fact self-reversal of the whole molecular transition band. This is actually comparable to the absorption dips of self-reversed atomic transitions, like e.g. in case of the Tl 535 nm line.
Besides the just discussed absorption in the green and red region, it has to be mentioned that there are additional absorption processes in the blue spectral range. It is known that the Tl 378 nm line as well as TlI on the red wing of the Tl 378 nm line do in fact cause strong absorption [12]. As fig. 3 shows, the increase of power results in a complete vanishing of the spectral emission below 400 nm. A more detailed discussion of this - also for other rare-earth iodides - is given in [16].
In the introduction it has been outlined that molecular emission causes substantial arc constriction. This constriction must lead to a loss of wall stabilization. In case that it is impossible to decrease the diameter of the discharge vessel accordingly, to reestablish wall stabilization, an electronically supported acoustic stabilization is required to ensure safe operation of such an arc. This means that a lamp of this type can not be operated at a conventional ballast, but needs a special ECG. The discharges investigated for this contribution were driven by a current amplifier which supplied a rectangular waveform with an amplitude modulation as proposed in [17].

PHOTOMETRIC DATA OF A HG-FREE HID DOMINATED BY MOLECULAR EMISSION

The results of the previous sections shall now be analyzed with the focus on photometric data. A series of spatially integrated spectra measured in an Ulbricht sphere is plotted in fig. 5. The measurements have been done for different power levels between 70 and 270 W. The corresponding photometric data derived from these spectra are displayed in fig. 6.

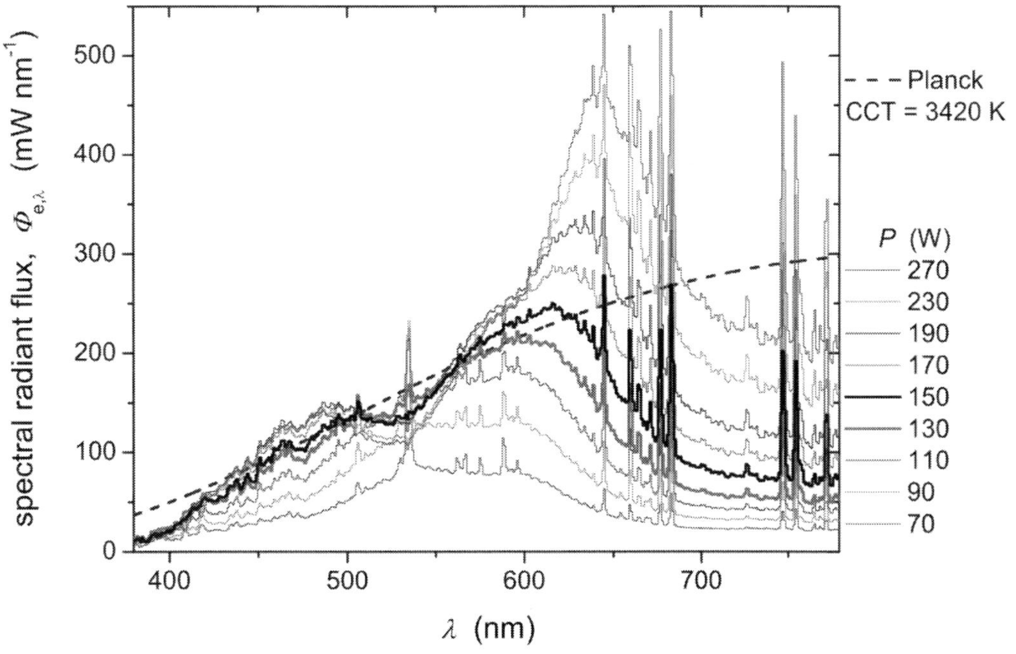

Figure 5. Spectral radiant flux measured in an Ulbricht sphere for different power levels between 70 and 270 W. For the bold traces at 130 W (gray) and 150 W (black) as well as the Planck curve (dashed line) see discussion in text.

From figure 5 it becomes obvious that with increasing power the dominant wavelength of the spectra shifts from rather green towards red. This "shift" can be related to ΔC values shown in fig. 6a. As known a ΔC value for a given spectrum defines the distance of its chromaticity coordinates from the Planck curve and in this way it is able to reflect the general color impression of the emission. For light sources which should meet the requirements for general lighting applications the corresponding ΔC values have to be small, to allow for the impression of a really white emitting light source without color fault. For the spectra of fig. 5 small ΔC values are achieved for power levels of about 140 W (see fig. 6a). Having this in mind the reason for the special bold marking of the traces for 130 and 150 W is becoming transparent. They enclose the spectrum at the operating point of the lamp being investigated.

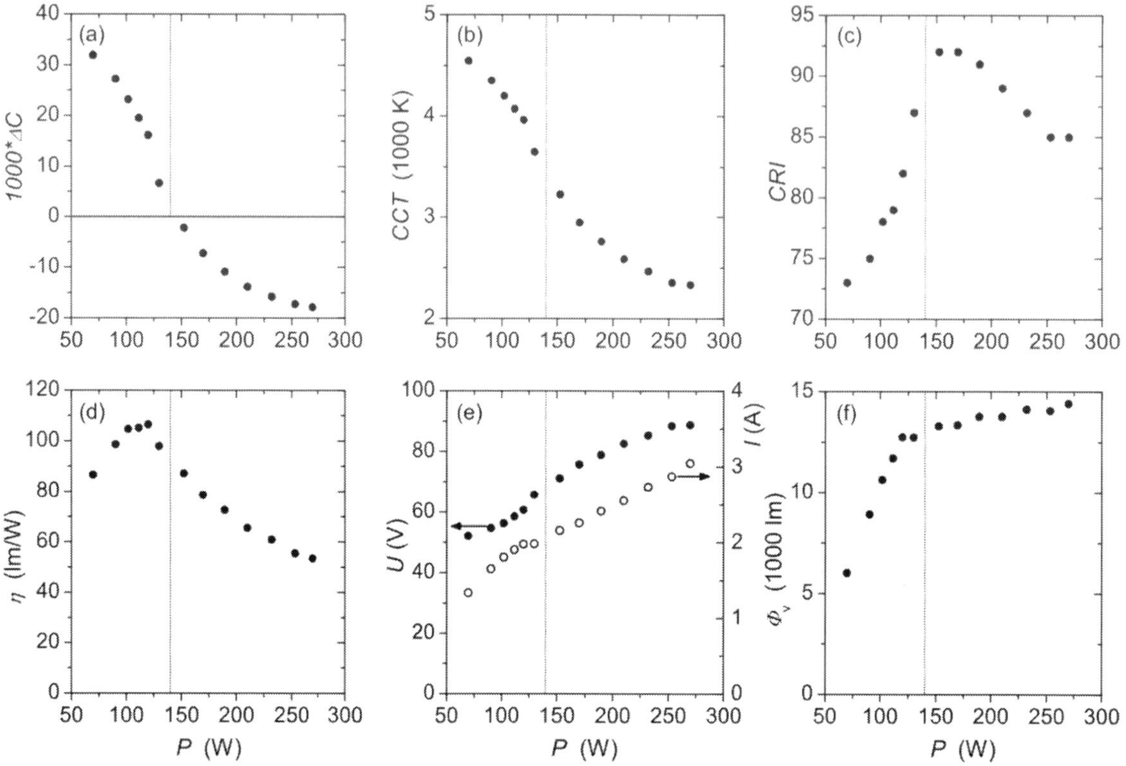

Figure 6. Photometric data derived from spectra in fig. 5 as well as electrical data of TmI-based discharge: (a) deviation from Planck, (b) correlated color temperature, (c) color rendering index, (d) efficacy, (e) discharge voltage and current, (f) luminous flux.

Next, in fig. 6b the correlated color temperatures for all spectral traces of fig. 5 are shown, which reflect the color "shift" of the dominant wavelength. At its operating point for the investigated lamp a color temperature of 3420 K has been achieved. To visualize the remaining spectral differences of the lamp's spectrum at the operating point from a blackbody radiation at 3420 K, the corresponding Planck curve has been included into fig. 5. The comparison of the Planck curve with the bold marked spectra shows that the latter render a Planck-like spectrum - at least up to 630 nm.

Bearing in mind the previous facts the behavior shown in fig. 6c is becoming obvious. As expected, at the operating point of 140 W the color rendering index, CRI, of the lamp is excellent and is reaching a value of about 90.

Further, fig. 6d shows the efficacies, η, for different power loads of the lamp. As demonstrated, the efficacy is going through a maximum of about 110 lm/W. Unfortunately, at maximum efficacy the color difference is still large ($\Delta C > 15 \cdot 10^{-3}$ in fig. 6a). Nevertheless, at the operating point a considerably high efficacy of 90 lm/W is still achieved.

In the introduction it has been mentioned that one important boundary condition for a technical realization of Hg-free discharge lamps is that it is possible to achieve a reasonably high operation voltage. Fig. 6e now reveals encouraging results. At the operating point of the lamp a discharge voltage of about 70 V at a current of about 2 A could be measured.

Finally, a few explanations on the emitted luminous flux should be given. Fig. 6f demonstrates that the luminous flux is rising steeply with increasing power until intermediate power levels of about 120 W are achieved. According to the earlier discussion, absorption processes are going to dominate the spectrum at higher lamp power loads. These absorption processes explain the insetting saturation of the luminous flux above 120 W.

One complicating factor has to be mentioned: Fig. 6a demonstrates that the dependence of the ΔC values on the power change is rather strong. The latter implies that the lamp will respond to small power changes rather sensitively with corresponding color changes. There are numerous reasons for this reaction. Power shifts would not only effect the temperature profile inside the arc, but also the evaporation of the salts due to this temperature profile change, which in turn directly influences the density of molecular radiators.

SUMMARY

High intensity discharges dominated by molecular radiation do in fact offer an alternative approach in replacing mercury. They are able to deliver competitive efficacy of about 90 lm/W, an excellent CRI of 90 and a CCT of 3420 K at the operating point near Planck. Moreover the operation voltages can reach 70 V.
The emission spectrum generated by HID lamps of this type differs significantly from those of conventional lamps. The reason for this is the dominance of molecular radiation processes. In comparison to conventional HID lamps atomic contributions are usually rather small. In the present case they amount to less than about 10% of the total intensity in the visible range.

REFERENCES

[1] G. Hartel, H. Schöpp, H. Hess, and L. Hitzschke. Radiation from an alternating current high-pressure mercury discharge: A comparison between experiments and model calculations. *J. Appl. Phys.*, 85(10):7076–7088, 1999.

[2] B. Schalk. *Funktionsprinzip des Entladungsplasmas quecksilberhaltiger Metallhalogenidlampen.* PhD thesis, TU München, 2007.

[3] B. Schalk, G. Hartel, M. Käning, and L. Hitzschke. Influence of power balance and radiation transport on temperature distributions of HID discharges. In *Proc. 11th Int. Symp. Sci. and Technol. of Light Sources*, page 323, 2007.

[4] K. Günther, T. Hartmann, and H. Sarroukh. Hg free ceramic automotive headlight lamps. In *Proc. 10th Int. Symp. Sci. and Technol. of Light Sources*, pages 219–220, 2004.

[5] M. Born and M. Strösser. Fundamental investigations of mercury-free automotive discharge lamps. *J. Phys. D: Appl. Phys.*, 40(13):3823–3828, 2007.

[6] R. Hilbig, A. Körber, J. Baier, and R. Scholl. Molecular discharges as light sources. In *Proc. 10th Int. Symp. Sci. and Technol. of Light Sources*, pages 75–84, 2004.

[7] M. Born. Untersuchungen zum Ersatz des Quecksilbers in keramischen Hochdruckgasentladungen mit Hilfe von metallischem Zink. ISBN: 3-89653-788-1, 2000.

[8] M. Born. Mercury free high pressure discharge lamps. In *Proc. 9th Int. Symp. Sci. and Technol. of Light Sources*, page 43, 2001.

[9] M. Born, H. Giese, and I. Niemann. Abschlußbericht BMBF-Projekt 13N8264, 2005.

[10] M. Ohnesorge. *Untersuchungen zur Hochtemperaturchemie quecksilberfreier Metallhalogenid-Entladungslampen mit keramischem Brenner.* PhD thesis, TH Aachen, 2005.

[11] L. Hitzschke, G. Hartel, A. Kloss, K. Stockwald, B. Schalk, and M. Käning. Abschlußbericht BMBF-Projekt 13N8261, 2005.

[12] M. Käning, B. Schalk, and H. Schneidenbach. Experimental determination of parameters for molecular continuum radiation of rare-earth iodides. *J. Phys. D: Appl. Phys.*, 40(13):3815–3822, 2007.

[13] L. Hitzschke, M. Käning, B. Schalk, St. Franke, R. Methling, H. Hess, H. Schöpp, and H. Schneidenbach. Patent WO2008/012228.

[14] St. Franke, R. Methling, H. Hess, H. Schneidenbach, H. Schöpp, L. Hitzschke, M. Käning, and B. Schalk. Mercury-free high-intensity discharge with high luminous efficacy and good colour rendering index. *J. Phys. D: Appl. Phys.*, 40(13):3836–3841, 2007.

[15] H. Schneidenbach and St Franke. Basic concepts of temperature determination from self-reversed spectral lines. *J. Phys. D: Appl. Phys.*, 41(14):144016, 2008.

[16] R. Methling, St. Franke, L. Hitzschke, M. Käning, and B. Schalk. Molecular Radiation from overcharged Hg-free HID. In *Proc. 12th Int. Symp. Sci. and Technol. of Light Sources and White LEDs And Solid State Lighting 3rd International Conference*, Eindhoven, Netherlands, 2010.

[17] J. A. Olsen, W. P. Moskowitz, P. B. Newell, and N. Brates. System for and method of operating a mercury free discharge lamp. *US-Patent 6,124,683*, 2000.

Ballasts and drivers for conventional and solid state light sources

J. Lester[1], W. Kaiser[2], A. El-Deib[3], F. Dawson[3]
[1] *Consultant, 83 Wood Drive, Essex, MA 01929, USA*
[2] *Depto. Engenharia de Energia e Automaçao Elétricas, Escola Politécnica da Universidade de Sao Paulo, CEP 05508-900 Sao Paulo, SP, BRAZIL*
[3] *Electrical and Computer Engineering Dept., University of Toronto, 10 King's College Road, Toronto, ON M5S3G4, Canada*
Contact: dawson@ele.utoronto.ca

ABSTRACT

The most commonly used ballasts and drivers for conventional discharge and solid state light emitting diode (LED) light sources are described in this review paper. Some of the light sources' critical properties are briefly discussed, for example, the impact of temperature on fluorescent lamp efficacy, acoustic resonance on the high intensity discharge (HID) lamp stability and junction temperature rise on the deterioration of the LEDs optical output power. These three light sources have different properties and therefore different types of ballasts and control methods are suggested for each type of light source. Voltage fed and current fed DC-AC resonant topologies for fluorescent lamps and HID lamps are reviewed. The drivers for the LEDs, which include the conventional DC-DC converters in addition to some newly proposed topologies, are also demonstrated.

GENERAL

Lighting ballasts and drivers have changed over the years as electronic components developed. Initially most lighting circuits were based on inductors, transformers, and capacitors and operated at line frequencies. These circuits tended to be large, heavy, and not very efficient. They had the advantages of reliability, low cost, and were largely environmentally insensitive. Cost, reliability, and environmental sensitivity were the main weaknesses of early electronic lighting circuits. Electronic circuits were developed in the 1950s and 1960s but did not begin to see common lighting use until the 1970s when the electronic components evolved to more acceptable levels. Today, operating costs and government regulations effectively ban the old magnetic circuits for general lighting.
The following factors, to name a few, impact the design of ballast and driver circuits for lamps:

- Low cost – There is plenty of competition in the ballast field. Sales margins are low. Every $0.01 US is important especially when a company sells many millions of ballasts a month.
- Simple – This keeps design time short and product costs low.
- Reliable – The cost to replace a ballast under warrantee is much higher than the profit made by the sale of the ballast. Parts count is minimized to increase reliability. Ballast companies work hard to keep their customers happy.
- Small – Available ballast space in fixtures keeps shrinking. Also shipping costs are lowered.
- Performance – There are requirements for input power factor, efficiency, input voltage range, compliance, electromagnetic immunity (EMI) and safety.

FLUORESCENT LAMPS

The initial concept of a fluorescent lamp evolved over nearly a 100 year period from the mid 1800s to the first lamps in production in 1938 [1]. High frequency ballasts for fluorescent lamps were suggested in the 1950s but reliable ballasts only started to sell well in the 1980s. Fluorescent lamps and ballasts work together as a system. It is possible to change one or the other to optimize the system. Lamp and ballast engineers should work together when a new lamp is being developed. In order to design a fluorescent lamp ballast, one needs to understand the basic properties of the lamp.

Lamp properties

The lamp voltage depends on many parameters such as the mercury vapour pressure, inert fill gas composition and pressure, lamp diameter, and lamp length. The lamp's starting and operating voltages can be somewhat independently controlled by adjusting these variables [2]. The lamp designer can lower the starting voltage if this will lower ballast cost and therefore system cost for example.

The amount of mercury vapour in the lamp controls the generated light efficiency. When designing a ballast, it is necessary to control the lamp's mercury vapour pressure. This can be done with a cold spot temperature controller. The temperature where the light production efficiency is highest for a simple mercury lamp is near 40 °C.

Some lamps must be operated at higher temperatures due to small fixtures or high lamp powers. These lamps use an amalgam of mercury and other metals. They will typically have two lumen peaks; one near 70 °C and one near 110 °C depending on the amalgam materials used. Amalgams are usually located in an exhaust tube.

Lamp efficacy, LPW (lumens per watt), increases as ballast frequency increases due to the following two primary reasons. Firstly, anode oscillations disappear as the operating frequency increases. Secondly, at low operating frequencies the ionization of the mercury gas dissipates when the arc current goes through a zero crossing so some energy has to be put into the discharge to re-ionize the mercury every half cycle of the line voltage. At operating frequencies above the ionization frequency, the gas stays ionized through the zero crossings. LPW can increase 15% if the lamp's operating frequency is well above the ionization frequency of the discharge. Several parameters impact the discharge's ionization frequency but of interest to ballast designers is the lamp's diameter. The ionization frequency decreases as the square of the lamp's diameter [2].

Fluorescent ballasts

Figure 1 shows a typical block diagram of the ballast for general lighting lamps. The power line connects to the Electromagnetic Interference (EMI) filter which may take up to ¼ of the area within the ballast. All electronic ballasts will have an EMI filter to satisfy government regulations.

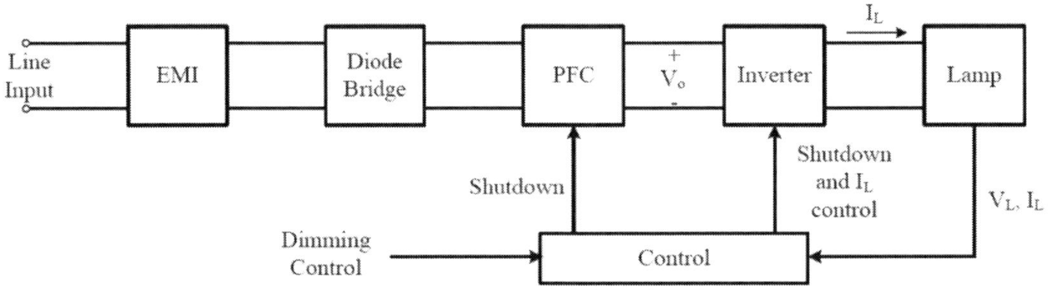

Figure 1. A typical block diagram of the ballast for general lighting lamps

The output voltage of this filter is rectified and converted to a pulsing DC voltage and then fed into a Power Factor Correction (PFC) stage. The PFC stage shapes the current waveform on the power line to appear as a sine wave. Power companies often charge more for their electricity when the ballast input current waveform is poor. Low wattage screw-in compact fluorescent lamps usually do not have a PFC stage in order to minimize cost.

The output of the PFC stage is a regulated DC voltage, V_o, that is fed to a DC voltage to AC current inverter stage to feed the lamp. A control circuit may or may not be present. The control circuit may monitor the lamp and shut down the ballast for an end of life (EOL) condition [3], a no-start condition, or may regulate the lamp's current. It may also provide dimming control, or external digital interface control with a lighting control system. It can also be used to sequence the inverter from the start mode to the run mode in sequence start ballasts. There are two common electronic ballast circuits in use today. These are discussed next.

Current fed parallel resonant ballast

Figure 2 shows the circuit of the current fed parallel resonant ballast. The two switches are on alternately and power a resonant circuit comprised of L_1 and C_1. The switching pattern causes a sine wave voltage to appear across C_1. That voltage is often stepped up to 600 to 1200 V by the circuit's transformer. Capacitor C_2 represents a high impedance that limits the current delivered to the lamp. The lamp's impedance is much lower than that of C_2 so that the inverter delivers essentially constant current, I_L to the lamp. In this example, the lamp is instant started by the high voltage appearing at the output of the transformer. The transformer also provides electrical isolation from the power line. This simple, low-cost circuit is used for cold cathode lamps in backlighting applications and has been popular for years ballasting T8 and T12 lamps [4, 5, 6].

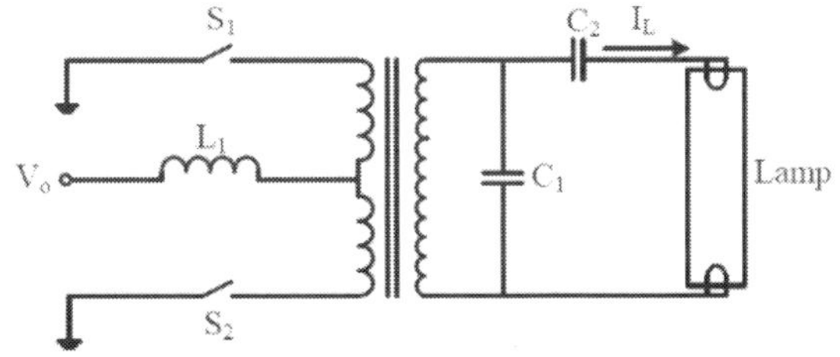

Figure 2. Circuit diagram of a current fed parallel resonant ballast

Voltage fed series resonant ballast

Figure 3 shows the circuit of the voltage fed series resonant ballast. This is probably the most popular ballast used today and has been analyzed by many [5-12]. It is used in 90% of the popular screw-in compact fluorescent lamps, many T4, T5, T8, and T12 ballasts and most of the electrodeless lamp ballasts on the market. There is nearly an infinite combination of L_1, C_1, and C_2 that can be selected to give the desired current to the lamp. There is one combination however that is optimal as discussed below. The two switches are on alternately feeding a square wave to inductor L_1. A control circuit can be used to drive the switches or feedback can be used to make the circuit self oscillate. The driving square wave can be modelled as a series of decreasing amplitude sine waves through Fourier analysis. The harmonics only contribute 8.8% to the RMS value of the square wave. Inductor L_1 and capacitors C_1 and C_2 form a second order filter that blocks the harmonics from reaching the lamp current I_L so the contribution of the harmonics to the lamp's current can be neglected. Therefore source voltage, V_s, driving L_1 can be modelled as a sine wave with the fundamental frequency.

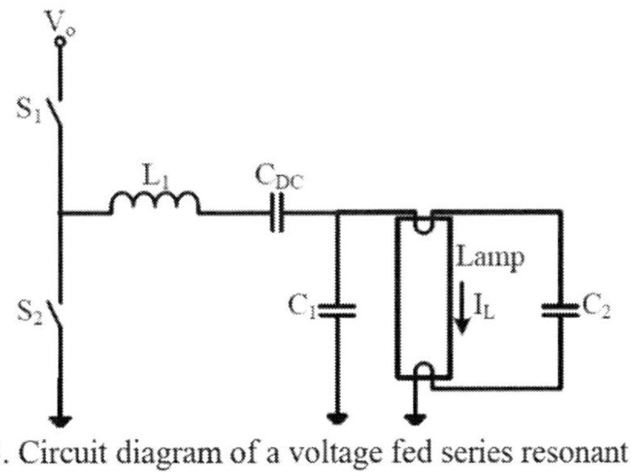

Figure 3. Circuit diagram of a voltage fed series resonant ballast

If this circuit is operated at the resonant frequency F_o of L_1 and the equivalent capacitance C_{eq} of C_1 and C_2, the current delivered to the lamp is constant and independent of the lamp load. When the lamp is off, the load on V_s is a series resonant circuit which at F_o is a short circuit. High current circulates in the circuit and a high voltage is generated across L_1 and C_{eq}. The high voltage across C_{eq} ignites the lamp. It can start and power an arc discharge lamp without control circuitry and steps up the voltage from V_s through resonant action. This boosting can be typically 3 to 5 times V_s depending on the loss in the resonant loop.

The power conversion efficiency of this ballast depends on the ratio Q of the lamp voltage to the source voltage V_s. This circuit works best with higher voltage lamps where Q is larger than one, which is often the case. The operating frequency, based on the resonance of LC, is always greater than the natural resonant frequency of the total RLC circuit causing the switches to turn on under zero voltage conditions thereby reducing the switching loss. If the operating frequency drops below the natural RLC frequency, the switches will be driving a capacitive load with associated high current spikes. This should be avoided during all phases of the lamp's operation.

HIGH INTENSITY DISCHARGE (HID) LAMPS

The acronym HID (High-Intensity Discharge) designates three types of high-pressure (near or above 100 kPa), discharge lamps: mercury vapour, metal halide and sodium. The light-producing element of these lamp types is a well-stabilized arc discharge contained within a refractory envelope (a translucent or transparent fused quartz or alumina arc tube) with wall loading in excess of 3 W/cm^2. The electrodes consist of a high fusion point metal support coated with an electron emissive material. The high operating temperature and pressure require an outer enclosure to provide thermal isolation and protect users from the rupture of the arc tube. The space within the two enclosures is evacuated and may contain a getter to absorb residual contaminants.

High-pressure discharge lamp operation issues

The starting voltage specifications to initiate the electrical discharge (plasma) in high-pressure lamps are much higher than those for low-pressure lamps. Moreover, the voltage required to force ignition depends on the initial temperature inside the discharge tube. That is the reason why it is much more difficult to initiate a discharge in a lamp that was initially operational and was turned off momentarily (for example during a power outage) and then turned on again – an event called "hot strike ignition voltage". This voltage is in the range of tens of kilovolts. There is no economic approach to address this issue and the majority of manufacturers recommend that the lamp be allowed to cool down for a specific period of time before ignition is reattempted.

The use of an electronic ballast for HID lamps faces a further problem namely the acoustic resonance phenomenon. The common explanation of the physical background for the acoustic resonance phenomenon is that when a HID lamp is operated with periodic power, the current is modulated and the intensity of the discharge and therefore the pressure of the filling gas inside the arc tube changes with the same frequency as the power. The fluctuation of the pressure, which originates in the vicinity of the discharge, propagates through the gaseous medium inside the arc tube.

Based on the theory of acoustic streaming, the physical causes of acoustic instabilities have been proposed. If the power frequency is at or close to an eigen-frequency of the arc discharge tube, travelling pressure waves will appear and distort the arc discharge path [13-16]. These waves reflect against the discharge tube walls and at certain frequencies standing pressure waves, called acoustic resonance, occur. Acoustic resonance with sufficient acoustic energy may cause undesired effects in HID lamps. Such effects range from a nearly imperceptible arc movement to extreme displacement of the discharge causing flicker, lamp voltage rise and even cracking of the discharge tube [17]. The critical frequency values depend on distance between electrodes, tube geometry and gas composition and pressure. Also, lamp aging leads to changes in lamp resonant frequencies since filling gases and lamp electrodes change with time.

HID lamp ballast

The traditional AC mains frequency ballasts are electromagnetic devices with built in igniter that limit the current flow to the lamp but do not change the frequency of the input power. Electronic ballasts usually change the frequency of the power from the standard mains frequency to a higher value. The basic idea of most ballasts is to reduce the power harmonics below the acoustic resonance excitation threshold. According to the output current frequency, the HID electronic ballasts can be classified in four categories as shown in table 1 [18]. Some manufacturers [19-21] suggest DC operation for small wattage metal halide lamps. The electronic ballast in this case is a constant output current chopper with an external igniter.

A ballast with a low frequency square-wave output should ideally be a current source that produces only a dc power component and thus cannot excite acoustic resonance phenomena. In practice, a square wave current source is impossible to achieve and a trapezoidal waveform with fast rise and fall times is produced instead. The first significant power harmonic occurs at twice the operating frequency that lies below the lowest first order acoustic resonant mode and whose energy is below the threshold level value that can excite an acoustic resonance. Practical designs impose converter output frequencies between 200 and 400 Hz with the lowest order power harmonic frequencies between 400 and 800 Hz. Due to the topology of electronic ballasts for such an application, there is a high frequency ripple in the current or voltage waveform. The ripple content with sufficient energy at the proper frequency is able to excite acoustic resonance in HID lamps [22]. The threshold value for high frequency ripples that excites an acoustic mode and destabilizes the lamp is still an open question for the industry and standard organizations [23].

High frequency (HF) ballasts need some strategies to make them resistant to acoustic resonance. Several lamp supply strategies are found in the literature:

i) Establishing HF operation in a free acoustic resonance zone [24]. The tuning process requires the predetermination of free acoustic resonance zones that depends on lamp model and is affected by lamp aging;

ii) Adopting adequate modulation strategies.

There are several papers in the literature describing different approaches: Injection of a white noise component along with the main switching frequency, spreading the input energy over a wide spectrum so as to minimize the magnitude of the input energy in a certain frequency range [25]; Using random switching schemes with amplitude limitation of each power harmonic, maintaining constant total power delivered to the lamp [26]; Injection of approaching frequencies signals confined within a selective band to avoid acoustic resonance under high frequency conditions [27].

The Very High Frequency (VHF) ballasts operate in the MHz range where acoustic resonance is not detectable. There are some papers in the literature reporting ballasts operating resonance free under these conditions [28-30].

One or more of the following factors motivates the move towards electronic ballasts: decreasing the ballast weight and volume, dimming feature, lumen maintenance over lamp lifetime by power regulation which can also counter the effect of arc voltage changing with lamp life and ballast shut down at lamp failure.

HID lamp ballast architectures

The literature contains a variety of HID lamp electronic ballast topologies with more or less complexity. Table 2 summarizes the general approaches for high pressure lamp ballasts although there are small differences between various HID lamp types.

Type 1 ballast has a resonant full-bridge inverter with an external igniter. A capacitor prevents a DC component in the lamp and the inductor limits lamp current. The lamp is shunted by an LC filter that ensures that high frequencies bypass the lamp. Type 1 ballast with internal resonant igniter has a half bridge LCC [31] inverter designed to operate at two frequencies. Prior to ignition, the lamp appears as an open circuit and the operating frequency is adjusted to ensure ignition. Current is sensed in the lamp and is used in a feedback system to maintain constant power through the lamp. After ignition, the impedance of the lamp drops and the switching frequency is adjusted to maintain a specific lamp power. Type 2 ballast has an asymmetrical bridge LCC inverter and its operation is quite similar to those described under the type 1 ballast category with internal igniter.

Type 3 ballast uses a controlled current source that is inverted to produce the low frequency square wave ac current. The external igniter normally uses a SIDAC. The SIDAC behaviour is similar to an avalanche diode: below a certain threshold voltage the device is an open circuit and above the threshold voltage, the SIDAC breaks down and provides a low impedance path for the previously charged capacitor.

Table 2 HID Lamp Electronic Ballast Architectures

Type 1: Half or full bridge resonant inverter	External igniter	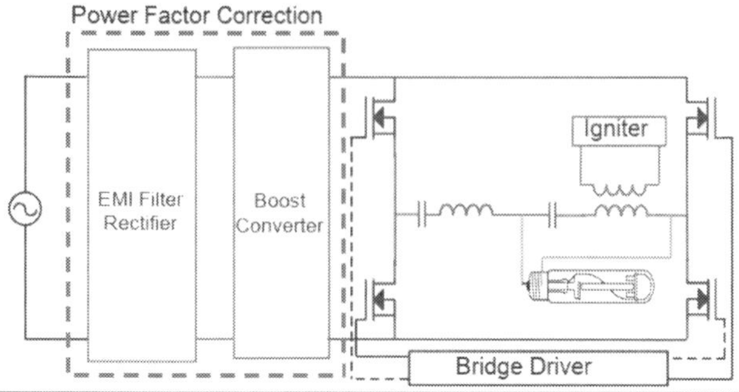
	Integrated resonant igniter.	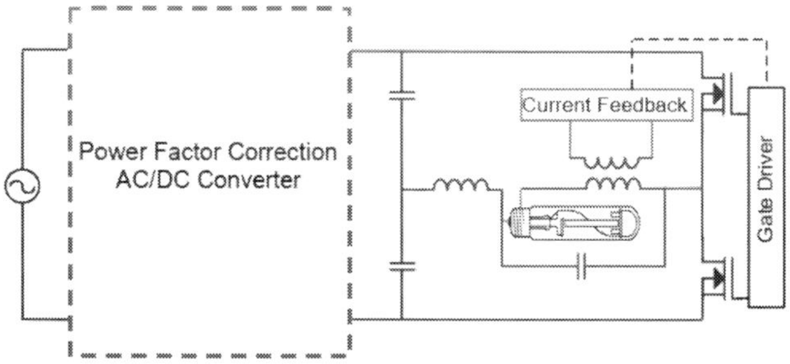
Type 2: Asymmetrical bridge resonant inverter with integrated resonant igniter		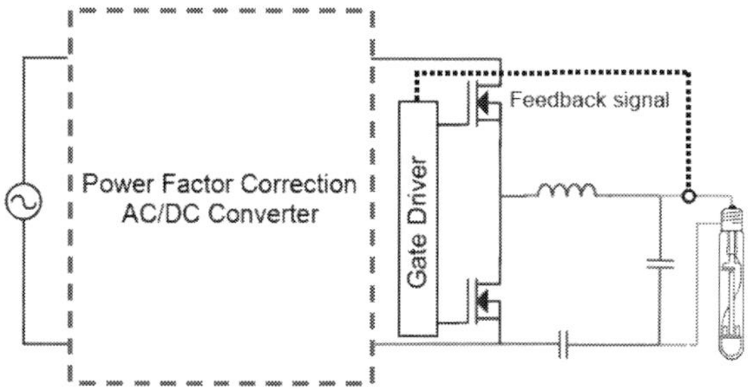
Type 3: Square wave current source and external igniter		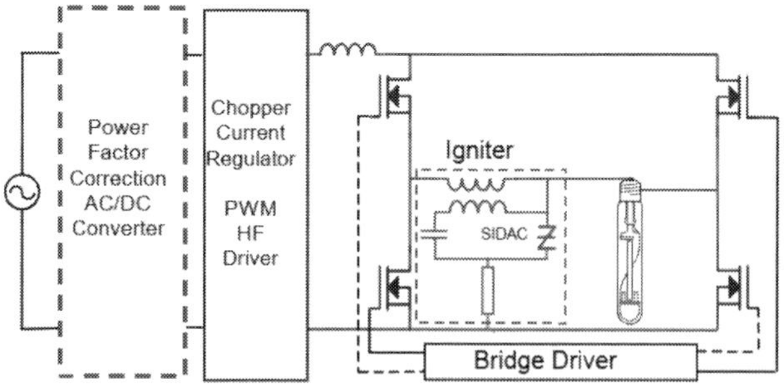

LIGHT EMITTING DIODE (LED) LAMPS

The light emitting diodes (LEDs) are attractive light sources due to their high efficiency, long life time and low maintenance requirements. One of the main obstacles facing the LED as a light source is the degradation of its optical output due to the increase of the junction temperature. The temperature increase is the result of the conduction losses in the LED. The junction temperature also affects the wavelength produced by the LED and therefore the light colour generated also depends on the junction temperature. A 10°C change in the junction temperature results in a change in the colour point of Δuv=0.005, as noted in [39]. Therefore different control techniques have been proposed to stabilize the colour produced by the LED [32, 33, 34].

Effect of the input current waveform

The luminous flux and colour produced by the LED are related to the instantaneous value of the input current. Therefore if LED dimming is required, the LED performance in terms of efficacy and colour will change. The current waveforms suggested in the literature include the following:

1. Constant current
2. Constant current with an imposed ripple of a specific frequency and amplitude
3. Pulsed width modulated (PWM) current with controlled duty cycle, frequency and peak current.

In [35], a comparison was performed between the previously mentioned current waveforms in terms of the LED efficacy. The result of this experimental study indicates that the pure DC current produces the highest efficacy over the whole dimming range. The DC current with linear ripple produced efficacies which are very close to the DC current when the ripple percentage was under 20% of the average current.

The PWM pulsed current waveform achieves the same efficacy as the DC current at high values of duty ratio corresponding to high output luminous flux. The PWM current produces a constant efficacy over the whole range of duty ratios.

The LED manufacturers always produce their datasheets based on a pure DC current fed to the LED which gives the best performance. However, for the power electronics driver it is too expensive to design a driver that can produce pure DC current waveform because lower ripple requires larger energy storage elements impacting the power density and cost of the system. Therefore, the selection of the current waveform to be used should be based on maximizing the efficiency of the whole system, increasing the power density and satisfying the required lifetime with the minimum cost.

LED lamp driver architectures

The LED should be operated as a current driven device because the optical output of the LED is determined by the input current. There have been many approaches suggested in the literature to supply this current. The drivers for LED applications can be categorized as belonging to one of the following two main types; non-isolated drivers and isolated drivers [36].

Non-isolated drivers do not contain galvanic isolation and therefore this category is suitable for battery connected LEDs, for example, emergency lighting systems. On the other hand, the isolated drivers include a transformer that performs the required galvanic isolation for powering from the AC grid. The isolated drivers are preferably operated at high switching frequencies to minimize the size of the transformer used for isolation. Also, isolated drivers must be connected to the AC mains through a power factor correction circuit (PFC) to comply with the power factor and the total harmonic distortion (THD) requirements.

The non-isolated drivers that have been suggested include most of the conventional DC-DC converters like Buck [33], Boost [40], Buck-boost [42] and Cuk [41] converters. A block diagram for these drivers is given in table 3. There are many options for LED drivers because the choice of the suitable driver depends on matching the input power source and the requirements of the LED as a load. If the power source is the AC mains, the input voltage has to be stepped down to the required voltage level of the LED string. In the case of stepping down the voltage level, the Buck converter is the suitable driver but for battery systems it might be required to step up the voltage therefore the Boost converter would be the better choice.

The main issue with these conventional converters is the presence of the electrolytic capacitor at the output. This electrolytic capacitor has a short life time compared to the LED lifetime. Therefore the electrolytic capacitor has to be oversized to meet the required lifetime of the system. Another disadvantage of having an electrolytic capacitor is its large size and therefore the power density of the driver will be low which is not permitted in some applications like the backlighting of LCD screens. Yet, another factor that should be considered in the design of the LED driver is the cost as it should be minimized as much as possible to maintain the advantage of a low cost lighting system.

The flyback converter is an example of an isolated driver that was proposed in [31], [36] and [37]. As discussed above, pulsed current waveforms can also be used to drive the LEDs. The flyback converter can be operated without a capacitor at the output to produce current pulses in the LED load. This allows for direct control of the LED current without adding a series sense resistance to the circuit. Another isolated converter that can directly control the LED current is the series resonant converter as suggested in [36] and [38]. The load current in the series resonant converter has the shape of sinusoidal pulses and the average current of these pulses can be controlled by the switching frequency. The block diagram for the isolated drivers is also shown in table 3.

A different concept for driving a number of LED strings is to use a multilevel converter which includes a main DC bus and a number of capacitors connected in series [39]. Each of these capacitors feeds a modular driver that supplies current to an LED string. The advantage of this approach is to have individual control over the different LED strings. Instead of connecting all the LEDs in series they are divided into strings and each string is driven by a separate driver to increase the system's reliability. Also this configuration can be used in the RGB LED systems, where each LED group can be controlled independently to control the chromaticity coordinates of the produced light.

A number of papers have investigated different methods for integrating the power factor correction PFC and the driver required to produce the current for the LED string in order to enhance the efficiency of the system and to minimize its cost.

Another topology was suggested in [43] where buck and flyback converters were integrated to drive the LEDs directly from the AC mains as shown in table 3. The buck converter performs the PFC and the flyback converter delivers the required current to the LEDs. Both converters are operated in DCM to provide the PFC action and fast output dynamics which is important to achieve dimming operation using the PWM approach. This integrated converter is capable of working under a wide range of input AC voltages however the efficiency of the converter has a maximum of 80% at an input voltage of about 150Vrms.

Table 3 LED Lamp Electronic Driver Architectures

Non isolated LED Driver (Buck, Boost, Buck-Boost, Cuk)	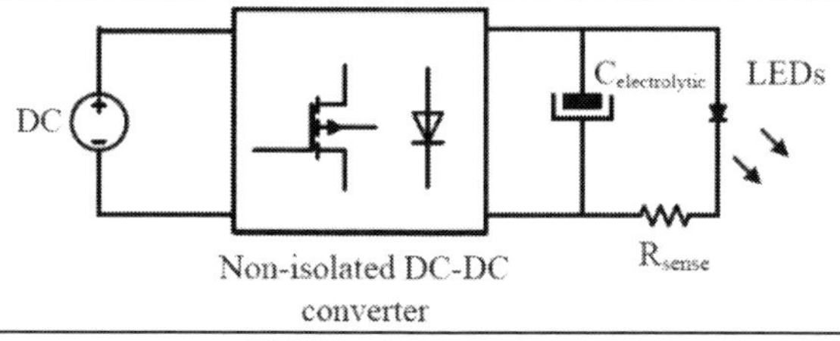
Isolated LED Driver (Flyback, series resonant)	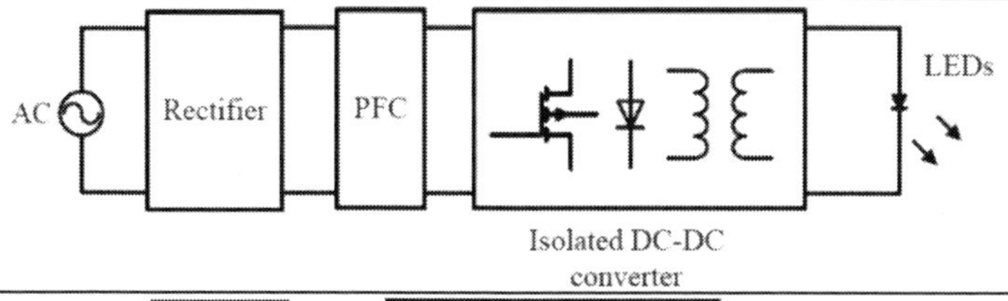
Integrated LED Driver (Integrated Buck-Flyback)	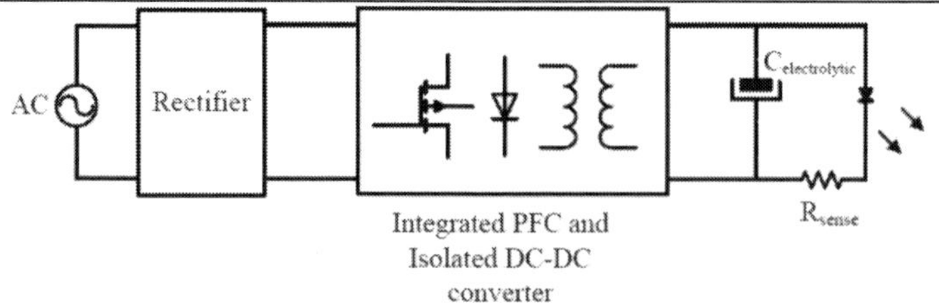

CONCLUSION

The most popular ballasts and drivers for the conventional and solid state lamps have been reviewed in this paper. The technology for the conventional lamps ballasts has matured. On the other hand, the drivers for the solid state lamps still face a number of challenges and therefore most of the research efforts have been directed towards this area. The challenge still facing all types of light source ballasts is finding a sufficiently accurate model for the lamp that can be used in the design process of the ballast. This model should demonstrate the important features of the lamp under different operating conditions and concurrently be suitable for integration within simulation software packages used for the ballasts' design especially in terms of short simulation times.

REFERENCES

[1] Inman G., "Electric Discharge Lamp", US Patent 2,259,040, Oct. 14, 1941

[2] Waymouth J. F., "Electric Discharge Lamps", Cambridge, MA, MIT Press, 1978.

[3] Lester J. N., "Fluorescent Lamp End-Of-Life Shutdown Sense Methods", Journal of the Illuminating Engineering Society, Vol. 31, No. 2, pp. 78-85, Summer, 2002

[4] Bairanzade M.. "Electronic Lamp Ballast Design", Motorola Application Note AN1543, 1995.

[5] Baxandall P. J., "Transistor Sine-Wave LC Oscillators", IEE (London) International Convention on Transistors and Associated Semiconductor Devices, pt. B, suppl. 16, May 1959, pp. 748-758.

[6] Chudobiak W. and Page D. F., "Frequency and Power Limitations of Class-D Transistor Amplifiers", IEEE Journal of Solid-State Circuits, Vol sc-4, No. 1, Feb. 1969, pp. 2537

[7] Kaufman M. and Seidman A.H., Handbook of Electronics Calculations, McGraw-Hill, P7.1, 1979.

[8] Bairanzade M.. "The Electronic Control of Fluorescent Lamps". Motorola Application Note AN1049, 1990.

[9] Kazimierczuk M. K. and Szaraniec W., "Electronic Ballast for Fluorescent Lamps", IEEE Transactions on Power Electronics, Vol. 8, No. 4, Oct. 1993, pp. 386-395.

[10] El-Hamamsy S. A., "Design of High-Efficiency RF Class-D Power Amplifier". IEEE Transactions on Power Electronics, Vol. 9, No. 3, May 1994, pp. 297-308.

[11] Nilssen O. K., "Electronic Ballast Circuit for Fluorescent Lamps", US Patent 4,503,363, March 5, 1985

[12] Nerone L. R., "A Mathematical Model of the Class D Converter for Compact Fluorescent Ballasts", IEEE Transactions on Power Electronics, Vol. 10, No. 6, Nov. 1995, pp. 708-715.

[13] F. Afshar, "The Theory of Acoustic Resonance and Acoustic Instability in HID Lamps", LEUKOS, vol 5, no.1, pp. 26-38, July 2008.

[14] H. Witting, "Acoustic resonances in cylindrical high-pressure arc discharges", Journal of Applied Physics, vol. 49, no.5, pp. 2680-2683, 1978.

[15] H. P. Stormberg, R. Schäfer, "Excitation of acoustic instabilities in discharge lamps with pulsed supply voltage", Lighting Research & Technology, vol. 15, no.3, pp.127-132, 1983.

[16] J. W. Denneman, "Acoustic resonances in high frequency operated low wattage metal halide lamps", Philips Journal of Research, vol. 38, no. 415, pp. 263 - 272, 1983.

[17] P. M. Morse, K. U. Ingard, Theoretical Acoustics, McGraw-Hill Book Company, 1968.

[18] M.W. Fellows, "A study of the high intensity discharge lamp-electronic ballast interface", 38th IAS Annual Meeting Conference Record, vol. 2, pp. 1043 – 1048, 2003.

[19] J.M. Davenport, R. L. Hansler, "Direct current operation of low wattage metal halide lamps", Journal of the Illuminating Engineering Society, pp. 274-282, Oct. 1984.

[20] General Electric Co., "32W HALARC metal halide lamp application guidelines, HD-1502, Feb. 1987.

[21] R. J. Wood, "Low watt metal halide apparatus", U.S. patent 5 291 100, March 1994.

[22] F. Afshar, "Allowed Level of the Power Ripple in Low Frequency Square-Wave Ballasts of Metal-Halide Lamps", LEUKOS, vol 3, no.2, pp. 143-157, October 2006.

[23] J. Olsen, P. W. Moskowitz, "Detrimental effect of a small amount of ripple in a metal halide system", 40th IEEE IAS Annual Meeting Conference Records, vol. 3, pp. 1581–1587, 2005.

[24] E Rasch and E. Statnic, "Behavior of metal halide lamps with conventional and electronic ballasts," Journal of the Illuminating Engineering Society, pp. 88–96, summer 1991.

[25] L. Laskai, P. N. Enjefi, I. J. Pitel, "White-noise modulation of high-frequency high-intensity discharge lamp ballasts", IEEE Transactions on Industry Applications, vol. 34, no. 3, pp. 597-605, 1998.

[26] W. Yan, Y.K.E Ho, S.Y.R Hui, "Stability study and control methods for small-wattage high-intensity-discharge (HID) lamps", IEEE Transactions on Industry Applications, vol. 37, no. 5, pp. 1522-1530, Sept.-Oct. 2001.

[27] L. Chhun, P. Maussion, S. Bhosle, G. Zissis, O. Durrieu, "HPS lamp control by signal injection of approaching frequency to avoid acoustic resonances: Application to an LCC circuit", 13th European Conference on Power Electronics and Applications, 2009.

[28] M. Gulko, S. Ben-Yaakov, "A MHz electronic ballast for automotive-type HID lamps", IEEE Power Electronics Specialists Conference, pp. 39-45, 1997.

[29] M. H. Ohsato, Q. Mao, H. Ohgochi, T. Shimiw, G. Kimura, H. Takagi, "Megahertz operation of voltage-fed inverter for HID lamps using distributed constant line", IEEE Transactions on Industry Applications, vol.34, no. 4, pp. 747-751, 1998.

[30] R. Redl, J. D. Paul, "A new high frequency and high efficiency electronic ballast for HID Lamps: topology, analysis, design, and experimental results", Proceedings of the IEEE Applied Electronics Conference, vol.1, pp. 486-492, 1999.

[31] R.M. Nelms, T.D. Jones, M.C. Cosby Jr., "A comparison of resonant inverter topologies for HPS lamp ballasts", 28th IEEE IAS Annual Meeting Conference Records, vol. 3, pp. 2317-2322, 1993

[32] Muthu, S.; Schuurmans, F.J.P.; Pashley, M.D., "Red, green, and blue LEDs for white light illumination," Selected Topics in Quantum Electronics, IEEE Journal of , vol.8, no.2, pp.333-338, Mar/Apr 2002

[33] Bernitz, F.; Schallmoser, O.; Sowa, W., "Advanced Electronic Driver for Power LEDs with Integrated Colour Management," Industry Applications Conference, 2006. 41st IAS Annual Meeting. Conference Record of the 2006 IEEE , vol.5, no., pp.2604-2607, 8-12 Oct. 2006

[34] Ackermann, B.; Schulz, V.; Martiny, C.; Hilgers, A.; Zhu, X., "Control of LEDs," Industry Applications Conference, 2006. 41st IAS Annual Meeting. Conference Record of the 2006 IEEE , vol.5, no., pp.2608-2615, 8-12 Oct. 2006

[35] Sauerlander, G.; Hente, D.; Radermacher, H.; Waffenschmidt, E.; Jacobs, J., "Driver Electronics for LEDs," Industry Applications Conference, 2006. 41st IAS Annual Meeting. Conference Record of the 2006 IEEE , vol.5, no., pp.2621-2626, 8-12 Oct. 2006

[36] van der Broeck, Heinz; Sauerlander, Georg; Wendt, Matthias, "Power driver topologies and control schemes for LEDs," Applied Power Electronics Conference, APEC 2007 - Twenty Second Annual IEEE , vol., no., pp.1319-1325, Feb. 25 2007-March 1 2007

[37] Yuequan Hu; Huber, L.; Jovanovic, M.M., "Single-Stage Flyback Power-Factor-Correction Front-End for HB LED Application," Industry Applications Society Annual Meeting, 2009. IAS 2009, pp.1-8, 4-8 Oct. 2009

[38] Yu-Kang Lo; Kuan-Hung Wu; Kai-Jun Pai; Huang-Jen Chiu, "Design and Implementation of RGB LED Drivers for LCD Backlight Modules," Industrial Electronics, IEEE Transactions on , vol.56, no.12, pp.4862-4871, Dec. 2009

[39] Qingcong Hu; Zane, R., "LED Drive Circuit with Series Input Connected Converter Cells Operating in Continuous Conduction Mode," Applied Power Electronics Conference and Exposition, 2009. APEC 2009. Twenty-Fourth Annual IEEE , vol., no., pp.1511-1517, 15-19 Feb. 2009

[40] Chun-Yu Hsieh; Ke-Horng Chen, "Boost DC-DC converter with charge-recycling (CR) and fast reference tracking (FRT) techniques for high-efficiency and low-cost LED driver," Solid-State Circuits Conference, 2008. ESSCIRC 2008. 34th European , vol., no., pp.358-361, 15-19 Sept. 2008

[41] de Britto, J.R.; Demian, A.E.; de Freitas, L.C.; Farias, V.J.; Coelho, E.A.A.; Vieira, J.B., "A proposal of Led Lamp Driver for universal input using Cuk converter," Power Electronics Specialists Conference, 2008. PESC 2008. IEEE , vol., no., pp.2640-2644, 15-19 June 2008

[42] Jianwen Shao, "Single Stage Offline LED Driver," Applied Power Electronics Conference and Exposition, 2009. APEC 2009. Twenty-Fourth Annual IEEE , vol., no., pp.582-586, 15-19 Feb. 2009

[43] Gacio, D.; Alonso, J.M.; Calleja, A.J.; Garcia, J.; Rico-Secades, M., "A Universal-Input Single-Stage High-Power-Factor Power Supply for HB-LEDs Based on Integrated Buck-Flyback Converter," Applied Power Electronics Conference and Exposition, 2009. APEC 2009. Twenty-Fourth Annual IEEE , pp.570-576, 15-19 Feb. 2009

Compact Fluorescent Lamp End-Of-Life: Behaviour Analysis and Possible Solutions

Libo Wu, Gus Bandyopadhyay, Yang Hu
Lighting Technology for Asia, Siemens Limited China Corporate Technology,
No.1 Pudong Avenue, Shanghai, China, 200120
Contact: libo.wu@siemens.com

ABSTRACT

Self-ballasted Compact Fluorescent Lamp behaviour at End of Life (EOL) is investigated for voltage and temperature trends. Different solutions by electronics protection and by burner/plastic structure are reviewed and discussed. A detection and protection circuit is introduced which has advantages of low cost, small size and reliable protection. Experimental results based on IEC standards are included to verify the feasibility of the proposed solution.

Key words:
Compact Fluorescent Lamp, End-Of-Life, overheating, protection, burner voltage

1. CFLI EOL BEHAVIOR

Fluorescent lamps including self-ballasted Compact Fluorescent Lamps (CFLi) exhibit EOL behaviour when one or both the electrodes are depleted of its emissive material. The cathode fall voltage rises from around 12 volts to 50 volts, or higher. In electronic ballast with approximately constant current output characteristic, considerable heat is generated around the cathode. This may cause localized heating of the glass burner envelope in the vicinity of the cathode, resulting in rare cases, melting/charring of the plastic housing cap. Safety concerns in CFLi, arising from this localized overheating, have been eliminated by the industry by specifying non-flammable V-0 plastic[1], as well as through good ballast design and judicious selection of ballast components. Furthermore, a number of tests for EOL potential in CFLi have been defined and adopted by the industry[2] as part of lamp qualification process.

For this study, a T2 full spiral 23W/120V/60Hz CFLi lamp type is adopted for evaluation. Figure 1 shows experimental curves of the burner voltage and the plastic junction temperature during EOL. At 25℃ ambient, in the last 3 ~ 5 minutes operation, the burner voltage increases from 160 V RMS to 250 V RMS due to high cathode fall voltage of one filament. The temperature curves show that the temperature of one junction reaches more than 220℃, while another junction keeps around 150℃.

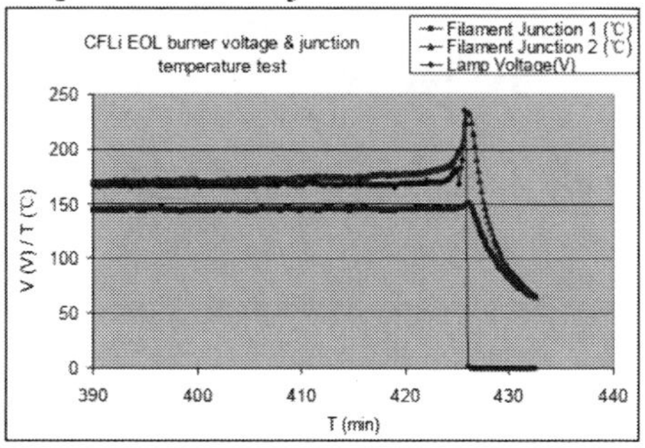

Figure 1: EOL behaviour of CFLi, Ta = 25℃

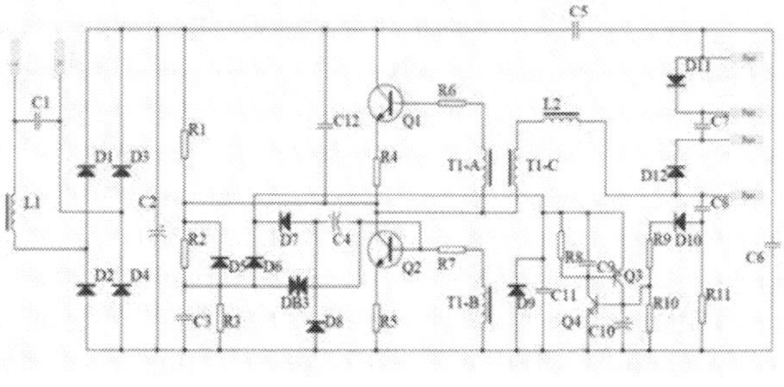

Figure 2: Schematic of the ballast of CFLi prototype with discrete component EOL protection

2. PROTECTION METHODS REVIEW

Some types of electronic ballasts[3], [4] have been designed with additional circuitry to sense the rise in burner voltage, or the asymmetric rectifier effect occurring on cathode depletion, and shut down the system. Some lighting controlling ICs are also designed with built-in EOL protection function. With

the integrated EOL protection function, it is easy to realize the high burner voltage detection or burner voltage DC component detection.
Some other approaches include adding a metal hydride deposit into the burner to realize arc quenching at the end of lamp life[5], or adding low thermal conductivity material to separate the heat from the cathode and the burner to the plastic cap of the lamp housing[6].

3. A BURNER VOLTAGE DETECTION AND PROTECTION CIRCUIT

For application in most common type of self-oscillating CFLi, cost effectiveness will be a crucial factor for successful EOL solutions. By comparing different circuits from cost and performance point of view, one EOL circuit that detects the high burner voltage and shuts down oscillation, is selected for further investigation. The ballast with burner voltage detection is shown in Figure 2. There are around 10~15 additional electronics components, of which most are low cost SMT types.
IEC asymmetric power test[2] has been done for the CFLi samples with the proposed EOL protection circuit. The ballast is protected before the asymmetric power exceeds 4W, as shown in Figure 3.

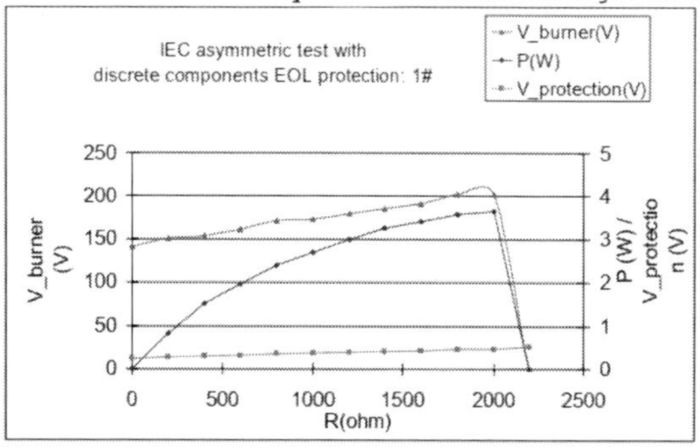

Figure 3: IEC asymmetric test with discrete components EOL protection

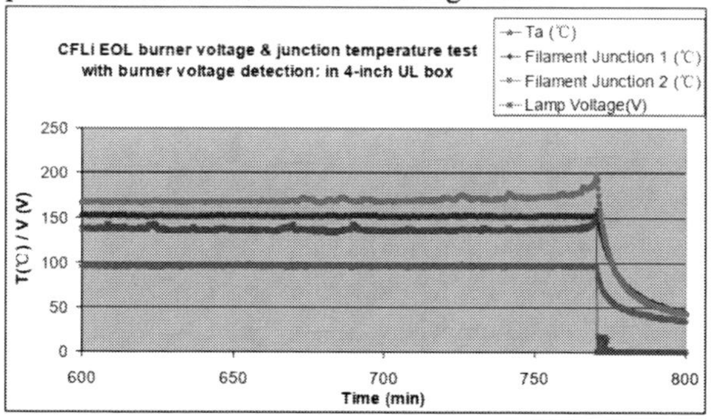

Figure 4: EOL behaviour of CFLi, with EOL, in 4-inch UL box

4. EXPERIMENTAL RESUTS & CONCLUSIONS

The experimental data and curves are recorded and shown as Figure 4. Test results show that when the lamp approaches the life end, one of the filaments (with lower emitter) will be overheated by the high cathode voltage fall. The burner voltage rises in the same time. The protection threshold of the burner voltage could be set by resistor parameters, and in the prototypes it is around 200V.
Self-oscillating CFLi has overheating issue near one or both filament regions when high cathode voltage fall occurs. Low emitter (1% emitter) lab prototypes with the proposed burner voltage detection EOL protection are built and tested for EOL performance evaluation in 4-inch UL boxes. Test results show that the circuit could effectively protect CFLi from overheating.

REFERENCES

[1] UL1993. Self-Ballasted Lamps and Lamp Adapters;
[2] IEC standard 61347-2-3;
[3] James N. Lester. Fluorescent Lamp End-Of-Life Shutdown Sense Methods. Journal of the illuminating engineering society, Vol. 31, No. 2, Pages 78-85, Summer 2002;
[4] Mihail S. Moisin. Ballast having a lamp end of life circuit. US patent, 6,127,786, 1998;
[5] John W. Shaffer. Fluorescent lamp with end of life arc quenching structure. US patent, 5,705,887, 1996;
[6] Albert Cavallaro. End-of-life protection for compact fluorescent lamps. US patent, 6,906,465, 2002.

ACKNOWLEDGEMENT

The authors gratefully acknowledge the support and advice of Mr. Rene Twardzik (OSRAM, China) regarding the EOL sensing circuit.

Energy Saving Medium Wattage Ceramic Metal Halide Lamps

J. Tu, A. Dunaevsky, R. Gibson, T. Steere, K. Graham
HID Lamp Development lab, Philips Lighting Company, 7265 Route 54, Bath, New York, USA
Contact: jim.tu@philips.com

ABSTRACT
The two technologically advanced energy saving medium wattage ceramic metal halide lamps have been developed for new lighting installations or retrofit applications. The system studies indicate that the new injection molded ceramic metal halide lamp has 25% - 50% energy savings compared with the quartz metal halide lamps. The novel retrofit ceramic lamp is suitable to direct replace probe-start or pulse start quartz metal halide lamps and save up to 18% energy. Both lamps are dimmable to realize additional energy savings with less color shift than the existing lamps.

INTRODUCTION

Lighting consumes 19% of worldwide electricity or 2650 billion MWh per annum.[1] Development of energy efficient lamps is very important to reduce greenhouse emissions. Furthermore, improvement of dimming properties and light quality is desired in many applications including new lighting systems and direct retrofit of the existing systems. In this paper we will present the two energy saving medium wattage (MW) ceramic metal halide lamps that can be used in new systems or retrofit applications.

ENERGY SAVING CERAMIC METAL HALIDE LAMP FOR NEW INSTALLATIONS

To reduce energy consumption for new installations, a family of new medium wattage (MW) injection molded ceramic metal halide lamps with enclosed or open fixture-rating and two color temperatures (3000K and 4200K) has been developed (Fig. 1).[2] Recent system studies indicate that the new ceramic lamps can save 25% to 50% energy compared to MW quartz metal halide lamps (Table 1). The energy savings for the lamps stem from high efficacy up to 123 lumens per watt, superior lumen maintenance (Fig. 2), efficient low frequency square wave electronic ballast, compact light source, and improved optical efficiency. Furthermore, the lamps are dimmable to 50% power with less than 900K color shift (Fig. 3).

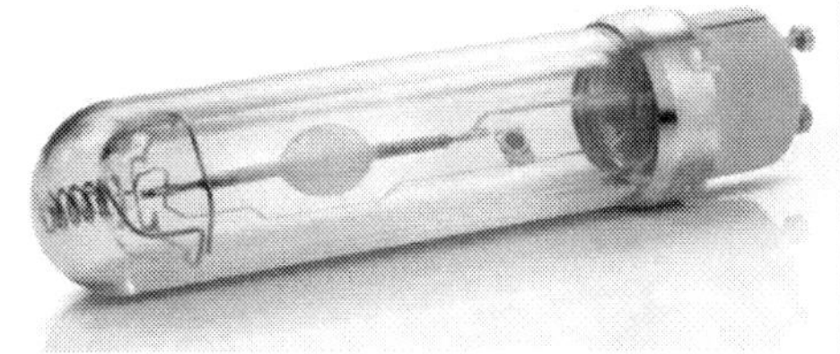

Fig.1. New open fixture-rated ceramic lamp

System	MH400W/U vs. new 210W	MH400W/PS vs. new 315W
QMH system watt	458W	452W
New system watt	225W	338W
Energy saving, watt	233W	114W
Energy saving, %	50.9%	25.2%

Table 1. System savings for new MW ceramic lamps

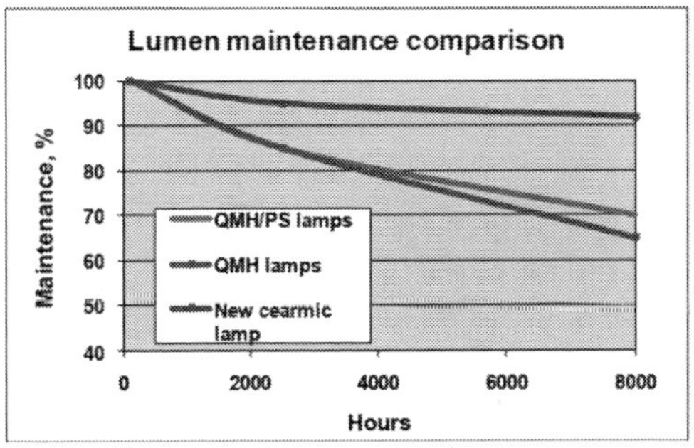

Fig. 2. Lumen maintenance comparisons.

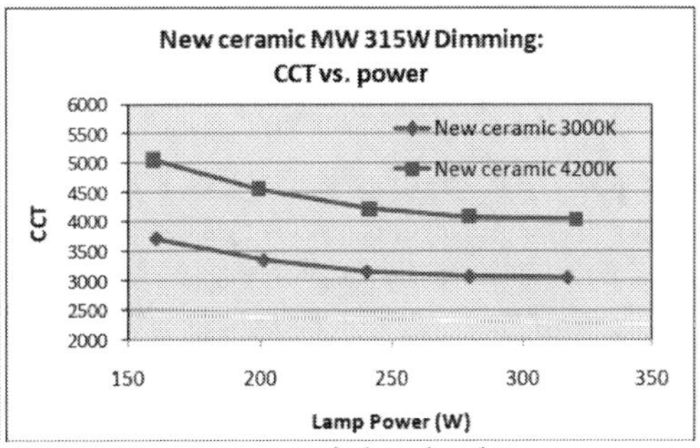

Fig.3. CCT shift under dimming

ENERGY SAVING CERAMIC METAL HALIDE LAMP FOR EXISTING SYSTEMS

In order to achieve energy savings for existing lighting systems, the technically advanced retrofit MW ceramic metal halide lamps (Fig. 4) have been recently developed. With optimized arc tube design, Ne-Ar penning starting gas, and new chemical fillings with low power factor, the novel lamps are suitable to direct retrofit probe-start or pulse-start quartz metal halide lamps and operate below the normal wattage to save energy. Recent system studies indicate that 80-watt and 45-watt can be saved when replacing QMH 400W and 250W lamps with new retrofit ceramic 330W and 205W lamps, respectively (Figure 6). The total energy savings over lifetime are 1600 kw-hr and 900 kw-hr. The lamps are also dimmable and the color temperature shift under 70% dimming is less than 400K (Fig. 5). The novel retrofit lamps are open-fixture rated and provide excellent lumen maintenance and superior color properties over quartz metal halide lamps.[3] Furthermore, the life time for the novel ceramic 205W lamp is two times longer than the existing quartz MH250W lamp.

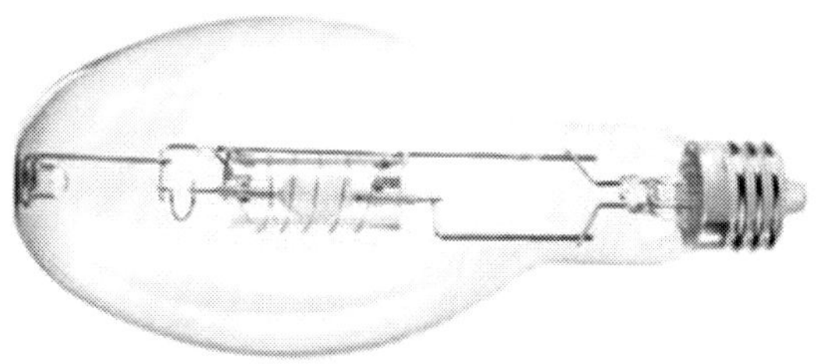

Fig. 4. Novel retrofit ceramic MW lamp

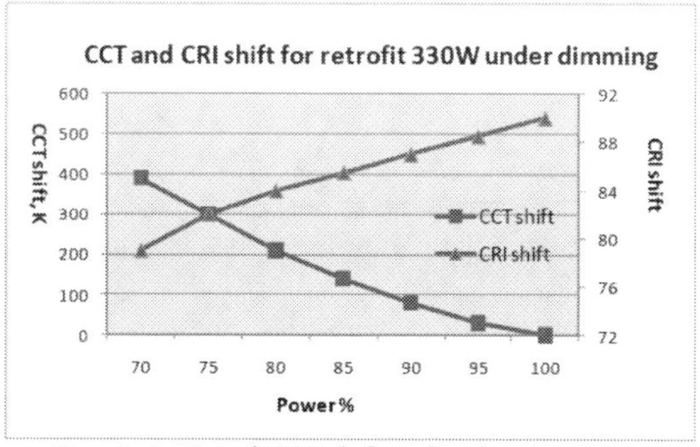

Fig. 5. CCT and CRI shift under dimming

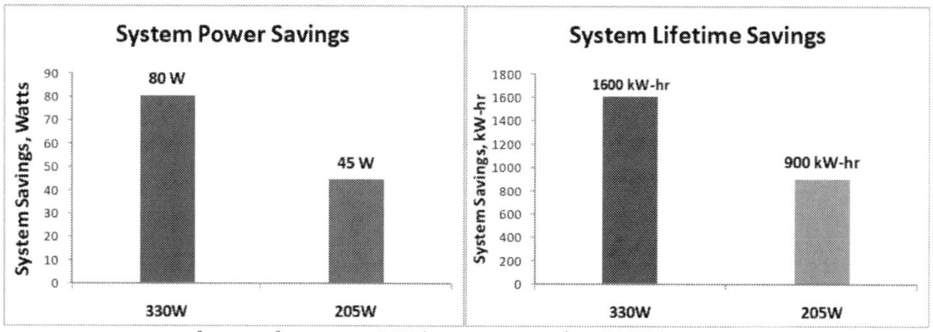

Fig. 6. System energy savings for retrofit ceramic MW lamp compared to QMH lamps

CONCLUSIONS

The two technologically advanced medium wattage ceramic metal halide lamps can be used in new lighting installations or retrofit applications. The system studies indicate that they have significant energy savings. The lamps are dimmable to realize additional energy savings with less color shift than the existing lamps. Moreover, they have superior color properties and performance.

REFERENCES:

1. G. Zissis and S. Kitsinelis, State of art on the science and technology of electrical light sources: from the past to the future, J. Phys D: Appl. Phys. **42** (2009) 173001 (16pp)
2. J. Tu, et al, New advancements in medium wattage ceramic metal halide lamps, LEUKOS, Vol. **5**, No. 3, Jan. 2009
3. T. Steere, R. Gibson, J. Tu, Energy saving ceramic metal halide lamps suitable for probe-start and pulse-start systems, IES Conference paper, Seattle, November, 2009

Thermal management and performance of thermal interface materials in new compact LED lamp-luminaire systems

H. Weiss[1], R. Hüttinger[1], S. Levchuk[2], G. Mitic[2]
[1] *Osram GmbH, Research Europe, Werner-von-Siemens-Str. 6, 86159 Augsburg, Germany*
[2]*Siemens AG, Corporate Technology, Otto-Hahn-Ring 6, 81739 Munich, Germany*
Contact: herbert.weiss@osram.de

ABSTRACT

For compact LED lamps exceeding a luminous flux of 1000 lm the retrofit approach is no longer practical. New concepts are therefore investigated which allow to remove the heat from the LEDs directly into the luminaire where it can be easily dissipated by larger cooling structures. Suitable geometries of cap holder systems are identified and thermal interface materials are examined with respect to their performance and maintenance under extreme conditions.

INTRODUCTION

A proper thermal management of LED lamps is crucial in order to reach the specified life times of up to 50.000 h as well as to maintain the spectral composition of the emitted light. Conventional light sources such as incandescent and halogen incandescent have very high operating temperatures and a large part of the heat is radiated away by infrared radiation. By contrast, since LEDs emit "cold" light an LED lamp relies on efficient heat removal from the LED by thermal conduction and subsequent convection and radiation cooling via additional cooling structures. In order to enter the market by making use of existing lighting units the retrofit approach is currently by far the most common. Here, the cooling structures are part of the lamp construction. However, when a luminous flux of 1000 lm or higher is required for a compact lamp these structures become too bulky and the retrofit approach is no longer applicable. In this case a powerful alternative is to remove the heat into the luminaire which can provide the heat transfer to the ambient atmosphere much more effectively. This allows for much smarter lamp designs. Two ways of assembling the lamp-luminaire system are possible:

1) The interface between lamp and luminaire is designed as an unlockable cap-holder system.
2) The LED arrangement is permanently fixed to the luminaire and not designed to be easily exchanged without any tool.

Both ways require the development of new fixture systems.

LED LAMP CAP HOLDER SYSTEMS

Various conventional and newly developed cap-holder systems including several plug, threaded, cylindrical and conical fittings with and without thermal interface materials (TIMs) have been experimentally investigated with regard to their overall thermal conductivities. It turned out that the required heat transfer is best realised by two plane surfaces which are held in close contact by a pressure of >1 bar. The thermal interface of both surfaces depends on the surface roughness and the parallelism. Therefore residual air gaps must be filled by TIMs.

THERMAL INTERFACE MATERIALS

For both fix and unlockable cap-holder systems TIMs are essential parts of the heat transfer path. Manufacturer datasheets do rarely contain information relevant for LED lamp operating conditions which are needed to estimate the long-term maintenance of TIMs. Changes of properties with pressure, temperature, humidity as well as with cycling of these variables are unknown. An experimental setup has been built which allows to measure the thermal impedance and to calculate the thermal conductivity of TIMs conforming to standard ASTM D5470 [1]. Special care was taken to account for the thermal expansion of the TIM which varies strongly from one product to another. A high thermal expansion coefficient leads to an increase in material thickness with increasing temperature thus lowering the thermal conductivity when pressure is held constant. In the case of an assembly where TIMs are held between two rigidly fixed surfaces a high thermal expansion coefficient results in a significantly increasing pressure with temperature thus increasing the thermal conductivity. High-performance TIM foils which are available on the market have been identified by these measurements. In the case of unlockable cap-holder systems TIMs must withstand at least several tens of lamp exchanges. That means that also the mechanical characteristics of the foil are important, i.e., they must not take damage at tensile and shear load under pressure. Therefore mechanical reliability tests were carried out. Depending on the type of TIM the life span of the tested samples varied from 3 to 70 mechanical cycles at a pressure load of 2 bars. If the repetitive shear load is combined with thermal cycles partial sticking of the elastomer matrix of the foil at the test socket and subsequent tearing off was observed.

Storage of selected TIMs at elevated temperatures up to 150°C did not influence the thermal conductivity. Storage for 100 h at a temperature of 95°C and at an enhanced relative humidity of 95% resulted in a 10% increase of the thermal resistance. Temperature cycling between 0° and 85° showed a strong effect with an increase of the thermal resistance by 20% after 3000 cycles.

SUMMARY

Compact LED lamps exceeding a luminous flux of 1000 lm need efficient heat transfer paths which remove the heat into larger cooling structures of the luminaire. Conventional cap-holder systems cannot provide the necessary heat transfer interface. The most efficient heat transfer is realised by two plane surfaces which are held in close contact by a pressure of >1 bar with a TIM in between the surfaces. First lifetest results suggest that high-performance TIMs are available which withstand extreme LED lamp operating conditions. Also unlockable cap-holder systems can be realised without degradation of the TIM by frequent lamp changes.

This work was partially supported by the German BMBF FKZ 13N9399.

REFERENCES

[1] ASTM Standard D5470 – 06, "Standard Test Method for Thermal Transmission Properties of Thermally Conductive Electrical Insulation Materials," ASTM International, West Conshohocken, PA, 2006, DOI: 10.1520/D5470-06, www.astm.org.

Drivers for OLEDs

J. Jacobs, Ch. Büttner
Philips Business Center OLED Lighting (BCOL), Philipsstraße 8, 52068, Aachen, Germany
Contact: joep.jacobs@philips.com

ABSTRACT

The work focuses on issues relate to driving OLEDs. An overview of important OLED characteristics required for the design of drivers is given. Additionally, a simplified equivalent circuit is presented, which may be used for calculations and simulations. Based on the characteristics and the equivalent circuit useful driver recommendations are given.

INTRODUCTION

Organic light emitting diodes (OLEDs) are expected to become important light sources in the future. At the Light and Building in 2010, many OLED lamps were presented by different manufacturers. In contrast to point source inorganic LEDs, OLEDs are large area device. The control and driving of OLEDs has many similarities with that of inorganic LEDs, but there are some differences. Both will be addressed in this work.

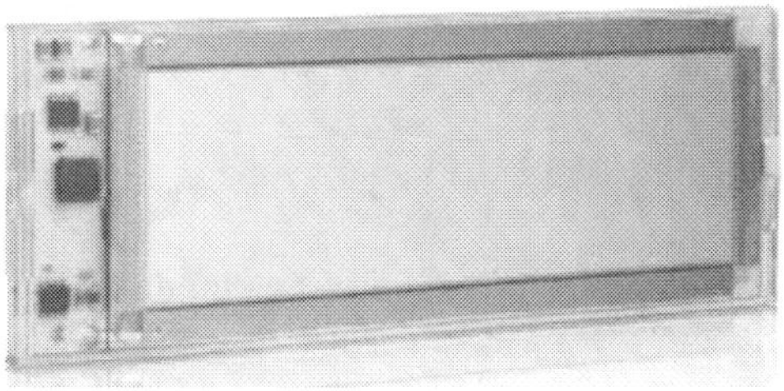

Figure 1: Picture of a Philips Lumiblade Module OLED with integrated driver [1]

OLED CHARACTERISTICS

IVL-characteristic

Typical examples of the OLED luminance of a small molecule OLED as a function of the forward voltage and current for different temperatures are depicted in Figure 2 [2]. The quasi-linear relationship between luminance and current requires a less accurate current control to obtain a desired luminance error compared to the non-linear luminance-voltage characteristic. In general, this means that a current control is cheaper and easier to realize. Additionally, the IL-characteristic does not change much with temperature. This also results in a preference for current control.

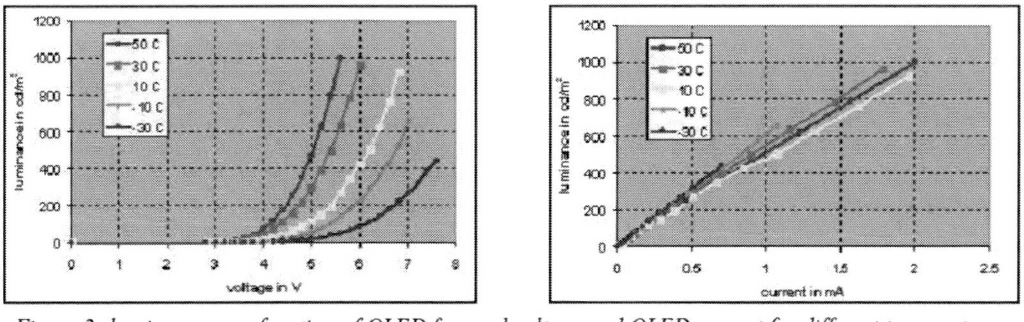

Figure 2: luminance as a function of OLED forward voltage and OLED current for different temperatures

Colour point variation

In the presentation, also the influence on the colour point and spectrum of the OLED will be addressed [3].

EQUIVALENT OLED CIRCUIT

The components of the equivalent circuit can be explained with the help of the schematic. One of the electrodes, through which the light goes, is transparent and mostly made of ITO. The resistivity of ITO is relatively high and is expressed by the ITO resistance R_{ITO} in the equivalent circuit. The combination of thin organic layers (few hundred nanometers) and large area device (large than 10 square centimetres) results in a relatively high intrinsic capacitance $C_{OLED} \approx 200pF/mm^2$. The diode represents the non-linear behaviour comparable with inorganic LEDs.

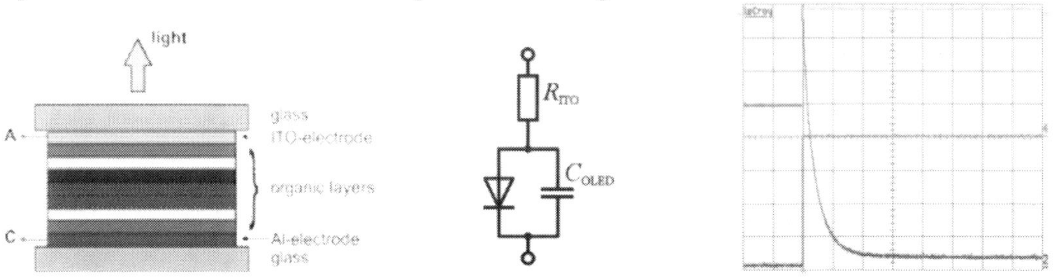

Figure 3: Schematic cross section (left), equivalent electrical circuit (middle) and current through OLED (pink) in case a voltage is applied instantly, i.e. at turn-on

The large internal OLED capacitance causes large current spikes in the case a voltage is applied. This is the reason why voltage PWM should be avoided (cf. Figure 3 – right).

OLED DRIVER

AM versus PWM dimming

In current control mode both dimming techniques, amplitude modulation (AM) and pulsed width modulations can be applied, but there are some differences: The efficacy (in lm/W) of OLEDs is lower at higher currents. Hence, the OLED efficacy is lower for PWM dimming [2]. The internal capacitance of the OLED C_{OLED} causes a delay in the voltage decay, which makes it extremely difficult and/or expensive to control the light output precisely.

Drivers for inorganic LEDs

Many drivers and driver ICs developed for the use of inorganic LEDs can be used for OLEDs as well, but not all drivers are suitable. Especially drivers having a current overshoot as depicted in Figure 4 are not suitable. Examples will be given in the presentation.

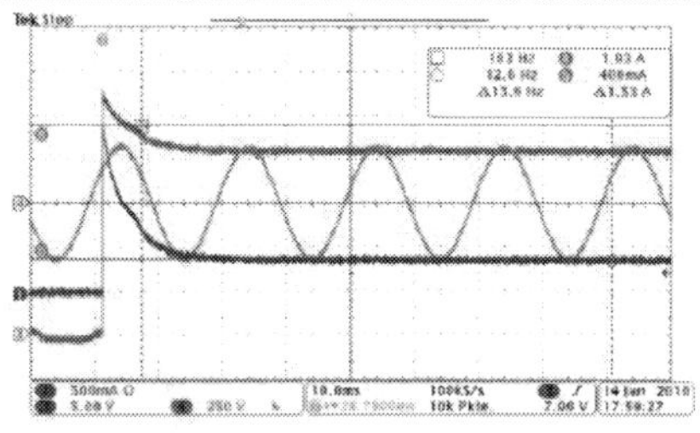

Figure 4: Example of a turn-on: U_{grid} (green), U_{OLED} (pink), I_{OLED} (blue)

Driver integration

Integrating the driver in the OLED lamp enables the inclusion of special functions. Examples are protection circuits, e.g. over current, over temperature and short circuit protection, dimming (AM and/or PWM), communication and compensation techniques. An example of an OLED with integrated driver is shown in Figure 1. Details on architecture and methods will be presented in the presentation.

REFERENCES

[1] Information and leaflet of the Philips Lumiblade module, www.lumiblade.com

[2] J. Jacobs, D. Hente, E. Waffenschmidt: "Drivers for OLEDs", IEEE IAS 2007, New Orleans, Louisiana, USA, 2007

[3] J. Jacobs, C. Singer, D. Hente, H.-P. Loebl: "Dynamic Colour Point Simulation of OLEDs", IEEE IAS 2008, Edmonton, Canada, 2008

Green joining of ceramics

D. Dudik[1], G. Kuenzler[1], M. Yee[1]
[1]*GE Lighting, 1975 Noble Rd., Cleveland, OH 44112 USA*
Contact: david.dudik@ge.com

ABSTRACT

A novel method for joining of green ceramics has been developed using the softening of the injection-molding (IM) binder in a hot fluid. This results in a hermetically joined ceramic body suitable for use in ceramic metal halide (CMH) lamps. The joining process enables the use of highly complex part shapes with minimal additional process equipment.

BACKGROUND

One of the more complex processes in the manufacture of polycrystalline alumina (PCA) CMH lamps is the fabrication of the ceramic arctube. The arctube must be high purity, translucent, hermetic, free of defects, and possess suitable geometry for optimal lamp performance. In many cases, the arctube is fabricated of separate components, often by different processes. These include injection molding, pressing, and extrusion. These disparate parts must be joined together in a hermetic and highly reliable manner. Often, variation in exact part dimensions from these separate processes causes defects and low production yields. Current joining processes include differential shrinkage[1], melt-bonding[2], and others, but these require multiple process steps, handling, and specialized equipment to prepare and assemble the individual parts into the final ceramic arctube. Although differential shrinkage can be used to produce successful hermetic bonds, the yield variation from batch to batch can be significant, caused by variation in the precise mating part dimensions that produces stress and cracking during the sintering step.

A need exists for a simple and inexpensive manner of bonding ceramic components together. It is known that green ceramic parts may be mechanically assembled, then placed into a debindering oven for removal of binders or other process chemicals. The debindered parts are placed into a high temperature furnace for sintering and densification. Yield issues involving cracking of the finished parts is discovered only after this step. A novel method of joining the green parts without application of mechanical stress has been investigated, and is shown to be a simple, low-cost alternative to current ceramic joining processes.

APPLICATION

The novel green joining process[3] uses tightly fit injection-molded parts that are submerged in a hot fluid, which softens the IM binder, and applies a hydrostatic force. As the part softens, the small air gap between the parts is removed, and sufficient mechanical strength is imparted to the joint to allow for handling. The joined assembly is then debindered and sintered as one piece using normal processes. Subsequent surface finishing, assembly, and handling operations on the individual injection molded parts are no longer required.

An important element of the "green joint" is a taper fit geometry between the two parts. Figure 1 shows a schematic of a green joint arctube design. The male and female tapers are matched at a carefully selected angle, chosen to maximize the sintered part yield. If the taper angle is too shallow from vertical (>45°), the bond interface will not be completely removed during subsequent steps. If the angle is too steep (<10°), the tapered sections become very thin and present challenges to both molding and handling. The taper drives a horizontal force between the two surfaces that closes the gap and removes internal voids, while maintaining axial alignment. Mold surface finishes of A1 or better were shown to produce sufficient surface finish in the parts that prevented voids along the joint interface. Therefore, assembly could be performed immediately with no further surface operations.

The green joining step is straightforward and requires minimal equipment and tooling. The assembled parts are lowered into a fluid bath, which is at a temperature higher than the softening point of the binder. The taper joint itself must be submerged into the bath, but it is not necessary to submerge the entire assembly. The assembly submerged long enough to soften the binder and join the two parts, but not so long that the binder flows, causing deformation or collapse. In instances where the lower end of the assembly is closed, a small amount of hydrostatic fluid pressure assists the joining process. In a typical example, parts containing a wax binder with a softening point between 40-60°C are lowered into a hot water bath for softening, followed by immersion in a "warm" water bath for controlled solidification. With properly chosen temperatures, such treatment will not introduce cracks or flaws from thermal shock or displacement. This process can also be applied to other binder systems, fluids, and ceramic parts using a similar technique.

Figure 2 shows an SEM image of a cross-section of a typical green joint after sintering to full density. The taper joint interface has been completely eliminated, and a void-free, monolithic ceramic joint has been created. Identical parts assembled in a similar manner that do not go through the fluid bath joining process exhibit large regions of porosity, and are not hermetic. The fluid bath joining process performs the critical step of removing the air gap between the two parts and bringing the ceramic particles into close contact.

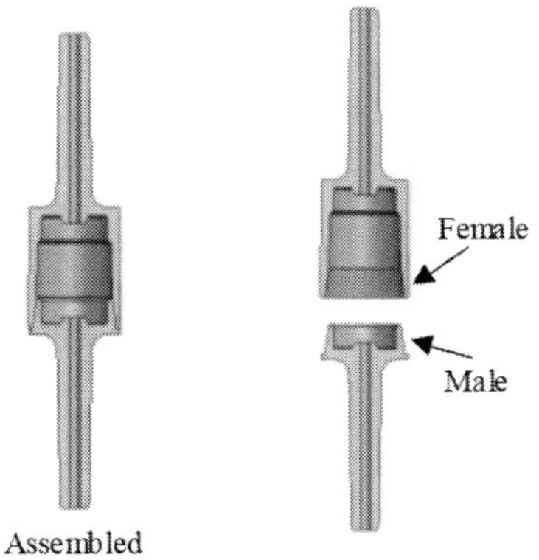

Figure 1. Green Joined Arctube Design

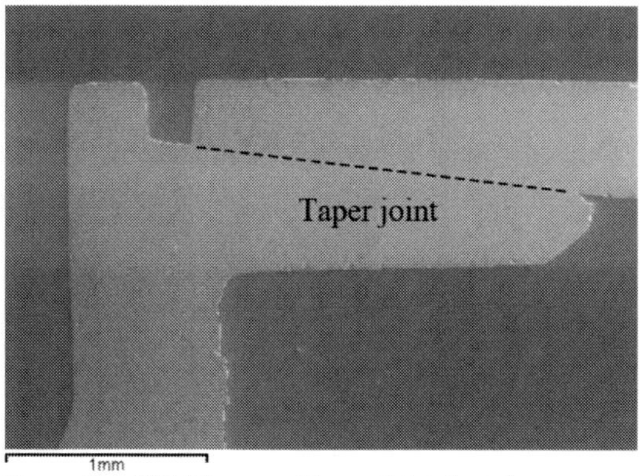

Figure 2. – SEM Image – Green Joint Arctube

CONCLUSIONS

Multiple benefits are realized by the implementation of the green joining process. Most importantly, the parts are joined in a nearly stress-free state, with no mechanical forces acting on the parts during the subsequent process steps that may cause cracking. Since no stress is applied, the possible shape complexity is greatly enhanced. Curved, rounded, and non-symmetric shapes that would otherwise require different processes are now possible. Arctubes requiring three or more parts from different fabrication techniques can now be created with only two injection-molded parts. In addition, the number of process steps is reduced, since the part is fully assembled in the green state prior to binder removal. The added step of hot fluid joining is simple and inexpensive in comparison to current alternatives.

REFERENCES

[1] German, R., Bose, A., *Injection Molding of Metals and Ceramics*, MPIF, (1997).
[2] Zaslavsky, *et al.*, US Patent No. 6620272, (2003).
[3] Kuenzler, *et al.*, US Patent No. 2009/013459A1, (2007).

Migration from Electromagnetic Driver to Electronic Driver for MP Lamps

B. Himaras, G. Fang, A. Cojocaru, C. Sheculski
Trojan Technologies, 3020 Gore Road, London, Ontario N5V 4T7, Canada
Contact: gfang@trojanuv.com

ABSTRACT

This paper provides insight of the migration from an electromagnetic driver to an electronic lamp driver for medium pressure (MP) mercury vapour germicidal lamps in the range of 9 to 12kW. The benefits of using an electronic lamp driver are discussed, especially better UV maintenance over the lifetime of the lamps.

INTRODUCTION

An electronic lamp driver has a number of benefits for the end user. The lamps in this study are medium pressure mercury arc lamps with the electric input power ranging from 9kW to 12kW, used in the TrojanUVSwift™30 and TrojanUVSwift™24 systems. The lamps provide the necessary UV light for water disinfection and environmental contamination treatment. The benefits of the application of electronic lamp drivers translate directly to monetary savings for the customer and improved lamp and driver performance over the life of the system.

A search for lamp driver vendor was performed that led to the evaluation of different driver topologies, high frequency sinusoidal and medium frequency square wave power supplies. The high frequency sinusoidal lamp driver that was evaluated had an operating frequency up to 100 kHz at full power while the medium frequency square wave power supply had an operating frequency of 400 Hz. One study [1] showed the effect of these driving waveforms on the UVC output of a MP lamp. The waveforms of lamp voltage and current when driven by the 45 kHz sinusoidal lamp driver at 100% power are shown in Fig. 1, and a photo of the lamp arc is shown in Fig. 2. Reference [2] also provides some detail performance of the lamp driver.

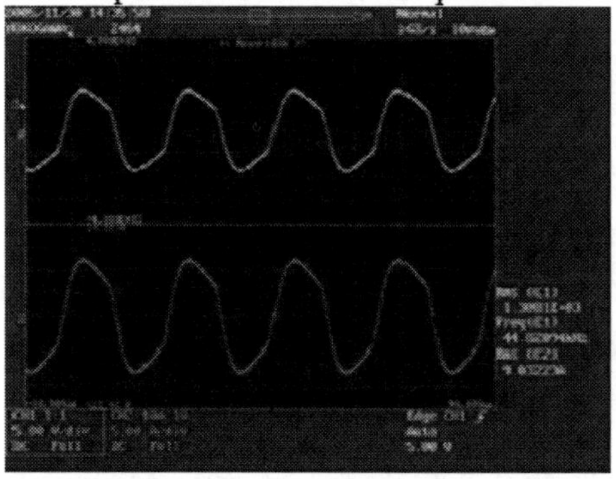

Fig. 1. MP lamp voltage and current when driven by 45kHz sinusoidal lamp driver at full power.

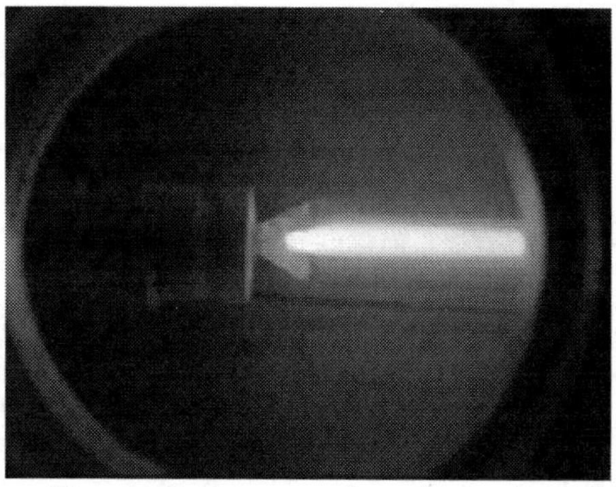

Fig. 2. Lamp arc when driven by 45kHz sinusoidal lamp driver at full power.

BENEFITS OF ELECTRONIC LAMP DRIVER

The lamp driver efficiency is defined as output electrical power divided by input electrical power. The efficiency of the electronic lamp driver is approximately 96%. Compared with the efficiency of electromagnetic driver at about 94%, the heat loss decreases by one third. If the panel is in a climate controlled environment then the heat load removed by the air conditioning system will be reduced accordingly.

The power factor of electronic lamp drivers is approximately 99%, while the power factor of a typical electromagnetic driver is approximately 93%. The increase of the power factor indicates that the driver draws less current for a given output. This reduces the peak demand in apparent power and the total current consumed, which could have a large impact on electricity costs.

The electronic lamp drivers can provide stepless power level control from 30% to 100%, while some of the electromagnetic drivers provide limited power level controls, such as 60%, 80% and 100% power levels. With an electronic lamp driver it is possible to control the lamp power to match the UV dose requirement precisely, resulting in electrical savings.

LAMP PERFORMANCE IMPROVEMENT

In addition to the electricity savings, the electronic lamp driver improves the lamp performance. The Trojan electronic driver operates at a high frequency (>45 kHz) compared with that of the electromagnetic driver at 50/60 Hz. A lamp powered by a high frequency electronic lamp driver would not have physical extinguishment of the plasma at zero crossing of the arc cycle [3]. The electronic lamp driver provides less temperature variation inside a MP lamp which could result in less stress on the electrodes, although the arc extinguishment is complicated not only by the time variation of arc properties through out the arc cycle, but also by the possible effects of convection and non-equilibrium of plasma [4].

The selected shape of lamp waveform is that based on the sinusoidal lamp driver with frequency range above 45 kHz. This frequency range accommodates the feature of lamp dimming. The electronic lamp driver operation with both 12 kW and 9 kW MP lamps becomes possible due to both lamps having the same electric power density. The lamp operating frequencies above 45 kHz was carefully chosen to avoid exciting any acoustic resonance of the MP lamps.

The lamp performance improvement by migration of electromagnetic driver to electronic lamp driver has been demonstrated by the end of lamp life (EOLL) performance. For EOLL testing, a calibrated GHz-Optik radiometer with a UVC detector and a cosine diffuser was used for the output measurement. The radiometer detector was mounted on a revolving fixture 87cm from the centre of the lamp arc. The lamp was measured normal to the midpoint of the lamp over 300 degrees around the measured lamp. The mean output of each measured lamp was obtained by averaging the output distribution measured at each revolving angle. The output of the 100h reference lamps was used as a basis to compare with the aged lamps to determine a relative output.

The lamp driven by the electronic lamp driver had significantly less lamp end darkening than the lamp driven by the electromagnetic driver at the same age, as shown in Fig. 3. This could confirm the fact of less stress on the electrodes. The lamp output at 10,000h was above 90% (EOLL UV maintenance) compared with 100h reference lamps.

Fig. 3. End darkening of lamp driven by electronic lamp driver (left) and a lamp driven by electromagnetic driver (right) at approximately 2,500h.

CONCLUSIONS

The migration from an electromagnetic driver to an electronic driver for MP mercury arc lamps has resulted in benefits of electricity savings for customers and better UV maintenance over the life of the system. The lamp waveform and the operating frequency were considered in the selection of the topology for a state of the art electronic lamp driver.

REFERENCES

[1] O. Paleka. B. Himaras, C. Sheculski, J. Robinson, and M. Sasges, "Effect of the driving waveform on the UVC output of a MP mercury-vapour lamp," IUVA 2nd Congress, Karlsruhe (2004).

[2] A. Voronov, V. Adam, G. van Eerden, T. Telgenhof, "Renaissance of medium pressure lamps driven by high frequency lamp driver", IUVA 5th Congress, Amsterdam (2009).

[3] Roger Philips, Sources and Application of Ultraviolet Radiation, Academy Press, London (1983).

[4] J.J. Lowke, R.J. Zollweg and R.W. Liebermann, "Theoretical description of ac arcs in mercury and argon", J of Applied Physics, 46 (2), 650-660 (1975).

Fabrication of Light Emitting Photonic Crystals as the Wavelength Converter for White Light Emitting Diodes

Sung Wook Kim, Byoung-Hwa Kwon, Minwon Suh, Dong Seok Kang, and Duk Young Jeon*
Department of Materials Science and Engineering, Korea Advanced Institute of Science and Technology, 335 Gwahangno, Yuseong-gu, Deajeon 305-701, Republic of Korea
**Contact: dyjeon@kaist.ac.kr*

ABSTRACT

In this study, we suggest light emitting photonic crystals (LEPCs) which are fabricated by self assembly of fluorescent polystyrene spheres as wavelength converter for WLEDs. The WLEDs composed of InGaN blue LED, green and red LEPCs were emitting white light with color rendering index of 80 and CIE color coordinates of (0.29, 0.36). The LEPCs showed the possibility as wavelength converter in order to generate white light with InGaN blue LED.

GENERAL

White light emitting diodes (WLEDs) have been attracted particularly as an alternative solid-state lighting source [1]. WLEDs have a good performance (e.g., high efficiency, long life-times, low power consumptions, and fast response times). There are several methods to fabricate WLEDs [2-3]. Combination of a blue LED and one or more inorganic phosphor (yellow phosphor or green-red phosphor) has been widely used to generate white light. Recently as high efficient organic dyes are developed, hybrid type of WLEDs is attempted [4]. In hybrid WLEDs, organic dyes are used as wavelength converters which have the same role as inorganic phosphor. Organic dyes can be easily processed from solution and designed to emit across the whole visible spectrum [5]. However, organic dyes are weak so that it can be affected by heat and photons generated during operation of LEDs. The heat and photons induce thermal quenching and photobleaching as well as degradation of molecular structure. In this study, we suggest light emitting photonic crystals (LEPCs) which are fabricated by self assembly of fluorescent polystyrene spheres as wavelength converter for WLEDs. Firstly, we synthesized fluorescent polystyrene spheres contained organic dyes (green dye: Coumarin 6 and red dye: Rhodamine B) by using microemulsion process. The organic dyes are imbedded into polystyrene sphere during polymerization of styrene monomer in micelles. In order to fabricate LEPCs, a few drop of colloidal suspension of the fluorescent polystyrene spheres was spread onto a pre-cleaned glass substrate and dried at room temperature for overnight. Figure 1 shows schematic diagram of WLED A and B composed of InGaN based blue LEDs and LEPCs layer on the surface of glass substrate. The photonic crystal layers self-assembled into a close-packed face-centered-cubic structures were shown in SEM (scanning electron microscope) images of Figure 1. Figure 2 (a) and (b) show electroluminescence (EL) spectrum of the WLED A with green LEPCs and EL spectra of the device at injection currents ranging between 10 and 50 mA. It can be seen that the EL intensities of the device were increased linearly with increasing currents. At the same time, no visible spectral shifts were observed. However, the WLED A was only able to generate white light with poor color rendering index (CRI) of 70 due to lack of red emission. In order to improve CRI value, we synthesized and mixed red fluorescent polystyrene spheres with the green fluorescent polystyrene spheres. Figure 2 (c) shows EL spectrum of the WLED B with LEPCs composed of green and red fluorescence polystyrene spheres. The WLED B emitted white light with CRI of 80. The CIE color coordinates of WLED A and B are (0.22, 0.37) and (0.29, 0.36) were shown in figure 2 (d). Figure 3 show wavelength conversion efficiency of each type of WLEDs. Wavelength conversion efficiency of case II and III were dramatically decreased with increasing operation time of the devices around 40% for 10min due to directly exposure of heat and photons of the blue LED. On the other hand, that of case I was decreased about only 20% for 1hr and then the degradation rate was slowly saturated. Because the organic dyes which were encapsulated by polystyrene could be protected from the heat and photon of the blue LED as well as moisture and oxygen. From the above results, The LEPCs which were fabricated by self assembly of fluorescent polystyrene spheres showed the possibility as wavelength converter in order to generate white light with InGaN blue LED.

Figures and equations

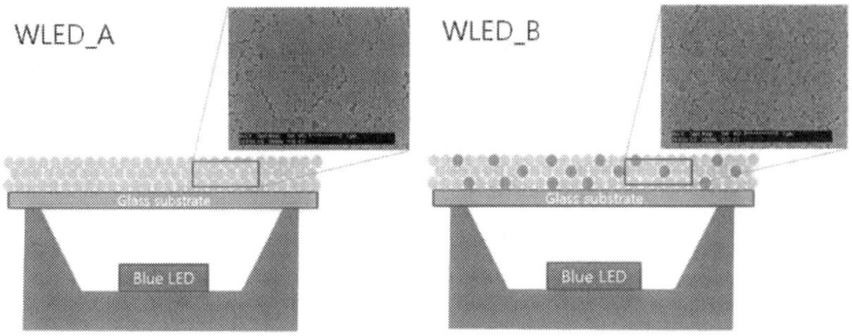

Figure 1. WLED A and WLED B composed of InGaN based blue LEDs and LEPCs wavelength converters on the surface of glass substrate.

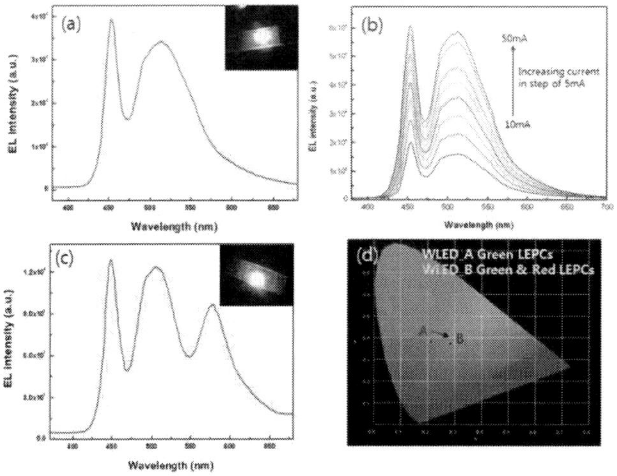

Figure 2. (a) Emission EL spectrum of WLED A with the green LEPCs. (b) EL spectra of WLED A at injection currents between 10 and 50 mA. (c) Emission EL spectrum of WLED B (d) CIE (1931) chart showing the coordinates of WLED A and WLED B.

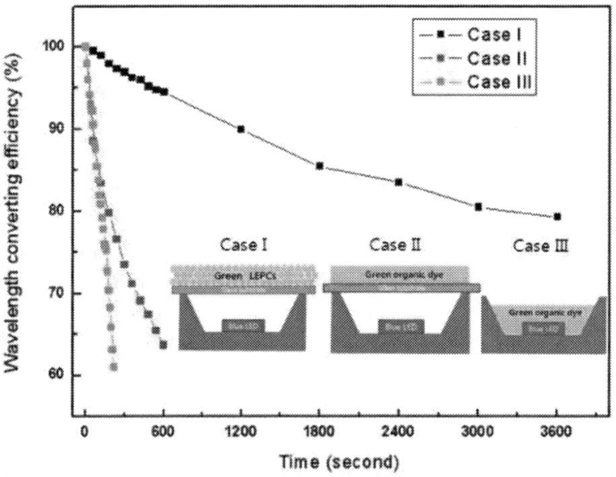

Figure 3. Wavelength conversion efficiency of each type of WLEDs

REFERENCES

[1] S. Nakamura, T. Mukai, and M. Senoh, Appl. Phys. Lett., 64, 1687 (1994).
[2] H. S. Jang, H. Yang, S. W. Kim, and D. Y. Jeon, et. al., Adv. Mater., 20, 2696-2702, 2008.
[3] J. K. Sheu, S. J. Chang, and R. K. Wu, et. al., IEEE Photon. Technol. Lett., 15, 18 (2003).
[4] S. Guha, R. A. Haight, N. A. Bojarczuk, and D. W. Kisker, J. Appl. Phys., 82 (8), 1997.
[5] G. Heliotis, P. N. Stavrinou, and M. D. Dawson, et. al., Appl. Phys. Lett., 87, 103505, 2005.

A HID PSpice lamp model including starting process

Deyan Lin[1, 2], Wei Yan[1], S. Y. R. Hui[1]
[1]*Electronic and Engineering Department, City University of Hong Kong, Kowloon, Hong Kong S.A.R, China*
[2]*School of Physics & Information Engineering, Jianghan University, Wuhan, Hubei, 430056 China*
Contact:ee114868@cityu.edu.hk

ABSTRACT

This paper presents a high intensity discharge (HID) lamp model which can predict the electrical characteristics of a HID lamp in both starting and steady operating states. The model can be implemented into a PSpice circuit model. The simulation results show a good agreement with the experimental waveforms.

I. The modelling ideas and assumptions

There are several HID lamp models which can simulate the electrical characteristics of the lamp over a wide frequency range [1-3]. However, there is no such a lamp model that can simulate both the starting and the steady operating state in the same lamp model. From extensive experimental observations on the starting process of HID lamps, we found that there is a time constant exists during the lamp warm-up process which determines the lamp resistance as shown in Fig. 1. It suggests that if one knows the lamp initial resistance value at the beginning of the starting process and the resistance value in the steady state, one can predict the dynamic lamp resistance value during the whole operating (in both starting and steady state) process according to the following formula:

$$R_{lamp} = R_{lamp\_start} * (1-p) + R_{lamp\_steady} * p \qquad (1)$$

where p is the normalized proportional coefficient according to the capacitor voltage of a 1st order capacitor charging circuit with the same time constant of the lamp, R_{lamp}, R_{lamp_start} and R_{lamp_steady} are the dynamic lamp resistance, the lamp resistance at the beginning of the starting process and the lamp resistance in the final steady state, respectively. R_{lamp_start} and R_{lamp_steady} can be obtained using the model proposed in [2] based on the experimental data shown in Fig. 2. The parameters for start point model and the steady state model are listed in table 1.

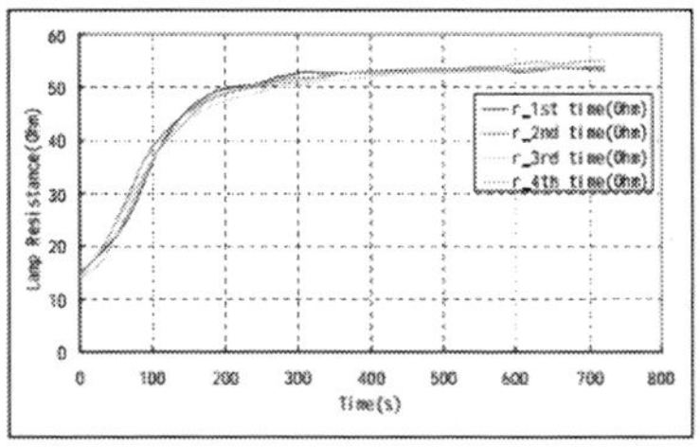

Fig. 1 Experimental curves of lamp resistance vs time during warm-up state (implemented by V_{rms}/i_{rms}, lamp voltage rms value divided by lamp current rms value)

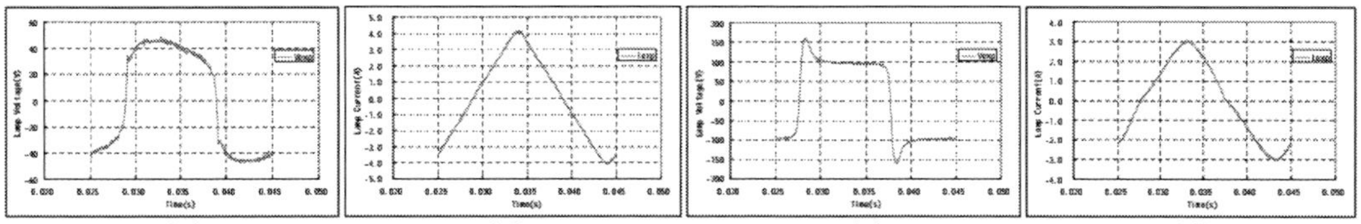

(a) (b) (c) (d)

Fig.2 The experimental waveforms of lamp voltage and lamp current: (a)lamp voltage at starting point (b)lamp current at starting point(c) lamp voltage at steady state (d) lamp current at steady state

Table 1 The parameters for start point model and steady state model

	a1	*a2*	*a3*	*a4*	*a5*	*a6*	*a7*
Starting point	14611.98	159166.3	3.543561	0.064685	2267.949	0.107997	0.964062
Steady state	39533.71	129770.7	2.670859	0.062697	5144.948	0.413432	0.835749

II. Proposed HID lamp model and the simulation results

The HID lamp model based on energy balance equations[2] including the starting process can be implemented into a Pspice lamp model as shown in Fig.3. The simulated lamp voltage and lamp current waveforms during the entire starting progress are shown in Fig.4. Where Fig.4 (a) is the entire waveforms, Fig.4 (b) is enlarged starting point curves and (c) is enlarged steady state waveforms. Good agreement have been achieved between experimental waveforms and the simulation results at 50 Hz.

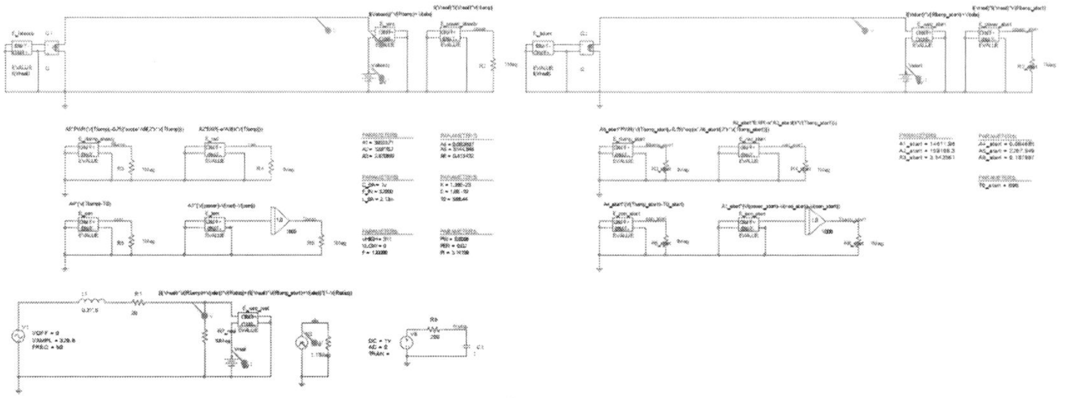

Fig. 3 Proposed Pspice lamp model

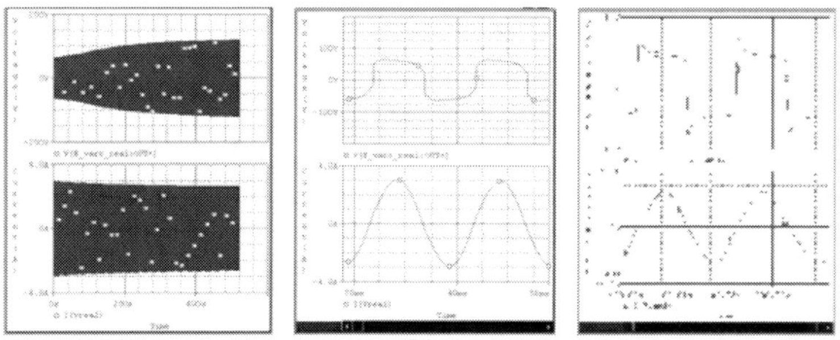

(a) (b) (c)

Fig. 4 Simulated waveforms (upper waveforms for voltage and lower waveforms for current): (a) entire waveforms; (b) waveforms at starting point; (c) waveforms at steady state.

III. Conclusion

A HID lamp model which can predict the lamp electrical characteristics in both starting and steady state was proposed in this paper. Both simulated lamp voltage and lamp current waveforms from the beginning of the warm-up process to the final steady state agree with the experimental waveforms very well. The model can be used to predict the transient process of the starting process in a real HID lamp system. Further work will be done to extend the model to high frequency to get a more flexible model for HID lamp ballast designer.

IV. References

[1] R. Osorio, M. A. Oliver, M. Ponce, M. Juarez, V. H. Olivares, S. E. Pinto, R. Katebi, and M. J. Grimble, "Thermal Dynamic Model for HID Lamps with the Outer-Bulb Effects," in *Electronics, Robotics and Automotive Mechanics Conference, 2006*, 2006, pp. 197-202.

[2] Y. Wei and S. Y. R. Hui, "A universal PSpice model for HID lamps," *Industry Applications, IEEE Transactions on*, vol. 41, pp. 1594-1602, 2005.

[3] C. Blanco, J. C. Anton, A. Robles, F. J. Ferrero, J. C. Campo, M. Gonzalez, and G. Zissis, "A Discharge Lamp Model Based on Lamp Dynamic Conductance," *Power Electronics, IEEE Transactions on*, vol. 22, pp. 727-734, 2007.

Further discussion on the mechanism of electrode change of UHP lamp driven by different current waveforms

Weiqiang Zhang[1, 2], Jianping Ying[2], Shanduan Zhang[1], Rongqing Liang[1]
[1] *Institute for Electric Light Sources, Fudan University, 220 Handan Road, Shanghai 200433, China*
[2]*Delta Power Electronics Center, 238 Min-Xia Road, Pudong, Shanghai, 201209, China*
Contact: zhang.weiqiang@deltaww.com.cn

ABSTRACT

Maintenance of sharp tips at the front ends of tungsten electrodes in AC-operated UHP lamp is very important in order to keep stable arc position. This paper shows the different electrode shapes in a UHP lamp driven by three different current waveforms. The mechanism of electrode tip change is discussed based on the tested results and analysis of different pulse position in the current waveform of an AC-operated lamp. The key factor to maintain a sharp electrode tip may be generating more tungsten ions before the phase transition from anode to cathode.

ELECTRODE SHPAES UNDER DIFFERENT CURRENT WAVEFORMS

Figure 1 shows three different current waveforms of tested AC-operated UHP lamp. Figure 1(a) is a standard square wave current (Mode I). Figure 1(b) has superimposed current pulse at the end of each half period of square waveform (Mode II), which is invented by Philips Company [1]. Figure 1(c) has superimposed current pulse at the middle of each half period of square waveform (Mode III).

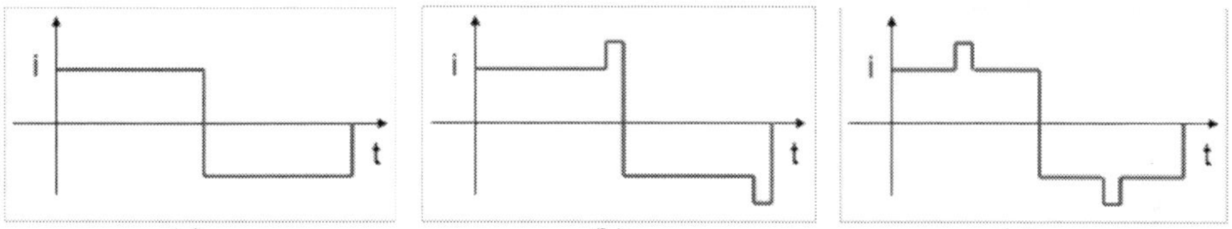

(a) (b) (c)

Figure 1 Current waveforms for AC-operated UHP lamp.

Figure 2 shows the electrode changes of one UHP lamp under different operated current waveforms. As shown in Figure 2(b), the lamp electrode front tip is still kept sharp to fix the arc position, which is operated in mode of Waveform II. However, the electrode tips in waveform Mode I become flat, as shown in Figure 2 (a), which means an unstable arc position. As shown in Figure 2(c), the lamp electrode front tip in waveform Mode III is sharper than the electrode shape show in Figure 2(a), but the arc jump still occur during operation.

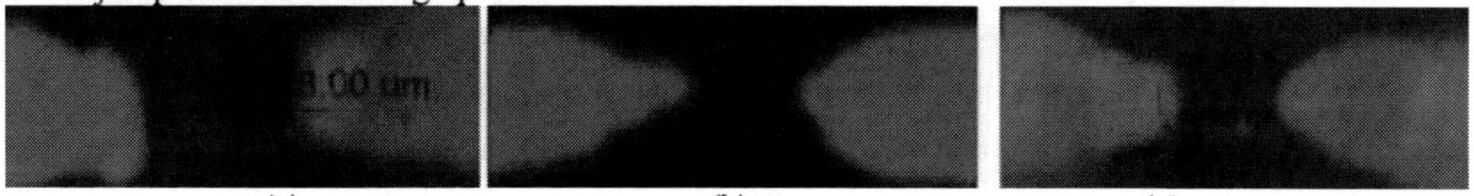

(a) (b) (c)

Figure 2 Photographs of electrodes operating in waveform Mode I, II and III.

FURHTER DISCUSSION ON ELECTRODE CHANGE

Although the detailed mechanism of the electrode change shown in Figure 2(a) is still not clear, the change process can be qualitatively described in the following: 1) Tungsten evaporate from a hot section near the tip of electrode; 2) Tungsten combine with halogen; 3) Tungsten halide decompose to tungsten atoms or ions in the hot gaseous phase near the electrode tip; 4) Due to diffusion of the tungsten atom which goes toward a low temperature section, a portion around the electrode tip happens; 5) When an electrode functions as a cathode, the tungsten ion is drawn toward electrode tip by electric field. This uncontrolled process results in the rough electrode front when the lamp is

operated by current waveform Mode I as shown in Figure 1(a) [1, 2].
Based on the tested result of Figure 2(b), it can be found that the additional current pulse is useful to maintain a sharp tip. The possible explanation is discussion in the following according to Figure 3. More tungsten ions, which are likely generated from the ionization of tungsten atoms in gas phase and the anode sheath due to the superimposed current pulse [3], are formed at the front of the electrode at step one. And then tungsten ions are drawn back to electrode tip by cathode fall voltage, which dominated the growth of electrode tip during operation and form a sharp tip.

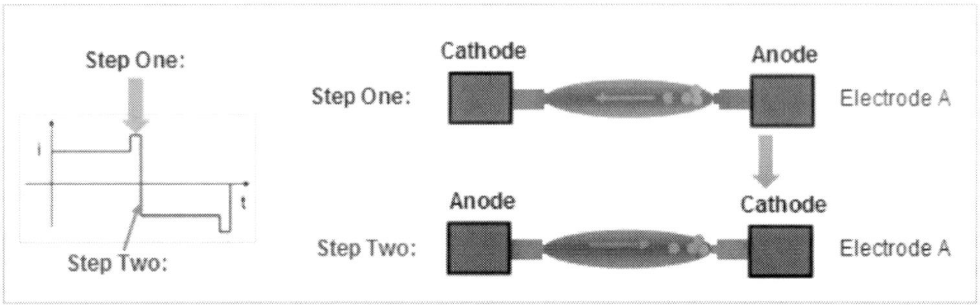

Figure 3 Tungsten ions (green dots) transportation driven by waveform Mode II.

However, when the pulse current is placed at the middle of half period, the sharp tip of electrode front will not be formed. The possible explanation is shown in Figure 4. More tungsten ions are formed at the front of the electrode at step one, which is same as Figure 3. But the density of tungsten ions in the vicinity of anode decreases at step two because both the current pulse disappears and the tungsten ions drift toward cathode. And then only a few tungsten ions are drawn back to electrode tip by cathode fall voltage at step three, which cannot maintain a sharp tip at the front of electrode.

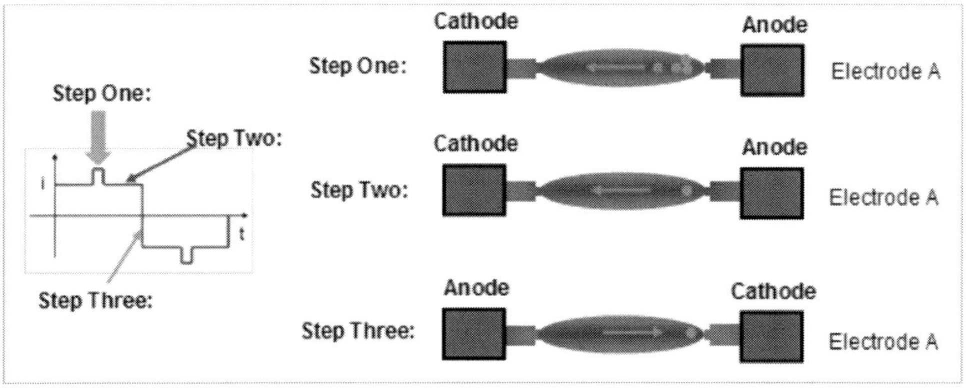

Figure 4 Tungsten ions (green dots) transportation driven by waveform Mode III.

CONCLUSION

Based on the above analysis, it can be found that generating more tungsten ions before the transition from anode to cathode phase is the likely key factor to maintain a sharp tip at the front of electrode. Tungsten atoms which return back from bulb wall and the tungsten layer in the anode sheath are used to generate enough ions by superimposed current pulse.

REFERENCES

[1] Gunther H. Derra High pressure lamp operating circuit with suppression of flicker, US5608294,
[2] Hirao T. Discharge lighting apparatus and projector, US2008/0315782A1
[3] M. Redwitz 2006 *J. Phys. D: Appl. Phys.* **39** 2160

Determination of rotational temperature of hydrogen molecule and hydroxyl radical in helium-hydrogen high-frequency electrodeless plasma

Z. Gavare[1], M. Zinge[1], J. Skudra[1]

[1] *Institute of Atomic Physics and Spectroscopy, University of Latvia, Skunu street 4, Riga, LV-1050, Latvia*
Contact: zanda.gavare@gmail.com

ABSTRACT

The goal of the present work was the investigation of the possibility to use intensity distribution of the Q-branch lines of hydrogen Fulcher-α (0-0) and (1-1) bands, and (A-X) (0-0) Q_1 branch of OH radical, to determine the plasma temperature of He+H_2 high-frequency electrodeless lamps. When determining the plasma temperature from Fulcher-α (0-0) and (1-1) Q lines, the rotational temperature of excited and ground states was analysed. Plasma temperature values determined from the rotational spectra of OH agree well with previous temperature measurements using Fulcher-α (2-2)Q lines.

INTRODUCTION

Such inductively coupled plasmas, like high-frequency electrodeless lamps (HFEDLs), are widely used in atomic absorption spectroscopy [1], where high intensity of emission and not-self-reversed spectral lines are of importance, and they are of particular interest in plasma-surface interaction studies due to the lack of electrodes, which usually cause impurities in plasma. One of the most important plasma parameters for the optimization of plasma conditions is plasma temperature.

It is well known that for the discharge gas temperature estimation in HFEDLs one can use the high-resolution spectroscopy method of the emission spectral line shape measurements using Fabry-Perot interferometer, and non-linear spectral line shape modeling [2].Since this method is very complicated due to the necessity to solve the incorrect inverse task, it is of great interest to find other methods for the determination of plasma temperature in HFEDLs. In the case of hydrogen containing plasmas, one of the commonly used techniques for the determination of gas temperature is based on the measurements of the intensity distribution in the rotational bands of hydrogen molecule, for instance (0-0), (1-1) and (2-2) Q-branches of Fulcher-α ($d^3\Pi_u^- \rightarrow a^3\Sigma_g^+$) electronic transition [3-5]. The determination of the gas temperature from measured intensity distribution in rotational bands is based on a certain theoretical model, which involves series of assumptions about the mechanism of the excitation-deexcitation of rotational-vibrational levels of hydrogen. According to this model, the translational gas temperature may be in equilibrium with rotational temperature determined from either upper or lower energy level depending on processes in plasma under study [4,5]. In each particular case, the relation between the rotational and translational temperatures needs to be verified. One way to establish the relationship between both temperatures is the comparison of the rotational temperature with that obtained by other methods.

It has been observed [6] that in some cases it not possible to register Fulcher-α Q-branches in the spectrum of plasma, although the hydrogen is present in the plasma. However, one can try to register the UV spectrum of OH radical ($A^2\Sigma^+ \rightarrow X^2\Pi$; v' = v'' = 0) at 306,4 nm. It is known that these bands can be observed even at very small amounts of hydrogen, and they can be used for the plasma temperature determination, too [7].

The Fulcher-α Q-branches and OH radical (A-X) (0-0)Q_1 branch have been used for the plasma temperature determination in different kinds of discharges; however for the diagnostics of helium-hydrogen high-frequency electrodeless discharge lamps (excited at about 100 MHz frequency) Fulcher-α (0-0) and (1-1) Q branches and (A-X) (0-0)Q_1 branch have not been used before.

MEASUREMENTS AND RESULTS

Plasma under study was several He-H_2 HFEDLs with total pressure 1 Torr (He:H_2 ratio was 9:1). The light from the lamp was imaged on the entrance slit of the spectrometer (*JobinYvon SPEX 1000M*, grating 1200 l·mm^{-1}, focal length 1 m) and detected by means of a charge-coupled device matrix detector (*2048 x 512 Thermoelectric Front Illuminated UV Sensitive CCD Detector, Simphony*). The emission spectra of the discharge was recorded in the wavelength range $\lambda = 600 - 630$ nm, containing (0-0), (1-1), and (2-2)Q branches of the Fulcher-α band system of the hydrogen molecule, and for the determination of rotational temperature of hydroxyl radical, emission spectra was recorded in the wavelength range $\lambda = 306 - 312$ nm, containing spectrum of Meinel transition ($A^2\Sigma^+ \rightarrow X^2\Pi$; $v' = v'' = 0$). Rotational temperature then was determined from semilogarithmic plots.

It was observed that rotational temperature obtained from the excited level energies is too low (300K-400K) to be approximated as plasma temperature [5], however rotational temperatures obtained from the ground level energies are in reasonable range to be equal to the plasma temperature. Rotational temperature determined from the (1-1)Q branch of Fulcher-α band agrees well with values from (2-2)Q branch, however the temperature obtained from lines of (0-0)Q branch are higher for about 20%.

The comparison of plasma temperatures determined from Fulcher-α (2-2)Q branch (obtained from ground level) with temperatures determined from OH Meinel transition showed that results of both methods agree well. The gas temperature ranging from 630 K to 740 K was obtained, depending on the applied current (80 – 200 mA). Relative uncertainty of the obtained temperature values using both methods is less than 10 %, for the method using OH rotational band the uncertainties were higher at 100 mA and 140 mA, being 11% and 12%, respectively. The difference between the results of both methods does not exceed 10 %, too.

The results of these measurements show that the methods of the gas temperature determination using H_2 Fulcher-α (0-0) and (1-1) Q-branches and (A-X) (0-0) Q_1 band of hydroxyl radical can be successfully used for the spectroscopic diagnostics of hydrogen-helium containing HFEDLs.

REFERENCES

[1] A.Ganeev et al., Spectrochim. Acta B 58 (2003) 879
[2] G. Revalde et al., J. Quant. Spectrosc. Radiat. Transfer 94 (2005) 311
[3] R. K. Garg et al., J. Phys. D: Appl. Phys. 4 (2008) 095206
[4] Yu. A. Lebedev and M. V. Mokeev, Plasma Phys. Rep.: Low-Temp. Plasma 29 (2003) 226
[5] Z. Gavare, G. Revalde, A. Skudra, International Journal of Spectroscopy 2010 (2010) 804506
[6] A. Skudra et al., Plasma Processes and Polymers 6 (2009) S183
[7] S. Pellerin et al., J. Phys. D: Appl.Phys. 29 (1996) 726

ACKNOWLEDGMENT

The work was partly supported by ESF project “Spectrometric techniques for detection of heavy metal contaminants” (Nr. 2009/0210/1DP/1.1.1.2.0/09/APIA/VIAA/100).

A new Hg free DBD lamps for applications in medicine

E.A. Sosnin, S.M. Avdeev, V.F. Tarasenko
Optical Radiation laboratory, High Current Electronics Institute SB RAS, 2/3, Akademicheskii ave., 634055, Russian Federation
Contact: badik@loi.hcei.tsc.ru

ABSTRACT

The discharge radiation sources for therapy of skin diseases existing in the market are analyzed. Their disadvantages are marked. As alternative to existing UV light sources it is offered to use DBD-driven lamps on working molecules N_2* and ·OH*.

BACKGROUND

It is well-known, what for erythema producing on skin the favorable wavelength range is (260–300) < λ < 400 nm. The wavelength at λ < 300 nm raise risk of skin carcinogenesis occurrence. UVA (315–400 nm) and UVB (280–315 nm) radiation is used for phototherapy of such skin diseases as psoriasis, neurodermatitis, occupational dermatosis and mycosis. For example, the spectrum of medical action of UVB radiation on psoriasis lays in a range of wavelengths 296-313 nm [1–3].
There are devices for UV phototherapy, which includes the luminescent lamp (for example, LER-40 (Russia) or TL-12 (Holland)) [1]. Such lamp represents a low pressure mercury lamp with short-wave radiation of mercury vapors converts to radiation useful to phototherapy. Serious lack of these devices is the presence of mercury in working mixtures. It makes their unsafe for operation in medical facilities and in household use (in a lamp crash case the rooms sanitization is needed). It demands the special and expensive procedure of recycling of the exhaust lamps [4]. Therefore, for example, in the EU since 2009 it is begun to phase out some types of mercury lamps from industrial lighting (EuP directive "Eco-Design Requirements for Energy-Using Products, 2005/32/EC" [5]).
There are devices for phototherapy, which includes the pulse discharge lamp (flash lamp) [6]. Complaint of such devices is their complexity: the specified lamps require the forced water cooling, special electro safety specification. Besides as the spectrum of such devices is wide, the special optical filters for selection of therapeutic spectrum part are needed.

RESULTS AND DISCUSSION

In current report as alternative to existing UV light sources for phototherapy we offer to use the lamps on working molecules N_2* and ·OH *, ensuring intensive radiation in ranges of lengths of waves 280-420 and 306-320 nm, accordingly. The power and spectral characteristics are presented.
Let us know that the N_2* molecules radiation is already used for phototherapy: the medical effect is confirm by action of radiation at λ = 337 nm on the affected parts of the skin [7]. But N_2-laser is relatively difficult-to-make and in service in comparison with lamps. Moreover the exposed skin surface does not exceed several mm in a diameter. Therefore we have many reasons to use spontaneous radiation of N_2* molecule, instead of lasing.
Our study of the N_2- and OH-lamps testify that these light sources have the following advantages:
1. Radiant exitance values up to several mW/cm^2. It is enough for minimal erythermal dose (and, accordingly, for medical effect) during an exposition of some minutes.
2. The application of dielectric barrier discharge provides spectrum stability and much greater service life of working gas mixture in comparison with electrode discharge lamps and N_2-laser.
3. It is possible to irradiate extended squares of skin and lamp size could be scaled.
4. N_2- and OH-lamps does not require the special water cooling and/or the composite construction, in comparison with flash lamps and N_2-laser.

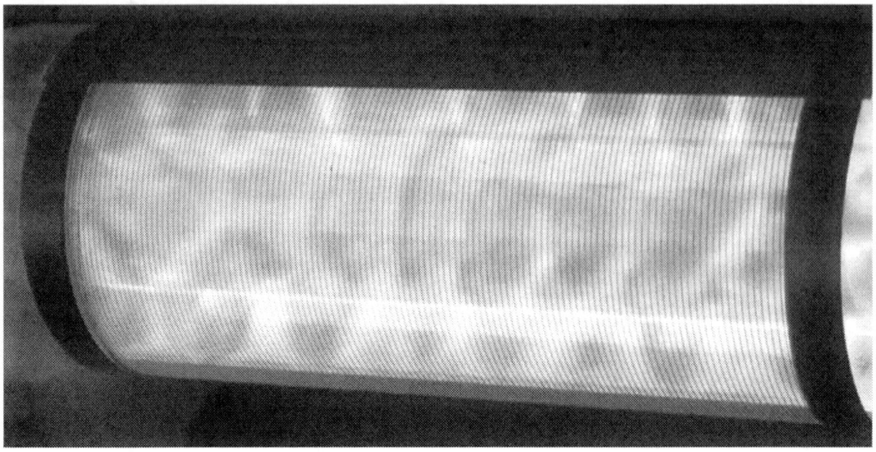

Figure 1. The view of discharge in DBD-driven N_2-lamp during operation.

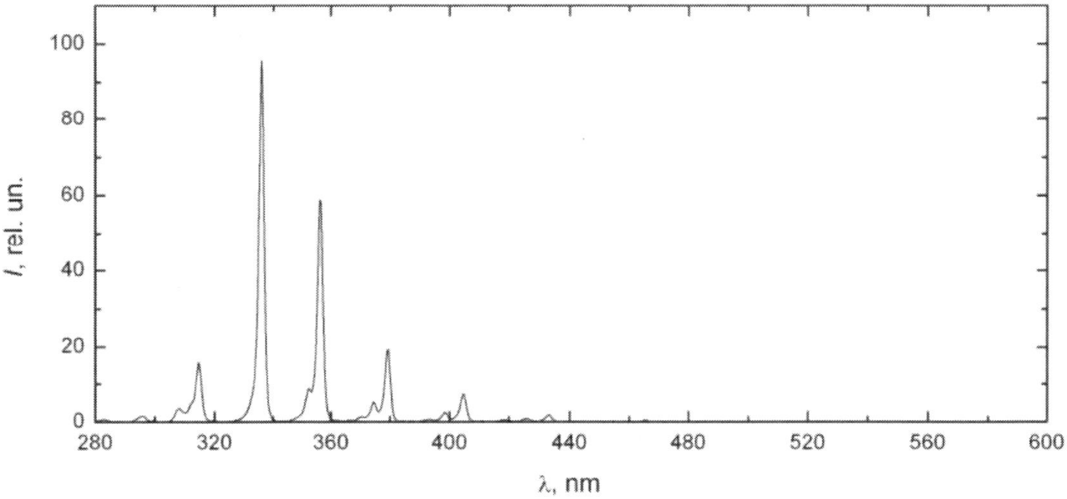

Figure 2. Spectra of N_2-lamp.

ASKNOLEGMENTS

This work was supported by the RFBR (Project No. 09-08-99080) and ISTC (Project No. 3583p).

REFERENCES

[1] Handbook of light sources (Ed. by Yu.B. Aizenberg), Moscow (2008) 951 p (Russian).
[2] J.A. Parrish, K.F. Jaencke, J. Invest. Dermatol., 76 (1981) 359.
[3] C.A. Cole, P.D. Forbes, R.E. Davies, Photochem. Photobiol., 43 (1986) 275.
[4] R. Wolsey, Lighting Futures, 3 (1998) 1.
[5] www.osram.com/incandescent
[6] Patent RU 2195342 C1 (A 61 N 5/06).
[7] Patent appl. US 2008/0172046 A1 (A 61 B 18/20).

Gas pressure characteristic of decay rate of Ne metastable atoms in cold cathode lamp

M. Goto and T. Arai
Department of Electrical and Electronic Engineering, Kanagawa Institute of Technology, 1030 Shimo-ogino, Atsugi 243-0292, Japan
Contact: miki@ele.kanagawa-it.ac.jp

ABSTRACT

We investigated the decay rate of Ne metastable atoms as a function of Ne gas pressure in cold cathode fluorescent lamps of 4mm inner diameter used for backlighting of LCD. The decay rate of Ne metastable atoms was deduced from the frequency characteristic of breakdown voltages measured. The trends of the decay rate of Ne metastable atoms as a function of gas pressure are similar to those obtained from calculated values (the loss rates by the diffusion and the de-excitation). The repetition frequency characteristic of the cathode voltages of lamp were measured with digital oscilloscope when the lamp was operated by DC pulse.

1. INTRODUCTION

The tendency of enlargement of a liquid crystal display (LCD) progresses and the high efficiency of the cold cathode fluorescence lamp (CCFL) is demanded increasingly. The greater part of total power dissipated in LCD is the power dissipated in the CCFL. The CCFL operated on low power becomes one problem. Previously, we investigated the breakdown voltage under varied frequency in the lamp of 4mm inner diameter used for LCD backlighting [1]. In this work, the decay rates of Ne metastable atoms were obtained through the frequency characteristic of breakdown voltages measured. Moreover the decay rates compared with the calculated values (the loss rates by ambipolar diffusion to the walls of the container and by de-exciting collisions with normal Ne gas atoms).

2. EXPERIMENTAL

The inner diameter of discharge tube used in this experiment is 4mm. The distance between electrodes is 53mm. The discharge tube is made of Pyrex glass. The material of electrode is a nickel. The electrode forms in the lamp are rods. The rod electrodes are covered with glass, but the tips (10mm) of those are exposed in plasma. The filling gas in the lamp was Ne. Ne was supplied to the discharge tube via mass flow controllers at flow rate of 5sccm. The pressure in the discharge tube was measured using a capacitance manometer and was varied from 6.7hPa to 80hPa. The vacuum system was performed by applying pumped using a turbomolecular pump. The cathode voltages of lamp were measured with high voltage probe (1000:1) connected to digital oscilloscope when the lamp was operated by DC pulse in varied repetition frequency from 1Hz to 10kHz.

3. RESULTS AND DISCUSSION

Figure 1 shows voltage waveforms when the DC pulse is supplied at repetition frequencies of 1Hz and 5kHz, Ne gas pressure of 53hPa. The sharp peak voltage is showing the breakdown voltage at cathode. The flat waveform is showing the maintaining voltage. τ is lifetime of Ne metastable atom, T is a period. Ne metastable atom doesn't exist for interval $T_1/2$ of 1Hz that is longer than lifetime τ.

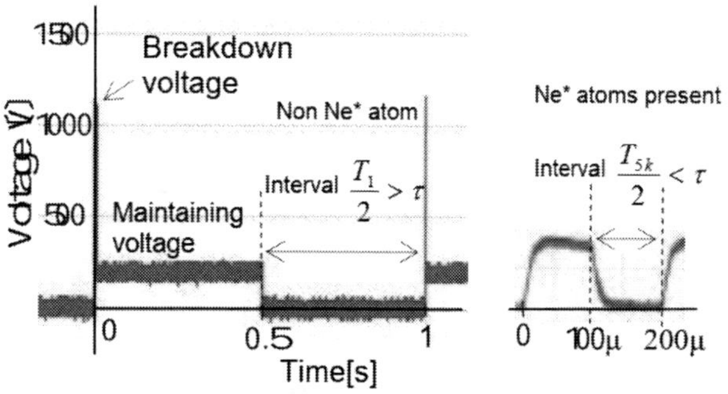

(a) 1Hz (b) 5kHz

Fig.1 Voltage waveforms

Figure 2 shows frequency

characteristics of breakdown voltage in Ne pressures of 6.7hPa, 13.3hPa and 53hPa. The measured breakdown voltages were almost constant at frequency region from 1Hz to 100Hz. At frequency of about 1kHz, the breakdown voltages decrease sharply, because the energy to breakdown is small and suffices above the frequency (1kHz) that corresponds to interval T/2 that are shorter than lifetime τ of Ne metastable atom. The interval of the Ne discharge plasma by DC pulse is a half cycle. Therefore, decay rates of Ne metastable atoms are two times frequency when breakdown voltages decrease sharply. Next, the decay rates of Ne metastable atoms have been estimated by calculation. The Ne metastable atoms are lost by diffusion to the walls of the container and by de-exciting collisions with normal Ne gas. The decay rate of Ne metastable atoms is given by

$$\frac{1}{\tau} = \frac{D_m}{\Lambda^2} + \nu_d ,$$

$$\nu_d = \sigma v N ,$$

where τ is the lifetime of Ne metastable atom, D_m is the diffusion coefficient[2], Λ is the diffusion length of the container, ν_d is the frequency of de-excitation, σ is the cross section[3] for the de-excitation of Ne metastable atoms by Ne atoms, v is the mean relative velocity of Ne metastable atoms and Ne atoms, and N is the density of Ne atoms.

Figure 3 shows the decay rates of Ne metastable atoms that were deduced from the frequency characteristic of breakdown voltages measured. A solid line shows calculated values of the decay rate of Ne metastable atoms. A dashed lines show calculated values of the diffusion loss rate and the destruction rate of Ne metastable atoms. The calculated decay rate and the measured decay rates are minimums at the Ne pressure of about 10hPa. The variations of these decay rates with gas pressure are similar.

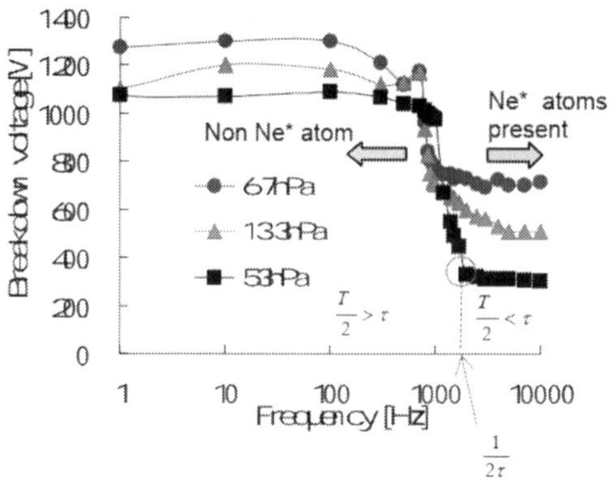

Fig.2 Frequency characteristics of breakdown voltages.

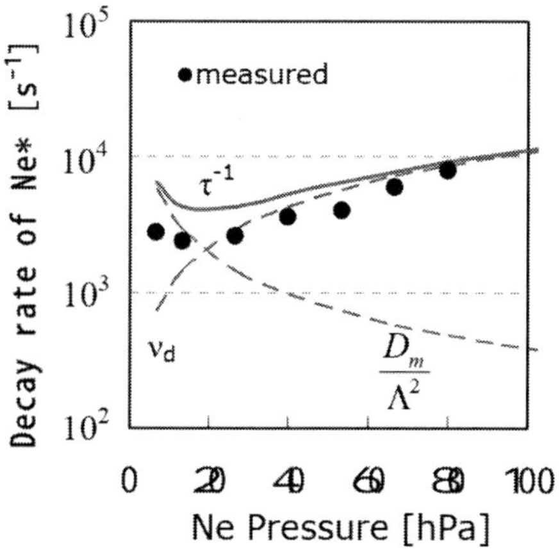

Fig.3 Decay rate of Ne metastable atoms as a function of Ne pressure.

4. CONCLUSIONS

The repetition frequency characteristic of the cathode voltages of lamp were measured by the operating DC pulse. The decay rate of Ne metastable atoms was calculated as total loss rates by the diffusion and the de-excitation. The calculated result is qualitatively good agreement with measured result.

5. REFERENCES

[1] M. Goto, T.Uehara, T.Shono and T.Arai, Proceedings of 28th ICPIG, Prague, Czech republic, 3p15-25, (2007) 1269.
[2] F.A.Grant, A.D.Krumbein: Phys. Rev., Vol.90, (1953) 59.
[3] M.A.Biondi: Phys. Rev. Vol.88, No.3 (1952) 660.

Evolution of the impurities pressure in discharge lamps using different getter materials

[1]A. Corazza, [1]V. Massaro, [2]S. Ceulemans, [3]A. Hendrikx

[1] *SAES Getters S.p.A., Viale Italia 77, Lainate (MI) 20020, Italy*
[2] *Philips Lighting, Mathildelaan 1, Eindhoven 5611 BD, The Netherlands*
[3]*Philips Innovative Applications NV; Steenweg op Gierle 417, Turnhout 2300, Belgium*

ABSTRACT

The evolution of the impurities level in the outer bulb of High Intensity Discharge (HID) lamps mounting different getter materials was studied. The vacuum degree achieved in lamps using the new Zr-Fe-Y alloy was much better than the residual pressure measured in lamps with the old Zr-Al getter.

INTRODUCTION

HID lamps have usually an outer bulb surrounding the arc tube in order to protect it against outside influences and to avoid contamination of the gas mixture in the burner from dangerous impurities.
Most of the outer jackets are under vacuum in order to allow the burner to achieve the proper thermal equilibrium without losses. In some discharge lamps the outer bulb is filled with pure inert gas, argon or nitrogen at a pressure of some hundred mbar: this filling gas is used to prevent non-passive failure and to avoid arcing between electrical current leads. Active gaseous impurities in the bulb can significantly contribute to thermal losses or can permeate through the burner walls, thus creating detrimental effects on the discharge: it is important to minimize the level of hydrogen, hydrocarbons and oxygenated gases in the outer bulb in order to assure optimal behaviour and performances during lamp life. The typically adopted solution is to mount, inside the jacket, a getter whose sorption action reduces gaseous contaminants below a "safe" level to obtain a higher quality device [1].
This paper will compare the evolution of impurities in the bulbs of HID lamps mounting two different getter materials: the old Zr-Al alloy (named St101®) and the new Zr-Fe-Y alloy (named St777) [2].

DESCRIPTION OF THE LAMPS AND MEASUREMENT PROCEDURE

Experimental 100W Metal Halide lamps mounting small getter tablets of Zr-Al alloy or of Zr-Fe-Y alloy were prepared: the getters were fixed in the vacuum outer bulb of the lamps in a position where the temperature achieved during operation was of about 300°-350°C.
After lamp production, the getters were activated by RF heating at a temperature of about 650°C for about 30 seconds. The lamps were connected to a vacuum system for Residual Gas Analyses (RGA) and the evolution of the residual pressure inside the outer bulbs was monitored during lamp operation.

PRESSURE EVOLUTION DURING FIRST HOURS OF LAMP OPERATION

In general, after thermal activation the getter is able to reduce the residual pressure in the outer bulb at a low level, but during the first period of operation of the lamp the initially high outgassing from inner components tends to generate a huge gas load, especially of water vapour, H_2, CO, CO_2 and CH_4, leading to a pressure increase even up to high values.
During this time it is important to exploit the action of an efficient

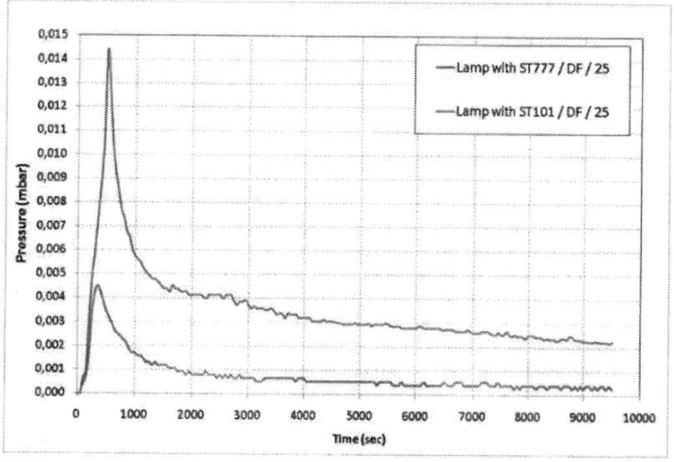

Fig.1: Pressure evolution in the outer bulb of 100W Metal-halide lamps with St777 and St101® getters in the first hours of operation

getter in order to cope with the released impurities and to reduce, in short time, the pressure at acceptably low values in order to avoid negative effects on lamp run-up behaviour and life. As shown in fig.1, the residual pressure in the outer bulbs of lamps mounting St101® and St777 is, after getter activation, in the low 10^{-4} mbar range; when the devices are switched-on and a high outgassing is induced, the Zr-Fe-Y alloy is more effective and much quicker in sorbing the released impurities than the Zr-Al: the St777 is able to promptly cope with the gas peak and to reduce the pressure below 10^{-3} mbar in less than 30 minutes, while the St101® needs many hours to achieve a comparable level. At the end, residual gases in both the bulbs were mainly CH_4 and Ar.

PRESSURE EVOLUTION AFTER AGING OF THE GETTER

Pressure evolution in the bulbs was also studied after dosing of 600 cc.mbar of hydrogen on the getters directly in lamp: investigations were carried out in order to simulate conditions typically occurring in aged lamps, by dosing H_2 quantities comparable to those possibly generated during some thousands of hours of operation. The graph of fig.2 puts well in evidence that the St777 getter was able to maintain always an optimum pressure in the range of 10^{-4} mbar in the bulb, whereas in the lamp with the St101 alloy the impurities level was significantly higher.

During lamp operation it was observed an increase of the pressure in the bulb related to the H_2 equilibrium pressure established on the getter at the actual working temperature: thanks to its excellent low H_2 equilibrium pressure characteristics the Zr-Fe-Y alloy was capable to keep a low H_2 pressure, while the H_2 pressure generated on the St101 material during lamp operation was in the high range of 10^{-3} mbar, getting close to critical values.

It is worth underling that H_2 partial pressures established on the getters in the bulbs were higher than the theoretical equilibrium pressures because, in both the cases, a portion of the getter alloy was "consumed" by the other impurities sorbed in lamp.

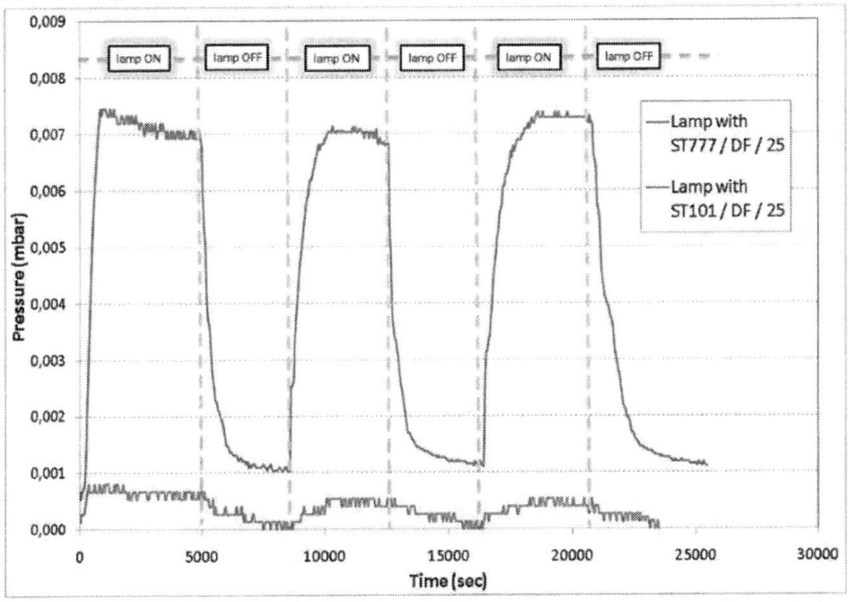

Fig.2: pressure evolution in the outer bulb of 100W MH lamps with St101® and St777 getters after some hours of operation and dosing of 600 cc.mbar of H_2

CONCLUSIONS

The use of getters based on the recently developed Zr-Fe-Y alloy in discharge lamps can provide significant improvements with respect to Zr-Al getters, both in terms of achievement of lower residual pressure during the first operation period and of maintenance of lower impurities level during lamp lifetime; this can result in a positive impact on the lamp performances during all the life.

REFERENCES

[1] C.Boffito, A.Corazza, Proc. of the 10th Int. Symp. on the Science and Technology of Light Sources, LS10 Toulouse (2004) 95.

[2] V.Massaro, A.Corazza, S.Giorgi, Proc. of the 11th Int. Symp. on the Science and Technology of Light Sources, LS11 Shanghai (2007) 203.

Phase diagram determination of the DyI_3-TlI binary system

Dietmar Kobertz[1], Klaus Hilpert[1*], Steve Hansen [2]

[1]*Forschungszentrum Jülich GmbH, 52425 Jülich, Germany*
[*]*In memoriam*
[2]*APL Engineered Materials, Inc., Urbana, Illinois 61802-7332 USA*
Contact: steve_hansen@adlt.com

ABSTRACT

Condensed phase equilibria of metal iodides are useful in assessing material properties such as vapor pressure and pellet fragility. A study was undertaken to determine the DyI_3-TlI phase diagram because of its importance to metal halide lighting and since no data on the system is available in the literature. The HoI_3-TlI system [1] has previously been determined. Differential Thermal Analysis (DTA) was used to determine the DyI_3-TlI phase diagram. Seventeen samples covering the complete composition range were investigated. Structure analysis by x-ray diffraction and determination of thermodynamic activities using Knudsen Effusion Mass Spectrometry (KEMS) assisted in the determination of the phase diagram. The system contained two eutectic reactions, $x_{\mathrm{TlI}} = 0.693$ at 434 ± 3°C and $x_{\mathrm{TlI}} = 0.865$ at 364 ± 3°C. A congruently melting compound $DyTl_3I_6$, (75 mole % TlI) also formed. The intermediate compound had a melting point of 443 ± 3°C. A solid-state transition of TlI was estimated at 177 ± 3°C and a solid-state transition in DyI_3 was estimated at 971°C.

EXPERIMENTAL

Samples were mixed together from the pure components (TlI 99.999%, DyI_3 99.99%, APL Engineered Materials, Inc) in a glove box and filled directly into baked-out quartz crucibles. A valve with an adapter to a vacuum pumping system facilitated connecting and sealing the crucible. The samples were subsequently sealed at a constriction in the crucible under vacuum at a pressure $< 10^{-6}$ bar. Ready-made compositions with 62.2 mole % TlI and 86.8 mole % TlI were also filled in the same way. A high TlI pressure, up to 10 bar [2], was expected due to the low boiling point of TlI. DTA measurements were done dynamically in a Netzsch STA 429 differential thermal analyzer on heating and cooling at rates of 5 K min^{-1} and 2 K min^{-1}. Alumina powder was placed in an open crucible as a reference cell to reduce base-line drift of the temperature difference signal. Thermodynamic activities in the solid state (20, 40, 62 and 72 mole % TlI) were measured using KEMS. Molybdenum cells with effusion orifices of 0.1 and 0.3 mm diameter were used. X-ray diffraction analysis was done on selected compositions.

RESULTS AND DISCUSSION

DTA measurements generally showed two peaks in the differential thermogram, both on heating and cooling for mixtures, and one peak for pure TlI at temperatures higher than 300°C. The temperature gap between the phase change temperatures varied from 7°C to 530°C depending on composition. A tendency toward a single phase change temperature occurred in the composition ranges between 65-70 mole % and 85-90 mole % TlI. A solid-state transition at 181°C attributed to TlI [3] occurred in compositions greater than 75 mole % TlI and was observed mainly only on heating. A solid-state transition in pure DyI_3, observed in a wide peak after signal separation, was measured at approximately 971°C.

A constant phase change temperature at 434°C occurred between 10 and 75 mole % TlI. In the same composition range, the liquidus decreased from 960 to 434°C and then increased again to 440°C. The minimum temperature was identical to the melting point of the eutectic composition of 69.2 mole % TlI. Beyond 75 mole % TlI, the behavior was similar but with an eutectic temperature at 364°C and a liquidus decreasing from 440 to 364°C and then increasing again to 430°C. The eutectic composition was found at 86.5 mole % TlI.

Vaporization studies of the four selected compositions in the range 0 to 75 mole % TlI were compared with those measured over pure TlI. A constant chemical activity of 0.37 ± 0.04 was determined for TlI beside the activity of one for DyI_3 in the same range. The TlI activity beyond 75 mole % determined by KEMS was 1 and the activity of solid DyI_3 was too low to measure. Both of these results support the assumption of a quasi-binary compound formed in the DyI_3-TlI system.

X-ray diffraction analysis of 62 mole % TlI showed an unknown structure. Taking all of the above information into account, the unknown structure was assigned to the compound (DyI_3·3TlI) $DyTl_3I_6$. Previously unknown in the literature, $DyTl_3I_6$ melts congruently at 443 ± 3°C. Fig. 1 shows the final construction of the DyI_3-TlI phase diagram. DyI_3 was reported to have a solid-state transition at 828°C [4] and 862°C [5].

CONCLUSIONS

The DyI_3-TlI phase diagram, determined by DTA and assisted by KEMS and x-ray diffraction, showed one intermediate phase ($DyTl_3I_6$) and two eutectic reactions.

REFERENCES

[1] A. K. Molodkin, E. F. Kudina, A. G. Dudareva and S. B. Nechitailov, Russ. J. Inorg. Chem., 32 (1988) 310.
[2] H. Schöpp, Elektrotechn. und Informatik, TU Berlin, Thesis, 2002.
[3] E. A. Secco, Thermochim. Acta, 342 (1999) 161.
[4] D. A. Johnson and J. D. Corbett, Rare Earth Elements, Book 1, Intl. Conf. of the National Center of Scientific Research, CNRS, Paris, France (1970) 429.
[5] A. K. Molodkin and A. G. Dudareva, Russ. J. Inorg. Chem., 31 (1986) 1603.

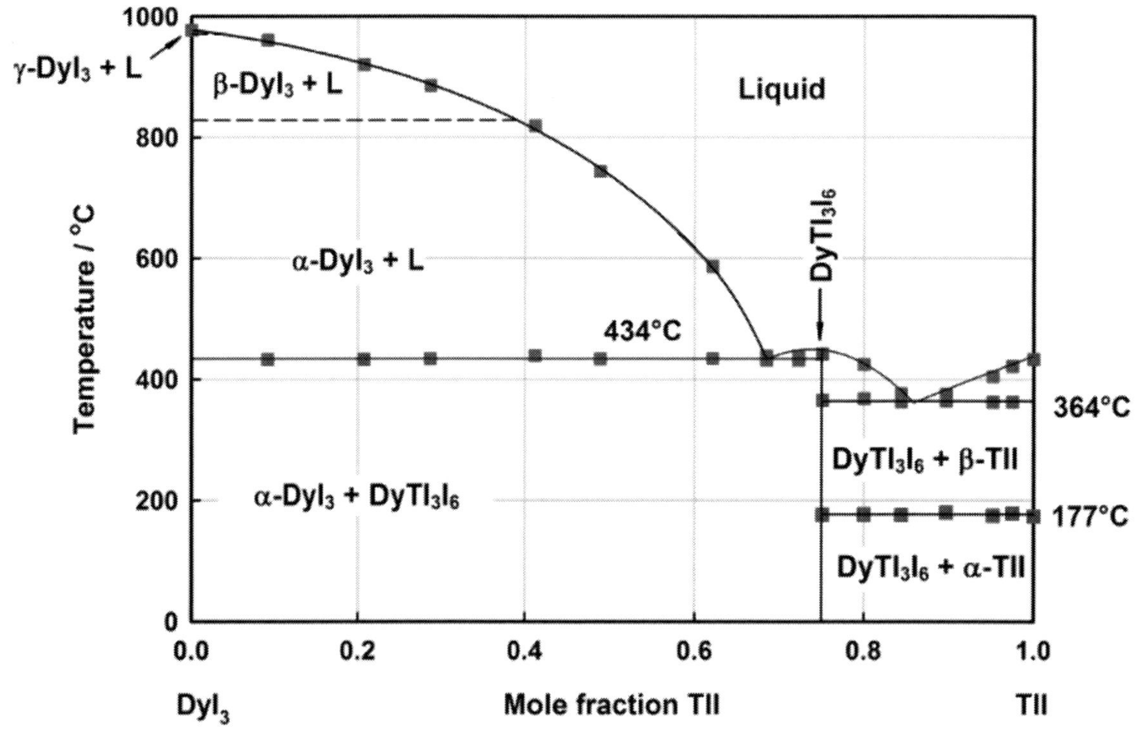

Figure 1. Measured phase diagram of the DyI_3-TlI system.

Hemispherical InGaN/GaN Light-emitting Diode for efficient fiber coupling

Ling Zhu, P.T Lai and H.W. Choi*

Department of Electrical and Electronic Engineering, University of Hong Kong, Pokfulam Road, Hong Kong

Contact: hwchoi@hku.hk

ABSTRACT

We demonstrate a novel blue InGaN/GaN LED structure capable of efficient coupling with plastic optical fiber(POF) through geometrical design. The blue LED chips were shaped into inverted truncated cones by laser micro-machining and further coated with metallic mirror onto its sidewalls. To further enhance the coupling efficiency, a half-ball lens was integrated forming a hemispherical structure. Based on this hemispherical LED, the coupling efficiency exceeded 51.7 %, which, to the best of our knowledge, is the highest reported value.

Although visible nitride-based light-emitting diodes (LED) are mainly developed for general lighting, other promising applications have been demonstrated recently, including fiber-coupled LED sources. The fiber enables the light to be channelled away from the source, offering a portable system with stable optical path and eliminating the need for bulky and lossy optics as with free-space transmission. At the output end, the signals can be easily focused and projected for imaging or detection purposes, as well as for data transmission. Compared with conventional fiber-coupled light sources such laser sources and super luminous diodes, our InGaN/GaN-based visible LEDs have advantages such as being bright, robust, reliable, responsive, economical and eye-safe, which may prove to be a cost-effective and valuable light source. To fulfill the needs for fiber-coupled applications, apart from requiring an efficient light source, its coupling efficiency with the fiber is a critical issue. Due to the divergent nature of LEDs, the coupling efficiency is typically low in conventional coupling configurations. To solve this problem, we have assembled a novel hemispherical LED for efficient coupling with POF, with a remarkable efficiency of 51.7%.

Our LEDs were shaped into truncated-conical structures (TC-LED) by laser micro-machining and coated with a metallic mirror layer on its inclined sidewall, which could suppress the divergence of emission pattern effectively. Then it was flip-chip bonded onto the package, as illustrated in Figure 2 (a) and (c), so that wire-bonds are not needed. Finally, a BK-7 half ball lens was bonded onto the surface of the exposed sapphire surface of the chip via liquid capillarity [1]. With the correct procedures, a strong capillary bond can be achieved. The SEM image in Figure 1 also reveals the close contact between sapphire and BK-7 lens, contributing to negligible light loss at the interface. The entire hemispherical LED assembly is shown in Figure 2 (b) and (d). Figure 2 (e) and (f) illustrate fiber-coupling arrangements with and without the hemispherical lens. When in an intimate contact, the maximum coupling efficiency of TC-LED and hemispherical LED was about 30.2% and 51.7% respectively.

The optical characteristics of the spherical LED was evaluated and compared with TC-LED. According to the experimental angular emission patterns in Figure 3 (a), the divergent half-angle γ of the spherical TC-LED is 36°, 9° less than that of reference LED with γ of 45° (γ is defined as the angle at which power intensity drops to $2^{½}$ of its peak value). The overall optical intensity of the spherical LED, obtained by integrating the angular intensities over 180°, is found to be 57.4% higher than that of the reference LED. The half-ball lens, an integral component in the hemispherical LED, played two vital roles for efficient coupling with the fiber. With a refractive index of 1.52, the lens can extract more light from the sapphire layer, resulting in higher light extraction efficiency than the TC-LED. At the same time, it focused the light into a convergent spot with negligible losses. The results have been verified by ray-tracing simulation using Trace Pro; the simulated emission profiles were also plotted in Figure 3 (b). The simulated pattern is consistent with the experimental data: the spherical LED with lens exhibits a 9° reduction (31° and 40° for them respectively) of divergence half-angle, a 131.9% increase of overall intensity.

Our fiber-coupled hemispherical LED can be demonstrated to exhibit higher output power, particularly under pulsed excitation. Figure 4 illustrates the measurement scheme. A 1 kHz input pulse train with a peak voltage of 8V with a pulse width of 3.75 μs was fed to the LED. The average power at the 10-feet POF fiber-coupled output was ~2.65 mW, measured with a calibrated Si detector. To determine the peak power, the output signal was collected with a photodetector connected to an oscilloscope to observe the impulse response. It was found that the width of the output pulse series was ~32.7 μs. Based on with the equation: $P_{average} / P_{peak} = \tau_{width} / T$, the peak output power of our spherical LED was evaluated to be ~200.4 mW. The measured optical pulse series are plotted in Figure 5. From this result, we can predict that the fiber-coupled spherical LED has great potential as a light source for micro-machining or data transmission over long range. The response time of LEDs can be reduced by dimensional downscaling [2].

Figure 1. SEM image showing the lens and sapphire interface.

Figure2. Optical images showing the TC-LED and hemispherical LED.

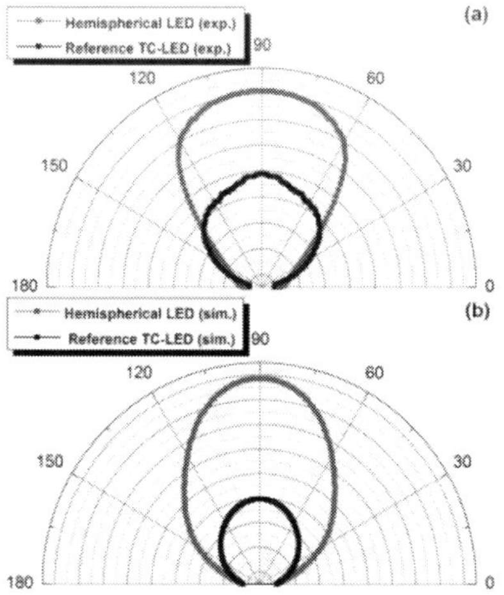

Figure 3. Experimental and simulated emission patterns.

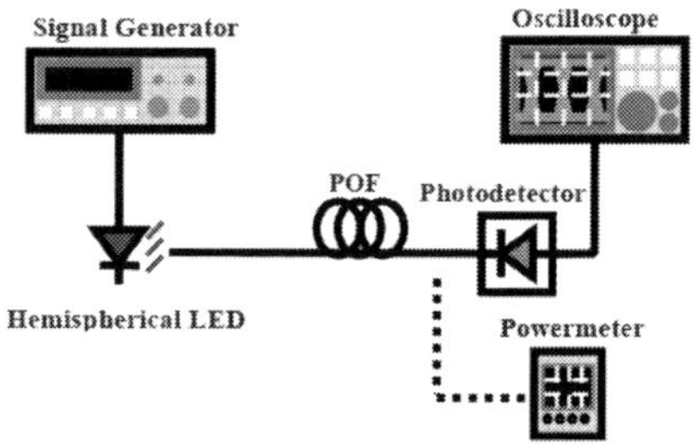

Figure 4. Schematic diagram of measurement setup.

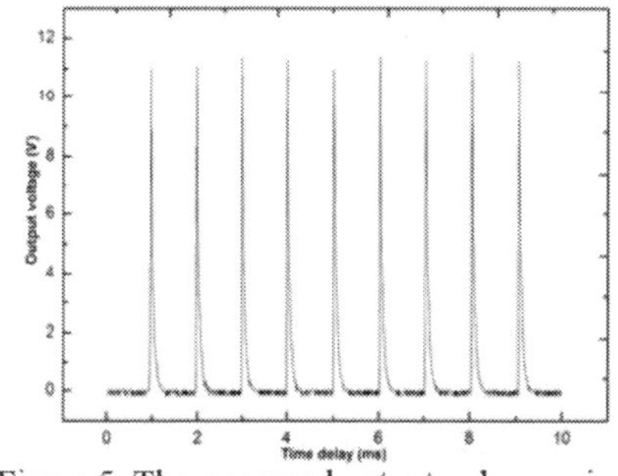

Figure 5. The measured output pulses series.

Reference:

[1] Z. L. Liau, Appl. Phys. Lett., **77**, 651, (2000)
[2] S. X. Jin, J. Shakya, J. Y. Lin, and H. X. Jiang, Appl. Phys. Lett. **78**, 3532 (2001)

Emission Enhancement of InGaN MQWs With Surface Plasmon Mediated Layers Ag/ITO

C. C. Lee[1], D. P. Cai[2], C. C. Chen[2], Bing- Jung Chen[3] , Jiun-Ting Lin[3], and, J. Y. Chang[2]

[1]*Optical Science Center, National Central University, Jhong-Li 32001, Taiwan*

[2]*Department of Optics and Photonics, National Central University, Jhong-Li 32001, Taiwan*

[3]*ALPHA CRYSTAL Technology Corporation, , Hsin-Chu 30075, Taiwan*

Abstract

Coupling of a InGaN/GaN multi-quantum well (MQW) and semitransparent metal layer is shown to result in dramatic enhancement of spontaneous emission rate by the surface plasmon effect in the optical spectral range. A six-pairs 17nm InGaN/GaN MQW is positioned 250nm away form its surface layer deposited with 80nm indium tin oxide(ITO) as the index matching layer of surface plasmon. Surface texture patterns(p=0.2~8μm) were defined on the ITO layer by e-beam lithography, then silver thin film was coated on these periodic patterns. We demonstrate the ITO (80nm)/Ag(50nm) film and ITO (80nm)/Ag(50nm) with grating period of 5μm exhibit significant enhancement about 44.8% and 52.5%, respectively.

1. INTRODUCTION

There have been various efforts to increase the extraction efficiency of Gallium nitride (GaN) based light emitting diodes (LED) [1-3]. Therefore, how to effectively extract light out of the material has become a major issue in device design. Surface plasmon has been used as a means to couple light out of the active layer [4, 5]. On the other hand, an enhanced rate of spontaneous emission recombination coupled into the surface plasmons in Ag/InGaN quantum well has been observed by photoluminescence and later confirmed by time-resolved photoluminescence measurement [6-9]. In this paper, we used flip-chip type InGaN/GaN LED, which was deposited with and without 80nm indium tin oxide(ITO) as the index matching layer of surface plasmon. We experimentally measured photoluminescence intensity (PL) of the fabricated structures and compared with the unprocessed samples, whilst still ensuring that most of the emission takes place into the surface plasmon (SP) mode. And the implication of these results for light extraction by using with surface texture to increase light scattering from light emission diode is discussed.

2. EXPERIMENTS AND RESULTS

An flip-chip type LED embedded with six-pairs InGaN/GaN multi-quantum well (MQW) grown by metallo-organic chemical-vapor deposition (MOCVD) was used in our experiments, which was deposited with 80nm indium tin oxide(ITO) as the index matching layer of surface plasmon. Surface texture patterns were defined on the surface layer of InGaN/GaN MQW by e-beam lithography, and Ag thin films with various thickness (15~100 nm) were evaporated on the top surface of these samples. The metal grating patterns were transferred into the top metal layer by using electron-beam evaporation and lift-off process. In this way, a direct comparison of the PL spectrum of the Ag-coated and the unprocessed parts of the sample could be made. Fig. 1(a) shows the schema of the sample structure.

The results reported here aimed to establish the best grating profile to produce the highest light-SP coupling efficiency in the case of simple square gratings of various periods. The grating-assisted coupling of light to SPPs was then investigated as a function of the grating period. As the PL peak of the Ag coated InGaN MQWs at 450 nm was normalized to 1, we demonstrate the ITO (80nm)/Ag(50nm) film and ITO (80nm)/Ag(50nm) with grating period of 5μm exhibit significant enhancement about 44.8% and 52.5%, as shown in Fig. 1(b). We showed how choice of patterned grating structures with ITO as the index matching layer of surface plasmon between active region and metal films can be used to optimize light emissions. We believe that this work provides another implication of the surface plasmon enhancement for highly efficient and high-speed solid state light emitters alternative to conventional light bulbs.

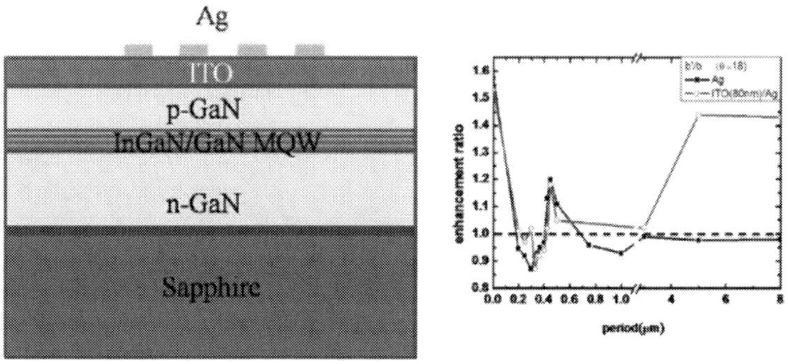

Fig. 1 (a) shows the schema of the sample structure, (b)The PL area intensity enhancement ratio of continuous silver film and various grating structures with periods from 0.2 to 8μm.

REFERENCES

[1] M. Yamada, T. Mitani, Y. Narukawa, S. Shioji, I. Niki, S. Sonobe, K. Deguchi, M. Sano, and T. Mukai, *Jpn. J. Appl. Phys.*, **vol. 41**, L1431-L1433, 2002.

[2] S. Nakamura, *Science*, **vol. 281**, pp956-961, 1998.

[3] S. Nakamura, M. Senoh, S. Nagahama, N. Iwasa, T. Yamada, T. Matsushita, H. Kiyoku, Y. Sugimoto, T. Kozaki, *Appl. Phys. Lett.* **vol. 72**, pp211-213, 1998.

[4] D. K. Gifford and D. G. Hall, *Appl. Phys. Lett.*, **vol. 81**, pp4315-4317, 2002.

[5] S. Wedge, J. A. E. Wasey, W. L. Barnes, and I. Sage, *Appl. Phys. Lett.*, **vol. 85**, pp182-184, 2004

[6] I. Gontijo, M. Boroditsky, E. Yablonovitch, S. Keller, U. K. Mishra, and S. P. DenBaars, "*Phys. Rev. B*, **vol. 60**, pp11564-11567, 1999.

[7] A. Neogi, C. W. Lee, H. O. Everitt, T. Kuroda, A. Tackeuchi, and E. Yablonovitch, *Phys. Rev. B*, **vol. 66**, pp153305, 2002.

[8] K. Okamoto, I. Niki, A. Shvartser, Y. Narukawa, T. Mukai, and A. Scherer, *Nat. Mater.* **vol. 3**, pp601-605, 2004.

[9] K. Okamoto, I. Niki, A. Shvartser, Y. Narukawa, T. Mukai, and Y. Kawakami, *Appl. Phys. Lett.*, **vol. 87**, pp071102, 2005.

Gas dependency of electrode temperature and emission of emitter material in a low-pressure fluorescent lamp

Y. Yamagata [1], M. Kai [2], S. Naito [1], K. Kamio [1], Y. Manabe [2], K. Uchino [1]
[1] *Interdisciplinary Graduate School of Engineering Sciences, Kyushu University, 6-1 Kasugakoen, Kasuga, Fukuoka 816-8580, Japan*
[2] *Lighting Company, Matsushita Electric Industrial Co., Ltd, 1-1 Saiwai, Takatsuki, Osaka 569-1193, Japan*
Contact: yamagata@ence.kyushu-u.ac.jp

ABSTRACT

In order to discuss the effect of ambient gas on the emitter material loss from the electrode of a low-pressure fluorescent lamp (FL), a LIF technique and a black-body radiation method have been applied to the Ba atom density measurement and the electrode temperature measurement, respectively. It is suggested that control of the hot spot temperature is most important for an effective and long-lived FL.

1. INTRODUCTION

An emitter material is coated onto the filament electrode in low-pressure fluorescent lamp (FL), which has some electric resistance. Sufficient thermionic electron emission from the electrode heated by discharge current itself leads a low sheath voltage resulting high efficiency operation of the FL. Ba atoms as one of the main source of emitter materials are evaporated and sputtered from the electrodes during FL operation. However, there are only a few studies describing the features of Ba atom loss from the electrode [1-3]. In those situations, we have clarified the mechanism of Ba atom loss in part, in which the loss is mainly due to thermal evaporation from a hot spot [4-6]. The electrode temperature and the Ba atom loss are strongly influenced by the discharge conditions. Therefore, in order to improve the life time and the efficiency of FL, it is strongly desired to clarify how the discharge condition influences the electrode temperature and the Ba atom loss in detail.

The aim of this work is to clarify the relationship between the electrode temperature and the Ba atom loss under various discharge conditions, then to contribute to make a design manual of much more effective and long-lived FL. To achieve this, we have applied laser-induced fluorescence (LIF) technique and black-body radiation method to measure the Ba atom loss and the electrode temperature, respectively [4-7]. In this paper, experimental results of measurements of the Ba atom loss and the electrode temperature of a FL in several gas mixtures and pressures are presented.

2. EXPERIMENTAL PROCEDURE

A commercially available FL (Panasonic, FL20SS) filled with a mixture of argon, krypton and mercury vapor was chosen. It consisted of a clear glass tube without any coating of fluorescent materials on the inner wall. The FL connected with a Cu-Fe ballast was continuously operated by AC (60 Hz) power supply. For Ba atom density measurement, LIF technique was used. Details of the experiment have been published previously [4-6]. Figure 1 shows an experimental setup for temperature measurement of the FL electrode using optical emission spectroscopy. Emission spectra (750–820 nm) from the electrode were detected through a lens, an optical chopper and a monochromator by a photomultiplier tube that has high quantum efficiency near 800 nm (Hamamatsu, R943-2). Electrode temperature was estimated by fitting a measured spectrum with a theoretical curve of Plank's law at a certain temperature. The optical detection system was calibrated by using a standard lamp. For the spatial-resolved measurements, the FL fixed on a

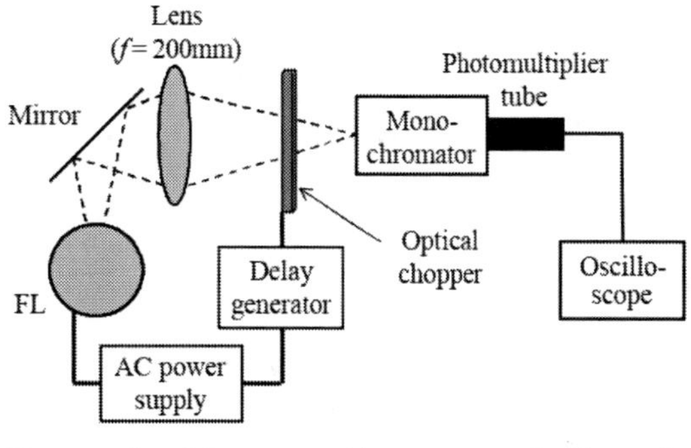

Figure 1 Schematic illustration of electrode temperature measurement system using black-body radiation method for a FL.

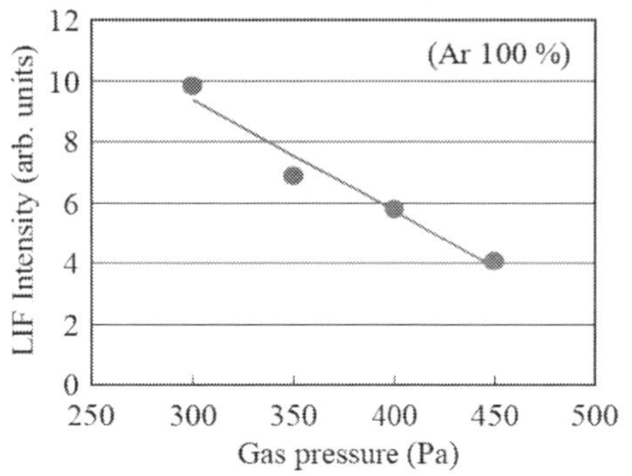

Figure 2 Temporally-averaged LIF intensity from Ba atoms in front of electrode as a function of gas pressure in a virgin FL operating at AC 60 Hz.

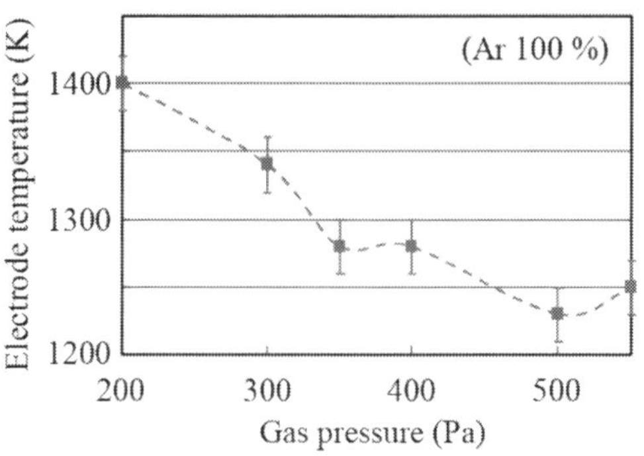

Figure 3 Temporally-averaged temperature of a hot spot as a function of gas pressure in a virgin FL operating at AC 60 Hz.

x-*y* translation stage was moved using micrometers without any change of the optical alignments. As for the time-resolved measurements, the laser pulse for LIF measurement or the optical chopper for temperature measurement was synchronized and varied with respect to the discharge voltage (V_d) using a delay generator. In order to increase the signal to noise ratio, the LIF signal or the optical emission signal was averaged over 600 times.

3. RESULTS AND DISCUSSION

Figure 2 shows the LIF intensity from the Ba atom in the vicinity of the electrode of FL filled with Ar 100 % as a function of the total pressure. The LIF signals were observed over the whole area in the direction of length of the electrode, and were temporally averaged. As can be seen in Fig. 2, the amount of the Ba atom loss from the electrode decreases with the pressure increase. Because the Ba atom loss is caused by the thermal evaporation from hot spot (HS) generated at the electrode edge connected to the power supply [4-6], the decrease of the Ba atom loss is considered to be caused by the vapour pressure suppression or the decrease of the HS temperature. The temporally-averaged HS temperature as a function of the total pressure is shown in Fig. 3. From the Fig. 3, the HS temperature decrease about 150 K is observed from 200 Pa to 500 Pa. From these experimental results and consideration, it is shown that the decrease of the Ba atom loss with the pressure increase is mainly caused by the decrease of the HS temperature.

4. CONCLUSION

The LIF technique and the black-body radiation method have been applied to Ba atom density measurement and electrode temperature measurement in FLs, respectively. The gas mixture and the total gas pressure affect strongly to the HS temperature and the resulting Ba atom loss from the HS. It is suggested that fine control of HS temperature by optimizing the discharge conditions is most important factor for an effective and long-lived FL.

REFERENCES

[1] A. K. Bhattacharya, J. Appl. Phys. **65** (1989) 4595.
[2] S Hadrath, M Beck, R C Garner, G Lieder, J Ehlbeck, J. Phys. D: Appl. Phys. **40** (2007) 163.
[3] K. Misono, M. Kando, J. T. Verdeyen, J. Illum. Engng. Inst. Jpn. **83** (1999) 819 *(in Japanese)*.
[4] A. Samir, G. Yamashita, Y. Yamagata, K. Uchino, T. Ueda and Y. Manabe, Jpn. J. Appl. Phys. **45** (2006) 8109.
[5] A. Samir, G. Yamashita, Y. Egashira, S. Shimada Y. Yamagata, K. Uchino, T. Ueda, Y. Manabe, Trans. Fund. Mater. IEEJ **A-127** (2007) 543.
[6] T. Ueda, A. Samir, Y. Egashira, G. Yamashita, S. Shimada Y. Yamagata, K. Uchino, Y. Manabe, Jpn. J. Appl. Phys. **46** (2007) 6828.
[7] Y. Yamagata, M. Kai, S. Naito, K. Tomita, K. Uchino, Y. Manabe, Thin Solid Films **518** (2010) 3449.

Psychological building blocks for dynamic road lighting: Understanding light's role in feelings of safety at night

Y.A.W, de Kort[1], A. Haans[1], L. Geerdinck[2], D. van Gennip[1], M. Horst[1], J. Servaes[1]
[1] Human-Technology Interaction, IE&IS, Eindhoven University of Technology, PO Box 513, 5600 MB Eindhoven, The Netherlands
[2]Philips Lighting, Mathildelaan 1, 5611 BD Eindhoven, The Netherlands
Contact: y.a.w.d.kort@tue.nl

ABSTRACT

Road lighting serves a host of functions at night, yet puts a substantial burden on our total energy consumption. Without exception, technical solutions to reduce the use of energy (e.g., new light sources, interactive road lighting) impact user perceptions and acceptance in ways that are not yet understood. In an experimental field study, conducted in our interactive road lighting testbed, we investigated pedestrians' preferences for various static road lighting scenarios, in particular as they relate to feelings of safety, and the psychological mechanisms behind them. The testbed enabled us to manipulate the light, keeping other factors constant, thus offering experimental rigor and control. Following a paired-comparison paradigm, fifty female participants rated three light distributions on two street segments according to perceived safety. They then assessed each light distribution for each street segment on psychological and perceptual attributes. Research is ongoing; results will be reported at the conference.

INTRODUCTION

Road lighting needs to provide light and safety for all users. In particular to those vulnerable to or fearful for personal attacks, roadway lighting is essential for experiencing safety and the freedom to go out at night. On the other hand, there is an increasing awareness for climate change and impending shortage of energy sources which accelerates the urge for energy use reduction. The total energy use of public lighting in the Netherlands nowadays is estimated to be 600.000 to 700.000 MWh a year. About 500.000 MWh per year is used by municipalities for the lighting of infrastructures (e.g. roads, footpaths) [1]. These conditions underline the clear and urgent need to develop more intelligent and efficient light sources and lighting systems for road lighting.

LEDs are a promising class innovation for energy conservation. Their efficiency is rising steadily and they offer new possibilities for interactive and adaptive lighting, by affording dimming for streets and/or periods with lower traffic densities, adjustment of the light output to contextual light conditions (e.g., snow), or intelligent light-on-demand scenarios. Such adaptation could result in reductions of energy use, but would require robust, accurate and intelligent sensing and recognition of use, type and number of user(s). Yet in addition to these technical challenges, a major challenge is to understand user perception and acceptance of such adaptive road lighting at night. Such understanding requires perception research for personal safety and investigations of the impact of various algorithms and scenarios on subjective personal safety and user behaviour.

People perceive light as the most important environmental feature in relation to perceived personal safety at night (e.g., [2,3,4]). Moreover, recent research (e.g., [2,5,6]) has demonstrated the importance of lighting as a determinant of feelings of safety. These studies also illustrate the complexity of relationships between lighting (illuminance, spectrum, e.g. [6,7]) and street characteristics (e.g., [2,3,4,5]). Few – if any – researchers have had the opportunity to manipulate road lighting, keeping all other factors (concealment, entrapment, etc. i.e., distal features) constant. The interactive road lighting testbed on the university campus allows us to do just that.

Rationale

In this project we focused on pedestrians as they are, in contrast to motorists, slower and hence more vulnerable. Pedestrians rely on lighting for several purposes, among which orientation and recognition of obstacles and faces [8]. As previous research has always considered constant and even light levels, it is as yet unknown how people would feel under dynamic lighting and what the optimal lighting scenario would look like: should it light mostly the direct surroundings of the pedestrian, or rather facilitate prospect farther away from him or her? This was the primary question for the current experiment. Before venturing into dynamic behaviour of street lamps, we first needed to investigate the optimal static patterns of light, so to speak as building blocks for dynamic scenarios.

METHOD

In a field experiment, participants saw and rated three lighting scenarios (5 poles evenly lit, 5 poles with levels decreasing from near to far, and 5 poles with levels increasing from near to far, see Figure 1). Since women have higher fears of crime at night (e.g., [2,3,4]), only females were recruited. Fifty females (not familiar with the campus) participated, their ages ranged between 18 and 30.

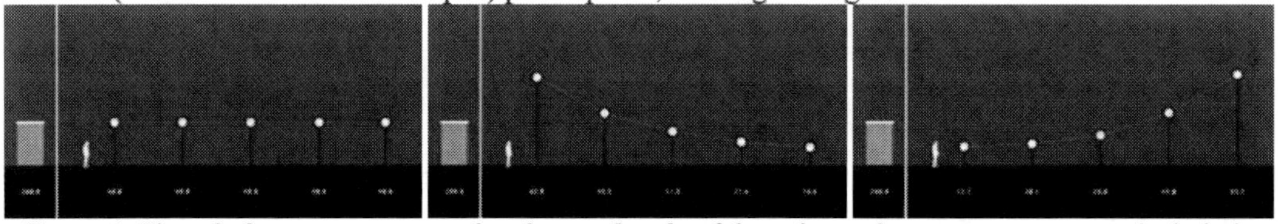

Figure 1. Three lighting scenarios used in the test (height of the poles in the picture represents the light level)

The first phase of the experiment followed a paired-comparison paradigm: two scenarios at a time were presented to participants on two street segments of the testbed, in all possible combinations (9 in total). For each pair, participants indicated on which of the two segments they would feel most comfortable/safe. In the second phase, all three scenarios were again presented to participants, on both street segments, but this time one by one. They rated every segment/scenario combination on a number of perceptual attributes and psychological experiences. Orders of the paired comparisons and scenarios for questionnaire ratings were randomised over participants.

RESULTS

The experiment is currently ongoing. Results will be available at the time of the conference.

ACKNOWLEDGEMENTS

This project is part of the ENSURE initiative, in which Philips and Eindhoven University of Technology collaborate in research on intelligent lighting solutions. We thank the Dutch Ministry of Economic Affairs for their financial support In particular, we thank Marco Haverlag, Gerrit Kroesen, Cagdas Atici, Tanir Ozcelebi, and Johan Lukkien for their input and assistance during the project.

REFERENCES

[1] SenterNovem en lokaal klimaatbeleid Beurskrant 2009. http://www.senternovem.nl
[2] Loewen, L. J., Steel, G. D., & Suedfeld, P. (1993). Perceived safety from crime in the urban environment. Journal of Environmental Psychology, 13, 323-331.
[3] Nasar, J. L., & Jones, K. M. (1997). Landscapes of fear and stress. Environment and Behavior, 29, 291-323. In bezit
[4] Nasar, J. L., Fisher, B., & Grannis, M. (1993). Proximate cues to fear of crime. Landscape and Urban Planning, 26, 161-178.
[5] Blöbaum, A., & Hunecke, M. (2005). Perceived danger in urban public space: The impact of physical features and personal factors. Environment and Behavior, 37, 465-486.
[6] Boyce, P. R., Eklund, N. H., Hamilton, B. J., & Bruno, L. D. (2000). Perceptions of safety at night in different lighting conditions. Lighting Research and Technology, 32, 79-91.
[7] Knight, C. (2009) Effect of lamp spectrum on perception of comfort and safety. In: de Kort, IJsselsteijn, Vogels, Aarts, Tenner, and Smolders, Proceedings of Experiencing Light 2009.
[8] Raynham, P. (2004). An examination of the fundamentals of road lighting for pedestrians and drivers. Lighting Research and Technology, 36, 307-313.

Transient Color stability of HID lamps

I. Francois[1] , J.T. van der Eyden[1] , W. Broeckx[1], S. Geudens[1]
[1] *GTD HID, Philips Lighting, Steenweg op Gierle 417, Turnhout, B-2300, Belgium*
Contact: joost.van.der.eyden@philips.com

ABSTRACT

This paper discusses the problem of "colour variation" in ceramic HID lamps, i.e. the phenomenon that the chromaticity coordinates of these lamps can vary in time in an unpredictable manner. The colour instability of a lamp can result in visible colour variation between lamps.
This paper describes the specification, the measurement methods and the description of the colour stability of a group of lamps.

TRANSIENT COLOR STABILITY IN HID LAMPS

The term transient colour (in)stability of a lamp describes the phenomenon that the chromaticity coordinates of that lamp varies in time; this variation can be on a time scale of minutes or hours, and the severity can change from minor changes in the colour point to steps of several hundred Kelvin. In the figure below a colour measurement of unstable (left) and stable (right) lamps is shown: the colour temperature is monitored as a function of time. The colour instabilities can be related to the movement of liquid salt droplets along the arc-tube wall and in the extended plugs of the burner.

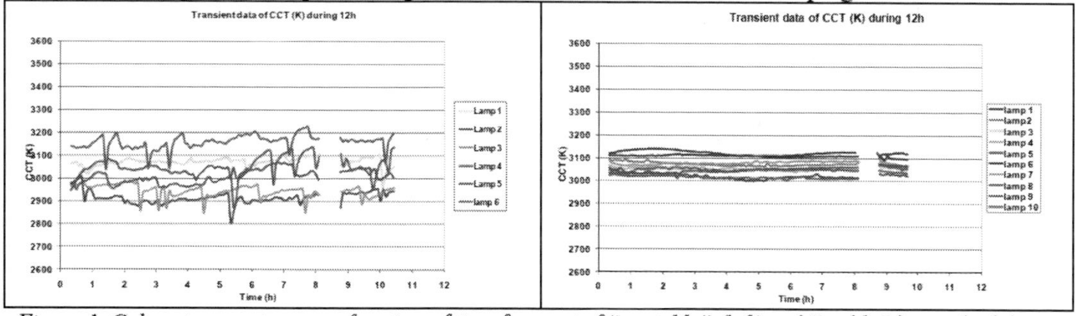

Figure 1. Colour temperature as a function of time for a set of "unstable" (left) and "stable" lamps (right).

A specification for the allowed colour variation between lamps was deduced from an appraisal test in which lamps with a well-defined colour variation were operated in a critical application (wall-washers on a white wall). The conclusion was that colour differences larger than 10 points in the x- and 6 points in the y-coordinate were considered as unacceptable by more than 30% of the observers. The Philips specification originating from this test (for shops and retail applications) is that 90% of the colour points of a population of lamps should be within a 3SDCM circle around the mean value.

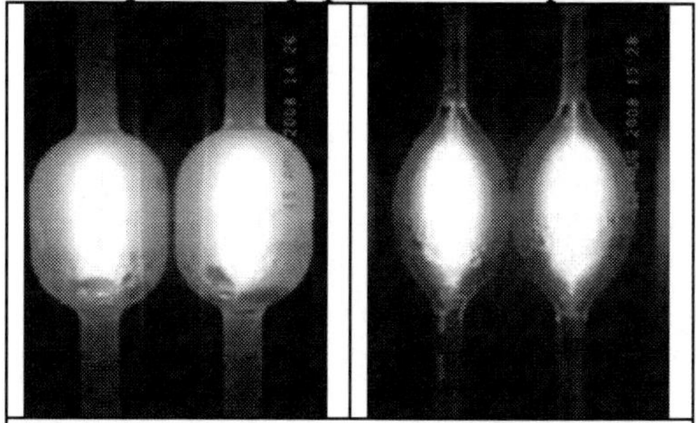

Figure 2. Projection images of an alumina and a YAG arc-tube.

MEASUREMENTS AND STATISTICS OF COLOR STABILITY

The standard integrating sphere measurements can be used to evaluate the color points of a set of lamps. This gives a clear indication of the spread in colorpoints. Since this represents only a single snapshot of the colorpoints after a standard stabilisation time of typically 10 minutes, these measurements are not adequate to evaluate the transient color stability phenomena. We have build up a special measurement setup consisting of 10 boxes for mounting 10 lamps. For each lamp the visible spectrum is measured using an OMA-spectrometer together with the electrical characteristics. Typically measurements are taken each 3 minutes over a period of 12 hours including one switching. In this way the characteristics of 10 lamps can be evaluated over time (figure 1). For better understanding of the mechanisms causing transient behaviour a setup is used consisting of an optical system to project the image of two sides of the lamp on a screen and a digital camera to take pictures, typically every minute. Some example pictures are shown in figure 2. With these measurements the movement of salt drops on the wall and the changes in arc attachement on the electrode can be monitored.

Color stability metric

To judge the transient color stability we need a quantitative metric. Color point differences can be measured using the MacAdam's method with the SDCM as unit of distance in color space. Since the SDCM distance defines a vector length in a 2-dimensional plane, the Rayleigh distribution is the appropriate distribution for the vector lengths. The Rayleigh distribution has a cummulative distribution function, $cdf(SDCM_{\max}|\sigma) = 1 - \exp\left(\frac{-SDCM_{\max}^2}{2\sigma^2}\right)$, which gives the percentage of colorpoints within $SDCM_{max}$. The distribution has one parameter, σ, which characterizes the spread of the colorpoints. The parameter σ, can be estimated from a sample of N colorpoints, taken from:

1. One lamp over time
2. A group of lamps over time
3. A group of lamps at a single instant in time

For each individual colorpoint (x_i, y_i) we can calculate the SDCM distance from the mean colorpoint $(\bar{x}, \bar{y})$ using: $SDCM_i = \sqrt{g_{11}(x_i - \bar{x})^2 + g_{12}(x_i - \bar{x})(y_i - \bar{y}) + g_{22}(y_i - \bar{y})^2}$.

From these values we can calculate the maximum likelihood estimator, $\hat{\sigma} = \sqrt{\frac{1}{2N}\sum SDCM_i^2}$. Depending on the sample type taken, we get a sigma value characterizing:

1. Color stability over time of one lamp, σ_{lamp}
2. Color stability of a group of lamps at a single instant in time, σ_{group}
3. Color stability over time of a group of lamps, σ_{total}

Since the transient colorpoints measurements are not absolutely calibrated, we cannot measure σ_{total} directly. We combine the transient measurements with integrating sphere measurements to get the color stability over time for a group of lamps, $\sigma_{total} = \sqrt{\sigma_{lamp}^2 + \sigma_{group}^2}$. From the value of σ_{total} we can directly deduce what percentage of the colorpoints of the total popluation is within 3 SDCM of the mean colorpoint using the relation shown in figure 3.

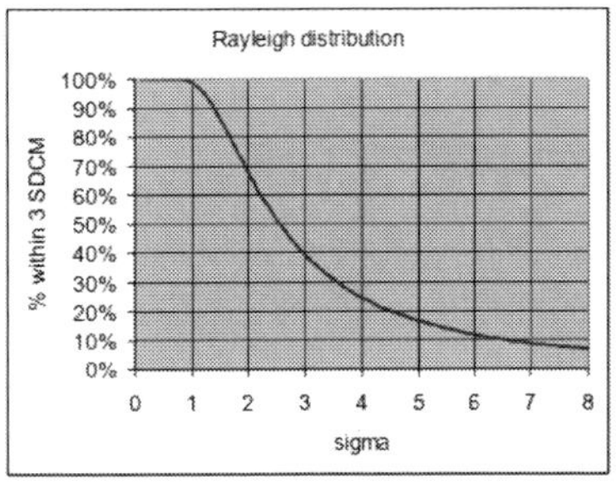

Figure 3.Relation between sigma value and percentage of colorpoints within 3 SDCM.

From this plot you can see that a sigma value below 1.4 corresponds to more than 90% of the colorpoints within 3 SDCM of the mean. Similarly a sigma below 2.3 corresponds to more than 90% within 5 SDCM. In figure 4. a validation of the color stability metric is demonstrated. Based on a sample of 10 lamps the estimated sigma total is 1.14. Measurements on a large group of 260 lamps shown a sigma total of 1.20.

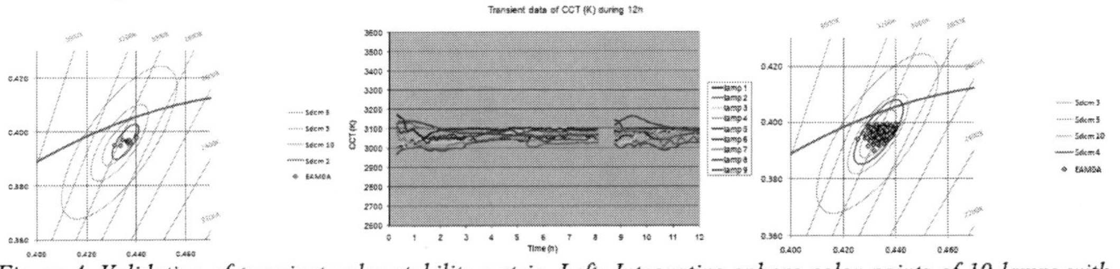

Figure 4. Validation of transient color stability metric. Left: Integrating sphere color points of 10 lamps with σ_{group} *=0.91. Middle: transient color stability measurement giving* σ_{lamp} *= 0.68. Calculated* σ_{total} *= 1.14. Right: Integrating sphere color points of 260 lamps giving sigma group 1.20.*

DBD-driven Xenon Iodide exciplex lamp operated with a $Kr/Xe/I_2$ mixture and attempts for the UV inactivation of *Saccharomyces cerevisiae*

M. M. Guivan[1], H. Dvořáková[2], H. Vojkovska[3], J. Slamova (Vrajova)[3,4], P. St'ahel[2], P. Slavíček[2], A. Brablec[2]

[1]*Department of Quantum Electronics, Uzhgorod National University, Pidgirna 46, Uzhgorod 88000, Ukraine*
[2]*Department of Physical Electronics, Masaryk University, Kotlářska 2, Brno 61137, Czech Republic*
[3]*Brno University of Technology, Faculty of Chemistry, Institute of Food Science and Biotechnology, Purkyňova 118, 61200 Brno, Czech Republic*
[4]*Technical Museum in Brno, Purkyňova 105a, Brno 612 00, Czech Republic*

Contact: m_guivan@rambler.ru

ABSTRACT

A dielectric barrier discharge (DBD)-driven XeI* excilamp operated with a $Kr/Xe/I_2$ mixture at the repetition frequency of AC sinusoidal voltage pulses of 100 kHz has been investigated. The developed lamp showed a good sterilization performance for the UV inactivation of *Saccharomyces cerevisiae*.

INTRODUCTION

The excilamps (excimer or exciplex lamps) based on mixtures containing xenon and iodine vapours emitting mainly due to $XeI(B^2\Sigma_{1/2} - X^2\Sigma_{1/2})$ at $\lambda = 253$ nm are considered to be the efficient sources in the UVC (200–290 nm) range. Recently, the sterilization action of a DBD-driven XeI* excilamp operated with a Xe/I_2 mixture has been reported for the inactivation of *Bacillus Subtilis* spores [1]. In this study, we report the investigation of DBD-driven XeI* excilamp operated with the $Kr/Xe/I_2$ mixture and using it for the UV inactivation of *Saccharomyces cerevisiae*.

RESULTS AND DISCUSSION

In these experiments a cylindrical DBD lamp in coaxial design made of two quartz tubes and electrodes, which were applied to the inner surface of the smaller tube and the outer surface of the outer tube, was used. An active length of the lamp was 10 cm. The discharge gap was 4.75 mm. For the external electrode a 0.1 mm nickel wire wrapped tightly on the outer tube with a 2 mm pitch providing transparency of 95% was used. To excite a working mixture a generator of high voltage AC sinusoidal pulses was used; peak-to-peak voltage up to 12 kV, $f = 1 - 125$ kHz. The voltage U applied to the lamp was measured using a high voltage probe (Tektronix P6015A). The waveforms of current I and charge Q were registered by means of a 50 Ω resistor or low-inductance 8 nF capacitor placed in series with the lamp and monitored by digital oscilloscope LeCroy WaveRunner 6100A (1 GHz, 10 GS/s). The radiation emitted by the discharge was registered and analyzed in the spectral range of 200–740 nm. The spectra were recorded by the Jobin Yvon FHR 1000 monochromator (grating 2400 grooves/mm, quartz optical fiber, the spectral resolution of about 8 pm) with the high-speed CCD detector. The radiation power of the lamp in absolute units was measured by a UV power meter Newport 1918-C equipped with a calibrated sensor head 918D-UV (200 – 1100 nm).

The yeast *Saccharomyces cerevisiae* CCM 8191 (ATCC 9763) has been chosen as a bio-indicator to evaluate the XeI* exciplex lamp microbial inactivation. The yeast was obtained in Czech collection of microorganisms (Brno Masaryk University – Faculty of Science). The culture of *Saccharomyces cerevisiae* was inoculated into wort solution and cultivated at 25°C for 24 hours. Afterwards the cell suspension in physiological solution containing cca $5{\cdot}10^6$ cells/ml was prepared and 100 μl of the cell suspension was inoculated on the sterilized substrate. The discs (with 3 cm diameter) of Whatman paper no. 1 were used as substrate. The paper samples with the cell suspension was let to dry overnight and then exposed to the XeI* excilamp. After the treatment the samples were immersed into 10 ml of sterile physiological solution with glass beads and placed overnight on the Horizontal Receptable Shaker in order to wash away the cells from the samples. The cell suspension was diluted and dispersed on wort agar plates. After 72 hours of incubation at 25° C the number of

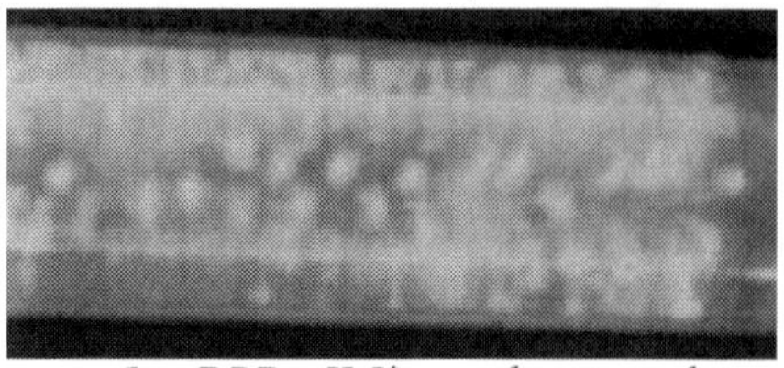

Figure 1: *DBD XeI* excilamp under AC excitation.*

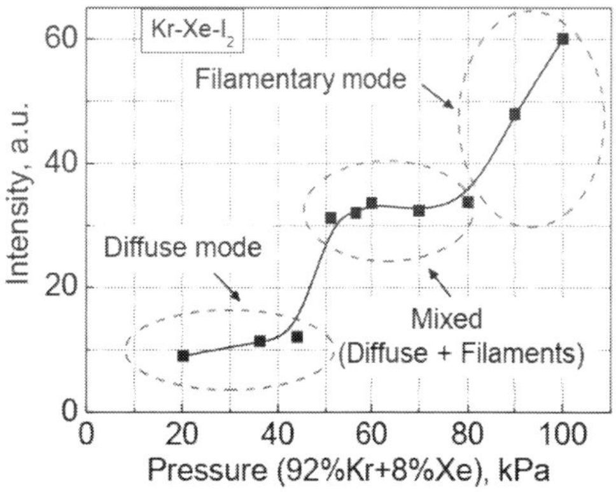

Figure 2: *UV radiation power as a function of (92%Kr + 8%Xe) pressure in DBD excilamp with the* $Kr/Xe/I_2 = p(Kr/Xe)/40$ *Pa mixture.*

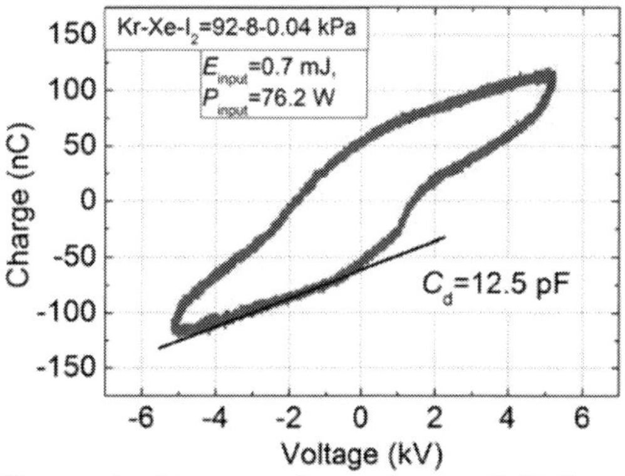

Figure 3: *Lissajous figure for a* $Kr/Xe/I_2$ = *92/8/0.04 kPa mixture.*

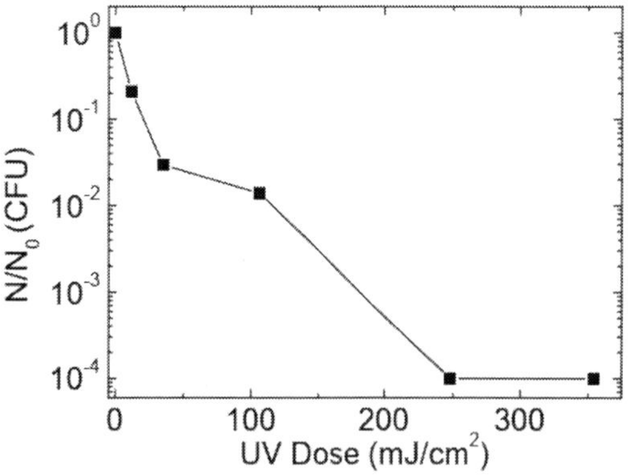

Figure 4: *Normalized number of CFU (Saccharomyces cerevisiae) as a function of the UV dose. Distance to the lamp was 5 cm.*

colony forming units (CFU) was counted.

Measurements of the UV–visible emission spectra of the DBD revealed the strong XeI*(*B*–*X*) exciplex radiation at λ_{max} = 253 nm. Weaker emissions of XeI*(*C*–*A*), XeI*(*B*–*A*) and I_2*(*D′*–*A′*) at λ = 265, 320 and 342 nm were also observed. Atomic iodine was represented in the spectrum by the emission at λ = 206.2 nm ($^2P_{3/2}$–$^2P°_{1/2}$). The DBD under AC excitation consisted of diffuse cylindrical microdischarges with a little wider base, forming a bright surface discharge on the internal tube (figure 1). The output excilamp power was measured as a function of the 92%Kr + 8% Xe pressure for the $Kr/Xe/I_2$ mixture with a fixed I_2 pressure $p(I_2)$=40 Pa. For a constant applied voltage, the UV output increased non-monotonically with gas pressure (figure 2). A jump in the UV power was observed when the discharge mode changed from a diffuse to mixed (diffuse + filaments) and then from the mixed to the filamentary. The average radiation power of 3.1 mW/cm^2 has been achieved with the $Kr/Xe/I_2$ = 92/8/0.04 kPa mixture at f = 100 kHz. The energy deposited into the gas-discharge plasma for one period and input power, calculated from the Lissajous figure, are equal to 0.7 mJ and 76 W accordingly (figure 3). In each half-cycle of the applied voltage U_0, the current waveform consisted of a displacement current and conductivity current peaks, with characteristics depending on the operating frequency and applied voltage. The peak current did not exceed 150 mA and their duration was in the range 0.8–1.1 μs.

In all experiments an increase in the exposure time and UV dose always resulted in CFU decrease (figure 4). The D-value (one order reduction) was about 20 mJ/cm^2. A four order reduction (99.99% inactivated microorganisms) was achieved at the UV dose of 250 mJ/cm^2.

SUMMARY

A detailed experimental study of the characteristics of a XeI* DBD excilamp under AC excitation has been carried out. UV radiation power density of 3.1 mW/cm^2 was obtained with the $Kr/Xe/I_2$ = 92/8/0.04 kPa mixture at f = 100 kHz. The developed lamp showed a good performance for sterilization application.

REFERENCES

[1] M.M. Guivan, T. Kamikozawa, H. Kurokawa, H. Motomura, K. Kadowaki, M. Jinno, IEEE Trans. Plasma. Sci., (2010) accepted for publication.

Electrode temperature measurement of mercury-free metal halide lamp for automotive HID headlamp operated with square-wave and DC current

M. Shido , Y. Itou , T. Serita , S. Yagi
R&D Dept., Koito Manufacturing Co., Ltd., Shimizu-ku Kitawaki 500, Shizuoka, 424-8764, Japan
Contact: mshido@koito.co.jp

ABSTRACT

Electrode temperature in mercury-free metal halide lamp for automotive HID headlamp with several kinds of electrode material and diameter was investigated and compared. The electrode temperature was measured pyrometrically along the electrode axis. The anode temperature operated with DC current increased from operated with square-wave drive and the proportion coefficient of the anode temperature to the lamp power operated with DC current were about the twice as operated with square-wave drive. In contrast, the cathode temperature operated with DC current decreased from operated with square-wave drive and that was almost independent of the lamp power.

INTRODUCTION

In the 1990s, new automotive HID headlamp started with the mercury containing metal halide lamp with quartz vessel, it was operated with AC current. Soon after, AC current drive was changed to low frequency square-wave drive by the demand for more compact, lightweight and lower cost HID system. In 2004, the mercury-free metal halide lamp for automotive headlamp was commercialized in the world's first.

Since DC current drive with the mercury containing metal halide lamp makes de-mixing and color separation of lamp arc according to cataphoresis phenomena, AC drive or low frequency square-wave drive has been used so far and thus the shapes of the lamp (electrodes and quartz vessel) are symmetry. However, the mercury-free metal halide lamp contains no mercury which vapor pressure is much higher than metal halides, de-mixing and color separation of the lamp arc are much smaller than mercury containing lamp even when operated with DC current as shown Figure 1. If the automotive HID system with DC current is possible, there are some advantages over existing square-wave drive HID system. For instance, the lamp life can be longer by means of the prevention of flickering at the end of life and more compact, lightweight and lower cost HID system can be realized by means of disappearance of EMI noise at the commutated and by removal of DC-AC converter of the ballast. This paper shows the electrode temperature measurement result and comparison of the mercury-free metal halide lamp with several kinds of electrode material and diameter which was operated with square-wave drive and DC current.

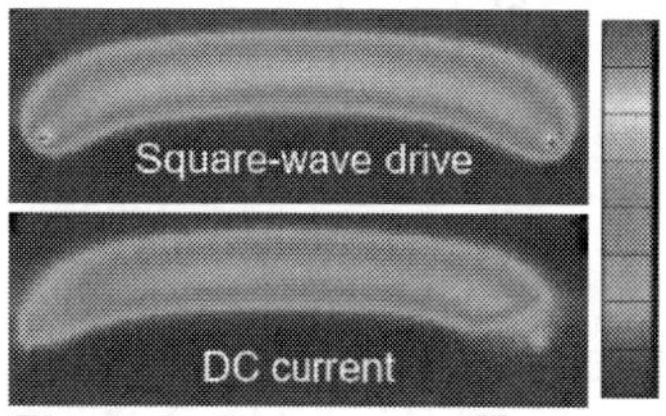

Figure 1: *Comparison of luminance distribution of lamp arc with square-wave drive and DC current.*

EXPERIMENTAL SETUP

Figure 2 shows the schematic diagram of electrode temperature measurement. The infrared radiation of electrode is measured using the gated CCD camera thorough the 980 nm narrow band filter. The relation between infrared radiation and temperature was calculated by the color temperature of the 100 W standard incandescent lamp. Figure 3 shows the spectrum distribution of the measured lamp. 980 nm is the wavelength which meets all the condition of the temperature measurement; sensitivity range of CCD camera, temperature range of electrode and relatively small emission range from the lamp arc.

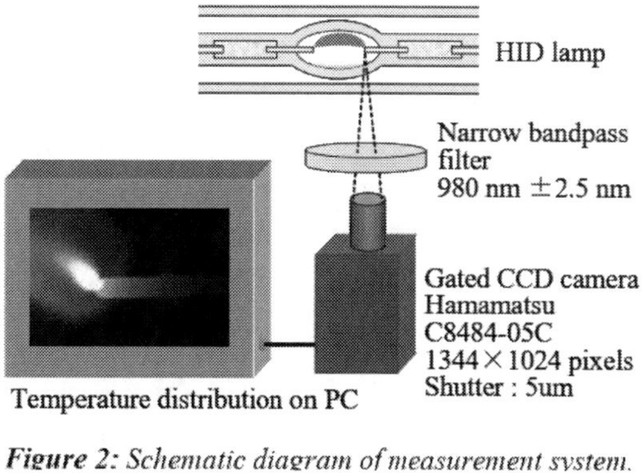

Figure 2: *Schematic diagram of measurement system.*

Table 1 shows three kinds of mercury-free metal halide lamps used for the experiment with different electrode material and diameter, the lamp power is 35 W and the distance between electrodes is approximately 4.0 mm and the lamp voltage is approximately 45 V. The electrode temperature was measured 2 seconds after switch ON (the run-up phase) and 5 minutes after switch ON (steady state), respectively. In the run-up phase, average run-up power up to 2 seconds was 65 W or 35 W. All the following electrode temperature data are at the cross point of center axis of the electrode and 0.5 mm apart from top of the electrode.

Table 1: *Specifications of measured samples.*

	Electrode material	Electrode diameter
Lamp:A	AKB-W (Pottasium dope tungsten)	0.31 mm
Lamp:B	ThO_2-W (Thorium dope tungsten)	0.31 mm
Lamp:C	ThO_2-W (Thorium dope tungsten)	0.38 mm

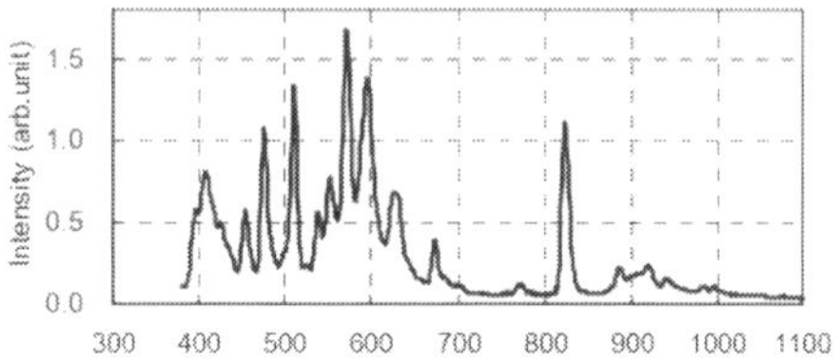

Figure 3: *Spectral distribution of measured lamp.*

RESULTS

Figure 4 shows the result of electrode temperature in the steady state operated with square-wave drive. The electrode temperature with square-wave drive changed with the material and diameter of the electrodes. Figure 5 shows the result of electrode temperature in the steady state operated with DC current. Since the cathode side vessel was covered with the metal halide filling and cathode electrode was hidden operated with DC current, no temperature data was acquired with the cathode of Lamp: B and Lamp: C. The anode temperature in Lamp: A operated with DC current decreased 50 K - 300 K from operated with square-wave drive, but the anode temperature in Lamp: B and Lamp: C operated with DC current increased 100 K – 500 K from operated with square-wave drive. In addition, the proportion coefficient of the anode temperature to the lamp power with DC current were about the twice as that with square-wave drive. In contrast, the cathode temperature in Lamp: A operated with DC current decreased compared with square-wave drive and that was almost independent of the lamp power.

Table 2 shows the result of electrode temperature in the run-up phase. In all the lamps, the anode temperature operated with DC current rose as compared to the square-wave drive. The 980nm emission spectra from the high pressure xenon gas in the run-up phase affect the measurement and the measurement data exceeded the melting temperature of the tungsten electrode, and actually top of the anode melted in Lamp: A and Lamp: B at the run-up power 65 W.

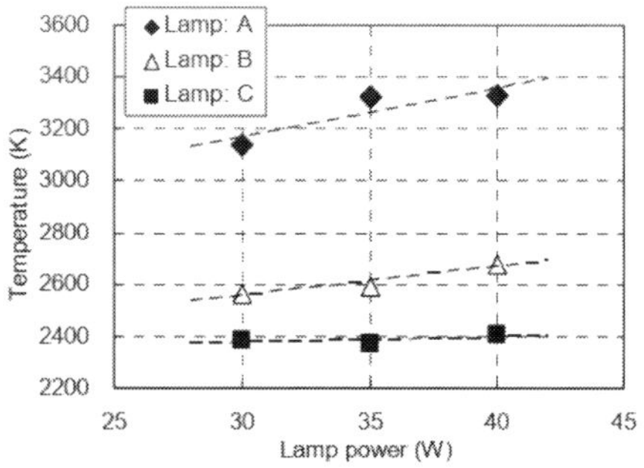

Figure 4: *Electrode temperature versus lamp power in the steady state operated with square-wave drive.*

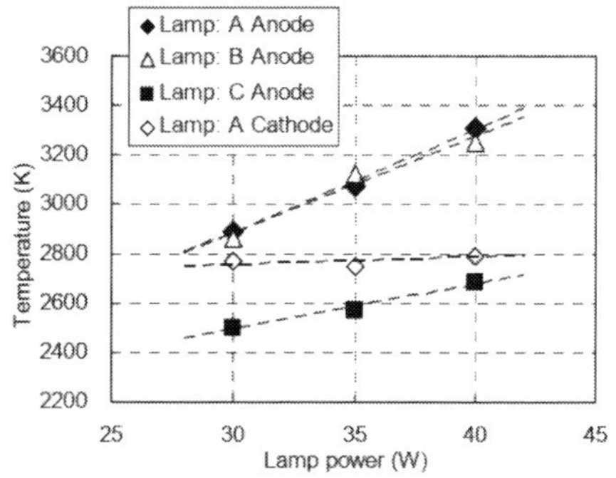

Figure 5: *Electrode temperature versus lamp power in the steady state operated with DC current.*

CONCLUSION

By choosing the material and the diameter of the anode electrode properly, the mercury-free metal halide lamp for automotive headlamp operated with DC current which have equivalent performance as operated with square-wave drive is possible.

Table 2: *Anode temperature in the run-up phase.*

Operating waveform	Squre-wave		DC current	
Run-up power	35 W	65 W	35 W	65 W
Lamp: A	3683 K	4653 K	2521 K	4952 K
Lamp: B	2754 K	3632 K	2816 K	4418 K
Lamp: C	2393 K	3332 K	2382 K	3922 K

REFERENCES

[1] M. Shido, T. Serita, M. Kando: Proc. of LS11, CP152, Shanghai, China (2007) 501.

Clarifying of physical processes in DBD-driven xenon iodide exciplex lamp by means of ICCD imaging, space-, time-resolved spectroscopy and IRLAS

Mykola M. Guivan[1], Hideki Motomura[2], Mikhail Lomaev[3], Masafumi Jinno[2]

[1]Department of Quantum Electronics, Uzhgorod National University, Pidgirna 46, Uzhgorod 88000, Ukraine
[2]Dep. of Electrical and Electronic Engineering, Ehime University, 3 Bunkyo-cho, Matsuyama 790-8577, Japan
[3]Institute of High Current Electronics, 2/3 Akademichesky Avenue, Tomsk 634055, Russia
Contact: m_guivan@rambler.ru

ABSTRACT

The space-, time-resolved optical emission spectroscopy, ICCD imaging, as well as infrared laser absorption spectroscopy (IRLAS) have been used to investigate the physical processes in the dielectric barrier discharge (DBD)-driven XeI* excilamp operated with a Xe/I_2 = 13.3/0.04 kPa mixture at the pulsed excitation.

RESULTS AND DISCUSSION

In this contribution to the research the results of investigation of the physical processes in a DBD-driven XeI* exciplex lamp (excilamp), as well as the peculiarities in dynamics of the electrical, optical, spatial and temporal characteristics have been presented. In these experiments a cylindrical DBD lamp in coaxial design made of two quartz tubes and electrodes, which were applied to the inner surface of the smaller tube and the outer surface of the outer tube, was used. An active length of the lamp was 10 cm. The discharge gap was 7.4 mm. For the external electrode a 0.1 mm nickel wire wrapped tightly on the outer tube with a 2 mm pitch providing transparency of 95% was used. To excite a working mixture a generator of high voltage unipolar or bipolar pulses was used; peak-to-peak voltage U_{pp} up to 8 kV, f = 10 – 80 kHz (unipolar), U_{pp} = 0 – 4.4 kV, f = 21.5 – 115 kHz (bipolar). The waveforms of voltage U, current I and charge Q were monitored by digital oscilloscope (Tektronix, 3034B). The radiation emitted by the discharge was registered and analyzed in the spectral range of 170–850 nm. The radiation power of the lamp in absolute units was measured by a UV power meter (Hamamatsu C8026) equipped with a calibrated H8025-254 sensor head. The discharge occurrence for a single filament and a lamp has been studied using a fast-speed ICCD camera (ANDOR iSTAR DH 720I). IRLAS measurements were realized by means of a diode laser (New Focus, Velocity Model 6316), monochromator (Jasco, CT-25TP), and PMT (Hamamatsu, R955).

The measurements with ICCD camera have revealed a diffuse distribution of radiation in VUV and UVC range, that completely different from that observed in the UVB, UVA, visible, and near IR (figure 1). It was observed a longer decay in the central part of a single filament at 823 nm compared to 828 nm, as well as the second maximum in the afterglow due to e/Xe$^+$ (Xe_2^+, Xe_3^+) recombination (figure 2(b)). It has been revealed that DBD appears in the lamp in the area with the highest electrical field strength and spreads uniformly simultaneously in two directions: along the whole active length of the lamp and across the gap from the anode to the cathode with velocity of ~185 km/s and ~80 km/s respectively (figure 3).

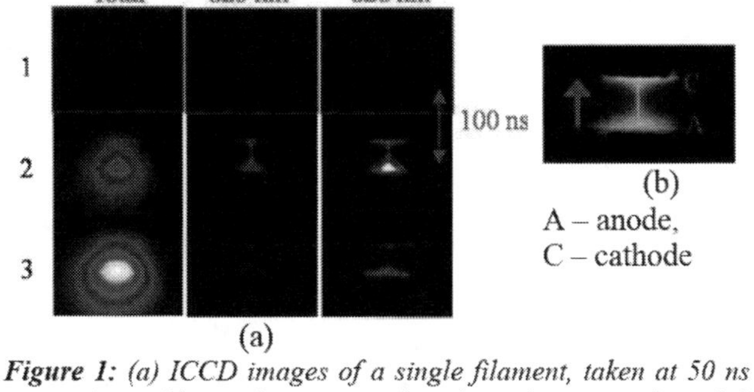

Figure 1: *(a) ICCD images of a single filament, taken at 50 ns gate pulse width: total emission taken through a quartz lens (left column), Xe* emission (823 nm – center column and 828 nm – right column) taken with the appropriate interference filters; (b) photo of a single filament, exposure time of 1/20 s.*

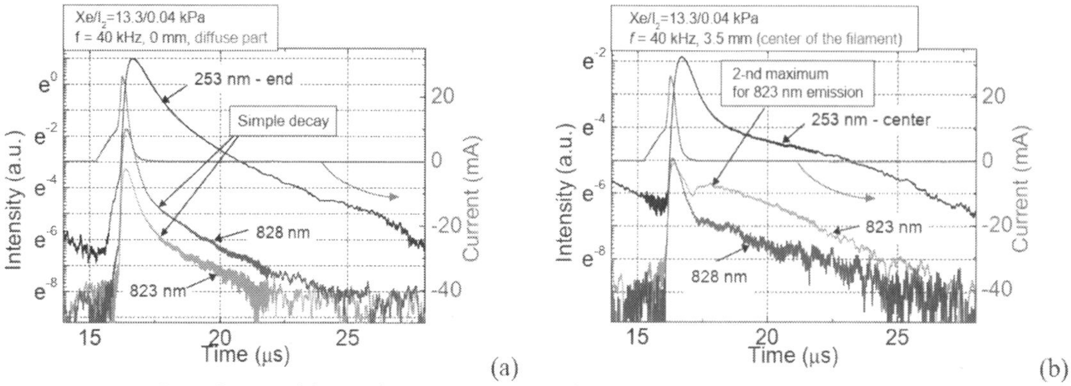

Figure 2: *Time dependences of the XeI*(B–X) (253 nm) and Xe* (823 and 828 nm) emission from the end (a) and center (b) of a single filament with the Xe/I$_2$ = 13.3/0.04 kPa mixture.*

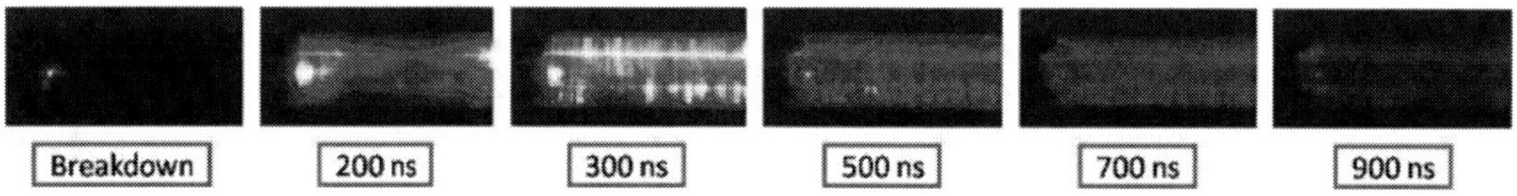

Figure 3: *ICCD images of XeI* excilamp, taken at 50 ns gate pulse width. U_{pp}=4.2 kV, f=60 kHz.*

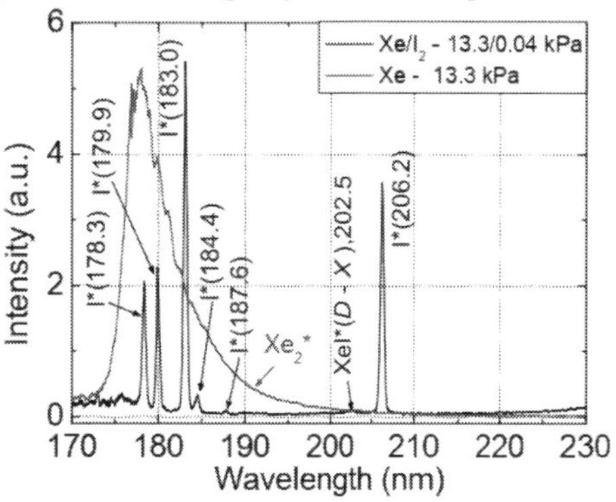

Figure 4: *VUV emission spectra of the XeI* and Xe$_2$* escilamps.*

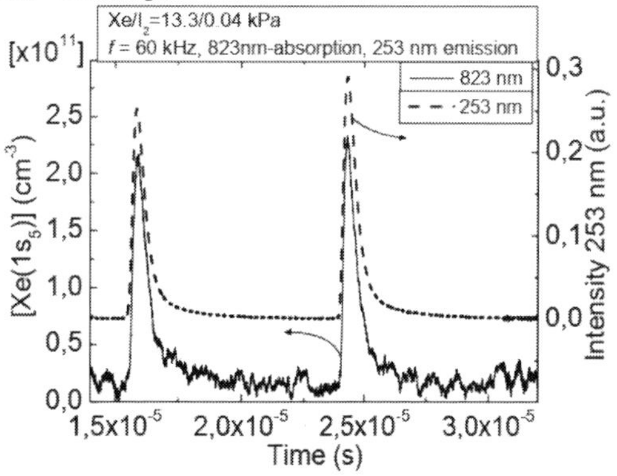

Figure 5: *Time dependences of the Xe*(1s$_5$) metastable density (left Y-axis) and XeI*(B–X) emission with the Xe/I$_2$ = 13.3/0.04 kPa mixture (right Y-axis).*

Measurements of the VUV-UV–visible emission spectra of the DBD have revealed atomic iodine emission in the range of 178-188 nm and traces of the XeI*(*D–X,* 202.5 nm) (figure 4) together with the known I*($^2P_{3/2}$–$^2P^{o}_{1/2}$, 206.2 nm), XeI*(*B–X,* 253 nm), XeI*(*C–A,* 265 nm), XeI*(*B–A,* 320 nm), and I_2*(*D′–A′*, 342 nm) radiation. Note that any Xe_2* VUV emission was not observed in the spectra of the XeI* excilamp contrary to the prediction suggested in [1]. The time evolution and absolute values of Xe(1s$_5$) metastables were measured by means of IRLAS at 823 nm and compared the obtained results with the time-resolved emission of XeI* at 253 nm (figure 5). An excellent coincidence has been observed and proved that the major part of xenon metastables is converted to the xenon iodide due to the Xe(1s$_5$) + I_2 → XeI(*B*) + I harpoon reaction.

SUMMARY

A detailed experimental study of the physical processes in the XeI* DBD excilamp under pulsed excitation has been carried out. The spatial intensity distribution in the VUV-UVC range is much more diffuse compared to other spectral ranges. It has been experimentally proved that the Xe(1s$_5$) metastable is the main precursor for the XeI(*B*) exciplex molecule.

ACKNOWLEDGMENT

This research has been supported by the JSPS, Grant-in-Aid for JSPS Fellows (06382).

REFERENCES

[1] R.J. Carman, B.K. Ward, R.P. Mildren, D.M. Kane, Proc. LS-11, Shanghai (China) (2007) 271.

Author Index

Author Index

D.

E.

F.

N.

O.

P.

R.

S.

T.

Z.